QUALITATIVE CHEMICAL SEMIMICROANALYSIS

В. Н. Алексеев

Курс качественного химического полумикроанализа

●

Издательство «Химия»
Москва

V. N. Alexeyev

Qualitative Chemical Semimicroanalysis

Edited by P. K. Agasyan, D. Sc. (Chem.)

Translated from the Russian
by
Artavaz Beknazarov

Mir Publishers Moscow

CBS PUBLISHERS & DISTRIBUTORS
4596/1-A, 11 Darya Ganj, New Delhi - 110 002 (India)

Revised from the 1973 Russian edition
First published 1975
Second printing 1980

На английском языке

ISBN : 81-239-0279-4

First Indian Edition : 1994
Reprint : 2004

This edition has been published in India by arrangement with Mir Publishers, Moscow

Publishing Director : **Vinod K. Jain**

Published by :
Satish Kumar Jain for CBS Publishers & Distributors,
4596/1-A, 11 Darya Ganj, New Delhi - 110 002 (India)
E-mail : cbspubs@del3.vsnl.net.in
Website : http://www.cbspd.com

Branch Office :
Seema House, 2975, 17th Cross, K.R. Road,
Bansankari 2nd Stage, Bangalore - 560070
Fax : 080-6771680 • E-mail : cbsbng@vsnl.net

Printed at :
Asia Printograph, Shahdara, Delhi - 110 032 (India)

Preface

The author of this book, Vladimir Nikolaevich Alexeyev, died in 1958. Fourteen years have now elapsed since the publication of the last edition of this textbook in 1962. In this interval, great developments have taken place in the theory and practice of analytical chemistry. In preparing this new edition we were aware of the necessity to revise and expand the text appropriately.

There is no need to mention the many changes introduced by the editors into this text. It should however be pointed out that substantial revisions have been made in all the chapters of the book. Modern conceptions of acids and bases, of the mechanism of certain types of reactions, etc., are presented.

What has been done by the editors in an attempt to make use of the theoretical achievements in the chemical science is possibly far from perfection. The editors will be grateful for any constructive criticism and useful suggestions which will doubtless be taken into account in a subsequent edition.

The editors wish to express their deep gratitude to their colleagues at Department of Analytical Chemistry, Moscow State University, and particularly to Academician I. P. Alimarin, for the perusal of the entire manuscript and for their valuable advice and comments for improvement of this text.

P. K. Agasyan
M. G. Tsyurupa

Contents

B. REACTIONS AND ANALYTICAL PROCEDURE FOR GROUP I CATIONS

Chapter 4. GROUP II CATIONS

A. THEORETICAL PRINCIPLES

B. REACTIONS AND ANALYTICAL PROCEDURE

FOR GROUP II CATIONS

Chapter 5. GROUP III CATIONS

A. THEORETICAL PRINCIPLES

B. REACTIONS AND ANALYTICAL PROCEDURE FOR GROUP III CATIONS

Chapter 6. CATIONS OF GROUPS IV AND V

A. THEORETICAL PRINCIPLES

B. REACTIONS AND ANALYTICAL PROCEDURE FOR GROUP IV AND V CATIONS

Chapter 7. VANADIUM, MOLYBDENUM AND TUNGSTEN

Chapter 8. ANIONS

A. GROUP I ANIONS

B. GROUP II ANIONS

C. GROUP III ANIONS

Chapter 9. GENERAL COURSE OF ANALYSIS
(Detection of Cations and Anions)

List of Symbols

a = Activity
C = Concentration
D = Distribution coefficient
e = Electron
E = Equilibrium oxidation-reduction potential
E^0 = Standard oxidation potential of a redox couple
f_a = Activity coefficient
h = Degree of hydrolysis
k = Rate constant
K = Ionization (dissociation) constant
K^0 = True or thermodynamic constant
K_a = Acid dissociation constant
K_b = Base dissociation constant
K_C = Concentration constant
K_{eq} = Equilibrium constant
K_h = Hydrolysis constant
K_{st} = Stability constant
K_{ins} = Instability constant
K_w = Ion (ionic) product of water
m = Detectable minimum (sensitivity)
M = Molarity
N = Normality
pH = Hydrogen ion exponent
pK = Logarithm of the reciprocal of the dissociation constant of an electrolyte; $\mathrm{p}K = \log (1/K)$
r = Ionic radius
R = Organic radical
R = Gas constant
S = Solubility
SP = Solubility product (constant)
T = Absolute temperature
v = Reaction rate
V = Volume of solution
z = Ionic charge
α = Degree of dissociation (degree of ionization)
γ = Side-reaction coefficient
λ = Equivalent electrical conductance
μ = Ionic strength

List of Tables

Chapter 1

Introduction

1.1. Subject and Methods of Analytical Chemistry

Analytical chemistry is a science concerned with development of the theoretical foundations and methods of chemical analysis. The practical objective of analytical chemistry is to establish the chemical composition of substances or their mixtures. In the first place, the qualitative composition of a substance is found, i.e., the problem is solved as to what elements, groups of elements, or ions this substance is composed of, and then the quantitative composition of the substance is determined: the relative proportions of the constituents in the substance are established.

The detection or identification of individual elements or ions entering into the composition of a substance constitutes the task of **qualitative analysis.**

The determination of the quantitative content of the constituents of the substance under study is the aim of **quantitative analysis.**

Qualitative analysis usually precedes quantitative analysis. One has to resort to qualitative analysis even in those cases when it is necessary to determine the percentage content of a certain component known to be present in a given substance. This is because the most convenient and efficient method for quantitative determination of the particular constituent can be chosen only when one knows what other elements or ions are present in the substance being studied.

On the other hand, a qualitative reaction is capable of providing information on the amount of a constituent in the sample under study: whether a large or small amount or traces of this constituent are present in the substance.

Analytical chemistry, and in particular qualitative analysis, is of enormous scientific and practical importance, being a science dealing with the various methods for investigating substances and their transformations. It also plays an important role in branches of science allied to chemistry—mineralogy, geology, physiology, microbiology, as well as in medicine, agriculture, and technology. In almost every scientific research work dealing in one way or another with chemical phenomena, the investigator has to make use of the methods of analytical chemistry.

Chemical analysis is highly important to the national economy; it is indispensable for chemical control of production in the most important sections of industry, for chemical investigation of soil, fertilizers, agricultural products, useful minerals, etc.

The tasks of analytical chemistry can be solved with the aid of a variety of methods, whether chemical, physical, or physico-chemical. These methods are continuously expanded and perfected. More advanced techniques of analysis are continuously designed, which require new procedures and new theoretical substantiation.

In **chemical methods** of qualitative analysis, the element or ion to be determined is converted into a compound having some specific properties, on the basis of which one can establish the identity of this compound. The chemical change involved is known as an **analytical reaction** and the substance responsible for this change is called a **reagent.**

Like any other experimental method, chemical methods of analysis have a definite field of application, beyond which they prove unsuitable. Thus, existing methods do not always satisfy the increasing demands of science and technology as to sensitivity, i.e., the suitability for detection and determination of traces of various impurities. Also, the speed with which analyses are carried out with the aid of chemical methods does not always satisfy the requirements of producing industries, where it is very essential that the results of an analysis be obtained in due time, so that there still remains a possibility of controlling the technological process in order to prevent spoilage. Therefore, along with chemical methods, an ever increasing use is being made of physical and physico-chemical analytical methods.

Physical methods of analysis rely on the measurement of any one parameter of a system, which is a function of the composition.

Spectrum analysis is concerned with study of emission spectra produced by substances introduced into the flame of a burner, electric arc, etc. Certain characteristic lines in the spectrum indicate the presence of some definite elements in the substance under study, and the quantitative content is estimated from the line intensities. This method is highly sensitive (about 10^{-6} to 10^{-8} gram of some elements can be detected) and, for the operations to be performed, requires little time and small amounts of substances.

Luminescence analysis is based on the observation of the luminescence emitted when the molecules of the material under study are excited by ultraviolet or corpuscular (cathodic rays, for example) radiation. Ultraviolet rays can be produced by a mercury quartz lamp with a light filter which transmits the invisible radiation of a bright mercury line of 3650 Å and stops visible radiation (which is essential in the observation of luminescence). Under the action of ultraviolet rays or other types of radiation, the substance under study emits a characteristic light. The colour and intensity of luminescence serve as analytical features which enable one to perform a qualitative and quantitative analysis of various substances. Luminescence analysis is even more sensitive than spectrum analysis (down to 10^{-10} gram and less of a substance can be detected). This method is successfully employed in a number of branches of science and technology (in analysis of organic and inorganic compounds, in biology, medicine, and agriculture, in investigations of food products and tanning agents, for grading glasses, in explorations of oil fields, etc.).

Among the **physico-chemical methods** used in qualitative analysis, mention should be made of **colorimetric techinques** which are based on the relationship between the colour intensity of a solution and the concentration of the coloured substance (ions or molecules), and also of the **chromatographic method.** In the latter method, the unknown solution is passed through a glass column packed with a powdered adsorbent (for example, alumina, Al_2O_3). Since the adsorbability (the degree of adsorption*) of different substances or ions differs essentially, they can be separated and detected in the column either by their characteristic colour or by treatment with reagents which form with them differently coloured compounds.

No strict boundary line can always be drawn between physical and physico-chemical methods of analysis. Sometimes, these methods are collectively named as **instrumental methods** since the measurements involved require instruments—precision devices enabling the values of definite parameters characterizing the properties of a substance to be measured with high accuracy. Instrumental methods of analysis are not discussed in detail in this textbook.

According to the amount of a substance used for analytical reactions, macro-, semimicro-, micro- and ultramicro-methods of qualitative analysis are distinguished.

In **macroanalysis**, relatively large quantities of substance (0.5-1.0 g) or solution (20-50 ml) are used. The reactions are carried out in ordinary test tubes (10-20 ml in capacity), beakers, or flasks. Precipitates are separated from solutions by filtration through filter paper.

The amount of substance used in **microanalysis** is as a rule about 1/100 of the quantity involved in macroanalysis work, i.e., a few milligrams of solid or some tenths of a millilitre of solution. Highly sensitive reactions are employed, which permit detection of individual constituents by the fractional method even if they are present in small amounts. Such reactions are carried out either by the microcrystalloscopic or by the drop method.

In the **microcrystalloscopic method**, the reactions are usually performed on glass microscope slides and the presence of a given ion (or an element) is deduced from the shape of the formed crystals examined under the microscope.

---

* Adsorption is defined as a process in which ions or molecules of a substance are concentrated from the surrounding medium in the surface layer of another substance.

Drop reactions, or **spot tests**, are reactions accompanied by a change of the colour of solution or by the formation of coloured precipitates. The tests are most commonly carried out on a strip of filter paper, drops of the unknown solution and of the reagents being applied to it in a definite sequence. The reaction produces a coloured spot on the paper, indicating the presence of a particular ion in the unknown solution. Spot tests can also be performed on special spot plates with indentations, on watch glasses, in porcelain crucibles or dishes, etc.

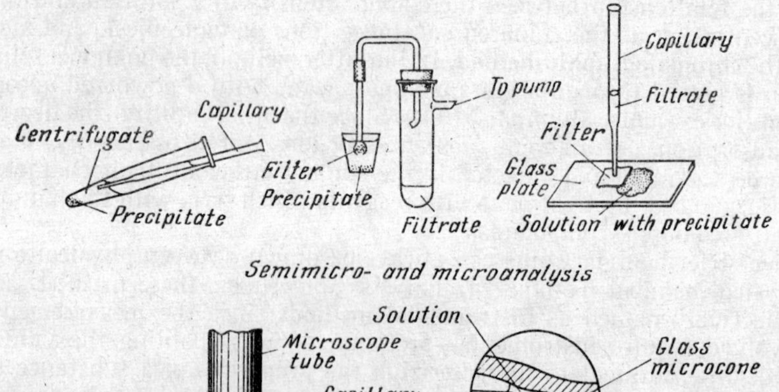

Fig. 1.1. Separation of a precipitate from solution in various methods of analysis.

Semimicroanalysis ranks between the macro- and the micro-method. The amount of substance used is about 1/20-1/25 of the amount taken for macroanalysis, which corresponds to about 50 mg of solid or 1 ml of solution.

In semimicroanalysis, the entire system of work used in macroanalysis with subsequent separation and identification of ions is retained but the operations are performed with small amounts of substance by means of special techniques and apparatus.

Semimicroanalysis has a number of advantages over macroanalysis, and gives equally reliable results with careful working.

Ultramicroanalysis is employed for investigation of less than 1 mg of substance. Almost all the analytical operations are performed under the microscope.

The differences in technique between the different methods of analysis are shown schematically in Fig. 1.1 in which a comparison of the various ways of separating precipitates from solution is made.

This book is concerned with methods of qualitative analysis of inorganic substances only. Analytical methods for organic substances constitute a specialized field and are therefore described in special books.

1.2. Methods of Carrying out Analytical Reactions

Analytical reactions can be carried out by "dry" and "wet" methods. In the former case, the unknown substance and the appropriate reagent are taken in the solid state and the reaction is usually performed by heating them to a high temperature; in the "wet" method, the reaction between the unknown substance and the reagent takes place in solution.

Dry reactions include reactions based, for example, on **flame colouration** by salts of certain metals. For example, when sodium salts are introduced on a platinum wire loop into the non-luminous flame of a gas burner, they colour the flame intense yellow; potassium salts give a violet colour, strontium salts—crimson, barium salts—green, etc. Under suitable conditions, the presence of these elements in unknown substances can be detected by these colours.

Dry reactions also include: (a) the formation of **coloured beads** (glasses) of sodium tetraborate (borax), $Na_2B_4O_7 \cdot 10H_2O$, or sodium-ammonium hydrogen phosphate ("microcosmic salt"), $NaNH_4HPO_4 \cdot 4H_2O$, when they are fused with salts of certain metals; and (b) the fusion of unknown substances with various fluxes such as mixtures of solid Na_2CO_3 and K_2CO_3, or Na_2CO_3 and KNO_3, etc.

These methods, known as **pyrochemical**, find wide application in the analysis of minerals and rocks.

Grinding or **trituration** of an unknown solid substance with a solid reagent is another method which is sometimes used. In this method proposed in 1898 by F. M. Flavitsky (1848-1917), use is made of reactions leading to the formation of coloured compounds of the elements being determined. For example, in order to detect cobalt, a few crystals of a solid, say $CoSO_4$, are ground on a porcelain plate with an approximately equal amount of solid reagent—ammonium thiocyanate, NH_4SCN. The mixture assumes a blue colour owing to the formation of a complex cobalt salt, $(HN_4)_2[Co(SCN)_4]$:

$$CoSO_4 + 4NH_4SCN \rightarrow (NH_4)_2[Co(SCN)_4] + (NH_4)_2SO_4$$

Such reactions can be accomplished with very small quantities of material, for which reason they are employed in semimicro- and microanalysis.

It should be pointed out that most reactions which take place during trituration involve water adsorbed from the air or present in a given compound in the form of crystallization water, so that they may be referred to as dry reactions with some reservation only.*

Dry reactions play a subsidiary part, and are generally used as preliminary tests.

Reactions carried out by the wet method, i.e., reactions taking place between substances in solutions, are most often used in qualitative analysis. For such reactions to proceed, the substance under test must first be dissolved. The usual solvent is water or an acid if the substance is insoluble in water. In the latter case, the substance is chemically converted into some water-soluble salt, for example:

$$CuO + H_2SO_4 \rightarrow CuSO_4 + H_2O$$

$$Fe(OH)_3 + 3HCl \rightarrow FeCl_3 + 3H_2O$$

$$CaCO_3 + 2HNO_3 \rightarrow Ca(NO_3)_2 + H_2O + CO_2\uparrow$$ **

In qualitative analysis, use is made only of reactions accompanied by *external effects*, i.e., easily recognizable changes which indicate that these reactions have actually taken place. The usual external effects are:

(a) changes in the colour of the solution;
(b) precipitation (or dissolution);
(c) evolution of gases.

Reactions accompanied by the formation of precipitates and changes in the colour of solutions are most commonly utilized.

In most cases, analysis of inorganic substances involves the use of aqueous solutions of salts, acids, or bases. These substances are known as electrolytes, i.e., in aqueous solutions they dissociate into ions. Therefore, wet reactions usually take place between simple or complex ions, and by using these reactions we detect the ions formed by elements and not the elements themselves***.

For example, chlorine is detected in HCl or in chloride solutions by the action of a solution of silver nitrate, $AgNO_3$. A characteristic curdy precipitate of AgCl is formed, and this is taken as an indication of the presence of chlorine:

$$HCl + AgNO_3 \rightarrow AgCl\downarrow + HNO_3$$

$$CaCl_2 + 2AgNO_3 \rightarrow 2AgCl\downarrow + Ca(NO_3)_2, \text{ etc.}$$

* For more detail concerning qualitative analysis by the grinding (trituration) method, see: Isakov, P. M., *Qualitative Chemical Analysis of Ores and Minerals by the Powder Grinding Method*, Gosgeoltekhizdat, Moscow, 1955 (in Russian); Hillebrand, W. F., G. E. F. Lundell, H. A. Bright, and J. I. Hoffman, *Applied Inorganic Analysis*, 2nd ed., J. Wiley and Sons, New York, 1953.

** An upward arrow indicates that the substance is given off as a gas; a downward arrow signifies precipitation.

*** Of course, when a particular ion has been identified, the presence of the corresponding element in the substance can be inferred.

In these equations all the salts, except the precipitated AgCl, are present in solution as ions. Accordingly, the last reaction may be written, for example, in the following form:

$$Ca^{2+} + 2Cl^- + 2Ag^+ + 2NO_3^- \rightarrow 2AgCl\downarrow + Ca^{2+} + 2NO_3^-$$

Since the Ca^{2+} and NO_3^- ions do not participate in the reaction, they may be omitted from the equation. So we have:

$$2Cl^- + 2Ag^+ \rightarrow 2AgCl\downarrow$$

or (after dividing through by 2):

$$Cl^- + Ag^+ \rightarrow AgCl\downarrow$$

This is the **ionic equation** for the reaction. The same ionic equation is obtained for the reaction between solutions of any other chloride and $AgNO_3$.

The ionic equation given above shows that the reaction consists essentially of the interaction of the Ag^+ and Cl^- ions present in solution, which form a precipitate of AgCl. If chlorine were present not in the form of Cl^- ions but as some other ions (such as chlorate, ClO_3^-) or undissociated molecules of non-electrolytes (for example, chloroform, $CHCl_3$), this reaction would have been impossible. Indeed, although compounds such as $KClO_3$ or $CHCl_3$ contain the element chlorine, AgCl is not precipitated by the action of silver nitrate. It is therefore evident that when using this reaction we detect the Cl^- ion and not the element chlorine. In the same way, the reagent for the Cl^- ion is not $AgNO_3$ but the Ag^+ ion present in solution. One can therefore replace $AgNO_3$ with Ag_2SO_4, CH_3COOAg, etc.:

$$2NaCl + Ag_2SO_4 \rightarrow 2AgCl\downarrow + Na_2SO_4$$

$$CaCl_2 + 2CH_3COOAg \rightarrow 2AgCl\downarrow + (CH_3COO)_2Ca$$

If an element forms ions which are in different oxidation states, each ion gives its own characteristic reactions. For example, the Fe^{3+} ion forms a reddish-brown precipitate of $Fe(OH)_3$ under the action of alkalis (i.e., OH^- ions), whereas the Fe^{2+} ion gives a dingy-green precipitate of $Fe(OH)_2$ with alkalis*. In the same way, Cu^{2+} and Cu^+, MnO_4^- and MnO_4^{2-}, etc., give different reactions.

Hence, the oxidation state of an element in the substance being analysed is usually established by qualitative analysis.

Since not elements but the ions formed by them are detected in wet reactions, it is possible in analysis of individual substances to establish their formulas even during qualitative tests. For instance, if we identify Na^+ and Cl^- ions by the appropriate reactions and find no other

---

* $Fe(OH)_2$ is white in colour. A dingy-green colour is produced due to the formation of $Fe(OH)_3$ when Fe^{2+} is partially oxidized to Fe^{3+} by the oxygen of the air.

ions in the unknown substance, we can obviously conclude that the substance is sodium chloride, NaCl. Similarly, if we detect only Fe^{3+} and SO_4^{2-} or Fe^{2+} and SO_4^{2-} ions, we may state that the formula of the substance is $Fe_2(SO_4)_3$ or $FeSO_4$, respectively, etc. On the other hand, quantitative as well as qualitative analysis is needed to establish even the simplest (empirical) formulas of organic compounds since they do not dissociate into ions.* Thus, compounds that undergo electrolytic dissociation are much easier to analyse than those which do not dissociate.

Electrolytic dissociation or ionization is important to analytical chemistry for another reason also. The number of the most important cations is relatively small, about 25. The number of the best known anions is almost about the same. The number of salts (neutral) which they form, however, exceeds 600. Since in solution all these salts are found to dissociate into ions to be detected in analysis, it will suffice to know the reactions of only 50 ions in order to identify any of these salts. In the case of un-ionized organic compounds, ionic reactions cannot be used, and therefore each such substance has to be identified from all its physical and chemical properties, which is a much more difficult undertaking.

1.3. Reaction Conditions. Sensitivity, Selectivity, and Specificity of Reactions

Any particular analytical reaction must be conducted under strictly definite conditions, otherwise the result of the reaction will be unreliable.

For example, precipitates which are soluble in acids cannot evidently form if the solution contains an excess of free acid. Analogously, precipitates soluble in alkalis cannot be formed in alkaline solution. If a given precipitate is soluble in both acids and alkalis, it can be obtained only in a neutral solution, etc.

It is clear from the above examples that one of the most important conditions for a reaction is a suitable medium which may be adjusted as required by addition of acid or alkali to the solution.

Another important condition is the *temperature* of the solution. If the solubility of a precipitate increases sharply with temperature, it will not separate from a hot solution; the reactions must be carried out at room temperature or sometimes the solution must even be cooled. Certain reactions proceed only on heating.

A very important condition for the occurrence of reactions is a *sufficiently high concentration* of a particular ion in solution; at very low concentrations, the reactions fail to proceed. The cause is not

* To establish the molecular formula, it is also necessary to determine the molecular weight of the substance.

difficult to see: any substance can be precipitated only when it is formed in solution at a concentration in excess of its solubility under given conditions. If the solubility of the substance is very low, it settles down even at a very small concentration of the ion to be detected; the corresponding reaction is said to be *sensitive*. Conversely, if the solubility of the compound formed is considerable, the sensitivity of the reaction is low and the reaction can be accomplished successfully only if the concentration of the ion to be detected is relatively high.

It is evident that we may extend similar reasoning to other types of reaction, not only to those involving the formation of precipitates. For example, the flame test (flame colouration reaction) for volatile sodium salts is quite sensitive since the flame is coloured even by extremely small amounts of such salts.

The **sensitivity** of a reaction is quantitatively characterized by two interrelated quantities—the detectable minimum and the dilution limit. The **detectable minimum** (m) is the least amount of a substance or ion, which can be detected by a given reaction under specified conditions.

Since this amount is very small, it is generally expressed in micrograms, i.e., in millionths of a gram, abbreviated to μg*: $1 \mu g = 0.000001$ gram $= 10^{-6}$ g.

The detectable minimum does not fully characterize the sensitivity of the reaction because not only the absolute amount but also the concentration in solution of the substance or ion to be detected is of importance. For this reason, the **dilution limit** is also commonly stated; this is the lowest concentration of the substance (or ion), at which it can be detected by a given reaction. The dilution limit is sometimes called the minimum concentration.

The dilution limit is expressed by the ratio $1/g$ (where g is the weight amount of solvent per 1 part by weight of the substance or ion to be detected).

To determine the sensitivity of a reaction experimentally, it is carried out repeatedly with solutions of gradually decreasing and accurately known concentrations of the test substance (or ion). These experiments are continued until the reaction no longer works; in this way it is possible to determine the lowest concentration at which the reaction is still reliable. If the concentration of the ion to be detected exceeds by far the dilution limit, the result of the reaction (for example, precipitation) is obtained very rapidly. As the concentration of the ion decreases, the rate of the reaction gradually falls, and at concentrations close to the dilution limit the reaction may give either a positive or a negative result. This is the region of *unreliable reaction*. As the concentration of the ion decreases further, the number of

---

* Formerly, the microgram was sometimes designated by the Greek letter γ (gamma).

cases when the reaction is negative increases and finally the reaction fails completely. The dilution limit is generally assumed to be the minimum concentration of the substance (or ion), at which half of the total number of tests give positive results. Knowing this concentration, one can easily calculate the value of the dilution limit and the detectable minimum.

Example. To determine the sensitivity of the reaction for the Ag^+ ion which proceeds according to the equation

$$2Ag^+ + CrO_4^{2-} \rightarrow Ag_2CrO_4 \downarrow$$

one litre of a solution of $AgNO_3$ containing 1 gram of Ag^+ (i.e., 1.57 g of $AgNO_3$) was prepared. It has been found that the reaction is positive when the solution is diluted 25-fold, but on further dilution it becomes unreliable. Find the detectable minimum and dilution limit for this reaction given that it is accomplished with a drop of solution 0.02 ml in volume.

Solution: The dilution limit is:

$$\frac{1}{g} = \frac{\text{mass of Ag}^+}{\text{mass of water}} = 1:25{,}000^*$$

The detectable minimum is found by proportion:

25,000 ml of solution contains 1 g Ag^+
0.02 ml of solution contains m g Ag^+

which gives

$$m = \frac{0.02}{25{,}000} = 0.0000008\ \text{g} = 0.8\ \mu\text{g}$$

Thus, the detectable minimum (expressed in micrograms) is related to the dilution limit by the expression

$$m = \frac{1}{g} V \cdot 10^6$$

where V is the volume of solution (in ml) taken for the reaction.

The reactions used to detect one and the same ion may differ greatly in sensitivity**; this is illustrated by the data of Table 1.1.

From Table 1.1 one can see that the reaction with $K_4[Fe(CN)_6]$ is the most sensitive since the amount of copper(II) detectable in solution is 1/50 of the amount detectable by the action of HCl and 1/10 of that identifiable by the use of NH_3.

---

* For highly dilute solutions the density of water may be assumed to be equal to unity.

** In qualitative analysis, use is generally made of reactions in which the detectable minimum does not exceed 50 micrograms and the dilution limit is 1 : 1000.

TABLE 1.1 **Sensitivities of Reactions* for Cu^{2+} Ions.**

Reagent	Compound formed	Reaction effect	Detectable minimum, μg	Dilution limit, l/g
HCl	$H[CuCl_3]$	Green solution	1	1 : 50,000
NH_3	$[Cu(NH_3)_4]Cl_2$	Blue solution	0.2	1 : 250,000
$K_4[Fe(CN)_6]$	$Cu_2[Fe(CN)_6]$	Brown precipitate	0.02	1 : 2,500,000

* The data in the table refer to tests carried out with drops of solution 0.05 ml in volume.

The sensitivity of a reaction may vary within wide limits, depending on a number of factors. For instance, it increases with increasing thickness of the solution layer used in the reaction; this enables detection of a slight turbidity or colour imperceptible in a thinner layer.

The reagent concentration is also of great importance: the sensitivity of the reaction generally increases (up to a certain limit) with increasing concentration of reagent. In many cases, the sensitivity rises with the reaction time, especially if non-electrolytes or weak electrolytes are involved. Moreover, as already pointed out, the important conditions for a reaction to be carried out are a suitable medium and temperature which also have a strong effect on the sensitivity of the reaction.

The sensitivity of reactions involving the formation of a precipitate can often be raised through addition to the solution of ethyl alcohol (ethanol) which lowers the solubility of inorganic compounds. The sensitivity can also be increased by shaking an aqueous reaction mixture with some organic liquid immiscible with water. The precipitate collects at the interface between the two liquid layers (it is "floated") and becomes thereby easily distinguishable.

By adding organic solvents, one can also enhance the sensitivity of reactions that involve no precipitation. For example, the addition of acetone, which is miscible with water, to an aqueous solution of a cobalt(II) salt in the presence of a thiocyanate ion can greatly increase the sensitivity of the reaction since this involves a decrease in the dissociation of the coloured complex ion $[Co(SCN)_4]^{2-}$:

$$[Co(SCN)_4]^{2-} \rightleftharpoons Co^{2+} + 4SCN^-$$

The equilibrium of this reaction is shifted to the left in the presence of acetone.

If this reaction is carried out by addition to the aqueous solution of an organic solvent immiscible with water, say isoamyl alcohol, the sensitivity increases for another reason too. The complex ion $[Co(SCN)_4]^{2-}$ is extracted into the organic solvent layer, and its blue colour becomes considerably more intense. Extraction processes play an important role in modern analytical chemistry since they permit increasing the sensitivity and selectivity of analytical methods.

In other cases, the sensitivity of a reaction can be raised by adding an extraneous ion. For example, a rather sensitive reaction for cobalt is the formation of a blue crystalline precipitate of $Co[Hg(SCN)_4]$. At a very low concentration of the Co^{2+} ion, however, no precipitate is formed. But if a small amount of Zn^{2+} ion is added to the solution, a pale blue precipitate begins to form immediately, consisting of $Zn[Hg(SCN)_4]$ and $Co[Hg(SCN)_4]$. This phenomenon is called **coprecipitation**; it will be discussed in greater detail in studying characteristic reactions for zinc(II) and cobalt(II). The phenomenon of coprecipitation can be used for detection and quantitative separation of traces of elements.

Highly sensitive reactions make it possible to detect very small quantities of the corresponding ion, thereby obviating the necessity of using large amounts of a substance or large volumes of solution. The importance of this circumstance is quite obvious.

It must be remembered, however, that the more sensitive the reaction the greater the danger of mistaking a slight contamination of the reagents with a given ion (traces) for the presence of the ion in the sample under study. To prevent this, it is necessary to confirm the purity of the reagents by a **blank experiment** (i.e., an experiment with reagents in the absence of the unknown substance). It is also obvious that cleanliness of the apparatus and carefulness in work are very essential, otherwise the results of the analysis will be vitiated.

Apart from the sensitivity, the **specificity** of a reaction is also very important in analysis.

A *specific reaction* for a given ion is one which enables its detection, under experimental conditions, in a mixture with other ions. An example is the detection of the NH_4^+ ion by the action of alkali on heating; this reaction is accompanied by evolution of ammonia which can easily be identified by its odour and other properties:

$$NH_4^+ + OH^- \rightarrow NH_3\uparrow + H_2O$$

Of the compounds generally encountered in analysis, only ammonium salts give ammonia under these conditions. The reaction with alkali is therefore specific of the NH_4^+ ion.

The number of known specific reactions, however, is not great. In analytical chemistry, one has to deal with reactions giving the same or similar effects with several and occasionally with many ions. Reactions that give a similar effect with only a limited number of ions are known as **selective**. The smaller the number of ions giving a positive test, the more selective is the reaction.

It is often found that the presence of certain extraneous ions, which do not react with a given reagent, nevertheless has an effect on the course of the reaction, reducing its sensitivity, altering the nature of the reaction products, etc. This influence is enhanced with increasing concentration of the extraneous ion, and at a certain **limiting ratio** of the concentrations of the two ions the reaction can no longer be used. For example, the microcrystalloscopic reaction for Pb^{2+} ions with KI, which involves the formation of characteristic PbI_2 crystals, is applicable only if the Cu^{2+} concentration in the unknown solution does not exceed the concentration of Pb^{2+} by more than 25 times. Hence, the limiting ratio is:

$$Pb^{2+} : Cu^{2+} = 1 : 25$$

In a number of cases, the interfering effects of extraneous ions in a reaction can be prevented by using **masking agents**, i.e., substances binding these ions into sufficiently stable complexes and thereby reducing considerably their concentration in solution. For example, the detection of Cd^{2+} ions as a characteristic bright yellow precipitate of CdS formed by the action of H_2S is interfered with by Cu^{2+} ions which give a black precipitate of CuS. If the reaction is carried out in the presence of KCN, the Cu^{2+} ion is bound into the complex ion $[Cu(CN)_4]^{3-}$ which does not react with hydrogen sulphide. In this way the Cu^{2+} is "masked" and does not interfere with detection of Cd^{2+}. As in the absence of Cu^{2+}, when H_2S is passed into the solution, a yellow precipitate of CdS is formed, indicating the presence of Cd^{2+}.

1.4. Fractional and Systematic Analysis

The application of specific reactions makes it possible to detect ions by the so-called **fractional method**, directly in separate portions of the unknown solution, without regard to the other ions present; the sequence in which individual ions are detected is unimportant.

If no sufficiently reliable specific reactions are available, and the specificity cannot be increased in any way, ions cannot be detected by means of the fractional method. In such cases, one has to plan a definite sequence of reactions for detection of individual ions, this being known as **systematic analysis**. Each given ion is detected only after all the ions which interfere with its detection (i.e., which react in a similar manner with the reagent used) have been detected and removed from solution.

To illustrate, let us consider the systematic analysis of a solution which is likely to contain Ca^{2+} and Ba^{2+} ions. The most sensitive reaction for the Ca^{2+} ion is the formation of a precipitate of calcium oxalate, CaC_2O_4, by the action of soluble salts of oxalic acid, $H_2C_2O_4$:

$$Ca^{2+} + C_2O_4^{2-} \rightarrow CaC_2O_4 \downarrow$$

This reaction is not specific of Ca^{2+} because a similar precipitate is also formed by Ba^{2+} with the reagent indicated. So one cannot start the analysis with detection of Ca^{2+} but must first find out whether the solution contains Ba^{2+}. The latter can be detected through the action of potassium chromate, K_2CrO_4. This results in the formation of a characteristic yellow precipitate of barium chromate:

$$Ba^{2+} + CrO_4^{2-} \rightarrow BaCrO_4 \downarrow$$

Since calcium chromate, $CaCrO_4$, is soluble in water, the presence of Ca^{2+} does not interfere with detection of Ba^{2+}; the Ba^{2+} ion can therefore be identified in a separate portion of the unknown solution. If no $BaCrO_4$ precipitate is formed and no Ba^{2+} ion is present, Ca^{2+} can be detected in another portion of the solution by the reaction with $(NH_4)_2C_2O_4$. Conversely, if Ba^{2+} is present, it must be completely removed from the solution before the test for Ca^{2+} is started. The Ba^{2+} ion can be picked up by the action of the same reagent, potassium chromate, the added amount of which must be such as to precipitate $BaCrO_4$ completely. To make certain that this aim has been achieved, a test is carried out. The precipitate formed is separated by centrifugation or filtration, and a small portion of the reagent is added to the centrifugate (filtrate). If no more precipitate settles down, there is no Ba^{2+} left in solution and Ca^{2+} can now be safely detected in it by the action of $(NH_4)_2C_2O_4$. In this case the formation of a white precipitate proves conclusively the presence of Ca^{2+}.

From what has just been said it becomes clear that in systematic analysis, apart from identification tests for individual ions, one has also to resort to **separation reactions** in order to separate them from one another. Separation reactions are mostly based on the solubility differences between analogous compounds of the ions being separated. An example is the process of separation of Ba^{2+} from Ca^{2+}, in which advantage is taken of the different solubilities of $BaCrO_4$ and $CaCrO_4$. Volatility differences may also be useful in some cases. For example, the NH_4^+ ion is separated from K^+, Na^+, and Mg^{2+} by evaporation of the solution, followed by ignition (calcination) of the dry residue. Ammonium salts are volatilized, while the non-volatile salts of potassium, sodium, and magnesium remain in the crucible used for the ignition.

An efficient method of separation is the extraction of ions from aqueous solutions with the aid of various organic solvents. At present this method finds wide application in analytical chemistry.

1.5. Classification of Cations into Analytical Groups. The Group Reagent. Classification of Cations and Mendeleyev's Periodic Law

In systematic analysis, ions are separated from complex mixtures into groups, advantage being taken of the same reaction of the ions within a group with a reagent which is called the **group reagent.**

A reagent used as a group reagent must meet the following requirements: (1) it must precipitate cations practically quantitatively (the cation concentration in solution after precipitation must not exceed 10^{-6} g-ion/litre); (2) the resulting precipitate must readily dissolve in acids so that the analysis can be continued; (3) the excess of reagent added must not interfere with detection of the ions left in solution.

These requirements are far from being satisfied by all reagents. For example, the sulphate ion fails to serve as a group reagent for cations of Group II (Ba^{2+}, Sr^{2+}, and Ca^{2+}) for two reasons: though SO_4^{2-} precipitates all the three cations in the form of $BaSO_4$, $SrSO_4$, and $CaSO_4$, the last-named compound is relatively well soluble in water, while the other two, $BaSO_4$ and $SrSO_4$, are practically insoluble in acids.

The phosphate ion cannot be used as a group reagent for cations of Group II either: though all the three phosphates ($BaHPO_4$, $SrHPO_4$, and $CaHPO_4$) are precipitated and are soluble in dilute hydrochloric and acetic acids, the excess of the phosphate ion interferes with the identification of sodium in the presence of one of the most sensitive reagents—uranyl acetate. For analogous reasons, one cannot make use of the oxalate ion as a group reagent for Group II cations. The best group reagent for Group II is the carbonate ion because it satisfies all the requirements indicated above: all the three cations are precipitated practically completely; the carbonates are readily soluble in acids, even in acetic acid; and, finally, an excess of the carbonate ion can be easily destroyed by the action of acid and is removed on boiling in the form of carbon dioxide:

$$BaCO_3\downarrow + 2CH_3COOH \rightarrow Ba^{2+} + 2CH_3COO^- + CO_2\uparrow + H_2O$$

That is why the carbonate ion is used in systematic analysis as a group reagent for separation of the cations of Group II from those of Group I (alkali metals and magnesium).

There exist several schemes for separation of cations into analytical groups. One of the most fitting and most commonly used is the classification based on the properties of sulphides. The separation by the so-called hydrogen-sulphide method is carried out as follows.

1. First, cations that give insoluble chlorides ($AgCl$, Hg_2Cl_2, and $PbCl_2$) are precipitated by addition of $2N$ HCl solution.

All the three cations indicated—silver(I), mercury(I), and lead(II)—form sulphides insoluble in water and in acids. One can however separate them more conveniently as the chlorides and examine the precipitates separately. Lead chloride is relatively well soluble in water. Therefore, part of Pb^{2+} remains in solution and is then precipitated as PbS by hydrogen sulphide. This group of cations is generally known as analytical **Group V** (the **silver group** or the **hydrochloric-acid group**).

The chlorides of the other cations dissolve well in water, with the exception of certain basic salts, for example, BiOCl and SbOCl. They can easily be removed from the precipitate by treating it 2 or 3 times with 2*N* HCl solution.

2. If after the separation of the precipitated chlorides the hydrochloric acid solution reacts with hydrogen sulphide at a definite acidity of the solution, this will result in the precipitation of cations whose sulphides are insoluble in dilute acids. To this category belong the cations of copper(II), cadmium(II), mercury(II), bismuth(III), lead(II), arsenic(III), arsenic(V), antimony(III), antimony(V), tin(II), and tin(IV). This group is known as **Group IV** or the **hydrogen-sulphide group.**

3. When the precipitated sulphides have been separated from the solution, the next group is separated through the action of the appropriate group reagent, namely, ammonium sulphide, $(NH_4)_2S$, in an ammoniacal medium. This brings down cations whose sulphides are soluble in acids but insoluble in water. To this group belong the cations of iron(II), iron(III)*, zinc(II), manganese(II), nickel(II), and cobalt(II). There will simultaneously precipitate the hydroxides of some cations, those of chromium(III), aluminium(III), titanium(IV), etc. The factors responsible for this phenomenon will be discussed in detail later in the book. This group is called **Group III** or the **ammonium-sulphide group.**

4. The separation of the precipitate of sulphides and hydroxides of Group III is followed by the precipitation of the cations of alkaline-earth metals by the action of $(NH_4)_2CO_3$ in an ammoniacal medium. This brings down the carbonates $BaCO_3$, $SrCO_3$, and $CaCO_3$. This group is known as **Group II** or the **ammonium-carbonate group.**

5. After the Group II carbonates have been separated, the solution still contains cations which, under the precipitation conditions, do not form water-insoluble carbonates. These are the cations of sodium, potassium, ammonium, and magnesium**. This group is classified as **Group I.** The cations of this group cannot be precipitated by any common reagent, i.e., they have no group reagent.

Apart from group reagents that precipitate ions as slightly soluble compounds, use is also made in some cases of group reagents that dissolve some of the constituents of the precipitate, leaving others undissolved. For example, from the precipitate of sulphides obtained by passing hydrogen sulphide into the acidified unknown solution, there is separated in systematic analysis a group of the ions of arsenic(III),

---

* When the cations of Group IV are precipitated by hydrogen sulphide, Fe(III) reduces to Fe(II).

** At a sufficient concentration of ammonium salts, magnesium carbonate does not precipitate. The ammonium ion is introduced into the solution together with the group reagents $(NH_4)_2S$ and $(NH_4)_2CO_3$. Therefore, the NH_4^+ ion is detected at the very beginning in a separate portion of solution.

arsenic(V), antimony(III), antimony(V), tin(II), and tin(IV), whose sulphides are soluble in $(NH_4)_2S_n$ (ammonium polysulphide) with the formation of special compounds—thiosalts (for example, $(NH_4)_3AsS_3$, $(NH_4)_3AsS_4$, etc.). Thus, ammonium polysylphide is a group reagent for this group (it is called the **arsenic subgroup** or the **ammonium-polysulphide subgroup**).

Note. It should be mentioned that copper(II) sulphide dissolves partly in ammonium polysulphide to give CuS_2^{2-}. If, instead of ammonium polysulphide use is made of sodium sulphide, Na_2S, mercury(II) will pass into solution as the HgS_2^{2-} anion. For more detail see Chapter 6.

Thus, the classification of ions in analytical chemistry is based on different solubilities of the compounds formed by them, which enables separation of some groups of ions from others. The general scheme for separating ions into five analytical groups is shown in Table 1.2.

TABLE 1.2. **Classification of Cations**

Sulphides soluble in water		Sulphides insoluble in water (or water-insoluble hydroxides are formed instead)		
Carbonates soluble in water	Carbonates insoluble in water	Sulphides (or hydroxides) soluble in dilute acids	Sulphides insoluble in dilute acids	
Group I	Group II	Group III	Group IV	Group V
K^+, Na^+, NH_4^+, Mg^{2+}, etc.*	Ba^{2+}, Sr^{2+}, Ca^{2+}, etc.	Al^{3+}, Cr^{3+}, Ti (IV), Fe^{3+}, Fe^{2+}, Mn^{2+}, Zn^{2+}, Co^{2+}, Ni^{2+}, etc.**	(a) Subgroup 1 (sulphides do not form thiosalts under the action of $(NH_4)_2S_n$): Cu^{2+}, Cd^{2+}, Bi^{3+}, Hg^{2+}, Pb^{2+}, etc.*** (b) Subgroup 2 (sulphides dissolve in $(NH_4)_2S_n$ to form thiosalts): As (V), As (III), Sb (V), Sb (III), Sn (IV), Sn (II), etc.	Chlorides insoluble in water and in dilute acids: Ag^+, Hg_2^{2+}, Pb^{2+}, etc.
No group reagent	Group reagent $(NH_4)_2CO_3$	Group reagent $(NH_4)_2S$	Group reagent H_2S in presence of HCl	Group reagent HCl

* Magnesium carbonate is soluble only in the presence of ammonium salts.
** The symbols Ti (IV), As (V), Sb (V), etc., indicate the oxidation state of the element regardless of the ionic form in which the element is present in solution.
*** As indicated above, copper is partially soluble in $(NH_4)_2S_n$ with the formation of CuS_2^{2-}

The use of group reagents offers considerable advantages since the involved analytical task is broken down into a number of simpler operations. If any of the groups is entirely absent, its group reagent gives no precipitate with the unknown solution. It is then pointless to carry out tests for the individual ions of this group, which saves a considerable amount of work, reagent, and time. Even if there are specific reactions for all ions, the systematic analytical scheme retains its value since it is more convenient and economic than fractional analysis of a substance of unknown composition. Conversely, if the composition of a substance is approximately known (as is often actually the case) and it is necessary only to establish the presence or absence of one or two impurities, fractional reactions are more advantageous.

The classification of cations adopted in this book is the one most widely used in the USSR. Groups are numbered differently in some countries. For example, cations that are precipitated by HCl are included in Group I; those precipitated by H_2S in Group II; Group III includes cations precipitated by $(NH_4)_2S$; Group IV consists of those brought down by $(NH_4)_2CO_3$; and Group V includes the cations K^+, Na^+, NH_4^+, and Mg^{2+}, which have no group reagent. This division of cations into groups corresponds to the sequence in which the groups are separated in systematic analysis. But the classification accepted in this textbook has a number of advantages: (1) the study of the material is facilitated by starting with alkali and alkaline-earth metals and then passing over to heavy metals; (2) the theoretical study of qualitative analysis is well coordinated with practical work; (3) the numbering is in better agreement with the numbering of the groups in Mendeleyev's periodic system.

Although the basis of the analytical classification of ions is purely practical at first sight, it is by no means accidental. Indeed, like all other properties, the solubility of the cation salts and hydroxides concerned, which lies at the basis of the analytical classification, is functionally tied up with the positions of the corresponding elements in the Periodic System. The relation becomes especially obvious if we examine the expanded form of the Periodic System (Table 1.3). In Table 1.3 the Roman numerals in the boxes indicate the oxidation states of the elements and the Arabic numbers signify their electronic structure (the number of electrons in the respective energy levels remaining after the atoms have lost their valency electrons).

With a few exceptions, the cations precipitated in the course of analysis as carbonates, hydroxides, and sulphides fit quite regularly into the Periodic System. It is precisely the cations forming slightly soluble sulphides that occupy the right-hand part of the table, starting with Group VI in the first half of the long periods and ending with the same group in the second half. Cations that do not form slightly soluble sulphides are in Groups I-V in the left-hand part of the table. The

TABLE 1.3. Analytical Groups of Cations and Mendeleyev's Periodic System of Elements

Periods \ Groups	I	II	III	IV	V	VI	VII	VIII	VIII	VIII	I	II	III	IV	V	VI	VII	O
1	I H																	He
2	I Li 2	II Be 2											B	C	N	O	F	Ne
3	I Na 8 2	II Mg 8 2	III Al 8 2											Si	P	S	Cl	Ar
4	I K 8 8 2	II Ca 8 8 2	III Sc 8 8 2	IV Ti 8 8 2	V V 8 8 2	III Cr 11 8 2	II Mn 13 8 2	II,III Fe 14 8 2	II Co 15 8 2	II Ni 16 8 2	I,II Cu 18 8 2	II Zn 18 8 2	III Ga 18 8 2	IV Ge 18 8 2	III,V As 18 8 2	VI Se 18 8 2	Br	Kr
5	I Rb 8 18 8 2	II Sr 8 18 8 2	III Y 8 18 8 2	IV Zr 8 18 8 2	V Nb 8 18 8 2	VI Mo 8 18 8 2	Tc	IV Ru 12 18 8 2	III Rh 14 18 8 2	II Pd 16 18 8 2	I Ag 18 18 8 2	II Cd 18 18 8 2	III In 18 18 8 2	II,IV Sn 18 18 8 2	III,V Sb 18 18 8 2	VI Te 18 18 8 2	I	Xe
6	I Cs 8 18 18 8 2	II Ba 8 18 18 8 2	III La 8 18 18 8 2	IV Hf 8 32 18 8 2	V Ta 8 32 18 8 2	VI W 8 32 18 8 2	VII Re 8 32 18 8 2	IV Os 12 32 18 8 2	III Ir 14 32 18 8 2	IV Pt 14 32 18 8 2	III Au 16 32 18 8 2	II,I Hg 18 32 18 8 2	I Tl 2 18 32 18 8 2	II Pb 2 18 32 18 8 2	III Bi 2 18 32 18 8 2	Po	At	Rn
7	Fr	II Ra 8 18 32 18 8 2	Ac	IV Th 8 18 32 18 8 2	Pa	VI U 8 18 32 18 8 2												
Pptd.	Not pptd	as carbonates	as hydroxides			as sulphides										Not precipitated		

- Elements forming Group I cations
- Elements forming Group II cations
- Elements forming Group III cations:
  - precipitated as hydroxides by $(NH_4)_2S$
  - precipitated as sulphides by $(NH_4)_2S$
- Elements forming cations of Groups IV and V
- Elements forming typical anions
- Inert gases

cations of the first and second analytical groups are in the same groups of the Periodic System. The cations of Group III precipitated by ammonium sulphide as the hydroxides occupy Groups III-V of the Periodic System in the left-hand side of the table. And, finally, Group III cations precipitated as sulphides by $(NH_4)_2S$ are mainly in the middle of the fourth long period.

From the electron shell structures of cations we see that the cations of analytical Groups I and II and those of Group III precipitated by $(NH_4)_2S$ as hydroxides have complete outer shells of two or eight electrons, like atoms of the inert gases. In contrast, the cations of analytical Groups III, IV, and V, which form sparingly soluble sulphides, have either complete outer shells of 18 electrons or incomplete shells transitional between 8-electron and 18-electron shells (in the middle of the long periods) or else they have $18 + 2$ electrons in the two outer layers.

Cations having incomplete outer shells and 8 electrons in the next lower level usually produce sulphides soluble in acids (for example, in HCl), i.e., they belong to analytical Group III. Cations with incomplete outer shells and 18 electrons in the next lower level form sulphides difficultly soluble in acids, i.e., they belong to analytical Groups V and IV.

Of course, there are some deviations from these regularities. But this is what could possibly be expected in view of the fact that the assignment of certain cations to their particular analytical groups is arbitrary and is occasionally dictated by rather fine peculiarities of precipitation revealed in the separation of the groups. For example, according to its position in the Periodic System, the Mg^{2+} cation should have been included in Group II; indeed, like $BaCO_3$, $SrCO_3$, and $CaCO_3$, magnesium carbonate is sparingly soluble in water; it dissolves however in solutions of ammonium salts and cannot be precipitated completely by $(NH_4)_2CO_3$. It is therefore preferable to separate Groups I and II in the presence of NH_4Cl so that the Mg^{2+} ion remains in solution with Group I. That is why it is assigned to Group I.

Another example is the Zn^{2+} cation. By its position in the Periodic System, this cation should be included in analytical Group IV rather than in Group III. Indeed, in distinction to all other cations of Group III, the Zn^{2+} cation is precipitated by hydrogen sulphide from moderately acid solution, like the cations of Groups IV and V. If the latter were precipitated at a hydrogen ion concentration equal to $0.01M$, the Zn^{2+} cation would have been included in Group IV instead of Group III. But, as the precipitation is effected in a strongly acid solution ($[H^+] \approx 0.3M$), the Zn^{2+} ion remains in solution with Group III. A number of other examples could be cited.

From the foregoing it follows that such deviations from regularity in the positions of individual cations in the Periodic System provide no ground for justifying the conclusion that there is no correlation between the Periodic System of the elements and the analytical classification of cations. Of course, the nature and character of this correlation has not yet been fully elucidated but its existence is undoubted. Formerly this correlation did not receive due attention. At present it is the subject of investigation by a number of analysts.*

It has been shown by N. I. Blok that the analytical classification of cations is intimately connected with the ionic potential, an energy characteristic, which is the ratio of the ionic charge z to the ionic radius r (z/r). The potentials of ions in the highest oxidation state are periodic functions of the atomic numbers of the elements. Ionic potentials are found to be closely associated with such important properties of ions and their compounds as the acidic or basic nature of oxides, the solubility of some compounds, their heat of dissolution, etc. And it is these properties that often lie at the basis of the analytical separation of cations.

K.B. Yatsimirsky maintains that another energy characteristic is also of great significance in this respect, namely, the ratio of the square of the charge of the ion to its radius (z^2/r). The dependence of the solubility of compounds on this ratio reveals itself quite distinctly in certain cases. The Periodic Law has made it possible for the chemical analyst to make use of the analogy in the properties of the elements in order to find new reactions and methods of determination of ions. For example, according to the position of the element germanium predicted by D. I. Mendeleyev, in Group IV of the Periodic System between silicon and tin, it may be concluded that the analytical reactions of germanium ions must show a similarity with the reactions of silicon and tin ions in the same oxidation state. Indeed, like Sn(IV), Ge(IV) is precipitated by hydrogen sulphide in acid solution as the sulphide GeS_2 which dissolves, like SnS_2, in solutions of NaOH, Na_2S, and other reagents to form thiosalts. Therefore, like Sn(IV), Ge(IV) must be assigned to the analytical arsenic subgroup. On the other hand, the anion GeO_3^{2-} formed by germanium(IV) could be expected to bear resemblance to the anion SiO_3^{2-}. Indeed, both these anions when reacted with ammonium molybdate, $(NH_4)_2MoO_4$, form the heteropolyacids $H_4[GeMo_{12}O_{40}]$ and $H_4[SnMo_{12}O_{40}]$, similar in structure and chemical properties. When these acids are reduced by the organic reagent benzidine, they give substances having a deep blue colour, this being used in analysis for identification of both ions.

A number of other examples could be quoted, illustrating a similar analogy in analytical reactions, based on the position of the elements in the Periodic System and used to develop new methods of analysis.

* Blok, N.I., *Qualitative Chemical Analysis*, Goskhimizdat, Moscow-Leningrad, 1952; Kreshkov, A. P., *Fundamentals of Analytical Chemistry*, vol. I, Khimiya, Moscow, 1970; Alimarin, I. P. and M.G., Tsyurupa, *A Hundred Years of the Periodic Law of Chemical Elements*, Nauka, Moscow, 1969; Yatsimirsky, K. B., *Zhurn. Analit. Khim.*, **24**, 1293 (1969) (in Russian).

1.6. Brief Historical Outline of Development of Analytical Chemistry

The origin and development of practical techniques of analytical chemistry called forth by the demands of the first chemical industries dates far back in antiquity. Originally, qualitative analysis consisted of the identification of natural compounds by their properties. Qualitative analysis originated as the art of assaying, which consisted in determining the purity of precious metals—gold and silver. The techniques were largely the same as those employed in the production of these metals.

As a science, analytical chemistry began to grow in the middle of the 17th century, when the British chemist Robert Boyle (1627-1691) introduced the concept of chemical elements as chemically indivisible constituents of complex substances, ultimately found in analysis. Boyle systematized all the previously known analytical reactions and suggested a number of new reactions, thereby laying the foundations of "wet" analysis. In particular, he was the first to use litmus and certain other vegetable dyes as indicators for detection of acids and bases.

An important advance in qualitative analysis was the separation of whole groups of metals (more precisely, their cations) from solution, introduced in the 18th century by the Swedish chemist and mineralogist T. O. Bergman (1735-1784). This was the origin of systematic analysis which has been used to the present day. Many other chemists, among whom special mention should be made of the German chemist K. R. Fresenius (1818-1897), worked in the 19th century on further development and improvement of the procedure.

Another method of qualitative analysis, microcrystalloscopic analysis, originated in Russia in the 18th century owing to the work of M. V. Lomonosov (1711-1765) and especially of T. E. Lowitz (1757-1804). The investigations of the Russian scientists in the field of utilization of crystal forms in analysis were brilliantly developed at the end of the 19th century in the classical works of the crystallographer E. S. Fedorov (1853-1919).

Apart from the microcrystalloscopic method, Lowitz made a number of important discoveries in the field of chemistry. The most significant is the discovery of adsorption phenomena which later found extensive application in the chromatographic method of analysis proposed by the Russian botanist M. S. Tsvett (1872-1919).

The founder of quantitative analysis as a scientific discipline is M. V. Lomonosov who was the first to use systematic weighing of the reacting substances and the products formed in chemical reactions. In 1756, Lomonosov experimentally confirmed the law of conservation of matter, which he had stated in general terms in 1748. Lomonosov's law is one of the most important laws of natural science and is the theoretical basis of quantitative analysis.

As a result of his experiments, Lomonosov was the first to find the true interpretation of the role of air in combustion and calcination* of metals, thereby refuting Boyle's view that the increase in the mass of a metal heated in air was due to the addition of "fire matter" to the metal (the so-called "phlogiston theory").

Subsequent development of the principal techniques of gas analysis and the isolation of a number of important gases (H_2, O_2, N_2, Cl_2) led to the overthrow of the phlogiston theory. A special credit in this respect goes to the French chemist Antoine Laurent Lavoisier (1743-1794) who gave birth to the "antiphlogiston chemistry" based on the oxygen theory of combustion. He was the first to determine the quantitative composition of air, water, and of a number of inorganic and organic substances.

A considerable contribution to the development of analytical chemistry in Russia was made by V. M. Severgin (1765-1826). Apart from analytical chemistry, he did a great deal of work on mineralogy and chemical technology. He is righteously considered to be the founder of chemical technology in this country. Severgin wrote a number of manuals on chemical analysis of various materials.

The application of methods of quantitative analysis led to the discovery, at the beginning of the 19th century, of stoichiometric laws (laws of constant composition, multiple proportions, and equivalents). The experimental verification of these laws due to the work of the English chemist John Dalton (1766-1844) marked the final enunciation of the atomic theory in chemistry. The advent of this theory stimulated further progress of quantitative analysis as it became necessary to determine more precisely the atomic weights of the elements. Much credit in this field is due to the famous Swedish chemist J. J. Berzelius (1779-1848) who set up quite accurate (for his time) values of the atomic weights of 45 elements, devised many new methods for quantitative determination, and improved earlier techniques. In particular, Berzelius developed a method for elementary analysis of organic compounds, which was subsequently refined by the German chemist Justus von Liebig (1803-1873) and other scientists. In 1824-1848, J. L. Gay-Lussac (1778-1850) worked up the titrimetric method of quantitative analysis, which was further extended in the middle of the 19th century.

In 1859, the German chemist R. W. Bunsen (1811-1899) and the German physicist G. R. Kirchhoff (1824-1887) developed the most remarkable physical method of qualitative analysis—spectrum analysis. Electrogravimetric analysis was also born in the 19th century.

The pioneering work on the electrolysis of a number of chemical compounds was carried out at the beginning of the 19th century by the

---

* Calcination is the process of conversion of metal into "metallic limes" (oxides) by ignition in the air.

Russian researcher V. V. Petrov (1761-1834) and the English chemist Humphrey Davy (1778-1829).

In 1871, there was published a classical manual on qualitative and quantitative analysis entitled *Analytical Chemistry*, written by N. A. Menshutkin (1842-1907), an eminent chemist and teacher. Up to 1931, the book had seen 16 editions of which 7 editions appeared after the death of the author. This book has been translated into German and English and exerted a great impact on the teaching of analytical chemistry in many countries. Menshutkin was the first to stress the necessity to relate the analytical properties of the elements to their position in the Periodic System.

The development of analytical chemistry during the second half of the 19th century and in the 20th century is associated with achievements in related branches of chemical science—inorganic, organic, and physical chemistry. First, mention must be made of the Periodic Law discovered by the Russian chemist D. I. Mendeleyev (1834-1907) and his Periodic System of the elements (1869) in which the elements are elegantly unified and a relationship is established between their properties and atomic weights. The Periodic System is one of the finest generalizations in chemistry and in all science; it still serves as the basis for understanding and further progress of chemistry.

For experimental confirmation of the Periodic System, it was necessary to determine more accurately the atomic weights of a number of elements, and this called for improvements in methods of quantitative analysis.

Nevertheless, the use of the Periodic System in analytical chemistry still remained inadequate until recently. As pointed out above, the analytical classification of ions was not correlated with the positions of the corresponding elements in the Periodic Table. Much attention has been devoted to this matter in recent years.

The rapid growth of organic synthesis had a great influence on the development of analytical chemistry. Some organic compounds proved to be valuable reagents for various ions; analysis is greatly facilitated and speeded up by the use of such reagents. A large number of organic compounds and the possibility of altering their composition and structure by a great variety of methods, thereby influencing their properties as analytical reagents, offer wide potentialities in analytical chemistry for discovery of highly sensitive and specific reagents for the various ions.

Organic reagents were first used as precipitants in analytical chemistry by the Russian scientists M. A. Il'insky (1856-1941) and L. A. Chugaev (1873-1922). Especially significant is the work of Chugaev who proposed in 1905 a reaction for the Ni^{2+} ion with dimethylglyoxime and set forth the problem of studying the analytical properties of compounds known as inner-complex salts. This problem still remains to be among the most topical in analytical chemistry.

Organic reagents are used most widely in spot tests, colorimetric analysis, and in photometry. The spot-test method was developed by the Russian chemist N. A. Tananaev (1878-1959) who applied it for fractional detection of ions.* Important studies into spot analysis were carried out by the Austrian chemist F. Feigl. (1891-).

On the basis of the spot-test method, Tananaev devised the "chipless" method of analysis of metals and alloys. In this method, the metal or alloy under study is dissolved by applying onto its surface a suitable solvent, such as HNO_3, following which the resultant drop of the salt solution is transferred to a strip of paper where a spot test is carried out.

The rapid advancement of physical chemistry at the end of the 19th and the beginning of the 20th century also greatly contributed to the further development of analytical chemistry. An especially significant role in the theory of analysis was played by the theory of electrolytic dissociation, enunciated by the Swedish scientist Svante Arrhenius (1859-1927) in 1887. It was used in conjunction with the law of mass action by the German chemist Wilhelm Ostwald (1853-1932) for substantiating theoretically a number of analytical reactions and experimental techniques, which had formerly been mainly empirical. Also important to analysis were the investigations of the German physical chemist Walther Nernst (1864-1941) who formulated the solubility product principle and worked up the theory of galvanic cells, and of the Russian scientist L. V. Pisarzhevsky (1874-1938) who revealed the nature of oxidation-reduction reactions as processes associated with electron transfer (1910-1914). Physical chemistry has also enriched analytical chemistry with the theory of oxidation-reduction potentials, which enables one to predict theoretically the direction of oxidation-reduction reactions, to calculate their equilibrium constants, to choose the most efficient oxidizing and reducing agents, and to solve a number of other very important problems.

Apart from valuable theoretical generalizations, the development of physical chemistry provided analysts with a number of new experimental methods and techniques. All these methods are of great significance and find extensive use in research and plant laboratories.

In many cases, physico-chemical methods of analysis have a number of valuable advantages over chemical methods, and therefore their use considerably increases the possibilities of qualitative and quantitative analysis. Chemical methods however still serve as a basis for the development of instrumental methods and therefore continue to play the leading role in analytical practice.

At the close of the 19th century, an essentially novel method was devised in Russia, mainly by N. S. Kurnakov (1860-1941) and his

* Tananaev, N. A., *Fractional Analysis*, Goskhimizdat, Moscow-Leningrad, 1950 (in Russian).

school. This method, known as physico-chemical analysis, consists of the determination of the chemical nature of complex systems (for example, alloys, solutions, and glasses) by plotting property-composition diagrams. In this way, one can find the most favourable conditions for chemical reactions, establish the composition of individual phases in heterogeneous systems, etc.

The demands of industry stimulated the growth of analytical chemistry even more than the progress of allied branches of science. The development of various industries called for rationalization of methods of chemical control of production, i.e., determination of the composition of raw materials, intermediates, and finished products. It is these industrial needs that had appeared to be the most important for the origin and advancement of analytical chemistry as a science. These problems, essential for production, are solved by a number of scientific research organizations, institutes, and plant laboratories. This has resulted in an unusually rapid growth of the amount of research work in the field of analytical chemistry.

Questions and Problems

1. In a microcrystalloscopic detection of the Ca^{2+} ion as CaC_2O_4, the test is positive even with 0.001 ml of 0.002*N* $CaCl_2$ solution.

Calculate: (a) the detectable minimum; and (b) the dilution limit for this reaction.

Answer: (a) 0.04 µg Ca^{2+}; (b) 1 : 25,000.

2. The dilution limit for the reaction for K^+ with a solution of $NaHC_4H_4O_6$ (the formation of a $KHC_4H_4O_6$ precipitate) is 1 : 1000.

What is the lowest concentration (normality) of a KCl solution, at which the K^+ ion can be identified by this reaction?

Answer: ca. 0.26*N*.

3. Solutions of calcium salts produce a white precipitate of $CaHPO_4$ when reacted with Na_2HPO_4.

Given that this precipitate is soluble in acids (including acetic acid) but is practically insoluble in alkalis and hot water, state whether this reaction can be carried out: (a) on heating; (b) in the presence of acids; and (c) in the presence of alkalis.

4. What conclusions can be drawn about the composition of an unknown solution if it does not give precipitates with HCl, H_2S, $(NH_4)_2S$, or with $(NH_4)_2CO_3$?

Chapter 2

Apparatus and Techniques

The amount of substance under study in semimicroanalysis is 1/20 of that used in macroanalysis (see Sec. 1.1). Thus, whereas the concentrations of the solution and reagents are the same in both cases, the volumes of precipitates formed in semimicroanalysis must also be 1/20 of those in the macro method. But analysis of such small quantities of precipitates obviously requires the use of special apparatus and techniques. For example, one cannot obtain such precipitates in the glassware employed in macroanalysis (test tubes, flasks, etc.) since the precipitate, spreading over the large surface area of the bottom, is difficult to distinguish. That is why in semimicroanalysis use is made of small conical test tubes (capacity 2-4 ml) in which even a small quantity of precipitate can be distinguished in the tip of the cone.

With small quantities of precipitates and small volumes of solutions, it is inconvenient to separate them by ordinary filtration which involves a large loss of precipitate and solution. In semimicroanalysis, centrifugation and special techniques of microfiltration are employed for the purpose.

Before the student starts studying the theory and practice of semimicroanalysis, he must become familiar with the apparatus and techniques used for carrying out the operations involved in this method.

2.1. Reagent Racks and Laboratory Equipment

Reagent Racks. Since the consumption of reagents in semimicro method is very low, they are stored in bottles with a capacity of 10-15 ml. This makes it possible to place the major portion of reagents needed for laboratory work on a small rack, owing to which the analysis can be performed directly at the bench.

The rack has several shelves with slots for bottles containing test solutions and reagents. Glassware, crucibles, filters, indicator papers, etc., are stored in the drawers of the bench.

Figure 2.1 shows a reagent bottle*. It is equipped with an ordinary medicine dropper or a pipette, i.e., a glass tubing with a drawn-out

* Bottles are made round-bottomed so that they could not be put on the bench.

end; a rubber cap is attached to the upper end of the tubing; the liquid is drawn into the dropper by squeezing the rubber cap between the fingers. The diameters of the bottle neck and the dropper tubing must be chosen so that the lower end of the rubber cap serves as a stopper for tightly closing the bottle.

Reagents are taken from bottles only with the aid of droppers. The bottles themselves should never be taken from the rack, otherwise the

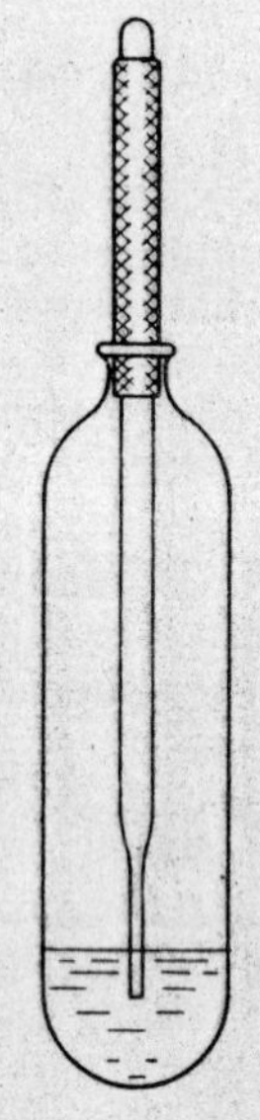

Fig. 2.1. Reagent bottle with a dropper.

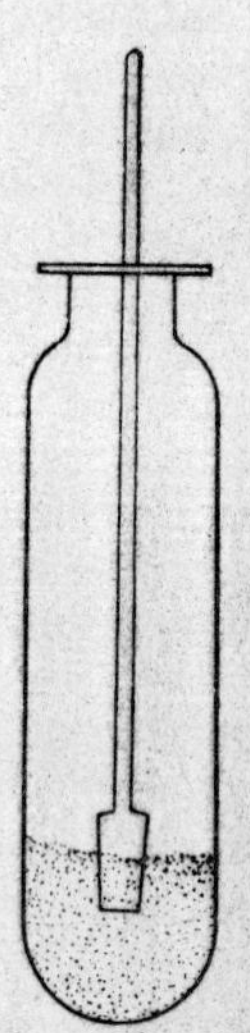

Fig. 2.2. Reagent bottle with a spatula.

order in which the bottles are arranged will be disturbed, and this will make it difficult to locate them. If the order remains unchanged, the analyst will become accustomed to it in a short time. Each rack must be supplied with a list of the test solutions and reagents, where they are numbered accordingly.

Bottles with solid reagents must be protected against the entrance of dust with a cardboard cover through which is passed a glass rod flattened at the end into a small spatula (Fig. 2.2). Reagents are taken from bottles with such a spatula.

All the bottles on the rack are fitted with labels containing formulas (or names) of the reagents. The labels must all be at the same height and covered with a layer of paraffin wax or Scotch tape to prevent damage to the inscription. When not in immediate use, the rack is covered with a special jacket to protect it from dust.

It is necessary to be careful so that the reagents are not contaminated during the work. To avoid this, the capillary tip of the dropper should never contact the inner sides of the vessel into which the reagent is poured. If this happens, the dropper must be rinsed thoroughly with distilled water before it is returned to the bottle. Droppers should never be put on the bench: a dropper may be either in the respective bottle or in the hand of the analyst.

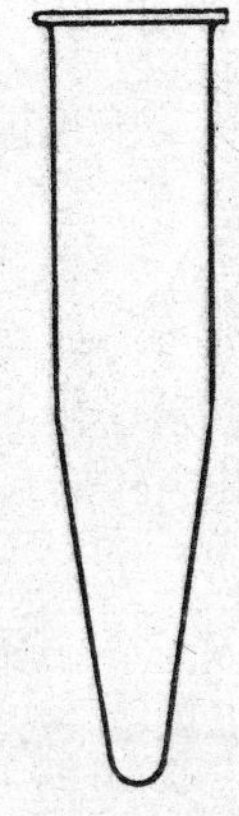

Fig. 2.3. Centrifuge tube.

The rack has about a hundred slots. This is quite enough for the number of reagents most commonly used in analysis. Reagents which are seldom used are kept in a special place. Bottles containing concentrated acids and ammoniacal solutions are stored in the hood, separately from the other reagents.

Laboratory Equipment. In semimicroanalysis, use is made of the following glass and porcelain vessels.

Test Tubes. Most operations in semimicroanalysis are carried out in conical or cylindrical test tubes having a capacity of 2-4 ml. Tubes with a cone-shaped bottom are most convenient for separation of precipitates from solutions by centrifuging (see below). That is why these tubes are called centrifuge tubes (Fig. 2.3). The particles of the precipitate collect in the point of the cone, forming a rather thick layer; its formation can easily be recognized, its colour conveniently identified, etc.

Tubes are placed into small racks which in the simplest case are wooden blocks with openings bored in them. Racks made of plastic materials are more convenient (Fig. 2.4).

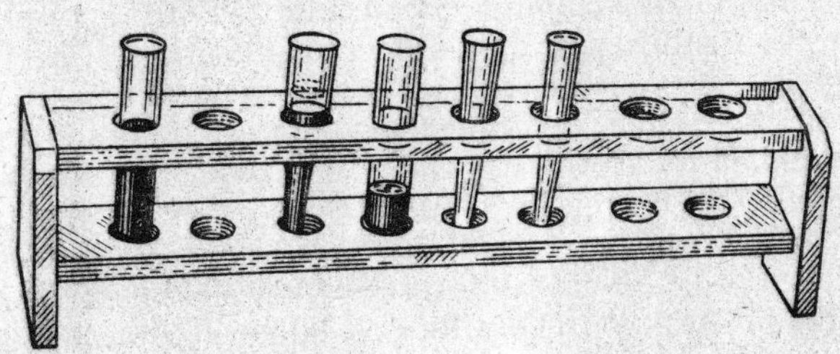

Fig. 2.4. Reagent rack.

Beakers. When relatively large volumes of solutions are used or when they are heated to boiling, use is made of small 10-ml beakers with a spout; in most cases they may be replaced with ordinary (macro) crucibles or with ordinary test tubes of a capacity of 10-15 ml.

Watch Glasses (Fig. 2.5). A watch glass is used for carrying out reactions. The "gas chamber" shown in Fig. 2.6 consists of two watch glasses in contact with each other; it is especially convenient when the reaction involves the evolution of a gas which is to be detected by the colour change of the reagent. In this case, a piece of filter paper moistened with the reagent is attached to the concave surface of the

Fig. 2.5. Watch glass.

Fig. 2.6. Gas chamber.

upper watch glass and the reaction mixture that evolves the gas to be detected is placed on the lower glass. Watch glasses must be dried for the gas chamber to operate reliably.

Drop Plates (*Spot Plates*). Spot plates are porcelain plates with indentations in which reactions are carried out (Fig. 2.7). They can be conveniently used for identification reactions for individual ions (with a drop of the test solution), which are accompanied by a colour change

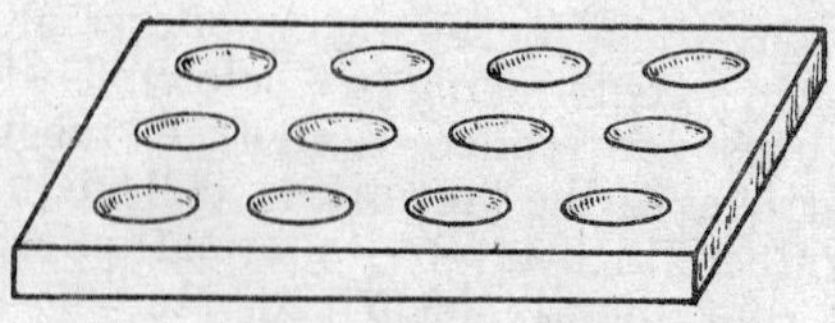

Fig. 2.7. Spot plate.

or by the formation of a coloured precipitate. A white spot plate is used to increase the visibility of dark-coloured precipitates; a black one is used to increase the visibility of light-coloured precipitates. Moreover, several reactions can be performed successively on a spot plate; it is washed more rapidly and easier than test tubes.

Dropping Pipettes. Pipettes similar to those in reagent bottles are used for taking samples of solution in the identification of ions, for separating the solution from the precipitate after centrifugation, for washing precipitates, etc. Therefore, a number of dropping pipettes must be in stock so that they are available for use. The capillary end of the pipette should be narrow and long enough so that it reaches the bottom of the centrifuge tube.

It is also necessary that a number of clean glass rods be available for stirring solutions in the course of the reaction.

Porcelain Equipment. Various types of porcelain vessel are employed to evaporate solutions, to ignite (calcine) precipitates, to fuse solid substances together, and other operations: small porcelain dishes (casseroles) (Fig. 2.8), ordinary (macro) crucibles (Fig. 2.9) or their covers, which are conveniently used instead of dishes, and, finally, micro-crucibles having a capacity of about 1 ml.

Washing of the Vessels. All the glass and porcelain vessels used in qualitative analysis must be scrupulously clean. The presence of

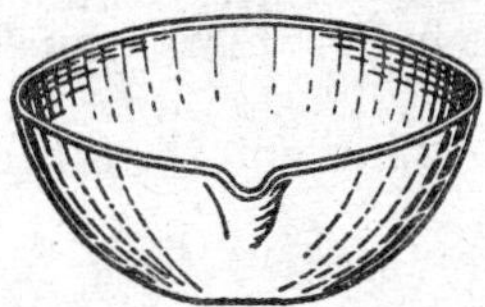

Fig. 2.8. Evaporating dish.

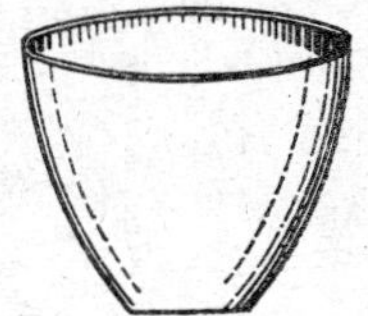

Fig. 2.9. Crucible.

even very small amounts of impurities often distorts the results of the reaction, which makes the student uncertain or leads to erroneous conclusions, and this in its turn entails waste of much time since the analysis has to be performed again. Therefore, very serious attention should be paid to the washing of all the glassware, porcelain vessels, droppers, and glass rods used for laboratory work.

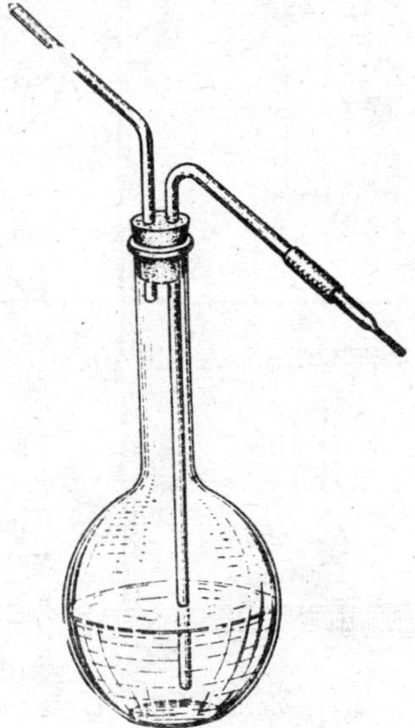

Fig. 2.10. Wash bottle.

Crucibles, test tubes, and other vessels should be emptied immediately upon the completion of the experiment. All solutions that are not needed should be poured out into a special vessel. The vessel is then put into a beaker with water and kept there until the washing is started; precipitates do not adhere to the glass and can easily be removed.

The vessels are first washed with tap water and cleaned with a test-tube brush of suitable size, after which the vessels are again washed several times and rinsed with distilled water. If close inspection reveals that the contaminants have not been removed completely, the vessel is filled with chromic acid mixture* (*Caution!*) and is allowed to stand for a few

* Chromic acid mixture is prepared in a *porcelain* beaker by mixing 16-20 g of technical-grade potassium bichromate, $K_2Cr_2O_7$, with 50 ml of water, after which concentrated sulphuric acid, H_2SO_4, is carefully added to bring the volume to about 100 ml.

minutes. When the chromic acid mixture is returned to its container, the vessel is washed several times repeatedly with running water and rinsed with distilled water.

Dropping pipettes are thoroughly washed after use. The glass tubing of pipettes is cleaned from time to time with chromic acid mixture.

Vessels should be washed at the bench. For this purpose each student is supplied with two wash bottles (Fig. 2.10), one with tap water and the other with distilled water.

Liquids are drained from vessels into a large glass beaker over which they are then washed.

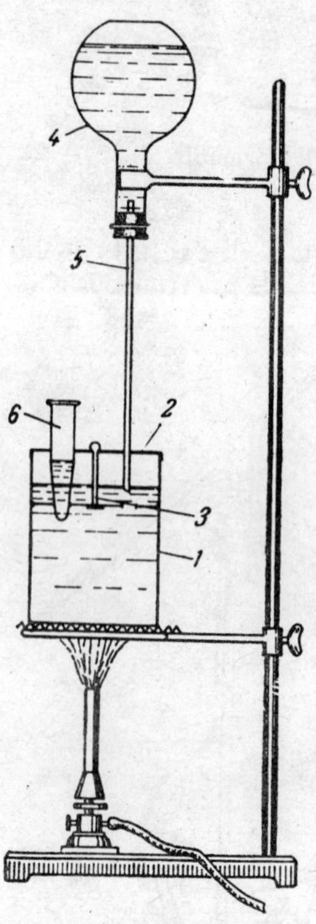

Fig. 2.11. Constant-level water bath:
1 — aluminium vessel; 2 — cover; 3 — insert; 4 — flask; 5 — glass tube; 6 — test tube.

2.2. Laboratory Techniques

Heating. As known, heating is required for carrying out many of the reactions used. Solutions contained in small test tubes should not be heated directly over a bare flame because the liquid may bump under the action of the vapour formed. Therefore, one must resort to the use of a water bath, i.e., a vessel with boiling water.

A water bath convenient for use is shown in Fig. 2.11. It is an aluminium vessel *1* of a capacity of 0.5 litre, fitted with a cover *2* and an insert *3* with openings for test tubes *6*. Since the level of water in the vessel is invisible, the bath is equipped with a special device for the water level to be kept constant. The device consists of an upturned flask *4* with water, into which is inserted (with the aid of a rubber stopper) a glass tubing *5* having a diameter of 11-12 mm.* The lower end of the tubing must be at the same height at which the water level in the bath is to be kept. On evaporation, the end of the tubing *5* is exposed and air gets into the flask, displacing from it an equal amount of water. Owing to this, the level of water in the bath is maintained constant until all the water in the flask is used up. When this happens, the flask must be filled with water again. The water bath is heated with a gas burner. Heating should be done so

* Water will not run out from the flask through too narrow a tubing; for water to flow out easily, the lower end of the tubing is beveled or widened into a small funnel.

that the water in the bath is slightly boiling. Strong boiling causes rapid evaporation of water. When solutions are heated in a water bath, they may be brought almost to boiling but they should not be boiled. If the solution is to be boiled, the reaction is conducted in a glass beaker or in a porcelain crucible by heating them over a gauze.

Evaporation. To concentrate solutions or to completely remove water from them, they are evaporated in porcelain dishes or crucibles. This operation is most safely carried out in a water or sand bath.

Evaporation can be speeded up by carrying it out over a gauze, but in such a case it must be continuously watched since:

(a) at the end of evaporation, the crystals formed are scattered by the vapour pressure, which leads to a loss and may distort the whole analysis;

(b) on further heating (ignition) of the dry residue after evaporation, many salts decompose to form oxides; the ignited oxides sometimes become insoluble in acids, which complicates the analysis considerably.

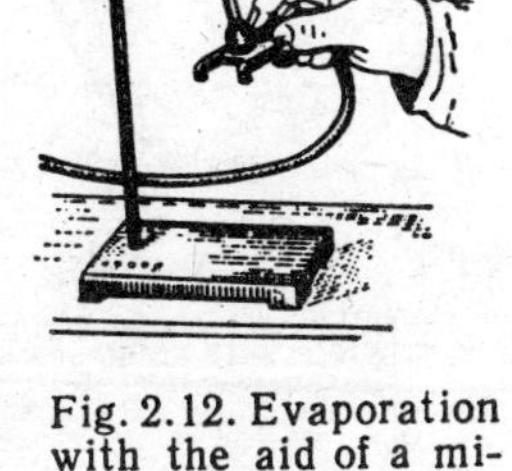

Fig. 2.12. Evaporation with the aid of a micro-burner.

Therefore, evaporation should be carried out with extreme care with the aid of a special micro-burner* (Fig. 2.12). The burner should be held in the hand and the flame moved all the time. The burner may also be put on the bench, not directly under the dish but beside it, at such a distance that the heat supplied through the gauze is enough for the liquid to boil gently.

In certain operations of semimicroanalysis (for example, when solutions containing H_2SO_4 and HNO_3 are to be evaporated), in order to avoid spilling out, heating can conveniently be carried out with the aid of an air bath (Fig. 2.13). This is a porcelain or nickel crucible *1* with a capacity of about 60 ml, on which is placed a copper or nichrome triangle *2* to support a small beaker or crucible *3* with the solution to be evaporated. By heating the bottom of crucible *1* with a burner, a sufficiently high and constant temperature of the air can be maintained inside the crucible.

Precipitation. This operation is most often carried out in centrifuge tubes. The required number of drops of the test solution are measured into the test tube and, if necessary, the tube is heated in a water bath, following which the specified number of drops of the reagent are added

* If no micro-burner is available, use may be made of an ordinary gas burner provided that its flame is made low.

with a pipette. Since solutions mix poorly in small test tubes employed in semimicroanalysis, it is necessary to stir the contents of the tube with a glass rod immediately after the solutions are mixed. It is still better to stir the contents after addition of each drop of the reagent. This provides a slower growth of crystals of the precipitate, which grow larger, are better separated from the solution upon centrifuging, and adsorb a lesser amount of foreign substances from the solution since they have a smaller surface area.

If the precipitate is likely to form a colloidal solution, it is useful to heat the contents of the tube in the water bath again for several minutes, after the reagent is added, in order to provide a better coagulation.*

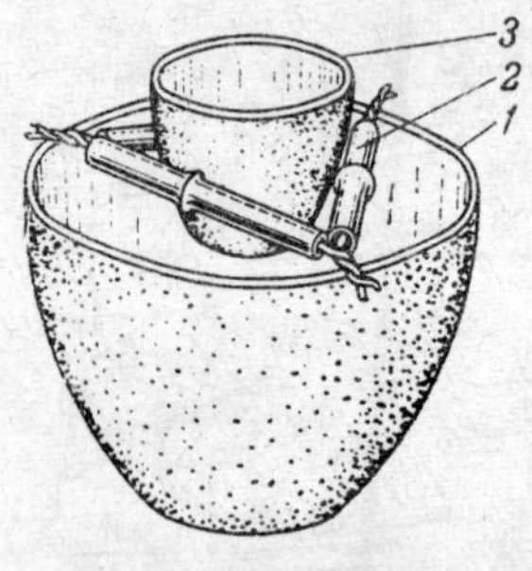

Fig. 2.13. Air bath:
1 — crucible; 2 — triangle; 3 — crucible with solution to be evaporated.

When the ions are separated by precipitation, the completeness of precipitation must necessarily be checked up. The test for the completeness of precipitation is ordinarily carried out as follows. The precipitate formed in the reaction is centrifuged (see below), and another drop of the precipitant is added to the clear supernatant liquid. If the solution remains clear, it means that the ion being separated has been precipitated completely. Otherwise, some more drops of the reagent are added to the solution, the latter is stirred, heated and, if required, centrifuged, following which the test for completeness of precipitation is repeated.

Centrifugation. While in macroanalysis the precipitate is separated from the solution by filtration, in semimicroanalysis use is most often made of centrifugation. This operation is based on the acceleration of the settling down of the particles of the solid phase under the action of the centrifugal force developed by rapid rotation of the centrifuge.

The centrifugal force throws the particles of the precipitate to the bottom of the test tube, as a result of which it quickly packs into a compact mass in the conical tip of the tube, and the solution above it (**centrifugate**) becomes clear.

Centrifuges are driven either manually or electrically. A manually operated centrifuge is shown in Fig. 2.14. The test tube with the mixture to be separated is placed in the cup attached to one arm of the centrifuge. To counterbalance, another test tube of the same size, containing approximately as much water as there is solution in the first tube, is placed in the opposite cup. Balancing the two arms in this way prevents a violent vibration which would ultimately destroy the centrifuge.

---

* For colloidal solutions and their coagulation, see Sec. 5.3.

The centrifuge is started slowly and then brought up gradually to the desired speed. After 1-1.5 min the handle is taken off and the centrifuge is allowed to stop on its own. By no means should the centrifuge be started with a jerk and stopped by the hand. If these precautions are not observed, the centrifuge may become faulty. The centrifuge must be well lubricated and rotate without knocking and rattling.

An electrically operated centrifuge (Fig. 2.15) rotates at a higher speed and develops a greater centrifugal force than a hand-operated centrifuge.

An electrical centrifuge is expensive and requires skilled and careful handling. The student must make certain that the tubes *1* placed in the two opposite cups *2* of the centrifuge are practically of the same mass. This is attained by weighing the tubes on a chemical balance and adding water by drops to the counterbalancing tube until it is

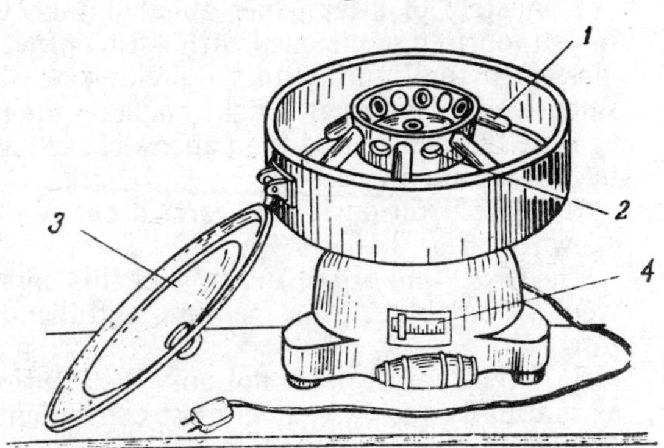

Fig. 2.14. Manual centrifuge.

Fig. 2.15. Electrical centrifuge
1 – tubes; *2* – cups; *3* – cover; *4* – rheostat handle.

of the same weight as the tube with the test mixture.

The current is switched on only after the centrifuge cover *3* is locked. When the centrifuge is started, the handle of the rheostat *4* is moved gradually and in 30 sec is returned in the same way to its initial position. When the centrifuge stops, the cover is taken off and the tubes are removed.

If anything is wrong during the operation of the centrifuge (unusual sound, violent vibration, or heating of the centrifuge), it should be immediately shut off and the matter reported to the laboratory instructor.

Transfer of Centrifugate. After the centrifugation is completed, the supernatant liquid or supernate (centrifugate) is removed by means of a pipette: the rubber bulb of the dropper is collapsed between the thumb and fingers and the capillary tip is immersed well beneath the surface of the liquid (be careful not to disturb the precipitate on the bottom of the tube). As the bulb is gradually released, the dropper fills with the liquid. Then the dropper is removed from the centrifuge tube, the liquid transferred to a clean tube, and the process repeated. If there is very little liquid in the centrifuge tube, it can conveniently be removed with a dropper tube (without the rubber bulb). The centrifuge tube is held in an inclined position and the capillary tip is carefully introduced into the liquid which is being drawn off due to capillarity. With a certain skill, one can remove almost all the liquid from the centrifuge tube without dispersing the precipitate. If the centrifugate turns out to be not quite clear, it is centrifuged again and the liquid is drawn off with a pipette.

Filtration. If centrifugation fails to be effective, which happens very seldom, filtration is resorted to. Filtration can be performed in different ways.

1. A strip of filter paper about 0.5 cm wide and 1-1.5 cm long is folded fourfold, moistened with a drop of water, and placed on a watch glass near the liquid with the precipitate. The tip of the pipette with smooth edges is put at a right angle on the paper strip, pressed lightly against the paper, and the paper strip is moved into the liquid which is drawn into the pipette.

2. Microfiltration is also carried out with the aid of the apparatus shown in Fig. 2.16.

The lower end of the tubing *2* of this apparatus is filled with cotton wool (or any other filter medium) and the air is sucked off through tubing *5**.

3. If the analyst needs not only the solution but also the precipitate, as is usually the case, it is most convenient to carry out filtration in the apparatus shown in Fig. 2.17.

Into a conical flask *1* with a side arm *2* is inserted, through a hole in a bung, a small funnel *3* with a filter *6* having a diameter of 2-3 cm. The funnel tubing gets into the opening of a centrifuge tube *4* which serves for collecting the filtrate (the filtered-off liquid). When the liquid to be filtered is transferred together with the precipitate to the filter, the air is drawn off from the flask through a rubber tubing *5*. This creates a vacuum in the flask, the filtration is speeded up, and the solution is sucked off well. To release part of the solution soaked up

* Suction is best done with mouth. If the solution being filtered evolve any harmful or objectionably smelling gas (e.g., H_2S), use is made of a water pump, the tubing *5* being connected with it. The stream of water should be very weak, otherwise the filter may be damaged.

by the filter, the latter is moistened with 3-4 drops of water and the solution is sucked off again.

Before the filter is inserted into the funnel, it is folded fourfold, after which one layer of paper is separated from the other three (Fig. 2.18). The resultant cone is put into the funnel, moistened with water, and pressed tightly against the glass with the finger so that no air is allowed to pass between the funnel and the paper. The upper edge of the filter must be below the edge of the funnel.

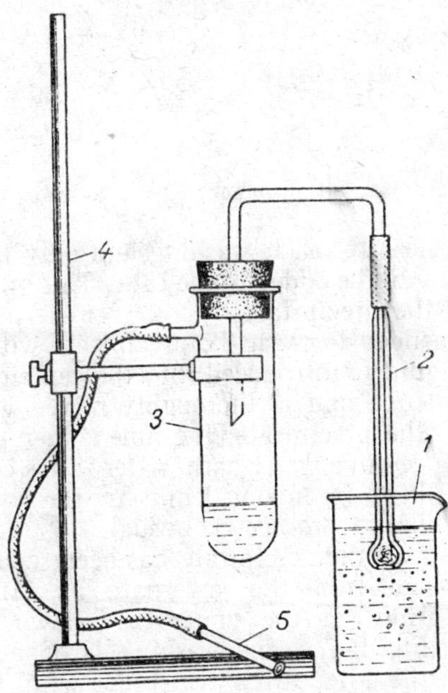

Fig. 2.16. Apparatus for microfiltration:
1 – beaker; 2 – tubing with filtering medium; 3 – test tube; 4 – stand; 5 – suction tubing.

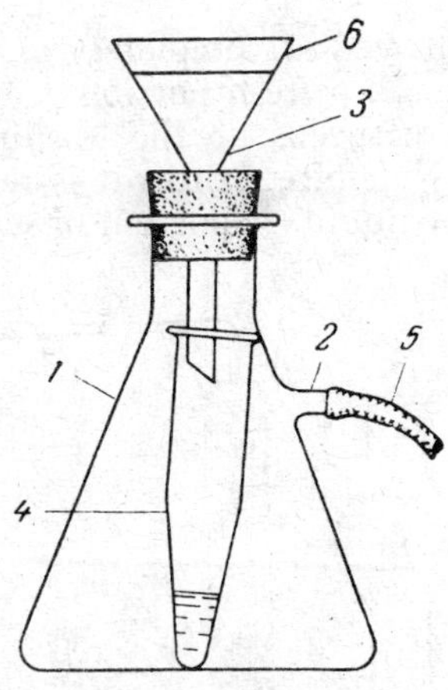

Fig. 2.17. Apparatus for filtration:
1 – flask; 2 – side arm; 3 – funnel; 4 – centrifuge tube; 5 – rubber tubing; 6 – filter.

Washing Precipitates. After the supernatant liquid has been removed, the precipitate is still wet with it and contains the ions that belong to the solution. For a complete separation to be achieved, the precipitate must be washed; it is usually washed in the same tube in which precipitation has been carried out. For this purpose, to the precipitate separated from the supernate is added 1-1.5 ml of distilled water (or any other wash liquid) and the solution is thoroughly mixed with a glass rod (otherwise the washing is hardly effective).

Then the solid is centrifuged to the bottom of the tube again and the supernate removed with a pipette and rejected. Often, even after a single washing, the amount of impurities left is found to be too small to interfere with tests to be made in the subsequent analysis.

In some cases, even very small quantities of impurity may be found to be interfering with the reactions to be carried out. In such cases, the washing is repeated several times. The supernatant liquid should be

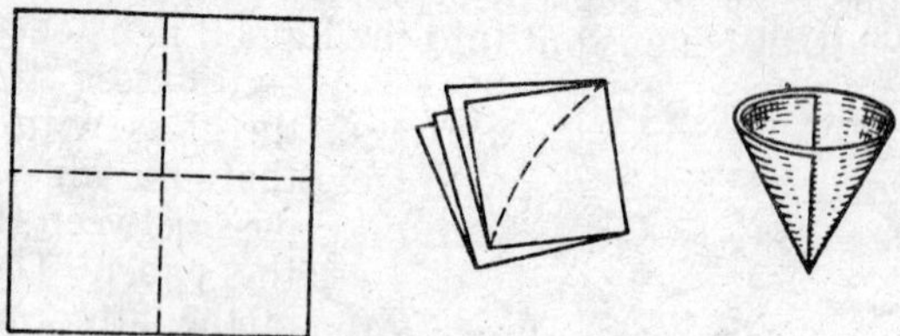

Fig. 2.18. Folding a filter paper.

removed as completely as possible after each washing and only then should a fresh portion of wash liquid be added to the tube. This procedure speeds up the washing of the precipitate.

Not infrequently it is recommended to wash the precipitate with a hot liquid. A portion of wash liquid is introduced into the centrifuge tube and is thoroughly mixed with the precipitate; the tube is then placed in a water bath. After it has been kept for about 1 min in the water bath, proceed as usual.

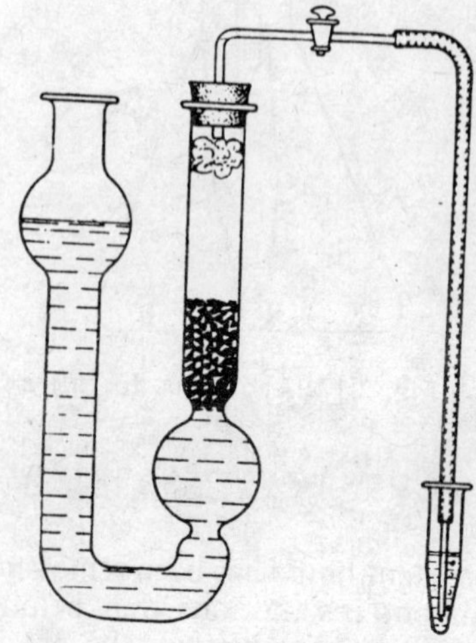

Fig. 2.19. Apparatus for preparing hydrogen sulphide.

If the precipitate has been separated from the supernate by filtration (in the apparatus shown in Fig. 2.17), the washing is done on the filter after the tube with the filtrate has been withdrawn from the flask. The wash liquids may be collected directly in the flask. Before a fresh portion of wash liquid is poured onto the filter, the previous one should be removed as completely as possible.

The washing of precipitates, especially in semimicro qualitative analysis, should by no means be neglected. A complete separation of ions cannot be effected without washing. Because of the incomplete separation of ions, the results of many reactions are found to be obscure, the analysis is erroneous, and it has to be repeated.

Precipitation with Hydrogen Sulphide. A convenient device for preparation of hydrogen sulphide is shown in Fig. 2.19. It is a U-shaped tube with two bulbs. Into the constricted part of the left arm is inserted a gauze (a porcelain or copper plate with holes) on which pieces of ferrous sulphide, FeS, are placed.

Through the left arm of the apparatus is introduced dilute (1 : 1) technical-grade hydrochloric acid which reacts with FeS to form H_2S:

$$FeS + 2HCl \rightarrow FeCl_2 + H_2S\uparrow$$

If the stopcock of the gas-delivery tubing is shut, the evolving hydrogen sulphide will force the acid into the left arm of the tube and the

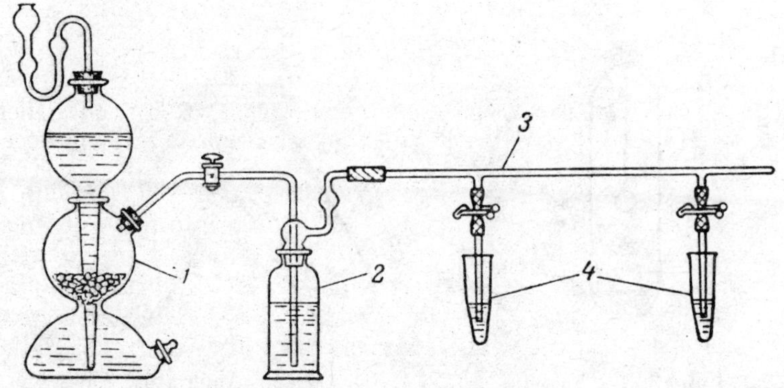

Fig. 2.20. Preparing hydrogen sulphide in Kipp generator and removal of the gas at several points:

1 – Kipp generator; *2* – wash bottle; *3* – gas delivery tube; *4* – test tubes with solution saturated with hydrogen sulphide.

reaction will be stopped. When the stopcock is opened, the reaction is resumed. To catch splashes, a piece of cotton wool or gauze is put into the upper portion of the right arm.

The rate of evolution of H_2S must be adjusted so that the bubbles of the gas passing through the test solution can be counted. Hydrogen sulphide should not be passed at a high rate because only part of it that has time to dissolve is effective.

Hydrogen sulphide is very poisonous! It must therefore be dealt with in a hood.

Hydrogen sulphide is introduced into the test solution through a glass tubing with a capillary tip which is pushed through the rubber tubing of the apparatus. Care must be exercised to see that the glass tubing is clean. It is best cleaned with hot concentrated HNO_3 (*Caution!*), after which it should be thoroughly washed with water.

When the experiment is run by a large group of students, instead of the apparatus described above, a Kipp-type H_2S generator (Fig. 2.20) is used, which is fitted with a gas-delivery tube *3* with side branchings, making it possible to collect the gas at several points. For the gas current to be conveniently controlled, it is necessary that the internal diameter of the branchings is sufficiently small (semi-capillary tubings).

When working with H_2S generators, the student must be extremely careful and see that *upon the completion of the analysis the stopcock (or clamp) in the gas-delivery tube is tightly shut!*

Another method of preparing hydrogen sulphide, which is especially convenient under semimicroanalysis conditions, consists in heating a mixture of paraffin wax (5 parts), sulphur (3 parts), and asbestos fibre (1 part). About 1 gram of the mixture is put into an ordinary (macro)

Fig. 2.21. Vessel for detection of gases formed during the reaction.

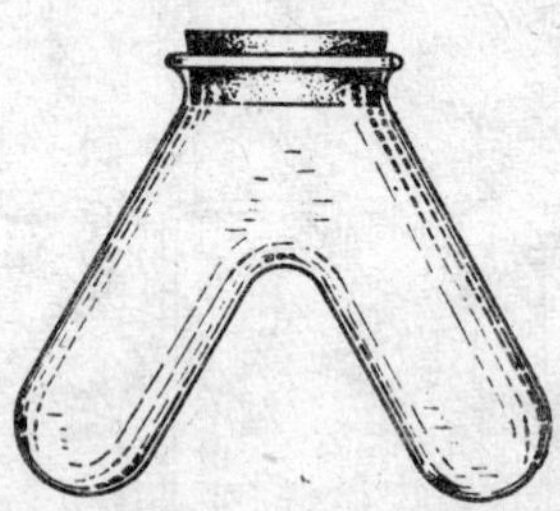

Fig. 2.22. Apparatus for detection of gases.

vessel fitted with a stopper, through which a gas-delivery tube is passed, secured in the stand clamp, and is heated gently. The evolving gas is bubbled through the test solution.* The reaction ceases to proceed as soon as heating is stopped. The rate of evolution of the gas is adjusted by changing the temperature. Only very small amounts of hydrogen sulphide gas escape into the air so that its disagreeable odour is not sensed even when no hood is used (this cannot be tolerated under the conditions of a study laboratory).

Work Involving Gases. To detect the gases evolved in the reaction, their interaction with appropriate reagents is observed. This is done in different ways. The simplest way is to place a drop of reagent in the loop of a copper or nichrome wire above the reacting mixture in the vessel (Fig. 2.21). Another way is to plug the vessel with a rubber bung with a pipette inserted into it; a drop of reagent is held in the capillary tip of the pipette.

---

* The hydrogen sulphide (H_2S) thus obtained is found to be contaminated with small amounts of CS_2 and paraffin vapours; these impurities do not interfere with the precipitation of sulphides.

Gases can be detected more conveniently by the use of the simple device shown in Fig. 2.22. The test solution and the reagent causing the evolution of gases are pipetted into one of the legs of the device. Into the other leg are put several drops of the reagent for detecting the gas, and then the apparatus is tightly stoppered. The evolving gas is allowed to get into contact with the reagent, causing the solution to change its colour (or to become turbid)..

In some cases, it is convenient to use the gas chamber described above (see Fig. 2.6) in order to detect the evolution of gases.

Other devices used for the same purpose will be described later in the book (see Figs. 8.1, 8.2, 8.3).

2.3. Reagents

The purity of reagents is of prime importance in semimicroanalysis. According to the degree of purity, the reagents sold are classified into six grades: (1) technical-grade (t.g.); (2) purified (pd); (3) pure (p); (4) analytically pure or pure-for-analysis (p.f.a.); (5) chemically pure (c.p.); (6) purissimum (puriss.). Purissimum reagents contain the least amount of impurity. For most analytical operations, use may be made of "pure-for-analysis" and "chemically pure" reagents.

The beginning student is frequently confused as to how much of reagent is to be used for carrying out a reaction; the amount taken is often too large and the reagent is wasted. As will be shown below, a large excess of reagent is often undesirable, causing the reaction to proceed in the unwanted direction.

To avoid such inefficient consumption and complications, solutions of definite normality are used in qualitative analysis.

The normality of a solution is defined as the number of gram-equivalents of solute per litre ol solution. Uninormal ($1N$) solutions contain 1 gram-equivalent of solute per 1 litre; seminormal solutions ($0.5N$) contain 0.5 gram-equivalent per litre of solution; decinormal solutions ($0.1N$) contain 0.1 gram-equivalent per litre, etc. In studying reactions and in performing analyses, use is most often made of $0.5N$ solutions of salts and $2N$ (dilute) solutions of acids and bases.

Let us recall that the *gram-equivalent of a substance is the number of grams of this substance chemically equivalent in a given reaction to 1 gram-atom (or gram-ion) of hydrogen.* For example, 1 gram-equivalent of HCl is equal to 1 gram-molecule (1 mole) of this substance, i.e., 36.5 g, since this amount of HCl contains 1 g-ion of hydrogen; the gram-equivalent of H_2SO_4 is equal to 1/2 mole (98 : 2 = 49 g); and that of H_3PO_4 to 1/3 mole (32.67 g).

The gram-equivalent of bases and salts is the quantity containing as many grams of metal as required to replace 1 g-ion of the hydrogen of the acid (or water).

Below are given the gram-equivalents of some substances:

KOH, NaOH, NaCl, KNO_3	1 mole
$Ca(OH)_2$, $Ba(OH)_2$, Na_2SO_4, $CuSO_4$	1/2 mole
$Al(OH)_3$, $FeCl_3$, $AlPO_4$, Na_3PO_4	1/3 mole
$Al_2(SO_4)_3$, $Ca_3(PO_4)_2$	1/6 mole

To find the gram-equivalent, we must divide the molecular weight (gram-molecule or mole): (1) of an acid by its basicity; (2) of a base by the charge on the metal ion (or by the number of hydroxyl radicals, OH); (3) of a salt by the number of ions of a metal and its charge.

It should however be noted that these rules cannot be applied without taking account of the reaction in which a given substance is taking part, since the gram-equivalent of a substance in different reactions may be different. For instance, as pointed out above, 1 g-eq of H_3PO_4 is equal to 1/3 mole; but this is valid only when in a particular reaction all the three hydrogen ions of ortho-phosphoric acid are replaced by the metal ions. In the reaction

$$H_3PO_4 + 2NaOH \rightleftharpoons Na_2HPO_4 + 2H_2O$$

ortho-phosphoric acid behaves as a dibasic acid, while in the reaction

$$H_3PO_4 + NaOH \rightleftharpoons NaH_2PO_4 + H_2O$$

it behaves as a monobasic acid. Therefore, 1 gram-equivalent of H_3PO_4 in the former reaction is equal to 1/2 of the mole, and in the latter, to one mole.

The gram-equivalents of oxidizing and reducing agents are calculated in a different way (see Sec. 5.13).

Gram-equivalents represent the quantities of substances needed for their interaction. Thus, any reaction consumes one and the same number of gram-equivalents of the two reacting substances involved. To neutralize 1 g-eq of HCl we need 1 g-eq of alkali; the precipitation of the Ag^+ ion from a solution containing 0.0015 g-eq of $AgNO_3$ requires exactly the same number of gram-equivalents of any soluble chloride, etc. Since equal volumes of solutions of the same normality contain the same number of gram-equivalents of the corresponding substances, for the reaction to take place between them we must take equal volumes of these substances, say, measure out the same number of drops.

If the solutions have different normalities, then the volume of a solution of higher normality must be proportionately smaller. It follows that the volumes of the solutions taken for a particular reaction are inversely proportional to their normalities. If we denote the volume and normality of one solution by V_1 and N_1, and those of the other by V_2 and N_2, we can write:

$$\frac{V_1}{V_2} = \frac{N_2}{N_1} \quad \text{or} \quad V_1 N_1 = V_2 N_2$$

In other words, the products of the volumes of solutions by their normalities are equal for both reacting substances.

Using these rules, one can easily calculate the volume of reagent needed for carrying out a given reaction. As an example, let us figure out how many drops of 0.1*N* $AgNO_3$ solution are required for precipitation of the Cl^- ions from 3 drops of 0.5*N* $CaCl_2$ solution. According to the above rule we have:

$$x \cdot 0.1 = 3 \cdot 05$$

which solves to

$$x = \frac{3 \cdot 0.5}{0.1} = 15 \text{ drops}$$

Since solutions of definite normality are convenient for use in semimicroanalysis work, we will describe two examples to illustrate the procedure of preparing such solutions.

Example 1. Prepare 250 ml of 0.5*N* solution of barium chloride. Barium chloride crystallizes with two molecules of water, i.e., has the formula $BaCl_2 \cdot 2H_2O$. Hence, its molecular weight is 244.3 and its gram-equivalent is 122.2 g. Thus, 1 litre of 0.5*N* solution contains 0.5×122.2 g, and 250 ml (i.e., 0.25 litre) contains $0.5 \times 122.2 \times \times 0.25 = 15.3$ g of $BaCl_2 \cdot 2H_2O$. Weigh out this amount of the salt and dissolve it in a small volume of hot water. Dilute the solution on cooling in a measuring cylinder or in a measuring flask (i.e., in a flask on whose long narrow neck a mark is made, indicating the level to which the flask should be filled) to bring up the volume to 250 ml.

Example 2. Prepare 1 litre of an approximately 2*N* solution of HCl from a concentrated acid of density 1.19 g/cm³.

The gram-equivalent of HCl is equal to its gram-molecular weight, i.e., 36.5 g. So you are to take $36.5 \times 2 = 73$ g HCl. Figure out what volume of concentrated hydrochloric acid will contain this number of grams of hydrogen chloride. Tables showing the densities of solutions of the most important acids and bases can be found in handbooks of chemistry. Using these tables, one can determine the percentage content of acids and bases from the corresponding densities. From tables we find that to the density 1.19 g/cm³ there corresponds 37.23 per cent HCl. We set up the proportion:

100 g of concentrated acid contains	37.23 g HCl
x g of concentrated acid contains	73 g HCl

which gives the required amount of concentrated acid

$$x = \frac{73 \cdot 100}{37.23} \approx 196 \text{ g}$$

Instead of weighing out such an amount of concentrated hydrochloric acid, it is more convenient to take its corresponding volume. As stated, each millilitre of concentrated hydrochloric acid weighs 1.19 g; evidently, the sought-for volume is equal to:

$$V = \frac{196}{1.19} \approx 165 \text{ ml}$$

Measure out the calculated volume of concentrated hydrochloric acid into a measuring cylinder (having a capacity of 200 ml), dilute it in the cylinder (or in a measuring flask) with water to make 1 litre, and mix the solution well.

In studying the questions relating to the theory of analysis, it is more convenient to use molar concentrations rather than normal concentrations of solutions. As known, *the molarity of a solution designated by M indicates the number of moles (i.e., gram-molecules) of solute per litre of solution.*

For example, 0.1 molar H_2SO_4 (also written 0.1*M* H_2SO_4) means a concentration corresponding to 0.1 mole of H_2SO_4 per litre of solution (i.e., to 9.8 g of H_2SO_4). Similarly, the concentrations of ions are expressed as the number of gram-ions per 1 litre of solution. If the concentration of CH_3COO^- ions is equal to 0.2 g-ion/litre, this means that one litre of solution contains $0.2 \times 59 = 11.8$ g of the ions.

2.4. Drop Reactions (Spot Tests)

The originator of spot analysis was N. A. Tananaev who, starting in 1920, devised a large number of spot tests, was the first to formulate the principle of fractional analysis by the spot-test method, and published the first book on this subject in which he generalized his work in the form of an original system of spot analysis.

About the same time, a systematic study of spot test methods was begun by the Austrian scientist F. Feigl.

In semimicro qualitative analysis, wide use is made of drop and microcrystalloscopic reactions. The student therefore must be acquainted with the specific features of these reactions and the procedures used for carrying them out.

In spot analysis, the reactions take place between drops of solutions. Solutions of reacting substances are taken in an amount from one to several drops. Spot tests are most often performed on filter paper; a coloured spot is formed on it, the colour of which points to the presence of a particular ion.

If the ion to be identified does not give a colour reaction, other types of reaction are used. In such cases, the tests are performed on spot plates which are occasionally employed for colour reactions as well.

The use of filter paper has a number of advantages; this is illustrated by the following example.

The Ni^{2+} ion is detected by the action of an organic reagent, dimethylglyoxime*. This reagent forms with Ni^{2+} a reddish precipitate of a complex salt.

The sensitivity of the reaction on a spot plate or in a test tube is characterized by the dilution limit 1 : 700,000; if the reaction is carried

* See Sections 5.11 and 5.23.

out on a filter paper, the dilution limit is 1 : 3,300,000, i.e., the sensitivity of the reaction increases almost by a factor of 5. This is accounted for by the fact that filter paper, being full of numerous very fine capillaries (through which the solutions spread), has a well-developed inner surface area and hence a high adsorptive power for dissolved substances.

This is easily confirmed by the following example. Place a drop of a dilute solution of an organic dye (e.g., ink) on a piece of filter paper and let it soak in. Note that all the dye is found to be concentrated in the centre of the moist spot surrounded by a colourless zone of water.

It is clear that, because of such adsorption of dissolved substances on the paper, their concentrations become considerably higher than those in the reaction of the same solutions taking place in a test tube or on a spot plate. So it is not surprising that the sensitivity of reactions carried out on filter paper is higher.

The advantages of using filter paper in drop reactions are more clearly seen when a separation process is required for detection of an ion. Suppose, for example, that a solution to be tested for Ni^{2+} contains Fe^{3+} which forms with NH_4OH (in the presence of which the reaction is carried out) a red-brown precipitate of $Fe(OH)_3$. The formation of this precipitate will evidently mask the red colour given by the Ni^{2+} ions and hence the Fe^{3+} ions must be separated.

To do this, a drop of a solution of Na_2HPO_4 (the first reagent applied to the paper is referred to as the **substrate** in spot analysis) is placed on a strip of filter paper. A drop of the unknown solution is then put into the centre of the moist spot formed. Spreading through the capillaries of the paper, the solution reacts with PO_4^{3-}, as a result of which Ni^{2+} and Fe^{3+} precipitate as sparingly soluble phosphates. The first to precipitate nearest to the centre of the spot will be the least soluble of the phosphates, ferric phosphate*, $FePO_4$. Nickel phosphate, being more soluble, will begin to precipitate only after the Fe^{3+} ions are completely precipitated with the solution moving in the direction of the periphery of the spot. Thus, the phosphate $Ni_3(PO_4)_2$ precipitates next to the phosphate $FePO_4$, nearer to the periphery of the spot. In this way, the Ni^{2+} ions are separated from the Fe^{3+} ions.

If we now treat the spot along its periphery with an ammoniacal solution of dimethylglyoxime, a red ring characteristic of Ni^{2+} will appear on the paper. Hence, the separation of Ni^{2+} from Fe^{3+} ions which interfere with its detection is effected owing to the filtering ability of the paper. In other words, a non-specific reaction becomes specific due to its occurrence on filter paper under appropriate conditions. Many other non-specific reactions can also be used in the same way for fractional detection of ions.

Procedure for Spot Tests. The filter paper used for spot tests is porous and relatively thick and soaks up solutions well. Since the

---

* For more detail, see Sec. 4.4.

mineral salts contained in ordinary filter paper may sometimes interfere with identification of ions, it is necessary to use "ashless filters" available in the market, i.e., filters made of paper freed of the major portion of mineral salts by treatment with acids (HCl or HF).

The procedure for applying drops of solutions to filter paper is as follows. The capillary tip of a pipette is dipped 1-2 mm into the solution and the liquid is allowed to rise into the pipette by capillarity. Then, with the pipette held in a vertical position, its tip is made to touch a strip of filter paper and is pressed lightly until a moist spot having a diameter of several millimeters is formed on the paper. Once this is done, the pipette is quickly withdrawn and the capillary with a solution of the corresponding reagent is applied to the centre of the moist spot, observing the same rules.

No drops of solution must be allowed to fall on the paper; the solution must be soaked up by the paper gradually when in contact with the capillary tip. Therefore, the amount of the liquid drawn into the pipette should not exceed the amount that can be held by capillary forces. No drop must hang from the capillary tip. Otherwise, it may fall to the side without reacting with the previously applied solution; as a result, the reaction becomes obscure and may lead to erroneous conclusions.

To avoid contamination of the reagent used in the reaction, the tip of the capillary must be brought into contact with clean paper surface before the pipette is returned to the bottle. In this way, the drop of reagent contaminated with the test solution is removed from the pipette. Then the pipette should be rinsed likewise with distilled water. If this rule is not obeyed, the reagents will soon become unusable.

The sequence in which the reagents are applied onto paper is often of great importance.*

2.5. Microcrystalloscopic Reactions

The first use of the microscope in chemical investigations dates back to 1744 when the great Russian scientist M. V. Lomonosov used it for studying the reaction of dissolution of iron in nitric acid. Lomonosov realized the importance of microscopic observations and employed the microscope extensively in his subsequent work.

In 1798, another eminent Russian chemist, T. E. Lowitz, who made a number of important discoveries in the field of chemistry, drew attention to the fact that the shape of microcrystals of salts, viewed under the microscope, is a very characteristic feature and can therefore be used for their identification.

---

* For a thorough study of spot analysis, see: Tananaev, N. A., *Drop Method*, 6th edn., Goskhimizdat, Moscow, 1954 (in Russian); Feigl, F., *Spot Tests in Organic Analysis*, 5th edn., Elsevier, Amsterdam, 1958.

Lowitz made about 100 sketches of various "salt deposits" and applied his discovery for analytical purposes. Lomonosov and Lowitz are thus the founders of microcrystalloscopic analysis. The method was later extended by P. N. Akhmatov who published in 1873 a paper *Application of the Microscope to Chemical Investigations.* He designed an instrument for measuring crystal angles (goniometer) and used it for analytical purposes. He also devised a number of microchemical methods for identification of various ions, in particular, cations of analytical Groups I and II. But it was not until later, in the 1890's, that microcrystalloscopic analysis found wide application.

The observation under the microscope of the shapes of crystals formed in reactions allows one to draw, rapidly and positively, a conclusion as to the presence of the corresponding ions in the solution being analysed. Crystals however assume their characteristic form only under definite crystallization conditions. Especially important is the fairly slow crystal growth observed when crystals grow from slightly supersaturated solutions. In this case, at first a relatively small number of minute crystal nuclei are formed; these gradually grow as the result of deposition of fresh amounts of the corresponding ions on their surface, which eventually results in the formation of large well-developed crystals. Conversely, if crystallization proceeds in a strongly supersaturated solution, numerous nuclei are formed at once, which cannot grow into large crystals. Furthermore, on rapid crystallization, the ions have no time to orient themselves properly relative to the neighbouring ions and, as a result, small crystals are formed, not very characteristic in form, which often give rise to concretions known as dendrites (tree-like formations).

The degree of supersaturation of a solution depends primarily on the concentration of the ion to be identified. To obtain good crystals, it is expedient to carry out precipitation from relatively dilute solutions. Moreover, the degree of supersaturation will evidently be the less, the higher the solubility of the compound formed. Not infrequently it proves useful to increase to a certain extent, by addition of a suitable reagent, the solubility of the substance formed. The crystallization conditions are improved but the sensitivity of the reaction is lowered. The slowest growth of crystals is observed when one of the reacting substances is taken in solid form. As the substance dissolves and diffuses slowly into the farther regions of the drop, there is enough time for growth of well-shaped crystals, particularly along the periphery of the drop. If both substances have to be taken in dissolved state, it is expedient not to simply mix the solutions but to place their drops side by side and connect them by means of a small fibre or glass thread so that the dissolved substances diffuse slowly from one drop into the other.

Another condition that favours slow crystallization is the slow evaporation of the drop, especially if the solubility of the substance is high. The solution is not allowed to dry up completely because, apart from the reaction product, all the other salts present in the solution would also crystallize. To avoid this, the evaporation is continued until the fringe of crystals is formed along the periphery of the drop.

In carrying out microcrystalloscopic reactions, one must remember that the presence of impurities may often drastically change the shape of the crystals formed. This is the main disadvantage of the method.

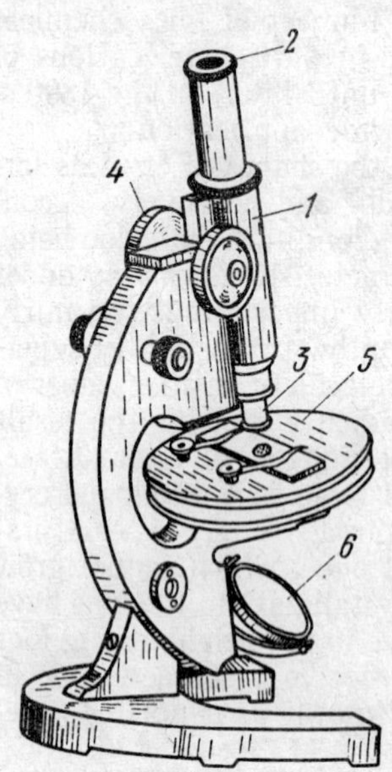

Fig. 2.23. Microscope:
1 – tube; 2 – eyepiece (ocular); 3 – objective; 4 – knob rack; 5 – stage; 6 – mirror.

The microcrystalloscopic technique is very simple. The reactions are carried out on microscope slides, which are rectangular glass plates (75 × 25 mm in size), under the conditions specified above. The reaction product is usually examined at a relatively low magnification (× 50-120) without using cover glasses (small square plates made of thin glass and used to cover the drop or any other object to be examined under the microscope).

The microscope (Fig. 2.23) consists of a tube *1* fitted with two lens systems: the upper, *2*, is known as the eyepiece (ocular), and the lower, *3*, is the objective.

The tube is moved in a vertical direction by means of a knob rack *4*. Under the tube is located a microscope stage *5*. Light rays reflected by a mirror *6* under the stage pass through an opening in the latter.

The first thing to be done is to adjust the mirror so that the field of vision is well illuminated; either daylight or artificial light may be used for the purpose. The slide with the specimen is placed on the stage (make certain that the underside of the slide is dry) and, with the microscope viewed from the side, the objective is adjusted to a distance of about 0.5 cm from the slide; then, with the left eye at the eyepiece (the right eye should be kept open), the tube is raised very slowly, the knob rack being turned until the specimen can be seen quite clearly. If you fail to do so, lower the tube again to a distance rather less than 0.5 cm between the objective and the slide and then raise again with the eye at the eyepiece.

Never lower the tube while looking through the eyepiece, as this would inevitably lead to wetting of the objective with the solution, which spoils the objective. If a drop of the solution wets the objective, it must immediately be wiped with a soft dry rag.

For high magnifications (used very rarely in microcrystalloscopic tests) the specimen must be covered with a cover glass, and the objective adjusted to about 0.5 mm from the specimen before the tube is raised.

Microcrystalloscopic reactions are in most cases highly sensitive. The detectable minimum in such reactions is usually tenths or hundredths of a microgram (μg); sometimes it is as low as ten-thousandths of a microgram. Therefore, cleanliness and carefulness in work are especially important in microcrystalloscopic tests. The reagents used must conform to high standards of purity.*

2.6. Chromatographic Separation of Substances

The chromatographic method worked out by M. S. Tsvett in 1903 belongs to the category of physico-chemical methods of analysis. It is based on the sorption of ions or molecules by various sorbents. The essence of this technique can be explained by reference to the following experiment. Into a glass tube (Fig. 2.24, *1*), on a piece of cotton, is placed a solid adsorbent, e.g., powdered alumina, Al_2O_3, following which 3 or 4 drops of a mixture of $CuSO_4$ and $CoSO_4$ solutions are pipetted into the chromatographic column thus formed. The Co^{2+} ions are sorbed on alumina considerably less strongly than the Cu^{2+} ions. Therefore, while the Cu^{2+} ions are completely sorbed in the top layer of the adsorbent, the Co^{2+} ions will pass into the underlying layer where they can be identified by the characteristic pink colour.**

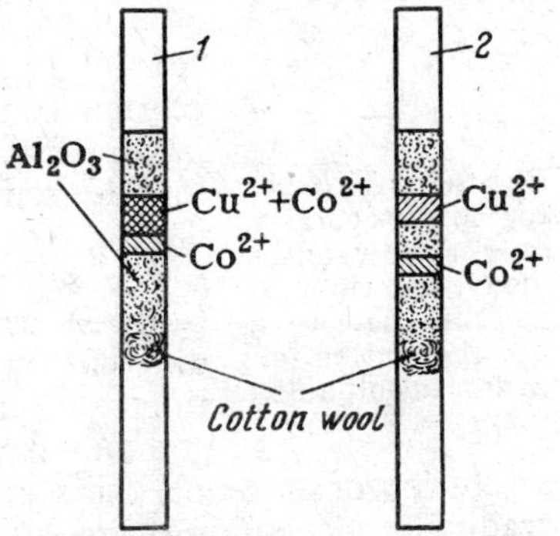

Fig. 2.24. Chromatographic separation and detection of copper and cobalt ions:

1 — primary chromatogram; *2* — ditto after washing with water.

Because of the different sorbability of the various ions or molecules of dissolved substances, they become separated when the solution is passed through the adsorbent column.

The mere introduction of the solution into the column is however insufficient for a complete separation of ions or dissolved substances. In the above example, the upper layer of the "primary chromatogram"

* For a thorough study of microcrystalloscopic analysis, see: Korenman, I. M., *Microcrystalloscopy*, Goskhimizdat, Moscow, 1955; Malyarov, K. L., *Qualitative Microchemical Analysis*, Izd. MGU, Moscow, 1951, (in Russian).

** The ability of ions to be sorbed depends on their charge and radius, and on the structure of the electronic shell.

contains also (in addition to the sorbed Cu^{2+} ions) a certain amount of Co^{2+} ions (see Fig. 2.24, *1*). To separate them completely, it is necessary to wash the column with several drops of water. The less strongly adsorbed Co^{2+} ions will travel to the bottom of the column more rapidly than the Cu^{2+} ions, and after a certain period of time a chromatogram is obtained, consisting of two spacially isolated coloured zones (bands): the upper one is the blue band of copper and the lower is the pink band of cobalt (see Fig. 2.24, *2*). The presence of Cu^{2+} and Co^{2+} ions in the solution is inferred from the characteristic colours of the bands on the chromatogram.

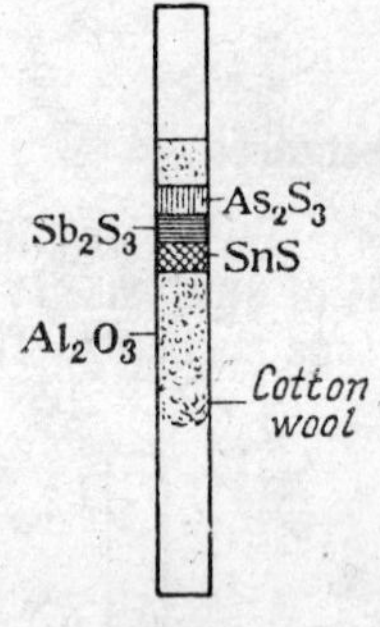

Fig. 2.25. Chromatographic separation of arsenic, antimony, and tin ions and their detection by the action of hydrogen sulphide.

If the ions being sorbed are colourless, the chromatogram has to be developed by passing a solution of an appropriate reagent, which forms differently coloured compounds with these ions, through the column. For example, when a solution containing As^{3+}, Sb^{3+}, and Sn^{2+} ions is passed through a column with Al_2O_3, the elution with a solution of HCl and the development with a solution of H_2S (or gaseous H_2S) result in a chromatogram consisting of three differently coloured bands (Fig. 2.25) the upper yellow (As_2S_3), the middle reddish-orange (Sb_2S_3), and the lower dark-brown (SnS) bands.

The selective sorption of dissolved substances (solutes) or ions may be used in analysis in a different way as well. For example, if the solution to be analysed is introduced into the column and eluted with a suitable solvent (e.g., water or a solution of an acid), the separate bands of the chromatogram will gradually move downward and emerge one after another from the column. The liquid leaving the column is collected in separate portions (fractions) and a so-called "liquid chromatogram" is produced. The least strongly adsorbed substances or ions are found to be contained in the first fractions and the most strongly adsorbed ones in the last fractions. In this way the complex mixture under study is separated into a series of simpler mixtures (or individual substances) which are much easier to analyse.

Also, the entire column of adsorbent can be pushed out of the tube and cut to separate the coloured bands; they are then treated with appropriate solvents, as a result of which the adsorbed substances pass into the solution.

Apart from various types of sorption, the chromatographic technique makes use of certain other physico-chemical phenomena as well.

The present-day classification of chromatographic techniques takes account of the specific features of atom-molecule interaction of the components to be separated and the column material. According to this classification, a distinction is made between: (1) **molecular chromatography**; and (2) **chemisorption chromatography**. The first includes absorption (molecular) chromatography on a solid adsorbent (the method described by Tsvett) and adsorption (partition) chromatography in which the role of adsorbent is played by the liquid phase present on the support material.

Chemisorption chromatography includes ion-exchange, precipitation, and other techniques. We shall discuss here some of the methods which are most important to qualitative analysis.

Adsorption Chromatography. This kind of chromatography is based on the selective sorption of the individual components of a mixture by a powdered adsorbent, and consists of passing the test solution through the column, eluting the column, and (if necessary) developing the resultant chromatogram.

Ion-Exchange Chromatography. This technique relies on exchange adsorption; when the solution is passed through the adsorbent, it takes up an equivalent amount of ions of the same charge, which were either previously adsorbed bv the adsorbent or present in its structure. Adsorbents capable of such an exchange of ions are called **ion-exchangers.** According to the charge of the ions exchanged, ion-exchangers are classified as cation-exchangers and anion-exchangers. So-called amphoteric ion-exchangers are also known; they are called ampholytes and can exchange both negatively and positively charged ions.

The ion-exchangers most commonly used are synthetic resins of various grades, e.g., Dowex, Amberlite, and also Permutite, sulphocarbon, and other compounds.* Cation-exchange resins contain active ionogenic groups: $—SO_3H$, $—COOH$, $—OH$, etc. These groups are firmly fixed to the polymeric skeleton of the cation-exchanger and only the hydrogen ions entering into their composition are mobile and capable of cation exchange. The cation-exchange process may be described by the following equation:

$$RH + Me^+ \rightleftarrows RMe + H^+$$

where R is the complex organic radical containing an active group and Me^+ is the cation of the salt.

The ion-exchange reaction is reversible. Therefore, when an acid is passed through a cation-exchanger, the absorbed Me^+ cations are displaced from it by the H^+ ions. Such a regeneration of the original H-cation-exchanger proceeds according to the scheme:

$$RMe + H^+ \rightleftarrows RH + Me^+$$

The active groups of anion-exchange resins are the groups $—NH_2$, $=NH$, $\equiv N$. Adding up water, they are converted into $—NH_3OH$, $=NH_2OH$, and $\equiv NHOH$ groups, respectively, which are capable of

* For more detail, see Osborn, G. H., *Synthetic Ion-Exchangers*, Chapman and Hall, London, 1961.

exchanging their hydroxyl ions for various anions. Consequently, the anion exchange may be represented as follows:

$$ROH + An^- \rightleftarrows RAn + OH^-$$

where An^- is the salt anion.

To recover the anion-exchanger, an alkaline solution is passed through it; for example,

$$RAn + OH^- \rightleftarrows ROH + An^-$$

In qualitative analysis of inorganic compounds, use is often made of ion-exchange chromatography. The above cases of chromatographic identification of copper, cobalt, arsenic, antimony, and tin compounds on a column with Al_2O_3 as an adsorbent involve cation-exchange. The aluminated alumina used as an adsorbent contains sorbed molecules of sodium aluminate, $NaAlO_2$, and is a Na-cation-exchanger capable of exchanging Na^+ ions for other cations, for example,

$$2[(Al_2O_3)_m \cdot AlO_2^-]Na^+ + Cu^{2+} \rightleftarrows [(Al_2O_3)_m \cdot AlO_2^-]_2Cu^{2+} + 2Na^+$$

It should be noted that when a uninormal solution of HNO_3 is passed through a layer of aluminated alumina, the following reaction takes place:

$$[(Al_2O_3)_m \cdot AlO_2^-]Na^+ + 2HNO_3 \rightleftarrows [(Al_2O_3)_m \cdot AlO^+]NO_3^- + Na^+ + NO_3^- + H_2O$$

In this reaction, the alumina changes from a cation-exchanger into an anion-exchanger because it now contains NO_3^- ions capable of being exchanged for other anions; for example,

$$3[(Al_2O_3)_m \cdot AlO^+]NO_3^- + PO_4^{3-} \rightleftarrows [(Al_2O_3)_m \cdot AlO^+]_3PO_4^{3-} + 3NO_3^-$$

Apart from being used for detection of cations, the cation-exchange process is successfully employed for removal of cations interfering with identification of anions. In such cases, the solution to be investigated is passed through a cation-exchanger column and eluted with water. The cations present in the solution are adsorbed by the cation-exchanger and the anions pass as the corresponding free acids into the solution to be analysed. Anions (e.g., PO_4^{3-}, $C_2O_4^{2-}$, etc.) that interfere with detection of cations can be removed from the solution in just the same way. To this end, the cations sorbed by the column (after it is eluted with water) are displaced by a solution of an acid, say HCl, following which the resulting solution is tested for cations.

Precipitation Chromatography. The technique of precipitation chromatography is based on the same principle of successive precipitation of sparingly soluble compounds that was previously used for fractional detection of ions in spot analysis.

In this technique, the solution under investigation is passed through a column with a support material (e.g., Al_2O_3) mixed with a suitable precipitant (or impregnated with its solution). The ions are precipitated as slightly soluble compounds which are formed in order of their increasing solubility. First (at the top of the column) the least soluble compound is formed, followed by the more soluble compound, etc. If these compounds are coloured differently, the resulting chromatogram will be completely analogous to those described above. For example, if a solution containing Pb^{2+} and Hg^{2+} ions is passed through a column with alumina impregnated with several drops of a solution of potassium iodide, KI, a chromatogram is obtained which consists of two zones: the upper orange-red (HgI_2) and the lower yellow (PbI_2).

V. B. Aleskovsky and Z. I. Kheifets suggested using ion-exchangers in precipitation chromatography instead of a mechanical mixture of a carrier and a precipitant; the ion-exchangers are charged with ions capable of forming precipitates with the ions of the unknown solution. For example, instead of a mixture of KI and a carrier, it is convenient to use an anion-exchanger containing iodide ions capable of being exchanged for other ions. When a solution containing Ag^+, Pb^{2+}, Hg^{2+}, etc., is passed through a column with such anion-exchanger, chromatograms are obtained, which show a definite arrangement of differently coloured zones of iodides of the corresponding metals. This procedure for obtaining precipitation chromatograms have certain advantages over the conventional technique.

Partition Chromatography. This technique is based on the distribution of solutes between two immiscible solvents. If an aqueous solution of bromine (Br_2) is mixed with chloroform, $CHCl_3$, then, since bromine dissolves in chloroform better than in water, it will pass into the chloroform layer which assumes, as a result of this, an intense orange colour. Conversely, the water solution will become paler. A state of equilibrium will eventually be established, at which the ratio of the bromine concentrations in the organic solvent (C_{org}) and in water (C_w) will become equal to a certain constant quantity D which is known as the **distribution coefficient (partition coefficient)**. This **distribution law** is expressed by the formula

$$\frac{C_{org}}{C_w} = D$$

The process of extraction described above, i.e., removal of a substance from the solution in a given solvent into the layer of another solvent, is very frequently employed in analysis. By extraction it is possible to concentrate the solute and thus to enhance the colour caused by its presence, which renders the corresponding reaction more sensitive. Extraction is sometimes resorted to if a given substance is found to be more stable in the organic solvent than in the aqueous solution. This process is also used for separation of some ions from others.

To obtain column partition chromatograms, the column is packed with an inert filling which firmly adsorbs a solvent (the nonmobile, or stationary phase). Into the column is introduced a solution of the substances to be separated in such a solvent (the mobile phase) which is immiscible with the first solvent. The solutes, in distributing themselves in the pores of the inert filling between the two solvents, proceed down the column: substances whose partition coefficient is higher (i.e., which dissolve in the mobile solvent better than the others) move down the column more rapidly than those for which the value of D is lower. A chromatogram is finally obtained, which contains spacially isolated bands of the components of the mixture being analysed.

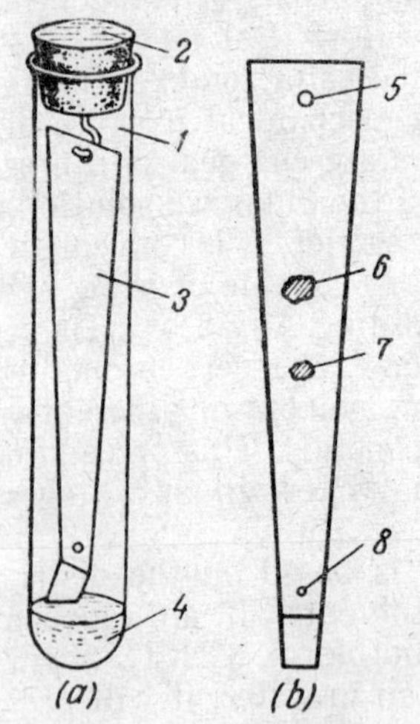

Fig. 2.26. Paper chromatography:
(a) paper-chromatographic device; (b) chromatogram with separated components; 1 — test tube; 2 — stopper with a hook for attachment of paper; 3 — strip of filter paper; 4 — solvent; 5 — hole for hook; 6 — zone of component A; 7 — zone of component B; 8 — point of solute application.

The support materials used in partition chromatography also include strips (or sheets) of filter paper containing no mineral substances (paper chromatography).

The apparatus used for carrying out the paper chromatographic technique is shown in Fig. 2.26. A drop of the aqueous solution to be analysed is placed on a strip of filter paper which is then suspended from a hook in a tall vessel. The bottom narrow end of the strip is dipped into a pool of mixed butanol (or some other organic solvent) and water. Being soaked by the paper, the solvent climbs up the strip by capillarity, carrying the constituents of the sample along with it at different speeds according to their partition coeficients. When the advancing front of the solvent approaches the top, the strip is removed, dried in air, and sprayed with a solution of a suitable reagent which forms differently coloured compounds with the ions to be identified.

Thin-Layer Chromatography. The technique of thin-layer chromatography (TLC) can be used for separation and analysis of minute quantities of numerous classes of compounds. A thin layer of an aqueous slurry of an inert adsorbent, such as alumina, silica gel, or cellulose, is evenly spread over the surface of a glass plate (about 10 × 10 cm in size). To ensure that the adsorbent adheres to the glass, a binder such as starch, plaster of paris, collodion, or a plastic dispersion is added (see Fig. 2.27). The plate is dried in an oven, as a result of which a thin, uniform layer of firmly bound adsorbent is left. (Ready-coated plates for TLC are also available.) The mixture to be analysed is applied near one end of the plate, and this end is dipped vertically into the solvent pool in a closed container. As the solvent rises (by capillary action, just as in paper chromatography) the mixture is resolved into very small discrete zones, or spots. After the solvent has moved some distance (which takes only 20 to 30 min), the plate is withdrawn and dried and the separated components are "visualized" by spraying the plate with a suitable reagent. The components can be recovered by scraping the patches of adsorbent off the plate and eluting the substances with an appropriate solvent. Two-dimensional thin-layer methods are also employed.

The chromatographic method of analysis has a number of highly valuable features. Separation of ions in this method is achieved by introducing the solution to be analysed into a column which is then eluted.

Fig. 2.27. Thin-layer chromatography.

Sometimes the detection of individual ions does not require the use of reagents at all, the ions being identified by theirinher ent characteristic colours. But even if the ions to be identified are colourless, one can often detect the presence of several ions at once by developing the chromatogram with an appropriate reagent. All this facilitates and accelerates the work.

The chromatographic method of separating mixtures is of great importance in science and technology. It finds wide application for purification of many organic compounds having a rather complex composition, say, vitamins, hormones, antibiotics. Even insignificant differences in the composition or structure of substances usually prove to be sufficient for these substances to be adsorbed differently by suitable adsorbents. For this reason, the chromatographic technique often appears to be applicable to mixtures which are very difficult to separate by any other method (for example, mixtures of rare-earth and transuranium elements). It should also be noted that in the chromatographic process the substances to be separated undergo no chemical change and are isolated in the form in which they were present in the original mixture; this is especially important in biochemical investigations.*

* For detailed familiarization with chromatography, see: Samuelson, O., *Ion Exchange in Analytical Chemistry*, John Wiley and Sons, New York, 1953; Pollard, F. H. and J. F. W. McOmie, *Chromatographic Methods of Inorganic Analysis*, Academic Press, New York, 1953; Stock, R. and C. B. F. Rice, *Chromatographic Methods*, Chapman and Hall, London, 1963.

2.7. Separation of Substances by Extraction

In analytical chemistry, wide use is made of extraction techniques for separation of the various substances, in particular, elements which are similar in their properties. Extraction is employed both in qualitative and quantitative analysis (much more frequently in the latter).

Extraction is a complicated physico-chemical process. In the simplest case, an individual substance (for example, Br_2) is distributed between two immiscible liquids (water and chloroform, see Sec. 2.6.). The distribution of a substance in two immiscible liquids (two phases) obeys the distribution law. For the given simplest case, this law can be expressed by the following mathematical formula:

$$\frac{C_{org}}{C_w} = D$$

where D is the distribution coefficient.

In the extraction of an individual substance (Br_2) which forms no chemical compounds with water or with an organic solvent, the process reduces to a purely physical phenomenon—mass transfer due to the operation of van der Waals forces. In this particular case, the distribution coefficient D coincides in value with the dynamic equilibrium constant, the dynamic equilibrium being established between the concentration of the substance in the water phase and that in the organic phase:

$$Br_2(H_2O) \rightleftharpoons Br_2(CHCl_3)$$

$$\frac{[Br_2]_{(CHCl_3)}}{[Br_2]_{(H_2O)}} = K_{eq} = D$$

In an ideal case where the equilibrium is established and the composition of the third component (Br_2 in this particular case) in the water and organic phases is the same (i.e., when no chemical interaction takes place between the two solvents and the solute), the distribution law may be formulated as follows. *At a given temperature the ratio of the concentrations of the third component in the two equilibrium phases is a constant quantity and does not depend on the concentration of the substance.* But in practice, as a rule, the solute is found to react either with one of the liquids or with both liquids. Moreover, in chemical analysis one has to deal not with simple substances but with complex compounds. Therefore, one has to take into account the composition of such complexes both in the water and in the organic phase, and all calculations of the distribution ratio and the equilibrium constant become complicated.

The separation of elements by extraction in the form of complexes, especially with organic reagents, is greatly influenced by the pH of the medium. It will suffice to quote a single example. The organic reagent dithizone gives stable complexes with more than 20 cations: Ag^{I}, Hg^{I}, Hg^{II}, Pd^{II}, Pt^{II}, Au^{I}, Au^{III}, Cu^{I}, Cu^{II}, Bi^{III}, In^{III}, Sn^{II}, Zn^{II}, Cd^{II}, Co^{II}, Pb^{II}, Ni^{II}, Fe^{II}, Mn^{II}, Tl^{I}, etc. By adjusting the pH of the solution, it is possible to separate the dithizonates of these elements. For example, at pH about 1 one can separate, using CCl_4 as an extracting agent, the dithizonates of Ag^{I}, Hg^{II}, Pd^{II}, Pt^{II}, Au^{III}, Cu^{II}, and Bi^{III} from the dithizonates of In^{III}, Sn^{II}, Zn^{II}, Cd^{II}, Co^{II}, Ni^{II}, Pb^{II}, Fe^{II}, Mn^{II}, Tl^{I}, etc., which are left in the aqueous solution. Then, by changing the pH of solution again, the other dithizonates can also be successively extracted.

Extraction processes are widely employed in present-day analytical chemistry:

(1) to increase the sensitivity and selectivity of reactions (see Sec. 1.3);

(2) to separate elements and to concentrate "traces";

(3) to study the instability constants of complex compounds;

(4) to study the state of substances in solution (to determine the type of ion or molecule in the form of which the element is present in solution; to establish the existence of polymeric ions or molecules in solution; to find out whether colloidal particles are formed or not, etc.).

A number of other aspects in the use of extraction techniques in anatytical chemistry could be cited, but they all refer mainly to quantitalive methods of analysis and are not discussed in this book.*

Questions and Problems

1. Calculate the normality of a solution of HCl containing 2.2 g of hydrogen chloride in 200 ml.

Answer: about 0.3*N*.

2. Calculate the normal and molar concentrations of a solution of sulphuric acid containing 5.88 g of H_2SO_4 in 300 ml.

Answer: ca. 0.4*N* and ca. 0.2*M*.

3. How many grams of KOH are contained in 200 ml of a 0.092*N* solution of it?

Answer: about 1.032 g.

4. How many grams of $ZnSO_4 \cdot 7H_2O$ are required to prepare 200 ml of 0.5 *N* solution?

---

* For more detail concerning extraction methods in analytical chemistry, see: Diamond, R. M. and D. G. Tacke, *Extraction of Inorganic Compounds*, Pergamon Press, 1965; Morrison, G. H. and H. Freiser, *Solvent Extraction in Analytical Chemistry*, John Wiley and Sons, New York, 1957; Berg, E. W., *Physical and Chemical Methods of Separation*, McGraw-Hill Book Co., New York, 1963.

Answer: about 14.38 g.

5. What volume of 56% H_2SO_4 having a density of 1.46 g/cm^3 must be taken to prepare 5 litres of a 2*N* solution of it?

Answer: about 600 ml.

6. How many millilitres of 2*N* Na_2CO_3 solution must be taken to prepare (by dilution with water) 250 ml of a 0.35*N* solution?

Answer: about 43.75 ml.

7. A saturated solution of $CaSO_4$ contains (at room temperature) 2 grams of salt in 1 litre. What are its molar and normal concentrations?

Answer: about 0.015*M*; about 0.03*N*.

8. How many grams of SO_4^{2-} ions are contained in 1 litre of solution if the concentration of these ions is equal to 0.2 g-ion/litre?

Answer: 19.2 g.

Chapter 3

Group I Cations

We have so far discussed the fundamental concepts of qualitative analysis and described the technique of semimicromethod. Now we shall pass over to the study of the characteristic reactions and analytical procedures for cations of individual groups. To acquire a proper understanding of the relevant material, it is necessary to know the fundamentals of the theory of qualitative analysis.

The most important theoretical questions are the theory of chemical equilibrium and the theory of solutions. These are considered in the present chapter.

A. THEORETICAL PRINCIPLES

3.1. The Law of Mass Action

The reactions used in analytical chemistry are mostly reversible, i.e., they proceed simultaneously in two mutually opposite directions.* Examples are the reactions used for identification and separation of Ba^{2+} and Zn^{2+} ions:

$$2Ba^{2+} + Cr_2O_7^{2-} + H_2O \rightleftharpoons 2BaCrO_4\downarrow + 2H^+$$

$$Zn^{2+} + H_2S \rightleftharpoons ZnS\downarrow + 2H^+$$

In equations for reversible reactions the equals sign is replaced by the double arrow ($\rightleftharpoons$) standing for reversibility. Of two simultaneously occurring reactions, the one proceeding to the right of the equation ($\rightarrow$) is known as the **forward reaction**, and the other which proceeds to the left ($\leftarrow$) is the **reverse reaction**.

We know from the course of general chemistry that reversible reactions result in the establishment of **chemical equilibrium** when the solution contains all the original substances (reactants) and all the compounds formed as the result of the reaction (reaction products).

---

* The classification of reactions as reversible and irreversible is arbitrary. Strictly speaking, every reaction is more or less reversible, but in many cases the reversibility observed is so slight that it may be ignored and the reaction may be considered practically irreversible.

Chemical equilibrium is attained because the rates of the forward and reverse reactions become equal due to changes in the concentrations of the reacting substances.

*The rate of chemical reactions is measured by the changes in the concentrations of the reactants in a unit time.**

Since any molecules or ions react chemically only when they collide, it follows that the more frequent are such collisions, the higher must be the reaction rate, other conditions being equal. The frequency of collisions between molecules or ions must obviously be the higher, the greater the concentration of each of them in solution. Suppose that the following reversible reaction proceeds in solution:

$$A + B \rightleftharpoons C + D$$

where A, B, C, and D are different substances present in the reaction mixture.

The rate of the forward reaction is proportional to the concentrations of substances A and B.

Let us recall that *the concentrations of the reactants involved and of the final products are usually expressed in moles per litre and designated by their formulas enclosed in square brackets.*

We can therefore write for the forward reaction:

$$v_1 = k_1[A][B] \tag{3.1}$$

Here v_1 represents the rate of the forward reaction, and k_1 is the proportionality coefficient, known as the **rate constant.**

If the concentrations [A] and [B] are taken to be equal to 1 mole/litre, then

$$v_1 = k_1$$

Thus, the rate constant can be defined as the rate at which a given reaction proceeds under specified conditions if the concentration of each of the reactants involved is equal to 1 mole/litre (or if the product of their concentrations is equal to unity).

Similarly, for the rate of the reverse reaction (v_2) we may write**:

$$v_2 = k_2[C][D] \tag{3.2}$$

As the reaction proceeds, the concentrations of substances A and B decrease and those of substances C and D increase. Consequently, the

---

* Since the rate of a chemical reaction is a variable (like the velocity of non-uniform motion), it may be expressed mathematically as the derivative of the concentration (C) with respect to time (t) with a minus sign because the value of C falls off with increasing t:

$$v = -\frac{dC}{dt}$$

** Since other substances are involved in this case, the rate constant (k_2) has a different meaning than that in the forward reaction (k_1).

rate of the forward reaction must gradually fall with time and that of the reverse reaction must increase from zero. The two reaction rates must eventually become equal.

The state of a system of reacting substances, at which the rates of the forward and reverse reactions become equal, is called the chemical equilibrium.

Once the state of chemical equilibrium is attained, the number of molecules of each of the substances (A, B, C, and D) that enter into the reaction per unit time is the same as the number of molecules of the substance formed as the result of the reverse reaction. Hence, at equilibrium the concentrations of all the substances in the reaction mixture cease to change as if the reaction has stopped; in fact, the two reactions continue but each counteracts the results of the other.

Chemical equilibrium is thus said to be a state of **dynamic equilibrium.**

We have already seen that $v_1 = v_2$ at equilibrium. Substituting into this equation the values of v_1 and v_2 from Eqs. (3.1) and (3.2), we obtain:

$$k_1[A][B] = k_2[C][D]$$

We now take the concentrations to one side of the equation and the constants k_1 and k_2 to the other. This gives:

$$\frac{k_1}{k_2} = \frac{[C][D]}{[A][B]}$$

But the ratio of the two constants, k_1/k_2, must also be a constant value. Denoting it as K, we finally get:

$$\frac{[C][D]}{[A][B]} = K \tag{3.3}$$

Similarly, we find that the equilibrium of the reaction

$$A + B + E \rightleftharpoons C + D$$

sets in at the moment when the concentrations of the reactants will satisfy the equation

$$\frac{[C][D]}{[A][B][E]} = K$$

If we assume that B and E are the same substance, the above equations will become:

$$A + 2B \rightleftharpoons C + D$$

$$\frac{[C][D]}{[A][B]^2} = K$$

In the general case, for the reaction

$$mA + nB \rightleftharpoons pC + qD$$

(where m, n, p, and q are the corresponding stoichiometric coefficients) the condition of the attainment of equilibrium is determined by the equation

$$\frac{[C]^p [D]^q}{[A]^m [B]^n} = K \tag{3.4}$$

This equation is the mathematical expression of one of the most important laws in chemistry, the **law of mass action***, which may be stated as follows: *At equilibrium the product of the concentrations of reaction products divided by the product of the concentrations of reactants is a constant quantity for a given reaction at a given temperature and is known as the* **equilibrium constant**. Each concentration is raised to the power corresponding to the appropriate coefficient in a balanced chemical equation.

Note. For more accurate calculations the equation for the law of mass action should be expressed in terms not of the concentrations but of the activities which are obtained by multiplying the concentration expressed in moles or gram-ions per litre by the corresponding coefficient called the "activity coefficient". This will be discussed in more detail in Sec. 3.6.

The physical meaning of the equilibrium constant K is clear from the fact that it is equal to k_1/k_2, i.e., it shows how many times the rate of the forward reaction at equal concentrations (namely, 1 mole/litre) and at a given temperature is greater than the rate of the reverse reaction. If the equilibrium constant is less than unity, the reverse reaction proceeds at a higher rate. It follows from the foregoing that the direction of the reaction can be deduced from the numerical value of K. If the constant K is very large, this is an indication that the forward reaction will go to near completion, while the reverse reaction will proceed but to a very negligible extent. In other words, the equilibrium will be shifted to the right. A very small value of K indicates that the reverse reaction predominates, so that the equilibrium is found to be strongly shifted to the left. And, finally, if the value of K is close to unity, the reaction is practically reversible. The equilibrium is established in this case at considerable concentrations of all the four substances involved in the reaction; none of these is converted sufficiently completely. We will show below that the state of equilibrium in a system depends not only on the value of K but also on the initial concentrations of the reactants. Conversely, *the equilibrium constant is independent of the initial concentrations of A and B*. It is in this sense that the constancy of K is to be understood. Thus, no matter what the concentrations of A and B may be, the ratio

$$\frac{[C]^p [D]^q}{[A]^m [B]^n}$$

* The law of mass action was first formulated by C. M. Guldberg and P. Waage in 1864—*Tr.*

at equilibrium is a constant value at a constant temperature. An increase in temperature will lead to the increase of the value of K. To put it otherwise, to each temperature there will correspond a definite value of the equilibrium constant of the reaction, just as any substance has a definite solubility at each temperature.

Suppose that at equilibrium in the reversible reaction

$$CO+H_2O \rightleftarrows H_2+CO_2$$

we add (without altering the total volume of the gaseous mixture) one of the original substances, say, the H_2O vapour. This will immediately result in the shift of the equilibrium toward the formation of the reaction products H_2 and CO_2. The reaction will proceed in this direction until the ratio $[H_2]\,[CO_2]/[CO]\,[H_2O]$ is again equal to the equilibrium constant of the reaction. Equilibrium will be set up this time at other concentrations of all the four substances.

In just the same way, addition of one of the reaction products, i.e., H_2 or CO_2, will temporarily increase the rate of the reverse reaction. Thus, addition of one of the reactants causes a *shift of chemical equilibrium*, i.e., a change in the concentrations of all the substances taking part in the reaction, which continues until the concentrations again satisfy the equation for the law of mass action. *The reaction in which the added substance is used up is the one to undergo an increase in its rate.*

An important conclusion can be drawn from the foregoing: *To achieve the most complete conversion of one of the starting substances (for example,* CO*), an excess of the reagent responsible for that conversion must be introduced into the reaction mixture* (H_2O in this particular case).

Example. It has been established experimentally that at 850°C, with equal initial concentrations of CO and H_2O vapour, an equilibrium sets in at the moment when the concentrations of all the four substances are equal. For instance, if the initial concentrations of CO and H_2O vapour in the gaseous mixture are equal to 1 mole/litre, then at equilibrium 0.5 mole of each remains and 0.5 mole each of H_2 and CO_2 are formed. Hence, the equilibrium constant of the reaction at the specified temperature is

$$K=\frac{0.5\times0.5}{0.5\times0.5}=1$$

From this one can easily calculate the equilibrium concentrations of each of the substances for any other initial concentrations of CO and H_2O.

Let us perform such a calculation for a case where 5 moles of H_2O is taken for 1 mole of CO (the total volume of the gaseous mixture being 1 litre).

Denote by x the number of moles of CO converted.

Since, according to the reaction equation, one molecule of CO reacts with one molecule of H_2O and one molecule each of H_2 and CO_2

are formed, the concentrations of these substances at equilibrium are:

$$[CO]=1-x \qquad [CO_2]=x$$
$$[H_2O]=5-x \qquad [H_2]=x$$

Substituting these values into the equilibrium-constant equation yields

$$\frac{xx}{(1-x)(5-x)}=1$$

which gives

$$x=\frac{5}{6}\approx 0.83 \text{ mole}$$

Thus, when a five-fold excess of H_2O is added, 83 per cent of the CO taken is converted, and with an equivalent amount of H_2O the conversion is only 50 per cent.

3.2. Equilibrium in Homogeneous Systems. Degree of Dissociation. Strong and Weak Electrolytes

The reactions in aqueous solutions, which are most often used in qualitative analysis, are ionic reactions (see Sec. 1.2). That is why it is very important for the analyst to know the extent to which the various electrolytes dissociate into ions in solution.*

The ratio of the fraction of the dissolved electrolyte that dissociates into ions, i.e., is ionized, to its total amount is called the **degree of dissociation** or **ionization** and is denoted by the Greek letter α (alpha).

For example, the statement that "the degree of dissociation (α) of acetic acid in 0.01 molar solution is 0.0419" means that at the stated concentration 4.19 per cent of the total amount of CH_3COOH dissociates into ions and 95.81 per cent is present in the form of undissociated molecules.

The degree of dissociation can be determined experimentally by a variety of methods, for example, from the electrical conductance of the solution, from the lowering of the freezing point of a solvent by a solute, etc.

According to the theory of electrolytic dissociation, the flow of current through a solution is effected by the movement of ions towards

* One should distinguish between two phenomena taking place during the dissolution of electrolytes in water. If a substance in the free state has a crystalline structure with heteropolar bonds (for example, NaCl), this substance undergoes dissociation in aqueous solution, i.e., ions that pass into solution are those which have previously entered into the composition of the crystal structure of the given substance. But if a substance in the free state has homopolar bonds (e.g., HCl), then, being in aqueous solution, this substance is found to be ionized, i.e., ions are formed during the dissolution; these ions have not been previously present in the crystal lattice of that substance.

the oppositely charged electrodes. The higher the concentration of ions in the solution and the greater the velocity of their movement, the higher the quantity of electricity passed through the solution and the higher its conductance*.

It has been found by experiment that with increasing dilution the so-called equivalent conductance (λ) of the solution gradually rises, approaching a limiting value (λ_∞). This limiting value, known as the **equivalent conductance at infinite dilution**, is reached in the case of strongly ionized (dissociated) electrolytes at a concentration of the order of 0.0001*M* and undergoes no change on further dilution.

The increase of λ is caused either by the increase of the degree of dissociation, when the solution is diluted, due to the disturbance of the equilibrium between the un-ionized molecules of the electrolyte and its ions, or by an increase in the velocity of movement of the ions.

In the classical theory of electrolytic dissociation proposed by S. Arrhenius in the period 1883-87, it is assumed that the velocity of the ions does not change with the concentration of the solution. Therefore, according to this theory, the increase of the equivalent conductance with dilution is attributed exclusively to the increase of a quantity proportional to it—the degree of dissociation (α), which tends to the limiting value $\alpha = 1$ corresponding to complete dissociation of the electrolyte.**

Hence, we have

$$\frac{\alpha}{1} = \frac{\lambda}{\lambda_\infty}$$

and

$$\alpha = \frac{\lambda}{\lambda_\infty}$$

The values of the degree of dissociation determined by various methods have been found to be fairly close, which led to the general acceptance of the Arrhenius theory. An outstanding role in the development and acceptance of this theory was played by the researches of I. A. Kablukov and V. A. Kistyakovsky.

Subsequent investigations have shown that the classical theory of Arrhenius is applicable only to weak electrolytes, i.e., to substances which dissociate to a slight degree.

---

* Electrical conductance is defined as the reciprocal of the resistance and is expressed in reciprocal ohms. If the resistance of a solution is 100 ohms, its electrical conductance will be equal to 0.01 ohm^{-1}. The equivalent electrical conductance of a solution is defined as the product of the specific conductance or conductivity (the electrical conductance of a column of solution 1 cm in length and with a cross section of 1 cm^2) by the solution volume (expressed in millilitres), containing 1 g-eq of the corresponding electrolyte.

** For other methods of determining the degree of ionization, see textbooks of inorganic chemistry.

For strong electrolytes, whose experimentally determined degree of dissociation is very high, it proved to be necessary to work out a new theory that would take into account the effect of the electrostatic attraction and repulsion forces operating between the ions on their velocity.

The main propositions of the theory of strong electrolytes will be discussed in Sec. 3.5. At this point, we merely note that all the methods mentioned above for determination of α provide a correct idea of the actual degree of dissociation of only weak electrolytes, the ions of which are so far apart in solution that the electrostatic forces may be disregarded. In the case of strong electrolytes, in whose solutions the concentration of ions is sufficiently high, the influence of these forces cannot be ignored.

The formula $\alpha = \lambda/\lambda_\infty$, which is derived without regard for this influence, gives not the true but the *apparent* degrees of dissociation of strong electrolytes. The *true* degree of dissociation of strong electrolytes is assumed at present to be approximately equal to 100 per cent on the basis of a number of considerations that will be discussed later in the book (see Sec. 3.5).

The degree of dissociation of electrolytes is of great importance in analysis since it characterizes the chemical activity of the respective substances. The validity of this is not difficult to see from the following facts. The determination of the conductance of uninormal solutions of HCl and CH_3COOH shows that hydrochloric acid is very strongly ionized, whereas the degree of dissociation of acetic acid is only 0.42 per cent. If equal pieces of metallic zinc are introduced into these solutions, then hydrogen will evolve vigorously in the HCl solution, while in the CH_3COOH solution the liberation of hydrogen is extremely slow. The same difference is observed in the rate of CO_2 evolution when solutions of these acids react with $CaCO_3$.

The difference between the solvent power of hydrochloric acid and that of acetic acid with respect to certain substances slightly soluble in water also indicates that their solutions of equal concentration differ in chemical activity. Indeed, there is a large number of substances which dissolve in hydrochloric acid solutions but are insoluble in solutions of acetic acid. Examples are $BaCrO_4$, CaC_2O_4, ZnS.

These facts can easily be interpreted from the standpoint of the theory of electrolytic dissociation. By the Arrhenius theory, the characteristic properties of acids are ascribed to H^+ ions. The fact that hydrogen ions are present in solutions of all acids is the reason why these solutions have a number of common properties, known as acidic properties. This being so, the acid must be the stronger the greater the amount of hydrogen ions formed by the acid, i.e., the higher its degree of dissociation.

The strongest acids are HCl, HBr, HI, HNO_3, and H_2SO_4. Their apparent degrees of dissociation are very high. They are followed by acids of medium strength, $H_2C_2O_4$, H_3PO_4, H_2SO_3, HF, and, finally, by weak acids which include CH_3COOH, H_2S, HCN, and others, whose degree of dissociation is negligibly small.

Bases, like acids, possess a number of properties in common, which depend on the presence of OH^- ions in their solutions. The strength of a base is determined by the concentration of OH^- ions formed in its solution, which depends on the degree of dissociation of the base. To the category of strong (chemically active) bases belong KOH and NaOH. An aqueous solution of ammonia, NH_4OH, is referred to as a weak base. In fact, NH_4OH is a strong base. The point is that in aqueous solutions of ammonia the molecules of NH_4OH are found to be almost completely decomposed:

$$NH_4OH \rightleftharpoons NH_3 + H_2O$$

The equilibrium of this reaction is strongly shifted to the right. The remaining NH_4OH molecules dissociate almost completely into ions:

$$NH_4OH \rightleftharpoons NH_4^+ + OH^-$$

But since this reaction gives a small quantity of OH^- ions (the concentration of NH_4OH molecules undecomposed in the first reaction is very low), ammonia in aqueous solutions* behaves as a weak base with a dissociation constant (K_{NH_4OH}) being approximately equal to 1.8×10^{-5}.

Let us now consider the specific features of the dissociation of di- and polybasic acids and also of polyacidic bases. As shown by investigations, the dissociation of such acids and bases proceeds by stages. For example, orthophosphoric acid, H_3PO_4, loses at first only one hydrogen ion:

$$H_3PO_4 \rightleftharpoons H^+ + H_2PO_4^- \quad \text{(ca. 28 per cent)}$$

The $H_2PO_4^-$ anions formed in this reaction release another H^+ ion:

$$H_2PO_4^- \rightleftharpoons H^+ + HPO_4^{2-} \quad \text{(ca. 0.12 per cent)}$$

And, finally, the HPO_4^{2-} ions give up the last hydrogen ion:

$$HPO_4^{2-} \rightleftharpoons H^+ + PO_4^{3-} \quad \text{(ca. 0.001 per cent)}$$

The figures in parentheses (relating to $0.1M$ H_3PO_4 solution) show that the highest degree of ionization corresponds to the first step. The $H_2PO_4^-$ anions behave as a much weaker acid, while the behaviour of the HPO_4^{2-} anions is like that of a very weak acid.

The reason is clear: it is more difficult for the positive H^+ ions to detach themselves from the negative $H_2PO_4^-$ ion than from the neutral

* Modern conceptions of the nature of acids and bases, which cover both aqueous and non-aqueous solutions, are expounded in Sec. 3.7.

H_3PO_4 molecule, and even more so, from the doubly charged HPO_4^{2-} ion. Because of the stepwise ionization, a moderately dilute solution of H_3PO_4 has a rather high content of $H_2PO_4^-$ anions, less HPO_4^{2-}, and very little PO_4^{3-} anions.

Other polybasic acids and polyacid bases dissociate in a similar way. In the case of bases, OH^- ions are split off by stages, for example,

$$Ca(OH)_2 \rightleftharpoons Ca(OH)^+ + OH^-$$

$$Ca(OH)^+ \rightleftharpoons Ca^{2+} + OH^-$$

Whereas acids and bases exhibit great variations in degree of dissociation, nearly all soluble salts (with a few exceptions) are strong electrolytes, i.e., their apparent degree of dissociation is high. This also applies to salts formed from weak acids and weak bases. For example, if an electric current is passed through solutions of acetic acid or ammonia of equal concentration, an electric lamp inserted into the circuit gives a dull light. Its light however becomes immediately bright once the solutions are mixed and the electrodes are dipped into the mixture. This is easily understood since the mixing of the slightly ionized compounds NH_4OH and CH_3COOH yields a well dissociated salt:

$$CH_3COOH + NH_4OH \rightleftharpoons CH_3COO^- + NH_4^+ + H_2O$$

It follows from the foregoing that *strong acids and bases, and nearly all soluble salts, are strong electrolytes. Weak acids and bases, water, salts such as $Hg(CN)_2$, $HgCl_2$, and some other substances are weak electrolytes.*

3.3. Dissociation Constants of Weak Electrolytes

Let us examine the mechanism of dissociation of weak electrolytes, using acetic acid as an example. The dissociation is caused by the attraction exerted by neighbouring water molecules on acetic acid molecules. As known, the water molecule has an asymmetric structure (Fig. 3.1). Owing to this, the "centres of gravity" of the negative and positive charges in the H_2O molecule do not coincide but lie at two different points in space. Such molecules are called **dipoles**. It is clear that, although the quantity of positive electricity is equal to that of negative electricity in such dipoles, they must exert a noticeable electrostatic effect on surrounding particles. For example, they must turn with their negative poles towards positive ions (or positive poles

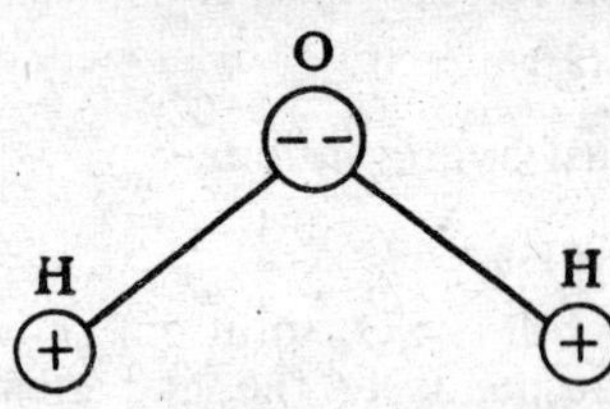

Fig. 3.1. Structure of the water molecule (schematic).

of dipoles), and with their positive poles towards negative ions. This orientation of water dipoles gives rise to forces of electrostatic attraction, which tend, as it were, to break the electrolyte molecule apart into its constituent ions and thereby greatly weaken the bonds between them. As a result, the energy of thermal motion of the molecules appears to be sufficient for the bonds to be broken, i.e., for the ions to be formed.

Dissociation results in the formation not of free ions but of compounds of ions with solvent molecules. Such compounds are collectively referred to as **solvated ions** or **solvates**; if the solvent is water, they

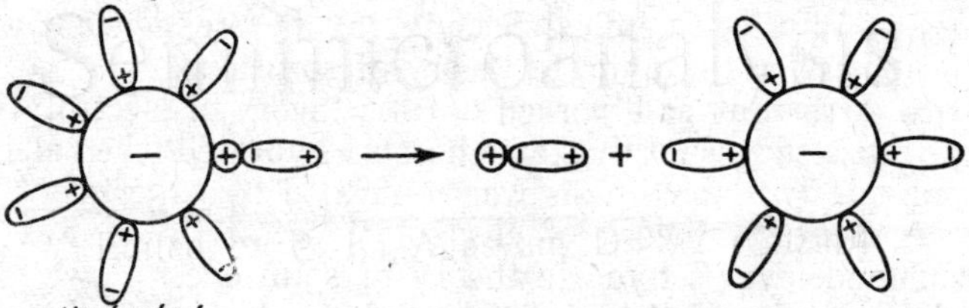

Hydrated molecule of CH_3COOH — *Hydroxonium ion* H_3O^+ — *Hydrated* CH_3COO^- *ion*

Fig. 3.2. Electrolytic dissociation of the CH_3COOH molecule (schematic).

are called **hydrated ions**. In a number of cases, the existence of solvates or hydrates has been demonstrated by experiment. It has been found, for example, that the free ion H^+ (i.e., the proton) cannot exist in aqueous solutions but combines with a water molecule to form a hydrate (H_3O^+) known as the **hydronium ion** (or **hydroxonium ion**):

$$H^+ + H_2O \rightarrow H_3O^+$$

The hydronium ion can in its turn be hydrated. For example, the existence of the particle $[H_9O_4]^+$ is known.

Electrolytic dissociation is shown schematically in Fig. 3.2 in which the ellipses represent water dipoles, the poles being marked accordingly, the small circles with plus signs are H^+ ions, and the large circles with minus signs are CH_3COO^- ions, both ions entering into the composition of undissociated molecules of acetic acid.*

The dissociation of CH_3COOH with the formation of H_3O^+ proceeds according to the equation

$$CH_3COOH + (n+1)H_2O \rightleftarrows H_3O^+ + [CH_3COO \cdot nH_2O]^-$$

* This interpretation of the ionization of CH_3COOH is simplified. In fact, as pointed out earlier (see Sec. 3.2), there are no H^+ and CH_3COO^- ions present in the molecules of CH_3COOH. These ions are formed when the CH_3COOH molecule interacts with water molecules. This is what constitutes the process of ionization.

The actual bearer of acidic properties is not the H^+ *ion as such but its hydrate, i.e., the hydronium ion.*

Further in the text we shall write (for the sake of simplicity) the H^+ ion instead of the hydronium ion H_3O^+ in reaction equations. Similarly, we shall not specify the presence of hydrate water in the formulas of all other ions, especially as its quantity is not exactly known and may vary depending on the conditions.

Thus, the process of dissociation (or ionization) should be regarded not as a purely physical process but rather as a chemical interaction between the solute and the solvent, leading to the formation of hydrates or solvates. This idea was first advanced in 1887, at a meeting of the Russian Physico-Chemical Society, by D. Mendeleyev who took it as the basis for the hydrate theory of solutions.

After S. Arrhenius had worked out his theory of electrolytic dissociation, it was suggested that ions become hydrated when electrolytes are dissolved. This hypothesis was confirmed by experiment. In this way S. Arrhenius's theory of electrolytic dissociation was tied up with D. Mendeleyev's hydrate theory of solutions.

The dissociation of the molecules of an electrolyte into ions is a reversible process; the resultant solvates of the oppositely charged ions collide with one another in solution and, being mutually attracted, may recombine to form solvated molecules. With hydrate water being disregarded, the dissociation of CH_3COOH may be represented by the following simplified equation:

$$CH_3COOH \rightleftarrows H^+ + CH_3COO^-$$

Like any reversible process, dissociation leads to chemical equilibrium and must obey the law of mass action.

We can therefore write:

$$\frac{[H^+][CH_3COO^-]}{[CH_3COOH]} = K_{CH_3COOH}$$

or, in general form,

$$\frac{C_{cat}C_{an}}{C_{mol}} = K \tag{3.5}$$

Here C_{cat} and C_{an} are respectively the concentrations of the cation and anion, and C_{mol} is the concentration of undissociated molecules of the electrolyte. The quantity K is the **dissociation constant** (or **ionization constant**) of the electrolyte. It is a measure of the tendency of an electrolyte to split up into ions. Indeed, as Eq. (3.5) shows, the greater the value of K, the higher must be the ion concentrations C_{cat} and C_{an}, i.e., the more highly ionized is the electrolyte.

We can easily arrive at another form of the equation for the dissociation constant if we note that in the case of a binary electrolyte, say, CH_3COOH, each molecule breaks down to form one cation (H^+) and one anion (CH_3COO^-). Hence, if the solution contains C moles of acetic acid per litre and its dissociation constant is α, the number of dissociated moles of CH_3COOH per litre of solution will be $C\alpha$ and so

$$C_{cat}=C_{an}=C\alpha$$

We can find the concentration of undissociated molecules of acetic acid (C_{mol}) by subtracting the number of moles dissociated ($C\alpha$) from its total concentration (C). It will equal

$$C_{mol}=C-C\alpha=C(1-\alpha)$$

Substituting the values of C_{cat}, C_{an}, and C_{mol} obtained into Eq. (3.5) gives

$$\frac{C\alpha C\alpha}{\alpha(1-\alpha)}=K$$

or

$$\frac{C\alpha^2}{1-\alpha}=K \tag{3.6}$$

This equation is an expression of the law known as **Ostwald's dilution law**. It establishes the relation between the degree of dissociation of a weak electrolyte and its concentration.

If the electrolyte is sufficiently weak and its solution is not too dilute, its degree of dissociation α is low and the value of $(1-\alpha)$ differs little from unity.

For this case we may write:

$$C\alpha^2 \approx K \quad \text{and} \quad \alpha \approx \sqrt{\frac{K}{C}} \tag{3.7}$$

From this it is seen that the degree of dissociation must increase as the solution is diluted (i.e., with decreasing C), which is found experimentally to be the case.

The dissociation constant of an electrolyte can easily be calculated by measuring its degree of dissociation α for any given concentration C and substituting the values into Eq. (3.6).

The results of such determinations for acetic acid are presented in Table 3.1.

TABLE 3.1. **Degree of Dissociation and Dissociation Constants of Acetic Acid at Different Concentrations***

Solution concentration C, mole/litre	Degree of dissociation α	Dissociation constant $K=\frac{C\alpha}{1-\alpha}$
0.2	0.00954 (0.954%)	1.83×10^{-5}
0.1	0.0136 (1.36%)	1.88×10^{-5}
0.01	0.0419 (4.19%)	1.83×10^{-5}
0.005	0.0585 (5.85%)	1.82×10^{-5}

* The calculation has been simplified, the activity coefficients being disregarded.

From this table one can see that, although the degree of dissociation of CH_3COOH increases with diminishing concentration of the acid, the dissociation constant remains unchanged (within the limits of experimental error). The fact that the quantity of K is independent of the solution concentration is the essence of the law of mass action. The data given in Table 3.1 serve as a good experimental confirmation of the validity of the law of mass action. At the same time, these data may be regarded as experimental proof of the concept of the existence of equilibrium between ions and undissociated molecules in solutions of weak electrolytes, which is what determines the behaviour of these electrolytes.

It should be noted that in the case of solutions of di- and polybasic acids there are several equilibria corresponding to the different dissociation steps, each with its own constant.

For example, for the two steps in the ionization of hydrogen sulphide (hydrosulphuric acid)

$$H_2S \rightleftarrows H^+ + HS^-$$

and

$$HS^- \rightleftarrows H^+ + S^{2-}$$

we accordingly have*:

$$K_{H_2S} = \frac{[H^+][HS^-]}{[H_2S]} = 8.9 \times 10^{-8}$$

and

$$K_{HS^-} = \frac{[H^+][S^{2-}]}{[HS^-]} = 1.3 \times 10^{-13}$$

Using the values of K, one can easily calculate the degrees of dissociation and ionic concentrations in solutions of various weak electrolytes, as shown in the following examples.

Example 1. Calculate the concentration of OH^- ions in $0.1N$ NH_4OH solution (at 25°C).

Solution. Write the equation for the dissociation constant of NH_4OH:

$$\frac{[NH_4^+][OH^-]}{[NH_4OH]} = K_{NH_4OH}$$

Denote the sought-for concentration of OH^- ions by x and note that according to the equation

$$NH_4OH \rightleftarrows NH_4^+ + OH^-$$

for each OH^- ion formed there appears one NH_4^+ ion in solution. Hence

$$[NH_4^+] = [OH^-] = x$$

* The numerical values given have been obtained for 25°C.

and

$$[NH_4OH] = C - x$$

where C is the total concentration of NH_4OH in the solution.

Consequently,

$$\frac{x^2}{C-x} = K_{NH_4OH}$$

Since the degree of dissociation of NH_4OH in not too dilute solutions is insignificant, the value of x is small as compared to C and it may therefore be neglected without loss of accuracy. Thus

$$\frac{x^2}{C} = K_{NH_4OH} \quad \text{and} \quad x = [OH^-] = [NH_4^+] = \sqrt{K_{NH_4OH}C} \qquad (3.8)$$

Substituting the numerical values of $K_{NH_4OH} = 1.76 \times 10^{-5}$ and $C = 0.1$ into Eq. (3.8), we get

$$x = [OH^-] = [NH_4^+] = \sqrt{1.76 \times 10^{-5} \times 0.1} = \sqrt{1.76 \times 10^{-6}} = 1.33 \times 10^{-3} \text{ g-ion/litre}$$

Example 2. Determine the degree of dissociation of NH_4OH in an $0.1N$ solution.

Solution. Substituting the numerical values of K_{NH_4OH} and C into formula (3.7), we have

$$\alpha = \sqrt{\frac{K_{NH_4OH}}{C}} = \sqrt{\frac{1.76 \times 10^{-5}}{0.1}} = \sqrt{1.76 \times 10^{-4}} = 1.33 \times 10^{-2} = 1.33 \text{ per cent}$$

If the concentration of the ions is already known (see Example 1), we can calculate the value of α as follows*:

$$\alpha = \frac{C_{ionized}}{C_{total}} = \frac{[OH^-]}{C} = \frac{1.33 \times 10^{-3}}{0.1} = 1.33 \times 10^{-2} = 1.33 \text{ per cent}$$

Example 3. Find the concentration of H^+ and HS^- ions in $0.1M$ H_2S solution (at 25°C).

Solution. Hydrogen ions are mainly formed as the result of the dissociation of H_2S at the first stage:

$$H_2S \rightleftarrows H^+ + HS^-$$

The dissociation at the second stage takes place to such a negligible extent that it may be disregarded without appreciable error. We shall therefore proceed from the equation

$$\frac{[H^+][HS^-]}{[H_2S]} = K_{H_2S} = 8.9 \times 10^{-8}$$

Reasoning as in Example 1, we obtain:

$$x = [H^+] = [HS^-] = \sqrt{K_{H_2S}C} = \sqrt{8.9 \times 10^{-8} \times 0.1} = \sqrt{89 \times 10^{-10}}$$
$$= 9.4 \times 10^{-5} \text{ g-ion/litre}$$

* The quantity $C_{ionized}$ is used to denote the molar concentration of NH_4OH which is present in solution in the form of ions (i.e., the ionized part of NH_4OH). Since each NH_4OH molecule yields one OH^- ion and one NH_4^+ ion, $C_{ionized} = [NH_4^+] = [OH^-]$.

3.4. Shift of Ionic Equilibria. The Common Ion Effect

As follows from the law of mass action, chemical equilibrium can be shifted in either direction by changing the concentration of one of the reacting substances. Such a shift of equilibrium can be effected in electrolytic dissociation as well.

If to a solution of acetic acid there is added one of its salts, a large quantity of CH_3COO^- ions will be introduced into the solution with the salt; these are known as **common ions** (i.e., the same as one of the ions of the given electrolyte). This will upset the equilibrium expressed by the formula*

$$K_{CH_3COOH} = \frac{[H^+][CH_3COO^-]}{[CH_3COOH]} = 1.74 \times 10^{-5}$$

The disturbed equilibrium will be restored as a result of the combination of the H^+ and CH_3COO^- ions with the formation of undissociated CH_3COOH molecules. Thus, *addition of a strong electrolyte containing a common ion* (CH_3COONa, CH_3COONH_4, *etc.*) *to a solution of a weak electrolyte* (CH_3COOH *in this case*) *decreases the degree of dissociation of the weak electrolyte*. The hydrogen ion concentration in the acetic acid solution is greatly diminished, as a result of which acetic acid behaves as a much weaker acid in the presence of its salt. For example, CH_3COOH reacts slowly with metallic zinc. But if to a solution of CH_3COOH is added a small amount of CH_3COONa or CH_3COONH_4, the evolution of hydrogen will almost entirely stop. In just the same way, it is easy to lower the concentration of OH^- ions in a solution of NH_4OH by addition of an ammonium salt which gives much more NH_4^+ ions than were present in the original solution.

This can easily be confirmed by the following experiment. To a solution of NH_4OH is added a drop of the indicator phenolphthalein which turns crimson in the presence of alkalis (i.e., in the presence of OH^- ions). After mixing, part of the coloured solution is left for comparison and to another portion is added a small amount of a solid ammonium salt, such as CH_3COONH_4, and the liquid is stirred to dissolve the added salt. The colour of the solution becomes much paler, indicating a considerable decrease of the OH^- ion concentration in the solution.

By using the equation for the dissociation constant, it is not difficult to figure out to what extent the concentration of a particular ion (e.g., OH^-) in a solution of a weak electrolyte will be changed by addition of a definite amount of a strong electrolyte with a common ion to that solution.

---

* This value of the thermodynamic ionization constant of acetic acid has been obtained with activity coefficients taken into account. Therefore, its absolute value differs somewhat from that given in Table 3.1 (see Sec. 3.3).

Example. By how many times will the concentration of OH^- ions decrease in 0.1*N* NH_4OH solution upon addition of solid NH_4Cl to 1*M* concentration?

Solution. Since in calculations made earlier (Sec. 3.3, Example 1) the concentration of OH^- ions in 0.1*N* NH_4OH solution was found to be equal to 1.33×10^{-3} g-ion/litre, it now remains to figure out what it will be in a mixture of NH_4OH and NH_4Cl. Let this concentration of OH^- ions be x.

The NH_4^+ ion concentration in the mixture is made up of two quantities: x g-ions formed in a litre by dissociation of NH_4OH and 1 g-ion formed in a litre by dissociation of NH_4Cl (as it is a strong electrolyte, it may be assumed to dissociate completely: each molecule of NH_4Cl gives one NH_4^+ ion). Hence

$$[NH_4^+] = (1+x) \text{ g-ion/litre}$$

To find the concentration of un-ionized NH_4OH molecules, we subtract the number of moles dissociated in 1 litre, i.e., x, from the total concentration of ammonia; hence, $[NH_4OH] = (0.1 - x)$ mole/litre. Substituting these values into the equation for the dissociation constant of NH_4OH, we obtain:

$$\frac{[NH_4^+][OH^-]}{[NH_4OH]} = \frac{(1+x)\,x}{0.1-x} = 1.76 \times 10^{-5}$$

Rearranging this expression, we can arrive at a complete quadratic equation solvable in a usual manner.

The problem however can be simplified if account is taken of the fact that the value of x is very small as compared with 1 and 0.1, and it may therefore be neglected without appreciable error.* We then have:

$$\frac{1x}{0.1} = 1.76 \times 10^{-5} \quad \text{and} \quad x = 1.76 \times 10^{-6} \text{ g-ion/litre}$$

Comparing this result with the above value (1.33×10^{-3}), we see that the addition of NH_4Cl decreases the OH^- ion concentration in the solution by a factor of

$$\frac{1.33 \times 10^{-3}}{1.76 \times 10^{-6}}$$

that is, approximately by 750 times.**

It has been found experimentally that this conclusion is qualitatively correct.

The possibility of lowering very strongly the concentration of a particular ion in solution by addition of an electrolyte with a common ion is extensively used in analytical practice.

---

* Small quantities may be neglected as compared with much larger ones provided only that they are added or subtracted but not multiplied or divided.

** This calculation is inaccurate because of the presence in solution of a strong electrolyte (NH_4Cl) which gives rise to strong electrostatic interaction forces operating between the ions (see Sec. 3.6); the effect of these interaction forces have not been taken into account in this case.

The equilibrium in the dissociation of a weak electrolyte can be shifted not only by increasing the concentration of one of its ions in solution but also by lowering it. For example, we can bind the OH^- ions of ammonia into un-ionized H_2O molecules by adding an acid to the solution, which will result in the ionization of new molecules of NH_4OH. Thus, un-ionized molecules play, as it were, the role of reserve which makes up, through dissociation, for the loss of ions in solution. Thus, *if one of the substances in chemical equilibrium is removed from the reaction space, the process that will predominate is the one in which new quantities of that substance are formed.*

It was shown in Sec. 1.2 that, since the reactions used in analytical chemistry take place between ions, it is more convenient to represent them in the form of ionic equations. For the reaction in question, the ionic equation is

$$NH_4OH + HCl \rightleftharpoons NH_4Cl + H_2O$$

Since NH_4OH and H_2O are weak electrolytes, they are present in solution mainly as undissociated molecules. Conversely, the strong electrolytes HCl and NH_4Cl are present in solution as ions. For example,

$$NH_4OH + H^+ + Cl^- \rightleftharpoons NH_4^+ + Cl^- + H_2O$$

and

$$NH_4OH + H^+ \rightleftharpoons NH_4^+ + H_2O$$

The last equation clearly shows that, as the result of the reaction, the un-ionized NH_4OH molecules disappear from the solution and NH_4^+ ions are accumulated in it. In the same way, the reaction between CH_3COOH and NaOH may be represented by the equation

$$CH_3COOH + OH^- \rightleftharpoons CH_3COO^- + H_2O$$

From all the foregoing it follows that, in setting up ionic equations, the formulas of all the strong electrolytes present in solution must be written in the form of ions and those of the weak electrolytes and slightly soluble substances which are precipitated, or of substances removed in the gaseous state, must be presented in the molecular form. The ions which remain unchanged are omitted from ionic equations.

3.5. The State of Strong Electrolytes in Solution

The dissociation of weak electrolytes leads to chemical equilibrium and can be characterized by a definite constant; *the dissociation of strong electrolytes, as shown by experiment, does not obey the law of mass action*. This means that if we determine the values of α for different concentrations of a strong electrolyte and substitute them into the equation

$$\frac{C\alpha^2}{1-\alpha} = K$$

the values of K obtained will also be different, as shown by the data of Table 3.2 (compare with the data given in Table 3.1).

TABLE 3.2. **Dissociation of KCl at 18°C**

Concentration C, moles/litre	Degree of dissociation α	Dissociation constant $K=\frac{C\alpha^2}{1-\alpha}$
2	0.712	3.52
1	0.756	2.34
0.5	0.788	1.46
0.1	0.862	0.538
0.01	0.942	0.153

For other strong electrolytes, the value of K is also different at different concentrations. Consequently, as distinct from weak electrolytes, *strong electrolytes have no dissociation constant.*

The fact that solutions of strong electrolytes do not obey the law of mass action is in sharp contradiction to the classical theory of electrolytic dissociation. This phenomenon could not be explained satisfactorily for a long time. The behaviour of strong electrolytes have come to be accounted for by the theory of strong electrolytes introduced by P. Debye and E. Hückel in 1923. Because of the mathematical complexity, only a very general outline of the theory is given here.

According to this theory, strong electrolytes, unlike weak electrolytes, dissociate virtually completely in solution. Apart from the fact that they do not follow the law of mass action, this view is confirmed by the optical and spectral investigations of solutions of strong electrolytes which have not revealed the presence in them of un-ionized molecules. This distinguishes them from solutions of weak electrolytes, in which the respective molecules are detected.

Note. In solutions of strong electrolytes, there may form ion associations or ion pairs, i.e., temporary formations consisting of two or more ions. These formations exist for a very short time and quickly break down under the influence of water dipoles. In accurate calculations however their existence cannot be ignored.

Finally, X-ray studies of crystals have shown that strong electrolytes in the solid state have ionic crystal lattices rather than molecular ones. This is an especially conclusive confirmation of the hypothesis of complete dissociation of strong electrolytes in solution. Indeed, if, for example, crystals of KCl are built up not of molecules but of K^+ and Cl^- ions, one can naturally suppose that, when the salt is dissolved, the solution will contain K^+ and Cl^- ions rather than KCl molecules. But if this is so, why determinations of electrical conductance, osmotic pressure, freezing-point depression or boiling-point elevation for solutions of these substances lead us to the conclusion that the dissociation of strong electrolytes is incomplete? The theory provides the following answer to this question.

The degree of dissociation of electrolytes is usually determined from the equivalent conductance of their solutions, which depends not only on the degree of dissociation but on the velocity of ions as well. Other conditions being equal, the quantity of electricity carried by a given amount of ions must be the greater the higher the velocity of the ions.

In the Arrhenius theory of electrolytic dissociation, it is assumed that the velocity of the ions is determined only by their nature and is independent of the solution concentration. Therefore, the Arrhenius theory ascribes the increase in the equivalent conductance of solutions with dilution exclusively to the increase in the degree of dissociation α.

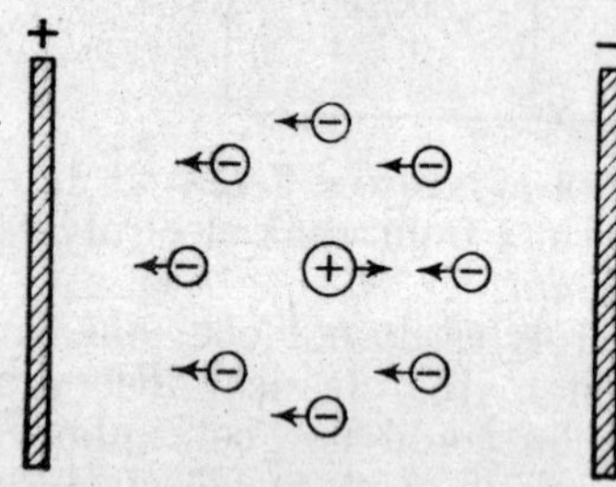

Fig. 3.3. Movement of ions in electrolysis (shown schematically).

It is difficult however to assume that the velocity of ions of a strong electrolyte is independent of their concentration in solution. Dissociation of strong electrolytes results in the formation of a large quantity of ions which (especially in more concentrated solutions) are so close to one another that electrostatic forces of attraction and repulsion must operate between them. But the action of these forces will diminish the velocity of the ions, and the nearer are the ions to one another, i.e., the higher the concentration of the solution, the greater is this effect.

Indeed, owing to the interionic forces, each ion appears to be surrounded by an **ionic atmosphere**, a spherical cloud of ions having a charge opposite to that of the central ion. When no electrostatic forces act on the solution and no diffusion, chemical reactions, and other similar processes occur in it, the ionic atmosphere is completely symmetrical and the attraction forces acting on the central ion are therefore mutually balanced. But if electrodes connected to a source of current are immersed in the solution, this symmetry will be continuously disturbed, as the central ion and its ionic atmosphere will move in opposite directions (Fig. 3.3). The ionic atmosphere left by the ion tends to hold it back, thereby slowing down its motion. All the ions are subjected to electrostatic attraction on their way to the respective electrodes because each ion, being pulled out of its ionic atmosphere, is found to get into the sphere of attraction of other ions, i.e., it becomes surrounded by a new ionic atmosphere, and so on. It is also apparent that the denser the ionic atmosphere, i.e., the higher the concentration (and also the charges) of the corresponding ions in solution, the greater must be the decrease in the mobility of the ions.

On this basis, the theory of strong electrolytes attributes the increase of the equivalent conductance with dilution not to an increase in the degree of dissociation of the electrolyte but to the increase in the velocity of its ions caused by the thinning of the ionic atmosphere and by the weakening of the interionic forces.

Thus, for solutions of strong electrolytes the ratio λ/λ_∞ is not a measure of their true dissociation into ions (which is assumed to be 100 per cent), but characterizes the effect of interionic forces on the ability of the ions to carry electricity. The value of α calculated from the electrical conductance of the corresponding solution is not the true but the **apparent degree of dissociation** for strong electrolytes. In this case, the value of α shows that, because of the retarding action of interionic forces, the corresponding electrolyte conducts an electric current so as if not all the molecules were dissociated into ions (as is actually the case) but only a fraction of them, equal to α. The same applies to other methods for determining α, in which not the true but the apparent degree of dissociation of a strong electrolyte is determined.

Later investigations have shown that the assumption of 100-percent dissociation of strong electrolytes is not quite rigorous. Certain facts indicate that solutions of strong electrolytes also contain some un-ionized molecules. Their number, however, is so small (incomparably less than was assumed by the Arrhenius theory) that their effect on the properties of the solutions is negligible. To simplify the reasoning, we shall assume that strong electrolytes are 100-percent dissociated.

3.6. Activity. Activity Coefficient

Interionic forces not only lower the conductance of solutions but also have an effect on the osmotic pressure, freezing point and boiling point of solutions, and on the ability of ions to participate in chemical interactions.

This ability is estimated in terms of **activity** (a). *The activity of an ion is its effective, apparent concentration in chemical reactions.* For example, if the activity of H^+ and Cl^- ions in 0.1*N* HCl solution is 0.0814, this means that these ions behave in chemical reactions as if their concentration were 0.0814 g-ion/litre and not 0.1 g-ion/litre.* It follows from the foregoing that the activity is expressed in the same units as the concentration, i.e., in moles or in gram-ions per litre. *The ratio of the activity to the true concentration of the ion is called the*

* The value of the activity given is the mean activity of these ions. It is equal to $a = \sqrt{a_c a_a}$, where a_c and a_a are the activities of the cation and anion, respectively.

activity coefficient (f_a). In the case under consideration

$$f_a = \frac{0.0814}{0.1} = 0.814$$

In a general form, denoting the activity by a and the ionic concentration by C, we have:

$$f_a = \frac{a}{C}$$

whence

$$a = f_a C$$

Thus, *the activity of an ion is the product of its concentration by the corresponding activity coefficient.*

The concept of the activity coefficient was first introduced by the Danish scientist N. Bjerrum in 1918. For not too concentrated solutions the activity coefficients are less than unity and $a < C$. For the reasons not considered here, in the case of concentrated solutions of strong electrolytes, the activity coefficients may assume values larger than unity. For example, in $2N$ and $3N$ solutions of HCl they are respectively equal to 1.02 and 1.32.

A further extension of the theory of strong electrolytes was the law of ionic strength discovered by the American scientists G. N. Lewis and M. Randall in 1921. This law states that the activity coefficient of a given electrolyte is the same in all dilute solutions of equal ionic strength. The **ionic strength** of a solution (μ) is defined as the half-sum of the products of the concentrations of all the ions present in solution by the square of the charge of the given ion:

$$\mu = \frac{1}{2}(C_1 z_1^2 + C_2 z_2^2 + \ldots + C_n z_n^2)$$

Here C_1, C_2, . . ., C_n stand for the concentrations of the individual ions present in solution, in g-ions/litre; and z_1, z_2, . . ., z_n are their charges.

Thus, the ionic strength of a solution, which is $0.1M$ with respect to HCl and $0.2M$ in $CaCl_2$, is equal to:

$$\mu = \frac{1}{2}(0.1 + 0.2 \times 2^2 + 0.5) = 0.7$$

Here 0.1 is the H^+ concentration, 0.2 is the Ca^{2+} concentration, and 0.5 is the Cl^- concentration; the value of z for H^+ and Cl^- is unity and for Ca^{2+} it is equal to 2.

Experiment shows that, in sufficiently dilute solutions of equal ionic strength, the activity coefficients of ions having equal charge are approximately the same.

The approximate values of mean activity coefficients are given in Table 3.3.

TABLE 3.3. **Activity Coefficients***

Ionic strength μ	Activity coefficient f_a			
	singly charged ions	doubly charged ions	triply charged ions	quadruply charged ions
0.001	0.96	0.86	0.73	0.56
0.005	0.92	0.72	0.51	0.30
0.01	0.89	0.63	0.39	0.19
0.05	0.81	0.44	0.15	0.04
0.1	0.78	0.33	0.08	0.01

* These are mean activity coefficients which are approximately equal for ions of equal charge. In fact, these coefficients differ somewhat from one another for different ions of equal charge since the ionic radius is an important factor here, which in its turn depends on the position of a given element in Mendeleyev's Periodic System.

The mathematical relationship between the ionic strength of a solution and the activity coefficient was found in 1923 by P. Debye and E. Hückel. Depending on the concentration of the solution, this relation is expressed by different formulas. For dilute solutions (0.01-0.05*N*) the following formula holds:

$$-\log f = 0.5z^2 \sqrt{\mu}$$

For more concentrated solutions (0.1-0.5*N*) the formula becomes complicated:

$$-\log f = 0.5z^2 \frac{\sqrt{\mu}}{1+\sqrt{\mu}}$$

In still more concentrated solutions, the dissolved electrolyte reacts with the solvent molecules, and ion associations, polymeric ions and molecules are formed; therefore, the calculation becomes still more involved. The empirical coefficients, the physical significance of which has not yet been conclusively explained, have to be introduced into the calculation formula.

For strong electrolytes, the ionic concentration C is calculated by proceeding from the assumption of complete dissociation of electrolytes. As regards the activity coefficient f_a, according to the theory of strong electrolytes, it is a measure of the effect of interionic forces on the ability of an ion to take part in chemical reactions. If f_a is less than unity, this means that the ion is confined in its movement; then $a < C$, i.e., the given amount of ions (C g-ion/litre) acts as if there were less ions (a g-ion/litre). If f_a is equal to unity, then $a = C$, i.e., the ion acts in accordance with its concentration in solution. For strong electrolytes, this is observed only in very dilute solutions (about $0.0001M$) in which the distance between the ions is so great that the forces operating between them have practically no effect. In the same way we may ignore interionic forces in not too concentrated solutions of weak electrolytes, where only a very small proportion of the molecules breaks down into ions. Therefore, for such solutions, we may also assume that $f_a = 1$ and $a = C$.

According to the activity concept, *not the concentrations of the ions but their activities must be included in the equation for the law of mass action.* In fact, activity coefficients were originally introduced as empirical factors to extend the law of mass action to cases where it was inapplicable in its conventional form. The physical meaning of these coefficients was later elucidated in the light of the theory of strong electrolytes. All the relevant questions are treated in detail in physical chemistry texts. Dilute solutions of weak electrolytes obey the law of mass action with sufficient accuracy. Conversely, the activity of ions of strong electrolytes generally differs greatly from their concentration, and the law of mass action in its usual form proves to be invalid. Nor can it be applied to weak electrolytes if the solution contains any extraneous strong electrolytes.*

From all that has been said about the effect of interionic forces, it follows that the activity coefficient depends not only on the concentration of the given electrolyte in solution but also on the concentration of all other electrolytes present in solution. The activity coefficient diminishes with increase of the ionic strength of the solution, which depends on the concentrations and charges of all the ions present.

Since there are usually several ions in solution, whose concentrations are not accurately known, the use of activity coefficients in calculations appears to be impossible. We shall therefore confine ourselves to illustrating their use by two or three examples; in all other cases, ionic activities will be replaced in calculations by ionic concentrations; for strong electrolytes, the calculations are based on the assumption of their complete dissociation in solution. This simplification is quite permissible since in qualitative analysis the calculations are usually made with the purpose of merely illustrating some theoretical aspects, and a high degree of accuracy is not needed. From the qualitative point of view, such simplified calculations are, in general, in

* Calculation of the ion concentrations for a weak electrolyte in the presence of a strong one with a common ion does not yield accurate results, as in the example given in Sec. 3.4, if the concentrations of ions are used instead of their activities. For the method of more accurate calculations, see the following page.

good agreement with experiment; for other cases, there will be given appropriate directions.*

Example. Let us calculate the activity of hydrogen ions (a_{H^+}) in a solution containing 0.1 mole of CH_3COOH and 0.05 mole of CH_3COONa in 1 litre.

Because of the presence of a strong electrolyte, CH_3COONa, the equation for the dissociation constant of acetic acid in its usual form

$$\frac{[H^+][CH_3COO^-]}{[CH_3COOH]} = K_{CH_3COOH}$$

fails to yield accurate results since K varies with the ionic strength of the solution. To obtain accurate results, it is necessary to substitute the activities for the concentrations:

$$\frac{a_{H^+}a_{CH_3COO^-}}{a_{CH_3COOH}} = K^0 \tag{3.9}$$

Here K^0 is the so-called **thermodynamic dissociation constant** of acetic acid, which is equal to 1.74×10^{-5} (at 25°C). In contrast to the ordinary (concentration) dissociation constant (K_C), the quantity K^0 remains constant in the presence of strong electrolytes.

Solving Eq. (3.9) for a_{H^+}, we get:

$$a_{H^+} = K^0 \frac{a_{CH_3COOH}}{a_{CH_3COO^-}} = K^0 \frac{[CH_3COOH]\, f_{CH_3COOH}}{[CH_3COO^-]\, f_{CH_3COO^-}}$$

The activity coefficient of un-ionized molecules (f_{CH_3COOH} in this particular case) is assumed to be equal to unity, and their concentration is taken to be practically equal to the total concentration (0.1*M*). The concentration of CH_3COO^- ions may be taken equal to the concentration of the salt (0.05*M*) since the dissociation of acetic acid results in the formation of a very small amount of these ions. The activity coefficient of the anions ($f_{CH_3COO^-}$) is found by calculating the ionic strength of the solution (μ). In this calculation, the small concentrations of H^+ and CH_3COO^- ions formed by the dissociation of CH_3COOH may be neglected, and only the concentrations of the Na^+ and CH_3COO^- ions resulting from the dissociation of the salt need to be taken into account. As these concentrations are equal to the total concentration of the salt, i.e., to 0.05 g-ion/litre, it follows that

$$\mu = \frac{1}{2}[0.05 \times 1^2 + 0.05 \times 1^2] = 0.05$$

---

* For more detail concerning the theory of strong electrolytes, see: Gerasimov, Ya. I. *et al.*, *Physical Chemistry*, vol. 1, Mir Publishers, Moscow, 1974; Harned, H. S. and B. B. Owen, *The Physical Chemistry of Electrolytic Solutions*, 3rd ed., Reinhold Publishing Corporation, New York, 1958. For calculations taking into account activity coefficients, see Laitinen, H. A., *Chemical Analysis*, McGraw-Hill Book Company, New York, 1960.

7—1026

With the ionic strength being equal to 0.05, the activity coefficient of singly charged ions is 0.81 (see Table 3.3). Hence

$$a_{H^+} = 1.74 \times 10^{-5} \times \frac{0.1}{0.05 \times 0.81} = 4.3 \times 10^{-5} \text{ g-ion/litre}$$

Thus, in reactions involving hydrogen ions, the solution behaves as if the hydrogen ion concentration were 4.3×10^{-5} g-ion/litre.

For more accurate calculations, in addition to electrostatic attraction forces operating between the ions, one should also consider the possibility of the occurrence of side reactions and the formation of ions of different composition.

For example, in calculating the concentration of S^{2-} ions in a saturated solution of H_2S at different hydrogen ion concentrations, one should take into account the fact that sulphur is present in the following forms: H_2S, HS^-, and S^{2-}. If the concentrations of these forms are denoted respectively by $[H_2S]$, $[HS^-]$, and $[S^{2-}]$, the total concentration of sulphur will then be:

$$C_S = [S^{2-}] + [HS^-] + [H_2S] \tag{3.10}$$

Since we are interested in the concentration of S^{2-}, we have to define the concentrations of $[HS^-]$ and $[H_2S]$ by proceeding from the corresponding dissociation constants of hydrogen sulphide (hydrosulphuric acid):

$$K_{H_2S} = \frac{[H^+][HS^-]}{[H_2S]} \quad \text{and} \quad K_{HS^-} = \frac{[H^+][S^{2-}]}{[HS^-]}$$

Rearranging the second constant with respect to $[HS^-]$, we have:

$$[HS^-] = \frac{[H^+][S^{2-}]}{K_{HS^-}} \tag{3.11}$$

The overall dissociation constant of hydrogen sulphide equals the product of the first and second constants:

$$K_{overall} = K_{H_2S} K_{HS^-} = \frac{[H^+]^2[S^{2-}]}{[H_2S]}$$

which gives

$$[H_2S] = \frac{[H^+]^2[S^{2-}]}{K_{H_2S} K_{HS^-}} \tag{3.12}$$

Substituting the values of $[HS^-]$ and $[H_2S]$ from Eqs. (3.11) and (3.12) into Eq. (3.10), yields

$$C_S = [S^{2-}] + \frac{[H^+][S^{2-}]}{K_{HS^-}} + \frac{[H^+]^2[S^{2-}]}{K_{H_2S} K_{HS^-}} = [S^{2-}] \left(1 + \frac{[H^+]}{K_{HS^-}} + \frac{[H^+]^2}{K_{H_2S} K_{HS^-}}\right)$$

whence

$$[S^{2-}] = C_S \left(\frac{1}{1 + \frac{[H^+]}{K_{HS^-}} + \frac{[H^+]^2}{K_{H_2S} K_{HS^-}}}\right) \tag{3.13}$$

The factor for C_S, which is enclosed in parentheses in Eq. (3.13), is a constant quantity at the given hydrogen ion concentration. Denoting it by γ, we have:

$$[S^{2-}] = C_S\gamma_S \text{ and } \gamma_S < 1 \tag{3.14}$$

For precise calculations, apart from the electrostatic coefficients f, coefficients γ taking into account the occurrence of side reactions must also be introduced into the expression for the equilibrium constant. In the general case, for the reaction

$$A + B \rightleftharpoons C + D$$

there are three constants.

The true or thermodynamic constant K^0 remains unaltered for solutions of any concentration, regardless of the ionic strength of the solution and the occurrence of side reactions:

$$K^0 = \frac{a_C a_D}{a_A a_B}$$

This constant depends on temperature and pressure only. The concentration constant K_C depends not only on the temperature and pressure but on the ionic strength of the solution as well and remains unchanged at a constant value of the ionic strength:

$$K_C = \frac{[C][D]}{[A][B]}$$

The relation between the thermodynamic and concentration constants is expressed by the following equation:

$$K^0 = \frac{[C][D]}{[A][B]} \cdot \frac{f_C f_D}{f_A f_B} = K_C \frac{f_C f_D}{f_A f_B}$$

If side reactions take place in the solution with the formation of ions of different composition (as in the case of hydrogen sulphide given above), the corresponding coefficients γ_A, γ_B, γ_C, and γ_D must be introduced.

If the total concentration of a given substance is denoted by C (C_A, C_B, C_C, C_D), then the so-called conditional equilibrium constant will be expressed by the equation

$$K_{cond} = \frac{C_C C_D}{C_A C_B}$$

The relation between the conditional and concentration constants is expressed as follows:

$$K_C = \frac{[C][D]}{[A][B]} = \frac{C_C C_D}{C_A C_B} \cdot \frac{\gamma_C \gamma_D}{\gamma_A \gamma_B} = K_{cond} \frac{\gamma_C \gamma_D}{\gamma_A \gamma_B}$$

The conditional constant depends also on the ionic strength of the solution and on the occurrence of side reactions. It remains constant provided only that the ionic strength of the solution is constant for a given definite concentration of the ion which is responsible for a side reaction (in the case of hydrogen sulphide—for a given definite concentration of hydrogen ions). The general relation between the three constants is given by the equation

$$K^0 = K_C \frac{f_C f_D}{f_A f_B} = K_{\text{cond}} \frac{\gamma_C \gamma_D}{\gamma_A \gamma_B} \cdot \frac{f_C f_D}{f_A f_B}$$

Since it is usually rather difficult to calculate the coefficients γ, they are seldom used and only the effect of the electrostatic attraction and repulsion forces operating between the ions is taken into account by introducing the activity coefficients f.

3.7. Modern Views on the Nature of Acids and Bases

The process of dissociation or ionization of substances in solutions is accounted for by the interaction between the solute and the solvent molecules (molecules of water in a particular case). This interaction results in the formation of compounds consisting of the solute ions and ions or molecules of the solvent. The classical theory of acids and bases fails to account for a number of phenomena which occur on dissolution of a given substance in various solvents. For instance, ammonium chloride dissociates in aqueous solutions into NH_4^+ and Cl^- ions, i.e., it behaves like other salts; at the same time, when dissolved in liquid ammonia, it exhibits all the typical properties of acids, including the ability to dissolve metals with evolution of hydrogen, though these solutions evidently cannot contain H^+ ions. Urea, $CO(NH_2)_2$, which is neutral in aqueous solutions, behaves as an acid in liquid ammonia and as a base in anhydrous acetic acid. Nitric acid, which is a very strong acid in aqueous solution, behaves as a base when dissolved in liquid HF or in anhydrous H_2SO_4. Many facts that contradict the theory of electrolytic dissociation could be cited.

Several theories have been put forward to account for all these contradictions. The most widely accepted was the protolytic theory of acids and bases proposed independently in 1923 by Brönsted in Denmark and by Lowry in England. They defined an acid as a substance that can donate a proton to another substance (a proton donor), and a base as a substance that can accept a proton from another substance (a proton acceptor). In these definitions the word "substance" is interpreted very broadly to mean any species of molecule or ion capable of undergoing reaction. If we designate a proton by the symbol H^+, then the relation between a given acid and the corresponding base may be represented by the equation

$$\text{Acid} \rightleftharpoons \text{Base} + H^+$$

Hence, we are led to the concept of a conjugate acid-base pair. This definition is much broader than the acid-base concept provided by the classical theory. For example, acetic acid will be an acid according to the Brönsted theory as well, since it is capable of giving up a proton:

$$\underset{\text{Acid}}{CH_3COOH} \rightleftharpoons \underset{\text{Base}}{CH_3COO^-} + H^+$$

Ammonia will be defined as a base in accordance with the Brönsted theory too, since it is capable of accepting protons:

$$\underset{\text{Base}}{NH_3} + H^+ \rightleftharpoons \underset{\text{Acid}}{NH_4^+}$$

According to Brönsted, however, not only molecules, but also ions capable of giving up protons, are acids:

$$NH_4^+ \rightleftharpoons NH_3 + H^+$$
$$HSO_4^- \rightleftharpoons SO_4^{2-} + H^+$$

In their turn, ions capable of accepting protons must be regarded as bases:

$$CH_3COO^- + H^+ \rightleftharpoons CH_3COOH$$
$$S^{2-} + H^+ \rightleftharpoons HS^-$$

Some ions or molecules may act both as an acid and as a base, depending on the conditions. For instance,

$$\underset{\text{Base}}{HSO_3^-} + H^+ \rightleftharpoons \underset{\text{Acid}}{H_2SO_3}$$
$$\underset{\text{Acid}}{HSO_3^-} \rightleftharpoons H^+ + \underset{\text{Base}}{SO_3^{2-}}$$

8-Hydroxyquinoline reacts as an acid in the equilibrium

$$C_9H_6N(OH) \rightleftharpoons C_9H_6N(O^-) + H^+$$

and as a base in the acid-base equilibrium

$$C_9H_6N(OH) + H^+ \rightleftharpoons C_9H_6NH^+(OH)$$

Since protons cannot exist independently in solution, the process of splitting-off of a proton from an acid must always be accompanied by the simultaneous process in which that proton is accepted by a base. For example, the reaction between acetic acid and ammonia may be represented as follows:

$$CH_3COOH \rightleftharpoons CH_3COO^- + H^+$$

$$NH_3 + H^+ \rightleftharpoons NH_4^+$$

As a result, the following neutralization reaction takes place:

$$\underset{Acid_1}{CH_3COOH} + \underset{Base_2}{NH_3} \rightleftharpoons \underset{Base_1}{CH_3COO^-} + \underset{Acid_2}{NH_4^+}$$

In an acid-base pair, to a weak acid there corresponds a strong base, and vice versa. To the weak acid, CH_3COOH, there corresponds a fairly strong base, CH_3COO^-, which adds up protons rather energetically. To the fairly strong base, NH_3, there corresponds a weak acid, NH_4^+. With all that has been said above we can arrive at an explanation of the processes that occur in solutions.

If the solvent is water, we should first of all emphasize its dual nature as an electrolyte. Water may act as an acid, giving up a proton:

$$H_2O \rightleftharpoons OH^- + H^+$$

or it may act as a base, accepting a proton:

$$H_2O + H^+ \rightleftharpoons H_3O^+$$

Combining these two equations*, we have:

$$2H_2O \rightleftharpoons OH^- + H_3O^+$$

Analogously, one could write:

$$2CH_3COOH \rightleftharpoons CH_3COOH_2^+ + CH_3COO^-$$

$$2NH_3 \rightleftharpoons NH_2^- + NH_4^+$$

Such compounds, which are capable of functioning as an acid and as a base simultaneously, are said to be **amphiprotic (amphoteric)**.

Depending on which tendency (to accept or to donate a proton) is predominating in a given solvent, it may exert a different effect on the ionization of the substance dissolved in it. It is clear that, other conditions being equal, acids must dissociate the more strongly, the more pronounced are the basic properties of the solvent, i.e., the greater its affinity for a proton. Conversely, bases dissociate the more strongly, the less pronounced is the proton affinity of the solvent, i.e., the more readily the solvent molecules give up protons.

Many examples could be cited to confirm the validity of these conclusions. For instance, water is a fairly strong basic solvent, having a relatively strong proton affinity. Accordingly, when strong acids such as $HClO_4$, H_2SO_4, HCl, and HNO_3, are dissolved in water, the protolytic equilibrium between the acid (HAn) and the solvent (H_2O)

$$\underset{Acid_1}{HAn} + \underset{Base_2}{H_2O} \rightleftharpoons \underset{Acid_2}{H_3O^+} + \underset{Base_1}{An^-}$$

---

* These equations are simplified: like other ions in aqueous solutions, hydronium ions are also hydrated; further in the text, formulas are given in simplified form.

is shifted almost completely to the right, which corresponds to complete dissociation of these acids in solution. The differences in their tendency to lose protons in aqueous solutions are completely masked by the greater tendency of this solvent to gain protons.

If a less basic solvent, such as ethyl alcohol, C_2H_5OH, is used instead of water, the difference in strength of the acids becomes noticeable. Indeed, in this solvent, nitric acid behaves as a moderately strong acid. In solvents of even lower proton affinity, such as acetone (CH_3COCH_3) or nitrobenzene ($C_6H_5NO_2$), all acids which are strong in aqueous solutions, with the exception of $HClO_4$ and H_2SO_4, become moderately strong.

Finally, in a solvent having an insignificant proton affinity and hence a high tendency to lose a proton, say, in anhydrous H_2SO_4, only perchloric acid, $HClO_4$, which has an even weaker proton affinity, continues to act as a strong acid. Nitric acid in this solvent exhibits basic properties as it gains protons from H_2SO_4 to form a nitronium ion $[NO_3H_2]^+$, which is a cationic acid similar in properties to ammonium and hydronium ions.* Organic acids, esters, alcohols, and other similar compounds are also bases in anhydrous H_2SO_4.

The reverse relations are observed in solvents with basic properties. An example is liquid ammonia, a solvent more basic than water, in which many acids, being very weak in aqueous solution, become quite strong.

Measurements carried out in solvents with acidic properties have revealed the following order of decreasing tendency to lose protons for acids which are equally strong in aqueous solution:

$$HClO_4 > H_2SO_4 > HCl > HBr > HI > HNO_3$$

The protolytic theory, which allows us to predict the behaviour of various substances in non-aqueous as well as aqueous solutions, is a much more general theory of acids and bases than the conceptions based on the Arrhenius theory of electrolytic dissociation. Furthermore this theory makes it possible to interpret, from a single viewpoint, processes which belong to entirely different types according to earlier views. These processes include neutralization reactions, the interaction between strong acids and salts of weak acids or between strong bases and salts of weak bases, the hydrolysis of salts, and the ionization of acids and bases. The protolytic theory** draws an analogy between protolytic reactions and oxidation-reduction reactions, the latter consisting in transferring electrons from atoms (or ions) of the reducing agent to atoms (or ions) of the oxidizing agent.

---

* Nitronium perchlorate, $[NO_3H_2]ClO_4$, has been isolated in the solid state.

** For more detail, see: Van der Werf, C. A., *Acids, Bases and the Chemistry of the Covalent Bond*, Reinhold, New York, 1961; Cotton, F. A. and G. Wilkinson, *Advanced Inorganic Chemistry*, Interscience (Wiley), New York, 1962.

3.8. Ionization of Water. The Hydrogen Ion Exponent

The reactions used in analytical chemistry take place most often in aqueous solution. Water is a weak electrolyte which ionizes, though very slightly, as follows:

$$H_2O \rightleftharpoons H^+ + OH^-$$

Therefore, in addition to the solute ions, the solution being analysed always contains H^+ and OH^- ions, a fact that must be considered by the analyst.

The degree of ionization of water is very low. For example, at 25°C only 1/10,000,000 (or 10^{-7}) mole of H_2O is dissociated per litre of water.*

Since a water molecule forms one H^+ ion and one OH^- ion on dissociation, the concentration of these ions in pure water is

$$[H^+] = [OH^-] = 10^{-7} \text{ g-ion/litre (at 25° C)}$$

As for other weak electrolytes, the ionization of H_2O molecules can be characterized by the corresponding ionization constant**:

$$\frac{[H^+][OH^-]}{[H_2O]} = K_{eq}$$

We transform this equation by transposing the quantity $[H_2O]$ to the right-hand side:

$$[H^+][OH^-] = K_{eq}[H_2O]$$

Since water is present almost entirely in the form of undissociated molecules, any change in the degree of ionization will have a very little effect on the concentration of these molecules. Therefore, the quantity $[H_2O]$ in the above equation may be regarded as practically constant.*** The ion product K_{eq} is also constant. Thus, the whole

---

* This value can be calculated from the electrical conductance of pure water.

** The electrical conductance method is used to determine the activity of ions in solution, and it would therefore be more correct to say that in 1 litre of pure water at 25°C:

$$a_{H^+} = a_{OH^-} = 10^{-7}$$

Since the concentration of H^+ and OH^- ions is extremely small (10^{-7}) and so is the ionic strength of the solution, the activity coefficients in the case under consideration are close to unity. Therefore, without appreciable error, for pure water (in the absence of strong electrolytes) we may express the quantity K_{eq} in terms of concentrations rather than activities.

*** The value of the ionization constant of water is not difficult to calculate. It is known that 1 litre of water weighs (at 25°C) 997 grams. Since the molecular weight of water is equal to 18.02, the number of moles of H_2O in 1 litre of water at 25°C is equal to

$$[H_2O] = \frac{997}{18.02} = 55.37 \text{ mole}$$

whence

$$K_{eq} = \frac{[H^+][OH^-]}{[H_2O]} = \frac{10^{-14}}{55.37} = 1,8 \times 10^{-16}$$

right-hand side of the equation, and consequently the product $[H^+][OH^-]$, must likewise be constant. This constant is known as the ion product of water and is designated as K_{H_2O} or K_w. Hence

$$[H^+][OH^-] = K_{H_2O} \quad (3.15)$$

At 25°C in pure water $[H^+] = [OH^-] = 10^{-7}$, and therefore at that temperature $K_w = 10^{-7} \times 10^{-7} = 10^{-14}$.

The quantity K_w increases rapidly with rise of temperature (see Table 3.4).

TABLE 3.4. **The Ion Product of Water (K_w) at Various Temperatures**

Temperature, °C	K_w	$\sqrt{K_w} = a_{H^+} = a_{OH^-}$	Temperature, °C	K_w	$\sqrt{K_w} = a_{H^+} = a_{OH^-}$
0	0.11×10^{-14}	0.33×10^{-7}	40	2.95×10^{-14}	1.70×10^{-7}
10	0.30×10^{-14}	0.54×10^{-7}	50	5.50×10^{-14}	2.34×10^{-7}
20	0.69×10^{-14}	0.83×10^{-7}	60	9.55×10^{-14}	3.09×10^{-7}
25	1.00×10^{-14}	1.00×10^{-7}	80	25.1×10^{-14}	5.01×10^{-7}
30	1.48×10^{-14}	1.20×10^{-7}	100	55.0×10^{-14}	7.41×10^{-7}

The meaning of the equation $[H^+][OH^-] = K_w$ is as follows: *No matter what the concentration of H^+ and OH^- is the ion product of water is always constant in aqueous solution at a given temperature; it is equal to 10^{-14} at 25°C.*

Note. Strictly speaking, it is the product of the activities, rather than that of the concenrations H^+ and OH^- ions, which is constant. Since the dissociation of H_2O is very slight, these activities in pure water are practically equal to the concentrations ($f_a = 1$). But in the presence of other electrolytes the values of $[H^+]$ and $[OH^-]$ differ markedly from their activities. For simplicity, however, we will use the approximate values of the ion product of water calculated from the above equation, especially as it will suffice for our purposes.

If an acid is added to pure water, the concentration of H^+ ions in solution will be raised above 10^{-7}, but the OH^- ion concentration will be reduced by approximately the same amount, to less than 10^{-7}. Therefore, in acid solution:

$$[H^+] > 10^{-7} > [OH^-]$$

Similarly, we find that in alkaline solution,

$$[OH^-] > 10^{-7} > [H^+]$$

and in neutral solutions,

$$[H^+] = [OH^-] = 10^{-7}$$

Hence, *any aqueous solution, whatever its reaction, must contain both H^+ and OH^- ions.* Since the concentrations of these ions are inversely proportional to each other, as it follows from Eq. (3.15)

$$[H^+] = K_w/[OH^-]$$

the reaction of any solution may be characterized quantitatively by indicating the H^+ ion concentration in it. Instead of the H^+ concentration, it is more convenient to use its negative logarithm denoted by the symbol pH and termed the **hydrogen ion exponent**:

$$pH = -\log[H^+]$$

More precisely, the pH is defined as the negative logarithm of the H^+ ion activity, i.e., $pH_a = -\log a_{H^+}$. If the pH is determined in a solution with a high concentration of ions, i.e., possessing a considerable ionic strength, the activity coefficient has to be introduced.

At $[H^+] = 10^{-4}$ the value of the hydrogen ion exponent is $-\log 10^{-4} = -(-4) = 4$. If $[H^+] = 5 \times 10^{-10}$, then the $pH = -\log 5 \times 10^{-10} = -(0.70 - 10) = 9.30$.*

It is obvious that the first solution (pH = 4) has an acid reaction because $[H^+]$ is greater than 10^{-7}; the reaction of the second solution (pH = 10) is alkaline bacause its $[H^+]$ is less than 10^{-7}.

Hence, in acid solutions pH < 7, in alkaline solutions pH > 7, and in neutral solutions pH = 7. It is not difficult to see that *the acidity of a solution increases with decrease of the pH and the alkalinity increases with increase of the pH*. For example, the reaction of a solution is more acidic at pH = 2 than at pH = 4 since the H^+ ion concentration in the former case (10^{-2}) is 100 times as high as in the latter (10^{-4}). Similarly, we find that at pH = 12 the reaction of a solution is more alkaline than at pH = 9. Indeed, in the first case the hydrogen ion concentration equals

$$\frac{10^{-14}}{[H^+]} = \frac{10^{-14}}{10^{-12}} = 10^{-2}$$

whereas in the second case it is equal to

$$\frac{10^{-14}}{10^{-9}} = 10^{-5}$$

that is, less by a factor of 1000.

All that has been said above may be schematically represented as follows:

pH	0 1 2 3 4 5 6	7	8 9 10 11 12 13 14
Reaction of solution	Acid ←	Neutral	→ Alkaline

* In calculating pH, the logarithms are usually rounded off to 0.01.

The extreme values of pH in this scheme roughly correspond to the hydrogen ion concentration of $1N$ HCl solution (pH ≈ 0) and of $1N$ NaOH solution (pH ≈ 14). Of course, more acid solutions with pH < 0 and more alkaline solutions with pH > 14 are also possible. But the acidity or alkalinity of such solutions are usually expressed in moles of acid or alkali per litre of solution.

Along with the hydrogen ion exponent, use is also made in some cases of the hydroxyl ion exponent:

$$\mathrm{pOH} = -\log[\mathrm{OH^-}]$$

If the equation

$$[\mathrm{H^+}][\mathrm{OH^-}] = 10^{-14}$$

is expressed in logarithmic form and the signs of the logarithms are then reversed, we will have

$$\mathrm{pH} + \mathrm{pOH} = \mathrm{p}K_w = 14 \text{ (at } 25^\circ\text{ C)}$$

where $\mathrm{p}K_w = -\log K_w = -\log 10^{-14} = 14$. From this it follows that in any aqueous solution the following relationship holds:

$$\mathrm{pH} = \mathrm{p}K_w - \mathrm{pOH} = 14 - \mathrm{pOH} \text{ (at } 25^\circ\text{ C)}$$

In solving various problems, it is often necessary to convert $[\mathrm{H^+}]$ to pH and vice versa. It is sometimes necessary to calculate the pH of a solution of an acid or a base, etc. We shall illustrate such calculations by the following examples.

Example 1. The hydrogen ion concentration in a solution is 5×10^{-4}. Calculate the pH and pOH of the solution.

Solution. We have

$$\mathrm{pH} = -\log[\mathrm{H^+}] = -\log 5 \times 10^{-4} = -(\log 5 + \log 10^{-4}) = -(0.70 - 4) = 3.30$$

$$\mathrm{pOH} = 14 - \mathrm{pH} = 14 - 3.30 = 10.70$$

Example 2. Carry out a similar calculation for a solution in which $[\mathrm{H^+}]$ is 4.5×10^{-11}.

Solution. We have

$$\mathrm{pH} = -\log 4.5 \times 10^{-11} = -(0.65 - 11) = 10.35$$

$$\mathrm{pOH} = 14 - 10.35 = 3.65$$

Example 3. What are the concentrations of $\mathrm{H^+}$ and $\mathrm{OH^-}$ ions in a solution with pH $= 4.87$?

Solution.

$$\log[\mathrm{H^+}] = -\mathrm{pH} = -4.87 = -(5 - 0.13) = -5 + 0.13$$

which gives

$$[\mathrm{H^+}] = 1.35 \times 10^{-5}; \ [\mathrm{OH^-}] = \frac{10^{-14}}{1.35 \times 10^{-5}} = 7{,}41 \times 10^{-8}$$

Example 4. Find the pH of a $0.003N$ HCl solution.

Solution. The acid HCl is known to be completely dissociated. And since every HCl molecule gives on dissociation one hydrogen ion, it follows that $[H^+] = 3 \times 10^{-3}$. Hence

$$pH = -\log 3 \times 10^{-3} = -(0.48 - 3) = 2.52$$

Example 5. Find the pH of a $0.05N$ NaOH solution.

Solution.

$$[OH^-] = C_{NaOH} = 5 \times 10^{-2}$$

Hence

$$pOH = -\log 5 \times 10^{-2} = -(0.70 - 2) = 1.30$$

$$pH = 14 - 1.30 = 12.70$$

Example 6. Derive a formula for calculating the pH of a solution of a weak monobasic acid (HAn) and use it to calculate the pH of $0.1\ M$ solution of formic acid, HCOOH, and of hydrosulphuric acid, H_2S.

Solution. From the equation for the ionization of the acid HAn

$$HAn \rightleftarrows H^+ + An^-$$

it is seen that the H^+ and An^- ions are formed in equal amounts. Hence, their concentrations in solution are equal.

The concentration of the un-ionized portion of the acid [HAn] is equal to the total concentration of the acid in solution minus the concentration of the ionized portion which can be expressed in terms of the concentration of one of the ions, say $[H^+]$.

Then $[HAn] = C_a - [H^+]$ and $[H^+] = [An^-]$.

Substituting these expressions into the formula for the dissociation constant of the acid gives

$$K_a = \frac{[H^+][An^-]}{[HAn]} = \frac{[H^+]^2}{C_a - [H^+]}$$

Solving this quadratic equation for $[H^+]$, we get

$$[H^+] = -\frac{K_a}{2} + \sqrt{\frac{K_a^2}{4} + K_a C_a}$$

If the degree of ionization of the acid is very low and the value of $[H^+]$ is less than 5 per cent of the value of C_a, the concentration of the un-ionized part of the acid may be assumed to be equal to the total concentration of the acid, i.e.,

$$[HAn] = C_a$$

The equation for the ionization constant will then have the form

$$K_a = \frac{[H^+]^2}{C_a}$$

and hence

$$[H^+]=\sqrt{K_aC_a} \qquad (3.16)$$

To convert $[H^+]$ to pH, we take logarithms and reverse the signs. We then have

$$-\log[H^+]=-\frac{1}{2}\log K_a-\frac{1}{2}\log C_a$$

But $-\log[H^+]$ is the pH of the solution in question. Similarly, $-\log K_a$ is denoted by pK_a and is called the **acid dissociation exponent**. Introducing this notation, we have

$$pH=\frac{1}{2}pK_a-\frac{1}{2}\log C_a \qquad (3.17)$$

We now use this equation to calculate the pH of a $0.1N$ solution of formic acid, HCOOH ($K = 1.8 \times 10^{-4}$). First, we calculate the value of pK:

$$pK_{HCOOH}=-\log 1.8\times 10^{-4}=-(0.25-4)=3.75.$$

Consequently,

$$pH=1.87-\frac{1}{2}\log 0.1=1.87+0.5=2.37$$

Similarly, for $0.1N$ acetic acid solution ($K = 1.74 \times 10^{-5}$), we find that pH = 2.88, etc.

Let us now calculate the pH of a saturated ($0.1M$) solution of H_2S. Hydrogen sulphide dissociates predominantly according to the equation

$$H_2S \rightleftarrows H^+ + HS^- \quad (K=8.9\times 10^{-8})$$

The second step in the dissociation is so slight that it may be ignored in solving the present problem. We can use the equation derived above for calculating the pH. Substituting the numerical values of K and C, we obtain

$$pK_{H_2S}=-\log 8.9\times 10^{-8}=-(0.95-8)=7.05$$

$$pH=3.52-\frac{1}{2}\log 0.1=3.52+0.5=4.02$$

Example 7. Derive a formula for calculating the pH of solutions of weak bases and use it to calculate the pH of a $0.1N$ NH_4OH solution ($K = 1.76 \times 10^{-5}$).

Solution. Using the equation for the ionization constant of a weak base and reasoning as in Example 6, we get the formula

$$pOH=\frac{1}{2}pK_b-\frac{1}{2}\log C_b$$

$$pH=pK_w-\frac{1}{2}pK_b+\frac{1}{2}\log C_b=14-\frac{1}{2}pK_b+\frac{1}{2}\log C_b \qquad (3.18)$$

For $0.1N$ solution of ammonia we have:

$$pK_{NH_4OH} = -\log 1.76 \times 10^{-5} = -(0.25-5) = 4.75$$

$$pH = 14 - 2.37 + \frac{1}{2} \log 0.1 = 11.13$$

The values of pK that may be required in similar calculations further in the book should be taken from special handbooks*.

3.9. Buffer Systems

If 0.01 mole of HCl is dissolved in 1 litre of pure water, a $0.01N$ hydrochloric acid solution will be obtained, with a hydrogen ion concentration of 10^{-2} g-ion/litre. Hence, the pH of the solution will decrease from 7 to about 2.

In the same way it is not difficult to figure out that, when 0.01 mole of sodium hydroxide, NaOH, is dissolved in 1 litre of pure water, the pH of the solution increases from 7 to about 12. Thus, addition of even relatively small amounts of a strong acid or base to pure water causes a drastic change of the pH.

Similar changes in the pH of a solution will be observed if pure water is replaced by solutions of NaCl, KNO_3, $CaCl_2$, and other salts formed by anions of strong acids and cations of alkali and alkaline-earth metals.

Very sharp changes of pH also occur when small amounts of a strong acid or base are added to *highly dilute* solutions of strong acids or bases, or to fairly concentrated solutions of weak acids and bases. For example, if 0.01 mole of HCl is added to 1 litre of $10^{-5}M$ HCl solution having pH 5, the total acid concentration in solution becomes equal to $10^{-5} + 10^{-2} \approx 10^{-2}$ M, and the pH of the solution decreases by 3 units to a value of 2. When 0.01 mole of NaOH is added to 1 litre of 10^{-5} M HCl solution, the concentration of NaOH in the solution becomes equal** to $10^{-2} - 10^{-5} \approx 10^{-2}$ M. Hence, the pOH of this solution will be about 2 and the pH close to 12. This means that the addition of sodium hydroxide changes the pH by 7 units.

The value of pH will change quite differently when small amounts of a strong acid or base are added to a mixture of a weak acid and one of its salts.*** Indeed, if to a litre of a solution containing, say,

---

* Lurie, Ju., *Handbook of Analytical Chemistry*, Mir Publishers, Moscow, 1975; Meites, L. (editor), *Handbook of Analytical Chemistry*, McGraw-Hill, New York, 1963.

** 10^{-5} mole of HCl will be neutralized by 10^{-5} mole of NaOH; thus, in order to find the concentration of the alkali left in excess in solution, it is necessary to subtract the amount used up for neutralization (10^{-5} mole) from the total amount (10^{-2} mole).

*** The amount of a strong acid or alkali added must not exceed 90 per cent of the total amount of the components present in the mixture.

acetic acid (CH_3COOH) and sodium acetate (CH_3COONa) in concentrations equal to $0.1M$, there is added 0.01 mole of HCl, the hydrogen ion concentration increases only slightly: the hydrogen ions do not remain free but combine with the acetate ions (CH_3COO^-) from the sodium acetate to form un-ionized acetic acid molecules.

Therefore, the pH of the solution is altered but to a relatively small extent. The corresponding calculation (see below) shows that the pH falls by only 0.09, from 4.76 to 4.67, whereas in the case of pure water the pH value decreases by 5 units.

If 0.01 mole of a strong base is added to 1 litre of the same mixture, the hydroxyl (or hydroxide) ions (OH^-) combine immediately with the hydrogen ions from the acetic acid. The hydrogen ion concentration however will not change considerably in this case either since the ratio of the concentrations of the acid and its salt will change very little. Consequently, the pH of the solution will undergo only a slight change (from 4.76 to 4.84).

And, finally, suppose that the solution in question is diluted a a hundredfold. It might seem that because of the great decrease of the acetic acid concentration the hydrogen ion concentration must also decrease considerably. But one should remember that the ratio of the concentrations of the acid and its salt remains nearly constant with dilution of the solution. The pH value therefore remains practically constant.*

Thus, the presence of a mixture of a weak acid and one of its salts in a solution has the effect of regulating the hydrogen ion concentration, thus reducing the influence of any possible factors which tend to alter the pH of the solution. Such a solution is said to be a **buffered solution**. The acid-salt mixture is called a **buffer mixture** or simply a a **buffer**. Buffers also include mixtures of weak bases and their salts, such as $NH_4OH + NH_4Cl$, and solutions of acid salts or of their mixtures with other acids or neutral salts ($NaH_2PO_4 + Na_2HPO_4$; $NaHCO_3 + Na_2CO_3$, etc.)

Finally, the buffering action is also exhibited by solutions of strong acids and bases if their concentrations are high enough. Of course, the mechanism of the buffering action in this case is entirely different. The point is that if the acid or base concentration is high, the quantities added must also be relatively large to cause any appreciable change of the pH of the solution. Addition of small amounts of acid or base has practically no effect on the pH. For example, if 0.01 mole of NaOH is added to a litre of $0.1M$ HCl solution, the hydrogen ion concentration is reduced to $0.1 - 0.01 = 0.09$ g-ion/litre. The pH of the solution is thus increased from 1 to 1.05, i.e., by only 0.05 unit.

---

* In accurate calculations, one should remember that, when the solution is diluted, the degree of ionization of the acid is somewhat changed even in the presence of its salt.

The pH of a buffered solution is not difficult to calculate.

Mixtures of Weak Acids and Their Salts. To illustrate, consider a mixture of a weak acid, CH_3COOH, and its salt, CH_3COONa. The hydrogen ion concentration, $[H^+]$, is determined from the equation for the ionization constant of acetic acid:

$$[H^+] = K_{CH_3COOH} \frac{[CH_3COOH]}{[CH_3COO^-]}$$

But CH_3COOH is a weak acid and is present in solution almost entirely in the form of un-ionized molecules. Besides, the ionization of the acid is strongly suppressed in the presence of a salt with a common ion. Therefore, the concentration of the un-ionized portion of the acid may be assumed to be equal to the total concentration of the acid in the solution, i.e., $[CH_3COOH] \approx C_a$.

On the other hand, the salt CH_3COONa is completely dissociated, while the dissociation of CH_3COOH is very slight; therefore, nearly all the CH_3COO^- anions present in solution result from the dissociation of the salt. Since each dissociated molecule of the salt gives one CH_3COO^- ion, evidently $[CH_3COO^-] \approx C_s$.

With all the foregoing taken into account, we have from the above equation:

$$[H^+] = K_a \frac{C_a}{C_s} \tag{3.19}$$

Taking logarithms and reversing the signs, we obtain:

$$-\log[H^+] = -\log K_a - \log \frac{C_a}{C_s}$$

or

$$pH = pK_a - \log \frac{C_a}{C_s} \tag{3.20}$$

where $pK_a = -\log K_a$ is the dissociation exponent of the acid (see Sec. 3.8).

Mixtures of Weak Bases and Their Salts. An example is a mixture of NH_4OH and NH_4Cl. In this case, from the equation for the ionization constant of NH_4OH we have:

$$[OH^-] = K_{NH_4OH} \frac{[NH_4OH]}{[NH_4^+]} = K_b \frac{C_b}{C_s}$$

from which we find that

$$pOH = pK_b - \log \frac{C_b}{C_s}$$

But, as shown in Sec. 3.8, pH + pOH = 14. Hence,

$$pH = 14 - pOH = 14 - pK_b + \log \frac{C_b}{C_s} \tag{3.21}$$

Let us now consider several examples to illustrate the use of these formulas in calculations.

Example 1. Calculate the pH of an acetate ($CH_3COOH + CH_3COONa$) buffer containing 0.1 mole of each of the compounds. Find out how the pH is changed by adding to 1 litre of the buffer solution: (a) 0.01 mole of HCl; (b) 0.01 mole of NaOH; (c) water to bring the solution to a hundredfold dilution.

Solution. Since for acetic acid pK = 4.76, we have:

$$\mathrm{pH} = 4.76 - \log \frac{0.1}{0.1} = 4.76$$

If 0.01 mole of HCl is added to 1 litre of this solution, then 0.01 mole of CH_3COONa will be converted to an equal number of moles of CH_3COOH. Hence,

$$\mathrm{pH} = 4.76 - \log \frac{0.11}{0.09} = 4.67$$

Similarly, when 0.01 mole of NaOH is added to 1 litre of the solution, an equal number of moles of CH_3COOH will be converted into CH_3COONa and the pH will be equal to:

$$\mathrm{pH} = 4.76 - \log \frac{0.09}{0.11} = 4.84$$

And, finally, the hundredfold dilution of the solution will result in

$$\mathrm{pH} = 4.76 - \log \frac{0.001}{0.001} = 4.76$$

Thus, calculations with the aid of Eq. (3.20) confirm the above statement that buffers are capable of maintaining pH values nearly constant on addition of small amounts of acids or bases and also on dilution of the solution.*

Example 2. What is the pH value of a ($NH_4OH + NH_4Cl$) buffer containing 0.1 mole of each of the compounds? Calculate the change of pH: (a) on addition of 0.01 mole of HCl to a litre of the mixture;

---

* To obtain more accurate results, calculations of the pH of buffer mixtures should be based on the formula

$$\mathrm{pH} = -\log a_{\mathrm{H^+}} = -\log[\mathrm{H^+}] f_{\mathrm{H^+}}$$

How the activity of H^+ ions in a mixture of a weak acid and its salt is calculated was shown in Sec. 3.6. Having found $a_{\mathrm{H^+}}$, we can calculate the pH value in a usual way. For example, the activity of hydrogen ions in 1 litre of a solution containing 0.1 mole of CH_3COOH and 0.05 mole of CH_3COONa has been found to be equal to 4.3×10^{-5} (see Sec. 3.6). Hence,

$$\mathrm{pH} = -\log 4.3 \times 10^{-5} = -(0.63 - 5) = 4.37$$

In calculations by the approximate equation (3.20), the result is pH 4.43. In the given example, the ionic strength of the solution is even larger because the concentration of CH_3COONa is 0.1 mole per litre.

(b) on addition of 0.01 mole of NaOH to a litre of the mixture; (c) on tenfold dilution of the mixture with water.

Solution. In accordance with Eq. (3.21), we have:

$$pH = 14 - 4.75 + \log \frac{0.1}{0.1} = 9.25$$

On addition of 0.01 mole of HCl, the value of C_b falls to 0.09*M* and that of C_s increases to 0.11*M*.

Therefore

$$pH = 14 - 4.75 + \log \frac{0.09}{0.11} = 9.15$$

When 0.01 mole of NaOH is added to 1 litre of the solution, we get:

$$pH = 14 - 4.75 + \log \frac{0.11}{0.09} = 9.33$$

Finally, the tenfold dilution of the solution yields*:

$$pH = 14 - 4.75 + \log \frac{0.01}{0.01} = 9.25$$

Example 3. Find out how the pH of a mixture of CH_3COOH and CH_3COONa will change as the result of a tenfold increase of the concentration of: (a) CH_3COOH; and (b) CH_3COONa.

Solution. We saw in Example 1 that at equal concentrations of the acid and its salt the pH of the mixture is equal to the p*K* of the acid, i.e., to 4.76. Suppose now that the concentration of CH_3COOH in the mixture is increased tenfold, i.e., to 1*M*. Then

$$pH = 4.76 - \log \frac{1}{0.1} = 4.76 - \log 10 = 4.76 - 1 = 3.76$$

If the salt concentration is increased tenfold, we get

$$pH = 4.76 - \log \frac{0.1}{1} = 4.76 + 1 = 5.76$$

For a mixture of NH_4OH and NH_4Cl it is easy to show that a tenfold increase of the NH_4OH concentration raises the pH by one unit, and a corresponding increase of the NH_4Cl concentration reduces the pH by the same amount.

The ability of buffer mixtures to maintain pH values almost constant is based on the fact that their individual components combine with

* As seen from Examples 1 and 2, the pH of buffer mixtures must not, it might seem, vary on dilution of the solution. In fact, it varies somewhat, this depending on the variation of the activity coefficients of the ions due to the diminution of the ionic strength of the solution on dilution. Taking activity coefficients into account, one can easily calculate the degree of the pH variation on dilution. For instance, in Example 2, the pH value increases approximately by 0.07.

the hydrogen or hydroxyl ions of acids or bases introduced into the solution (or resulting from the reaction). Of course, this ability is limited, which depends on the concentrations of the components of the buffer mixture. For example, if to a litre of a 0.1*N* ammonium buffer (i.e., a mixture containing NH_4OH and NH_4Cl in concentrations equal to 0.1*N*) there is added more than 0.1 mole of HCl or NaOH, the pH of the solution will change drastically in either case, because the amounts of NH_4OH or NH_4Cl available are insufficient to pick up the hydrogen or hydroxyl ions. An excess of the added strong acid or base remains in solution, which causes a sharp change of the pH. Calculations using Eq. (3.21) (see Sec. 3.9) and experiment show that the pH begins to change considerably long before the corresponding component of the buffer mixture has been completely used up in the reaction (Table 3.5).

TABLE 3.5. **Effect of Added Base or Acid on the pH of Ammonium Buffer Mixtures**

To 1 litre of buffer solution is added 1 litre of NaOH solution, mole/litre	pH		To 1 litre of buffer solution is added 1 litre of HCl solution, mole/litre	pH	
	0.1*N* buffer	1*N* buffer		0.1*N* buffer	1*N* buffer
—	9.25	9.25	—	9.25	9.25
0.01	9.33	9.25	0.01	9.16	9.24
0.02	9.42	9.26	0.02	9.07	9.23
0.05	9.72	9.29	0.05	8.77	9.20
0.10	11.12	9.33	0.10	5.12	9.16
0.50	13.30	9.72	0.50	0.70	8.77
1.00	13.65	11.62	1.00	0.35	4.62

The following important conclusions can be made on the basis of the data presented in Table 3.5.

1. Every buffer mixture maintains a practically constant pH value only up to a certain limiting amount of added acid or base, i.e., it has a definite buffer capacity.

The **buffer capacity** is defined as the limiting amount of a strong acid or base of definite concentration (expressed in mole/litre or in g-eq/litre) needed to change the pH of the buffer solution by not more than 1 unit.

2. The maximum buffer capacity is observed in those solutions which contain equal concentrations of a weak acid and its salt, or a weak base and its salt.

3. The buffer capacity of a solution is the greater the higher is the concentration of the components of the buffer mixture. This is clear from a comparison of the values presented in Table 3.5 for buffer mixtures.

For example, if the pH value is to be kept constant to within 0.5 unit for a 0.1*N* buffer mixture, to a litre of the mixture one may add not more than 0.05 mole of a strong acid or base.

In the case of a 1*N* buffer mixture, one may add 0.5 mole of a strong acid or base to 1 litre of the solution. Addition of 0.05 mole of one of these substances will change the pH value by only 0.04. For a 0.1*N* buffer solution this change will amount to 0.48.

4. With increasing additions of acid or base, the resistance of a a buffer solution to changes in pH gradually diminishes.

For instance, the changes in pH caused by additions of 0.01 mole of NaOH or HCl to a litre of a 0.1*N* buffer mixture are as follows (on the average):

in the 0.00-0.02 mole range	0.085 pH unit
in the 0.02-0.05 mole range	0.10 pH unit
in the 0.05-0.10 mole range	0.28 pH unit

Thus, the **capacity** of buffer mixtures must be taken into account when they are used in analysis. For example, if we wish to conduct a reaction, which is accompanied by the accumulation of hydrogen ions in solution, at a certain, practically constant, pH, it is not sufficient to merely introduce a buffer mixture of the required pH into the solution; we must ensure that its buffer capacity is high enough. Only at a sufficiently high capacity of the buffer mixture will the pH of the solution be maintained practically constant. In other words, it is necessary that the components of the buffer mixture be present in solution in suitable and sufficiently high concentrations.

Buffer solutions find wide application in analytical practice for carrying out various reactions.

The most commonly used buffer mixtures are listed in Table 3.6.

TABLE 3.6. **Most Commonly Used Buffer Mixtures**

Buffer	Composition	pH*
Formate	Formic acid, HCOOH, and sodium formate, HCOONa	3.8
Benzoate	Benzoic acid, C_6H_5COOH, and sodium benzoate, C_6H_5COONa	4.2
Acetate	Acetic acid, CH_3COOH, and sodium acetate CH_3COONa	4.8
Phosphate	Sodium dihydrogen phosphate, NaH_2PO_4, and disodium hydrogen phosphate, Na_2HPO_4	6.6
Ammonium	Ammonium hydroxide, NH_4OH, and ammonium chloride, NH_4Cl	9.2

* The pH values given refer to cases where the concentrations of both components of the buffer are equal.

It is also possible, in accordance with Eqs. (3.20) and (3.21) (see this section), to change the pH by altering the concentrations of the components of a buffer. The concentration of one of the components should not be more than ten times that of the other to ensure that the change of the buffer capacity is not too great, i.e., that the buffer action is sufficiently effective. But since a tenfold change of one of the concentrations causes a unit change of pH (see Example 3, Sec. 3.9), it is clear that the efficiency range of the buffering action extends by one pH unit to either side of the values given above. For example, it is possible to prepare various ($NH_4OH + NH_4Cl$) buffers capable of maintaining the pH between 8.2 (with 10 times as much NH_4Cl as NH_4OH) and 10.2 (with a tenfold excess of NH_4OH). The range of efficiency of the acetate buffer lies between pH = 3.7-5.7.

3.10. Determination of pH in the Course of Qualitative Analysis

Since the pH of the medium is important in most analytical reactions, it often has to be determined in the course of analysis. There are various methods for pH determination. The most accurate methods are rather complicated and require special apparatus; they are hardly suitable for qualitative analysis, where the accuracy exceeding one pH unit is not needed (it may often be even lower). That is why the only method employed in qualitative analysis is the colorimetric technique based on the use of reagents which change colour depending on the hydrogen ion concentration. Such reagents (e.g., litmus) are called **acid-base indicators**. Indicators are weak organic acids or bases whose undissociated molecules and ions have different colours.* Litmus contains the so-called azolitmin whose undissociated molecules are red and anions blue in colour.

If we designate an indicator acid by HInd and its anions by Ind^-, the dissociation of litmus in solution may be represented by the following scheme:

$$\underset{\text{red}}{\text{HInd}} \rightleftharpoons H^+ + \underset{\text{blue}}{\text{Ind}^-}$$

When litmus is dissolved in water (or in any neutral solution), its HInd molecules and the simultaneously present Ind^- anions confer an intermediate (violet) colour on the solution. If a small amount of an acid is added to this solution, the equilibrium will be shifted to the left. In other words, the hydrogen ions introduced with the acid will combine with all the Ind^- anions present in solution to form un-ionized HInd molecules and the solution will turn red. If to a neutral

* The theory of acid-base indicators is expounded in more detail in textbooks of quantitative analysis.

litmus solution there is added an alkali, its OH^- ions will combine with the hydrogen ions of the indicator to give un-ionized H_2O molecules. As a result, the reaction, which is the ionization equilibrium of the indicator, will be shifted to the right, i.e., towards the formation of more Ind^- anions, and the solution will turn blue.

The colour change of other indicators is accounted for analogously. For example, phenolphthalein has colourless un-ionized molecules, whereas its anions are red:

$$\underset{\text{colourless}}{HInd} \rightleftharpoons H^+ + \underset{\text{red}}{Ind^-}$$

Consequently, phenolphthalein should become red with alkalis and colourless with acids.

Addition of an acid or alkali evidently changes the pH of the solution, this being responsible for the colour change of the indicators in question. Studies have shown that the colour change takes place within certain pH limits characteristic of each particular indicator. This pH range is called the **transition range**. For example, at any pH below 5, litmus has precisely the same red colour. Above pH 5, its colour begins to change appreciably and becomes bright blue at pH 8. No matter how much the pH is then raised, i.e., no matter how much more alkali is added to the solution, the same blue colour is retained as at pH 8. The pH interval, which lies between 5 and 8, is the transition range of litmus. Other indicators have other transition ranges.* For example, the transition range of phenolphthalein is pH 8-10; at pH $\leqslant 8$ this indicator is colourless and at pH $\geqslant 10$ it acquires an intense-red colour. Within this range, the originally colourless solution becomes progressively redder, reaching maximum intensity at pH 10.

The simplest method of determining the reaction of a solution is to test it with litmus solution or litmus paper. Litmus paper is a strip of filter paper impregnated with a litmus solution which has been previously coloured by addition of a very small amount of acid (red litmus paper) or alkali (blue litmus paper). If a drop of the test solution turns blue litmus paper red, the reaction of the solution is acid (pH $\leqslant$ 5). A solution having an alkaline reaction (pH $\geqslant$ 8) turns red litmus paper blue. If neither paper changes colour by the action of an unknown solution, the latter is regarded as neutral (pH $\approx$ 7).

Such an approximate determination of the reaction of a solution is sufficient. In a number of cases, however, it is very important to have a more accurate idea of the pH of a solution. This can be achieved by using not one but several indicators whose transition ranges correspond to different pH values.

---

* This depends on the differences in the value of the ionization constants of different indicator acids (or bases).

Apart from litmus, other indicators are also used for pH determinations (see Table 3.7).

TABLE 3.7. **Some Mixed Acid-Base Indicators**

Indicator	pH range	Colour change (acid → alkali)
Thymol blue*	1.2-2.8	red → yellow
Thymol blue	8.0-9.6	yellow → blue
Methyl orange	3.0-4.6	pink → orange**
Methyl red	4.2-6.3	red → yellow
Phenol red	6.8-8.0	yellow → red
Phenolphthalein	8.3-10.0	colourless → red

* Thymol blue has two transition ranges. At pH 1.2 and less, its colour is red; when the pH increases up to 2.8, it becomes yellow and remains as such up to pH 8.0. With further increase of pH, the indicator gradually turns blue and retains its blue colour beginning from pH 9.6.

** At small concentrations of the indicator the colour of methyl orange at pH ⩾ 4.6 is yellow.

In addition to the indicators listed in Table 3.7, the following two may also be used:

Methyl violet	0.5-3	green → violet; blue at pH 1.5
Thymolphthalein	9.4-10.6	colourless → blue

Comparing the action of a test solution on various indicators (or on papers impregnated with their solutions), one can easily determine the pH of the solution with an accuracy sufficient for qualitative analysis.

Suppose, for example, that the following colours were produced when a drop of an unknown solution was mixed with drops of indicator solutions on a spot plate:

Indicator	Colour
Thymol blue	Yellow
Methyl orange	Orange (or yellow)
Methyl red	Yellow
Phenol red	Yellow
Phenolphthalein	Colourless

These results allow one to make the following conclusions.

The yellow colour of thymol blue indicates that the pH of the solution lies between 2.8 and 8.0. That no red colour is produced with phenolphthalein confirms that the pH does not exceed 8.0. The orange colour of methyl orange is evidence that pH $\geqslant$4.6, and the yellow colour of methyl red is an indication that pH $\geqslant$6.3. But since phenol red turned yellow, the pH must be below 6.8. Thus, the pH value lies within the range 6.3-6.8, i.e., pH $\approx$ 6.5. It may be seen from this that it is not always necessary to make use of all the indicators. For example, in this particular case, tests with methyl orange and phenolphthalein are unnecessary. It is most convenient to start tests with thymol blue because, if it gives a red colour (pH $\leqslant$ 1.2, strongly acid solution) or a blue colour (pH $\geqslant$9.6, strongly alkaline solution), further tests would not be needed.

If an indicator gives a colour intermediate between the lower and the upper limit of its transition range, this means that the pH of the solution lies within the range of that indicator. For example, if, instead of the yellow or blue colour, thymol blue gives a green colour formed by optical addition of yellow and blue, we may conclude that the pH is intermediate between 8.0 and 9.6, i.e., about 9. The same inference may be made if phenolphthalein gives a pale-pink colour rather than bright red.

Instead of solutions, use may be made of a corresponding indicator paper onto which a small amount of the test solution is applied with a glass rod, the colour formed being observed.

The use of paper requires little solution and is therefore more convenient for semimicroanalysis. The colours produced are however less bright and besides they are somewhat changed due to the adsorption of solutes by the paper.

Instead of the range of indicators described above, it is more convenient to use so-called **universal indicators**; these are mixtures of individual indicators which change colour over a wide range of pH.

For example, one of the universal indicators is a mixture of five different indicators, which changes colour in a pH interval of 2-10 as follows:

pH	Colour	pH	Colour
2 (and below) . .	red → pink	7	yellow → green
3	red → orange	8	green
4	orange	9	blue → green
5	yellow → orange	10 (and above) . .	violet
6	lemon → yellow		

The indicator may be used in the form of either a solution, a drop being mixed on a spot plate with 1-3 drops of the test solution, or an indicator paper impregnated with the universal indicator solution and dried. A colour scale is supplied with the paper, indicating the colours produced at different pH values. The determination of pH with the aid of the colour scale is very simple. If no colour scale is available, it is useful to have a series of buffer solutions, testing them with the universal indicator at the same time as the unknown solution.

Using more sophisticated techniques, one can determine the pH of solutions colorimetrically to within about 0.2 pH unit. The procedure used is described in textbooks of quantitative analysis.

In interpreting the results of pH determinations one should take into account that distilled water used for preparing the solutions is not absolutely pure: it contains CO_2 absorbed from the air and also substances, such as Na_2SiO_3, extracted from the glass of containers. As a result, the pH of distilled water determined experimentally does not agree with the theoretical value (7.0); usually it is somewhat lower (5-6). If this water is used to prepare solutions of neutral salts, such as NaCl or KNO_3, the pH of the solutions will be found to equal not 7 but that of the water (i.e., 5-6).

To prepare solutions of these salts with a strictly neutral reaction, either water neutralized to pH 7 or water freed from CO_2 by prolonged boiling must be used. To prevent the re-absorption of CO_2 on cooling, the containing vessel is stoppered after boiling with a stopper through which is inserted a tube with soda lime (a calcined mixture of NaOH and CaO). While passing through the tube, the air is purified from CO_2 which is absorbed by the NaOH and CaO to form the solid compounds Na_2CO_3 and $CaCO_3$. Solutions of salts should be protected from CO_2 in the same way if their pH value is to be kept at the theoretical value of 7. It should also be remembered that the pH of a solution may rise on prolonged storage because of soluble silicates being extracted from the glass by water; these silicates show an alkaline reaction due to hydrolysis. To prevent this, the container walls are coated inside with paraffin wax.

All these precautions, which complicate the work considerably, have rarely to be exercised in practice because in most cases no special advantage is gained. One should merely remember that the pH of neutral solutions determined by experiment may not correspond to the theoretical pH 7.

Questions and Problems

1. Find the concentrations of H_2 and CO_2 at which the equilibrium of the reversible reaction $CO + H_2O \rightleftharpoons H_2 + CO_2$ sets in if the initial concentrations of CO and H_2O vapour are 1 mole and 10 moles per litre, respectively. Calculate the percentage of CO converted to CO_2 ($K = 1$).

Answer: about 0.9*M*; 90 per cent.

2. In which direction will the equilibrium of the reaction $N_2 + 3H_2 \rightleftharpoons 2NH_3$ be shifted if the concentration of (a) hydrogen and (b) ammonia in the gas mixture is increased?

3. Will the equilibrium in the above reactions be influenced by compression of the gas mixture to half the original volume? If so, in which direction will it be shifted?

4. What is the degree of ionization of formic acid in $0.5N$ solution, given that the hydrogen ion concentration in it is 0.01 g-ion/litre?

Answer: $\alpha = 0.02$ (2 per cent).

5. Do the degree of dissociation and the dissociation constant of hydrocyanic acid, HCN, change when its solution is diluted with water?

6. Using the data of Problem *4*, calculate the ionization constant of formic acid and its degree of dissociation in $1N$ solution.

Answer: $K \approx 2 \times 10^{-4}$; $\alpha = 0.014$ (1.4 per cent).

7. What is the concentration of OH^- in $0.05N$ solution of NH_4OH ($K = 1.8 \times 10^{-5}$) containing 0.1 mole of NH_4Cl per litre? How much does the presence of NH_4Cl reduce this concentration? An approximate calculation is required and therefore the activity coefficients of the ions may be disregarded.

Answer: $[OH^-] = 9 \times 10^{-6}$ g-ion/litre; in the presence of 0.1 mole of NH_4Cl per litre the hydroxyl ion concentration is reduced by a a factor of about 105.

8. What is the ionic strength of solutions containing in 1 litre: (a) 0.1 mole of KCl; (b) 0.1 mole of K_2SO_4; (c) 0.1 mole of $MgSO_4$; (d) 0.1 mole of $AlCl_3$; (e) 0.01 mole of K_2SO_4 + 0.01 mole of $Al_2(SO_4)_3$?

Answer: (a) 0.1; (b) 0.3; (c) 0.4; (d) 0.6; (e) 0.18.

9. What are the activities of Ca^{2+} and Cl^- ions in $0.02N$ $CaCl_2$ solution?

Answer: $a_{Ca^{2+}} = 0.0054$; $a_{Cl^-} = 0.0172$.

Hint. To solve this problem, first calculate the ionic strength of the solution and then proceed from it to find the activity coefficients of Ca^{2+} and Cl^- by interpolation of the data of Table 3.3.

10. Calculate the activity of OH^- ions in $0.05N$ NH_4OH solution, which also contains NH_4Cl at $0.1N$ concentration, and compare the result with the activity of the same ions in the absence of the ammonium salt. How does the presence of the ammonium salt influence the value of a_{OH^-}? Compare the result with the value found by approximate calculation (see Problem *7*).

Answer: $a_{OH^-} = 1.17 \times 10^{-5}$; the presence of the ammonium salt in $0.1N$ concentration lowers a_{OH^-} by a factor of about 80.

Hint. Assume the ionization (thermodynamic) constant of NH_4OH to be equal to 1.8×10^{-5}; the activity coefficients of the ions resulting from the ionization of ammonium hydroxide in the absence of NH_4Cl are to be taken equal to unity.

11. What is the pH of solutions containing: (a) 2×10^{-4} g-ion of H^+ per litre; (b) 0.008 g-ion of OH^- per litre?

Answer: (a) 3.7; (b) 11.9.

12. Two solutions have the following pH: (a) 2.63; (b) 12.45. Find the H^+ and OH^- concentrations in them.

Answer: (a) $[H^+] = 2.34 \times 10^{-3}$ g-ion/litre; $[OH^-] = 4.27 \times 10^{-12}$ g-ion/litre; (b) $[H^+] = 3.54 \times 10^{-13}$ g-ion/litre; $[OH^-] = 2.82 \times$ $\times 10^{-2}$ g-ion/litre.

13. What are the pH values of: (a) $0.005N$ HCl solution; (b) $0.015N$ KOH solution?

Answer: (a) 2.30; (b) 12.18.

14. Calculate the pH of: (a) $0.02N$ CH_3COOH solution ($K_{CH_3COOH} = 1.74 \times 10^{-5}$); (b) saturated (at 20°C) 3.8×10^{-2} M solution of CO_2 ($K_{H_2CO_3} = 4.5 \times 10^{-7}$); (c) $0.05N$ NH_4OH solution.

Answer: (a) 3.22; (b) 3.89; (c) 10.98.

15. How will the pH of HCl and KOH solutions change if they are diluted hundredfold with water?

16. Figure out the changes in the pH of a solution containing $10^{-5}N$ NaOH, which are brought about by the addition of: (a) 0.001 mole of NaOH; and (b) 0.001 mole of HCl to 1 litre of the solution.

Answer: (a) it will increase by about 2; (b) it will decrease by about 6.

17. A solution contains 0.056 mole of NH_4OH and 0.1 mole of NH_4Cl per litre. What is the pH of this solution and how will it change on addition of 0.001 mole of (a) NaOH and (b) HCl to 1 litre of the solution? (Compare the answer with the result of the previous problem.)

Answer: pH 9; (a) will increase by 0.01; (b) will decrease by 0.01.

18. Calculate the pH of buffer mixtures containing: (a) $0.01M$ CH_3COOH and $0.01M$ CH_3COOK; (b) $0.01M$ CH_3COOH and $0.5M$ CH_3COOK; (c) $0.5M$ CH_3COOH and $0.01M$ CH_3COOK ($K_{CH_3COOH} = 1.74 \times 10^{-5}$).

Answer: (a) 4.76; (b) 6.46; (c) 3.06.

19. To prepare a formate buffer, 100 ml of $23N$ formic acid solution was mixed with 3 ml of $15N$ NH_4OH solution. Calculate the pH of the mixture ($K_{HCOOH} = 2 \times 10^{-4}$).

Answer: pH 2.05.

20. What changes are produced in the pH of (a) $0.2N$ acetate buffer and (b) $0.1N$ acetate buffer by addition of 0.05 mole of HCl or NaOH to 1 litre of the buffers?

Answer: (a) decreased by 0.22 with HCl, increased by the same amount with NaOH; (b) decreased by 0.48 with HCl, increased by the same amount with NaOH.

21. What conclusions may be drawn from the answers to Problem *20* about the relative buffer capacities of $0.2N$ and $0.1N$ acetate buffers?

22. What must be the ratio of the concentrations of formic acid, HCOOH, and its salt, HCOONa, in a solution of formate buffer for its pH to be equal to 2.25? Would the buffer capacity of this solution be sufficiently effective?

Answer: $C_a : C_s = 31.6$.

23. How many millilitres of $0.5N$ CH_3COONa solution must be added to 100 ml of $2N$ CH_3COOH solution to prepare a buffer mixture with pH 4.0?

Answer: 74 ml.

B. REACTIONS AND ANALYTICAL PROCEDURE FOR GROUP I CATIONS

3.11. General Characteristics of Group I

This group includes the ions of ammonium, potassium, sodium, magnesium, and some rare-earth elements. Most of the salts formed by these ions are readily soluble in water. Of special importance to analysis is the solubility of their sulphides, hydroxides, carbonates, and chlorides in water since it distinguishes Group I from all the other groups (see Table 1.2, Sec. 1.5). Owing to the solubility of these salts, Group I cations are not precipitated by the group reagents for the other groups, i.e., HCl, H_2S, $(NH_4)_2S$, and $(NH_4)_2CO_3$, and remain in solution when cations of the other groups are separated as slightly soluble salts. There is no group reagent that would precipitate all cations of Group I. Since cations of the other groups interfere with the detection of K^+, Na^+, and Mg^{2+}, these ions are detected in systematic analysis only after all other cations have been removed completely from solution.

Potassium and sodium belong to Group I of the Periodic System of the elements and are highly active alkali metals. They decompose water very rapidly even at room temperature with evolution of hydrogen and formation of the corresponding hydroxide, for example,

$$2Na + 2H_2O \rightarrow 2NaOH + H_2\uparrow$$

The alkali-metal hydroxides are the strongest alkalis. In aqueous solution, they are completely ionized. Therefore, sodium and potassium salts of strong acids undergo no hydrolysis* and their solutions show a neutral reaction; this means that the pH of such solutions is determined by the pH of water used for their preparation. Potassium or sodium salts of weak acids are more or less strongly hydrolysable and their solutions give an alkaline reaction ($pH > 7$).

Ammonium hydroxide, NH_4OH, a solution of ammonia in water, behaves as a weak base. Its degree of ionization in $1N$ solution is about 0.4 per cent. Solutions of ammonium salts of strong acids give an acid reaction ($pH < 7$). For example, the pH of $1N$ NH_4Cl solution is 4.6.

The cations K^+, Na^+, and Mg^{2+} (and also Li^+, Rb^+, and Cs^+ belonging to Group I) have completed shells of eight or two electrons, similar to the shells of atoms of the inert gases. The Mg^{2+} ion differs from the other Group I cations; it belongs to Group II of the Periodic System and is intermediate between analytical Groups I and II, so that it can be included in either of these groups. The Mg^{2+} cation resembles

---

* The hydrolysis of salts is considered in Sections 5.5 and 5.6.

Group II cations in that its basic carbonate* $(MgOH)_2CO_3$ is slightly soluble in water. It however dissolves in the presence of ammonium salts and since the precipitation of Group II cations by ammonium carbonate is carried out in the presence of NH_4Cl, the ion Mg^{2+} is found in the course of analysis to be not in the precipitate but in solution together with Group I. It is therefore more conveniently studied with Group I.

All the Group I cations are colourless in aqueous solution.

In studying the properties of ions and characteristic reactions, one should pay attention to the conditions under which these reactions are carried out (temperature, ionic concentration, the action of acids and bases on a precipitate, i.e., its solubility, etc.). The factors responsible for the solubility of precipitates in acids and bases are discussed in greater detail in Chapter IV after the solubility product principle is studied. Special consideration in the discussion of the properties of Group I cations will be given to the ability of precipitates to dissolve in acids and alkalis for the optimum conditions to be chosen. Therefore, after a precipitate is formed, the student should test the action of various acids (mineral and organic) on it and also study the solubility of the precipitate in alkalis. The corresponding directions are given elsewhere in the text where characteristic reactions of the ions are considered.

3.12. Reactions of the Potassium Ion

There are very few precipitation reactions for identification of the K^+ cation. The most important ones are the following.

1. Tartaric acid. $H_2C_4H_4O_6$, and its acid sodium salt, sodium hydrogen tartrate ($NaHC_4H_4O_6$), produce a white crystalline precipitate of potassium hydrogen tartrate, $KHC_4H_4O_6$, in solutions of potassium salts.

Tartaric acid is a relatively weak organic acid. Its molecule has six hydrogen atoms but only two of these atoms can split off as H^+ ions, so that the acid is dibasic.** It is capable of forming both neutral and acid salts. The neutral potassium and sodium tartrates, $K_2C_4H_4O_6$ and $Na_2C_4H_4O_6$, and sodium hydrogen tartrate, $NaHC_4H_4O_6$, are

* Let us recall that basic salts are compounds which contain hydroxyl groups, apart from metal ions and acid residues. Usually, basic salts result from the hydrolysis of neutral salts, e.g., $MgCl_2 + H_2O \rightleftharpoons MgOHCl + HCl$ (for more detail see Section 5.5).

** This is accounted for by the presence of two carboxyl groups (—COOH) in the molecule of tartaric acid

```
HOOC   COOH
   |   |
 H—C———C—H
   |   |
  HO   OH
```

readily soluble in water. On the other hand, the solubility of potassium hydrogen tartrate, $KHC_4H_4O_6$, is relatively low*, this being used for detection of the K^+ cation:

$$KCl + NaHC_4H_4O_6 \rightarrow KHC_4H_4O_6 \downarrow + NaCl$$

Other potassium salts react analogously; for example,

$$K_2SO_4 + 2NaHC_4H_4O_6 \rightarrow 2KHC_4H_4O_6 \downarrow + Na_2SO_4$$

Certainly this might be expected since the reaction involves only K^+ and $HC_4H_4O_6^-$ ions. The other ions may be omitted from the equation. We then have the ionic equation for the reaction:

$$K^+ + HC_4H_4O_6^- \rightarrow KHC_4H_4O_6 \downarrow$$

Study this reaction experimentally as follows. Pipette 4 or 5 drops of a solution of a potassium salt (KCl or KNO_3) into a conical test tube and add to it the same amount of a solution of $NaHC_4H_4O_6$. Stir the contents of the tube with a glass rod and, if a precipitate does not form at once, scratch the side of the tube gently to induce crystallization. The point is that not infrequently supersaturated solutions are formed in precipitation reactions, which fail to form a precipitate for a relatively long time. Scratching, shaking, and similar mechanical operations speed up the formation of precipitates.

Having obtained a precipitate of $KHC_4H_4O_6$, study its properties.

1. First confirm that the precipitate is crystalline. To do this, close the top of the test tube, turn it upside down, and examine the small crystals of $KHC_4H_4O_6$ adhering to the bottom and sides. Amorphous precipitates (and also those consisting of very fine crystals) readily separate from the glass wall.

2. Test the action of acids and bases on the precipitate. To do this, shake the contents of the test tube and place a drop of the turbid liquid into each of the three indentations of a spot plate (or into three test tubes). Add a drop of a strong acid (HCl, HNO_3, or H_2SO_4) to the first, a drop of an alkali (NaOH or KOH) to the second, and a drop of a weak acid (CH_3COOH) to the third. The precipitate dissolves in the first two cases. The reactions that take place may be represented by the following equations:

$$KHC_4H_4O_6 \downarrow + H^+ \rightarrow H_2C_4H_4O_6 + K^+$$

$$KHC_4H_4O_6 \downarrow + OH^- \rightarrow C_4H_4O_6^{2-} + H_2O + K^+$$

3. To the portion of the precipitate left in the test tube add 2 drops of distilled water and heat it in a water bath, stirring the contents of

* Among the hydrotartrates of Group I cations, ammonium hydrogen tartrate is also slightly soluble. The tartrates of cations of other groups are mostly slightly soluble in water but dissolve readily in the presence of $Na_2C_4H_4O_6$ or $K_2C_4H_4O_6$ due to the formation of soluble complex compounds.

the tube with a glass rod. As a result, the precipitate dissolves too. Hence, the solubility of $KHC_4H_4O_6$ increases with rise of temperature.

Thus, the $KHC_4H_4O_6$ precipitate is a white crystalline substance which is soluble in hot water and in strong acids and bases but insoluble in weak acids. It should be noted that $KHC_4H_4O_6$ dissolves also in a large quantity of cold water and the sensitivity of the reaction is therefore relatively low (dilution limit 1 in 1000).

These properties of the precipitate determine the test conditions. For example, no precipitate of $KHC_4H_4O_6$ separates from hot solution and from solutions having a strongly acid or strongly alkaline reaction. To be sure, to 2 or 3 drops of a potassium salt add several drops of a strong acid or base and try to detect K^+ in the resulting solution. Can you succeed in achieving this?

Evidently, no precipitate of $KHC_4H_4O_6$ can be formed either if the concentration of K^+ in solution is low.

On the basis of the results of all the experiments, one can conclude that the reaction conditions for detection of K^+ with the aid of $NaHC_4H_4O_6$ are as follows:

(a) the concentration of the K^+ ion must be high enough;

(b) the test must be carried out in the cold;

(c) the unknown solution must give a weakly acid or neutral reaction (pH 4-7).

It may happen in the course of analysis that the conditions have to be adjusted for the reaction to proceed. For instance, if the solution is too dilute, it is concentrated by evaporation. If the solution shows an alkaline or acid reaction, it is neutralized until the required pH value is obtained. Alkaline solutions can be neutralized by any of the acids available in the laboratory, most preferably acetic, since an excess of this acid would not interfere with the precipitation of $KHC_4H_4O_6$. Strongly acid solutions can be neutralized by NaOH; KOH can by no means be used for the purpose because this would introduce the ion which is being detected. Instead of NaOH, it is more convenient to use sodium acetate, CH_3COONa, which when reacted with the strong acid present in the test solution forms weak acetic acid:

$$CH_3COO^- + H^+ \rightleftharpoons CH_3COOH$$

If sodium acetate is introduced into the solution in some excess, an acetate buffer mixture will be obtained with pH $\approx$ 5 (see Sec. 3.9). The precipitation of $KHC_4H_4O_6$ in this medium proceeds sufficiently completely. The same buffer mixture is produced when a small excess of free acetic acid is added to the alkaline test solution.

It has already been pointed out that instead of $NaHC_4H_4O_6$, as a reagent for K^+, we can use tartaric acid because its first dissociation step results in the formation of $HC_4H_4O_6^-$ ion which is capable of precipitating the potassium ion:

$$H_2C_4H_4O_6 \rightleftharpoons H^+ + HC_4H_4O_6^-$$

The reaction however is accompanied in this case by the formation of a strong acid:

$$K^+ + H_2C_4H_4O_6 \rightleftharpoons KHC_4H_4O_6\downarrow + H^+$$

which makes it reversible. Therefore, in carrying out this reaction, apart from the reagent, one must add to the test solution sodium acetate which reacts with the acid to form an acetate buffer mixture.

2. Sodium Cobaltinitrite, $Na_3[Co(NO_2)_6]$. This reagent, also known as sodium hexanitritocobaltate (III), produces in solutions of potassium salts a yellow precipitate of the predominant composition $K_2Na[Co(NO_2)_6]$. The reagent used is a complex cobalt(III) salt which dissociates to give $[Co(NO_2)_6]^{3-}$ ions; it is these ions that combine with K^+ and Na^+ ions:

$$2K^+ + Na^+ + [Co(NO_2)_6]^{3-} \rightarrow K_2Na[Co(NO_2)_6]\downarrow$$

The reaction can be conducted in a test tube or on a spot plate. To a drop of a potassium salt solution is added 1 or 2 drops of the reagent solution, and if no precipitate separates, the solution is allowed to stand for a while. A distinct bright yellow precipitate is formed. This precipitate dissolves in acids to a considerably lesser extent than $KHC_4H_4O_6$, but in order to render the reaction more sensitive, sodium acetate should be added to the acid solutions in this case too; the pH of the solution must in this case be within the limits 4-5.

The precipitate is insoluble in acetic acid since it is weaker than nitrous acid in the composition of $[Co(NO_2)_6]^{3-}$:

$$K_{HNO_2} = 5.1 \times 10^{-4}$$

$$K_{CH_3COOH} = 1.74 \times 10^{-5}$$

The complex $[Co(NO_2)_6]^{3-}$ is sufficiently stable and the quantity of NO_2^- ions, which pass into the solution from this complex, is very small. The small concentration of hydrogen ions resulting from the dissociation of acetic acid is insufficient for the formation of a stronger acid, nitrous acid. That is why the precipitate fails to dissolve in acetic acid.

The precipitate dissolves in strong mineral acids to give very unstable weak nitrous acid which decomposes readily. At the same time cobalt(III) is reduced to cobalt(II).

All these processes can be represented by a series of reaction equations:

$$K_2Na[Co(NO_2)_6]\downarrow + 6H^+ \rightarrow 6HNO_2 + Co^{3+} + Na^+ + 2K^+$$

$$2Co^{3+} + 2Cl^- \rightarrow 2Co^{2+} + Cl_2\uparrow$$

$$2Co^{3+} + HNO_2 + H_2O \rightarrow 2Co^{2+} + NO_3^- + 3H^+$$

$$2HNO_2 \rightarrow H_2O + NO\uparrow + NO_2\uparrow$$

These reactions can easily be accounted for by referring to the standard oxidation-reduction potentials of the corresponding redox couples:

$$E^0_{Co^{3+}/Co^{2+}} = 1.84 \text{ volt}$$

$$E^0_{Cl_2/2Cl^-} = 1.36 \text{ volt}$$

$$E^0_{NO_3^-/HNO_2} = 0.94 \text{ volt}$$

Thus, cobalt(III) is a stronger oxidizing agent than chlorine and nitric acid, and therefore chloride is oxidized to chlorine and nitrous acid to nitric acid. *No alkaline reaction* (pH > 7) *must take place* since under the action of alkalis the reagent undergoes decomposition and a dark-brown precipitate of $Co(OH)_3$ is formed:

$$[Co(NO_2)_6]^{3-} + 3OH^- \rightarrow Co(OH)_3\downarrow + 6NO_2^-$$

The reaction in question is more sensitive than the reaction with $NaHC_4H_4O_6$ (dilution limit about 1 in 13,000). Therefore, this reaction is used in practice for detection of the potassium ion. It must be remembered however that $Na_3[Co(NO_2)_6]$ decomposes relatively rapidly on standing and its brown colour changes to pink (the colour of the Co^{2+} ion). If the solution has turned pink, it is unsuitable for detection of the potassium ion.

The NH_4^+ ion gives an analogous yellow precipitate of $(NH_4)_2Na[Co(NO_2)_6]$. This compound differs from the corresponding potassium compound in that it decomposes on heating. The reaction should therefore be carried out on heating.

To eliminate the interfering effect of the NH_4^+ ion, to the solution is added a solution of formaldehyde which reacts with NH_4^+ to form hexamethylenetetramine (urotropin), $(CH_2)_6N_4$. The reaction conditions are described at the end of this chapter (see Fractional Reaction for K^+). The other cations of Group I and those of Group II do not interfere with this reaction.

3. Chloroplatinic Acid, $H_2[PtCl_6]$. This acid produces in fairly concentrated solutions of potassium salts a yellow crystalline precipitate of potassium chloroplatinate:

$$2K^+ + [PtCl_6]^{2-} \rightarrow K_2[PtCl_6]\downarrow$$

When viewed under the microscope, the crystals of the precipitate appear as octahedra. Similar precipitates are formed by NH_4^+ ions (and also by Cs^+ and Rb^+ ions).

4. Perchloric acid, $HClO_4$. The acid produces a white precipitate of potassium perchlorate:

$$K^+ + ClO_4^- \rightarrow KClO_4\downarrow$$

The solubility of the precipitate is increased considerably by heating and can be greatly reduced by addition of ethyl alcohol.

5. Microcrystalloscopic Test. Put a drop of a solution of a potassium salt on a glass slide and evaporate it to dryness on a microburner or on the edge of an electric hot plate (not over the heating coil). Cool the slide, treat the dry residue with a drop of a specific reagent for K^+, which has the formula $Na_2PbCu(NO_2)_6$, and after one minute examine the crystals formed under the microscope. Characteristic cubic crystals

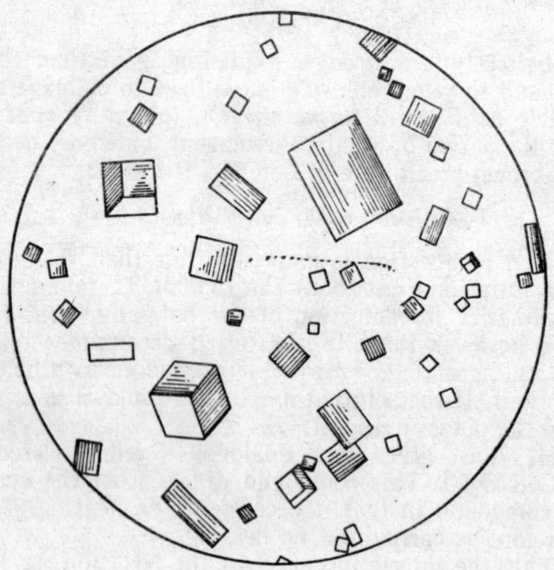

Fig. 3.4. Crystals of $K_2PbCu(NO_2)_6$.

of the triple nitrite $K_2PbCu(NO_2)_6$, black or brown in colour, are formed (Fig. 3.4). The dilution limit is 1 in 6600. The NH_4^+ ion forms similar crystals and therefore interferes with the reaction (*check this!*).

If the NH_4^+ ion is present, do as follows. Dry a drop of the test solution on a glass slide and treat it several times with concentrated nitric acid, evaporating to dryness each time. The ammonium ion will be broken down:

$$2NH_4NO_3 \rightarrow N_2\uparrow + 2NO\uparrow + 4H_2O$$

$$NH_4NO_3 \rightarrow N_2O\uparrow + 2H_2O$$

Dissolve then the dry residue in a drop of distilled water and carry out the reaction for potassium in accordance with the procedure described above.

The ions Na^+ and Mg^{2+} form no crystals with this reagent.

6. Drop Reaction. To identify the potassium ion by the drop method, use may be made of an organic reagent known as dipicrylamine

(Poluektov's reagent), $[C_6H_2(NO_2)_3]_2NH$, which forms a finely divided crystalline orange-red precipitate of its potassium salt with the K^+ ion. To carry out the reaction, a drop of a neutral solution of the potassium salt is placed on a filter paper and a drop of the reagent is added at once. An orange-red spot is formed on the paper, which remains unchanged when moistened with 1-2 drops of 2*N* hydrochloric acid.* When present at a high concentration, the NH_4^+ ion also gives a precipitate and therefore interferes with the reaction.

7. Flame Test. Take a glass rod with a sealed-in thin platinum wire** the end of which is bent into a loop. First make certain that the wire is clean and does not colour the non-luminous flame of the gas burner. If there appears a colour, moisten the wire with a solution of HCl and ignite until the colour disappears.

Now touch a finely powdered potassium salt, preferably KCl, with the red-hot wire. Then introduce the wire with the adhering salt grains into the non-luminous part of the flame; note the characteristic pale violet flame. A solution of a potassium salt may be used instead of the dry salt but in this case the test may be less effective. Sodium salts strongly interfere with the test because they colour the flame intense yellow, which covers up the violet colour of the flame. If sodium salts are present, the reaction is carried out and potassium is identified by viewing the flame through a special prism of blue glass (see Fig. 3.6).

3.13. Reactions of the Sodium Ion

Most sodium salts are readily soluble in water. Therefore, there are very few reactions for Na^+ based on the formation of slightly soluble compounds.

1. Potassium dihydrogen antimonate, KH_2SbO_4. The potassium salt of orthoantimonic acid, H_3SbO_4, produces in solutions of sodium salts a white crystalline precipitate of sodium dihydrogen antimonate:

$$Na^+ + H_2SbO_4^- \rightarrow NaH_2SbO_4 \downarrow$$

For the reaction to be carried out, to 2 or 3 drops of a solution of a sodium salt add an equal volume of the reagent solution and scratch the sides of the test tube with a glass rod. Make sure that the precipitate is crystalline (a feature characteristic of NaH_2SbO_4). To this end, leave the solution for some time so that the precipitate is formed. Then stopper the tube and turn it upside down. Large crystals of cubic form will appear at the sides of the tube.

The following conditions must be fulfilled in carrying out this reaction.

(1) The concentration of the sodium salt must be high enough. Since the solubility of NaH_2SbO_4 is rather high and it tends to form

* The reaction can also be carried out in a test tube or on a spot plate.

** A nichrome wire may also be used.

supersaturated solutions, the precipitation may be delayed or may not occur at all in dilute solutions. The sensitivity of the reaction is therefore low. Accordingly, dilute solutions or solutions of unknown concentration must first be concentrated by evaporation. If no precipitate is formed at once, the test tube must be left to stand for some time and only after this may one conclude whether Na^+ is absent.

(2) The test solution must have a neutral reaction (pH $\approx$ 7). In an acid medium (pH $<$ 7) the reagent is decomposed to form a white amorphous precipitate of metaantimonic acid, $HSbO_3$. The reaction proceeds according to the equation

$$H_2SbO_4^- + H^+ \rightarrow HSbO_3\downarrow + H_2O$$

An alkaline medium tends to dissolve the NaH_2SbO_4 precipitate with the formation of the neutral salt Na_3SbO_4 soluble in water. For the reaction to be accomplished successfully, acid solutions are neutralized by KOH and alkaline solutions by CH_3COOH.

It should be kept in mind that, if the solution is neutralized not quite accurately, an amorphous precipitate of $HSbO_3$ may separate out in the absence of Na^+. One should know how to distinguish it from the crystalline NaH_2SbO_4 precipitate. The formation of an amorphous precipitate must not be taken as an indication of the presence of Na^+ in the solution.

(3) The test must be carried out in the cold* since the solubility of NaH_2SbO_4 (like that of $KHC_4H_4O_6$) is greatly increased with rise of temperature.

2. Microcrystalloscopic Reaction with Uranyl Acetate.** Put a a drop of the sodium salt test solution on a glass slide and evaporate it carefully to dryness. Treat the dry residue after cooling with a drop of a solution of uranyl acetate, $(CH_3COO)_2UO_2$, and after several minutes examine under the microscope the crystals of $CH_3COONa\cdot(CH_3COO)_2UO_2$ (sodium uranyl acetate). They have the form of regular tetrahedra or octahedra (Fig. 3.5).

The reaction is very sensitive (the dilution limit is 0.8 μg of Na^+). In the presence of Mg^{2+}, there separate out crystals of a triple salt of the composition $CH_3COONa\cdot(CH_3COO)_2Mg\cdot3(CH_3COO)_2UO_2\cdot9H_2O$, which resemble the crystals formed by sodium.

The reaction may also be carried out with zinc uranyl acetate or nickel uranyl acetate which form in neutral or acetic acid solutions of sodium salts a yellowish crystalline precipitate corresponding to the formula $(CH_3COO)_2Me\cdot CH_3COONa\cdot3(CH_3COO)_2UO_2\cdot9H_2O$ or $NaMe(UO_2)_3(CH_3COO)_9\cdot9H_2O$.

The shape of the crystals is the same as in Fig. 3.5.

---

* The test tube is immersed in a beaker with cold water.

** Uranyl is the ion UO_2^{2+}; in salts it is a cation.

This reaction is much more sensitive than the reaction with potassium dihydrogen antimonate (the detectable minimum is 12.5 μg in 1 drop of solution). Moreover, it is specific of sodium. The presence of 20-fold amounts of K^+, NH_4^+, Mg^{2+}, Ca^{2+}, Sr^{2+}, Ba^{2+}, Al^{3+}, Fe^{3+}, Mn^{2+}, Zn^{2+}, Co^{2+}, Ni^{2+}, Cd^{2+}, Bi^{3+}, Cu^{2+}, Pb^{2+}, and Hg^{2+} does not interfere with identification of sodium. It is for this reason that zinc uranyl acetate is used for detection of an impurity of Na^+ in salts of

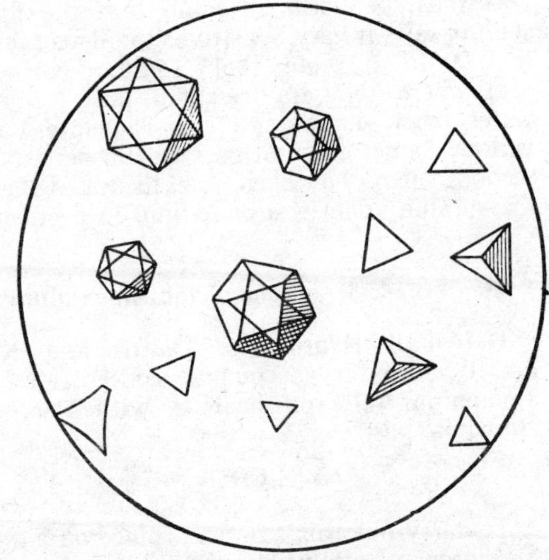

Fig. 3.5. Crystals of $CH_3COONa \cdot (CH_3COO)_2UO_2$.

zinc, manganese, iron, lead, copper, etc. If large amounts of K^+ are present, needle-shaped crystals of $CH_3COOK \cdot (CH_3COO)_2UO_2$ are formed. Therefore, if the test solution contains a high concentration of K^+ (if an abundant precipitate is formed in the test for K^+), the solution should be diluted with water to two or three times the original volume before the test for Na^+ with zinc uranyl acetate and a drop of the diluted solution taken for the reaction.

The solution used for identification of Na^+ should be as nearly neutral as possible. Acid solutions are preliminarily neutralized with ammonia or zinc oxide. In the latter case, excess zinc oxide is removed by centrifuging and a drop of the centrifugate is used for the reaction.

3. Flame Test. Volatile sodium salts impart a bright-yellow colour to the flame. This is a very characteristic test for sodium and it is carried out like the analogous test for potassium.

Mix solutions of potassium and sodium salts, dip a platinum (or nichrome) wire into the mixture, and repeat the flame test. The violet colour characteristic of potassium cannot be seen in this case. By viewing the flame through blue glass or through a solution of indigo placed in a special "indigo prism" (Fig. 3.6) or a bottle, one can identify potassium even in the presence of sodium bacause the blue glass or indigo solution absorbs the yellow rays of Na.

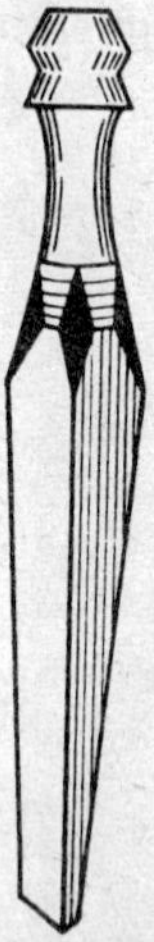

It must be borne in mind that the flame test for sodium salts is very sensitive: the detectable minimum is 0.0001 μg. Note that sodium compounds occur everywhere (they are present in dust, are extracted by water from glassware, etc.). Therefore, only a bright-yellow flame colouration, which persists for several seconds, may be taken as evidence of the presence of measurable quantities of sodium in a given substance.

Fig. 3.6. Indigo prism.

3.14. Reactions of the Ammonium Ion

1. Alkali Hydroxides, NaOH and KOH. These alkali hydroxides decompose concentrated solutions of ammonium salts on heating with the evolution of ammonia gas:

$$NH_4^+ + OH^- \rightarrow NH_3 \uparrow + H_2O$$

To carry out this reaction, place on a watch glass 1 or 2 drops of a solution of an ammonium salt and 3 or 4 drops of a solution of alkali and heat the solution. The escaping ammonia gas may be identified not only by its characteristic odour but in other ways as well.

(1) Cover the watch glass, on which the test solution (a mixture of the solutions of the ammonium salt and alkali) has been put, with another *dry* one to whose middle convex surface a piece of moist filter paper impregnated with a solution of phenolphthalein is attached. In the gas chamber thus prepared (see Fig. 2.6), the indicator paper turns red after several minutes because the escaping ammonia gas dissolves in water to form an alkali (NH_4OH).

To speed up the evolution of ammonia, the reaction mixture may be warmed slightly by placing it over the opening in a water bath.

Instead of filter paper impregnated with a phenolphthalein solution, use may be made of a piece of red litmus paper which turns blue in the escaping ammonia gas; in this case however the change of the colour is less distinct.

(2) The escaping ammonia may also be detected by allowing it to come in contact with a piece of filter paper moistened with a solution of mercurous nitrate, $Hg_2(NO_3)_2$, which is turned black under the

action of NH_3:

$$2[Hg_2]^{2+} + 4NH_3 + H_2O \rightarrow \left[O\langle^{Hg}_{Hg}\rangle NH_2 \right]^+ + 2Hg\downarrow + 3NH_4^+$$

2. Nessler's Reagent. This reagent, which is a solution of the complex salt $K_2[HgI_4]$ in KOH, produces in solutions of ammonium salts a characteristic reddish-brown precipitate.

The ionic equation of the reaction is

$$NH_4^+ + 2[HgI_4]^{2-} + 4OH^- \rightarrow \left[O\langle^{Hg}_{Hg}\rangle NH_2 \right] I\downarrow + 7I^- + 3H_2O$$

Very small amounts (traces) of ammonium salts give a yellow-orange precipitate instead of a red-brown one. The reaction is extremely sensitive. For this reaction to be accomplished, to a drop of a dilute solution of an ammonium salt put on a spot plate, there is added 1 or 2 drops of a solution of Nessler's reagent (the reagent is taken in excess as compared with the amount of the present ammonium salt since the precipitate formed is soluble in excess salt).

Nessler's reagent contains an alkali; and alkalis when reacted with most of the cations produce precipitates of slightly soluble hydroxides, many of which are coloured. The formation of such coloured precipitates may cover up the colour of the precipitate formed in the presence of NH_4^+. Hence, the reaction is less specific of NH_4^+ than the reaction with NaOH and KOH. The reaction with Nessler's reagent however is sufficiently sensitive if it is carried out in a gas chamber with a drop of this reagent being placed on the top glass instead of phenolphthalein paper. The NH_3 expelled by the action of the alkali is absorbed by the drop, causing a characteristic red-brown precipitate to be formed.

This reaction may be carried out in a different way too: the metal hydroxides are precipitated in a test tube (by the action of an excess of NaOH or KOH on 2 or 3 drops of the solution), after which the precipitate is centrifuged and Nessler's reagent is added to the centrifugate separated from the precipitate.

3. The Behaviour of NH_4^+ towards the Action of Reagents for K^+ and Na^+. Having studied the most important qualitative reactions of the NH_4^+ ion, we should find out how it behaves with the reagents used for detection of K^+ and Na^+. This is necessary for understanding the systematic procedure for analysis of Group I cations.

(a) $NaHC_4H_4O_6$, $Na_3[Co(NO_2)_6]$, and $Na_2PbCu(NO_2)_6$ give with NH_4^+ the corresponding precipitates indistinguishable in appearance from those formed by the potassium ion. Confirm this experimentally.* Thus, the presence of NH_4^+ interferes with detection of K^+.

(b) The reaction of KH_2SbO_4 with solutions of ammonium salts may lead to the formation of a white amorphous precipitate of $HSbO_3$ because these solutions show an acid reaction (due to hydrolysis). The precipitate is formed as a result of the following series of reactions:

$$NH_4^+ + H_2O \rightarrow NH_4OH + H^+$$
$$KH_2SbO_4 + H^+ \rightarrow H_3SbO_4 + K^+$$
$$H_3SbO_4 \rightarrow HSbO_3 \downarrow + H_2O$$

Hence, the presence of NH_4^+ interferes with detection of Na^+ by this reaction too.

Using a microcrystalloscopic reaction with zinc uranyl acetate, one can however detect Na^+ in the presence of NH_4^+.

4. Separation of the Ammonium Ion. The NH_4^+ ion interferes with the identification of K^+ and Na^+ (in reaction with KH_2SbO_4). Therefore: (a) the presence of NH_4^+ in the solution should be established before the test for K^+ and Na^+ is made; (b) if the NH_4^+ ion has been detected, it must be removed completely from the solution.

To remove the ammonium ion from the solution, advantage is usually taken of the volatility of its salts on heating.**

The solution, from which the NH_4^+ ion is to be removed, is acidified with concentrated nitric acid or hydrochloric acid and evaporated to dryness in an evaporating dish (or crucible), after which the dry residue is ignited until the NH_4^+ ion is completely removed (test with Nessler's reagent).

The reaction with NH_4Cl proceeds according to the equation

$$NH_4Cl \rightleftarrows NH_3 \uparrow + HCl \uparrow$$

From this equation, it is seen that this reaction is reversible. At a high temperature, NH_4Cl decomposes to gaseous NH_3 and HCl. These gases, volatilizing from the high-temperature region, recombine to form NH_4Cl. It is small crystals of this salt that give rise to the white fume which evolves on ignition of ammonium chloride.

Potassium and sodium salts volatilize at much higher temperatures than ammonium salts. Experiment shows that the ignition of KCl and NaCl at 800°C for half an hour does not cause an appreciable loss. If however these salts are ignited at 1000°C (when working with semimicroanalytical amounts), they are found to be almost completely volatilized. From this it follows that, if the NH_4^+ ion is to be removed, a too strong and prolonged ignition must be avoided.

---

* A sufficiently concentrated (saturated) solution of NH_4Cl should be used for the reaction with $NaHC_4H_4O_6$ to take place.

** There is a method for analysis of Group I cations, according to which the ammonium ion can be removed by combining it with formaldehyde (see Sec. 3.16).

The residue left after the ignition is dissolved in water; the resulting solution contains all the ions involved, with the exception of NH_4^+.

If the ammonium salt is derived not from hydrochloric acid but from a non-volatile acid such as H_3PO_4, only ammonia is volatilized on heating while the acid is left. It is clear that no white fume is formed in this case. Neither is it formed from NH_4NO_3 and other ammonium salts formed by those acids which are oxidizing agents and whose decomposition takes place in a different way; for example, in the case of NH_4NO_3, the reaction proceeds according to the equation

$$NH_4NO_3 \rightarrow 2H_2O + N_2O \uparrow$$

But, whatever ammonium salt is taken, it will either volatilize on heating or decompose with the formation of volatile products, so that the aim of the operation—removal of NH_4^+—will be achieved in one way or another.

As pointed out above, when the potassium ion is to be detected, the ammonium ion can be bound into a molecule of the organic compound $(CH_2)_6N_4$ (hexamethylenetetraamine or urotropin) by adding a solution of formaldehyde, CH_2O:

$$4NH_4^+ + 6CH_2O \rightarrow (CH_2)_6N_4 + 6H_2O + 4H^+$$

The reaction goes to near completion provided that the concentration of hydrogen ions formed is being reduced. This is accomplished by the action of sodium carbonate solution which precipitates, in addition, Mg^{2+} and the cations of all other groups (if they are present in the solution) in the form of the corresponding carbonates (or hydroxides). The method can also be employed for identification of K^+ with the aid of the fractional technique.*

3.15. Reactions of the Magnesium Ion

The reactions of the Mg^{2+} cation differ considerably from those of the other Group I cations.

1. Alkali Hydroxides. These hydroxides produce with Mg^{2+} a white *amorphous* precipitate of magnesium hydroxide, $Mg(OH)_2$:

$$Mg^{2+} + 2OH^- \rightarrow Mg(OH)_2 \downarrow$$

The reaction may be used for separation of Mg^{2+} since the hydroxides of the remaining metals of Group I are well soluble in water.

Magnesium hydroxide is soluble in acids and solutions of ammonium salts (confirm this experimentally) and so in the presence of NH_4^+ the magnesium ion cannot be separated completely. Sometimes, when the solution contains NH_4^+, $Mg(OH)_2$ may not be precipitated at all. The following experiments provide a confirmation of this.

---

* The method is described in Sec. 3.16.

(a) To 2 drops of a solution of a magnesium salt add 2 drops each of a saturated solution of NH_4Cl and 2*N* ammoniacal solution; then check whether a precipitate is formed.

(b) Repeat the same experiment with the only difference that 2 drops of water is added instead of the saturated NH_4Cl solution (to make the concentrations of magnesium and ammonium ions equal in both experiments).

It has been shown in Sec. 3.4 that the presence of NH_4Cl has a strong repressing action on the dissociation of NH_4OH, thereby reducing the concentration of OH^- in the solution. As a result, no precipitate of $Mg(OH)_2$ will be formed in the presence of NH_4Cl. This phenomenon will be discussed in greater detail in Sec. 4.3. The formation of a $Mg(OH)_2$ precipitate can be prevented by addition of an ammonium salt; this will be used repeatedly in further experiments.

2. **Disodium Hydrogen Phosphate, Na_2HPO_4.** This compound (also called disodium phosphate) forms with magnesium salts in the presence of NH_4OH and NH_4Cl a white *crystalline* precipitate of magnesium ammonium phosphate, $MgNH_4PO_4$:

$$Mg^{2+} + HPO_4^{2-} + NH_4OH \rightarrow MgNH_4PO_4 \downarrow + H_2O$$

The ammonium salt is added so that no precipitate of $Mg(OH)_2$ be formed by the action of ammonia. This reaction is more conveniently carried out in acid solution by gradually lowering its acidity with the aid of NH_4OH added drop by drop. The solubility of $MgNH_4PO_4$ will be decreased in this case and conditions will thereby be provided for slow crystallization, which will result in the formation of a well developed crystalline precipitate.

The $MgNH_4PO_4$ precipitate is readily soluble in mineral acids:

$$MgNH_4PO_4 + 3H^+ \rightarrow Mg^{2+} + NH_4^+ + H_3PO_4$$

The dissolution also takes place under the action of acetic acid. In this case, the reaction proceeds according to the equation

$$MgNH_4PO_4 + CH_3COOH \rightarrow Mg^{2+} + NH_4^+ + CH_3COO^- + HPO_4^{2-}$$
$$K_{HPO_4^{2-}} = 4.4 \times 10^{-13}$$

or

$$MgNH_4PO_4 + 2CH_3COOH \rightarrow Mg^{2+} + NH_4^+ + 2CH_3COO^- + H_2PO_4^-$$
$$K_{H_2PO_4^-} = 6.2 \times 10^{-8}$$

No orthophosphoric acid can be formed since this acid is strong ($K_{H_3PO_4} = 7.6 \times 10^{-3}$). It is stronger than acetic acid ($K_{CH_3COOH} = 1.74 \times 10^{-5}$) which cannot give enough H^+ in the solution for the formation of un-ionized H_3PO_4 molecules. To produce well-shaped crystals of $MgNH_4PO_4$, the reaction should be conducted by observing certain conditions.

Mix 1 or 2 drops of a solution of a magnesium salt in a test tube with 2 or 3 drops of 2*N* HCl and 1 or 2 drops of a solution of Na_2HPO_4. Then add 2*N* NH_4OH solution drop by drop, mixing the contents of the test tube after each drop. At first the ammonia neutralizes the acid to give NH_4Cl required to prevent the precipitation of $Mg(OH)_2$. When the neutralization is complete, a characteristic *crystalline** precipitate of $MgNH_4PO_4$ begins to settle down. It is necessary to continue adding ammonia until a distinct odour is felt or a weakly alkaline reaction with respect to phenolphthalein takes place (the pink colour of the paper corresponding to pH ≈ 9). A strongly alkaline reaction should be avoided since at pH > 10, instead of $MgNH_4PO_4$, a precipitate of $Mg_3(PO_4)_2$ is formed, which is not quite characteristic of this compound. In dilute solutions, the precipitate of $MgNH_4PO_4$ forms slowly. Scratching the sides of the tube with a glass rod speeds up its formation. If no precipitate separates out, the contents of the tube is allowed to stand for some time; only after this may a conclusion be made concerning the presence or absence of Mg^{2+}.

3. **Soluble Carbonates.** The CO_3^{2-} ion forms with Mg^{2+} a white *amorphous* precipitate of basic magnesium carbonate, $(MgOH)_2CO_3$, which is but slightly soluble in water, in distinction to Na_2CO_3, K_2CO_3, and $(NH_4)_2CO_3$:

$$2Mg^{2+} + 2CO_3^{2-} + H_2O \rightarrow (MgOH)_2CO_3\downarrow + CO_2\uparrow$$

The precipitate is soluble in acids and solutions of ammonium salts and therefore, if $(NH_4)_2CO_3$ is taken instead of Na_2CO_3 as a reagent, the precipitation of Mg^{2+} cannot be complete due to the occurrence of the reaction

$$(MgOH)_2CO_3\downarrow + 4NH_4^+ + H_2O \rightarrow 2Mg^{2+} + 4NH_4OH + CO_2\uparrow$$

If the concentration of the ammonium salt is increased by adding it to the test solution, no precipitate of $(MgOH)_2CO_3$ will be formed.

As already pointed out (see Sec. 3.11), the action of $(NH_4)_2CO_3$ on Mg^{2+} in the presence of NH_4Cl allows one to assign Mg^{2+} to Group I cations with which it is found to be present in systematic analysis.

4. **8-Hydroxyquinoline (Oxine), C_9H_7ON.** This reagent produces in ammoniacal solutions of magnesium salts a greenish-yellow *crystalline* precipitate of magnesium hydroxyquinolate, $Mg(C_9H_6ON)_2$, which is a slightly soluble inner-complex compound.

The structural formula of 8-hydroxyquinoline is

OH
N

* In the absence of NH_4OH, the ion Mg^{2+} gives with Na_2HPO_4 a non-characteristic *amorphous* precipitate of $MgHPO_4$.

The structural formula of its compound with Mg^{2+} is

$O - Mg^{2+}/2$
$\uparrow$
N

The reaction is carried out as follows. To a drop of the magnesium salt solution add a drop each of NH_4Cl and NH_4OH solutions and then a drop of 5-percent alcoholic or ammoniacal solution of the reagent. The ions Na^+, K^+, NH_4^+, and also Ba^{2+}, Sr^{2+}, and Ca^{2+} do not interfere with the reaction.

The reaction should be conducted on heating to avoid the precipitation of 8-hydroxyquinoline itself (this is what often happens on addition of an alcoholic solution of 8-hydroxyquinoline to an aqueous solution of a magnesium salt).

5. Organic Dyes. Magnesium hydroxide is capable of adsorbing certain organic dyes, in which case their colour in the adsorbed state appears to be different from that in solution. This circumstance mav be used to advantage for detection of Mg^{2+}. *p*-Nitrobenzene-azo-resorcinol*, for example, may be used as a reagent. It has a red or red-violet colour in alkaline medium; when adsorbed by magnesium hydroxide, the reagent imparts to it a dark-blue colour. The reaction is carried out as follows. To a drop of a solution of a magnesium salt add 1 or 2 drops of an alkaline solution of the reagent. Depending on the amount of Mg^{2+}, either a blue precipitate is formed or the red-violet colour of the solution changes to blue. If the solution assumes a yellow colour, this is evidence that the test solution is too acid. In this case, several drops of alkaline solution must be added to it. The ions Ca^{2+}, Sr^{2+}, Ba^{2+}, and also Al^{3+} and Mn^{2+} do not interfere with the reaction. If large amounts of ammonium salts are present, they have to be removed. The reaction enables detection of 0.5μg of Mg^{2+}

* The structural formula of *p*-nitrobenzene-azo-resorcinol is

$O_2N-C_6H_4-N{=}N-C_6H_3(HO)-OH$

This reagent is known as magneson I. Use is often made at present of *p*-nitrobenzene-azo-α-naphthol or magneson II. The latter, too, forms with Mg^{2+} in alkaline solution a blue adsorption complex. The structural formula of a precipitate of magneson I and Mg is

$O_2N-C_6H_4-N{=}N-C_6H_3(HO)-O-MgOH$

p-Nitrobenzene-azo-resorcinol colours hydroxides of nickel, cobalt, and cadmium as well.

. Instead of this reagent, use may be made of quinalizarin* which in the presence of small amounts of magnesium gives a blue colour in alkaline solution. In the presence of considerable amounts of magnesium, the reagent is absorbed by the precipitated magnesium hydroxide and imparts to it a deep-blue colour. In the absence of magnesium, the solution acquires a red-violet colour.

6. Drop Reaction (Spot Test). This reaction is based on the fact that magnesium hydroxide when partly dissolved in water produces pH $\approx$ 10 at which phenolphthalein has a bright-red colour.

The reaction is carried out as follows. Put a drop of the test solution and a drop of 2*N* ammoniacal solution on a strip of phenolphthalein paper. The paper turns red, which depends both on the presence of excess NH_4OH and the formation of magnesium hydroxide:

$$Mg^{2+} + 2NH_4OH \rightarrow Mg(OH)_2 \downarrow + 2NH_4^+$$

Then, holding the paper over the flame, dry it carefully. The ammonia and water volatilize and the red colour disappears. If the paper is now treated with a drop of water, then in the presence of Mg^{2+} it will become red again due to the partial dissolution of $Mg(OH)_2$; in the absence of Mg^{2+}, the paper will remain colourless.

The reaction under consideration enables Mg^{2+} to be detected in the presence of the ions K^+, Na^+, Ca^{2+}, Sr^{2+}, and Ba^{2+}. The cations of Groups III, IV, and V must be preliminarily removed. Neither must the NH_4^+ ions be present because they prevent the precipitation of $Mg(OH)_2$.

7. Microcrystalloscopic Reaction. To identify Mg^{2+} by microcrystalloscopic means, advantage is also taken of the reaction of formation of $MgNH_4PO_4$, which has been considered above. Figure 3.7 shows the crystals of this salt formed on slow crystallization from dilute solutions. On rapid crystallization from more concentrated solutions or from solutions containing a large amount of ammonium salts, the crystals separate out in the form of stars or dendrites, i.e., tree-like formations (Fig. 3.8).

---

* The structural formulas of quinalizarin, 1,2,5,8-hydroxyanthraquinone (I), and its compound with Mg^{2+}(II) are as follows:

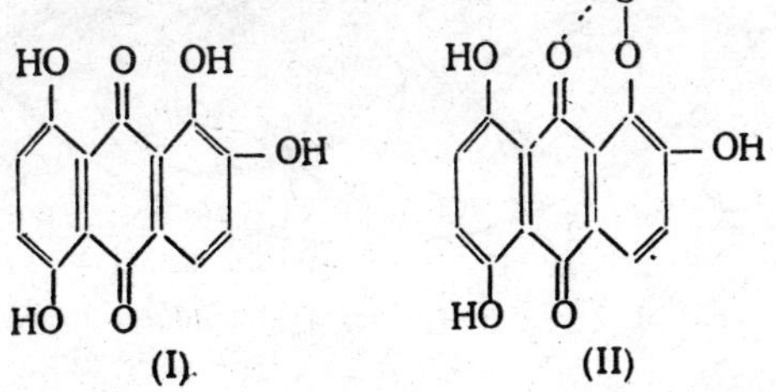

(I) (II)

Place a drop of an ammoniacal solution of a magnesium salt, which contains a little NH_4Cl, on a glass slide. Then put a small crystal of $Na_2HPO_4 \cdot 12H_2O$ into the drop and examine the crystals formed under the microscope. The dilution limit is 1 in 83,000. Potassium, sodium, and ammonium salts do not interfere with the reaction.

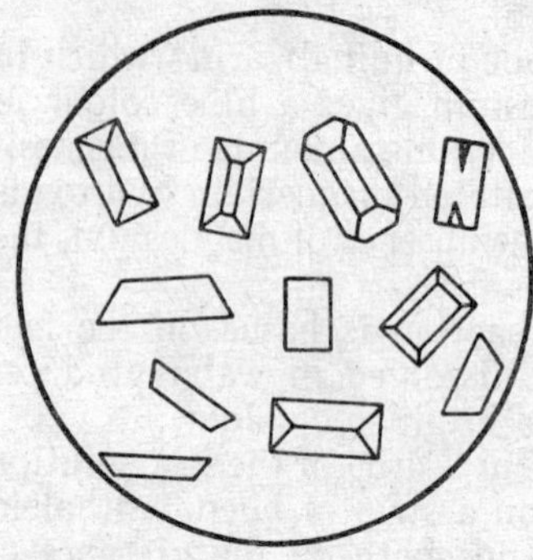

Fig. 3.7. Crystals of $MgNH_4PO_4 \cdot 6H_2O$ formed on slow crystallization.

8. The Action of Reagents for K^+ and Na^+ on Mg^{2+}. Now that we have familiarized ourselves with the most important reactions of the magnesium ion, let us consider the action of the reagents used for detection of the ions K^+ and Na^+ on Mg^{2+}.

(1) The compounds $NaHC_4H_4O_6$ and $Na[Co(NO_2)_6]$ do not produce precipitates with Mg^{2+}. Thus, Mg^{2+} does not interfere with detection of K^+.

(2) The compound KH_2SbO_4 forms with Mg^{2+} a white *crystalline* precipitate of $Mg(H_2SbO_4)_2$, which is indistinguishable in appearance

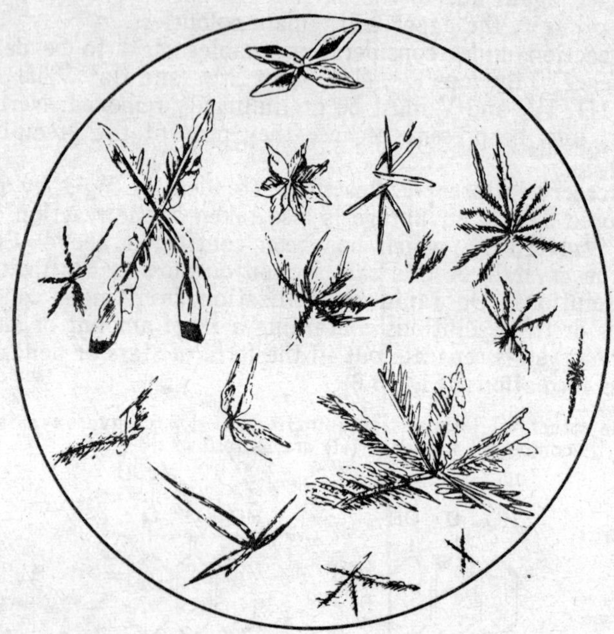

Fig. 3.8. Crystals of $MgNH_4PO_4 \cdot 6H_2O$ formed on rapid crystallization.

from NaH_2SbO_4. Hence, the magnesium ion must be preliminarily separated, for the sodium ion to be identified by reaction with KH_2SbO_4. It is more convenient to use a microcrystalloscopic reaction for Na^+ which is not interfered with by Mg^{2+}

3.16. Analysis of Mixtures of Group I Cations

A comparison of the behaviour of the cations of Group I with respect to the most important reagents is given in Table 3.8. The boldface type indicates the most characteristic reactions or properties of the cations, on which is based their identification or separation. A dash signifies the absence of a precipitate (or colour).

Table 3.8 is a spectacular illustration of the conclusions reached in our study of the reactions of Group I cations: (a) the NH_4^+ ion interferes with the identification of K^+ and Na^+ (by the reaction with KH_2SbO_4); (b) other Group I cations do not interfere with the detection of NH_4^+ and Mg^{2+}; (c) the potassium ion can be identified in the presence of Na^+ and Mg^{2+}; (d) the NH_4^+ ion can be removed by evaporation of the solution followed by the ignition of the residue, while Mg^{2+} is removed by precipitation as $Mg(OH)_2$ by the action of alkali.

The analitycal procedure for mixtures of Group I cations is described in detail below.

When starting analysis, one must bear in mind that the objective of analysis is not only to identify the elements (or ions) present in a given substance but also to find out which are the main constituents of the substance and which are present as impurity. In the case of wet reactions, the last-named objective is easily achieved by reference to the amount of precipitate formed or to the colour intensity of the ions or compounds produced.

Flame colouration reactions (flame tests) do not make it possible to draw such conclusions; but they must be carried out in order to choose the most suitable methods for qualitative analysis of substances.

Therefore, although in analysis of Group I cations the Na^+ and K^+ ions could be identified by the fractional technique (with the aid of flame tests), one must not confine oneself to such reactions alone. The cations Na^+ and K^+ must of necessity be detected by wet reactions, the flame test being used as a confirmatory test only. Moreover, one should not place too much reliance on the flame test for another reason too, namely, that in some cases (for example, in detection of Na^+) it can easily lead to errors because of its extremely high sensitivity.

Before the identification of individual ions, it is necessary to approximately determine the pH of the unknown solution since a suitable pH is one of the most important conditions for analytical reactions to be accomplished. Besides, such an estimate provides some information regarding the composition of the test solution. For instance, a weakly acid reaction points to the possible presence of NH_4^+ and Mg^{2+}

TABLE 3.8. **Reactions of Group I Cations**

Reagent	Cation			
	K^+	Na^+	NH_4^+	Mg^{2+}
$NaHC_4H_4O_6$	White crystalline precipitate: **$KHC_4H_4O_6$**	—	White crystalline precipitate: $NH_4HC_4H_4O_6$	—
$Na_3[Co(NO_2)_6]$	Yellow crystalline precipitate: **$K_2Na[Co(No_2)_6]$**	—	Yellow crystalline precipitate: $(NH_4)_2Na[Co(NO_2)_6]$	—
KH_2SbO_4	—	White crystalline precipitate: **NaH_2SbO_4**	White amorphous precipitate: $HSbO_3$	White crystalline precipitate: $Mg(H_2SbO_4)_2$
$(CH_3COO)_2Zn + (CH_3COO)_2UO_2 + + CH_3COOH$	—*	Yellowish crystalline precipitate: **zinc-sodium-uranyl acetate****	—	—
KOH, NaOH	—	—	**NH_3** evolved	White amorphous precipitate: **$Mg(OH)_2$**

Nessler's reagent $K_2[HgI_4] + KOH$	—	—	Red-brown precipitate: $\left[O \langle {}^{Hg}_{Hg} \rangle NH_2 \right] I$	White amorphous precipitate: $Mg(OH)_2$
$Na_2HPO_4 + NH_4OH + NH_4Cl$	—	—	—	White crystalline precipitate: $MgNH_4PO_4$
8-Hydroxyquinoline	—	—	—	Greenish-yellow precipitate: $Mg(C_9H_6ON)_2$
p-Nitrobenzene-azoresorcinol or quinalizarin	—	—	—	Blue precipitate or colouration
$(NH_4)_2CO_3 + NH_4OH + NH_4Cl$	—	—	—	—
Action of heat on salts	Slightly volatile	Slightly volatile	Volatile	Slightly volatile
Flame colouration	Violet	Yellow	—	—

* At a high concentration of K^+ there may separate out a precipitate of $CH_3COOK \cdot (CH_3COO)_2UO_2$.
** $CH_3COONa \cdot (CH_3COO)_2Zn \cdot 3(CH_3COO)_2UO_2 \cdot 9H_2O$.

In the case of a strongly acid reaction (pH ≈ 2), one has to admit the presence in solution of acid salts, such as $NaHSO_4$, or free acids. Finally, if the reaction is strongly alkaline, this indicates the presence either of free alkalis or of potassium or sodium salts of weak acids (K_2CO_3 or Na_2CO_3). The result of the pH determination and the relevant conclusions must be entered into the analytical report and presented to the instructor together with the other results obtained.

1. Determination of the pH. The pH of a solution is assayed with the aid of a range of indicators or the universal indicator.

2. Identification of NH_4^+. The analysis should be started with the detection of ammonium ion, the presence of which interferes with identification of K^+ and Na^+; these ions, on the other hand, do not interfere with detection of NH_4^+ since, on heating with an alkali, ammonia can be formed under the conditions of analysis only from an ammonium salt.

To detect NH_4^+, treat one or two drops of the test solution in a gas chamber with 2-4 drops of 2*N* NaOH solution in the presence of moist phenolphthalein paper (or red litmus paper) and heat in a water bath.

If the phenolphthalein paper becomes red (or the red litmus paper turns blue), this indicates the presence of ammonium.*

Another test can also be made simultaneously. Add 2 or 3 drops of Nessler's reagent to 1 drop of the test solution. The formation of a red-brown precipitate** confirms the presence of NH_4^+.

3. Identification of Mg^{2+}. For this purpose, use may also be made of the fractional method since the other cations of Group I do not interfere with detection of Mg^{2+}. To 2 or 3 drops of the test solution add 2*N* HCl solution drop by drop until the reaction is acid and 1 or 2 drops of Na_2HPO_4 solution. Then add a solution of NH_4OH drop by drop with stirring until the reaction is alkaline or the distinct odour of ammonia appears. If no precipitate is formed at once, the solution is allowed to stand for some time. A white *crystalline* precipitate of $MgNH_4PO_4$ indicates the presence of Mg^{2+}. Instead of this reaction, a drop reaction or a microcrystalloscopic reaction with a drop of the solution can be carried out for Mg^{2+}, or else the magnesium ion can be detected by reaction with quinalizarin or 8-hydroxyquinoline (see Sec. 3.15).

4. Removal of NH_4^+. If NH_4^+ is present***, it must be removed before the identification of K^+. To do this, place 12-15 drops of the test solution into a microcrucible or into a small evaporating dish, acidify

* Use may be made of paper moistened with a solution of $Hg_2(NO_3)_2$ which turns black by the action of ammonia.

** In addition to $K_2[HgI_4]$, Nessler's reagent contains KOH and therefore it gives with magnesium salts a white precipitate of $Mg(OH)_2$.

*** In the systematic analysis, after the precipitation of the cations of Group II from solution, the NH_4^+ ion is found to be introduced with $(NH_4)_2CO_3$, NH_4OH, and NH_4Cl.

with concentrated HNO_3 or HCl, evaporate to dryness, and then heat until the ammonium salts are completely removed. To make sure that they have been completely removed, collect a little dry deposit from the crucible (or dish) into a conical test tube and add 2 or 3 drops of distilled water and 2 or 3 drops of Nessler's reagent. The formation of an orange-brown precipitate or a yellow colour indicates that not all NH_4^+ has been removed. In this case, treat again the dry residue in the crucible (or dish) with concentrated HNO_3 or HCl, evaporate, and ignite it. This operation is repeated several times until the test for NH_4^+ is negative. Having removed all the NH_4^+ present, transfer the solution into a conical test tube and if it contains a turbidity or precipitate, the MgOHCl formed is separated by centrifuging and discarded.* The centrifugate is examined as described below.

The removal of ammonium in analysis of Group I cations can be achieved by converting it into $(CH_2)_6N_4$. Then the potassium ion is detected by one of its characteristic reactions (see Sec. 3.12); in this case, Na^+ must be detected in a separate portion of the solution by a microcrystalloscopic reaction (see below).

5. Identification of K^+. After the NH_4^+ is broken down and the dry residue is dissolved, the pH of the test solution may change and it must therefore be checked. To 1 or 2 drops of the unknown solution add 2 or 3 drops of a solution of $Na_3[Co(NO_2)_6]$ and (if the reaction is acid) 1 or 2 drops of a solution of CH_3COONa. If no precipitate is formed, let the solution stand for some time. A yellow precipitate of $K_2Na[Co(NO_2)_6]$ indicates the presence of K^+.

A drop of the solution may also be evaporated to dryness on a slide. In this case, cool the dry residue, treat it with a drop of a solution of $Na_2PbCu(NO_2)_6$, and after 1 or 2 minutes examine the precipitate under the microscope. The formation of black cubic crystals of $K_2PbCu(NO_2)_6$ proves the presence of K^+.**

6. Identification of Na^+. The procedures used are as follows.

(1) Evaporate a drop of the unknown solution to dryness on a glass slide and after it is completely cooled treat the dry residue with a drop of a solution of uranyl acetate, $(CH_3COO)_2UO_2$. The formation of $CH_3COONa \cdot (CH_3COO)_2UO_2$—characteristic crystals having the form of tetrahedra or octahedra when viewed under the microscope (see Fig. 3.5)—points to the presence of Na^+. The reaction may also be carried out by using zinc or nickel uranyl acetate.

In carrying out the reactions, it should be kept in mind that the K^+ ions when present in a high concentration produce crystals of

* This precipitate is formed by the hydrolysis of Mg^{2+} when a solution containing $MgCl_2$ is evaporated:

$$Mg^{2+} + Cl^- + H_2O \rightarrow MgOHCl\downarrow + H^+$$

** Let us recall that both these reactions are also induced by ammonium and if it is not removed completely the K^+ will be detected in error.

$CH_3COOK \cdot (CH_3COO)_2UO_2$ too. When examined under the microscope, this precipitate appear as needles (Fig. 3.9) and it is easily distinguished from a precipitate of the corresponding sodium salt. If the reaction is performed with zinc uranyl acetate in the presence of large amounts of K^+, the solution should first be diluted with water two- or threefold and the reaction carried out with a drop of the diluted solution. In doubtful cases, a drop of the liquid together with the precipitate may be examined under the microscope. Crystals of zinc uranyl acetate have the shape of octahedra, like those shown in Fig. 3.5.

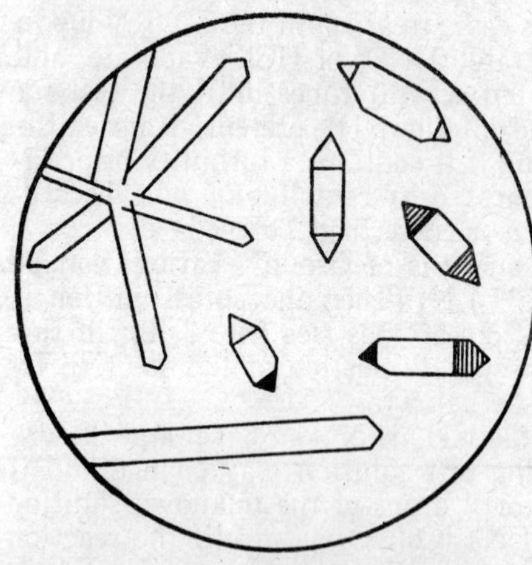

Fig. 3.9. Crystals of $CH_3COOK \cdot (CH_3COO)_2UO_2$.

(2) If the reagents specified above are not available in the laboratory, the Na^+ ion can be identified by carrying out the reaction with KH_2SbO_4 in a strictly neutral medium (pH 7), in which case a glass rod is used for scratching the sides of the vessel and the solution is cooled. A white *crystalline* precipitate of NaH_2SbO_4 is an indication of the presence of Na^+.

It must however be remembered that this reaction is inapplicable in the presence of Mg^{2+} because this ion forms a precipitate of $Mg(H_2SbO_4)_2$ which resembles NaH_2SbO_4. Hence, if Mg^{2+} is detected in the test according to Procedure *3*, it must be completely removed.

To do this, introduce an excess of KOH (until a strongly alkaline reaction occurs)* into the solution, from which the NH_4^+ ion has been removed, and heat it for several minutes. The $Mg(OH)_2$ precipitate formed is centrifuged and discarded.

The centrifugate containing Na^+ together with excess KOH must first be neutralized and concentrated. Add 2*N* HCl solution to the centrifugate until the reaction is acid, evaporate the solution to dryness in a crucible, and heat slightly the dry residue to remove excess HCl. After cooling dissolve the contents of the crucible in 3 drops of water (not more), transfer the resulting fully transparent solution to a test tube**, check its pH, and detect the Na^+ ion by reaction with KH_2SbO

* First, a solution of KOH is added until phenolphthalein paper turns red (pH $\approx$ 10), followed by addition of another 2 or 3 drops of KOH.
** If the solution is not quite transparent, it is to be centrifuged.

7. Confirmatory Tests for K^+ and Na^+. The conclusions reached regarding the presence of K^+ and Na^+ should be confirmed by a flame test, watching the flame with the naked eye (detection of Na^+) and through indigo solution or blue glass (detection of K^+).

If no NH_4^+ is detected, the presence of Mg^{2+} can be confirmed by a drop reaction (see Sec. 3.15).

8. Identification of K^+ and Na^+ by the Fractional Method. It is known that if the test solution contains the cations of Groups II-V, they are precipitated successively from solution. Group I cations are then detected in the solution left after the separation. Separation reactions are always associated with losses due to coprecipitation and other factors (see Sec. 5.4). Therefore, the concentrations of Group I cations greatly decrease by the moment they are being detected. Since the reactions for K^+ and Na^+ are slightly sensitive, it is clear that these ions may sometimes be "lost" during the systematic course of analysis. To avoid errors, it is expedient, if K^+ and Na^+ have not been detected in the systematic analysis, to try to detect them by the fractional method.

First Method. (1) *Fractional Detection of K^+.* To 2 or 3 drops of the unknown solution add, if NH_4^+ is present, an equal volume of 40-percent formaldehyde solution and 1 drop of phenolphthalein. Then add a solution of Na_2CO_3 drop by drop to the appearance of a persistent red colour corresponding to $pH \geqslant 10$. If a precipitate of carbonates and hydroxides is formed in this case, heat the solution for 1 or 2 minutes, separate the precipitate by centrifuging, and discard it. Acidify the centrifugate with $2N$ CH_3COOH until the red colour disappears and try to detect K^+ in it by reaction with $Na_3[Co(NO_2)_6]$.

When this reaction is being carried out, the solution should be heated. If the NH_4^+ ion has not been removed completely and a precipitate of $(NH_4)_2Na[Co(NO_2)_6]$ has been formed, then on heating it will decompose, while the $K_2Na[Co(NO_2)_6]$ precipitate will remain unchanged.

(2) *Fractional Detection of Na^+.* To another portion of the test solution add a solution of potassium carbonate and, being sure that the precipitation is complete, heat the carbonate and hydroxide precipitate, separate it by centrifuging, and discard it. Acidify the centrifugate slightly with $2N$ acetic acid ($pH \approx 4$-5).

After this try to detect Na^+ by a microcrystalloscopic reaction. In this method, it is necessary to use chemically pure Na_2CO_3 and K_2CO_3 containing no impurities of the ion being detected.

Second Method. Another method for analysis is based on the decomposition of ammonium salts under the action of bleaching powder:

$$2NH_4Cl + 3Ca\begin{matrix}\diagup OCl \\ \diagdown Cl\end{matrix} + Ca(OH)_2 \rightarrow 4CaCl_2 + 5H_2O + N_2\uparrow$$

Since bleaching powder contains $Ca(OH)_2$ in addition to ClO—Ca—Cl, it precipitates most of the cations as the hydroxides or basic salts. The ions Ca^{2+}, Sr^{2+}, Ba^{2+}, and Mg^{2+} left in the solution do not interfere with detection of K^+ by the action of $Na_3[Co(NO_2)_6]$ and of Na^+ by the action of zinc uranyl acetate.

To 15-20 drops of the unknown solution add about 1 gram of bleaching powder and 3 ml of water. Stir thoroughly the contents of the test tube with a glass rod until bubbles of the gas (N_2) cease to evolve. To make sure that the removal of the ammonium salts is complete, add to the solution a small amount of bleaching powder (no gas bubbles must evolve), and then separate the precipitate from the solution and discard it. Acidify the solution with acetic acid and boil it to remove the chlorine being evolved (*In the hood*!). After this try to detect in separate portions of the solution the potassium ion by the reaction with $Na_3[Co(NO_2)_6]$ and the sodium ion by a microcrystalloscopic reaction.*

* It should be kept in mind that the iodide (I^-) ions (introduced in the systematic course of analysis of the cations of all five groups in order to reduce As^V to As^{III}) interfere with the reaction since they cause the decomposition of $Na_3[Co(NO_2)_6]$. Therefore, if the I^- ions are present, it is necessary to evaporate to dryness several drops of the solution to be used for detection of K^+, with 1 or 2 drops of 6*N* HNO_3 solution, and to heat slightly the dry residue. The iodide ions are oxidized by HNO_3 and iodine is formed, which volatilizes on evaporation. The dry residue is dissolved in 2-3 drops of water and tested for K^+ as described above.

Chapter 4

Group II Cations

In studying the cations of Group II and those of the succeeding groups we shall deal with numerous cases involving the precipitation of individual ions and of whole groups of ions in the form of slightly soluble compounds. To control this most important operation of qualitative analysis consciously, it is necessary to familiarize with the theory of the process of precipitation and also with the theoretical principles on which is based the opposite process—the dissolution of precipitates. It is these theoretical principles that are discussed below.

A. THEORETICAL PRINCIPLES

4.1. Equilibrium in Heterogeneous Systems. The Solubility Product

In Chapter 3 we considered some examples of ionic equilibria in electrolyte solutions which are homogeneous systems. In analytical practice we also deal with more complex heterogeneous systems. As examples may be cited: a system consisting of a saturated solution in contact with a precipitate of the corresponding substance; two immiscible liquids; a liquid and a gas in equilibrium with that liquid.

The individual components of a heterogeneous system separated by interfaces are known as its phases. For example, the precipitate and the saturated solution are different phases of a heterogeneous system.

It is this type of heterogeneous systems that we are concerned with in precipitation reactions carried out during analysis; study of such heterogeneous systems is therefore of great importance to analytical chemistry, and we shall consider them in detail.

Suppose that a slightly soluble salt, say AgCl, is brought into contact with water. The ions Ag^+ and Cl^-, which the crystals of this salt are composed of, are attracted by the nearest water dipoles and gradually break loose from the crystal surfaces and pass into solution in the form of hydrated ions. This process is accompanied by the opposite process of precipitation of AgCl from solution. While moving in the solution, the hydrated ions Ag^+ and Cl^- encounter AgCl crystals and are attracted by the oppositely charged lattice ions, as a result of which they occasionally become dehydrated and redeposited on the crystal surfaces (Fig. 4.1). The occurrence of these opposing processes invariably leads to a state of dynamic equilibrium, in which the number of Ag^+ and Cl^- ions supplied in unit time from unit area of the solid phase into the solution exactly equals the number of these ions deposited from solution. When the condition of equilibrium is reached, the Ag^+ and Cl^- ions cease to pass into solution just as the amount of

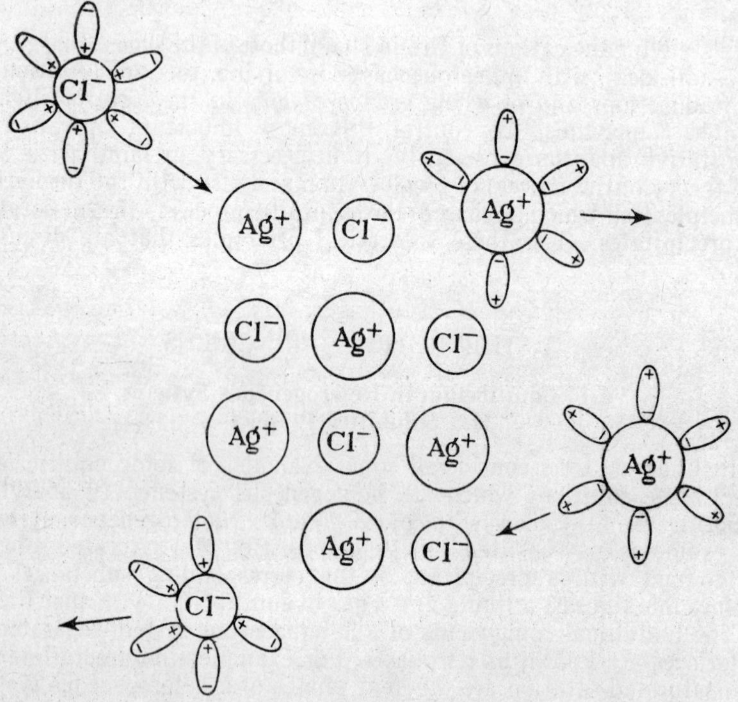

Fig. 4.1. Schematic diagram showing the dissolution of AgCl.

solid ceases to decrease, and the solution becomes saturated. We can therefore say that *a saturated solution is one which is in dynamic equilibrium with the corresponding solid phase*.

To derive the general law governing this equilibrium, let us write expressions for the rates at which the Ag^+ and Cl^- ions go into solution (v_1) and deposit on the AgCl crystal surfaces (v_2).

The rate v_1, i.e., the number of Ag^+ and Cl^- ions passing into solution from unit area of the solid phase in a unit time, is proportional to the number of these ions per unit area of the solid surface. It may be assumed that this number of ions remains constant with time.

Consequently, the rate of dissolution may be regarded as a constant quantity at a constant temperature:

$$v_1 = k_1$$

The rate v_2 of the process of deposition of Ag^+ and Cl^- ions on the AgCl crystal surface depends on the number of collisions between the ions Ag^+ and Cl^- and unit area of the solid surface per unit time. If no interionic forces existed in solution, which tended to retard the movement of ions (and hence to decrease the number of collisions), the rate v_2 would be proportional to the concentration of Ag^+ and Cl^- ions in solution. For these forces to be taken into account, one has evidently to replace the concentrations of ions with their activities. Consequently,

$$v_2 = k_2 a_{Ag^+} a_{Cl^-}$$

where k_2 is the proportionality coefficient, i.e., a constant quantity at constant temperature.

At equilibrium $v_2 = v_1$, i.e.,

$$k_2 a_{Ag^+} a_{Cl^-} = k_1$$

$$a_{Ag^+} a_{Cl^-} = \frac{k_1}{k_2} = SP_{a,\ AgCl} \tag{4.1}$$

Thus, the product of the ionic activities of a given slightly soluble electrolyte in its saturated solution is a constant quantity at a constant temperature. Since this constant is a measure of the ability of a substance to dissolve in the solvent concerned, it is called the **solubility product** and denoted by the symbol SP_a*.

The values of ionic activities in the expression for the solubility product are raised to powers equal to the number of ions participating in the process of dissociation; for example,

$$SP_{a,\ Ca_3(PO_4)_2} = a^3_{Ca^{2+}} a^2_{PO_4^{3-}}$$

Strictly speaking, the assumption that the number of Ag^+ and Cl^- ions per unit surface of the solid phase remains constant is not quite true. The point is that in the course of precipitation the crystals always adsorb those of the common ions (Ag^+ and Cl^-) which are present in solution in excess. And since the amount of adsorbed ions depends on their concentration in solution, it follows that the amount of Ag^+ and Cl^- ions per unit surface of the solid phase varies somewhat with change of concentration. It is not difficult however to show that this change does not affect the final result.

Let us consider in more detail the equilibrium conditions for each of the ions separately.

1. The rate of transfer of Ag^+ ions into solution (v_1) is proportional to the number of these ions per unit surface area of solid phase (n_1),

* The symbols K_{sp}, K_s and L are also sometimes used.

i.e.,

$$v_1 = k_1 n_1$$

The rate of deposition of Ag^+ ions is proportional to:

(a) the number of collisions of Ag^+ ions with unit area of solid phase in unit time, and hence to a_{Ag^+};

(b) the number of Cl^- ions (that attract Ag^+ ions) per unit surface area (n_2).

Thus,

$$v_2 = k_2 n_2 a_{Ag^+}$$

At equilibrium $v_1 = v_2$, i.e.,

$$k_1 n_1 = k_2 n_2 a_{Ag^+} \qquad (4.2)$$

2. Analogously, for Cl^- ions the rate of their transfer (v_3) into solution is

$$v_3 = k_3 n_2$$

and the rate of their deposition (v_4) on the surface of solid phase

$$v_4 = k_4 n_1 a_{Cl^-}$$

Under the conditions of equilibrium $v_3 = v_4$ or

$$k_3 n_2 = k_4 n_1 a_{Cl^-} \qquad (4.3)$$

To derive the general equilibrium equation for the system as a whole from the expressions obtained for the individual ions, (4.2) and (4.3), we multiply them termwise:

$$k_1 n_1 k_3 n_2 = k_2 n_2 a_{Ag^+} k_4 n_1 a_{Cl^-}$$

The quantities n_1 and n_2 present on both sides of the equation cancel out. Consequently, whatever the amounts of Ag^+ and Cl^- ions per unit area of the AgCl crystals, and whatever their variations during precipitation, the final result will not be affected. Indeed, if we transpose the constants k_2 and k_4 to the left-hand side of the equation, we shall have:

$$\frac{k_1 k_3}{k_2 k_4} = a_{Ag^+} a_{Cl^-}$$

If we now designate the ratio of the constants $k_1 k_3 / k_2 k_4$ by $SP_{a,\ AgCl}$, we obtain Eq. (4.1) given above.

Since the ionic activities are equal to ionic concentrations multiplied by the activity coefficients, the expression for $SP_{a,\ AgCl}$ may be written as follows:

$$SP_{a,\ AgCl} = [Ag^+][Cl^-] f_{Ag^+} f_{Cl^-}$$

But since the solubility of AgCl is very slight, the ion concentrations in its saturated solution is very low and the interionic forces practically play no part. Therefore, without appreciable error, f_{Ag^+} and f_{Cl^-} may be assumed to be equal to unity. This yields an approximate equation

$$SP_{a,\ AgCl} \approx [Ag^+][Cl^-]$$

which is generally used in practice.

In the case of readily soluble electrolytes, the value of f is not equal to unity and the activities of the ions cannot be replaced with their concentrations. Neither can this be tolerated with respect to slightly soluble electrolytes, such as AgCl, if other strong electrolytes are also present in solution in high concentrations. In calculations based on the solubility product principle, however, there is no need for a high degree of accuracy and the use of activities is inconvenient (and sometimes impossible), for which reason they are normally replaced by concentrations (even in the presence of foreign electrolytes), with the attending error being neglected.

Thus, the thermodynamic constant $SP_a = a_{Cat^+}a_{An^-}$ and the concentration constant $SP_C = [Cat^+][An^-]$ are connected by the following relation:

$$SP_a = SP_C f_1 f_2$$

If the occurrence of side reactions is possible during the precipitation of a sparingly soluble compound, the coefficient γ should be introduced (see Sec. 3.6). For instance, when AgCl is precipitated by excess HCl, the AgCl may dissolve with the formation of a complex compound:

$$Ag^+ + HCl \rightarrow AgCl\downarrow + H^+$$

$$AgCl\downarrow + HCl \rightarrow AgCl_2^- + H^+$$

The total concentration of silver C_{Ag^+} is composed of two components: (1) the concentration of silver ions $[Ag^+]$ that pass into solution due to the solubility of the AgCl precipitate; and (2) the concentration of silver entering into the composition of the complex $[AgCl_2]^-$. Thus,

$$C_{Ag^+} = [Ag^+] + [AgCl_2^-]$$

Hence $[Ag^+]$ will be less than the total silver concentration: $[Ag^+] < C_{Ag^+}$; $[Ag^+] = C_{Ag^+}\gamma_{Ag^+}$, the value of γ being less than unity. We shall call the product of C_{Ag^+} by C_{Cl^-} a **conditional solubility product**. All the three constants are interrelated as follows:

$$SP_a = a_{Cat^+}a_{An^-} = SP_C f_1 f_2 = SP_{con} f_1 f_2 \gamma_1 \gamma_2$$

It is not always easy to calculate the coefficients γ, and therefore they are taken into account only in those cases where very precise estimations are required. If side reactions take place but to a very insignificant degree, the electrostatic coefficient f alone will suffice.

Thermodynamic constants of the solubility product, SP_a, can be found in handbooks of chemistry. Therefore, if it is required to calculate the concentrations of ions in solution, expressed in gram-ions per 1 litre, the activities of the ions are first determined from the corresponding thermodynamic constant and then the sought-for concentrations are found by dividing these activities by the activity coefficients.

According to the present-day views, $S_{\text{Cat, An}} \neq a_{\text{Cat}^+}$ (or a_{An^-}), where $S_{\text{Cat, An}}$ is the solubility, as the solution under analysis invariably contains un-ionized molecules of a slightly soluble compound and also ion pairs and polymeric ions. Therefore, in the simplest case (if we disregard the presence of ion pairs and polymer ions), we have:

$$K = \frac{a_{\text{Cat}^+} a_{\text{An}^-}}{a_{\text{Cat, An}}}$$

that is, $a_{\text{Cat}^+} a_{\text{An}^-} = K a_{\text{Cat, An}} = SP_a$.

But since $a_{\text{Cat, An}}$ is very negligible in value, it is assumed here that $SP_a = a_{\text{Cat}^+} a_{\text{An}^-} = S^2$.

Even with the assumptions indicated, the solubility product principle is of great significance in analytical chemistry. It enables one to elucidate numerous, rather sophisticated problems arising in analytical practice. The conclusions reached with the aid of the solubility product principle are nearly always in qualitative agreement with experiment. From now on we shall use the solubility product rule in its simplified form, i.e., without taking activity coefficients and other coefficients into account, resorting to them only when it is absolutely necessary.

To find the solubility product of a slightly soluble electrolyte, its solubility is determined at a specified temperature by any suitable method. Knowing the solubility, one can easily calculate the solubility product required. We shall illustrate this by the following examples.

In solving the problems given below it is necessary to know that, if for a given slightly soluble compound SP_a is less than $n \cdot 10^{-7}$, the activity coefficients may be ignored; if the value of SP_a is greater than $n \cdot 10^{-7}$, the activity coefficients must be taken into account in calculations. Examples *3* and *5* are solved with account taken of the activity coefficients; in other examples, they may be ignored.

Further in the text, in a number of cases involving slightly soluble compounds ($SP_a < n \cdot 10^{-7}$), the solubility product will be designated as SP with the subscript omitted.

Example 1. Calculate $SP_{a,\,\text{AgCl}}$ if it is known that at 25°C a saturated solution of AgCl contains 1.86×10^{-3} g of salt per 1 litre.

Solution. First we find the molar concentration of the saturated solution of AgCl. Considering that the molecular weight of AgCl is 143.3, we have:

$$C_{\text{AgCl}} = \frac{1.86 \times 10^{-3}}{143.3} = 1.3 \times 10^{-5}\ M$$

Since each gram-molecule (mole) of AgCl gives on dissolution 1 g-ion each of Ag^+ and Cl^-, the concentrations of these ions in the saturated AgCl solution will be the same. Hence,

$$SP_{C,\,AgCl} = [Ag^+][Cl^-] = 1.3 \times 10^{-5} \times 1.3 \times 10^{-5} = 1.78 \times 10^{-10} \approx SP_{a,\,AgCl}$$

Example 2. Find the solubility product (SP_C) of silver chromate given that one litre of a saturated solution at 25°C contains about 2.2×10^{-2} g of the salt.

Solution. According to the equation

$$C_{Ag_2CrO_4} = \frac{2.2 \times 10^{-2}}{331.8} = 6.5 \times 10^{-5}\ M$$

Since each gram-molecule (mole) of Ag_2CrO_4 dissolves to give 1 g-ion of CrO_4^{2-} and 2 g-ion of Ag^+, we have:

$$[CrO_4^{2-}] = 6.5 \times 10^{-5} \text{ g-ion/litre}$$
$$[Ag^+] = 1.3 \times 10^{-4} \text{ g-ion/litre}$$

whence

$$SP_{C,\,Ag_2CrO_4} = [Ag^+]^2[CrO_4^{2-}] = [1.3 \times 10^{-4}]^2 \times 6.5 \times 10^{-5} =$$
$$= 1.1 \times 10^{-12} \approx SP_{a,\,Ag_2CrO_4}$$

Example 3. From the solubility of calcium sulphate ($CaSO_4$), which is approximately equal to 1 g/litre (at room temperature), calculate its solubility product constant (SP_a).

Solution. The solubility of $CaSO_4$ is so high that the activity coefficients of the ions cannot be assumed to be equal to unity. To find the activity coefficients, let us first calculate the ionic strength of the solution. To do this, the molar concentration of a saturated solution must first be calculated:

$$C_{CaSO_4} = \frac{1}{136} = 7.4 \times 10^{-3}\ M$$

Hence,

$$[Ca^{2+}] = [SO_4^{2-}] = 7.4 \times 10^{-3} \text{ g-ion/litre}$$

$$\mu = \frac{1}{2}\{[Ca^{2+}]2^2 + [SO_4^{2-}]2^2\} = \frac{1}{2}[7.4 \times 10^{-3} \times 4 + 7.4 \times 10^{-3} \times 4] = 0.03$$

From Table 3.3 (see Sec. 3.6) we find:

$$f_{Ca^{2+}} \approx f_{SO_4^{2-}} \approx 0.5$$

Thus,

$$SP_{a,\,CaSO_4} = [Ca^{2+}][SO_4^{2-}]\,f_{Ca^{2+}} f_{SO_4^{2-}} = (7.4 \times 10^{-3} \times 0.5)^2 = 1.4 \times 10^{-5}$$

The result obtained differs from the tabulated value ($SP_a = 2.37 \times 10^{-5}$). This is accounted for by the fact that the data of Table 3.3 are the mean activity coefficients which are actually somewhat different for different ions. If we calculate the solubility product of calcium sulphate without taking activity coefficients into account, we shall have 5.5×10^{-5}, i.e., the value that greatly differs from the experimentally found value.

It should be noted that although the solubility of silver chloride (1.3×10^{-5}) is lower than that of silver chromate (6.5×10^{-5}), the solubility product of AgCl (1.78×10^{-10}) is greater than that of Ag_2CrO_4 (1.1×10^{-12}). This is explained as follows: in calculating the solubility product (SP_a) of Ag_2CrO_4 one has to square the small value—the concentration of Ag^+ (1.3×10^{-4}), which is not needed when $SP_{a,\ AgCl}$ is calculated. From this it follows that the relative solubility of compounds can be judged by the values of their solubility products only when their formulas are of the same type, e.g., AgCl, $BaSO_4$, $CaCO_3$, or Ag_2CrO_4, Ag_2CO_3, HgI_2. Evidently, it is also easy to solve the reverse problem—to calculate the solubility of an electrolyte if the solubility product is known.

Example 4. Calculate the solubility of $Mg(OH)_2$ at 25°C if the solubility product $SP_a = 6.0 \times 10^{-10}$.

Solution. Let the required solubility (in moles per litre) be x. When passed into solution, these x moles of $Mg(OH)_2$ will give rise to the following ion concentrations:

$$[Mg^{2+}] = x \quad \text{and} \quad [OH^-] = 2x$$

that is,

$$SP_{C,\ Mg(OH)_2} = [Mg^{2+}][OH^-]^2 = x[2x]^2 = 4x^3 \approx 6.0 \times 10^{-10}$$

whence

$$x \approx \sqrt[3]{150 \times 10^{-12}} \approx 5.3 \times 10^{-4} \text{ mole/litre}$$

To convert it to the solubility expressed in grams per 1 litre (y), the value found is multiplied by the molecular weight of $Mg(OH)_2$ equal to 58.31. The result is

$$y \approx 5.3 \times 10^{-4} \times 58.31 \approx 3.1 \times 10^{-2} \text{ g/litre}$$

Example 5. Given that the solubility product of silver bromate, $AgBrO_3$, at 25°C is 5.5×10^{-5}, find its solubility.

Solution. The value of the solubility product of $AgBrO_3$ is relatively large so that activity coefficients cannot be disregarded. To calculate the ionic strength of the solution, we first calculate the approximate values of ionic concentrations from the simplified formula

$$SP_{C,\ AgBrO_3} = [Ag^+][BrO_3^-]$$

Since $[Ag^+] = [BrO_3^-]$, we have

$$[Ag^+] = [BrO_3^-] \approx \sqrt{5.5 \times 10^{-5}} \approx 7.4 \times 10^{-3} \approx 0.007 \text{ g-ion/litre}$$

Hence,

$$\mu = \frac{1}{2}(0.007 \times 1^2 + 0.007 \times 1^2) = 0.007$$

From Table 3.3 (see Sec. 3.6) we find:

$$f_{Ag^+} = f_{BrO_3^-} = 0.90$$

We can now find a more accurate value for the solubility of $AgBrO_3$. Designating it by x and remembering that $[Ag^+] = [BrO_3^-] = x$, we have:

$$[Ag^+][BrO_3^-]f_{Ag^+}f_{BrO_3^-} = x^2 \cdot 0.90^2 = 5.5 \times 10^{-5}$$

which gives

$$x = \frac{\sqrt{5.5 \times 10^{-5}}}{0.90} = \frac{7.4 \times 10^{-3}}{0.90} = 8.2 \times 10^{-3} \text{ mole/litre}$$

Converting this to the solubility in grams per litre, we have:

$$y = 8.2 \times 10^{-3} \times 235.8 = 1.93 \text{ g/litre}$$

Had we not taken the activity coefficients into account in calculating the solubility, we would have found:

$$y = 7.4 \times 10^{-3} \times 235.8 = 1.74 \text{ g/litre}$$

4.2. Effect of Common Ions on Solubility. The Salt Effect

According to the solubility product principle, at the moment when the product of the activities (or, more simply, the concentrations) of the ions (the **ion concentration product** or simply **ion product**) of a slightly soluble electrolyte reaches the value of its solubility product at a given temperature, the solution becomes saturated with respect to that electrolyte.

If the ion product is less than the solubility product, the solution is said to be unsaturated.

If however the ion product exceeds the solubility product, the solution becomes supersaturated and part of the solute must sooner or later separate out as a solid phase (precipitate).

Indeed, increasing the electrolyte ion concentration causes an increase in the rate at which the ions are deposited on the solid phase surface. When the ion product becomes equal to the solubility product, the rates of dissolution and precipitation are equilibrated. But if the ion product increases further and goes beyond the value of the solubility product, the rate of precipitation will prevail over the rate of dissolution and a precipitate will separate out from the solution. For instance, for AgCl we find:

(a) in unsaturated solution

$$[Ag^+][Cl^-] < SP_{AgCl}$$

(b) in saturated solution

$$[Ag^+][Cl^-] = SP_{AgCl}$$

(c) in supersaturated solution

$$[Ag^+][Cl^-] > SP_{AgCl}$$

An unsaturated solution of an electrolyte can be made saturated and even supersaturated by adding to it an electrolyte with a common ion.

Indeed, if to an unsaturated solution of AgCl there is gradually added HCl or KCl, the ion product, initially less than $SP_{C,\,AgCl}$ (i.e., ca. 1.78×10^{-10} at 25°C), gradually reaches and then exceeds this value. Accordingly, the unsaturated solution becomes saturated and then supersaturated.

From what has been said it follows that *the solubility of a slightly soluble electrolyte falls off when a strong electrolyte with a common ion is added to its solution.*

If a more soluble salt, such as $KClO_3$, is taken instead of AgCl, the common ion effect will be more pronounced. For example, if to a saturated solution of $KClO_3$ there is gradually added, on continuous stirring, a concentrated solution of KCl or $NaClO_3$, a white crystalline precipitate of $KClO_3$ will form after some time.

Thus, in full accordance with the solubility product rule, the solubility of $KClO_3$ in the presence of KCl or $NaClO_3$ appears to be less than in pure water. The same is observed in other analogous cases.

Experiment shows, however, that the solubility of electrolytes is influenced not only by salts with a common ion but also by salts having no such ions. In the latter case, however, the solubility is usually increased. The phenomenon described is called the **salt effect**. For example, the solubility of $PbSO_4$ in the presence of KNO_3, $NaNO_3$, and other salts increases, the more so the higher the concentration of the corresponding salt.

Since nc ions capable of combining with Pb^{2+} or SO_4^{2-} are introduced on addition of KNO_3 or $NaNO_3$ (because $Pb(NO_3)_2$ and K_2SO_4 are strong electrolytes), the increase of the solubility of $PbSO_4$ cannot be accounted for by the solubility product principle in its simplified form: $[Pb^{2+}][SO_4^{2-}] = SP_{C,\,PbSO_4}$.

It can easily be explained if use is made of the exact equation

$$[Pb^{2+}][SO_4^{2-}]f_{Pb^{2+}}f_{SO_4^{2-}} = SP_{a,\,PbSO_4}$$

more conveniently written in the form

$$[Pb^{2+}][SO_4^{2-}] = \frac{SP_{a,\ PbSO_4}}{f_{Pb^{2+}} f_{SO_4^{2-}}}$$

Indeed, the quantity $SP_{a,\ PbSO_4}$ in this equation is strictly constant. The activity coefficients $f_{Pb^{2+}}$ and $f_{SO_4^{2-}}$ are usually decreased upon addition of any ions because of the increasing ionic strength of the solution. Therefore, the product $[Pb^{2+}][SO_4^{2-}]$ and, hence, the solubility of lead sulphate must increase.

The physical cause of this increase of the solubility is clear: the Pb^{2+} and SO_4^{2-} ions, being restricted in their motion by interionic forces, collide less frequently with the surfaces of the $PbSO_4$ crystals

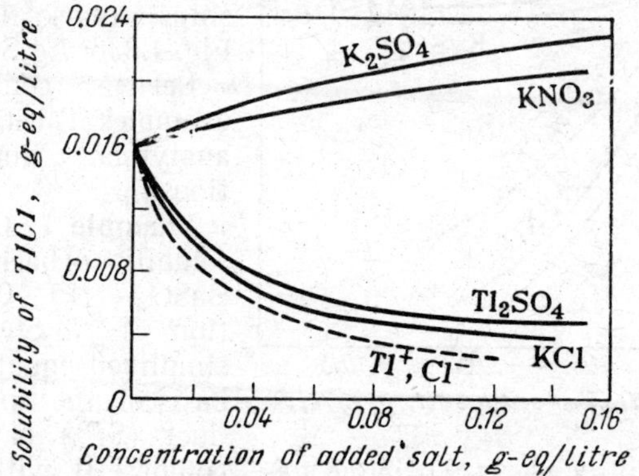

Fig. 4.2. Solubility of TlCl in the presence of salts with foreign (uncommon) and common ions.

when KNO_3 or $NaNO_3$ is introduced into the solution. As a result, the transfer of the Pb^{2+} and SO_4^{2-} ions into the solution proves to be more intensive than the process of their precipitation, following which equilibrium sets in at a higher concentration of the ions in solution; the solubility of $PbSO_4$ is increased.

Since the activity coefficients of ions also decrease when electrolytes with a common ion are introduced into the solution, the salt effect must show up in this case too. It is however usually masked by the opposing effect of the common ion on solubility and may be disregarded when we consider the phenomenon qualitatively. Quantitatively, the decrease of the solubility due to the salt effect proves to be less than that calculated from the solubility product rule in its simplified form.

The foregoing can be illustrated by a graph (Fig. 4.2) showing the effect of various electrolytes on the solubility of thallium chloride,

TlCl. In the presence of Tl_2SO_4 and KCl, the solubility of TlCl is decreased but to a lesser extent than it follows from calculations by the simplified equation without regard for the activity coefficients (the dashed line). Conversely, in the presence of salts containing no common ion, the solubility appears to be increased (the salt effect). In this case K_2SO_4, which forms the divalent ion SO_4^{2-}, is more effective than KNO_3 which gives only univalent ions.* The same relationships are also illustrated by Fig. 4.3 for Ag_2SO_4 solutions.

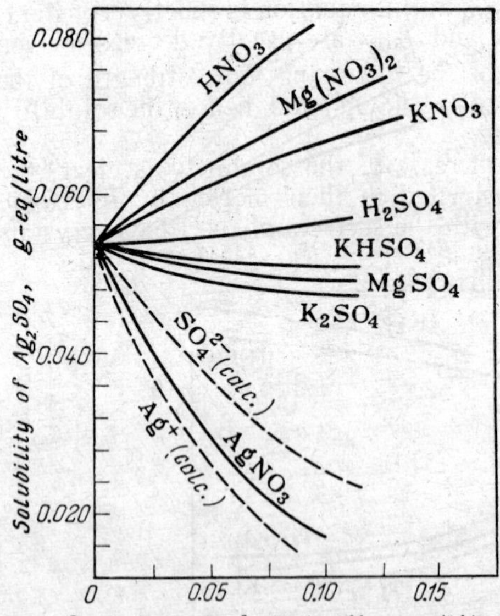

Fig. 4.3. Solubility of Ag_2SO_4 in the presence of salts with foreign (uncommon) and common ions.

Let us consider some examples illustrating these analytically important relationships.

Example 1. Calculate the solubility of barium sulphate, $BaSO_4$, in $0.01M$ solution of Na_2SO_4, using the simplified equation, on the basis of the solubility product (SP_a) of $BaSO_4$ and compare it with its solubility in water.

Solution. Denote the sought-for solubility of barium sulphate by x. These x moles of $BaSO_4$ give x g-ion of Ba^{2+} and SO_4^{2-} per litre. But the SO_4^{2-} ions are also formed as the result of the dissociation of Na_2SO_4. Taking into account that each gram-molecule (mole) of this salt gives 1 g-ion of SO_4^{2-}, we may write:

$$[Ba^{2+}] = x \quad \text{and} \quad [SO_4^{2-}] = 0.01 + x$$

Substituting these values into the simplified equation for $SP_{a,\ BaSO_4}$, we get:

$$SP_{a,\ BaSO_4} = 1.1 \times 10^{-10} \approx x(0.01 + x)$$

* Detailed information on the calculation of solubility and solubility products for various compounds can be found in: Komar', N. P., *Fundamentals of Qualitative Chemical Analysis*, vol. I, Kharkov, 1955; Nadeinsky, B. P., *Theoretical Interpretation and Calculations in Analytical Chemistry*, Moscow, 1956.

The theoretical basis of dissolution processes is formulated in a monograph by Yatsimirsky, K. B., *Thermochemistry of Complex Compounds*, Moscow, 1951.

The calculation can be considerably simplified if it is taken into account that the solubility of $BaSO_4$ is very low and, hence, the value of x may be ignored as compared with 0.01. Having done this, we obtain:

$$x \cdot 0.01 \approx 1.1 \cdot 10^{-10}; \quad x \approx 1.1 \cdot 10^{-8} \text{ mole/litre}$$

The solubility of $BaSO_4$ in water is equal to:

$$y \approx \sqrt{1.1 \times 10^{-10}} \approx 1.05 \times 10^{-5} \text{ mole/litre}$$

Thus, the presence of 0.01 mole of Na_2SO_4 in one litre must lower the solubility of $BaSO_4$ approximately by a factor of $1.05 \times 10^{-5} : 1.1 \times 10^{-8}$, i.e., by 950 times.

Example 2. Perform the same calculation with the activity coefficients of the ions being taken into account.

Solution. First calculate the ionic strength of the solution. Since the solubility of $BaSO_4$ is very low, the ionic strength is determined by the concentration of Na_2SO_4 alone.

The concentrations of the ions given by the salt Na_2SO_4 are equal to:

$$[SO_4^{2-}] = 10^{-2} \text{ g-ion/litre} \quad \text{and} \quad [Na^+] = 2 \times 10^{-2} \text{ g-ion/litre}$$

Hence,

$$\mu = \frac{1}{2}([Na^+]1^2 + [SO_4^{2-}]2^2) = \frac{1}{2}(2 \times 10^{-2} + 4 \times 10^{-2}) = 0.03$$

To the ionic strength found, there correspond the values of the activity coefficients $f_{Ba^{2+}} = f_{SO_4^{2-}} = 0.54$ (see Table 3.3, Sec. 3.6).

Denoting the required solubility of $BaSO_4$ by x and taking into account what has been said in Example *1*, we have:

$$x \cdot 0.01 \cdot 0.54^2 = 1.1 \cdot 10^{-10}$$

which solves to

$$x = \frac{1.1 \cdot 10^{-10}}{10^{-2} \cdot 0.54^2} = 3.8 \cdot 10^{-8} \text{ mole/litre}$$

Thus, the solubility of $BaSO_4$ is decreased by the action of Na_2SO_4 by approximately 280 times and not by a factor of 950 as found from the simplified solubility product formula.

Example 3. Calculate the solubility of $BaSO_4$ in $0.1M$ KNO_3 solution.

Solution. Find the ionic strength of the solution. In practice, it is determined exclusively by the concentration of KNO_3 and is equal to:

$$\mu = \frac{1}{2}(0.1 \times 1^2 + 0.1 \times 1^2) = 0.1$$

Hence,

$$f_{Ba^{2+}} = f_{SO_4^{2-}} = 0.33$$

Denoting the sought-for solubility of $BaSO_4$ by x, we have:

$$x^2 \cdot 0.33^2 = 1.1 \cdot 10^{-10}$$

whence

$$x = \frac{\sqrt{1.1 \cdot 10^{-10}}}{0.33} = 3.2 \cdot 10^{-5} \text{ mole/litre}$$

Thus, the solubility of $BaSO_4$ in the presence of 0.1 mole of KNO_3 per litre increases from 1.05×10^{-5} to 3.2×10^{-5} mole/litre, i.e., approximately by 3 times (the salt effect).

4.3. Formation of Precipitates. Effect of Various Factors on Completeness of Precipitation

The solubility product rule enables us to predict the effect of common ions on the solubility of precipitates. It is especially useful in those cases when a precipitate is formed or dissolved as a result of a chemical reaction.

Suppose that a solution of a lead salt is mixed with a soluble chloride, say, NaCl. The Pb^{2+} ions collide with Cl^- ions and the formation of a $PbCl_2$ precipitate becomes possible. The precipitation of lead chloride depends on whether the solution becomes supersaturated with respect to the salt. It will become supersaturated only when the product of the activities of Pb^{2+} and Cl^- exceeds the solubility product (SP_a) of $PbCl_2$ which is equal to about 1.6×10^{-5} (at 25°C). One can thus predict whether a particular salt will precipitate under given conditions.

Example 1. Will a precipitate of $PbCl_2$ form when equal volumes of $0.1M$ solutions of $Pb(NO_3)_2$ and NaCl are mixed?

Solution. When the solutions are mixed, the concentration of each of the substances decreases to $5 \times 10^{-2}M$. The concentrations of Pb^{2+} and Cl^- immediately after mixing the solutions will also be equal to $5 \times 10^{-2}M$. Since the solubility of $PbCl_2$ is rather high, the activities of the ions greatly differ from the concentrations. It is therefore necessary to calculate the ionic strength of the solution, to find the corresponding activity coefficients for this ionic strength, and to insert them into the solubility-product expression for $PbCl_2$.

Since

$$[Pb^{2+}] = 5 \times 10^{-2}; \quad [NO_3^-] = 0.1; \quad [Na^+] = 5 \times 10^{-2}; \quad [Cl^-] = 5 \times 10^{-2}$$

it follows that

$$\mu = \frac{1}{2}[5 \times 10^{-2} \times 4 + 0.1 + 5 \times 10^{-2} + 5 \times 10^{-2}] = 0.2$$

With the ionic strength $\mu = 0.2$, $f_{Pb^{2+}} = 0.24$ and $f_{Cl^-} = 0.70$. Now we find the product of the activities of Pb^{2+} and Cl^- in the solution:

$$[Pb^{2+}] \times 0.24 \times [Cl^-]^2 \times [0.70]^2 = 5 \times 10^{-2} \times 0.24[5 \times 10^{-2}]^2 \times [0.70]^2 = 1.47 \times 10^{-5}$$

The solubility product (SP_a) of $PbCl_2$ is equal to 1.6×10^{-5} and under the given conditions $PbCl_2$ will not form a solid phase because $1.47 \times 10^{-5} < 1.6 \times 10^{-5}$.

Thus, if equal volumes of $0.1M$ solutions of lead nitrate and sodium chloride are mixed, no precipitate of lead chloride will be formed. If the same experiment is repeated with a $1M$ solution of sodium chloride, then (on shaking) an abundant white precipitate of $PbCl_2$ will form.

From the above example it is clear that the concentration of the precipitant (NaCl in this case) is very important in precipitation reactions. The greater the concentration, the higher is usually the sensitivity of the reaction.*

From what has been said above we can draw the following conclusion: *any slightly soluble electrolyte is precipitated only if the product of the activities of its ions in solution exceeds its solubility product* SP_a *at a given temperature.*

It should however be kept in mind that the formation of a precipitate may begin not immediately after the corresponding solutions are mixed (the formation of a supersaturated solution) but after a more or less prolonged period of time. As known, scratching with a glass rod, shaking, and other similar mechanical operations do accelerate precipitation. A stronger effect is gained by introducing into the solution a small crystal of the compound to be precipitated (the so-called *seed*).

Once started the precipitation continues until the ion product, which gradually decreases, becomes equal to the solubility product of the precipitate. When this takes place, a dynamic equilibrium will be set up between the precipitate and the solution and no further precipitation is observed.

Since there exist no substances completely insoluble in water, the solubility product is never equal to zero. Thus, even very difficultly soluble compounds, such as sulphides of heavy metals, have measurable, though extremely small, solubility products.** It follows from this that no precipitation can ever be quite complete. Part of the ions being precipitated, corresponding to the solubility product of the precipitated compound, will always remain in solution. Sometimes the amount left in solution is so small that it cannot interfere with the subsequent analysis. In such cases, the precipitation may be regarded as practically complete. Not infrequently, however, cases are encountered where the precipitation cannot be considered complete, and the analyst has to take steps to achieve a practically complete precipitation.

---

* The limitation of this rule will be discussed later in the book.

** The solubility and solubility products of such very difficultly soluble compounds are generally determined by electrochemical methods which are treated in courses of physical chemistry.

The completeness of precipitation is an exceptionally important factor since only in the case of practically complete precipitation it is possible to separate ions in the course of analysis and to eliminate the interfering action of the ions being removed in the subsequent operations. In view of this, we shall consider in more detail the factors on which the completeness of precipitation depends.

1. Solubility of the Compound to Be Precipitated. It is evident that, under given conditions, the less soluble is the compound in the form of which a given ion is precipitated, the more complete is the separation. For instance, the Pb^{2+} ion can be precipitated as the chloride, sulphate, or sulphide; the solubilities and solubility products of these salts are as follows:

	$PbCl_2$	$PbSO_4$	PbS
Solubility product	1.6×10^{-5}	1.6×10^{-8}	2.5×10^{-27}
Solubility, mole/litre	1.6×10^{-2}	1.26×10^{-4}	5×10^{-14}

Comparison of these figures shows that the most soluble is the chloride and the least soluble is the sulphide. Hence, lead can be precipitated most completely as PbS and least completely as $PbCl_2$. This can easily be confirmed by experiment. Mix 10 ml of 1*N* $Pb(NO_3)_2$ solution with 10-11 ml of 1*N* NaCl solution in a large test tube or flask, shake it vigorously, and filter off the $PbCl_2$ precipitate formed. Test the filtrate for completeness of precipitation (by adding several drops of NaCl) and divide it into two parts. Add a little $(NH_4)_2S$ solution to one part and a little Na_2SO_4 solution to the other. A black precipitate of PbS will form in the first case and a white precipitate of $PbSO_4$ in the second. The latter is usually formed only after shaking (precipitation from supersaturated solution).

It is also easy to show that the precipitation of Pb^{2+} as $PbSO_4$ is more complete than when the ion is precipitated in the form of PbS. To 10 ml of 1*N* $Pb(NO_3)_2$ solution add 10 ml of 1*N* Na_2SO_4 solution; filter off the precipitated $PbSO_4$, test for completeness of precipitation, and add $(NH_4)_2S$ solution to the filtrate. The liquid in the test tube will turn dark brown owing to the formation of a small amount of PbS.*

These experiments clearly show the different degrees of precipitation attained when a given ion is converted into slightly soluble compounds having different solubility products. One should however remember that the more informative quantity is the solubility of the precipitate expressed in moles per litre rather than the solubility product. Indeed, as pointed out above (see Sec. 4.1), the solubility

* The lead sulphide prepared in this experiment is usually in the form of a colloidal solution. If a small amount of NH_4Cl solution (an electrolyte-coagulant) is added to it and the solution is heated to boiling, the PbS will coagulate and precipitate as black flakes.

of AgCl ($1.3 \times 10^{-5}M$) is less than that of Ag_2CrO_4 (6.5×10^{-5} M), although the solubility product of silver chloride (1.78×10^{-10}) is greater than that of silver chromate (1.1×10^{-12}). For this reason, the Ag^+ ion is precipitated more completely as AgCl than as Ag_2CrO_4, under given conditions.

2. Amount of Precipitant Added. The precipitation of any ion may be complete provided that the amount of precipitant added is not less than the equivalent amount, i.e., not less than the amount required in accordance with the reaction equation. But if the solubility product of the precipitate is not small enough, addition of the equivalent amount of precipitant does not usually result in a practically complete precipitation. The solubility product principle shows that the precipitation becomes more complete if a certain excess of precipitant is introduced. Indeed, since during the precipitation of $PbSO_4$ the ion concentration product $[Pb^{2+}][SO_4^{2-}]$ is nearly permanent and approximately equal to the solubility product (SP_a) of $PbSO_4$, it follows that the increase of the SO_4^{2-} concentration caused by addition of excess precipitant (Na_2SO_4 or H_2SO_4) will lead to a corresponding decrease of the Pb^{2+} concentration in solution, i.e., it will make the precipitation of the lead ion more complete.*

Thus, if the reaction used for separating an ion involves the formation of a compound, which is appreciably soluble in water, the precipitant is taken in some excess, i.e., more precipitant is taken than required by the reaction equation.

On the basis of the equation

$$a_{Pb^{2+}}a_{SO_4^{2-}} = SP_{a,\ PbSO_4}$$

one might expect that the more H_2SO_4 we add, the more complete must be the precipitation of Pb^{2+}. But this is not the case in reality. Addition of a very large excess of sulphuric acid does not decrease the solubility of the precipitate at all but rather increases it considerably due to the formation of HSO_4^- ion with a dissociation constant of $K_{HSO_4^-} = 1.2 \times 10^{-2}$:

$$PbSO_4\downarrow + 2H^+ \rightarrow Pb^{2+} + 2HSO_4^-$$

Analogously, in the reaction

$$Hg^{2+} + 2I^- \rightarrow HgI_2\downarrow$$

the addition of too large an excess of I^- causes all the precipitate to dissolve due to the formation of the complex ion HgI_4^{2-}:

$$HgI_2\downarrow + 2I^- \rightarrow HgI_4^{2-}$$

---

* This is a direct consequence of the law of mass action.

Precipitates of amphoteric hydroxides can also dissolve in an excess of precipitant—an alkali:

$$Al(OH)_3\downarrow + OH^- \rightarrow [Al(OH)_4]^-$$

But even if no acid salts and complex or other soluble compounds are expected to form, the use of too large an excess of precipitant is not reasonable. The point is that an excess of precipitant not only lowers the solubility of the precipitate, due to an increase in the concentration of the precipitant ion, but also tends to raise the solubility by enhancing the salt effect. With a small excess of precipitant the solubility of the precipitate decreases, and when too large an excess of precipitant is used the reverse phenomenon takes place—the solubility of the precipitate increases. For these reasons, the amount of precipitant generally used is not more than 50 per cent excess; sometimes even less is used.

3. The pH of the Solution. Let us consider in more detail the effect of the pH of a solution in the precipitation of slightly soluble hydroxides and also of salts of weak and strong acids.

(a) *Precipitation of Slightly Soluble Metal Hydroxides.* In this case, the precipitant is the OH^- ion whose concentration is the higher the greater is the pH of the solution. Whether the hydroxide is precipitated and whether the precipitation is practically complete, depends on the OH^- ion concentration. It is clear that the more soluble is the hydroxide, the higher is the OH^- concentration needed for its solubility product to be exceeded, i.e., the higher is the pH value required to start its precipitation.

On the basis of the solubility product, one can approximately calculate the pH value at which the precipitation of a particular hydroxide begins and ends. Let us perform such a calculation for magnesium hydroxide ($SP = 6 \times 10^{-10}$). If the concentration of Mg^{2+} in solution is 1×10^{-2} g-ion/litre, then, using the equation

$$[Mg^{2+}][OH^-]^2 = SP_{Mg(OH)_2} = 6 \times 10^{-10}$$

one can easily calculate the OH^- concentration at which the solubility product is reached and precipitation of $Mg(OH)_2$ begins. It is equal to:

$$[OH^-] = \sqrt{\frac{SP_{Mg(OH)_2}}{[Mg^{2+}]}} = \sqrt{\frac{6 \times 10^{-10}}{1 \times 10^{-2}}} = 2.5 \times 10^{-4} \text{ g-ion/litre}$$

Hence

$$pOH = -\log 2.5 \times 10^{-4} = -(0.4 - 4) = 3.6$$

$$pH = 14 - pOH = 14 - 3.6 = 10.4$$

Thus, the precipitation of magnesium hydroxide from a $0.01M$ solution of its salt begins at pH $\approx$ 10.4.

The precipitation of an ion may be considered complete only when its concentration in solution decreases to $10^{-5} - 10^{-6}$ g-ion/litre.

At a concentration of Mg^{2+} ions equal to 10^{-6} g-ion/litre, the value of $[OH^-]$ must be

$$[OH^-]=\sqrt{\frac{6\times10^{-10}}{10^{-6}}}\approx 2.5\times10^{-2}\ \text{g-ion/litre}$$

This corresponds to pOH $\approx$ 1.58 and pH $\approx$ 12.42.

If pH $>$ 12.42, then the precipitation of Mg^{2+} will be still more complete (i.e., the concentration of the Mg^{2+} ion left in solution will become less than 10^{-6} g-ion/litre). Conversely, at pH $<$ 12.42 the precipitation of Mg^{2+} will be incomplete and at pH $\leqslant$ 10.4 no Mg^{2+} will precipitate at all. Thus, the complete precipitation of Mg^{2+} in the form of a hydroxide is attained provided the pH $\geqslant$ 12.42.

If such a calculation is made for a much less soluble compound—ferric hydroxide, $Fe(OH)_3$ ($SP_a = 3.2\times10^{-38}$), we will find that its precipitation from a 0.01*M* solution of a ferric salt begins even at pH = 2.2 and becomes practically complete at pH $\geqslant$ 3.5.

The results of analogous calculations for a number of other slightly soluble hydroxides*, the formation of which is generally used in analysis, are listed in Table 4.1. This table gives pH values at which: (a) the precipitation of hydroxides begins when the concentrations of the ions are C = 1 g-ion/litre and $C = 1\times10^{-2}$ g-ion/litre; (b) practically complete precipitation is attained, corresponding to $C = 1\times10^{-6}$ g-ion/litre. The calculated pH values at the start of precipitation for C = 1 g-ion/litre are compared with the experimentally found values. As seen, the theoretically calculated values coincide satisfactorily in most cases with the experimental values in spite of the fact that the ionic activity coefficients were not taken into account and that the solubility products of some of the hydroxides were not determined with sufficient accuracy. This suggests that the pH values (not yet determined experimentally) for practically complete precipitation of the hydroxides must also be close to the calculated values listed in Table 4.1.

Using the data of Table 4.1, one can predict some circumstances important for analysis. To illustrate, let us consider the following example. As shown above, the precipitation of $Mg(OH)_3$ from a 0.01*M* solution of a magnesium salt begins at pH 10.4. Since the pH of 0.1*M* NH_4OH solution equals 11.13 (see Example 7 in Sec. 3.8), this solution must give a precipitate of Mg^{2+}, which is confirmed by experiment. If however to the NH_4OH solution is added an equivalent amount of NH_4Cl, then an ammonium buffer is formed, whose pH can be calculated from the formula (see Sec. 3.9):

$$pH = 14 - pK_b + \log\frac{C_b}{C_s}$$

* Table 4.1 gives also the formulas of the hydrous oxides of a number of elements, which correspond more accurately to the composition of the compounds formed.

TABLE 4.1. **Precipitation pH Values for Certain Hydroxides**

Formula of compound being precipitated	Solubility product	Start of precipitation			Complete precipitation
		$C = 1$ g-ion/litre		$C = 1 \cdot 10^{-2}$ g-ion/litre	$C = 1 \cdot 10^{-6}$ g-ion/litre
		exp.*	calculated		
$Ag_2O(Ag^+, OH^-)$**	1.6×10^{-8}	—	6.2	8.2	12.2
$Mg(OH)_2$***	6×10^{-10}	—	9.4	10.4	12.42
$Mn(OH)_2$	4.5×10^{-13}	—	7.83	8.83	10.83
$Cd(OH)_2$	2.2×10^{-14}	6.8	7.18	8.18	10.18
$Fe(OH)_2$	1×10^{-15}	6.6	6.51	7.51	9.51
$Co(OH)_2$ (blue)	6.3×10^{-15}	6.8	6.9	7.9	9.9
$Co(OH)_2$ (pink)	2×10^{-15}	—	6.65	7.65	9.65
$Ni(OH)_2$	2×10^{-15}	6.7	6.65	7.65	9.65
$Zn(OH)_2$	7.1×10^{-18}	5.8	5.43	6.43	8.43
$Cu(OH)_2$	5×10^{-20}	4.2	4.34	5.34	7.34
$Cr(OH)_3(Cr_2O_3\cdot xH_2O)$	6.3×10^{-31}	—	3.9	4.25	5.9
$Al(OH)_3(Al_2O_3\cdot xH_2O)$	1×10^{-32}	3.6	3.32	4.0	5.32
$Fe(OH)_3(Fe_2O_3\cdot xH_2O)$	3.2×10^{-38}	1.6	1.5	2.2	3.5
$Sn(OH)_2$	6.3×10^{-27}	—	0.9	1.9	3.9
$Bi(OH)_3(Bi_2O_3\cdot xH_2O)$	3.2×10^{-40}	—	0.83	1.5	2.83
$Sb_2O_3(2Sb^{3+}, 3OH^-)$	4×10^{-42}	—	0.2	0.87	2.2

* The experimentally found pH values refer, strictly speaking, to the activities of the corresponding ions ($a = 1$) and not to their concentrations ($C = 1$).

** Some elements whose oxidation state $\geqslant 3+$ and also silver, mercury, and others are precipitated as hydrous oxides and not as hydroxides.

*** All the values of SP_a are given for freshly precipitated hydroxides. After "aging" the hydroxides change into other modifications, which are less soluble — they have a lower value of SP_a. Processes taking place during the aging of precipitates are considered in Chapter 5.

Applying this formula to the case under consideration, we obtain

$$pH = 14 - 4.76 + \log\frac{0.1}{0.1} = 9.24$$

Since the result obtained (9.24) is less than the pH at which precipitation of $Mg(OH)_2$ begins (10.4), one has to conclude that a precipitate of $Mg(OH)_2$ will not form by the action of the buffer. This conclusion was confirmed in our study of the reactions of the Mg^{2+} ion (see Sec. 3.14).

But if we add a mixture of NH_4OH and NH_4Cl to a solution of $FeCl_3$, a precipitate of $Fe(OH)_3$ will be formed because the precipitation of this compound is complete even at pH $\geqslant$ 3.5. This is also easily confirmed by experiment.

It is not difficult to understand why $Fe(OH)_3$ is precipitated while $Mg(OH)_2$ is not. This may evidently be explained as follows. The common NH_4^+ ions of NH_4Cl so strongly depress the degree of ionization of NH_4OH, and hence the OH^- concentration, that the solubility product of magnesium hydroxide (6×10^{-10}) is not reached. On the other hand, the solubility product of $Fe(OH)_3$, which is considerably less (3.2×10^{-38}), is found to be greatly exceeded even at the OH^- concentration (1.8×10^{-5} g-ion/litre) present in the mixture of NH_4OH and NH_4Cl. Therefore, $Fe(OH)_3$ is precipitated even in the presence of NH_4Cl.

The different effect of ammonia on Mg^{2+} and Fe^{3+} ions in the presence of an ammonium salt can be used to separate these ions. Such separations based on the adjustment of the pH are often used in analysis and are of great practical importance.

As shown above 0.1*M* NH_4OH solution partially precipitates Mg^{2+} in the absence of NH_4Cl. Complete precipitation of $Mg(OH)_2$ however is achieved at pH $\geqslant$ 12.42, but the pH of 0.1*M* NH_4OH is 11.13 and therefore the precipitation of Mg^{2+} by ammonia cannot be complete. This becomes especially obvious if we note that one of the products of the reaction

$$Mg^{2+} + 2NH_4OH \rightarrow Mg(OH)_2\downarrow + 2NH_4^+$$

is NH_4^+ ions. The accumulation of these ions in solution leads to the formation of a buffer mixture and the pH of the solution gradually falls off and becomes much less than the value required for $Mg(OH)_2$ to be precipitated. Accordingly, the precipitation of Mg^{2+} by ammonia is rather incomplete.

If an alkali such as potassium hydroxide is added instead of NH_4OH to a solution of a magnesium salt, it will be easy, with a small excess of alkali, to bring the pH value to 12.42 and hence to achieve practically complete precipitation of the Mg^{2+} ion.

This example shows that, under given conditions, the greater the degree of dissociation of the precipitant used the more complete is the precipitation of the ion.

(b) *Precipitation of Slightly Soluble Salts of Weak Acids.* The pH of the solution is also very important in the precipitation of various slightly soluble salts of weak acids, such as carbonates, sulphides, sulphites, phosphates, arsenates, chromates, oxalates, and tartrates. The precipitants used are the anions of the corresponding acids: CO_3^{2-}, S^{2-}, SO_3^{2-}, PO_4^{3-}, etc. Since these acids are weak (i.e., have a low tendency to dissociate into ions), their anions combine with the H^+ ions present in solution to form the anions HCO_3^-, HS^-, HSO_3^-, HPO_4^{2-}, $H_2PO_4^-$ and the un-ionized molecules of H_2CO_3, H_2S, H_2SO_3, and H_3PO_4.

Thus, the concentrations of CO_3^{2-}, S^{2-}, SO_3^{2-}, PO_4^{3-}, and other anions greatly depend on the hydrogen ion concentration, decreasing with increase of the latter, i.e., with decrease of the solution pH. If this is so, it then follows that whether a given salt is precipitated and whether its precipitation is complete or not also depends on the pH.

As in the case of hydroxides, the pH required for practically complete precipitation of a slightly soluble salt of a weak acid depends primarily on the solubility product constant of that salt. The lower is the solubility product constant, the less is the concentration of the precipitating ion needed to exceed it; accordingly, the higher is the H^+ concentration (i.e., the lower is the pH of the solution) at which the precipitation of the salt is practically complete. For example, a fairly complete precipitation of cadmium sulphide, CdS, ($SP = 7.9 \times 10^{-27}$) by hydrogen sulphide is achieved even in a strong acid solution ($pH \geqslant 0.5$), whereas manganese sulphide*, MnS, whose solubility product constant is 2.5×10^{-10}, is precipitated only in alkaline or neutral medium ($pH \geqslant 7$).

Apart from the solubility product constant, the ionization constants of a corresponding weak acid play an important role. The lower these constants, the greater the decrease of the concentration of the precipitating ion because of its combination with H^+ ions and the higher the pH value required for practically complete precipitation of the salt. For instance, since carbonic acid ($K_1 = 4.5 \times 10^{-7}$, $K_2 = 4.8 \times 10^{-11}$) is much weaker than oxalic acid ($K_1 = 5.6 \times 10^{-2}$, $K_2 = 5.4 \times 10^{-5}$), a higher pH value is required to precipitate Ca^{2+} as $CaCO_3$ than as CaC_2O_4, though the solubility product constants of these salts are of the same order of magnitude ($SP_{CaCO_3} = 4.8 \times 10^{-9}$ and $SP_{CaC_2O_4} = 2.3 \times 10^{-9}$). This conclusion is borne out by experiment and calculations—whereas an alkaline solution ($pH > 9$) is required for $CaCO_3$ to be completely precipitated, the precipitation of CaC_2O_4 is complete already at $pH \approx 4$.

Generally, calculation of the pH value required for practically complete precipitation of a slightly soluble salt is a comparatively simple task. When slightly soluble salts of weak acids are to be precipitated, the precipitants commonly used are soluble salts of such acids, which show an alkaline reaction due to hydrolysis. Therefore, the pH of the solution is predetermined to a considerable extent by the presence of excess precipitant in solution. In this case, the problem should be tackled in a different way. Having found the pH developed by the excess of precipitant, we must calculate how completely the particular ion is precipitated at that pH.

Let us consider, as an example, the precipitation of the cations of Group II by the group reagent, ammonium carbonate, $(NH_4)_2CO_3$. Being a salt of a weak acid (H_2CO_3) and a weak base (NH_4OH), am-

---

* The solubility product is given for a modification of MnS having a pale-pink colour. For a green-coloured modification, the solubility product is 2.5×10^{-13}.

monium carbonate is strongly hydrolysed in aqueous solution.* It is known that the hydrolysis of salts of di- and polybasic acids proceeds by stages.

In ionic form the equations for the hydrolysis are:

$$NH_4^+ + CO_3^{2-} + H_2O \rightleftharpoons HCO_3^- + NH_4OH$$

$$NH_4^+ + HCO_3^- + H_2O \rightleftharpoons H_2CO_3 + NH_4OH$$

It is also known that the hydrolysis proceeds the faster the lower the degree of ionization of the hydrolysis products. The ion HCO_3^- formed at the first stage of hydrolysis is ionized much less ($K_{HCO_3^-} = 4.8 \times \times 10^{-11}$) than H_2CO_3 ($K_{H_2CO_3} = 4.5 \times 10^{-7}$), which is formed at the second stage. Therefore, the hydrolysis of $(NH_4)_2CO_3$ is incomparably greater at the first stage than at the second. About 90 per cent of all the CO_3^{2-} ions are converted into HCO_3^- ions (see Sec. 5.6). The second stage of hydrolysis is slight and may be neglected without appreciable error. If in the precipitation of Group II cations the excess of $(NH_4)_2CO_3$ is $0.1M$, then, considering what has just been said, one may assume, as a first approximation, that practically all the CO_3^{2-} ions (0.1 g-ion/litre) are converted into HCO_3^- ions whose concentration is thus close to the value $[HCO_3^-] \approx 0.1$ g-ion/litre. On the other hand, the almost complete hydrolysis of $(NH_4)_2CO_3$ at the first stage signifies that the solution is a mixture of nearly equivalent amounts of free ammonia and an ammonium salt (NH_4HCO_3) and is therefore an ammonium buffer with pH ≈ 9.2. This pH value corresponds to $[H^+] \approx 6.6 \times 10^{-10}$ g-ion/litre. Inserting the values found for $[HCO_3^-]$ and $[H^+]$ into the equation for the K_2 of carbonic acid, we obtain:

$$K_{HCO_3^-} = \frac{[H^+][CO_3^{2-}]}{[HCO_3^-]} \approx \frac{6 \times 10^{-10}[CO_3^{2-}]}{10^{-1}} \approx 4.8 \times 10^{-11}$$

whence

$$[CO_3^{2-}] \approx \frac{4.8 \times 10^{-11} \times 10^{-1}}{6 \times 10^{-10}} = 0.8 \times 10^{-2}$$

Knowing this concentration of CO_3^{2-} ions, one can easily find the concentration of the metal ions in solution from the solubility product constants of the carbonates upon completion of precipitation. For example,

$$[Ba^{2+}] = \frac{SP_{BaCO_3}}{[CO_3^{2-}]} \approx \frac{5.1 \times 10^{-9}}{0.8 \times 10^{-2}} = 6.4 \times 10^{-7} \text{ g-ion/litre}$$

Thus, the precipitation of Ba^{2+} is practically complete. Since the values of the solubility product constants of $CaCO_3$ (4.8×10^{-9}) and

---

* The hydrolysis of salts is discussed in detail in Sec. 5.5.

$SrCO_3$ (1.1×10^{-10}) are even less than the solubility product of $BaCO_3$, the ions Ca^{2+} and Sr^{2+} are precipitated even more completely at the given pH.

(c) *Precipitation of Slightly Soluble Salts of Strong Acids.* Slightly soluble salts such as AgCl, AgBr, and AgI are precipitated by the anions of strong monobasic acids: Cl^-, Br^-, I^-, etc. These anions evidently cannot combine with H^+ ions because HCl, HBr, and HI dissociate practically completely in solution. Hence, the completeness of precipitation of their silver salts is almost independent of the pH. Sometimes the effect of excess acid in solution has to be taken into account for the only reason that the salt effect is enhanced or the cations of the precipitated salt are bound by the acid anions into a complex, as a result of which the solubility of the precipitate is increased.

The situation is somewhat different in the case of slightly soluble sulphates since H_2SO_4 dissociates completely only at the first stage, i.e., with formation of H^+ and HSO_4^- ions. Although the second dissociation stage

$$HSO_4^- \rightleftharpoons H^+ + SO_4^{2-}$$

proceeds to a very considerable extent ($K_{HSO_4^-} = 1.2 \times 10^{-2}$), it is not complete. Therefore, at not too low SO_4^{2-} concentrations and sufficiently high concentrations of H^+, these ions combine partly to form HSO_4^- anions. Hence, the pH of the solution is not entirely immaterial in this case. The precipitation of slightly soluble sulphates from strongly acid solutions must be somewhat less complete than from neutral or alkaline solutions. This circumstance has to be taken into account in qualitative analysis only in the precipitation of more soluble sulphates—$CaSO_4$, $SrSO_4$, and $PbSO_4$. The least soluble of these, $BaSO_4$, is precipitated rather completely even from strongly acid solutions.

(d) *Adjustment of the pH*. As shown above, for a sufficiently complete separation of ions to be achieved, it is necessary to adjust the pH of the solution. This is done by adding suitable acids, bases, or buffers to the test solution. The acid or base is added drop by drop, the solution being stirred continuously and tested with appropriate indicator papers or universal indicator. Sometimes an indicator is introduced into the solution and the addition of acid or base is continued until the colour of the solution is changed by a further drop.

In selecting the acid or base, one must be sure that the ions introduced it will not interfere with detection of the ions in solution and will not cause any complications in the subsequent analysis. For example, if the K^+ ion is to be detected, acid solutions must not be neutralized with KOH or NH_4OH; use must of necessity be made of NaOH. Conversely, for the Na^+ ion to be detected, the acid must be neutralized with KOH rather than with NaOH. In the same way, alkaline solutions that may contain Group II cations or Pb^{2+} should not be neutralized with sulphuric acid because this acid precipitates slightly soluble sulphates of these cations, which complicates the analysis.* In many cases, it is not sufficient to adjust the pH to the required value before precipitation; one must also take into consideration that the pH may vary as a result of the reaction involved in the precipitation. This occurs when slightly soluble salts are precipitated by the action of the corresponding free acids:

$$K^+ + H_2C_4H_4O_6 \rightarrow KHC_4H_4O_6\downarrow + H^+$$
$$Ca^{2+} + H_2C_2O_4 \rightarrow CaC_2O_4\downarrow + 2H^+$$
$$Zn^{2+} + H_2S \rightarrow ZnS\downarrow + 2H^+$$

As seen from the above equations, these reactions involve the formation of H^+ ions, so that the pH of the solution gradually falls as the precipitation proceeds. As a result, the precipitation is incomplete unless appropriate steps are taken to prevent this decrease of pH. As known, buffers are capable of resisting the change of the pH. Therefore, for the pH to be maintained at a specified level, it is necessary to add a suitable buffer to the solution. For instance, the first two of the above reactions are conducted in the presence of the acetate buffer whose pH $\approx$ 4.7; under these conditions, the K^+ and Ca^{2+} ions are practically completely precipitated because of the H^+ ions being bound by the CH_3COO^- anions from sodium acetate. Analogously, the precipitation of Zn^{2+} by hydrogen sulphide can be effected in the presence of the formate buffer (formic acid, HCOOH, and its salt—formate) with pH $\approx$ $\approx$ 2; under these conditions, zinc sulphide is almost completely precipitated and, at the same time, the precipitation of other sulphides of Group III is prevented. This procedure can be used for fractional detection of Zn^{2+} in the presence of all the other cations of Group III.

The use of buffers is convenient for another reason too. They produce a definite pH despite the presence in solution of small amounts of acid or alkali. But every buffer has a definite capacity, being capable of binding a definite, but not too large, amount of H^+ and OH^- ions. If either of these ions is found to be present in a very large excess, the pH of the solution will be far from the expected value and will not remain constant (see Sec. 3.9).

From this it follows that in strongly acid or strongly alkaline solutions either most of the acid or alkali must be neutralized before the buffer is added, or use must be made of buffers possessing a high buffer capacity, i.e., with high enough concentrations of both components. Only then will the buffer be effective.

Occasionally, to strongly acid solutions there is added a certain amount of a dry salt of a weak acid (CH_3COONa), and to strongly alkaline solutions, a dry salt of a weak base (NH_4Cl). In this way, the solution is buffered to an appropriate pH value.

* In some cases where there is already an SO_4^{2-} ion in solution but its amount is not enough for complete precipitation of the sulphates, it is recommended that 2*N* sulphuric acid be added until the slightly soluble sulphates are completely precipitated.

4.4. Fractional Precipitation

In analytical practice, cases are not infrequently encountered, where the reagent forms slightly soluble compounds not with one but with two or more ions present in solution. For example, this takes place when group reagents are used. On the basis of the solubility product principle, one can easily understand how the precipitation of the ions proceeds in such cases.

Suppose, for example, that to a mixture of KCl and K_2CrO_4, in which the concentration $[Cl^-] = [CrO_4^{2-}] = 10^{-1}$ g-ion/litre, there is added drop by drop a solution of $AgNO_3$. When the Cl^- and CrO_4^{2-} ions collide with Ag^+, the following reactions take place:

$$Ag^+ + Cl^- \rightarrow AgCl\downarrow$$
$$2Ag^+ + CrO_4^{2-} \rightarrow Ag_2CrO_4\downarrow$$

Which of the two salts will precipitate first? Evidently the one whose solubility product is reached first (i.e., at a lower concentration of Ag^+). The Ag^+ concentrations required to start precipitation of AgCl and Ag_2CrO_4 are easily calculated from their solubility product constants:

$$SP_{AgCl} = [Ag^+][Cl^-] = 1.78\times10^{-10}$$
$$SP_{Ag_2CrO_4} = [Ag^+]^2[CrO_4^{2-}] = 1.1\times10^{-12}$$

Denoting the Ag^+ concentrations as $[Ag^+]_{AgCl}$ and $[Ag^+]_{Ag_2CrO_4}$, we can write:

$$[Ag^+]_{AgCl} = \frac{SP_{AgCl}}{[Cl^-]} = \frac{1.78\times10^{-10}}{10^{-1}} = 1.78\times10^{-9} \text{ g-ion/litre}$$

$$[Ag^+]_{Ag_2CrO_4} = \sqrt{\frac{SP_{Ag_2CrO_4}}{[CrO_4^{2-}]}} = \sqrt{\frac{1.1\times10^{-12}}{10^{-1}}} = 3.3\times10^{-5} \text{ g-ion/litre}$$

Hence, the first to precipitate is AgCl, the less soluble salt, whose solubility product is reached earlier than that of Ag_2CrO_4.*

This is easy to confirm by experiment. Mix a few drops of KCl and K_2CrO_4 solutions of equal molar concentration in a small test tube and add $AgNO_3$ solution to the mixture drop by drop, stirring the contents of the tube thoroughly after each addition. First a white curdy precipitate of AgCl is formed, followed by a brick-red precipitate of Ag_2CrO_4.

When does the precipitation of Ag_2CrO_4 begin? This can easily be calculated by using the solubility product principle. Indeed, the product of the concentrations $[Ag^+]$ $[Cl^-]$ remains constant throughout the precipitation and is equal to the solubility product of AgCl. As the precipitation of Cl^- proceeds, the Ag^+ concentration in solution will

* The solubility of AgCl and Ag_2CrO_4 is discussed in Sec. 4.1.

increase. There will, in the long run, come a moment when the solubility product of Ag_2CrO_4 is attained and both salts are precipitated together (in relative amounts determined by their concentrations in solution at that instant). This occurs when the following condition is fulfilled:

$$\frac{SP_{AgCl}}{[Cl^-]}=\sqrt{\frac{SP_{Ag_2CrO_4}}{[CrO_4^{2-}]}}$$

Rearranging this expression, we obtain:

$$\frac{\sqrt{[CrO_4^{2-}]}}{[Cl^-]}=\frac{\sqrt{SP_{Ag_2CrO_4}}}{SP_{AgCl}}=\frac{\sqrt{1.1\times10^{-12}}}{1.78\times10^{-10}}\approx 5.8\times10^3$$

Since the concentration of CrO_4^{2-} in solution is known to be 10^{-1} g-ion/litre, it is easy to find the Cl^- concentration that must be present at this instant:

$$\frac{\sqrt{10^{-1}}}{[Cl^-]}=5.8\times10^3$$

that is,

$$[Cl^-]\approx\frac{0.32}{5.8\times10^3}=5.7\times10^{-5}\text{ g-ion/litre}$$

Thus, the precipitation of CrO_4^{2-} will begin only after the Cl^- ions are almost completely removed from solution. It is apparent from this that the precipitation of ions from their mixture takes place in the same sequence in which the solubility products of slightly soluble compounds formed by the action of a given reagent are reached. Such precipitation is called **fractional precipitation.** In fractional precipitation, the first to precipitate is usually the least soluble of the compounds formed in the reaction, followed by the precipitation of the compound whose solubility product is satisfied next, etc. The principle of fractional precipitation is extensively used in spot analysis and in precipitation chromatography (see Sections 2.4 and 2.6).

The sequence of precipitation of ions depends not only on the solubility of the reaction products but also on the concentrations of the corresponding ions in solution. This sequence can be reversed by changing appropriately the concentrations of ions of a slightly soluble electrolyte. For example, if the concentrations of Cl^- and CrO_4^{2-} were chosen so that $[Ag^+]_{Ag_2CrO_4}<[Ag^+]_{AgCl}$, i.e., so that

$$\frac{\sqrt{SP_{Ag_2CrO_4}}}{\sqrt{[CrO_4^{2-}]}}<\frac{SP_{AgCl}}{[Cl^-]}$$

the first to precipitate would be Ag_2CrO_4 and not AgCl.

4.5. Dissolution of Precipitates

The above examples demonstrate that the solubility product principle is used to solve the various problems associated with the precipitation of ions in the form of slightly soluble compounds. But it is no less useful for solving the reverse problem, equally important in analysis, i.e., to choose the conditions required for dissolution of slightly soluble precipitates.

Suppose that a precipitate of $Mg(OH)_2$ is to be dissolved in order to detect Mg^{2+} in solution. The solution in contact with this precipitate is saturated with $Mg(OH)_2$, which means that the ion product $[Mg^{2+}][OH^-]^2$ is equal to the solubility product constant of $Mg(OH)_2$, i.e., to 6×10^{-10} (at 25°C).

If a small amount of an acid is added to this solution, the hydrogen ions of the acid combine with the hydroxyl ions present in solution to form un-ionized H_2O molecules. As a result, the ion product $[Mg^{2+}][OH^-]^2$ becomes less than the solubility product of $Mg(OH)_2$. The solution then becomes unsaturated with respect to magnesium hydroxide whose precipitate begins to dissolve, i.e., the Mg^{2+} and OH^- ions pass into solution.

This evidently continues until the ion concentration product $[Mg^{2+}][OH^-]^2$ again equals the solubility product constant of $Mg(OH)_2$, following which equilibrium is established. But if the addition of the acid is continued, this equilibrium is continuously disturbed and more and more of the precipitate dissolves. The dissolution process goes on until all the precipitate is dissolved.

Hence, in order to dissolve a precipitate, one of the ions it supplies into the solution must be bound by an ion that forms with it a slightly dissociated compound or a gaseous substance.

The same conclusion can be reached on the basis of the following considerations. Since the ion concentration product $[Mg^{2+}][OH^-]^2$ must remain almost constant *at equilibrium*, and be equal to the solubility product constant of $Mg(OH)_2$, any change in the concentration of one of the ions in solution will cause a change in the concentration of the other ion. Thus, in order to reduce the concentration of Mg^{2+}, i.e., to precipitate it more completely, it is necessary to add an excess of the precipitant, i.e., to increase the hydroxyl ion concentration.

But if we are to increase the Mg^{2+} concentration in solution, i.e., to dissolve the $Mg(OH)_2$ precipitate, the hydroxyl ion concentration should be lowered by binding these ions into un-ionized molecules of a weak electrolyte.

The reaction which takes place when acids act on $Mg(OH)_2$ may be represented by the following equations:

$$Mg(OH)_2\downarrow \rightleftharpoons \underset{\text{(solution)}}{Mg(OH)_2} \rightleftharpoons Mg^{2+} + 2OH^-$$

$$2OH^- + 2H^+ \rightleftharpoons 2H_2O$$

$$Mg(OH)_2\downarrow + 2H^+ \rightleftharpoons Mg^{2+} + 2H_2O$$

Other slightly soluble bases dissolve in acids similarly; for **example,**

$$Mn(OH)_2\downarrow + 2H^+ \rightleftharpoons Mn^{2+} + 2H_2O$$

$$Fe(OH)_3\downarrow + 3H^+ \rightleftharpoons Fe^{3+} + 3H_2O$$

The hydroxyl ions can also be bound by introducing NH_4^+ ions into the solution, a relatively slightly ionized compound (NH_4OH) being formed. Therefore, a number of hydroxides dissolve not only in acids but in solutions of ammonium salts as well:

$$Mg(OH)_2\downarrow + 2NH_4^+ \rightleftharpoons Mn^{2+} + 2NH_4OH$$

$$Mn(OH)_2\downarrow + 2NH_4^+ \rightleftharpoons Mn^{2+} + 2NH_4OH$$

The compound NH_4OH however is ionized much more strongly than H_2O and therefore the decrease of the hydroxyl ion concentration in solution is not as great as when an acid is added. Hence, the solvent power of ammonium salts on bases is much weaker than that of acids and only certain more soluble bases do dissolve in solutions of ammonium salts. Conversely, the least soluble bases, such as $Fe(OH)_3$, $Al(OH)_3$, and $Cr(OH)_3$, which dissolve readily in strong mineral acids, are insoluble in solutions of ammonium salts and acetic acid.

Acids slightly soluble in water can evidently be dissolved by binding the hydrogen ions supplied by them to the solution by the hydroxyl ions of any base. For example,

$$H_2WO_4\downarrow + 2OH^- \rightleftharpoons WO_4^{2-} - 2H_2O$$

Slightly soluble salts of weak acids dissolve in acids because the hydrogen ions of the acid combine with the salt anions to form unionized molecules of the corresponding weak acid, for example,

$$KHC_4H_4O_6\downarrow + H^+ \rightleftharpoons H_2C_4OH_4O_6 + K^+$$

Here the $HC_4H_4O_6^-$ anions supplied to the solution by the precipitate combine with H^+ ions to form molecules of tartaric acid, $H_2C_4H_4O_6$, which dissociates to a slight extent. As the concentration of these anions in solution diminishes, the K^+ concentration increases due to the dissolution of the $KHC_4H_4O_6$ precipitate. A precipitate of CaC_2O_4 dissolves in acids in just the same way:

$$CaC_2O_4\downarrow + H^+ \rightarrow Ca^{2+} + HC_2O^-$$

$$HC_2O_4^- + H^+ \rightarrow H_2C_2O_4$$

Precipitates of sulphides which are slightly soluble in water dissolve in a similar way. For instance,

$$ZnS\downarrow + H^+ \rightarrow Zn^{2+} + HS^-$$

$$HS^- + H^+ \rightarrow H_2S\uparrow$$

Conversely, slightly soluble salts of strong acids, such as $BaSO_4$, $SrSO_4$, AgCl, AgI, and others, do not dissolve in acids because no compounds of low degree of dissociation can form in this case.*

We should also consider the dissolution of precipitates of acid salts. It has been shown above that a salt such as $KHC_4H_4O_6$ must dissolve in acids. But it dissolves in bases as well. This is accounted for by the fact that the $HC_4H_4O_6^-$ ions supplied by the $KHC_4H_4O_6$ precipitate to the solution partly dissociate:

$$HC_4H_4O_6^- \rightleftarrows H^+ + C_4H_4O_6^{2-}$$

The OH^- ions introduced with the base combine with the H^+ ions, as a result of which the dissociation equilibrium is shifted to the right and the concentration of $HC_4H_4O_6^-$ is reduced. Consequently, the K^+ concentration increases and the precipitate dissolves if enough alkali is present:

$$KHC_4H_4O_6 \downarrow + OH^- \rightarrow K^+ + C_4H_4O_6^{2-} + H_2O$$

In all the cases considered above the ions are bound during the dissolution of precipitates as a result of the formation of undissociated molecules of a weak electrolyte (water, weak acids, NH_4OH, etc.). Another factor that may be responsible for the binding of the ions is the formation of sufficiently stable complex ions, which also results in the dissolution of precipitates.** For example, silver chloride is insoluble in acids but it dissolves very readily in aqueous solution of ammonia since Ag^+ ions that pass into solution are bound by NH_3 molecules to yield complex $[Ag(NH_3)_2]^+$ ions. The resultant decrease in the Ag^+ concentration disturbs the equilibrium between the solution and the precipitate and the latter passes into solution:

$$AgCl \downarrow + 2NH_3 \rightarrow [Ag(NH_3)_2]^+ + Cl^-$$

In just the same way, a precipitate of HgI_2 dissolves readily in KI because its Hg^{2+} ions that pass into solution form complex $[HgI_4]^{2-}$ ions in the presence of an excess amount of I^- ions formed as the result of the dissociation of KI:

$$HgI_2 \downarrow + 2I^- \rightarrow [HgI_4]^{2-}$$

---

* It should however be noted that, because of the binding of SO_4^{2-} ions by the hydrogen ions of the acid with the formation of HSO_4^- anions, the solubility of slightly soluble sulphates increases somewhat under the influence of acids (see Sec. 4.3).

** Complex compounds are more fully discussed in Sections 5.8 through 5.12. It should be noted here that complex ions of these substances, e.g., $[Ag(NH_3)_2]^+$, though being capable of ionizing with formation of the corresponding simple ions (Ag^+), generally have a rather low degree of ionization. The formation of such complex ions therefore affects the course of the dissolution of precipitates, just as the formation of un-ionized molecules of weak electrolytes.

Thus, the introduction of common iodide ions does not lower the solubility of the HgI_2 precipitate at all but, on the contrary, causes its complete dissolution.

It has been pointed out above that the possibility of formation of complexes is one of the most important reasons why a large excess of precipitant should not be added in separation reactions.

Finally, the dissolution of precipitates is very often associated with the occurrence of the various oxidation-reduction processes. For example, sulphides such as CoS, NiS, CuS, Bi_2S_3, which are insoluble in HCl and H_2SO_4, dissolve readily on heating in concentrated HNO_3. The dissolution of CuS proceeds according to the equation

$$3CuS\downarrow + 2NO_3^- + 8H^+ \rightarrow 3Cu^{2+} + 3S\downarrow + 4H_2O + 2NO\uparrow$$

This reaction consists essentially in the following: the S^{2-} ions supplied to the solution by the precipitate are oxidized to free sulphur by nitric acid. As a result, the S^{2-} concentration in solution decreases, while the concentration of Cu^{2+} increases, i.e., the CuS precipitate is dissolved.

Such cases where the precipitate insoluble in HCl dissolves in HNO_3 are encountered not infrequently. Therefore, when substances slightly soluble in water are to be dissolved, they are first acted upon by HCl. If the precipitate does not dissolve in it, use is made of concentrated HNO_3 and, finally, of aqua regia (a mixture of 1 volume of concentrated HNO_3 and 3 volumes of concentrated HCl), the oxidizing action of which is even stronger than that of nitric acid.

Sometimes, for a precipitate to be dissolved, it is necessary to reduce one of its elements. For example, a precipitate of manganese(IV) hydroxide, $MnO(OH)_2$, does not dissolve in H_2SO_4, but when KNO_2 or $NaNO_2$ is added to the solution, Mn(IV) will be reduced to Mn(II) whose compounds dissolve well:

$$MnO(OH)_2\downarrow + NO_2^- + 2H^+ \rightarrow Mn^{2+} + NO_3^- + 2H_2O$$

Of course, there is no point in using oxidizing (or reducing) agents unless one of the components of the precipitate can be oxidized (or reduced). It would be obviously quite useless to try to dissolve $BaSO_4$ in HNO_3 or in aqua regia because the elements Ba and S of this substance are in their highest oxidation states.

Some additional information on the dissolution of slightly soluble compounds is given in Sections 5.2 and 9.4.

4.6. Interconversion of Slightly Soluble Compounds

In analytical chemistry, we often have to deal with reactions in which one slightly soluble compound, e.g. Ag_2CrO_4, is converted by the action of an appropriate reagent, say KCl, into another slightly soluble compound, AgCl in this case.

The course of such reactions is explained as follows. The solution over the Ag_2CrO_4 precipitate contains Ag^+ ions which interact with Cl^- ions when KCl is added. Since AgCl is less soluble than Ag_2CrO_4, the Ag^+ concentration appears to be sufficient for the solubility product constant of silver chloride to be exceeded, and AgCl is precipitated. But the removal of Ag^+ makes the solution unsaturated with respect to Ag_2CrO_4 which therefore dissolves. The resulting Ag^+ ions react at once with Cl^- ions, etc. It is obvious that all the processes represented by the scheme

$$\begin{array}{rcl} Ag_2CrO_4\downarrow & \rightleftharpoons & 2Ag^+ + CrO_4^{2-} \\ 2KCl & \rightarrow & 2Cl^- + 2K^+ \\ & & \downarrow \\ & & 2AgCl\downarrow \end{array}$$

would lead to complete conversion of Ag_2CrO_4 into AgCl if enough KCl is present.

This can easily be demonstrated by experiment. If a solution of $AgNO_3$ is added to a few drops of a K_2CrO_4 solution, a red precipitate of Ag_2CrO_4 will form. Then add an excess of KCl solution to the test tube and shake its contents. The red precipitate (Ag_2CrO_4) turns white (AgCl).

Various slightly soluble compounds can likewise be converted into even less soluble compounds under given conditions. For example, a precipitate of $SrSO_4$ ($SP = 3.2 \times 10^{-7}$) is readily converted to $SrCO_3$ ($SP = 1.1 \times 10^{-10}$) by the action of Na_2CO_3.

This fact is of great practical importance since $SrSO_4$, which is a slightly soluble salt of a strong acid, is insoluble in acids (see Sec. 4.5). Therefore, in order to dissolve strontium sulphate, it is first converted into the carbonate, i.e., a salt soluble in acids.

If we compare the value of the solubility product constant of $BaSO_4$ ($SP = 1.1 \times 10^{-10}$) with that of $BaCO_3$ ($SP = 5.1 \times 10^{-9}$), it may seem that it is impossible to convert barium sulphate into the carbonate. But this is not actually the case. A precipitate of $BaCO_3$ is formed whenever the solubility product constant is found to be exceeded, i.e., provided that

$$[Ba^{2+}][CO_3^{2-}] > SP_{BaCO_3} \quad \text{or} \quad [Ba^{2+}] > \frac{SP_{BaCO_3}}{[CO_3^{2-}]}$$

But the Ba^{2+} concentration is determined in this case by the solubility of $BaSO_4$ and is equal to

$$[Ba^{2+}] = \frac{SP_{BaSO_4}}{[SO_4^{2-}]}$$

The conversion of $BaSO_4$ into $BaCO_3$ must take place on condition that

$$\frac{SP_{BaSO_4}}{[SO_4^{2-}]} > \frac{SP_{BaCO_3}}{[CO_3^{2-}]}$$

that is, if

$$\frac{[CO_3^{2-}]}{[SO_4^{2-}]} > \frac{SP_{BaCO_3}}{SP_{BaSO_4}}$$

Substituting the numerical values of the solubility product constants into this expression, we obtain

$$\frac{[CO_3^{2-}]}{[SO_4^{2-}]} > \frac{5.1 \times 10^{-9}}{1.1 \times 10^{-10}} \quad \text{or} \quad \frac{[CO_3^{2-}]}{[SO_4^{2-}]} > 50$$

Thus, for $BaSO_4$ to be converted into $BaCO_3$, the concentration of CO_3^{2-} in solution must be 50 times as great as that of SO_4^{2-}. Since the SO_4^{2-} concentration in a saturated solution of $BaSO_4$ is very low (ca. 10^{-5} g-ion/litre), this condition is easily satisfied.

It is not difficult to see however that the reaction under consideration does not still go to completion. Indeed, from the equation

$$BaSO_4\downarrow + CO_3^{2-} \rightarrow BaCO_3\downarrow + SO_4^{2-}$$

it is clear that as the reaction proceeds the CO_3^{2-} concentration decreases, whereas the SO_4^{2-} concentration increases, and, as soon as the ratio of these concentrations becomes equal to 50, equilibrium is established and further conversion of $BaSO_4$ into $BaCO_3$ is stopped.

If however we decant the liquid from the precipitate and act on the latter with a fresh portion of Na_2CO_3 solution, the reaction of formation of $BaCO_3$ will be resumed. By repeating the reaction several times one can achieve a complete conversion of $BaSO_4$ into $BaCO_3$.

Thus, although the conversion of $BaSO_4$ into the more soluble compound $BaCO_3$ is possible, the reaction proceeds with difficulty and, in the case of a single treatment with the reagent (Na_2CO_3), it does not even go to completion. It is evident that the possibility of the conversion is itself based on the fact that the solubility product constant of the sulphate does not differ greatly from that of the carbonate (by a factor of 50). In the case of the reaction

$$AgI\downarrow + Cl^- \rightarrow AgCl\downarrow + I^-$$

the solubility product constant of the compound formed (AgCl) is approximately 1,000,000 times as great as the solubility product constant of the original compound (AgI). This reaction will not proceed even if the AgI precipitate is repeatedly treated with a KCl solution. The reverse reaction-conversion of AgCl into AgI—proceeds very smoothly and goes to near completion even after a single treatment of the AgCl precipitate with a KI solution. This is well illustrated by the following example. If we precipitate AgCl, separate it from the mother liquor, and mix it with KI solution, the precipitate will turn yellow because of the conversion of AgCl into AgI.

Interconversion of slightly soluble compounds can be successfully used for separation of ions, especially in fractional detection of ions.

In certain cases, owing to their higher selectivity, solid reagents slightly soluble in water have considerable advantages as precipitants over those readily soluble in water. For example, if we add a soluble cadmium salt (e.g., $CdSO_4$) to a solution containing S^{2-} and SO_3^{2-} ions, both these anions will be precipitated because they form slightly soluble cadmium salts. If we replace $CdSO_4$ with slightly soluble $CdCO_3$, which is more soluble than CdS but less so than $CdSO_3$, only S^{2-} ions will precipitate, but not SO_3^{2-}. This method is employed for separation of the indicated anions.

In the same way, one can separate Cu^{2+} from Cd^{2+} through the action of solid PbS on a sulphuric acid solution containing these ions. The reactions involved proceed according to the equations

$$PbS\downarrow + Cu^{2+} \rightarrow CuS\downarrow + Pb^{2+}$$

$$Pb^{2+} + SO_4^{2-} \rightarrow PbSO_4\downarrow$$

The Cd^{2+} ions remain in solution. If H_2S were used, the sulphides of both metals would be precipitated.

Reactions involving the conversion of some slightly soluble compounds into compounds even less soluble under given conditions can be used for separation of ions by another method. Not infrequently it proves to be possible to dissolve the ions of any one element by the action of a suitable reagent on a mixture of slightly soluble substances present in the precipitate, and then to detect these ions in the resultant solution.

For instance, the anion of thiocyanic acid (HSCN) is detected by the action of $FeCl_3$, with which it forms the intensely-red ferric thiocyanate, $Fe(SCN)_3$:

$$3SCN^- + Fe^{3+} \rightarrow Fe(SCN)_3$$

The presence in solution of iodide ions interferes with this reaction as they are oxidized by ferric ions to form iodine which also imparts a red-brown colour to the solution, similar to the colour of $Fe(SCN)_3$:

$$2I^- + 2Fe^{3+} \rightarrow I_2 + 2Fe^{2+}$$

For I^- to be separated from SCN^-, the test solution is acidified with HNO_3 and a solution of $AgNO_3$ is added. The I^- and SCN^- ions precipitate as AgI and AgSCN, respectively. The precipitate is separated from the solution, washed, and shaken with a solution of KBr. Since the solubility product constant of AgBr (5.3×10^{-13}) is less than that of AgSCN (1.1×10^{-12}), silver thiocyanate is converted into silver bromide:

$$AgSCN\downarrow + Br^- \rightarrow AgBr\downarrow + SCN^-$$

The ions SCN^- that pass from the precipitate into solution can be detected by the reaction with $FeCl_3$. Since AgI has a much lower solubility product constant (8.3×10^{-17}) than AgBr, the action of KBr does not practically convert AgI into AgBr, and hence the I^- ions do not pass into solution. Thus, the I^- ions do not interfere with the detection of SCN^-.

The choice of a reagent for separating only a given ion from the precipitate can be based on a comparison of the solubility product constants of the corresponding salts provided that the formulas of the precipitates formed are of the same type, as in the preceding example. If the precipitates are compounds of different types, say, $Ag_4[Fe(CN)_6]$ and AgBr, or AgI and Ag_2S, then in selecting the reagent we must calculate the concentration of the precipitating ion required to start the precipitation of each of the given ions taken in equal concentrations (for example, in unimolar concentrations):

$$[Ag^+]_{AgBr} = \frac{SP_{AgBr}}{1} = 5.3 \times 10^{-13}$$

$$[Ag^+]_{Ag_4[(FeCN)_6]} = \sqrt[4]{\frac{SP_{Ag_4[Fe(CN)_6]}}{1}} = \sqrt[4]{1.5 \times 10^{-41}} = 0.63 \times 10^{-10}, \text{ etc.}$$

The substances to be precipitated are arranged in a series which shows not only the sequence of their precipitation but also the sequence of interconversion of slightly soluble compounds. For example, the following series is obtained for silver salts:

$$Ag_3[Fe(CN)_6] \rightarrow AgCl \rightarrow Ag_4[Fe(CN)_6] \rightarrow AgCN \rightarrow AgSCN \rightarrow AgBr \rightarrow AgI \rightarrow Ag_2S$$

If the silver salts in this series are acted on, say, by a solution of KI, those on the left of AgI will be converted to AgI, and the corresponding anions, i.e., $[Fe(CN)_6]^{3-}$, Cl^-, $[Fe(CN)_6]^{4-}$, CN^-, SCN^-, and Br^-, will pass into solution. For instance,

$$Ag_3[Fe(CN)_6]\downarrow + 3I^- \rightleftharpoons 3AgI\downarrow + [Fe(CN)_6]^{3-}$$

Conversely, Ag_2S, which stands on the right of AgI in the series and which is precipitated earlier, will not practically be converted to silver iodide and no noticeable amounts of S^{2-} ions will pass into solution.

Similar series can be written for any other slightly soluble compounds, for example, for barium and lead sulphates, for the sulphides of various metals, etc.

4.7. Direction of Exchange Reactions

One of the most important factors determining the direction of an exchange reaction is evidently the solubility of the compounds formed. Indeed, if the solubility product of one of the compounds is sufficiently low and is found to be considerably exceeded, the reaction will proceed towards practically complete precipitation of that compound from solution. If the reaction involves the formation of several different slightly soluble compounds, the least soluble of these will usually precipitate first, followed by the remaining ones in order of their increasing solu-
bility

Another factor governing the direction of exchange reactions is the formation of any slightly dissociated compounds. This also refers to the reaction of neutralization which consists essentially of the binding of H^+ and OH^- ions:

$$H^+ + OH^- \rightleftarrows H_2O$$

Another consequence of the formation of slightly dissociated substances is the displacement of weak acids from their salts by strong acids or of weak bases by strong alkalis. For example, if a solution of sodium acetate, CH_3COONa, is acted on by a solution of HCl, the greater part of H^+ ions of hydrochloric acid will immediately be bound by CH_3COO^- ions into un-ionized molecules of weak acetic acid:

$$CH_3COO^- + H^+ \rightleftarrows CH_3COOH$$

This phenomenon is often used in practice for reducing the acidity of a solution, for example, when we deal with the formation of precipitates soluble in strong acids but insoluble in acetic acid ($KHC_4H_4O_6$, CaC_2O_4, $BaCrO_4$, ZnS, etc.). The reaction is carried out in the presence of sodium acetate. If sodium acetate is taken in some excess, acetate buffer results, which maintains the pH value practically constant and sufficient for complete precipitation of such salts.

The ability of strong acids to interact with salts of weak acids is occasionally employed to neutralize acid solutions; use is generally made of carbonates such as Na_2CO_3, K_2CO_3, and $CaCO_3$, whose interaction with hydrogen ions is accompanied by the formation of free H_2CO_3 which breaks down into the anhydride and water:

$$2H^+ + CaCO_3\downarrow \rightarrow Ca^{2+} + CO_2\uparrow + H_2O$$

The decrease in the solution concentration of ions, which is needed for exchange reactions to proceed, may be caused not only by the formation of slightly soluble or slightly dissociated compounds but also by the formation of compounds that are volatile under given conditions. For example, in the reaction

$$NaCl + H_2SO_4 \rightarrow NaHSO_4 + HCl\uparrow$$

which is carried out on prolonged heating or in the absence of water, HCl is removed in the form of a gas.

Similarly, when nitrate solutions are evaporated with H_2SO_4, HNO_3 is evolved, which is removed from solution in vapour form:

$$Cu(NO_3)_2 + H_2SO_4 \rightarrow CuSO_4 + 2HNO_3\uparrow$$

This method of removing NO_3^- is used in all cases where the presence of an oxidizing agent in the solution is undesirable.

It is evident that in order to obtain an acid from one of its salts by this method, the salt must be treated with an acid less volatile than the one to be formed. Sulphuric acid is commonly used as a relatively poorly volatile acid.

By taking advantage of the differences in volatility, one can sometimes displace a stronger acid from one of its salts by a weaker one. For example, when a solution of sodium chloride is evaporated with boric acid, H_3BO_3, which is one of the weakest acids, the latter, being involatile, displaces HCl from the salt. A similar phenomenon based on the low solubility of one of the substances formed is observed in the reaction

$$CuCl_2 + H_2S \rightarrow CuS\downarrow + 2HCl\uparrow$$

From what has been said it follows that for exchange reactions to take place in solutions of electrolytes it is necessary that certain ions be removed from solution as the result of the formation of slightly soluble, slightly dissociated, and readily volatile substances.

If no compound having such properties is formed, the reaction fails to proceed. This becomes especially clear if we try to write an ionic equation for the reaction

$$NaCl + KNO_3 \rightleftharpoons NaNO_3 + KCl$$

Indeed, all the members on the left-hand side of the ionic equation are found in this case to be absolutely identical with those on the right-hand side of the equation:

$$Na^+ + Cl^- + K^+ + NO_3^- = Na^+ + NO_3^- + K^+ + Cl^-$$

that is, no reaction occurs in solution.

Questions and Problems

In solving the problems the student should remember the following. The values of thermodynamic solubility product constants are given in special handbooks. If the value of SP_a is less than $n \times 10^{-7}$, then in the absence of soluble strong electrolytes it may be assumed that $SP_a \approx SP_C$, and the calculation may be made with activity coefficients being neglected. If SP_a is greater than $n \times 10^{-7}$, activity coefficients should be taken into account.

1. The solubility of $CaCO_3$ is equal to 0.0069 g/litre. Calculate the solubility product constant of $CaCO_3$.

Answer: $SP_C \approx SP_a = 4.8 \times 10^{-9}$.

2. The solubility of $Fe(OH)_3$ equals 2×10^{-8} g/litre. Determine the solubility product constant of $Fe(OH)_3$.

Answer: $SP_C \approx SP_a = 3.2 \times 10^{-38}$.

3. From the solubility of $MgNH_4PO_4$, which is equal to 8.6×10^{-3} g/litre, find its solubility product constant.

Answer: $SP_C \approx SP_a = 2.5 \times 10^{-13}$.

4. From the solubility of $AgNO_2$, equal to 2.22 g/litre, calculate $SP_{a,\, AgNO_2}$. Take activity coefficients into account.

Answer: $SP_a = 1.7 \times 10^{-4}$ (averaged activity coefficients are taken for calculation).

5. $SP_{a,\ PbSO_4} = 1.6 \times 10^{-8}$; $SP_{a,\ PbI_2} = 1.1 \times 10^{-9}$. Calculate the solubilities of these compounds in moles and grams per litre. Why is the salt with a lower value of SP_a more soluble?

Answer: (a) the solubility of $PbSO_4$ is 1.3×10^{-4} mole/litre or 3.9×10^{-2} g/litre; (b) the solubility of PbI_2 is 6.49×10^{-4} mole/litre or 21.9×10^{-1} g/litre.

6. From the solubility product constant of $Pb_3(PO_4)_2$ (7.9×10^{-43}) calculate the solubility of this salt in grams per litre and find the concentrations of Pb^{2+} and PO_4^{3-} ions in a saturated aqueous solution.

Answer: 1.2×10^{-6} g/litre; $[Pb^{2+}] = 4.5 \times 10^{-9}$; $[PO_4^{3-}] = 3 \times 10^{-9}$

7. Figure out by how many times the solubility of AgCl ($SP_a = 1.78 \times 10^{-10}$) in $0.01N$ NaCl solution is lower than its solubility in pure water.

Answer: by about 730 times.

8. Solve the preceding problem, taking into account the activity coefficients of the ions.

Answer: by about 600 times.

9. By how many times is the solubility of CaC_2O_4 ($SP_a = 2.3 \times 10^{-9}$) lower in $0.01M$ $(NH_4)_2C_2O_4$ solution than in pure water? Perform calculations: (a) with the activity coefficients being neglected; and (b) with the activity coefficients taken into account.

Answer: (a) by about 200 times; (b) by about 66 times.

10. How is the solubility of CaC_2O_4 affected by the presence in solution of KCl in a concentration of $0.1N$?

Answer: the solubility is increased by about 3 times.

11. Figure out whether a precipitate of $PbCl_2$ ($SP_a = 1.6 \times 10^{-5}$) is formed when $0.05N$ $Pb(NO_3)_2$ solution is mixed with an equal volume of $0.05N$ HCl solution. Use the exact formula.

Answer: it is not formed.

12. Figure out whether a precipitate is formed when equal volumes of $0.001N$ solutions of $SrCl_2$ and K_2SO_4 are mixed. The value of SP_a for $SrSO_4$ is equal to 3.2×10^{-7}.

Answer: a precipitate is not formed.

13. The solubility of $CaSO_4$ is 1 g/litre. A saturated solution of calcium sulphate is mixed with an equal volume of a solution of ammonium oxalate, $(NH_4)_2C_2O_4$, which contains 0.0248 g of this salt per litre. Determine the ion concentration product $[Ca^{2+}][C_2O_4^{2-}]$ at the moment of mixing and find out whether a precipitate of CaC_2O_4 is formed.

Answer: $[Ca^{2+}][C_2O_4^{2-}] = 3.7 \times 10^{-7}$; $SP_{a,\ CaC_2O_4} = 2.3 \times 10^{-9}$; the precipitate is formed.

14. The concentration of S^{2-} in aqueous solutions of hydrogen sulphide is equal to 1.2×10^{-13} g-ion/litre.

Taking this and the solubility product constants of lead chloride and lead sulphide into account, figure out whether a precipitate of PbS is formed when gaseous H_2S is passed into a saturated solution of

$PbCl_2$. $SP_{a,\ PbS} = 2.5 \times 10^{-27}$; $SP_{a,\ PbCl_2} = 1.6 \times 10^{-5}$ Take activity coefficients into account.

Answer: the precipitate is formed.

15. How many grams of Pb^{2+} will be left in 100 ml of solution if a solution of $Pb(NO_3)_2$ is acted on: (a) by an equivalent amount of H_2SO_4; (b) by H_2SO_4 taken in such an excess that its concentration in the solution becomes equal to 0.01 mole/litre upon completion of precipitation? The value of SP_a for $PbSO_4$ is 1.6×10^{-8}.

Answer: (a) 2.4×10^{-3} g Pb^{2+}; (b) 3.4×10^{-5} g Pb^{2+}.

16. At what value of pH will the precipitation of aluminium hydroxide begin from a $0.1M$ solution of its salt? At what pH value will the concentration of Al^{3+} ions left in solution be equal to 1×10^{-5} g-ion/litre? The value of SP_a for $Al(OH)_3$ is 1×10^{-32}

Answer: (a) at pH = 3.7; (b) at pH = 5.

17. At what pH value will Ca^{2+} be practically completely precipitated by the action of $(NH_4)_2C_2O_4$ if this salt is added in such an excess that the concentration of $HC_2O_4^-$ ions formed as a result of hydrolysis becomes equal to 1×10^{-2} g-ion/litre?

Hint. First, using the equation for the solubility product constant of the compound being precipitated (the value of SP_a for CaC_2O_4 is 2.3×10^{-9}), find the concentration of $C_2O_4^{2-}$ in solution required for the precipitation of Ca^{2+} to be practically complete ($[Ca^{2+}] = 10^{-6}$). Then, from the value of $[C_2O_4^{2-}]$ found and the value of $[HC_2O_4^-]$ given in the problem, calculate the sought-for concentration of hydrogen ions and the corresponding pH value, using the equation for the second-stage ionization constant of oxalic acid ($K_{HC_2O_4} = 5.4 \times 10^{-5}$).

Answer: pH = 3.64.

18. What is the minimum pH value at which the precipitation of Sr^{2+} by ammonium carbonate could be practically complete if the excess of added $(NH_4)_2CO_3$ is $0.1M$? Assume the hydrolysis of $(NH_4)_2CO_3$ at the first stage to be complete, neglecting the second stage of hydrolysis. Compare the result obtained with the pH value produced by an excess of $(NH_4)_2CO_3$ in solution and, on the basis of this comparison, find out whether practically complete precipitation of Sr^{2+} would be achieved (see the preceding problem). The value of SP_a for $SrCO_3$ is 1.1×10^{-10}.

Answer: pH = 7.36; the precipitation will be practically complete.

19. What will be the concentration of Ca^{2+} in solution upon completion of the precipitation effected by the action of $(NH_4)_2C_2O_4$ if the final concentration of the precipitant is $0.01M$ and the precipitation is carried out under the following conditions: (a) at the end of precipitation pH equals 4; (b) at the end of precipitation pH equals 1?

Hint. The excess of reagent present in solution yields 0.01 g-ion of $C_2O_4^{2-}$ per litre of solution. But, combining with hydrogen ions, $C_2O_4^{2-}$ is partly converted into $HC_2O_4^-$ ions and $H_2C_2O_4$ molecules.

Hence,

$$[C_2O_4^{2-}]+[HC_2O_4^-]+[H_2C_2O_4]=0.01 \quad (4.4)$$

Then, proceeding from the given concentration of hydrogen ions (assuming that pH is constant throughout the precipitation) and using the equations for both ionization constants of oxalic acid, express the concentrations $[HC_2O_4^-]$ and $[H_2C_2O_4]$ in terms of $[C_2O_4^{2-}]$:

$$K_{H_2C_2O_4}=\frac{[H^+][HC_2O_4^-]}{[H_2C_2O_4]}=5.6\times10^{-2}$$

$$K_{HC_2O_4^-}=\frac{[H^+][C_2O_4^{2-}]}{[HC_2O_4^-]}=5.4\times10^{-5}$$

Substituting the expressions obtained for $[HC_2O_4^-]$ and $[H_2C_2O_4]$ into Eq. (4.4), we solve it for $[C_2O_4^{2-}]$. Substituting the thus obtained value of $[C_2O_4^{2-}]$ into the formula for the solubility product constant of CaC_2O_4 ($SP_{a,\ CaC_2O_4}= 2.3 \times 10^{-9}$), calculate the sought-for concentration of Ca^{2+} ions in solution.

Answer: (a) 6.57×10^{-7} g-ion/litre; (b) 1.2×10^{-3} g-ion/litre.

20. Under what conditions and why does not the pH value practically affect the completeness of precipitation of slightly soluble salts?

21. A solution containing Cl^- and I^- ions is made to react with a solution of $AgNO_3$. Which of the salts, AgCl or AgI, is formed first if the concentrations of these anions are equal to 0.01 g-ion/litre? At what concentration ratio will both these salts precipitate simultaneously? The value of SP_a for AgCl is 1.78×10^{-10} and that for AgI is 8.3×10^{-17}.

Answer: both these salts will precipitate simultaneously at a concentration ratio of $[Cl^-]:[I^-] \approx 2{,}000{,}000$.

22. A solution containing 1 g-ion of Ba^{2+} and 0.01 g-ion of Ca^{2+} per litre is made to react with a solution of $(NH_4)_2C_2O_4$. Which of the cations will precipitate first and what is the percentage of it that will be precipitated by the time when the precipitation of the other cation begins?

The value of SP_a for BaC_2O_4 is 1.1×10^{-7} and that for CaC_2O_4 is 2.3×10^{-9}.

Answer: the first to precipitate is Ba^{2+}; by the time when the precipitation of Ca^{2+} begins, Ba^{2+} will have been precipitated by 52 per cent.

23. A solution containing Ba^{2+} and Ca^{2+} in equal concentrations is made to react with a mixture of equal volumes of $(NH_4)_2C_2O_4$ and $(NH_4)_2SO_4$ solutions of equal normality. What compounds will be precipitated as the result of the reaction?

24. State the condition for conversion of a precipitate of $BaCrO_4$ to $BaCO_3$ by the action of Na_2CO_3 solution. What is the ratio of the concentration of CO_3^{2-} to that of CrO_4^{2-} at which equilibrium is established?

Answer: $[CO_3^{2-}]:[CrO_4^{2-}] = 42.5$.

B. REACTIONS AND ANALYTICAL PROCEDURE FOR GROUP II CATIONS

4.8. General Characteristics of Group II

Analytical Group II includes the following cations: Ca^{2+}, Sr^{2+}, and Ba^{2+}. In contrast to most of Group I cations, these ions form a number of slightly soluble salts with various anions. For example, the sulphates, phosphates, oxalates, and carbonates of Group II cations are slightly soluble. In the form of which salts should Group II be separated from Group I? The answer to this question is determined by a number of considerations (see Sec. 1.5). First, the salts must be as insoluble as possible, i.e., they should have the lowest possible solubility product constants. For example, Group II cannot be precipitated sufficiently completely in aqueous solution in the form of sulphates since one of them ($CaSO_4$) has a rather high solubility product constant (2.37×10^{-5})*. Furthermore, the salts of a strong acid—sulphates—are practically insoluble in acids and it would therefore be relatively difficult to dissolve them after the separation from Group I.

It is also difficult to separate Group II in the form of phosphates and oxalates because the PO_4^{3-} and $C_2O_4^{2-}$ ions introduced into the solution would complicate the subsequent analysis.

The most suitable procedure is to separate Group II cations from the cations of Group I in the form of carbonates—$CaCO_3$, $SrCO_3$, and $BaCO_3$. Indeed, the solubility product constants of these salts are sufficiently small (of the order of 10^{-9}), and therefore all the cations of Group II can be precipitated practically completely. The dissolution of the precipitates obtained, which is required for further analysis of Group II, is simple because, unlike sulphates, carbonates are salts of a weak acid and are readily soluble in acids. And, finally, the excess of precipitating CO_3^{2-} ions is easily removed from the solution when the latter is acidified, because the carbonic acid formed decomposes into CO_2 and H_2O.

From all the foregoing it follows that the most important analytical property of Group II cations, which is used to separate them from Group I cations, is the practical insolubility of the carbonates $CaCO_3$, $SrCO_3$, and $BaCO_3$ in water.

Conversely, the sulphides** of Group II cations, like those of Group I, are soluble in water, a feature which distinguishes Group II from Groups III, IV, and V.

---

* It should however be noted that addition of ethyl alcohol (ethanol) strongly reduces the solubility of $CaSO_4$; this is sometimes made use of in analysis (see Sec. 4.13).

** It would be more correct to say "hydrosulphides" since the dissolution of the sulphides CaS, SrS, and BaS is associated with their conversion into soluble hydrosulphides and hydroxides; for example,

$$CaS + H_2O \rightarrow Ca^{2+} + HS^- + OH^-$$

Group II carbonates are precipitated when the ionic concentration product $[Me^{2+}][CO_3^{2-}]$ in a solution exceeds the solubility product of $MeCO_3$ (where Me^{2+} is any cation of Group II). The only group reagent that can be used for Group II is ammonium carbonate, $(NH_4)_2CO_3$, since the use of Na_2CO_3 or K_2CO_3 would introduce Na^+ and K^+ ions; it is clear that in the latter case it is impossible to establish whether these ions were present in the solution being analysed. The introduction of ammonium ion does not result in error because ammonium is identified in a separate portion of the solution before the Group II reagent is added.

All the metals forming the cations of Group II belong to the second group of the Periodic System; they are in the first half of the long periods, i.e., in the even rows. These metals are known as the alkaline-earth metals; they decompose water vigorously at room temperature:

$$Ca + 2H_2O \rightarrow Ca^{2+} + 2OH^- + H_2\uparrow$$

The chemical activity increases in the series from Ca to Ba. Many other properties, such as the solubility of the salts and the basic properties of the hydroxides, change in the same direction.

Since the basic properties of Group II metals are strongly pronounced their salts formed by strong acids are hardly hydrolyzable and show an almost neutral reaction in solution (pH $\approx$ 6.)

Like Group I cations, Ca^{2+}, Sr^{2+}, and Ba^{2+} have complete outer shells of 8 electrons. In water solutions, these ions are colourless.

4.9. Action of the Group Reagent

The precipitation of Group II cations by the group reagent $(NH_4)_2CO_3$ is carried out under conditions which provide a complete separation from Group I cations. One of the most important conditions for complete precipitation of slightly soluble salts of weak acids, such as the carbonates of Group II cations, is the appropriate adjustment of the solution pH.

The pH value is largely predetermined by the presence in solution of an excess of $(NH_4)_2CO_3$ which is almost completely hydrolyzed (see Sec. 4.3) in solutions according to the equation

$$NH_4^+ + CO_3^{2-} + H_2O \rightleftharpoons HCO_3^- + NH_4OH$$

As a result, ammonium carbonate solution is essentially a mixture of approximately equivalent amounts of NH_4OH and the ammonium salt NH_4HCO_3, i.e., it is an ammonium buffer with pH $\approx$ 9.2. At this pH the precipitation of all the three carbonates of Group II cations is practically complete (see Sec. 4.3). The ions K^+, Na^+, and NH_4^+, which form soluble carbonates, remain in solution.

Although basic magnesium carbonate, $(MgOH)_2CO_3$, and also magnesium hydroxide, $Mg(OH)_2$, are slightly soluble, they are not precipitated at pH ≈ 9.2. For example, as shown above (see Sec. 4.3), the precipitation of $Mg(OH)_2$ begins at pH > 10.4 and becomes complete at pH 12.42. The conditions for precipitation of $(MgOH)_2CO_3$ are about the same.

Thus, Group II cations are completely separated from the cations of Group I if the precipitation is effected at pH ≈ 9.2. At pH < 9.2 the precipitation of Group II cations will be incomplete, and at pH > 10.4 there will be observed a partial precipitation of the slightly soluble compounds of magnesium.

The reason why the buffer capacity of the solution has to be taken into account is that ammonium carbonate solution is a mixture of equivalent quantities of NH_4OH and NH_4HCO_3 (i.e., HCO_3^-). Hence, the reaction involved in the precipitation of Group II cations by ammonium carbonate takes place according to the equation

$$Me^{2+} + HCO_3^- + NH_4OH \rightarrow MeCO_3 \downarrow + NH_4^+ + H_2O$$

From this equation it is seen that NH_4OH takes part in the reaction and that NH_4^+ ions accumulate in the solution. Evidently, accumulation of NH_4^+ ions must also take place when ammonium carbonate reacts with the acids present in the solution:

$$2H^+ + HCO_3^- + NH_4OH \rightarrow CO_2 \uparrow + 2H_2O + NH_4^+$$

The pH of the ammonium buffer is determined by the formula

$$pH = 14 - pK_{NH_4OH} + \log \frac{[NH_4OH]}{[NH_4^+]} = 9.24 + \log \frac{[NH_4OH]}{[NH_4^+]}$$

As a result of the accumulation of NH_4^+ ions in the solution, the ratio of the concentrations of NH_4OH and NH_4^+ decreases and becomes less than unity. Accordingly, the pH of the solution falls to below 9.24.

If this decrease of pH is more or less considerable, i.e., if the buffer capacity of the solution is insufficient, the precipitation of Group II cations will be incomplete. The buffer capacity however increases with the concentration of the components of the buffer (see Sec. 3.9). For example, the buffer capacity of $1N$ ammonium buffer is much larger than that of $0.1N$ ammonium buffer (see Table 3.5), and that of $0.01N$ buffer is still lower. For this reason, even relatively large changes in the concentrations of NH_4OH and NH_4^+ have a very little effect on the pH of $1N$ buffer, but they can bring about a very sharp change in the pH of $0.1N$ or $0.01N$ ammonium buffer.

Thus, it is not enough to introduce into the solution a buffer mixture (e.g., ammonium carbonate solution) having the required pH; it is also necessary that its buffer capacity be sufficient. For this purpose, before Group II cations are precipitated by ammonium carbonate, to the solution is added a solution of ammonia until an odour appears (or until the reaction is alkaline) and then a little NH_4Cl.

Ammonium hydroxide neutralizes free acids, but the solution becomes so alkaline that this may result in the precipitation of Mg^{2+}

Addition of NH_4Cl depresses the pH to the required value (pH $\approx$ 9), which is checked with the aid of indicator paper. At the same time, the buffer capacity of the solution is so increased due to the introduction of ammonia and ammonium chloride that the consumption of NH_4OH and the accumulation of NH_4^+ ions during the precipitation can no longer cause any appreciable change in the solution pH.

The temperature of the solution is another factor that should be mentioned among those influencing the completeness of precipitation of Group II cations. The point is that, when ammonium carbonate is stored in the solid state, it undergoes partial decomposition with the liberation of water and the formation of ammonium carbamate (a salt of carbamic acid, NH_2COOH):

$$(NH_4)_2CO_3 \rightleftharpoons NH_2COONH_4 + H_2O$$

Therefore, commercial ammonium carbonate contains a large amount of carbamate which reduces considerably the concentration of CO_3^{2-} in the solution obtained. This however is easily avoided by precipitating Group II from a solution heated up to 80°C since an increase in the temperature results in the equilibrium of the above reaction being shifted to the left, i.e., towards the conversion of ammonium carbamate into the carbonate. Heating is advantageous for another reason too, namely it speeds up the conversion of the originally amorphous carbonate precipitate into a crystalline one.

Hence, the precipitation of Group II cations by the group reagent should be carried out in the presence of ammonium hydroxide and ammonium chloride (at pH = 9.2) from a solution heated to 80°C. The compounds precipitated are $CaCO_3$, $SrCO_3$, and $BaCO_3$, while Group I cations remain in solution together with the added ammonium salts.

To study the action of the group reagent, to 2 or 3 drops of solutions of $BaCl_2$, $SrCl_2$, and $CaCl_2$ is added a drop each of 2*N* solutions of NH_4OH and NH_4Cl and, after the contents of the test tubes is heated in a water bath, a solution of $(NH_4)_2CO_3$ is introduced. This results in the formation of white precipitates of $BaCO_3$, $SrCO_3$, and $CaCO_3$. For example,

$$CaCl_2 + (NH_4)_2CO_3 \rightarrow CaCO_3\downarrow + 2NH_4Cl$$

Being salts of weak carbonic acid, which decomposes with the evolution of a gas (CO_2), the carbonates of Group II cations dissolve readily in HCl, HNO_3, and CH_3COOH. For example,

$$CaCO_3\downarrow + 2CH_3COOH \rightarrow Ca^{2+} + 2CH_3COO^- + H_2O + CO_2\uparrow$$

4.10. Reactions of the Barium Ion

1. Potassium Dichromate, $K_2Cr_2O_7$. This reagent forms with Ba^{2+} a yellow precipitate of $BaCrO_4$ and not a precipitate of $BaCr_2O_7$ as might be expected. This can be accounted for as follows. Apart from $Cr_2O_7^{2-}$, the solution of $K_2Cr_2O_7$ contains also a small amount of CrO_4^{2-}

which results from the interaction of $Cr_2O_7^{2-}$ with water:

$$Cr_2O_7^{2-} + H_2O \rightleftharpoons 2HCrO_4^- \rightleftharpoons 2H^+ + 2CrO_4^{2-}$$

$$K_{HCrO_4^-} = \frac{[H^+][CrO_4^{2-}]}{[HCrO_4^-]} = 3.2 \times 10^{-7}$$

The concentration of CrO_4^{2-} however is sufficient for the solubility product constant of $BaCrO_4$ to be exceeded before the solubility product constant of $BaCr_2O_7$ is reached. It is for this reason that the compound precipitated is $BaCrO_4$:

$$2CrO_4^{2-} + 2Ba^{2+} \rightarrow 2BaCrO_4 \downarrow$$

Summing up the two equations, we obtain the general equation for the reaction under consideration:

$$Cr_2O_7^{2-} + 2Ba^{2+} + H_2O \rightarrow 2BaCrO_4 \downarrow + 2H^+$$

The $BaCrO_4$ precipitate is soluble in strong acids and insoluble in acetic acid. Since a strong acid is obtained here as the result of the reaction itself, this reaction does not run to completion. Complete precipitation of Ba^{2+} can however be achieved if, apart from $K_2Cr_2O_7$, to the solution is added CH_3COONa to replace the strong acid by a weak one—acetic acid (see Sec. 4.7), in which $BaCrO_4$ does not dissolve:

$$CH_3COO^- + H^+ \rightarrow CH_3COOH$$

Excess CH_3COONa is required so that part of it remains unused, i.e., so that an acetate buffer is formed, which, despite the formation of H^+ ions in the reaction, maintains a constant value of pH (ca. 5) sufficient for complete precipitation of $BaCrO_4$.

The ions Sr^{2+} and Ca^{2+} form no precipitates with $K_2Cr_2O_7$ and do not interfere with the identification of Ba^{2+}. The reaction in question is used not only for identification of Ba^{2+} but for separation of this ion from Ca^{2+} and Sr^{2+} as well.

A precipitate of $BaCrO_4$ is also formed when solutions of barium salts are made to react with potassium chromate. But K_2CrO_4 gives with Sr^{2+} a yellow precipitate of $SrCrO_4$, which differs from $BaCrO_4$ only by being soluble in acetic acid. Therefore, in order to prevent the formation of $SrCrO_4$, the reaction is conducted in the presence of CH_3COOH or preferably in an acetate buffer. The hydrogen ions of acetic acid bind CrO_4^{2-} anions into $HCrO_4^-$, and the concentration of CrO_4^{2-} in solution is decreased to such an extent that the ion product $[Sr^{2+}][CrO_4^{2-}]$ does not reach the value of the solubility product of $SrCrO_4$, and no precipitate is formed. Conversely, the solubility product of less soluble barium chromate is found to be exceeded even in the presence of CH_3COOH which therefore does not interfere with its precipitation.

2. **Sulphuric Acid and Soluble Sulphates (SO_4^{2-} Ions).** These produce with Ba^{2+} a white precipitate of $BaSO_4$:

$$Ba^{2+} + SO_4^{2-} \rightarrow BaSO_4 \downarrow$$

Barium sulphate is a slightly soluble salt of a strong acid and is insoluble in acids; $BaSO_4$ is dissolved by converting it into $BaCO_3$ which is then dissolved in acid:

$$BaSO_4\downarrow + CO_3^{2-} \rightarrow BaCO_3\downarrow + SO_4^{2-}$$
$$BaCO_3\downarrow + 2CH_3COOH \rightarrow Ba^{2+} + 2CH_3COO^- + H_2O + CO_2\uparrow$$

In practice, this conversion is accomplished, for example, by heating $BaSO_4$ with a saturated solution of sodium carbonate (*wet method*).

But since $BaSO_4$ is less soluble than the $BaCO_3$ formed in the reaction, this reaction is reversible (see Sec. 4.6) and can be brought to completion only when the precipitate is repeatedly treated with a solution of Na_2CO_3. In this treatment the liquid (containing the SO_4^{2-} ions formed in the reaction) is decanted from the precipitate and is replaced with a fresh portion of the Na_2CO_3 solution, etc.

Barium sulphate can also be converted into the carbonate by fusing it in a crucible with a five- or sixfold amount of a mixture of Na_2CO_3 and K_2CO_3 (*dry method*). The reason why a mixture of carbonates is used is that it has a lower melting temperature than the corresponding pure compounds. After the mass in the crucible is melted, it is allowed to cool, placed into an evaporating dish, covered with water, and heated. The Na_2SO_4 and K_2SO_4 formed in the reaction dissolve in water, together with excess Na_2CO_3 and K_2CO_3, while $BaCO_3$ remains in solution. The precipitate is separated from the solution, washed thoroughly*, and dissolved in acid.

3. **Ammonium Oxalate, $(NH_4)_2C_2O_4$.** This reagent, which is the ammonium salt of oxalic acid, produces with Ba^{2+} a white precipitate of BaC_2O_4 (barium oxalate):

$$Ba^{2+} + C_2O_4^{2-} \rightarrow BaC_2O_4\downarrow$$

The precipitate is soluble in HCl and HNO_3; when heated, it also dissolves in concentrated CH_3COOH.

4. **Disodium Hydrogen Phosphate, Na_2HPO_4.** This compound produces with Ba^{2+} a white precipitate of $BaHPO_4$ (barium phosphate):

$$Ba^{2+} + HPO_4^{2-} \rightarrow BaHPO_4\downarrow$$

which is soluble in HCl, HNO_3, and CH_3COOH.**

If the reaction is conducted in the presence of alkalis or ammonia solution, HPO_4^{2-} is converted into PO_4^{3-} and a neutral salt is precipitated:

$$HPO_4^{2-} + OH^- \rightarrow PO_4^{3-} + H_2O$$
$$3Ba^{2+} + 2PO_4^{3-} \rightarrow Ba_3(PO_4)_2\downarrow$$

* Otherwise, when the precipitate is dissolved, part of the Ba^{2+} ions will be reprecipitated by the SO_4^{2-} ions that have not been removed by washing.

** That $BaHPO_4$ dissolves in acetic acid is accounted for by the formation of $H_2PO_4^-$ ($K_{H_2PO_4^-} = 6.2 \times 10^{-8}$; see Sec. 3.15).

The action of acids on this precipitate is the same as on $BaHPO_4$.

5. **Flame Test.** Volatile barium salts, for example, $BaCl_2$ and $Ba(NO_3)_2$, impart a yellow-green colour to the colourless flame.

4.11. Reactions of the Strontium Ion

1. **Sulphuric Acid and Soluble Sulphates.** These produce a white precipitate of $SrSO_4$ which is practically insoluble in acids (*Explain why!*); the precipitate can be dissolved in the same way as $BaSO_4$. Because of the low solubility of $SrCO_3$ as compared with that of $SrSO_4$, strontium sulphate is converted into the carbonate much more readily than $BaSO_4$ (see Sec. 4.6).

2. **Gypsum Water.** A saturated aqueous solution of gypsum, $CaSO_4 \cdot 2H_2O$, also forms with Sr^{2+} a precipitate (turbidity) of $SrSO_4$. The explanation for this fact is that, although gypsum is slightly soluble in water, its solubility product constant (2.37×10^{-5}) is considerably greater than that of $SrSO_4$ (3.2×10^{-7}). Therefore, the concentration of SO_4^{2-} in a saturated solution of $CaSO_4$ is found to be sufficient for the solubility product constant of $SrSO_4$ to be exceeded. But since the concentration of SO_4^{2-} is very low, the reaction results in the formation of only a slight turbidity of $SrSO_4$, which appears *not immediately* (the formation of a supersaturated solution). Heating of the solution speeds up the formation of the turbidity.

Since the solubility product constant of $BaSO_4$ (1.1×10^{-10}) is lower than that of $SrSO_4$, barium salts give with gypsum water a turbidity of $BaSO_4$, which is formed *at once*. Hence, the reaction with $CaSO_4$ can be used for detection of Sr^{2+} only after the separation of Ba^{2+}.

3. **Disodium Hydrogen Phosphate, Na_2HPO_4.** This reagent produces with Sr^{2+} a white precipitate of strontium hydrogen phosphate:

$$Sr^{2+} + HPO_4^{2-} \rightarrow SrHPO_4\downarrow$$

which is soluble in acids (including acetic acid).

4. **Ammonium Oxalate, $(NH_4)_2C_2O_4$.** The reagent precipitates Sr^{2+} in the form of a white precipitate of SrC_2O_4:

$$Sr^{2+} + C_2O_4^{2-} \rightarrow SrC_2O_4\downarrow$$

which is soluble in mineral acids and when heated it also dissolves in concentrated acetic acid.

5. **Identification of Sr^{2+} and Ba^{2+} by Spot Test with Sodium Rhodizonate*, $Na_2C_6O_6$.** A drop of a neutral solution of a strontium salt

* The structural formula of sodium rhodizonate is

```
              O
              ‖
NaO — C — C — CO
      ‖   |
NaO — C — C — CO
          ‖
          O
```

is placed on filter paper and a drop of an aqueous solution of sodium rhodizonate is added to it. A red-brown spot of precipitated strontium rhodizonate, SrC_6O_6, appears on the paper. When the spot is moistened with a drop of dilute HCl, the precipitate dissolves and the spot disappears.

With rhodizonate the Ba^{2+} ion produces a similar red-brown precipitate of barium rhodizonate, BaC_6O_6, which however can easily be distinguished from strontium rhodizonate because HCl turns it pink-red due to conversion into acid barium rhodizonate.

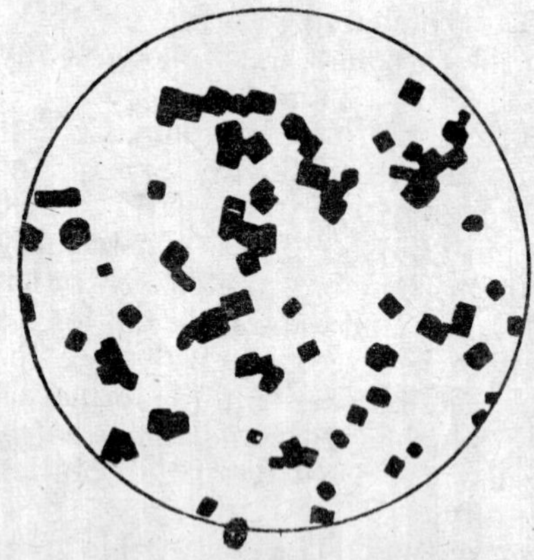

Fig. 4.4. Crystals of $K_2SrCu(NO_2)_6$.

The reaction is very convenient for fractional detection of Sr^{2+} and Ba^{2+} in the presence of all the cations of Groups II and I (no cations of other groups must be present).

Place a drop of the neutral solution to be tested on a strip of paper and moisten it with a drop of a rhodizonate solution. Apply a drop of the rhodizonate solution next to it and compare the colours. If they are the same, neither Ba^{2+} nor Sr^{2+} is present in solution. If the spot is turned red-brown, moisten it with a drop of 0.5*N* HCl solution.* If the spot becomes colourless, it may be concluded that Ba^{2+} is absent and Sr^{2+} is present. But if the spot turns red (or, more precisely, pink), this is an indication that Ba^{2+} is present. To detect Sr^{2+} in this case, the following test is performed. Moisten a strip of paper with a solution of K_2CrO_4 and place a drop of the unknown solution in the centre of the spot. This precipitates Ba^{2+} as $BaCrO_4$ which does not react with the rhodizonate, whereas strontium chromate reacts with the rhodizonate because of the higher solubility.

To make certain that the precipitation is complete, moisten the spot once again with a solution of K_2CrO_4 and treat it with a drop of the rhodizonate. In the presence of Sr^{2+}, the spot readily assumes a reddish-brown colour; if Sr^{2+} is absent, the spot turns orange (the colour of the control spot produced when a drop of the reagent is placed on the paper).

---

* It should be kept in mind that at small amounts of Ba^{2+} the spot may disappear because of the larger amount or higher concentration of HCl, as a result of which the Ba^{2+} ion will be missed. In dubious cases it is useful to carry out a parallel reaction for Ba^{2+} with a solution of K_2CrO_4.

6. Microcrystalloscopic Reaction. Evaporate a drop of the test solution of a strontium salt to dryness on a glass slide. Dissolve the dry residue in a drop of a 2-percent solution of $Cu(NO_3)_2$ and evaporate to dryness again. After it is cooled, add 2 drops of a 0.05*N* solution of CH_3COOH and introduce small crystals of KNO_2 into the resultant clear solution. After the lapse of several minutes (or sooner with gentle warming), small blue-green cubes of the triple nitrite $K_2SrCu(NO_2)_6$ are formed along the edges (Fig. 4.4). The dilution limit is 1 in 10,000.

The ions Ba^{2+} and Ca^{2+} produce green crystals whose shape differs from that of crystals formed by the strontium salt.

7. Flame Test. Volatile strontium salts impart a characteristic crimson colour to the colourless flame.

4.12. Reactions of the Calcium Ion

1. Ammonium Oxalate, $(NH_4)_2C_2O_4$. This compound forms with Ca^{2+} a white precipitate of calcium oxalate:

$$Ca^{2+} + C_2O_4^{2-} \rightarrow CaC_2O_4 \downarrow$$

which is soluble in mineral acids but insoluble in acetic acid.

This most important qualitative reaction of Ca^{2+} is interfered with by the presence of Ba^{2+} and Sr^{2+} which produce analogous precipitates with $(NH_4)_2C_2O_4$.

2. Soluble Sulphates (SO_4^{2-} ions). These sulphates produce a white precipitate of $CaSO_4$ only in relatively concentrated solutions of calcium salts:

$$Ca^{2+} + SO_4^{2-} \rightarrow CaSO_4 \downarrow$$

The precipitate is soluble in $(NH_4)_2SO_4$ because of the formation of a complex salt, $(NH_4)_2[Ca(SO_4)_2]$. Gypsum water gives no turbidity with solutions of calcium salts (in contrast to Ba^{2+} and Sr^{2+}).

3. Potassium Ferrocyanide, $K_4[Fe(CN)_6]$. This compound, which is also known as potassium hexacyanoferrate(II), makes it possible to identify Ca^{2+} in the presence of Sr^{2+} and also to separate these cations.

The reaction may be represented by the equation

$$Ca^{2+} + 2NH_4^+ + [Fe(CN)_6]^{4-} \rightarrow Ca(NH_4)_2[Fe(CN)_6] \downarrow$$

The reaction is carried out at pH ≈ 9, i.e., in the presence of ammonium buffer. Therefore, in carrying out this reaction, to 1 or 2 drops of the unknown solution add one drop each of solutions of NH_4Cl and NH_4OH, heat, and add an approximately equal volume of a saturated solution of $K_4[Fe(CN)_6]$ for precipitation to be effected. A white crystalline precipitate of $Ca(NH_4)_2[Fe(CN)_6]$ is formed, which is insoluble in acetic acid. This enables one to distinguish it from $SrCO_3$ which may form in the presence of $(NH_4)_2CO_3$ contained in ammonia used

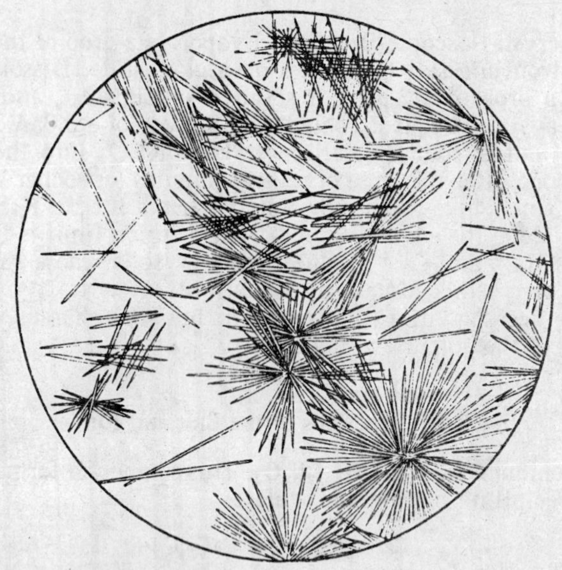

Fig. 4.5. Crystals of $CaSO_4 \cdot 2H_2O$ formed from dilute solutions.

for the reaction. The Ba^{2+} ion when present in a considerable concentration can also produce a precipitate with $K_4[Fe(CN)_6]$. That is why this reaction should not be used for identification of Ca^{2+} in the presence of Ba^{2+}.

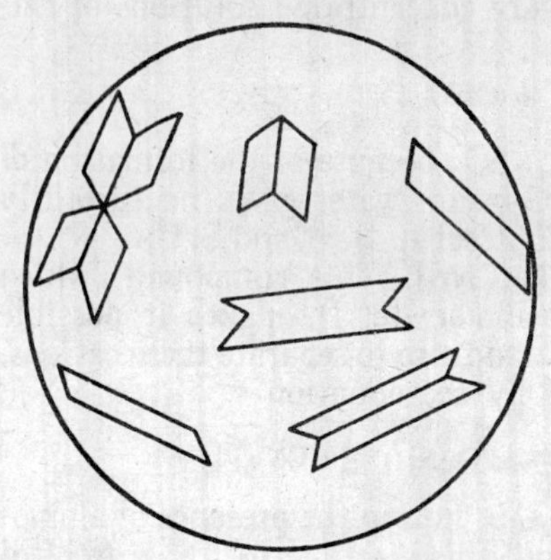

Fig. 4.6. Crystals of $CaSO_4 \cdot 2H_2O$ formed from concentrated solutions.

4. Disodium Hydrogen Phosphate, Na_2HPO_4. This phosphate produces a white precipitate of $CaHPO_4$:

$$Ca^{2+} + HPO_4^{2-} \rightarrow CaHPO_4 \downarrow$$

which is soluble in acids, including acetic acid.

5. Microcrystalloscopic Reaction. Place a drop of a calcium salt solution on a glass slide and add to it a drop of $2N$ H_2SO_4 solution. Evaporate it slightly until a fringe appears around the drop. The resulting crystals of gypsum, $CaSO_4 \cdot 2H_2O$, assume various shapes (see Figs. 4.5 and 4.6), depending on the concentration of Ca^{2+} in solution. The dilution limit is 1 in 25,000. This reaction can generally be used to identify Ca^{2+} in the presence of Ba^{2+} and Sr^{2+} since barium and strontium sulphates are precipitated in the form of small crystals which can easily be distinguished from characteristic large crystals of gypsum.

6. Flame Test. Volatile calcium salts impart a brick-red colour to the flame.

4.13. Analysis of Mixtures of Group II and I Cations

The effects of the most important reagents on Group II cations are compared in Table 4.2. The results of the action of the most important reagents on Group II cations are given in Table 4.2. As seen from the table, the carbonates, phosphates, oxalates, and sulphates of Group II cations are slightly soluble in water. Slightly soluble among the chromates are $BaCrO_4$ and $SrCrO_4$; $CaCrO_4$ dissolves well. The solubility of the sulphates increases in the following sequence: Ba, Sr, Ca.

TABLE 4.2. **Reactions of Group II Cations**

Reagent	Cation		
	Ba^{2+}	Sr^{2+}	Ca^{2+}
$(NH_4)_2CO_3$	White precipitate: $\mathbf{BaCO_3}$	White precipitate: $\mathbf{SrCO_3}$	White precipitate: $\mathbf{CaCO_3}$
Na_2CO_3 and K_2CO_3	$BaCO_3$	$SrCO_3$	$CaCO_3$
Na_2HPO_4	White precipitate: $BaHPO_4$	White precipitate: $SrHPO_4$	White precipitate: $CaHPO_4$
H_2SO_4 and soluble sulphates	White precipitate: $\mathbf{BaSO_4}$	White precipitate: $\mathbf{SrSO_4}$	White precipitate*: $\mathbf{CaSO_4}$
$CaSO_4$	White precipitate: $\mathbf{BaSO_4}$, formed at once	White precipitate: $\mathbf{SrSO_4}$, formation delayed	—
$(NH_4)_2C_2O_4$	White precipitate: BaC_2O_4	White precipitate: SrC_2O_4	White precipitate: $\mathbf{CaC_2O_4}$
K_2CrO_4	Yellow precipitate: $BaCrO_4$	Yellow precipitate: $SrCrO_4$	—
$K_2Cr_2O_7 + CH_3COONa$ or $K_2CrO_4 + CH_3COOH$	$\mathbf{BaCrO_4}$	—	—
$K_4[Fe(CN)_6] + NH_4OH + NH_4Cl$	—**	—	White precipitate: $Ca(NH_4)_2[Fe(CN)_6]$
Flame colouration	Yellow-green	Crimson-red	Brick-red

* The precipitate separates out only from not too dilute solutions.
** A precipitate may form at a high concentration of Ba^{2+}.

Based on the reactions studied, one may outline several different methods for analysis of mixtures of Group II cations; here we will consider the simplest one.

Apart from a description of the course of analysis, schemes are also given (see Table 4.3) for the operations to be carried out during analysis and the separations of ions involved.

The student must not, of course, work from the scheme alone, without having thoroughly studied the respective material in the description of the analytical procedure. This would inevitably lead to errors since the scheme does not specify the conditions and techniques for carrying out individual operations. Moreover, the individual operations are explained in the description; if they are not understood, the work becomes mechanical.

The description of the analytical procedure and the scheme in Table 4.3 allow for the possible presence of Group I cations together with Group II cations in the unknown solution. The solution may also contain a precipitate in which may be present compounds soluble in acids (such as Group II carbonates, oxalates, and phosphates, magnesium hydroxide, etc.) and also the acid-insoluble sulphates of barium, strontium, and calcium.

(a) Procedure for Analysis of a Mixture of Group II and I Cations

1. Preliminary Tests. The analysis is started with preliminary test with separate portions of the unknown solution.

(1) *Determination of the solution pH* with the aid of a range of indicators or a universal indicator (see Sec. 3.10).

(2) *Identification of NH_4^+* by the action of alkali in a gas chamber or by means of Nessler's reagent (see Sec. 3.14). Let us recall that NH_4^+ will be introduced later into the solution with the group reagent and therefore it must be detected preliminarily in a separate portion of the solution.

Note. If Mg^{2+} is present in the solution, $Mg(OH)_2$ is precipitated by the action of alkali. The same effect is shown by Nessler's reagent which contains alkali. Moreover, if NH_4^+ is being detected by both reactions, the carbonates of Group II cations are precipitated since alkalis (NaOH and KOH) usually contain Na_2CO_3 or K_2CO_3 as impurity (due to absorption of CO_2 from the air). The detection of NH_4^+ is not interfered with by the formation of these precipitates.

(3) *Test for Group II sulphates* (to be carried out only in the presence of a precipitate)*. Shake up the precipitate, transfer 2 or 3 drops of the turbid liquid into another test tube, and add 3 or 4 drops of 2*N* HCl

---

* If no precipitate is present, the analysis (after NH_4^+ has been detected) is carried out according to Procedure *4* and the succeeding ones.

solution. If the precipitate does not dissolve, it means that the sulphates of Group II cations are present.

2. Dissolution of the Precipitate and Separation of Group II Sulphates. (1) *In the absence of sulphates*, place 12-15 drops of the test solution together with the precipitate into a conical test tube and dissolve the precipitate in as small a volume of $2N$ HCl solution as possible with thorough stirring and heating.

(2) *In the presence of sulphates*, to 12-15 drops of the unknown solution add 4 or 5 drops of $2N$ H_2SO_4 solution and, after heating for 2-3 minutes in a water bath, centrifuge and wash the undissolved precipitate. Sulphuric acid is added to achieve a dual purpose: to dissolve all the precipitates soluble in acid and, on the other hand, to postprecipitate the sulphates of barium, strontium and, partly, calcium. A large excess of sulphuric acid should, of course, be avoided to prevent the dissolution of the sulphates with the formation of hydrosulphates.

At a sufficiently large amount of SO_4^{2-} in solution, the ions Ba^{2+} and Sr^{2+} are precipitated and Ca^{2+} is always left partly in solution because of the considerable solubility of $CaSO_4$. For this reason, in the analysis of the sulphate precipitate, Ca^{2+} should not be sought for in it.

3. Investigation of the Sulphate Precipitate. To dissolve the sulphates of barium, strontium, and calcium, they must first be converted into the corresponding carbonates by the action of Na_2CO_3. To do this, heat the precipitate for a few minutes in a boiling water bath (or, what is better, boil in an evaporating dish or crucible) with 20 drops of a saturated solution of Na_2CO_3, stirring thoroughly from time to time. Centrifuge, discard the centrifugate, and treat the precipitate analogously with Na_2CO_3. Repeat the operation 2 or 3 times, wash the precipitate thoroughly (2 times) with hot distilled water, and dissolve on heating in 3-5 drops of $2N$ CH_3COOH solution.

The precipitate must be washed thoroughly to remove SO_4^{2-}. If this is not done, on further treatment of the precipitate with acetic acid, part of the cations will be reprecipitated as sulphates.

If the precipitate does not dissolve completely in CH_3COOH, the undissolved residue (part of $BaSO_4$) is centrifuged and discarded since, under the conditions specified above, the amount of Ba^{2+} that will inevitably pass into solution is quite sufficient for identification.

A more complete conversion of $BaSO_4$ into $BaCO_3$ is achieved by fusing the precipitate (which must be preliminarily dried carefully in a crucible or an evaporating dish) with a mixture of Na_2CO_3 and K_2CO_3. The fusion can conveniently be done in the loop of a platinum or nichrome wire. The bead resulting from the fusion is put (together with the wire) into a test tube with water and heated in water bath until the fused mass is converted into fine powder which (after it is separated from the solution by centrifuging and washed) is examined as described below.

The centrifugate is tested for the presence of Ba^{2+} and Sr^{2+} as described in Procedures *7-9*.

Note. If it is not required to establish which of the sulphates of Group II were present in the precipitate, the centrifugate can more conveniently be mixed with the solution obtained according to Procedure *6* instead of being examined separately.

4. Test for the Presence of Group II Cations. Before starting to analyse the centrifugate (see Procedure *2*), or the original solution in the absence of sulphates, one has to find out whether there are Group II cations in it. A test for their presence is also necessary when the unknown solution contains a precipitate of sulphates since there may be calcium ions in solution. It is expedient to perform this test not with all the solution but with a separate portion of it. To 2 drops of the solution add 2*N* NH_4OH solution until the reaction is alkaline and a drop of NH_4Cl solution, mix with 2 or 3 drops of a solution of $(NH_4)_2CO_3$, and heat. The formation of a precipitate indicates the presence of Group II cations. In this case, further investigation is carried out according to Procedure *5*. If Group II is absent, the solution is investigated as described in Sec. 3.16.

5. Separation of Group II Cations from Group I Cations. First of all it is necessary to separate Group II cations from the cations of Group I because the ions Ca^{2+}, Sr^{2+}, and Ba^{2+} form precipitates with some of the reagents used in the analysis of Group I (for example, with Na_2HPO_4, $NaHC_4H_4O_6$, etc.).

Adjust the reaction of the test solution to pH ≈ 9 using universal indicator or phenolphthalein paper (it must turn pink but not bright-red in colour). To do this, add to the solution a few drops each of NH_4OH and NH_4Cl solutions* and determine pH. If pH < 9, add more NH_4OH. If pH > 9, NH_4Cl or HCl must be added. When the required medium is produced, the solution is heated and 8-10 drops of $(NH_4)_2CO_3$ solution are added to it. Then the contents of the test tube is stirred thoroughly and heated for 1-2 minutes in a water bath. The precipitate is centrifuged and, without being separated from the solution, is tested for completeness of precipitation by adding 1 drop of $(NH_4)_2CO_3$ solution.

When the cations of Group II are completely precipitated, the precipitate is separated from the solution and washed once with hot water**.

---

* If the solution is strongly acidic and for the acid to be neutralized, one has to add a few drops of an ammoniacal solution; no ammonium chloride should be added before the test for the pH of the solution since it results from the neutralization reaction.

** If the precipitation of Group II cations is carried out after Group III has been precipitated by its group reagent and $(NH_4)_2CO_3$ gives no precipitate, it is expedient to perform the test described on page 207 before a decision is made as to the absence of Group II cations.

The precipitate of $CaCO_3$, $SrCO_3$, and $BaCO_3$ is examined in accordance with Procedure *6*; the centrifugate containing Group I cations, excess $(NH_4)_2CO_3$, and other compounds of ammonium is investigated as described in Sec. 3.16.

6. Dissolving the Precipitated Group II Carbonates. Dissolve the precipitated carbonates ($BaCO_3$, $SrCO_3$, and $CaCO_3$) on heating in 8-10 drops of $2N$ CH_3COOH solution. If the resulting solution is slightly turbid*, centrifuge it and discard the precipitate; then transfer the absolutely clear centrifugate into another test tube and examine it in accordance with Procedure 7.

7. Detection and Separation of Ba^{2+}. First it is necessary to detect Ba^{2+} in the resultant solution by the action of $K_2Cr_2O_7$ (or K_2CrO_4) and, since Ba^{2+} interferes with identification of Sr^{2+} and Ca^{2+}, it must be eliminated from the solution. To this end, to a drop of the solution add a drop each of CH_3COONa and $K_2Cr_2O_7$ solutions. The formation of a yellow precipitate of $BaCrO_4$ indicates the presence of Ba^{2+}.

If the Ba^{2+} ion has been identified, add CH_3COONa to the solution to bring the pH up to 4-5, following which add a solution of $K_2Cr_2O_7$, drop by drop, until the supernatant assumes an orange-yellow colour. The formation of this colour points to the presence of an excess of $Cr_2O_7^{2-}$ in the solution and hence to the completeness of precipitation of Ba^{2+}.

After the $BaCrO_4$ precipitate is heated for 1-2 minutes, it is separated by centrifuging and discarded.

The centrifugate is investigated in different ways, depending on the presence (or absence) of Sr^{2+} in the solution.

If Sr^{2+} is likely to be present, examine the centrifugate in accordance with Procedures *8-10*. If it is absent (the test for the presence of Sr^{2+} is carried out according to Procedure *9*); the investigation is accomplished as directed in Procedure *11*.

8. Precipitation of Sr^{2+} and Ca^{2+}. First it is necessary to remove the excess of potassium dichromate added for the precipitation of Ba^{2+} because it has a bright colour and will interfere, in subsequent analysis, with detection of Sr^{2+}. For this purpose, add solid Na_2CO_3 to the solution until the reaction is alkaline (pH > 10) and heat the contents of the test tube for 2-3 minutes in a water bath, after which separate the precipitated $SrCO_3$ and $CaCO_3$ by centrifuging, wash it, and investigate as described below.

The centrifugate, which contains chromates and excess Na_2CO_3, is not examined.

9. Detection and Separation of Sr^{2+}. Dissolve the precipitated $SrCO_3$ and $CaCO_3$ in 4-6 drops of $2N$ CH_3COOH and, if the solution is not absolutely transparent, separate the precipitate (turbidity) by centrifuging and discard.

---

* A turbidity may form if CH_3COOH contains an impurity of sulphates or H_2SO_4. This turbidity is usually either $BaSO_4$ or $SrSO_4$.

To 1 or 2 drops of the centrifugate add 2 or 3 drops of a saturated solution of $CaSO_4$ and heat in a water bath. The appearance of a white turbidity of $SrSO_4$ after some time indicates the presence of Sr^{2+}.

If no turbidity is present, heat for another 10 minutes and only after this may a conclusion be made concerning the presence or absence of Sr^{2+}. If a turbidity appears at once upon the action of $CaSO_4$, this is an indication that the separation of Ba^{2+} is incomplete. Whether Sr^{2+} is present or not remains unsolved in this case.

If Sr^{2+} is present, it is precipitated from the entire solution by adding an excess (8-10 drops) of a solution of $(NH_4)_2SO_4$ and a few drops of NH_4OH and heating for 10-15 minutes. The ion Sr^{2+} is precipitated as $SrSO_4$, while Ca^{2+}, whose sulphate has a much higher solubility product constant, remains largely in solution. This is also favoured by the formation of a complex compound with $CaSO_4$ under the action of $(NH_4)_2SO_4$ (see Sec. 4.12).

After the precipitate is separated by centrifuging and washed, confirmatory tests for Sr^{2+} may be carried out. The centrifugate is to be examined as described in Procedure *10*.

Note. To confirm the presence of Sr^{2+} in the precipitate by a flame test, the precipitate is first separated from the small amount of $CaSO_4$ present as impurity by washing it off twice with $(NH_4)_2SO_4$ solution on heating and stirring. Then the $SrSO_4$ is converted by heating with a solution of Na_2CO_3 into $SrCO_3$, which is separated from the liquid, washed, and dissolved in 1-2 drops of HCl solution. The resulting $SrCl_2$ solution is investigated by means of a flame test. The crimson-red colour of the flame proves the presence of Sr^{2+}. Alternatively, the $SrCO_3$ may be dissolved in CH_3COOH and a drop of the solution tested for Sr^{2+} by a microcrystalloscopic reaction (see Sec. 4.11).

10. Identification of Ca^{2+}. After the $SrSO_4$ is separated, add to the centrifugate 4 or 5 drops of a solution of $(NH_4)_2C_2O_4$ and heat. A white precipitate of CaC_2O_4 indicates the presence of Ca^{2+}.

To confirm the test, the precipitate is centrifuged and dissolved in as small a volume of 2*N* HCl as possible. A drop of the unknown solution is used for carrying out a microcrystalloscopic reaction for Ca^{2+} with H_2SO_4 (see Sec. 4.12). This reaction usually makes it possible to identify Ca^{2+} in the presence of Ba^{2+} and Sr^{2+} since the crystals of gypsum, $CaSO_4 \cdot 2H_2O$, have a very characteristic shape, while $BaSO_4$ and $SrSO_4$ produce much smaller crystals.

When use is made of the reaction of Ca^{2+} with $K_4[Fe(CN)_6]$, which is not interfered with by Sr^{2+}, the separation of strontium may be dispensed with. It is only necessary to test the precipitate settled down by the action of $K_4[Fe(CN)_6]$ for solubility in CH_3COOH, otherwise it would be easy to detect Ca^{2+} in error. The Ba^{2+} ions when present in a sufficient concentration can also give a precipitate with $K_4[Fe(CN)_6]$. In the presence of Ba^{2+}, this reaction should not be used for identification of Ca^{2+} (see Sec. 4.12).

A drop of the resulting solution is used for carrying out a microcrystalloscopic reaction for Ca^{2+} (as described in Sec. 4.12). The formation of characteristic crystals of gypsum (see Figs. 4.5 and 4.6) is proof of the presence of Ca^{2+} ions.

(b) Detection of Group II Cations in Separate Portions of Solution (Fractional Method)

Considerable losses of Group II cations occur during the systematic analysis of cations. There are several factors responsible for these losses. First, when the cations of Group III are precipitated by the group reagent $(NH_4)_2S$ in the presence of NH_4OH, the cations of Group II may precipitate as carbonates since these reagents usually contain $(NH_4)_2CO_3$ as impurity as a result of the absorption of CO_2 from the air. Second, Ba^{2+} and Sr^{2+} may also separate out due to the oxidation of S^{2-} to SO_3^{2-}, $S_2O_3^{2-}$, and SO_4^{2-} by the air oxygen. And, finally, there takes place the coprecipitation of Group II cations with the hydroxides of the triply charged cations of Group III.

If the carbonates of Group II cations are not precipitated by the action of $(NH_4)_2CO_3$ in the systematic course of analysis, one has to take into account what has been said above and try to detect Group II cations in separate portions of the original solution. Use may be made of the following procedure.

1. Identification of Ca^{2+}. To 2 drops of the solution under test with the precipitate (if there is one) being in suspension, add 1 drop of 4*N* H_2SO_4 solution and 3 drops of ethyl alcohol (or 6 drops of methyl alcohol) to reduce the solubility of $CaSO_4$. In order to speed up the precipitation of $CaSO_4$, scratch the sides of the test tube with a glass rod. After 3-5 minutes separate the precipitate by centrifuging, test it for completeness of precipitation, and separate from the solution. Shake the precipitate for some time with 5 or 6 drops of water (a considerable amount of $CaSO_4$ will be dissolved). Centrifuge the undissolved residue, evaporate a drop of the centrifugate slightly on a glass slide, and examine the crystals formed around the edges under the microscope. If Ca^{2+} is present, one can see needle-shaped crystals of gypsum, $CaSO_4 \cdot 2H_2O$, usually in clusters (see Fig. 4.5).*

2. Identification of Ba^{2+} and Sr^{2+}. These ions can be detected in different ways.

(1) Take 15-20 drops of the unknown solution and add HCl to precipitate the cations of Group V (see Sec. 6.19, Procedure *2*). To the cen-

---

* Since the solubility product constant of CaC_2O_4 (2.3×10^{-9}) is less than that of $CaSO_4$ (2.37×10^{-5}), $(NH_4)_2C_2O_4$ may be added to the solution left. After some time (sooner on stirring with a glass rod) the solution will become turbid because of the precipitation of CaC_2O_4. The turbidity does not dissolve when a drop of concentrated acetic acid is added but it disappears under the action of HCl.

TABLE 4.3. **Scheme of Analysis of Mixtures of Group II and I Cations**

<table>
<tr><td colspan="4">1. Preliminary tests: (a) determine the pH of the solution; (b) identify NH_4^+; (c) test for sulphates of Group II cations.</td></tr>
<tr><td colspan="4">2. Dissolve the precipitate in HCl or separate sulphates of Group II cations by the action of H_2SO_4.</td></tr>
<tr><td rowspan="4">Precipitate 2a: $BaSO_4$, $SrSO_4$ ($CaSO_4$).
3. Convert sulphates into carbonates by the action of Na_2CO_3, wash, dissolve in CH_3COOH. Add the liquid to the solution obtained in Procedure 6, or test separately for Ba^{2+}, Sr^{2+}, and Ca^{2+} (see Procedures 7-10).</td><td colspan="3">Centrifugate 2a: cations of Groups II and I (HCl, H_2SO_4).
4. Test for Group II cations by the action of $(NH_4)_2CO_3$ in the presence of NH_4OH and NH_4Cl at pH $\approx$ 9.</td></tr>
<tr><td colspan="3">5. Separate Group II by the action of the same reagents from the whole solution.</td></tr>
<tr><td colspan="2">Precipitate 5a: $BaCO_3$, $SrCO_3$, $CaCO_3$.
6. Dissolve in CH_3COOH.
7. Detect Ba^{2+} and separate it by the action of $K_2Cr_2O_7$ in the presence of CH_3COONa (at pH $\approx$ 5).</td><td rowspan="2">Centrifugate 5a: Mg^{2+}, K^+, Na^+, excess $(NH_4)_2CO_3$ and other compounds of ammonium. Investigate as for Group I (see Sec. 3.16).</td></tr>
<tr><td>Precipitate 7a: $BaCrO_4$. Not to be investigated.</td><td>Centrifugate 7a: Sr^{2+}, Ca^{2+} ($Cr_2O_7^{2-}$).
8. Precipitate Ca^{2+} and Sr^{2+} by the action of excess Na_2CO_3 (at pH $>$ 10).</td></tr>
<tr><td colspan="2">Precipitate 8a: $SrCO_3$, $CaCO_3$.
9. Dissolve in CH_3COOH, detect Sr^{2+} by the action of $CaSO_4$, and separate it by the action of excess $(NH_4)_2SO_4$.</td><td colspan="2">Centrifugate 8a: K_2CrO_4, Na_2CO_3. Not to be investigated.</td></tr>
<tr><td colspan="2">Precipitate 9a: $SrSO_4$. Confirmatory test for Sr^{2+}.</td><td colspan="2">Centrifugate 9a: Ca^{2+} and excess $(NH_4)_2SO_4$.
10. Identify Ca^{2+} by the action of $(NH_4)_2C_2O_4$, dissolve in HCl, and perform a confirmatory test for Ca^{2+}.</td></tr>
</table>

trifugate obtained after the separation, add H_2SO_4. This will precipitate $BaSO_4$, $SrSO_4$, $PbSO_4$, and partly $CaSO_4$ which may be present in the original sample; in this case the precipitate is examined as indicated below.

When you have made certain that the precipitation is complete, separate the precipitate by centrifuging, wash, and heat it with 10-15 drops of 6*N* NaOH solution. The $PbSO_4$, which forms the plumbite Na_2PbO_2, passes into solution. When the undissolved residue has been centrifuged off and washed, convert barium and strontium sulphates into the carbonates as described in Procedure *3* and, with the resulting precipitate of $BaCO_3$, $SrCO_3$, and, possibly, $CaCO_3$ being dissolved in 3-5 drops of 2*N* acetic acid, test the solution for Ba^{2+} and Sr^{2+} in the usual way (Table 4.3).

(2) If the original sample has sulphate ions, then the ions Ba^{2+}, Sr^{2+}, Pb^{2+}, and, partly, Ca^{2+} are present in the precipitated sulphates. In this case, one has first of all to add 2*N* H_2SO_4 until the precipitation of the sulphates of Ba^{2+}, Sr^{2+}, and Pb^{2+} is complete (most of Ca^{2+} will remain in solution). Thus, the precipitate will contain the chlorides of Group V and the sulphates of Ba^{2+}, Sr^{2+}, and Pb^{2+}. The full analysis of this precipitate is described in the systematic procedure for all the cations of Groups I-V.

Chapter 5

Group III Cations

A. THEORETICAL PRINCIPLES

5.1. Precipitation of Sulphides

In contrast to the sulphides of the cations of Groups I and II, which are well soluble in water, the cations of Groups III, IV, and V form, under the action of sulphide ions, compounds which are practically insoluble in water. On this difference is based the classification of cations presented in Table 1.2 (see Sec. 1.5). Hence, in the systematic analysis, the analytical Groups III and IV are separated from Groups I and II by the action of sulphide ions.* But sulphide ions are the anions of very weak hydrosulphuric acid, H_2S, which dissociates according to the equations

$$H_2S \rightleftharpoons H^+ + HS^-$$

$$HS^- \rightleftharpoons H^+ + S^{2-}$$

Since H_2S is a weak acid, the S^{2-} ions introduced into the solution are tied up by the hydrogen ions present to form HS^- ions and undissociated H_2S molecules. As a result, the concentration of S^{2-} very strongly depends on the hydrogen ion concentration, i.e., on the pH of the solution. The relation between these concentrations can easily be established by proceeding from the equations for the two ionization constants of H_2S:

$$K_{H_2S} = \frac{[H^+][HS^-]}{[H_2S]} = 8.9 \times 10^{-8}; \; pK_{H_2S} = 7.05$$

$$K_{HS^-} = \frac{[H^+][S^{2-}]}{[HS^-]} = 1.3 \times 10^{-13}; \; pK_{HS^-} = 12.89$$

Multiplying the two equations termwise and cancelling $[HS^-]$ in the numerator and denominator, we get:

$$K_{H_2S\,(overall)} = \frac{[H^+]^2[S^{2-}]}{[H_2S]} = 1.2 \times 10^{-20} \qquad (5.1)$$

---

* Group V is preliminarily separated as the chlorides AgCl, Hg_2Cl_2, and $PbCl_2$ (because of the relatively high solubility $PbCl_2$ is precipitated incompletely).

and

$$[S^{2-}] = \frac{1.2 \times 10^{-20}[H_2S]}{[H^+]^2} \qquad (5.2)$$

Equation (5.2) shows that the concentration of S^{2-} is inversely proportional to the square of the concentration of hydrogen ions in solution. Thus, if the pH decreases by one unit, i.e., if the $[H^+]$ increases tenfold, the S^{2-} concentration will decrease by a factor of 100. From this it follows, for example, that the S^{2-} concentration in a solution with $[H^+] = 10^{-1}$, i.e., with pH = 1, must be $(10^8)^2$, i.e., 10^{16} times lower than in a solution with $[H^+] = 10^{-9}$, i.e., with pH = 9.

The relation between the concentration of S^{2-}, the concentration of H^+, and the initial concentration of hydrogen sulphide, C_S, is given by formula (3.13) in Sec. 3.6:

$$[S^{2-}] = C_S \left[\frac{1}{1 + \frac{[H^+]}{K_{HS^-}} + \frac{[H^+]^2}{K_{H_2S}K_{HS^-}}} \right]$$

or

$$[S^{2-}] = \frac{K_{HS^-}K_{H_2S}C_S}{[H^+]^2 + K_{H_2S}[H^+] + K_{HS^-}K_{H_2S}} \qquad (5.3)$$

If this is so, the precipitation of a sulphide and also the completeness of its precipitation must largely depend on the pH of the solution.

For example, at small values of pH, i.e., in more or less strongly acid solutions, only those sulphides can be precipitated which have very small values of solubility product constant since the very minute concentrations of S^{2-} ions that are present in acid solutions are enough for their solubility product constants to be exceeded. Conversely, in neutral and alkaline solutions, not only these least soluble sulphides but also those with higher solubility product constants will be precipitated. It is clear from this that, by regulating the pH of the solution in the precipitation by S^{2-} ions, one can achieve a separation of cations.

It is this principle on which is based the separation of Group IV from Group III and the precipitation of Group III in systematic analysis.

Let us compare the solubility product constants of some sulphides of Group III and IV cations:

	Group III	Group IV
Most soluble sulphides . . .	$SP^*_{MnS} = 2.5 \times 10^{-10}$	$SP_{CdS} = 7.9 \times 10^{-27}$
Least soluble sulphides . .	$SP^{**}_{ZnS} = 2.5 \times 10^{-22}$	$SP_{HgS} = 1.6 \times 10^{-52}$

* What is meant here is a modification of MnS having a pink colour. There is also a green modification of MnS ($SP = 2.5 \times 10^{-13}$).

** $SP_{ZnS} = 2.5 \times 10^{-22}$ corresponds to β-ZnS (wurtzite), in the form of which ZnS begins to precipitate. On standing zinc sulphide changes into a less soluble form, α-ZnS (sphalerite), whose solubility product constant is 1.6×10^{-24}.

From the data given it is seen that the sulphides of Group IV cations are much less soluble than those of Group III cations and therefore they can be precipitated at lower values of pH, i.e., in more acid solutions.

Let us now pass over from qualitative conclusions to quantitative calculations of the pH values required for separation of Groups III and IV from each other and from the cations of Groups II and I.

First we calculate the pH at which the sulphides of Group IV cations are completely precipitated. To do this, it is sufficient to find the pH value needed for complete precipitation of CdS since, if it is precipitated sufficiently completely, all the other sulphides of Group IV cations, which are less soluble, will be precipitated even more completely at that pH. If hydrogen sulphide is passed into the solution for precipitation to take place, the concentration of H_2S will then be equal to its solubility, i.e., to about 1×10^{-1} mole/litre at room temperature.

Thus, using Eq. (5.1), we get:

$$[H^+]^2[S^{2-}] = 1.2 \times 10^{-20}[H_2S] = 1.2 \times 10^{-21} \tag{5.4}$$

Solving Eq. (5.4) for $[H^+]$, we obtain:

$$[H^+] = \sqrt{\frac{1.2 \times 10^{-21}}{[S^{2-}]}} \tag{5.5}$$

On the other hand, if, as previously, the precipitation is assumed to be complete when the concentration of the precipitated ion Cd^{2+} in solution has decreased to 10^{-6} g-ion/litre, we may write:

$$[S^{2-}] = \frac{SP_{CdS}}{[Cd^{2+}]} = \frac{7.9 \times 10^{-27}}{10^{-6}} = 7.9 \times 10^{-21} \text{ g-ion/litre}$$

Substituting the value obtained for $[S^{2-}]$ into Eq. (5.5), we have:

$$[H^+] = \sqrt{\frac{1.2 \times 10^{-21}}{7.9 \times 10^{-21}}} = 0.39 \text{ g-ion/litre}$$

To this value of $[H^+]$ there corresponds the following value of pH:

$$pH = -\log 0.39 = 0.41$$

Thus, the complete precipitation of Cd^{2+} and, hence, of all the other cations of Group IV, is achieved at pH $\gg$ 0.41.

Since the reaction with H_2S liberates hydrogen ions, which gradually decrease the pH of the solution, the precipitation is started at a somewhat larger pH value, i.e., at pH 0.5, which corresponds to $[H^+] \approx \approx 0.3$ g-ion/litre.

Under these conditions, all the cations of Group III will remain in solution, which is not difficult to see from the following calculation.

Let the concentration of Zn^{2+} in solution be 1×10^{-2} g-ion/litre. The concentration of S^{2-} at $[H^+] = 0.3$ g-ion/litre is equal to:

$$[S^{2-}] = \frac{1.2 \times 10^{-21}}{(0.3)^2} = 1.3 \times 10^{-20} \text{ g-ion/litre}$$

Hence, the ion concentration product for ZnS will be:

$$[Zn^{2+}][S^{2-}] = 10^{-2} \times 1.3 \times 10^{-20} = 1.3 \times 10^{-22}$$

The value obtained (1.3×10^{-22}) is lower than the solubility product constant for ZnSβ, which is equal to 2.5×10^{-22}. It might seem that at such pH of the solution containing Zn^{2+} in a high concentration, zinc sulphide must separate out as a precipitate. This however does not actually occur since when zinc sulphide is precipitated by hydrogen sulphide:

$$Zn^{2+} + H_2S \rightleftharpoons ZnS\downarrow + 2H^+$$

hydrogen ions accumulate in the solution all the time.

When, at $[H^+] = 0.3$ g-ion/litre, hydrogen sulphide is passed into a mixture of cations of Groups IV and III, the first to precipitate are the least soluble sulphides of Group IV cations (for example, Hg^{2+}, Cu^{2+}, Pb^{2+}, Bi^{3+}, and others). This is accompanied by an even greater increase of the hydrogen ion concentration; it is not difficult to show, by way of an analogous calculation, that in the presence of at least two species of cations of Group IV at a concentration equal to 10^{-1} g-ion/litre no precipitation of zinc sulphide will take place.*

It is obvious that all the other (more soluble) sulphides of Group III cations are not precipitated by hydrogen sulphide at pH 0.5. Of course, the cations of Groups II and I, whose sulphides are soluble in water, do not precipitate either. Thus, at pH 0.5, the cations of Group IV can be separated from those of Groups III, II, and I by the action of H_2S.

The centrifugate obtained after the separation of Group IV cations is used to precipitate the sulphides of Group III cations. Let us calculate at what pH this operation must be carried out. Group III cations can be precipitated either by H_2S or by $(NH_4)_2S$. In the former case, the calculation is performed in just the same way as for precipitation of Group IV sulphides. In this case, however, the calculation has to be based on the solubility product constant for the most soluble sulphide of Group III cations, i.e., manganese sulphide ($SP = 2.5 \times 10^{-10}$).

* If, for example, the precipitation of 0.1 g-ion of Zn^{2+} per litre took place at $[H^+] = 0.3$ g-ion/litre, 0.2 g-ion/litre of H^+ would form and the total concentration of hydrogen ions in solution at the end of precipitation would then become equal to 0.5 g-ion/litre. But this is insufficient for all the Zn^{2+} to be held in solution. Calculations show that in this case about 50 per cent of Zn^{2+} (ca. 0.05 g-ion/litre) is precipitated if no Group IV cations are present.

Assuming that the precipitation of Group III cations is complete provided the concentration of Mn^{2+} decreases to 10^{-6} g-ion/litre, we shall arrive at the required concentration of S^{2-}:

$$[S^{2-}] = \frac{SP_{MnS}}{[Mn^{2+}]} = \frac{2.5 \times 10^{-10}}{10^{-6}} = 2.5 \times 10^{-4} \text{ g-ion/litre}$$

From this, using Eq. (5.5), we obtain:

$$[H]^{+} = \sqrt{\frac{1.2 \times 10^{-21}}{[S^{2-}]}} = \sqrt{\frac{1.2 \times 10^{-21}}{2.5 \times 10^{-4}}} = 2.2 \times 10^{-9} \text{ g-ion/litre}$$

Hence,

$$\text{pH} = -\log 2.2 \times 10^{-9} = 8.7$$

Thus, the complete precipitation of Group III cations by hydrogen sulphide is achieved at pH 8.7.

One should not however conclude that Group III cations can be precipitated by passing hydrogen sulphide into a solution containing no free acid: even without being acidified, solutions of salts of Group III cations show an acid reaction due to hydrolysis. Besides, being an acid, hydrogen sulphide, when dissolved in water, lowers pH approximately to 4 and, when Group III sulphides are precipitated, hydrogen ions accumulate:

$$Me^{2+} + H_2S \rightleftharpoons MeS\downarrow + 2H^{+}$$

Therefore, in contrast to Group IV sulphides, the sulphides of Group III are not usually precipitated when hydrogen sulphide is passed into the solution.

Thus, in order to achieve complete precipitation of all the cations of Group III by the action of hydrogen sulphide, it is necessary to introduce into the solution a buffer that would maintain the pH practically constant (pH $\geqslant$ 8.7). Of those commonly used in practice, the ammonium buffer with pH $\approx$ 9.24 (see Table 3.5 in Sec. 3.9) is the only one suitable for this purpose.

To precipitate Group III cations, use is more often made, instead of H_2S, of a solution of ammonium sulphide, $(NH_4)_2S$, which is actually the group reagent for Group III.

Just as in the separation of Group II cations (see Sec. 4.9), the question as to the significance of pH required to precipitate Group III cations by ammonium sulphide appears to be predetermined because in aqueous solutions $(NH_4)_2S$ is almost completely hydrolysed*:

$$NH_4^{+} + S^{2-} + H_2O \rightleftharpoons HS^{-} + NH_4OH$$

Hence, the solution of $(NH_4)_2S$ is in fact a mixture of equivalent amounts of ammonium hydroxide and an ammonium salt (NH_4HS),

* For a detailed mathematical treatment, see Sec. 5.6, Example 5.1.

i.e., like solutions of $(NH_4)_2CO_3$, it is itself an ammonium buffer (pH 9.24). Let us now find out how complete is the precipitation of Group III cations at this pH. To this value of pH there corresponds a hydrogen ion concentration equal to 5.8×10^{-10} g-ion/litre. If the excess of $(NH_4)_2S$ in solution was equal to 0.1*M*, then, since this salt is practically completely converted into NH_4HS as a result of hydrolysis, we may write: $[HS^-] = 0.1$ g-ion/litre. Substituting the values of $[H^+]$ and $[HS^-]$ into the equation for the K_{HS^-} of hydrosulphuric acid, we obtain:

$$K_{HS^-} = \frac{[H^+][S^{2-}]}{[HS^-]} = \frac{5.8 \times 10^{-10}[S^{2-}]}{10^{-1}} = 1.3 \times 10^{-13}$$

which solves to

$$[S^{2-}] = 2.3 \times 10^{-5} \text{ g-ion/litre}$$

Now we find the concentration of Mn^{2+} ions that remain in solution after the precipitation is complete:

$$[Mn^{2+}] = \frac{SP_{MnS}}{[S^{2-}]} = \frac{2.5 \times 10^{-10}}{2.3 \times 10^{-5}} \approx 1.1 \times 10^{-5} \text{ g-ion/litre}$$

Thus, Mn^{2+} precipitates practically completely. All the other sulphides of Group III cations must be precipitated even more completely at the given pH because the solubility product constants of these compounds are even lower than the solubility product of MnS.* So, the separation of Groups III and IV and also of Group III from Groups II and I is based on the difference between the solubility products of the sulphides formed by the cations of these groups, and is achieved by proper adjustment of the solution pH.

One would think that there is, in this case, a certain contradiction to the data given above, according to which, for the concentration of manganese to be decreased to 10^{-6} g-ion/litre, i.e., $[Mn^{2+}] = 10^{-6}$, it is necessary that the pH be equal to 8.7. And the result obtained here is: $[Mn^{2+}] = 10^{-5}$ g-ion/litre at pH 9.24. This is accounted for by the fact that, for the concentration of S^{2-} to be equal to 2.3×10^{-5} at pH 8.7, it is necessary that the concentration of HS^- in solution be approximately equal to 4 g-ion/litre. This is unattainable on pasage of H_2S unless a pH of 8.7 is produced by adding an appropriate buffer. It is also necessary that hydrogen sulphide be passed for a sufficiently prolonged period of time, otherwise the precipitation of MnS would not be complete. The concentration of S^{2-} must in this case be calculated by using a more exact Eq. (5.3) since at pH 8 the hydrogen sulphide present in solution will be mainly in the form of HS^- ions.

---

* The conditions under which Group III cations are precipitated by the group reagent are discussed in greater detail in Sec. 5.15.

5.2. Action of Acids on Sulphides

The sulphides of Group III and IV cations behave differently when acted on by acids. This is very important for analysis.

As pointed out in Sec. 4.5, in order to dissolve a precipitate, we must tie up one of the ions it gives up to the solution. The binding of the ions of a salt is usually achieved by the action of acids, whose hydrogen ions combine with the anions of the salt to form a corresponding weak (slightly ionized) acid.

If a salt of a weak acid is represented by the formula MeAn and the acid dissolving it by the formula HX, then the processes occurring when the precipitated salt is dissolved may be schematically written as follows:

$$\begin{array}{ccl} \text{Precipitate} & & \text{Solution} \\ \text{MeAn} & \rightleftharpoons & Me^{+} + An^{-} \\ \text{HX} & \rightleftharpoons & X^{-} + H^{+} \\ & & \qquad\;\; \updownarrow \\ & & \qquad \text{HAn} \end{array}$$

The weak acid HAn formed in the reaction ionizes, though to a slight degree, and gives up the same An^- anions, the binding of which results in the dissolution of the precipitate. The given process therefore is reversible and leads in the long run to equilibrium. This equilibrium will evidently set in when the concentration of the anions, $[An^-]$, satisfies simultaneously the following two equations:

$$[Me^+][An^-] = SP_{MeAn} \tag{5.6}$$

$$\frac{[H^+][An^-]}{[HAn]} = K_{HAn} \tag{5.7}$$

Solving Eq. (5.6) for $[An^-]$, substituting its value into Eq. (5.7), and rearranging the latter, we obtain:

$$[Me^+][HAn] = \frac{[H^+]SP_{MeAn}}{K_{HAn}} \tag{5.8}$$

From the ionic equation of the reaction

$$MeAn + H^+ \rightleftharpoons Me^+ + HAn$$

it is clear that, when one mole of the MeAn precipitate dissolves, one gram-ion of Me^+ and one mole of HAn appear in the solution. Thus, designating the solubility of the precipitate in the acid by S (in moles per litre), we may write:

$$S = [Me^+] = [HAn]$$

From this and Eq. (5.8) we have:

$$S = \sqrt{\frac{[H^+]SP_{MeAn}}{K_{HAn}}} \tag{5.9}$$

Thus, the solubility of the precipitate of a slightly soluble salt in acids must be the higher: (a) the greater the solubility product constant of the precipitate; (b) the lower the ionization constant of the weak acid formed (HAn); and (c) the higher the concentration of hydrogen ions in solution (i.e., the lower the pH of the solution).

These relationships hold for all slightly soluble salts, sulphides included. If the sulphides of the cations of Groups III and IV are acted on by the same acid, at the same concentration, the reaction product will be the same slightly ionized compound (H_2S); and the difference in the behaviour of the sulphides concerned may be explained only by the differences between their solubility product constants. Since the solubility products of Group IV sulphides (see Sec. 5.1) are much lower than those of Group III sulphides, the former should be expected to be much less easily soluble in acids.

In fact, the sulphides of Group III cations dissolve well in dilute acids (HCl and H_2SO_4), whereas those of Group IV are practically insoluble in them. Group IV sulphides, evidently, yield so few S^{2-} ions into the solution that equilibrium sets in even if very small amounts of H_2S have been formed, i.e., when a negligibly small quantity of sulphide has been dissolved. Conversely, in the case of Group III sulphides, whose solubility product constants are incomparably higher, the formation of H_2S proceeds further, so that when sufficient acid is present, all the precipitate dissolves (this is also aided by the removal of hydrogen sulphide in gaseous form). This explanation is well supported by calculation.

Calculation of the solubility of sulphides in acids is complicated by the fact that the ionization of hydrosulphuric acid takes place by steps. The reverse process of binding of S^{2-} ions by hydrogen ions, which occurs when sulphides dissolve in acids, is also stepwise in character: HS^- ions are formed first, which add up hydrogen ions and are partly converted into un-ionized H_2S molecules. If the formation of both reaction products (i.e., HS^- and H_2S) is taken into account, then, in calculating the solubility of sulphides in acids, we obtain a quadratic equation which is rather difficult to solve*. But if the value of the solubility product of the precipitate is small and the K_1 of the corresponding weak acid is much greater than K_2 (as in the case of H_2S), then it may be assumed, without appreciable error, that in the presence of excess H^+ all the HS^- ions are practically completely converted into H_2S, i.e., that the reaction with doubly charged cations proceeds according to the equation

$$MeS + 2H^+ \rightleftharpoons Me^{2+} + H_2S$$

---

* Komar', N. P., *Fundamentals of Qualitative Chemical Analysis*, vol. I, Izd. Kharkovskogo Universiteta, Kharkov, 1955, page 255 (in Russian).

With this assumption, the solubility of the sulphide can be calculated from the equation

$$\frac{[H^+][S^{2-}]}{[H_2S]} = K_{H_2S\ (total)}$$

or

$$[H_2S] = \frac{[H^+]^2[S^{2-}]}{K_{H_2S\ (total)}} \qquad (5.10)$$

where

$$K_{H_2S\ (total)} = K_{H_2S}K_{HS^-} = 1.2 \times 10^{-20} \text{ (see Sec. 5.1)}$$

The concentration of S^{2-} anions is related to the concentration of Me^{2+} cations by the equation

$$[Me^{2+}][S^{2-}] = SP_{MeS}$$

Solving this equation for $[S^{2-}]$ and substituting the value obtained into Eq. (5.10), we get:

$$[Me^{2+}][H_2S] = \frac{[H^+]^2 SP_{MeS}}{K_{H_2S\ (total)}} \qquad (5.11)$$

From the equation of the reaction

$$MeS + 2H^+ \rightleftarrows Me^{2+} + H_2S$$

it follows that the dissolution of each mole of the MeS precipitate requires 2 g-ion of H^+ (i.e., 2 mole of a strong acid such as HCl); this dissolution is accompanied by the formation of 1 g-ion of Me^{2+} and 1 mole of H_2S. Hence, denoting the initial concentration of the solvent acid by C, we may write*:

$$S = [Me^{2+}] = [H_2S] \quad \text{and} \quad [H^+] = C - 2S$$

Substituting the values of $[Me^{2+}]$, $[H_2S]$, and $[H^+]$ into Eq. (5.11) yields:

$$S^2 = (C - 2S)^2 \frac{SP_{MeS}}{K_{H_2S\ (total)}}$$

Taking square roots of both sides of the equation and solving it for S, we find:

$$S = \frac{C\sqrt{\dfrac{SP_{MeS}}{K_{H_2S\ (total)}}}}{2\sqrt{\dfrac{SP_{MeS}}{K_{H_2S\ (total)}}} + 1}$$

* The volatilization of H_2S is neglected here; this is permissible if the concentration of the acid is such that at equilibrium the solution is found to be unsaturated with respect to H_2S (i.e., if the amount of H_2S formed is less than 0.1 mole/litre).

Dividing the numerator and denominator of the right-hand side of the equation by

$$\sqrt{\frac{SP_{MeS}}{K_{H_2S\,(total)}}}$$

we have:

$$S=\frac{C}{2+\sqrt{\frac{K_{H_2S\,(total)}}{SP_{MeS}}}} \tag{5.12}$$

On the basis of Eq. (5.12) one can draw the following conclusions:

1. If the solubility product of MeS is high as compared with $K_{H_2S\,(total)}$ (just as for most of the sulphides of Group III), then the quantity under the root sign is small. In such a case, the solubility of a sulphide approaches a maximum value* corresponding to the reaction equation, i.e., $C/2$.

From Table 5.1 one can see that, although Eq. (5.12) is approximate, the calculated results are in full *qualitative* agreement with experimental data. Table 5.1 is a good illustration of the effect of the solubility product of sulphides on their solubility in HCl.

If the calculation of the solubility of a sulphide in an acid is based on the concentration of hydrogen ions at equilibrium rather than on the initial concentration of the acid, then the following equation for S,

TABLE 5.1. Solubilities of Selected Sulphides in 0.1*N* HCl Solution

Sulphide	Solubility product	Solubility of sulphide	
		mole/litre	in per cent of maximum possible solubility ($C/2$)
MnS (pale-pink)	2.5×10^{-10}	5×10^{-2}	100.0
MnS (green)	2.5×10^{-13}	ca. 5×10^{-2}	ca. 100.0
FeS	5×10^{-18}	4.9×10^{-2}	97.8
ZnS (alpha)	1.6×10^{-24}	1.2×10^{-3}	2.3
ZnS (beta)	2.5×10^{-22}	1.14×10^{-2}	22.7
NiS (alpha)	3.2×10^{-19}	4.6×10^{-2}	91.3
NiS (gamma)	2×10^{-26}	1.3×10^{-4}	0.26
CoS (alpha)	4×10^{-21}	3.7×10^{-2}	54.0
CoS (beta)	2×10^{-25}	4.1×10^{-4}	0.83
CdS	7.9×10^{-27}	7×10^{-5}	0.16
PbS	2.5×10^{-27}	4.6×10^{-5}	9.3×10^{-2}
CuS	6.3×10^{-36}	2.3×10^{-9}	4.6×10^{-6}

* Since 2 mole of HCl is to be utilized for 1 mole of MeS to be dissolved, C mole of the acid will dissolve $C/2$ mole of the sulphide if 100 per cent of HCl is used up for the reaction.

derived by A. Babko, is obtained:

$$S = \sqrt{SP_{MeS}\left(1 + \frac{[H^+]}{K_{HS^-}} + \frac{[H^+]^2}{K_{H_2S}K_{HS^-}}\right)} \quad (5.13)$$

In deriving this equation, account was taken of the stepwise binding of S^{2-} ions by hydrogen ions. This equation however is not exact either because it does not take account of the activity coefficients of the ions.

When precipitates are "aged" (ripened), i.e., stored under the mother liquor, their solubility usually decreases. This is explained mainly by two reasons: (a) the aging changes the hydration of precipitates, which leads to a decrease of the solubility; (b) precipitates are gradually converted, with time, into other less soluble modifications (for example, α-NiS is transformed to γ-NiS; α-CoS changes into β-CoS, etc.; see Table 5.1)*.

On the other hand, the solubility of sulphides is sometimes found to be increased because of the formation of complex ions. For instance, a precipitate of CdS dissolves in HCl of sufficient concentration since a rather stable complex ion, $[CdCl_4]^{2-}$, is formed.

Similar phenomena are observed not only in the case of the sulphides of the cations in question but also for other slightly soluble electrolytes. Thus, in Sec. 4.5 it was pointed out that magnesium hydroxide ($SP = 6 \times 10^{-10}$), manganese hydroxide ($SP = 4.5 \times 10^{-13}$), ferrous hydroxide ($SP = 1 \times 10^{-15}$), and others, are readily soluble in solutions of ammonium salts because of the OH^- ions being bound into the weak base NH_4OH. At the same time, ferric hydroxide, $Fe(OH)_3$ ($SP = 3.2 \times 10^{-38}$), as well as aluminium and chromium hydroxides, whose solubility products are very small, does not dissolve in solutions of ammonium salts. Use is made of this fact in analysis for separation of Group III cations into two subgroups.

If however a hydroxide is acted on by an acid instead of an ammonium salt, it dissolves readily since the reaction involves the formation of H_2O which dissociates to a much smaller extent than NH_4OH. In this case, the decisive factor is evidently the difference in the value of the ionization constants of the reaction products.

Different solvents act differently on sulphides. For example, the sulphides of Group IV are insoluble in HCl and readily soluble when heated with HNO_3 which oxidizes S^{2-}:

$$S^{2-} + 2NO_3^- + 8H^+ \rightarrow S\downarrow + 2NO\uparrow + 4H_2O$$

This reaction is based on the oxidation of the sulphide ion:

$$S^{2-} - 2e \rightleftharpoons S$$

* Other factors have also been suggested to account for the existence of various modifications of certain slightly soluble compounds, for example, the change of the oxidation state of an element, the formation of basic salts.

The equilibrium constant of this reaction may be written thus:

$$K' = \frac{[S]}{[S^{2-}]}.$$

But the concentration of free sulphur, [S], in its saturated solution is a constant quantity. Hence,

$$[S^{2-}] = \frac{[S]}{K'} = K$$

The equilibrium is established when $[S^{2-}]$ becomes equal to the value of K which is very small. The oxidation of S^{2-}, therefore, also takes place at concentrations at which the S^{2-} ions are no longer tied up by hydrogen ions. Thus, the factor responsible for the different behaviour of Group IV sulphides under the action of HCl and HNO_3 is the difference in the equilibrium constants of reactions, as a result of which the S^{2-} ions are removed from the solution.

Other oxidizing agents may also be used to dissolve slightly soluble sulphides. For instance, the modifications γ-NiS and β-CoS, which are insoluble in HCl, dissolve readily under the action of hydrogen peroxide in a weak acid such as acetic:

$$NiS\downarrow + H_2O_2 + 2CH_3COOH \rightarrow Ni^{2+} + 2CH_3COO^- + S\downarrow + 2H_2O$$

The effect of the hydrogen ion concentration on the solubility of precipitates in acids has already been repeatedly discussed. For example, the precipitates of $KHC_4H_4O_6$, $BaCrO_4$, CaC_2O_4 are well soluble in strong (mineral) acids and practically insoluble in a weak (acetic) acid which produces a much lower concentration of hydrogen ions. Such phenomena will be encountered elsewhere in the text.

The effect of this factor is observed only when the solubility product constant of the precipitate is relatively small or K_{HAn} is large. For example, in contrast to $BaCrO_4$ ($SP = 1.2 \times 10^{-10}$), $SrCrO_4$ ($SP = 3.6 \times 10^{-5}$) dissolves not only in strong acids but in acetic acid as well. Analogously, although the solubility product of calcium oxalate is close to that of calcium carbonate, the former (CaC_2O_4) is insoluble in acetic acid, while the latter ($CaCO_3$) dissolves readily in this acid; it is because the ionization constants of carbonic acid are much lower than those of oxalic acid.

5.3. Colloidal Systems

The process of precipitation is very often complicated by the formation of colloidal solutions. As a result of this colloid formation, the corresponding slightly soluble substances (e.g., the sulphides of Group III cations) cannot be separated from solution either by centrifugation or by filtration. The aim of the entire operation—separation of ions—is therefore not achieved.

To prevent the formation of colloidal solutions or to destroy them if they are formed, certain measures have to be taken. An understanding is therefore required of the nature of colloidal solutions and of their most important properties in relation to analytical chemistry.

If the minute particles of a substance are distributed in any other substance, a **disperse system** is formed. The fine particles in such a system are collectively called the **dispersed phase**, and the substance in which these particles are distributed is known as the **dispersion medium.** Disperse systems differ among themselves primarily in **degree of dispersion**, i.e., particle size. The smaller the particles the higher is the degree of dispersion, and vice versa. If the particles of the dispersed phase are larger than 100 nm*, the corresponding disperse system is called a **suspension** in those cases when we deal with solid particles, and an **emulsion** if the particles are liquid. An example of a suspension system is water containing suspended clay particles. Milk with very minute fat particles being in a suspended state is an example of an emulsion. If the particle size is in the range from 100 nm to 1 nm, disperse systems are called **colloidal solutions** or **sols****; if the size of particles is less than 1 nm, the corresponding solutions are termed **true solutions**, or simply solutions.

In true solutions, the dispersed particles are separate molecules or ions of dissolved substances (solutes). They are too small to be visible under the microscope with any available magnification. True solutions are homogeneous to the naked eye. Conversely, suspensions and emulsions, and also colloidal solutions, in which the dispersed particles consist of many molecules***, are characterized by various degrees of distinct heterogeneity. The heterogeneity of suspensions and emulsions manifests itself primarily in the form of opaqueness and turbidity. It shows up even more distinctly when suspensions and emulsions are viewed under the microscope, with the aid of which the individual particles of the dispersed phase can be detected.

Smaller particles of colloidal solutions cannot be seen under the ordinary microscope. In most cases, however, their presence can be detected with the aid of a so-called *ultramicroscope*. In the ultramicroscope, the object under study is illuminated from the side and not from below as in the ordinary microscope, so that the light rays

* This is an abbreviation for nanometre (1 nm = 1 mμ = 10Å = $= 10^{-9}$ metre).

** Strictly speaking, it is not quite correct to refer to such disperse systems as colloidal solutions. The point is that, at present, solutions are taken to mean only thermodynamically stable disperse systems; sols, however, are more or less unstable, as will be shown later. But since the term "colloidal solution" is one which has been widely used up to now, we shall use it along with the term "sol".

*** The molecules of compounds such as proteins, starch, etc., are so large that their solutions exhibit all the properties of colloidal systems. Hence, the particles in such systems are found to be the individual molecules of high-molecular-weight compounds (polymers).

do not reach the observer's eye and the field of vision appears dark. When scattering the incident light rays, the colloidal particles reflect light to the observer and become visible under the ultramicroscope as tiny dancing specks of light, which are in continuous random movement (Brownian movement).*

The large particles of suspensions and emulsions usually fail to pass through the pores of filter paper or clay filters, through which smaller particles of colloidal solutions pass easily. Such filters therefore cannot be used to separate the dispersed phase from the dispersion medium in colloidal solutions. **Membranes** however exist in nature or can be made artificially, which have pores so small that even the dispersed particles of colloidal solutions cannot pass through them. Membranes are made of collodion, parchment, or cellophane. The molecules and ions present in true solutions pass easily even through such filters.

Colloidal solutions can be produced by grinding particles to colloidal size by any suitable method (for example, with the aid of special "colloid mills").

Dispersion, as a method of preparing colloids, can be replaced by the reverse process—condensation in which colloidal particles are built up or condensed from the individual molecules or ions of dissolved substance. This may take place in chemical reactions accompanied by the formation of slightly soluble compounds. If large particles are formed in this process, they separate out as a precipitate under the influence of gravity. But if the experimental conditions chosen are such (for example, a sufficiently low concentration of the reacting substances in solution) that the particles cease to grow when they have reached the size limits indicated above (1-100 nm), a colloidal solution results.

It was originally thought that colloidal solutions could be formed only by a certain class of substances which were therefore named colloids in contrast to crystalloids which failed to form such solutions. It has since been found however that under suitable conditions even typical crystalloids, such as NaCl, may form colloidal solutions. Thus, one should now speak of the colloidal and crystalline states of substances rather than of colloids and crystalloids.

To produce a sufficiently stable colloidal solution, it is necessary to prevent the dispersed particles from collecting in aggregates and from settling down to the bottom of the container under gravity. But in colloidal solutions there exist forces favouring such aggregation of particles, for example, the forces of molecular attraction. The same effect is produced by the surface tension of the liquid, which tends to reduce the surface of contact between the liquid and the particles to a minimum. For a given total volume of particles, their total surface area decreases with increasing particles size.

---

* This property of colloids is known as the Tyndall effect—*Tr.*

This can easily be proved by a simple calculation. A cube with an edge of 1 cm has a surface area of 6 cm^2. Now if this cube is subdivided into cubes with an edge of 0.1 cm, the total surface area (the number of small cubes being 10^3) will become equal to $6 \times 0.1^2 \times 10^3 = 60$ cm^2, i.e., 10 times as large as before the subdivision. With cubes of colloidal size, i.e., cubes having edges of 0.0001 mm, or 0.00001 cm, the total surface is 600,000 cm^2.

Despite all this, typical colloidal solutions may fail to form a precipitate for a very long time. Evidently, there are some factors which prevent aggregation of colloidal particles.

One such factor is the presence of like electric charges on colloidal particles. Under the action of these charges, the colloidal particles undergo mutual repulsion which counteracts their aggregation into larger particles.

The existence of such charges can be easily demonstrated by passing an electric current through a colloidal solution. Many colloidal particles, say, those of metals or metal sulphides, move towards the anode, i.e., they are negatively charged. Other particles, such as the particles of metal hydroxides, move towards the cathode, i.e., they bear a positive charge.

The charges of colloidal particles arise in most cases as a result of adsorption of ions by the particles (nuclei) from solution, most often common ions, i.e., ions contained in the crystal lattice of the particles. For example, when a sol of arsenous sulphide is prepared by the action of H_2S on a solution of arsenite (the salt of arsenious acid, H_3AsO_3), particles are formed from a large number (*m*) of molecules of As_2S_3, which adsorb a certain amount (*n*) of HS^- anions from solution and become therefore negatively charged.

These negatively charged particles attract from solution the corresponding number (*n*) of positively charged hydrogen ions (gegenions or counter-ions), a part of which (*x*) remains in the liquid phase, forming the so-called diffuse layer; the rest (*n* — *x*) are adsorbed by the particle surface and move together with the particle. This adsorption layer consists, as it were, of two layers of oppositely charged ions, and is known as the electrical double layer (it is also often called the Helmholtz double layer). The total charge that the particle bears is due to the fact that the sum of the charges on counter-ions in the adsorption layer does not completely compensate for the charges on the primarily adsorbed ions. The colloidal particle and the diffuse layer form a micelle. This may be schematically represented as follows:

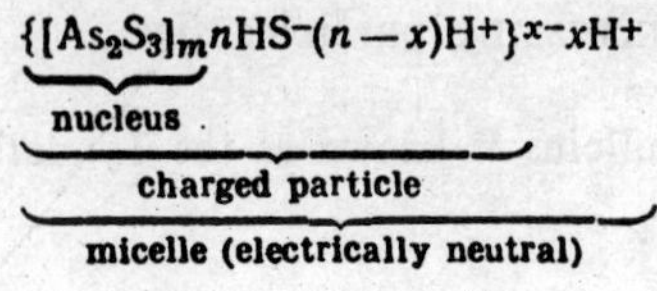

From this it follows that the particle is charged (negatively in this particular case) and the micelle is electrically neutral.

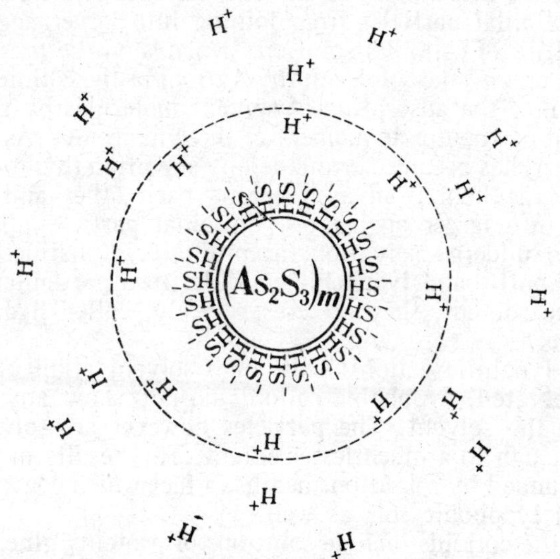

Fig. 5.1. Structure of the As_2S_3 micelle (shown schematically).

The structure of the As_2S_3 micelle is shown schematically in Fig. 5.1. The micelle of AgI formed in a solution containing a small excess of KI has an analogous structure:

$$\underbrace{\underbrace{\{\underbrace{[AgI]_m}_{\text{nucleus}} nI^-(n-x)K^+\}^{x-}}_{\text{charged particle}} xK^+}_{\text{micelle (electrically neutral)}}$$

If the same sol is formed in the presence of excess $AgNO_3$, its micelles have a structure represented by the following formula:

$$\{[AgI]_m nAg^+(n-x)NO_3^-\}^{x+} xNO_3^-$$

In both cases, the nucleus of the micelle, $[AgI]_m$, has the same chemical composition and crystal structure, but the charge on the particles is different since it results from the adsorption of I^- ions in the first case and of Ag^+ ions in the second, both being present in solution in some excess. Substances, which on being adsorbed impart a charge to the colloidal particle, so that the colloid becomes stable, are termed **stabilizers**. In the examples discussed, H_2S, KI, and $AgNO_3$ are stabilizers.

From what has been said above, it becomes clear that the formation of a sufficiently stable colloidal solution requires not only an appropriate degree of dispersion but also the presence of a stabilizer to prevent the colloidal particles from joining into larger aggregates.

The stability of colloidal solutions depends on the presence of electric charges on and the solvation (hydration) of the colloidal particles. **Solvation**, i.e., the adsorption of solvent molecules by the particles, is dependent on the dipole moment of these molecules. As a result, the colloidal particles become surrounded by solvation (hydration) sheaths which prevent them from approaching each other and hence from combining into larger aggregates. Colloidal particles differ in their tendency to undergo solvation; accordingly, a distinction is made between **lyophilic** and **lyophobic** colloids. In a particular case, when water is the solvent, they are, respectively, called **hydrophilic** and **hydrophobic**.

Whereas lyophilic colloids have high solvent affinity and become strongly solvated, lyophobic colloids do not show any appreciable affinity for the solvent. The particles however are solvated in this case too, though to a much lesser extent. This results in the particles being surrounded by solvation sheaths, a fact which contributes to the stability of lyophobic sols as well.

Hydrophilic colloids include solutions of proteins, glue, starch, etc.; examples of hydrophobic colloids are sols of metals, sulphides, and of various salts. The sols of metal hydroxides occupy an intermediate position. If to a hydrophobic sol is added a small amount of an electrolyte, the colloidal particles join together to form larger aggregates, i.e., the degree of dispersion (dispersity) of the colloid decreases. This process known as **coagulation** may end in **sedimentation**, i.e., the settling-down of the solid phase under the influence of gravity.

The phenomenon of coagulation is most simply explained by the fact that colloidal particles adsorb differently charged ions of a coagulant electrolyte, as a result of which the charges on the particles decrease and they become capable of uniting with one another (this is assisted by the simultaneously occurring process of destruction of solvation sheaths). In full conformity with this theory, the coagulating effect of various electrolytes increases rapidly with increasing charge on the electrolyte ion whose charge is opposite in sign to the charge on the colloidal particles. For example, for a sol of As_2S_3 the coagulating concentrations (i.e., the minimum concentrations causing the given sol to coagulate) of Al^{3+}, Ba^{2+}, and K^+ are in the ratio of 1:20:1000. For a sol of $Fe(OH)_3$, whose particles are usually charged positively, the charge on the anion of the coagulant electrolyte is of importance. In this case, the coagulating effect increases rapidly in the series: Cl^-, SO_4^{2-}, $[Fe(CN)_6]^{3-}$. Coagulation may also occur when the solution contains colloidal particles of opposite electric charges, say, when a sol of $Fe(OH)_3$ is mixed with a sol of As_2S_3.

The phenomenon of coagulation is rather complicated and such a simplified "adsorption theory" does not provide a more or less satisfactory explanation for a great variety of experimental data. Other, more sophisticated theories have been proposed, which are discussed in detail in courses of colloid chemistry. None of them, however, may be considered quite exhaustive.

For instance, as far back as 1930's, the electrostatic theory of coagulation was developed. Briefly, it consists in the following. The adsorption layer of a colloidal particle is a double electrical layer; it is characterized by the so-called zeta potential ζ. When a strong electrolyte is added, the double layer contracts, the value of the zeta potential decreases to a critical value, the sol becomes unstable and undergoes coagulation followed by the precipitation of a solid phase.

With all its merits, the electrostatic theory of coagulation has some limitations. The main drawback is the necessity to postulate the existence of the critical value of zeta potential at which the stability of the sol is disturbed.

The most important stability factor in the case of lyophilic colloids is the presence of solvent sheaths or shells around the particles. To induce coagulation, it is necessary that these solvent shells be broken down. They can be destroyed in various ways, in particular, by addition of electrolytes but in considerably higher concentrations than in the case of lyophobic colloids. At high concentrations, the electrolyte ions when being solvated wrench solvent molecules out of the colloidal particles, thereby conferring on them the properties of lyophobic colloidal particles. Since the electrolyte ions also discharge the particles, the colloid undergoes coagulation, in which case the phenomenon is called the **salting-out effect.**

The addition of a hydrophilic colloid to a lyophobic colloid sol greatly enhances its stability to the action of electrolytes. This phenomenon known as **colloid protection** consists in the adsorption of the hydrophilic colloid by the particles of the hydrophobic colloid, which thus acquire solvent shells.

Let us cite an example of colloid protection. If to a solution of $AgNO_3$ we add a "protective" colloid (gelatine), followed by a solution of HCl, the usual curdy precipitate of AgCl is not formed. Instead, the solution becomes turbid due to the formation of an AgCl sol. This sol does not coagulate even when excess precipitant is added.

Rise of temperature usually favours the coagulation of colloidal solutions. The adsorption of ions, which imparts a charge to the particles, is diminished and, besides, the solvent sheaths are broken down.

When a lyophilic colloidal solution is coagulated, it often becomes a jelly-like mass called a **gel** (for example, gelatine, starch, silicic acid, etc.).

As already pointed out above, the formation of colloidal solutions cause complications in analysis. Colloidal particles pass freely through the pores of a filter and do not precipitate; they cannot be separated either by filtration or by centrifugation. To make possible the separation of substances being in the colloidal state, it is necessary to induce the coagulation of the colloid. The role of a coagulant electrolyte is mainly played by the precipitant which is always added in a slight excess*. Coagulation may also be effected by addition to the solution of an electrolyte, usually ammonium salts, or by heating.

For coagulation to take place, the solution being analysed must not contain protective colloids. It is for this reason that organic compounds must be removed, this being achieved by treatment of the solution with a mixture of concentrated acids—HNO_3 and H_2SO_4 (to oxidize organic substances to CO_2 and H_2O).

The transition of substances to the colloidal state often takes place when precipitates are washed. This is accounted for by the fact that the electrolyte ions, which are added in order to coagulate the precipitate, are adsorbed by the surface of the precipitate and are then gradually washed out of it. The particles become charged again and begin to repel each other. As a result, a colloidal solution is formed, which passes through the filter and is transferred from the precipitate into the solution. To avoid this undesirable phenomenon called **peptization,** one has to wash certain precipitates with a dilute solution of an electrolyte (usually an ammonium salt) rather than with pure water. The ions of an ammonium salt when being adsorbed by the precipitate gradually replace the ions removed by washing and the peptization of the precipitate is thus prevented.**

5.4. Coprecipitation

When a precipitate separates out from a solution, it adsorbs both common ions and various foreign ions on its surface. Especially extensive is the adsorption by colloidal particles which have a large surface area. The analyst has to consider the phenomenon of adsorption, particularly in detecting small quantities (traces) of an impurity. In precipitation reactions, not infrequently such small amounts of impurities are completely carried down by the precipitate and cannot be detected in solution.

---

* A large excess of precipitant is harmful since it may cause re-charging of particles and their transition into the colloidal state.

** For a more detailed discussion of the properties of colloidal systems, the reader should consult the following books: Voyutsky, S. S., *A Course in Colloid Chemistry*, Khimiya, Moscow, 1964, (in Russian); Jirgensons, B. and M. E. Straumanis, *A Short Textbook of Colloid Chemistry*, Pergamon Press, London, 1954; Ward, A. G., *Colloids—Their Properties and Applications*, Blackie, London, 1948.

Which of the foreign ions are adsorbed by the precipitate in a particular case depends to a considerable extent on the conditions under which precipitation takes place.

Experiment shows that ions which enter into the composition of colloidal particles (lattice ions) are adsorbed first, the particles acquiring a charge of the corresponding sign. When being coagulated, such electrically charged particles carry down foreign, oppositely charged ions. For example, if ferric hydroxide is precipitated in an ammoniacal solution, the particles of $Fe(OH)_3$ adsorb OH^- ions, acquiring a negative charge, and therefore on coagulation carry down various cations, say, Mg^{2+}, Ca^{2+}, and, especially strongly, Co^{2+}, Mn^{2+}, Ni^{2+}, and other ions, a fact that must be taken into account in analysis of Group III cations.

Conversely, if ferric hydroxide is brought down in a weakly acid medium, the particles of $Fe(OH)_3$ are charged positively because of the adsorption of $Fe(OH)_2^+$ or $Fe(OH)^{2+}$. As a result, the precipitate adsorbs various anions (Cl^-, SO_4^{2-}, etc.) from the solution. The carrying down of foreign substances (impurities), which are not to be precipitated by a given precipitant under the conditions of the experiment, is called **coprecipitation.**

Coprecipitation may take place not only when impurities are adsorbed on the surface of particles but also when they are entrapped in the interstices and microcapillaries present in true crystals (internal adsorption). The mechanism of internal adsorption (or occlusion) is not difficult to visualize: the ions adsorbed on the surface of the growing crystals of the precipitate become isolated from the solution due to the deposition of new layers of the substance.

Coprecipitation is very often caused by the precipitation of so-called **mixed crystals**, i.e., crystals formed by two or more substances crystallizing in the same crystalline form (this is also known as solid solution formation). Substances capable of forming mixed crystals are called **isomorphous.** This type of coprecipitation is termed **isomorphous coprecipitation** in distinction to the **adsorption coprecipitation** considered above.

And, finally, coprecipitation may occasionally be caused by the formation of true chemical compounds by the precipitate and coprecipitated impurity. For instance, the coprecipitation of Zn^{2+} with manganese when the latter is precipitated as the acid* H_2MnO_3 is due to the formation of a slightly soluble zinc salt of this acid, $ZnMnO_3$. This type of coprecipitation however is encountered relatively seldom. More often one has to deal with adsorption or isomorphous coprecipitation.

Thus, two cases should be distinguished in coprecipitation: (a) the coprecipitated impurity is adsorbed on the surface of precipitate par-

* Its formula is often written as $MnO(OH)_2$. The more correct formula is $MnO_2 \cdot xH_2O$.

ticles (surface adsorption); and (b) the coprecipitated impurity is occluded within the particles, i.e., is entrapped in the interior of primary particles (internal adsorption, formation of mixed crystals, etc.).

In the former case, the adsorbed ions are in a state of dynamic equilibrium with the corresponding ions in solution (adsorption equilibrium). When the precipitate is being washed, this equilibrium is continuously disturbed and the adsorbed ions gradually pass from the precipitate surface into the solution. Hence, the impurities adsorbed on the surface can be removed from the precipitate by washing. Conversely, occluded impurities cannot be washed out of the precipitate. For these occluded particles to pass into solution, all the precipitate has to be dissolved.

If precipitation reactions involve an intensive coprecipitation of any ions, it is expedient, if possible, to detect them by fractional reactions until a separation is effected. The amount of coprecipitation can sometimes be materially reduced by replacing some ions with others. For example, if Ba^{2+} ions are to be precipitated by SO_4^{2-} ions in the presence of Fe^{3+} ions, which coprecipitate extensively with $BaSO_4$, the ferric ions should best be preliminarily reduced to ferrous ions which coprecipitate to a much smaller degree.

A very important and widely used measure (especially in quantitative analysis) to counteract coprecipitation is **reprecipitation.** Let us illustrate this process by the following example. Suppose that Ca^{2+} is precipitated as CaC_2O_4 in the presence of Mg^{2+}; in this case, a little MgC_2O_4 will also be coprecipitated. To remove MgC_2O_4, the precipitate formed is separated from the solution, washed slightly, and dissolved in HCl. A solution is thus obtained, which contains all the Ca^{2+} ions that were present in the original solution and a small portion of Mg^{2+} contained in it. If the acid is neutralized and the CaC_2O_4 is precipitated again, i.e., reprecipitated, then the second precipitation takes place at a much lower concentration of Mg^{2+} than the first. The CaC_2O_4 precipitate obtained will therefore be free from the MgC_2O_4 impurity.

Apart from coprecipitation, the analyst has to deal with the phenomenon of **delayed precipitation (postprecipitation)**. It consists in gradual settling-down of an impurity to the precipitate when the solution is allowed to stand.

For instance, when the cations of Group IV are separated from those of Group III by the action of hydrogen sulphide in an acid medium, a precipitate is formed, which does not contain zinc sulphide. If however the precipitate is separated from the solution some time after the precipitation, it is found to be contaminated with ZnS because the latter comes down later, after the main precipitate has been brought down.

The factor responsible for the postprecipitation of zinc sulphide is the adsorption of S^{2-} ions from the solution by the particles of Group IV sulphides. Because of this, the concentration of S^{2-} ions on the surface of the particles turns out to be higher than in solution, as a result of which the solubility product of ZnS is found to be exceeded by the product of the concentrations of S^{2-} and Zn^{2+}. It is for this reason that zinc sulphide settles down on the surface of the other precipitated sulphides, though in their absence, at the same pH value, this should not have taken place.

Coprecipitation and postprecipitation are serious obstacles in many analytical separations of ions. In certain cases, however, coprecipitation may prove to be a boon. Owing to coprecipitation, the ions present in trace amounts in solution can be concentrated, an advantage not infrequently used in analysis.

Let us consider an example. If a solution contains traces of Cu^{2+} ions, they will not be precipitated by H_2S. But if we add a little of a mercuric salt, the precipitate of HgS formed by the action of hydrogen sulphide will also entrain Cu^{2+} ions. When the HgS + CuS precipitate is ignited, the mercury will be removed (because of the decomposition of the HgO formed into Hg and O_2 which volatilize), and the CuS will be converted into CuO. After the cupric oxide is dissolved in a few drops of HCl, a solution results, in which Cu^{2+} can easily be detected.

The operation just described is known as carrier precipitation. The role of a carrier is played in this particular case by HgS on which Cu^{2+} ions are concentrated, this making it possible to detect even very small amounts of the ions. Carrier precipitation is extensively employed in quantitative analysis as well for determination of traces of various impurities, which is of great importance to a number of branches of industry and science. Use is made of a large variety of carriers, such as manganese dioxide, calomel (Hg_2Cl_2), some other salts, sulphur, etc.

5.5. Hydrolysis of Salts

The hydrolysis of salts plays an important part in analysis. For example, the pH of solutions of many salts depends on their hydrolysis. If such salts are to be used as precipitants, the excess of precipitant added determines to a considerable extent the pH of the solution and, hence, the completeness of precipitation. We have already encountered similar cases in discussing the precipitation of the cations of analytical Groups II and III by their group reagents (see Sections 4.9 and 5.1). Not infrequently it is also found that, as a result of the hydrolysis of the reagents, instead of slightly soluble salts, for example, Al_2S_3 and $Cr_2(CO_3)_3$, there are precipitated the hydroxides of the corresponding metals, $Al(OH)_3$ and $Cr(OH)_3$. All this shows that, in studying analytical chemistry, serious attention must be paid to the process of hydrolysis.

*Hydrolysis is defined as the interaction of the ions of a dissolved salt with the H^+ and OH^- ions of water.** Three types of hydrolysis are distinguished.

1. Hydrolysis of Salts Formed from Strong Bases and Weak Acids. Almost all salts, with rare exceptions, are strong electrolytes. This means that, regardless of what acid and what base are made to interact with each other to form salts, these salts dissociate in water practically completely.

When the salt of hydrocyanic acid and a strong base, say, KCN, is dissolved in water, this salt breaks down into K^+ and CN^- ions. Encountering OH^- ions in aqueous solution, the K^+ ions may give an un-ionized KOH molecule which immediately dissociates again into K^+ and OH^- ions under the action of water dipoles.

But when encountering H^+ ions, the CN^- ions form molecules of a slightly dissociated weak acid ($K_{HCN} = 6.2 \times 10^{-10}$).

This process may be schematically represented as follows:

$$\begin{array}{ccc} KCN \rightarrow K^+ + & CN^- \\ & + \\ H_2O \rightleftharpoons OH^- + & H^+ \\ & \uparrow\downarrow \\ & HCN \end{array}$$

In the course of the process, there are gradually accumulated OH^- ions and therefore the hydrolysis of the salt continues until the product of the concentration of the OH^- ions accumulated and the concentration of the H^+ ions resulting from the ionization of the acid HCN formed reaches the ion product of water ($K_w = 10^{-14}$).

Just as in all chemical reactions, the equilibrium established is reversible. But at equilibrium the concentration of OH^- ions must exceed considerably the hydrogen ion concentration. In other words, the solution of the salt KCN must show an alkaline (pH > 7) rather than a neutral reaction. Indeed, as experiment shows, the $0.1N$ solution has pH 11.1.

The occurrence of the alkaline reaction of the solution is especially distinct if the hydrolysis equation is written in the ionic form. Let us write the ionic equation of the reaction, taking into account that H_2O and HCN as weak electrolytes should be written in the molecular form and KCN and KOH in the form of ions:

$$K^+ + CN^- + H_2O \rightleftharpoons K^+ + OH^- + HCN$$

and, finally,

$$CN^- + H_2O \rightleftharpoons OH^- + HCN$$

Like KCN, a solution of CH_3COONa also has an alkaline reaction, which is seen from the hydrolysis equation:

$$CH_3COO^- + H_2O \rightleftharpoons CH_3COOH + OH^-$$

* Thus, hydrolysis is the reverse of neutralization.

With all the similarity between this reaction and the reaction considered above, there is however an essential difference between them too, which is accounted for by the fact that the ionization constant of CH_3COOH ($K = 1.74 \times 10^{-5}$) is much higher than that of HCN. Hence, the process of binding of hydrogen ions cannot in this case go so far as with HCN, and therefore at equilibrium the smaller portion of the salt is found to have been converted into the corresponding base and acid, i.e., the degree of hydrolysis of the salt* is lower. In view of the lower degree of hydrolysis of CH_3COONa, a 0.1*N* solution of this salt shows a less alkaline reaction (pH 8.88) than a solution of KCN of the same concentration (pH 11.1).

2. Hydrolysis of Salts Formed from Weak Bases and Strong Acids. In aqueous solutions of such salts, the OH^- ions of water are bound and hydrogen ions accumulate in solution. For example,

$$NH_4^+ + H_2O \rightleftharpoons NH_4OH + H^+$$

Thus, an NH_4Cl solution will give an acid reaction due to hydrolysis. Indeed, a 0.1*N* solution of this salt has a pH of 5.12.

We shall encounter this type of hydrolysis in studying the cations of Group III since the hydroxides formed by these cations are very weak bases. Solutions of the salts of Group III cations, which are formed by the anions of strong acids, show an acid reaction due to hydrolysis.

3. Hydrolysis of Salts Formed from Weak Bases and Weak Acids. In this case, both the hydrogen and hydroxyl ions of water are bound:

$$\begin{array}{ccccc} CH_3COONH_4 \rightarrow & NH_4^+ & + & CH_3COO^- \\ & + & & + \\ H_2O \rightleftharpoons & OH^- & + & H^+ \\ & \downarrow\uparrow & & \downarrow\uparrow \\ & NH_4OH & & CH_3COOH \end{array}$$

Since the ionization constants of NH_4OH (1.76×10^{-5}) and CH_3COOH (1.74×10^{-5}) are almost equal, the H^+ and OH^- ions must be bound (removed) to the same extent and therefore the reaction of the solution remains practically neutral (pH $\approx$ 7). It is not difficult however to see that hydrolysis does take place in this case too, and to a considerable extent at that (because of the fact that not one but two processes simultaneously shift the dissociation equilibrium of water). This can be recognized by the odour of the CH_3COONH_4 solution which smells both of acetic acid and of ammonia.

It is clear that the reaction of solutions of the salts in question will be neutral only if the base and acid formed in the hydrolysis have close ionization constants, i.e., are practically equal in strength. Solutions of salts such as $(NH_4)_2S$ or $(NH_4)_2CO_3$, which are formed by a base stronger than the acid, show a weakly alkaline reaction (pH $\approx$ $\approx$ 9.2).

* The method used for calculating the degree of hydrolysis is considered in Sec. 5.6.

4. Salts Formed from Strong Bases and Strong Acids. Salts of this kind undergo no hydrolysis since, when they are dissolved in water, neither H^+ nor OH^- (both resulting from the dissociation of water) can be removed.

Since no binding of the ions from water occurs, the pH cannot be changed either. Hence, the pH of solutions of the salts formed from strong bases and acids is the same as that of pure water (pH $\approx$ 7).

Thus, hydrolysis is caused by the shift of the dissociation equilibrium of water due to the formation of slightly ionized compounds. If such compounds do not result, i.e., if both the acid and the base which form a salt are strong, no hydrolysis takes place either. Conversely, the weaker the acid and base, the more strongly will be shifted the water equilibrium and the greater will be the degree of hydrolysis of the salt. What has just been said of hydrolysis is illustrated by Table 5.2.

TABLE 5.2. **Hydrolysis of Salts**

Salt-forming compounds		Hydrolysis	Reaction of solution
Base	Acid		
Strong	Weak	Occurs	Alkaline (pH $>$ 7)
Weak	Strong	Occurs	Acidic (pH $<$ 7)
Weak	Weak	Extensive	Depends on the relative strength of acid and base
Strong	Strong	Does not occur	Neutral (pH $\approx$ 7)*

* The reaction of the solution is determined by the pH of water used to prepare the solution.

In the case of di- and polybasic acids, the process of hydrolysis is complicated by the fact that the hydrolysis takes place by steps just as the ionization of the acid. For example, the hydrolysis of Na_2CO_3 may be represented by the following equations:

(1) for the first stage

$$CO_3^{2-} + H_2O \rightleftharpoons HCO_3^- + OH^-$$

(2) for the second stage

$$HCO_3^- + H_2O \rightleftharpoons H_2CO_3 + OH^-$$

Since HCO_3^- anions ($K_2 = 4.8 \times 10^{-11}$) are ionized to a much smaller extent than H_2CO_3 molecules ($K_1 = 4.5 \times 10^{-7}$), the first-stage hydrolysis proceeds to an incomparably greater extent than the second stage.

The hydrolysis of Na_3PO_4 involves three steps, the first step being the strongest:

$$PO_4^{3-} + H_2O \rightleftharpoons HPO_4^{2-} + OH^-$$

One must clearly understand that the stepwise hydrolysis of salts, say, Na_2CO_3, cannot be represented by the overall equation

$$CO_3^{2-} + 2H_2O \rightleftharpoons H_2CO_3 + 2OH^-$$

obtained by summing the equations for the individual steps of hydrolysis. Indeed, when the equations are added up, the common term HCO_3^- cancels out. This would be permissible if all the HCO_3^- ions formed at the first step were converted into H_2CO_3 in accordance with the second-stage hydrolysis equation. But in fact only a negligibly small amount of HCO_3^- is converted into H_2CO_3, while most of it remains in solution and is the main product of hydrolysis; and yet the overall equation given above does not contain HCO_3^- at all.

Thus, after the hydrolysis is completed, out of p g-ion of CO_3^{2-} there is left in solution $(p - x)$ g-ion of CO_3^{2-}, and $(x - y)$ g-ion of HCO_3^- and y mole of H_2CO_3 are formed.

As shown above, the hydrolysis of neutral salts of polybasic acids leads to the formation of anions of acid salts. Likewise, the hydrolysis of neutral salts of polyacid bases involves the formation of cations of basic salts which have metal and hydroxyl ions in their composition. For example, in the analysis of Group I cations, when a solution of $MgCl_2$ is evaporated, one of the Cl^- ions of this salt is replaced by the OH^- ion from water and the basic salt MgOHCl is formed:

$$MgCl_2 + H_2O \rightarrow MgOHCl + HCl\uparrow$$

When the hydrated aluminium ion is hydrolysed, the following reactions take place*:

$$Al(OH_2)_6^{3+} + H_2O \rightarrow Al(OH_2)_5OH^{2+} + H_3O^+ \quad \text{(1st stage)}$$
$$Al(OH_2)_5(OH)^{2+} + H_2O \rightarrow Al(OH_2)_4(OH)_2^+\downarrow + H_3O^+ \quad \text{(2nd stage)}$$
$$Al(OH_2)_4(OH)_2^+\downarrow + H_2O \rightarrow Al(OH_2)_3(OH)_3^+\downarrow + H_3O^+ \quad \text{(3rd stage)}$$

But even these reactions do not cover all the processes involved in the hydrolysis of aluminium salts. It has been found that, along with the monomer ions $Al(OH)^{2+}$ and $Al(OH)_2^+$, there also exist in solution the polymer ions $Al_2(OH)_2^{4+}$ and $Al_3(OH)_6^{3+}$. Moreover, complex ions such as $Al[(OH)_5Al_2]_n^{(n+3)+}$ and $Al_6(OH)_{15}^{3+}$ may also be expected to exist in solution.

Analogously, the hydrolysis of salts of Th(IV) evidently involves the formation of a number of polymeric complexes of the composition $Th[(OOH)Th]_n^{(n+4)+}$ (where n has various values which sometimes exceed 6).

---

* Since in hydrated metal ions the metal ion and water molecules are bound through oxygen atoms, it is adopted that the formula of water be written as OH_2.

When salts of Cu(II) undergo hydrolysis, the formation of the dimer $Cu_2(OH)_2^{2+}$ predominates.

The hydrolysis of salts is a very complicated multi-stage process, and this circumstance has to be taken into account in accurate calculations. For simplicity, however, in a number of cases in writing reaction equations, we shall make use of the formulas of unhydrated ions and molecules, for example, $Al(OH)_3$ instead of $Al_2O_3 \cdot nH_2O$.

As pointed out above, hydrolysis is a reversible process which leads to the establishment of equilibrium. In the majority of cases, only a small part of a salt is found to have been converted into new substances. If however both the acid and base that form a salt are very weak or if one of the substances formed in the hydrolysis accumulates in solution in such amounts that its precipitation or vaporization becomes possible, the hydrolysis will become irreversible and will lead to a practically complete decomposition of the salt by water. This is what happens in the hydrolysis of Al_2S_3:

$$Al_2S_3 + 6H_2O \rightarrow Al(OH)_3\downarrow + 3H_2S\uparrow$$

It is obvious that such salts can exist only in the absence of water.

In carrying out numerous analytical reactions, one has to consider the hydrolysis of the salts used as reagents because sometimes not the ions of the salt itself but the products of its hydrolysis enter into reaction. To illustrate, let us consider an example.

Most of Group III cations are precipitated by the group reagent $(NH_4)_2S$ in the form of the sulphides FeS, MnS, ZnS, etc. Under the same conditions, however, Al^{3+}, Cr^{3+}, and Ti^{4+} form, instead of sulphides, the corresponding hydroxides: $Al(OH)_3$, $Cr(OH)_3$, and $Ti(OH)_4$ (more precisely, the hydrated or hydrous oxides $Al_2O_3 \cdot xH_2O$, $Cr_2O_3 \cdot xH_2O$, and $TiO_2 \cdot xH_2O$). The cause will be clear if it is taken into account that ammonium sulphide forms on hydrolysis considerable quantities of ammonium hydroxide:

$$NH_4^+ + S^{2-} + H_2O \rightleftharpoons HS^- + NH_4OH$$

The hydroxyl ion concentration formed in the solution is found to be sufficient for the solubility products of the hydroxides of aluminium(III), chromium(III), and titanium(IV) to be exceeded, which results in their precipitation. Conversely, for other Group III cations, the sulphides whose solubility products are reached first are those which are much less soluble than the hydroxides. It is because of this that the sulphides and not the hydroxides of the corresponding cations are precipitated.

The reaction between $(NH_4)_2S$ and $AlCl_3$ may be represented by the following equations:

$$\begin{array}{c} 6NH_4^+ + 3S^{2-} + 3H_2O \rightarrow 3NH_4^+ + 3HS^- + 3NH_4OH \\ Al^{3+} + 3NH_4OH \rightarrow Al(OH)_3\downarrow + 3NH_4^+ \\ \hline Al^{3+} + 3S^{2-} + 3H_2O \rightarrow Al(OH)_3\downarrow + 3HS^- \end{array} \qquad (5.14)$$

It should be noted that this equation, the derivation of which is based only on the first stage of hydrolysis of $(NH_4)_2S$, does not reflect quite accurately the processes involved. In fact, part of the hydrosulphide formed undergoes hydrolysis:

$$NH_4^+ + HS^- + H_2O \rightarrow H_2S\uparrow + NH_4OH$$

Therefore, apart from the reaction given above, there also takes place the following reaction:

$$2Al^{3+} + 3S^{2-} + 6H_2O \rightarrow 2Al(OH)_3 + 3H_2S\uparrow \qquad (5.15)$$

Whether the reaction proceeds according to Eq. (5.14) or (5.15) depends largely on the conditions of the experiment: the solution pH and temperature. If the pH increases, HS^- is predominantly formed according to Eq. (5.14). If the pH decreases, the compound preferentially formed according to Eq. (5.15) is H_2S.

This process is also favoured by heating since rise of temperature results in an extensive vaporization of hydrogen sulphide which is thus removed from the reaction space.

Naturally, when the solution pH decreases to a very considerable extent, aluminium hydroxide may dissolve.

The formation of aluminium(III), chromium(III), and titanium(IV) hydroxides under the action of ammonium sulphide may also be attributed to the shift of the hydrolysis equilibrium of salts of these metals due to the binding of the acid formed (H^+ ions) by the S^{2-} ions from the sulphide. For example,

$$Al^{3+} + 3H_2O \rightarrow Al(OH)_3\downarrow + 3H^+$$

$$3S^{2-} + 3H^+ \rightarrow 3HS^-$$

$$\overline{Al^{3+} + 3S^{2-} + 3H_2O \rightarrow Al(OH)_3\downarrow + 3HS^-}$$

This explanation by no means contradicts the above interpretation of the mechanism of this reaction but rather supplements it. Indeed, as pointed out above, the cause of the precipitation of aluminium hydroxide is the fact that the OH^- concentration in solution is sufficient for the solubility product constant of $Al(OH)_3$ to be exceeded. But the concentration of OH^- ions results not only from the formation of NH_4OH on hydrolysis of ammonium sulphide but also from the binding of the acid formed when the ammonium salt is hydrolysed. But the removal of hydrogen ions is tantamount to the formation of hydroxyl ions since the product of their concentrations is always constant.

The shift of the hydrolysis equilibrium caused by the consumption of one of the products of hydrolysis for a certain reaction has to be taken into consideration in other cases too. If to a solution of $AlCl_3$ there is added an alkali in small amounts, then, as the hydrochloric acid formed by hydrolysis is neutralized, ever increasing amounts of the acid and of aluminium hydroxide are formed, the latter beginning to precipitate:

$$Al^{3+} + 3H_2O \rightarrow Al(OH)_3\downarrow + 3H^+$$
$$3H^+ + 3OH^- \rightarrow 3H_2O$$

It is not difficult however to understand that, as far as there is some $AlCl_3$, the solution will show an acid reaction (due to hydrolysis) as before. Thus, the acidity of solutions of the salts of Group III cations, which depends on the hydrolysis of these salts, cannot be eliminated by neutralization. But if one nevertheless speaks occasionally of the neutralization of such solutions, it is not the acid formed by hydrolysis but the one added to the solution that is meant in this case. The added acid can be neutralized as follows. To the solution is added, drop by drop, a solution of an alkali until a turbidity of the hydroxides appears, which still persists on mixing; this turbidity is dissolved in 1 or 2 drops of acid and the neutralization is considered complete. Though the resulting solution gives an acid reaction as before (pH < 7), the acidity of the solution is attributed to the acid that results from the hydrolysis.

Hydrolysis is intensified by removal of one of its products, but it can be suppressed by addition of such a product. Advantage is often taken of this possibility when solutions of readily hydrolysable salts are to be stored. For instance, if we prepare an aqueous solution of an iron salt, say, $Fe_2(SO_4)_3$, a precipitate of the basic salts will rapidly separate out from solution due to hydrolysis. If this solution is acidified, i.e., if H^+ (one of the hydrolysis products) is introduced into it, the reaction opposing hydrolysis will predominate, and the solution will be more stable on storage.

In just the same way, the hydrolysis of salts of weak acids (KCN, CH_3COONa, etc.) can be suppressed by addition of hydroxyl ions, which are one of the hydrolysis products in this case. The concentration of the anions of the salts (CN^-, CH_3COO^-) in solution increases.

From what has been said above, it becomes clear that the hydrolysis of salts can be repressed by adjustment of the solution pH.

Moreover, the hydrolysis equilibrium is influenced greatly by the temperature and the concentration of the salt solution.

5.6. Calculation of Hydrolysis Constant, Degree of Hydrolysis and pH of Solutions of Hydrolysable Salts

The degree of hydrolysis is defined as the number indicating the fraction hydrolysed of the total amount of the salt, i.e., the fraction converted by the action of H_2O into the corresponding acid or base (or into acid or basic salts). The degree of hydrolysis is not difficult to calculate by using the ionization-constant equations for the respective acid (or base) and the ion product of water.

Let us derive equations for calculating the hydrolysis constant K_h, the degree of hydrolysis h, and the concentration of hydrogen ions $[H^+]$ (and hence the pH value) in solutions of various types of hydrolysable salts.

1. Salts of a Monobasic Weak Acid and a Monoacid Strong Base. Let us denote a weak acid by HAn (where An^- is the anion of the acid) and a strong base by BOH (where B^+ is the cation of the base). The hydrolysis of the salt formed will involve only the anion because the cation gives no slightly ionized compound and remains in solution in the form of a hydrated ion.

The equation for the reaction of hydrolysis is written in this case as follows:

$$An^- + H_2O \rightleftharpoons HAn + OH^- \tag{5.16}$$

The equation for the equilibrium constant of this reaction is

$$K_{eq} = \frac{[OH^-][HAn]}{[An^-][H_2O]}$$

In dilute solutions water is present in a large excess and, since the ionization of water is negligibly slight, the concentration of un-ionized water molecules may be regarded as practically constant and unchanging with any shift of the hydrolysis equilibrium. Therefore, multiplying both sides of the equation by $[H_2O]$ yields:

$$K_{eq}[H_2O] = \frac{[OH^-][HAn]}{[An^-]}$$

The product K_{eq} $[H_2O]$ is a constant quantity and is called the hydrolysis constant K_h. Then

$$K_h = \frac{[OH^-][HAn]}{[An^-]} \tag{5.17}$$

The numerical value of K_h is easily found from the values of the ion product of water (K_w) and K_{HAn}. Indeed, from the equation for the ion product of water

$$[H^+][OH^-] = K_w$$

we have:

$$[OH^-] = \frac{K_w}{[H^+]}$$

Substituting this expression for $[OH^-]$ into Eq. (5.17), we get:

$$\frac{K_w[HAn]}{[H^+][An^-]} = K_h$$

or

$$K_h = \frac{K_w}{K_{HAn}} \tag{5.18}$$

Thus, combining Eqs. (5.17) and (5.18) gives:

$$\frac{[OH^-][HAn]}{[An^-]}=\frac{K_w}{K_{HAn}} \tag{5.19}$$

Using Eq. (5.19), we can calculate the degree of hydrolysis of the salt.

Let us assume that the concentration of the salt is C_{BAn} mole/litre. If the degree of hydrolysis is h, there are hydrolysed $C_{BAn}h$ moles of the salt, which results in the formation of $C_{BAn}h$ moles of HAn and $C_{BAn}h$ g-ions of OH^-. With this taken into account, we obtain from Eq. (5.19):

$$\frac{C^2_{BAn}h^2}{C_{BAn}-C_{BAn}h}=\frac{K_w}{K_{HAn}}$$

which gives

$$\frac{C_{BAn}h^2}{1-h}=\frac{K_w}{K_{HAn}} \tag{5.20}$$

If the value of h is small as compared with 1, then

$$C_{BAn}h^2 \approx \frac{K_w}{K_{HAn}}$$

and

$$h \approx \sqrt{\frac{K_w}{C_{BAn}K_{HAn}}} \approx \sqrt{\frac{K_h}{C_{BAn}}} \tag{5.21}$$

From Eq. (5.21) it is seen that the degree of hydrolysis is the greater:

(a) the higher the value of K_w, i.e., the higher the temperature of the solution;

(b) the lower the value of K_{HAn}, i.e., the weaker the acid whose salt undergoes hydrolysis;

(c) the smaller the value of C_{BAn}, i.e., the more strongly diluted is the solution.

The pH of solutions of such salts can be calculated from Eq. (5.19). From the equation for the hydrolysis (5.16) it follows that $[OH^-] = [HAn]$. Then Eq. (5.19) can be transformed into the following:

$$K_h=\frac{[OH^-]^2}{[An^-]}=\frac{K_w}{K_{HAn}} \tag{5.22}$$

If the degree of hydrolysis of the salt is low (less than 10 per cent), it may be assumed that the concentration of the unhydrolysed part of the salt $[An^-] \approx C_{BAn}$ (being a strong electrolyte, the salt dissociates practically completely in solution).

Equation (5.22) may be rewritten thus:

$$[OH^-]^2=\frac{K_w}{K_{HAn}}C_{BAn}$$

and

$$[OH^-]=\sqrt{\frac{K_w}{K_{HAn}}C_{BAn}}$$

Passing from $[OH^-]$ to $[H^+]=(K_w/[OH^-])$, we have:

$$[H^+]=\sqrt{\frac{K_wK_{HAn}}{C_{BAn}}}$$

and

$$pH=-\log[H^+]=-\frac{1}{2}\log K_w-\frac{1}{2}\log K_{HAn}+\frac{1}{2}\log C_{BAn}$$

Designating $-\log C_{BAn}$ by pC_{BAn}, $-\log K_{HAn}$ by pK_{HAn}, and $-\log K_w$ by pK_w, we obtain:

$$pH=\frac{1}{2}pK_w+\frac{1}{2}pK_{HAn}-\frac{1}{2}pC_{BAn} \quad (5.23)$$

If the degree of hydrolysis is great ($\geqslant$10 per cent), the value of h and the pH of the solution should be calculated by using a more exact equation.

The degree of hydrolysis is calculated from Eq. (5.20) by solving it as a full quadratic equation. The result is:

$$h=\frac{-K_w}{2C_{BAn}K_{HAn}}+\sqrt{\left(\frac{K_w}{2C_{BAn}K_{HAn}}\right)^2+\frac{K_w}{C_{BAn}K_{HAn}}} \quad (5.24)$$

On the other hand, considering that $[An^-]\neq C_{BAn}$ and $[An^-]=$ $=C_{BAn}-[HAn]=C_{BAn}-[OH^-]$, we substitute these values into Eq. (5.22):

$$K_h=\frac{[OH^-]^2}{C_{BAn}-[OH^-]}=\frac{K_w}{K_{HAn}}$$

We solve the quadratic equation for $[OH^-]$:

$$[OH^-]=\frac{-K_w}{2K_{HAn}}+\sqrt{\frac{K_w^2}{4K_{HAn}^2}+\frac{K_wC_{BAn}}{K_{HAn}}}$$

and find the pH as the difference:

$$pH=pK_w-pOH \quad (5.25)$$

We can equally successfully calculate the pH by using the quantities h and C_{BAn} since h is the ratio of $[OH^-]$, which results from the hydrolysis of the salt, to the concentration of this salt, C_{BAn}:

$$[OH^-]=C_{BAn}h \quad pOH=-\log(C_{BAn}h)$$
$$pH=pK_w-pOH=pK_w-p(C_{BAn}h) \quad (5.26)$$

2. Salts of a Monobasic Strong Acid and a Monoacid Weak Base. In this case only the cation of a weak base undergoes hydrolysis:

$$B^+ + H_2O \rightleftharpoons BOH + H^+$$

$$K_h = \frac{[H^+][BOH]}{[B^+]} = \frac{K_w}{K_{BOH}} \qquad (5.27)$$

whence

$$h \approx \sqrt{\frac{K_w}{C_{BAn}K_{BOH}}} \approx \sqrt{\frac{K_h}{C_{BAn}}} \qquad (5.28)$$

Let us recall that, just as in the previous case, this approximate formula can be used to calculate the degree of hydrolysis of only slightly hydrolysable salts ($h < 10$ per cent).

Similarly to the previous case, let us find equations for calculating $[H^+]$ and pH for salt solutions of this kind:

$$[H^+] = \sqrt{\frac{K_w C_{BAn}}{K_{BOH}}}$$

$$pH = \frac{1}{2}pK_w - \frac{1}{2}pK_{BOH} + \frac{1}{2}pC_{BAn} \qquad (5.29)$$

3. Salts of a Monoacid Weak Base and a Monobasic Weak Acid. Both the anion of a weak acid and the cation of a weak base undergo hydrolysis:

$$B^+ + An^- + H_2O \rightleftharpoons HAn + BOH$$

$$K_h = \frac{[HAn][BOH]}{[An^-][B^+]} \qquad (5.30)$$

According to the ionization constants of the acid and base, the ratios $[HAn]/[An^-]$ and $[BOH]/[B^+]$ are equal to:

$$\frac{[HAn]}{[An^-]} = \frac{[H^+]}{K_{HAn}} \quad \text{and} \quad \frac{[BOH]}{[B^+]} = \frac{[OH^-]}{K_{BOH}}$$

Substituting the data obtained into Eq. (5.30) yields:

$$K_h = \frac{[HAn][BOH]}{[An^-][B^+]} = \frac{[H^+][OH^-]}{K_{HAn}K_{BOH}} = \frac{K_w}{K_{HAn}K_{BOH}} \qquad (5.31)$$

We denote, as before, the total concentration of the salt by C_{BAn} and the degree of hydrolysis by h. Then the number of moles of the salt that undergo hydrolysis is $C_{BAn}h$. From each molecule of this salt there are formed, on hydrolysis, one molecule of an acid and one molecule of a base. So we have:

$$[HAn] = [BOH] = C_{BAn}h$$

The amount of unhydrolysed salt is:

$$C_{BAn} - C_{BAn}h = C_{BAn}(1-h)$$

Assuming this amount to be practically completely dissociated, we have as a first approximation:

$$[An^-] = [B^+] = C_{BAn}(1-h)$$

Substituting the values of the concentrations into Eq. (5.30) and taking into account Eq. (5.31), we obtain:

$$\frac{C_{BAn}hC_{BAn}h}{C_{BAn}(1-h)\,C_{BAn}(1-h)} = \frac{K_w}{K_{HAn}K_{BOH}} = K_h$$

which gives:

$$\frac{h^2}{(1-h)^2} = \frac{K_w}{K_{HAn}K_{BOH}} = K_h$$

Taking square roots of both sides of this equation gives:

$$\frac{h}{(1-h)} = \sqrt{\frac{K_w}{K_{HAn}K_{BOH}}} = \sqrt{K_h} \qquad (5.32)$$

and, finally,

$$h = \frac{\sqrt{\dfrac{K_w}{K_{HAn}K_{BOH}}}}{1+\sqrt{\dfrac{K_w}{K_{HAn}K_{BOH}}}} = \frac{\sqrt{K_h}}{1+\sqrt{K_h}} \qquad (5.33)$$

If $h \ll 1$, the value of h in the denominator of Eq. (5.32) may be ignored and h can be calculated from the following formula:

$$h = \sqrt{K_h} \qquad (5.34)$$

If the hydrolysis constant [see Eq. (5.31)] is small, hydrolysis does not go too far. The acidity of the solutions of such salts can be determined as follows.

In the case under consideration we may again assume that $[HAn] \approx \approx [BOH]$ and $[An^-] \approx [B^+] \approx C_{BAn}$. The calculation then can be carried out by using Eq. (5.31):

$$K_h = \frac{K_w}{K_{HAn}K_{BOH}} = \frac{[HAn]^2}{C_{BAn}^2} \qquad (5.35)$$

On the other hand, since

$$[HAn] = \frac{[H^+][An^-]}{K_{HAn}} = \frac{[H^+]C_{BAn}}{K_{HAn}}$$

it follows that, substituting this value of [HAn] into Eq. (5.35) and solving it for $[H^+]$, we obtain:

$$[H^+] = \sqrt{\frac{K_wK_{HAn}}{K_{BOH}}}$$

and

$$pH = \frac{1}{2}pK_w + \frac{1}{2}pK_{HAn} - \frac{1}{2}pK_{BOH} \qquad (5.36)$$

Equation (5.36) shows that, if the values of pK_{HAn} and pK_{BOH} are approximately equal, the solutions of hydrolysable salts give an almost neutral reaction (pH $\approx$ 7). If $pK_{HAn} < pK_{BOH}$, then the reaction of the medium is acidic (pH $<$ 7). If $pK_{HAn} > pK_{BOH}$, the reaction is alkaline (pH $>$ 7). To put it in other words, if the acid formed due to hydrolysis is stronger than the base, the solution shows a weakly acid reaction. Conversely, if the base is stronger than the acid, the reaction of the solution is weakly alkaline.

In such cases, the concentrations of the hydrolysis products, though being close, are not equal, i.e., $[HAn] \neq [BOH]$. Therefore, here we should not speak of the degree of hydrolysis of the salt; the hydrolysis of the cation and that of the anion must be treated separately.

But usually, if the degree of hydrolysis is not very great, then the presence in solution of a large concentration of B^+ and An^- of the salt represses the dissociation of the weak acid and weak base produced by hydrolysis. In view of this, it may be assumed in practice that $[BOH] \approx$ $\approx [HAn]$. This enables one to calculate h and pH from approximate equations (5.33) or (5.34) and (5.36).

Apart from the three cases considered, the hydrolysis of normal and acid salts formed by polybasic weak acids and strong or weak monoacid bases is also of importance to analytical chemistry (salts of the type Na_2S, $(NH_4)_2S$, NaHS, NH_4HS, Na_3PO_4, Na_2HPO_4, NaH_2PO_4 etc.).

The hydrolysis of salts formed by polyacid weak bases, for example, $FeCl_3$ and $Fe(CH_3COO)_3$, is complicated by the formation of slightly soluble compounds or colloidal solutions. Therefore, such cases of hydrolysis will not be discussed here.

4. Normal Salts Formed by Polybasic Weak Acid and a Monoacid Strong Base. Normal salts of this type, having the general formula B_nAn (where B^+ is the cation of a monoacid strong base, and $n \geqslant 2$), are hydrolysed by stages:

$$An^{n-} + H_2O \rightleftharpoons HAn^{(n-1)-} + OH^-$$

$$HAn^{(n-1)-} + H_2O \rightleftharpoons H_2An^{(n-2)-} + OH^-, \text{ etc.}$$

Usually, the successive ionization constants for polybasic weak acids are different, and therefore, as a first approximation, we may assume that the successive steps of hydrolysis may be neglected; the values of K_h, h, and pH for solutions of such salts can be calculated on the basis of the first step of hydrolysis according either to Eqs. (5.18), (5.21), and (5.23) or (5.26), or to Eqs. (5.18), (5.24), and (5.25) or (5.26).

5. Normal Salts Formed by a Polybasic Weak Acid and a Monoacid Weak Base. Normal salts of this kind, having the general formula B_nAn (where B^+ is the cation of a monoacid weak base and $n \geqslant 2$), are also hydrolysed by stages, both the cation and the anion undergoing hydrolysis:

$$B^+ + H_2O \rightleftharpoons BOH + H^+$$
$$An^{n-} + H_2O \rightleftharpoons HAn^{(n-1)-} + OH^-$$
$$HAn^{(n-1)-} + H_2O \rightleftharpoons H_2An^{(n-2)-} + OH^-, \text{ etc.}$$

In this case too, in view of what has been said in the previous example, we may ignore, as a first approximation, all the steps of hydrolysis of the anion, except the first step. On the other hand, since both the cation B^+ and the anion An^{n-} are hydrolysed and the H^+ and OH^- ions formed are bound into un-ionized molecules of water, it should be expected that the greater part of the ions of the salt undergoes hydrolysis. If $K_{HAn^{(n-1)-}}$ and K_{BOH} are sufficiently small and differ greatly from each other in magnitude, we can no longer neglect the fact that the hydrolysis proceeds by stages and goes rather far. Therefore, the quantities h and pH should be calculated using more exact equations, the solution of which is difficult.

Usually, in order to obtain a sufficiently approximate idea of the values of K_h, h, and pH for such salts, use may be made of the equations given below.

If the completely dissociated salt has the formula B_2An, we can write, neglecting the second hydrolysis step (in case K_h is small):

$$B_2An \rightleftharpoons 2B^+ + An^{2-} \quad \text{(dissociation)}$$
$$\left.\begin{aligned} B^+ + H_2O &\rightleftharpoons BOH + H^+ \\ An^{2-} + H_2O &\rightleftharpoons HAn^- + OH^- \end{aligned}\right\} \text{(hydrolysis)}$$

The overall hydrolysis process will be written thus:

$$B^+ + An^{2-} + H_2O \rightleftharpoons BOH + HAn^-$$

(since the H^+ and OH^- ions resulting from the hydrolysis of the cation and anion will again give an un-ionized water molecule).

Hence, the hydrolysis constant will be equal to

$$K_h = \frac{[BOH][HAn^-]}{[B^+][An^{2-}]} = \frac{[BOH][HAn^-]K_w}{[B^+][OH^-][An^{2-}][H^+]} = \frac{K_w}{K_{BOH}K_{HAn^-}} \quad (5.37)$$

and the degree of hydrolysis will be

$$h = \frac{[BOH]}{C_{B_2An}} \approx \frac{[HAn^-]}{C_{B_2An}}$$

As a first approximation,

$$[BOH] \approx [HAn^-] \approx hC_{B_2An}$$
$$[B^+] = 2C_{B_2An} - [BOH] = C_{B_2An}(2-h)$$

and

$$[An^{2-}] = C_{B_2An} - [HAn] = C_{B_2An}(1-h)$$

Taking into account that $h \ll 1$, one has:

$$K_h = \frac{h^2 C^2_{B_2An}}{C^2_{B_2An}(2-h)(1-h)} \approx \frac{h^2}{2}$$

and

$$h \approx \sqrt{2K_h} = \sqrt{2\frac{K_w}{K_{BOH}K_{HAn^-}}} \qquad (5.38)$$

The pH of the solution can be calculated either approximately using Eq. (5.36) or somewhat more accurately from the values of h and K_{HAn^-}, assuming that $[An^{2-}] = C_{B_2An} - [HAn^-]$:

$$[HAn^-] = \frac{[H^+](C_{B_2An} - [HAn^-])}{K_{HAn^-}} = \frac{[H^+]C_{B_2An}}{K_{HAn^-} + [H^+]} \approx hC_{B_2An}$$

Solving this equality for $[H^+]$ gives:

$$[H^+] = \frac{hK_{HAn^-}}{1-h} = \frac{\sqrt{2\frac{K_wK_{HAn^-}}{K_{BOH}}}}{1-\sqrt{2\frac{K_w}{K_{BOH}K_{HAn^-}}}}$$

Since the case under consideration is one in which $h \ll 1$, the second term in the denominator of this fraction may be neglected and the following written:

$$[H^+] = \sqrt{2K_w\frac{K_{HAn^-}}{K_{BOH}}}$$

$$pH = \frac{1}{2}pK_w + \frac{1}{2}pK_{HAn^-} - \frac{1}{2}pK_{BOH} - \frac{1}{2}\log 2 \qquad (5.39)$$

As seen, Eq. (5.39) differs little from Eq. (5.36).

But if the value of K_h found at the beginning is fairly large, then the h and pH must be calculated as follows, though to a certain degree of approximation too.

A salt of the type B_2An dissociates according to the equation

$$B_2An \rightleftharpoons 2B^+ + An^{2-}$$

But if K_h is sufficiently large, one then cannot assume that $[B^+] = 2C_{B_2An}$ and $[An^{2-}] = C_{B_2An}$. In this case, $[B^+] \neq C_{B_2An}$ and $[An^{2-}] \neq C_{B_2An}$.

From the foregoing it follows that, if the solution still contains unhydrolysed anions at a concentration $[An^{2-}]$, then the concentration of the hydrolysed anion is equal to $[HAn^-] = (C_{B_2An} - [An^{2-}])$; the concentration of the hydrolysed cation is the same, as a first approximation, i.e., $[BOH] = (C_{B_2An} - [An^{2-}])$. Hence, the unhydrolysed portion of B^+ cations is $[B^+] = 2C_{B_2An} - (C_{B_2An} - [An^{2-}]) = (C_{B_2An} + [An^{2-}])$.

Substituting all these expressions for the concentrations into Eq. (5.31) gives:

$$K_h = \frac{(C_{B_2An} - [An^{2-}])^2}{(C_{B_2An} + [An^{2-}])[An^{2-}]} = \frac{K_w}{K_{HAn^-} K_{BOH}} \qquad (5.40)$$

Since K_h is large, i.e., the hydrolysis goes too far, $[An^{2-}]$ is fairly small as compared with the total initial concentration of salt, C, and therefore $[An^{2-}]$ may be neglected. Equation (5.40) then will assume the following form:

$$K_h = \frac{C^2_{B_2An}}{C_{B_2An}[An^{2-}]} = \frac{C_{B_2An}}{[An^{2-}]} = \frac{K_w}{K_{HAn^-} K_{BOH}}$$

and, finally,

$$[An^{2-}] = C_{B_2An} \frac{K_{HAn^-} K_{BOH}}{K_w} = \frac{C_{B_2An}}{K_h} \qquad (5.41)$$

Having found the concentration $[An^{2-}]$ from Eq. (5.41), one can calculate all the other concentrations:

$$[B^+] = C_{B_2An} + [An^{2-}] = C_{B_2An} + \frac{C_{B_2An}}{K_h} = C_{B_2An} \frac{K_h + 1}{K_h} \qquad (5.42)$$

$$[BOH] \approx [HAn^-] = C_{B_2An} - [An^{2-}] = C_{B_2An} - \frac{C_{B_2An}}{K_h} = C_{B_2An} \frac{K_h - 1}{K_h} \qquad (5.43)$$

Hence, the degree of hydrolysis of the An^{2-} ions of this salt is:

$$h_{An^{2-}} = \frac{[HAn^-]}{C_{B_2An}} = \frac{C_{B_2An} - [An^{2-}]}{C_{B_2An}} = 1 - \frac{[An^{2-}]}{C_{B_2An}} = 1 - \frac{1}{K_h} \qquad (5.44)$$

and the degree of hydrolysis of the B^+ ions is given by:

$$h_{B^+} = \frac{[BOH]}{2C_{B_2An}} = \frac{C_{B_2An} - [An^{2-}]}{2C_{B_2An}} = 0.5 - \frac{1}{2K_h} \qquad (5.45)$$

The pH of the solution of the salt B_2An with a concentration of C_{B_2An} mole/litre can be calculated using one of the following equalities:

$$\frac{[HAn^-]}{[An^{2-}]} = \frac{[H^+]}{K_{HAn^-}} \quad \text{or} \quad \frac{[BOH]}{[B^+]} = \frac{[OH^-]}{K_{BOH}}$$

From the first equality we have:

$$[H^+] = K_{HAn^-} \frac{[HAn^-]}{[An^{2-}]}$$

Substituting the value of $[HAn^-]$ from Eq. (5.43) and of $[An^{2-}]$ from Eq. (5.41) yields

$$[H^+] = K_{HAn^-}(K_h - 1)$$

which gives:

$$pH = pK_{HAn^-} + p(K_h - 1) \qquad (5.46)$$

From the second equality we obtain:

$$[OH^-] = K_{BOH} \frac{[BOH]}{[B^+]}$$

Substituting the values of [BOH] from Eq. (5.43) and of $[B^+]$ from Eq. (5.42) gives:

$$[OH^-] = K_{BOH} \frac{(K_h - 1)}{(K_h + 1)}$$

from which one has:

$$pOH = pK_{BOH} + p(K_h - 1) - p(K_h + 1)$$

and

$$pH = pK_w - pOH = pK_w - pK_{BOH} - p(K_h - 1) + p(K_h + 1) \qquad (5.47)$$

In a more accurate calculation, especially when the hydrolysis constant is not very large, the $[An^{2-}]$ should be found by using Eq. (5.40) taking into account the quantity $[An^{2-}]$ enclosed in parentheses and solving this equation as a full quadratic equation for $[An^{2-}]$:

$$[An^{2-}] = \frac{-C_{B_2An}(2 + K_h)}{2(K_h - 1)} + \sqrt{\frac{C^2_{B_2An}(2 + K_h)^2}{4(K_h - 1)^2} + \frac{C^2_{B_2An}}{(K_h - 1)}} \qquad (5.48)$$

Further calculation is carried out as described above.

6. Acid Salts Formed by a Polybasic Weak Acid and a Monoacid Strong Base. The hydrolysis of acid salts of this type proceeds by stages if the basicity of a weak acid is equal to or larger than 2.

Let us consider, as an example, a salt of the composition B_2HAn (where B^+ is the cation of a monoacid strong base).

In a solution of the salt containing the anion of a polybasic weak acid, there must also be present the ions or molecules formed as a result of the addition of hydrogen ions or ionization. In this particular case, apart from the anion of the salt itself, HAn^{2-}, the solution also contains H_2An^- and An^{3-} ions and H_3An molecules. The concentrations of all these ions and also of the un-ionized molecules of the acid must satisfy all the ionization constants of the given weak acid.

According to what has just been said, the hydrolysis of HAn^{2-} may be written in the following form:

$$HAn^{2-} + H_2O \rightleftharpoons H_2An^- + OH^-$$

and

$$H_2An^- + H_2O \rightleftharpoons H_3An + OH^-$$

On the other hand, the ionization of HAn^{2-} gives H^+ and An^{3-} ions. As a first approximation, we may neglect the secondary hydrolysis (it is slight as compared with the first stage of hydrolysis) and hence the ionization of H_3An molecules (since the concentration of H_3An is small as compared with that of H_2An^- and the more so with the concentration of HAn^{2-}). When the equilibrium between the hydrogen and hydroxyl ions is reached, the hydrolysis of ions of the type HAn^{2-} may finally be represented as follows:

$$\begin{array}{c} HAn^{2-} + H_2O \rightleftharpoons H_2An^- + OH^- \\ HAn^{2-} \rightleftharpoons H^+ + An^{3-} \\ \hline 2HAn^{2-} \rightleftharpoons H_2An^- + An^{3-} \end{array} \quad (5.49)$$

So the concentration of these ions must satisfy the two ionization constants ($K_{HAn^{2-}}$ and $K_{H_2An^-}$) of the weak acid H_3An:

$$K_{H_2An^-} = \frac{[H^+][HAn^{2-}]}{[H_2An^-]} \quad \text{and} \quad K_{HAn^{2-}} = \frac{[H^+][An^{3-}]}{[HAn^{2-}]}$$

From this equation we have:

$$\frac{[H^+]}{K_{H_2An^-}} = \frac{[H_2An^-]}{[HAn^{2-}]} \quad \text{and} \quad \frac{K_{HAn^{2-}}}{[H^+]} = \frac{[An^{3-}]}{[HAn^{2-}]} \quad (5.50)$$

The hydrolysis constant of the salt is expressed in accordance with Eq. (5.49) as follows:

$$K_h = \frac{[H_2An^-][An^{3-}]}{[HAn^{2-}]^2} = \frac{[H_2An^-][An^{3-}]}{[HAn^{2-}][HAn^{2-}]} \quad (5.51)$$

Substituting into the right-hand side of Eq. (5.51), in place of the ratios

$$\frac{[H_2An^-]}{[HAn^{2-}]} \quad \text{and} \quad \frac{[An^{3-}]}{[HAn^{2-}]}$$

the corresponding values from Eq. (5.50), we obtain:

$$K_h = \frac{[H_2An^-][An^{3-}]}{[HAn^{2-}][HAn^{2-}]} = \frac{[H^+]}{K_{H_2An^-}} \frac{K_{HAn^{2-}}}{[H^+]} = \frac{K_{HAn^{2-}}}{K_{H_2An^-}} \quad (5.52)$$

Equation (5.49) shows that the hydrolysis of two HAn^{2-} ions involves the formation of one H_2An^- ion and one An^{3-} ion. Hence, if x ions of HAn^{2-} are hydrolysed, $x/2$ of H_2An^- and $x/2$ of An^{3-} are obtained. The amount left unhydrolysed is $(C - x)$ of HAn^{2-} ions.

Substituting these values into Eq. (5.52), one gets:

$$K_h = \frac{K_{HAn^{2-}}}{K_{H_2An^-}} = \frac{\left(\frac{x}{2}\right)^2}{(C_{B_2HAn} - x)^2} = \frac{x^2}{4(C_{B_2HAn} - x)^2}$$

If the value of x is small as compared with that of C_{B_2HAn}, which is what usually occurs, it may be omitted from the denominator. Then we finally obtain:

$$K_h = \frac{K_{HAn^{2-}}}{K_{H_2An^-}} = \frac{x^2}{4C^2_{B_2HAn}} \quad (5.53)$$

Solution of Eq. (5.53) for x enables one to calculate the degree of hydrolysis of acid salts of polybasic acids:

$$x=\sqrt{4C^2_{B_2HAn}K_h}=2C_{B_2HAn}\sqrt{K_h}=2C_{B_2HAn}\sqrt{\frac{K_{HAn^{2-}}}{K_{H_2An^-}}}$$

and

$$h=\frac{x}{C_{B_2HAn}}=2\sqrt{K_h}=2\sqrt{\frac{K_{HAn^{2-}}}{K_{H_2An^-}}} \qquad (5.54)$$

The pH of a solution of a salt of the B_2HAn type can be found from the values of $K_{HAn^{2-}}$ and $K_{H_2An^-}$.

According to Eq. (5.49), the concentrations of the products of hydrolysis of the HAn^{2-} ions of the acid salt B_2HAn are equal: $[H_2An^-] = [An^{3-}]$. Hence, the expression for K_h [see Eq. (5.52)] may be rewritten as follows:

$$K_h=\frac{[H_2An^-]^2}{[HAn^{2-}]^2}=\frac{K_{HAn^{2-}}}{K_{H_2An^-}}$$

From the ionization constant of the H_2An^- ion we have:

$$\frac{[H_2An^-]}{[HAn^{2-}]}=\frac{[H^+]}{K_{H_2An^-}}$$

Hence,

$$K_h=\frac{[H_2An^-]^2}{[HAn^{2-}]^2}=\frac{[H^+]^2}{(K_{H_2An^-})^2}=\frac{K_{HAn^{2-}}}{K_{H_2An^-}}$$

From this equation it follows that

$$[H^+]=\sqrt{K_{H_2An^-}K_{HAn^{2-}}}$$

and

$$pH=\frac{1}{2}pK_{H_2An^-}+\frac{1}{2}pK_{HAn^{2-}} \qquad (5.55)$$

7. Acid Salts Formed by a Polybasic Weak Acid and a Monoacid Weak Base. If we take, as an example, an acid salt of the composition BHAn (where B^+ is the cation of a monoacid weak base), both the cation B^+ and the anion HAn^- undergo hydrolysis. The hydrolysis of the anions of polybasic weak acids proceeds in a complicated way since all the ions (HAn^-, An^{2-}, B^+, H^+, and OH^-) and un-ionized molecules of the acid, base, and water (H_2An, BOH, and H_2O) present in solution must be in equilibrium with one another.

Depending on which of the ionization constants (of the acid H_2An or of the base BOH) is lower, hydroxyl or hydrogen ions will predominate in solution. This means that, depending on the magnitude of these ionization constants, one of the ions (HAn^- or B^+) will be hydrolysed to a larger extent.

In the simplest case, it may be assumed that HAn^- and B^+ are hydrolysed to an equal extent (neglecting the ionization of HAn^-, i.e., the formation of An^{2-}). Then

$$\begin{array}{c} HAn^- + H_2O \rightleftarrows H_2An + OH^- \\ B^+ + OH^- \rightleftarrows BOH \\ \hline HAn^- + B^+ + H_2O \rightleftarrows H_2An + BOH \end{array}$$

Here one can, to a fairly large degree of approximation, calculate K_h and h from formulas (5.31) and (5.33) or (5.34), i.e., regard the anion HAn^- as the anion of a monobasic weak acid. The pH of a solution of such a salt cannot however be calculated from formula (5.51) because the pH obtained will be the same irrespective of whether a strong or a weak base forms the given acid salt. For example, we would obtain the same pH value for solutions of NaHS and NH_4HS, which is not in accord with reality.

To a certain degree of approximation, we can calculate the pH of a solution of a salt of the type BHAn (where B^+ is the cation of a weak base) by making use of the degree of hydrolysis and the ionization constant of a weak acid or a weak base.

If the pH is calculated from the degree of hydrolysis and the ionization constant of a weak acid, we obtain:

$$\frac{h_{HAn^-}}{1-h_{HAn^-}} = \frac{[H_2An]}{[HAn^-]} = \frac{[H^+]}{K_{H_2An}}$$

which gives

$$[H^+] = \frac{h_{HAn^-} K_{H_2An}}{(1-h_{HAn^-})}$$

and

$$pH = pK_{H_2An} + ph_{HAn^-} - p(1-h_{HAn^-}) \tag{5.56}$$

If the calculation is based on the degree of hydrolysis and the ionization constant of a weak base, we have:

$$\frac{h_{B^+}}{1-h_{B^+}} = \frac{[BOH]}{[B^+]} = \frac{[OH^-]}{K_{BOH}}$$

whence

$$[OH^-] = \frac{h_{B^+} K_{BOH}}{(1-h_{B^+})} \qquad pOH = pK_{BOH} + ph_{B^+} - p(1-h_{B^+})$$

and, finally,

$$pH = pK_w - pOH = pK_w - pK_{BOH} - ph_{B^+} + p(1-h_{B^+}) \tag{5.57}$$

It is also possible to calculate pH (though somewhat less accurately) by formula (5.36), using the quantities pK_w, pK_{H_2An}, and pK_{BOH}.

In all the calculations given above several assumptions were made. For instance, in calculating h and pH, the concentrations of the hydrolysis products were assumed to be equal, though the ionization constants of the weak acid and weak base obtained by hydrolysis differed from each other. Furthermore, we neglected the stepwise hydrolysis of the anion of the weak polybasic acid resulting from the hydrolysis.

There exists a method which allows one to calculate, unambiguously and quite accurately, the values of h and pH for solutions of hydrolysable salts independently of their type. To this end, use is made of the general rule which states that all solutions of electrolytes (both weak and strong) are electrically neutral; in other words, the sum of the concentrations of all positively charged particles is equal to the sum of the concentrations of negatively charged particles in solution. As an example, we may consider the hydrolysis of the salt B_3An formed from a tribasic weak acid and a monoacid weak base, for example, $(NH_4)_3PO_4$; it is assumed that no polymerization of the ions or molecules takes place.

A solution of such a salt, no doubt, contains the ions An^{3-}, HAn^{2-}, H_2An^-, B^+, H^+, and OH^- and also H_3An, BOH, and H_2O molecules.

In conformity with the rule of electroneutrality of electrolyte solutions,

$$[H^+]+[B^+]=[OH^-]+[H_2An^-]+[HAn^{2-}]+[An^{3-}] \quad (5.58)$$

Transposing $[OH^-]$ to the left-hand side and $[B^+]$ to the right-hand side of Eq. (5.58) and expressing $[OH^-]$ through $K_w/[H^+]$, we get:

$$[H^+]-\frac{K_w}{[H^+]}=[H_2An^-]+[HAn^{2-}]+[An^{3-}]-[B^+] \quad (5.59)$$

On the other hand, denoting the initial concentration of the salt by C_{B_3An}, we obtain another equality:

$$C_{B_3An}=[H_3An]+[H_2An^-]+[HAn^{2-}]+[An^{3-}]=[B^+]+[BOH] \quad (5.60)$$

To solve Eq. (5.59) for $[H^+]$, it is necessary to find the concentrations of An^{3-}, HAn^{2-}, H_2An^-, and B^+ from the respective ionization constants:

$$K_{H_3An}=\frac{[H^+][H_2An^-]}{[H_3An]} \qquad K_{H_2An^-}=\frac{[H^+][HAn^{2-}]}{[H_2An^-]}$$

$$K_{HAn^{2-}}=\frac{[H^+][An^{3-}]}{[HAn^{2-}]} \qquad K_{BOH}=\frac{[B^+][OH^-]}{[BOH]}=\frac{[B^+]K_w}{[BOH][H^+]}$$

Multiplying the first three equalities together, we get:

$$K_{H_3An}K_{H_2An^-}K_{HAn^{2-}}=\frac{[H^+]^3[An^{3-}]}{[H_3An]}$$

which gives

$$[H_3An]=\frac{[H^+]^3[An^{3-}]}{K_{H_3An}K_{H_2An^-}K_{HAn^{2-}}}$$

Multiplying the second and third equalities together, we have:

$$K_{H_2An^-}K_{HAn^{2-}} = \frac{[H^+][An^{3-}]}{[H_2An^-]}$$

whence

$$[H_2An^-] = \frac{[H^+]^2[An^{3-}]}{K_{H_2An^-}K_{HAn^{2-}}}$$

The concentrations of HAn^{2-} and BOH are found from the third and fourth equalities, respectively:

$$[HAn^{2-}] = \frac{[H^+][An^{3-}]}{K_{HAn^{2-}}} \quad \text{and} \quad [BOH] = \frac{[B^+]K_w}{K_{BOH}[H^+]}$$

Substituting the expressions for $[H_3An]$, $[H_2An^-]$, $[HAn^{2-}]$, and $[BOH]$ into Eq. (5.60) and solving it for $[An^{3-}]$ and $[B^+]$, we obtain:

$$[An^{3-}] = \frac{K_{H_3An}K_{H_2An^-}K_{HAn^{2-}}C_{B_3An}}{[H^+]^3 + K_{H_3An}[H^+]^2 + K_{H_2An^-}K_{HAn^{2-}}[H^+] + K_{H_3An}K_{H_2An^-}K_{HAn^{2-}}} \tag{5.61}$$

$$[B^+] = \frac{K_{BOH}[H^+]C_{B_3An}}{K_{BOH}[H^+] + K_w} \tag{5.62}$$

If we denote the denominator in Eq. (5.61) by the letter α, this equation may be rewritten thus:

$$[An^{3-}] = \frac{K_{H_3An}K_{H_2An^-}K_{HAn^{2-}}C_{B_3An}}{\alpha} \tag{5.61a}$$

Inserting the expression for $[An^{3-}]$ from Eq. (5.61a) into the equations for the concentrations of H_2An^- and HAn^{2-}, we find that

$$[H_2An^-] = \frac{[H^+]^2K_{H_3An}C_{B_3An}}{\alpha} \tag{5.63}$$

$$[HAn^{2-}] = \frac{[H^+]K_{H_3An}K_{H_2An^-}C_{B_3An}}{\alpha} \tag{5.64}$$

The values of $[An^{3-}]$, $[HAn^{2-}]$, $[H_2An^-]$, and $[B^+]$ calculated from Eqs. (5.61a), (5.62), (5.63) and (5.64) are inserted into Eq. (5.59), which is solved for $[H^+]$, and the pH of a solution of the given electrolyte is then found.

The value of $[H^+]$ is used to calculate the value of α [the denominator in Eq. (5.61)] and to find the degree of hydrolysis of the salt B_3An from the following formula:

$$h = \frac{C_{B_3An} - [An^{3-}]}{C_{B_3An}} = \frac{\alpha - K_{H_3An}K_{H_2An^-}K_{HAn^{2-}}}{\alpha} \tag{5.65}$$

Interpretation of Hydrolysis from the Standpoint of the Protolytic Theory. According to the present-day concepts of acids and bases, which are in accord with the Brönsted acid-base theory, there is no fundamental difference between the ionization constant and the hydrolysis constant.

As known, water is an ampholyte and is capable of protonation:

$$H_2O + H^+ \rightleftharpoons H_3O^+$$

and of giving up a proton:

$$H_2O \rightleftharpoons H^+ + OH^-$$

The overall equation is

$$2H_2O \rightleftharpoons H_3O^+ + OH^-$$

For the protonation of water molecules to take place, the solution must contain a stronger acid capable of giving up its proton:

$$\underset{\text{Acid}_1}{CH_3COOH} + \underset{\text{Base}_2}{H_2O} \rightleftharpoons \underset{\text{Base}_1}{CH_3COO^-} + \underset{\text{Acid}_2}{H_3O^+}$$

Conversely, for a proton to be split off by a water molecule, it is necessary that a stronger base be present:

$$NH_3 + H_2O \rightleftharpoons NH_4^+ + OH^-$$

or

$$\underset{\text{Base}_1}{CH_3COO^-} + \underset{\text{Acid}_2}{H_2O} \rightleftharpoons \underset{\text{Acid}_1}{CH_3COOH} + \underset{\text{Base}_2}{OH^-}$$

From the foregoing we may conclude that aqueous solutions contain the following conjugated pairs:

$$\text{Acid/Base} \quad \text{and} \quad H_3O^+/H_2O \quad \text{or} \quad H_2O/OH^-$$

In a general form we may write:

$$\text{Acid}_1 + H_2O \rightleftharpoons \text{Base}_1 + H_3O^+ \qquad (5.66)$$

and

$$\text{Base}_2 + H_2O \rightleftharpoons \text{Acid}_2 + OH^- \qquad (5.67)$$

When Acid_1 reacts with Base_2, we have:

$$\text{Acid}_1 + \text{Base}_2 \rightleftharpoons \text{Base}_1 + \text{Acid}_2$$

For example,

$$CH_3COOH + NH_3 \rightleftharpoons CH_3COO^- + NH_4^+$$

The equilibrium constant (K_{eq}) of reaction (5.66) is given by

$$K_{eq} = \frac{[\text{Base}][H_3O^+]}{[\text{Acid}][H_2O]}$$

and

$$K_{eq}[H_2O] = K_a = \frac{[\text{Base}][H_3O^+]}{[\text{Acid}]} \qquad (5.68)$$

The greater the value of K_a, the stronger the acid and the higher the concentration of H_3O^+. In a conjugated acid-base pair, to a strong acid there corresponds a weaker base, and conversely, to a weak acid there corresponds a stronger base (there is here a full analogy with redox couples).

In fact, the constant K_a corresponds to the classical concept of the ionization constant of the acid K_{HAn}.

Analogously, for Eq. (5.67) we have:

$$K_b=\frac{[\mathrm{Acid}][\mathrm{OH^-}]}{[\mathrm{Base}]} \tag{5.69}$$

It is also known that $K_w = [H_3O^+][OH^-]$, whence $[OH^-] = (K_w/[H_3O^+])$. Substituting this value of $[OH^-]$ into Eq. (5.69), we obtain:

$$K_b=\frac{[\mathrm{Acid}]K_w}{[\mathrm{Base}][\mathrm{H_3O^+}]}$$

From Eq. (5.68) it is clear that

$$\frac{[\mathrm{Acid}]}{[\mathrm{Base}][\mathrm{H_3O^+}]}=\frac{1}{K_a}$$

whence

$$K_b=\frac{K_w}{K_a} \tag{5.70}$$

This value of K_b corresponds to the hydrolysis constant in the classical interpretation. On the other hand, the constant K_b embraces acids and bases by means of a single common characteristic—the quantity K_a.

For example, if a given acid is stronger than H_3O^+, the equilibrium in Eq. (5.66) is shifted completely to the right, and the solution practically contains only its conjugated weak base. Conversely, if the base is stronger than OH^-, the equilibrium in Eq. (5.67) is also found to have been shifted to the right, and only its conjugated weak acid is present in the solution. For example,

$$\mathrm{HCl+H_2O \longrightarrow Cl^- + H_3O^+}$$
$$\mathrm{NaOH+H_2O \longrightarrow Na^+(H_2O)+OH^-}$$

If it is assumed that $-\log[H_3O^+] = \mathrm{pH}$, this quantity is calculated in different cases in the following way.

For strong acids and bases:

$$\mathrm{pH=p[H_3O^+]=p}C_a$$
$$\mathrm{pOH=p[OH^-]=p}C_b$$

or

$$\mathrm{pH=p}K_w-\mathrm{p[OH^-]=p}K_w-\mathrm{p}C_b$$

This follows from the fact that, as shown above, strong acids and strong bases are completely converted into H_3O^+ and OH^-, respectively.

For weak acids (when the acid is not completely converted into H_3O^+), we have, according to Eq. (5.68):

$$[H_3O^+] = K_a \frac{[Acid]}{[Base]}$$

But [Base] = $[H_3O^+]$ according to Eq. (5.66), and [Acid] $\approx C_a$ since K_a is small. Hence,

$$H_3O^+]^2 = K_aC_a \quad \text{and} \quad [H_3O^+] = \sqrt{K_aC_a}$$

$$pH = \frac{1}{2}\,pK_a + \frac{1}{2}\,pC_a \qquad (5.71)$$

Equation (5.71) may also be used to calculate the pH of salts formed by the reaction between a strong acid and a weak base, say NH_4Cl.

Since salts dissociate completely, we have:

$$NH_4Cl \longrightarrow \underset{\text{Acid}}{NH_4^+} + \underset{\text{Base}}{Cl^-}$$

Chloride ions are a very weak base and cannot affect the pH of the solution. Therefore, the pH of the medium is controlled by the acid NH_4^+

$$NH_4^+ + H_2O \rightleftarrows NH_3 + H_3O^+ \qquad (5.72)$$

$$K_{NH_4^+} = \frac{[NH_3][H_3O^+]}{[NH_4^+]} \qquad (5.73)$$

From Eq. (5.72) it follows that $[NH_3] = [H_3O^+]$; on the other hand, since $K_{NH_4^+}$ is small, we may assume that $[NH_4^+] \approx C_s$. Substituting these data into Eq. (5.73) gives:

$$[H_3O^+]^2 = K_{NH_4^+}C_s$$

$$[H_3O^+] = \sqrt{K_{NH_4^+}C_s}$$

and

$$pH = \frac{1}{2}\,pK_{NH_4^+} + \frac{1}{2}\,pC_s \qquad (5.74)$$

As can be seen, Eq. (5.74) is identical to Eq. (5.71). Eq. (5.74) can be transformed if we take into account that

$$K_a = \frac{K_w}{K_b} \quad \text{[see Eq. (5.70)]}$$

whence

$$pK_a = pK_w - pK_b$$

Considering that $pK_a = pK_{NH_4^+}$ and $pK_b = pK_{NH_3}$ and replacing $pK_{NH_4^+}$ in Eq. (5.74) by its value, we get:

$$pH = \frac{1}{2}\,pK_w - \frac{1}{2}\,pK_{NH_3} + \frac{1}{2}\,pC_s \qquad (5.75)$$

For weak bases (when a base is not completely converted into OH^-) and salts formed from a weak acid and a strong base, the pH is calculated as follows.

For a weak base, according to Eq. (5.67), we have:

$$K_{eq}=\frac{[\text{Acid}][\text{OH}^-]}{[\text{Base}][\text{H}_2\text{O}]}$$

and

$$K_{eq}[\text{H}_2\text{O}]=K_b=\frac{[K_a][\text{OH}^-]}{[\text{Base}]} \quad (5.76)$$

If K_b is small, then [Base] $\approx C_s$ and [Acid] = $[\text{OH}^-]$; Eq. (5.76) upon rearrangement gives:

$$[\text{OH}^-]^2=K_bC_b \quad \text{and} \quad [\text{OH}^-]=\sqrt{K_bC_b}$$

Taking into account that $[\text{OH}^-] = K_w/[\text{H}_3\text{O}^+]$, we obtain:

$$\frac{K_w}{[\text{H}_3\text{O}^+]}=\sqrt{K_bC_b} \quad \text{and} \quad [\text{H}_3\text{O}^+]=\frac{K_w}{\sqrt{K_bC_b}}$$

On the other hand, $K_b = K_w/K_a$, and, consequently,

$$[\text{H}_3\text{O}^+]=\sqrt{\frac{K_w^2K_a}{K_wC_b}}=\sqrt{\frac{K_wK_a}{C_b}}$$

and

$$\text{pH}=\frac{1}{2}\text{p}K_w+\frac{1}{2}\text{p}K_a-\frac{1}{2}\text{p}C_b \quad (5.77)$$

For instance, sodium acetate is dissociated completely:

$$\text{CH}_3\text{COONa} \longrightarrow \underset{\text{Base}}{\text{CH}_3\text{COO}^-}+\underset{\text{Acid}}{\text{Na}^+}$$

Na^+ is a very weak acid and does not affect the pH of the solution. Hence, with account taken only of the effect of CH_3COO^-, we have:

$$\text{CH}_3\text{COO}^-+\text{H}_2\text{O} \rightleftarrows \text{CH}_3\text{COOH}+\text{OH}^-$$

$$K_{\text{CH}_3\text{COO}^-}=\frac{[\text{CH}_3\text{COOH}][\text{OH}^-]}{[\text{CH}_3\text{COO}^-]}=K_b$$

If K_b is small, then $[CH_3COO^-] \approx C_s$ and $[CH_3COOH] = [OH^-]$. We thus obtain:

$$[\text{OH}^-]^2=K_{\text{CH}_3\text{COO}^-}C_s$$

$$[\text{OH}^-]=\sqrt{K_{\text{CH}_3\text{COO}^-}C_s}$$

$$[\text{H}_3\text{O}^+]=\frac{K_w}{\sqrt{K_{\text{CH}_3\text{COO}^-}C_s}}=\frac{K_w}{\sqrt{K_bC_s}}$$

But $K_b = (K_w/K_a)$, and hence

$$[\text{H}_3\text{O}^+]=\sqrt{\frac{K_w^2K_a}{K_wC_s}}=\sqrt{\frac{K_wK_a}{C_s}}$$

and

$$\text{pH}=\frac{1}{2}\text{p}K_w+\frac{1}{2}\text{p}K_a-\frac{1}{2}\text{p}C_s \quad (5.78)$$

Equation (5.78) is similar to Eq. (5.77), the only difference being that C_s has been substituted for C_b.

If the solution contains a mixture of a weak acid and one of its salts formed by reaction with a strong base, then a buffer is formed. The pH of such a solution is determined, in conformity with the classical concepts, by the equation

$$pH = pK_{HAn} + pC_a - pC_s \tag{5.79}$$

The same result is obtained in calculations based on the Brönsted theory of acids and bases.

Consider, for example, the calculation of the pH of a solution containing a mixture of CH_3COOH and CH_3COONa:

$$\underset{\text{Acid}}{CH_3COOH} + H_2O \rightleftarrows \underset{\text{Base}}{CH_3COO^-} + H_3O^+$$

$$K_a = \frac{[CH_3COO^-][H_3O^+]}{[CH_3COOH]} \tag{5.80}$$

The salt CH_3COONa is completely dissociated:

$$CH_3COONa \longrightarrow \underset{\text{Base}}{CH_3COO^-} + \underset{\text{Acid}}{Na^+}$$

Na^+ is a very weak acid and does not influence the pH of the solution; hence

$$\underset{\text{Base}}{CH_3COO^-} + H_2O \rightleftarrows \underset{\text{Acid}}{CH_3COOH} + OH^-$$

$$K_b = \frac{[CH_3COOH][OH^-]}{[CH_3COO^-]} = \frac{[CH_3COOH]K_w}{[CH_3COO^-][H_3O^+]} \tag{5.81}$$

From Eq. (5.80) it follows that

$$[H_3O^+] = K_a \frac{[CH_3COOH]}{[CH_3COO^-]}$$

From Eq. (5.81) we have

$$[H_3O^+] = \frac{K_w[CH_3COOH]}{K_b[CH_3COO^-]}$$

Since one and the same solution is concerned, $[H_3O^+]$ from Eq. (5.81) must be equal to $[H_3O^+]$ from Eq. (5.80):

$$[H_3O^+] = K_a \frac{[CH_3COOH]}{[CH_3COO^-]} = \frac{K_w}{K_b} \frac{[CH_3COOH]}{[CH_3COO^-]} \tag{5.82}$$

But $(K_w/K_b) = K_a$. Substituting this value into Eq. (5.82) yields:

$$[H_3O^+] = K_a \frac{[CH_3COOH]}{[CH_3COO^-]} \tag{5.83}$$

The acid CH_3COOH is a weak acid and is but slightly dissociated, and therefore the concentration of CH_3COO^- ions resulting from the dissociation of the acid may be neglected. The base CH_3COO^- is also weak and therefore we may ignore the concentration of CH_3COOH molecules formed by the reaction of CH_3COO^- with water. Thus, we may assume that

$$[CH_3COOH] \approx C_a \quad \text{and} \quad [CH_3COO^-] \approx C_s$$

Substituting these values into Eq. (5.83), we find:

$$[H_3O^+] = K_a \frac{C_a}{C_s}$$

and

$$pH = pK_a + pC_a - pC_s \tag{5.84}$$

We see that these equations are identical to Eq. (5.79) whose derivation is based on the classical concepts of buffer solutions.

Thus, the Brönsted theory permits us to consider acids, bases, salts, and the processes of their interaction with each other and with water molecules (i.e., the process of hydrolysis) from a unified point of view.

Example 1. Calculation of the K_h, h, and pH for solutions of salts formed by the interaction between weak monobasic acids and strong monoacid bases.

(1) For a $0.1M$ solution of sodium acetate, CH_3COONa:

$$K_{CH_3COOH} = 1.74 \times 10^{-5} (pK_{CH_3COOH} \approx 4.76)$$

The hydrolysis equation is:

$$CH_3COO^- + H_2O \rightleftharpoons CH_3COOH + OH^-$$

According to Eqs. (5.18) and (5.19) we have:

$$K_h = \frac{[CH_3COOH][OH^-]}{[CH_3COO^-]} = \frac{K_w}{K_{CH_3COOH}} = \frac{10^{-14}}{1.74 \times 10^{-5}} \approx 5.7 \times 10^{-10}$$

Since K_h is small, the extent of hydrolysis of the $0.1M$ solution of CH_3COONa can be calculated from Eq. (5.21):

$$h = \sqrt{\frac{K_h}{C_{CH_3COONa}}} = \sqrt{\frac{5.7 \times 10^{-10}}{10^{-1}}} \approx 7.6 \times 10^{-5} \text{ or } 0.0076 \text{ per cent}$$

According to Eq. (5.23) the pH of this solution is:

$$pH = \frac{1}{2} pK_w + \frac{1}{2} pK_{CH_3COOH} - \frac{1}{2} pC_{CH_3COONa} \approx 7.00 + 2.38 - 0.50 \approx 8.88$$

(2) For a $0.1M$ solution of potassium cyanide, KCN:

$$K_{HCN} = 6.20 \times 10^{-10} (pK_{HCN} = 9.21)$$

The hydrolysis equation is:

$$CN^- + H_2O \rightleftharpoons HCN + OH^-$$

Although in this case $K_{HCN} \ll K_{CH_3COOH}$, the values of K_h and h, as we shall see below, are sufficiently small, and therefore use may be made of Eqs. (5.18), (5.21), and (5.23):

$$K_h = \frac{[HCN][OH^-]}{[CN^-]} = \frac{K_w}{K_{HCN}} = \frac{10^{-14}}{6.2 \times 10^{-10}} \approx 1.6 \times 10^{-5}$$

$$h = \sqrt{\frac{K_h}{C_{KCN}}} = \sqrt{\frac{1.6 \times 10^{-5}}{10^{-1}}} \approx 1.3 \times 10^{-2} \approx 0.013 \text{ or } 1.3 \text{ per cent}$$

$$pH = \frac{1}{2} pK_w + \frac{1}{2} pK_{HCN} - \frac{1}{2} pC_{KCN} \approx 7.00 + 4.60 - 0.50 = 11.10$$

Example 2. Calculate the K_h, h, and pH for solutions of salts formed by monobasic strong acids and monoacid weak bases.

For a $0.1M$ solution of ammonium chloride, NH_4Cl:

$$K_{NH_4OH} = 1.76 \times 10^{-5} (pK_{NH_4OH} \approx 4.75)$$

The hydrolysis equation is:

$$NH_4^+ + H_2O \rightleftharpoons NH_4OH + H^+$$

From Eqs. (5.27), (5.28), and (5.29) it follows that

$$K_h = \frac{[NH_4OH][H^+]}{[NH_4^+]} = \frac{K_w}{K_{NH_4OH}} = \frac{10^{-14}}{1.76 \times 10^{-5}} \approx 5.6 \times 10^{-10}$$

$$h = \sqrt{\frac{K_h}{C_{NH_4Cl}}} = \sqrt{\frac{5.6 \times 10^{-10}}{10^{-1}}} \approx 7.4 \times 10^{-5} \text{ or } 0.0074 \text{ per cent}$$

$$pH = \frac{1}{2} pK_w - \frac{1}{2} pK_{NH_4OH} + \frac{1}{2} pC_{NH_4Cl} \approx 7.00 - 2.38 + 0.50 = 5.12$$

Example 3. Calculate the K_h, h, and pH for solutions of salts formed by monobasic weak acids and monoacid weak bases.

(1) For a $0.1M$ solution of ammonium acetate, CH_3COONH_4:

$$K_{CH_3COOH} = 1.74 \times 10^{-5} (pK_{CH_3COOH} \approx 4.76)$$
$$K_{NH_4OH} = 1.76 \times 10^{-5} (pK_{NH_4OH} \approx 4.75)$$

The hydrolysis equation is:

$$NH_4^+ + CH_3COO^- + H_2O \rightleftharpoons NH_4OH + CH_3COOH$$

According to Eqs. (5.31), (5.34), and (5.36) we have:

$$K_h = \frac{[NH_4OH][CH_3COOH]}{[NH_4^+][CH_3COO^-]} = \frac{K_w}{K_{CH_3COOH} K_{NH_4OH}} =$$

$$= \frac{10^{-14}}{1.74 \times 10^{-5} \times 1.76 \times 10^{-5}} \approx 3.2 \times 10^{-5}$$

$$h = \sqrt{K_h} = \sqrt{3.2 \times 10^{-5}} \approx 5.7 \times 10^{-3} \text{ or } 0.57 \text{ per cent}$$

$$pH = \frac{1}{2} pK_w + \frac{1}{2} pK_{CH_3COOH} - \frac{1}{2} pK_{NH_4OH} \approx 7.00 + 2.38 - 2.38 = 7.00$$

(2) For a $0.1M$ solution of ammonium cyanide, NH_4CN:

$$K_{NH_4OH} = 1.76 \times 10^{-5} \quad (pK_{NH_4OH} \approx 4.75)$$

$$K_{HCN} = 6.20 \times 10^{-10} \quad (pK_{HCN} \approx 9.21)$$

The hydrolysis equation is:

$$NH_4^+ + CN^- + H_2O \rightleftharpoons NH_4OH + HCN$$

Using Eqs. (5.31), (5.33), and (5.36), we can approximately calculate the values of K_h, h, and pH for the solution in question:

$$K_h = \frac{[NH_4OH][HCN]}{[NH_4^+][CN^-]} = \frac{K_w}{K_{NH_4OH}K_{HCN}} =$$

$$= \frac{10^{-14}}{1.76 \times 10^{-5} \times 6.20 \times 10^{-10}} \approx 0.912$$

$$h = \frac{\sqrt{K_h}}{1+\sqrt{K_h}} \approx \frac{\sqrt{0.912}}{1+\sqrt{0.912}} \approx 0.488 \text{ or } 48.80 \text{ per cent}$$

$$pH = \frac{1}{2}pK_w + \frac{1}{2}pK_{HCN} - \frac{1}{2}pK_{NH_4OH} \approx 7.00 + 4.60 - 2.38 \approx 9.22$$

In this case, since $pK_{HCN} \gg pK_{NH_4OH}$, one would think that the values of h and pH obtained are rather approximate. Accurate calculation by means of the equations given below (according to the electroneutrality of electrolyte solutions) however proves the opposite, and therefore we may make use of Eqs. (5.33) and (5.36) in order to calculate h and pH.

For example, according to the equation for the hydrolysis of the salt NH_4CN, the solution contains NH_4OH, HCN, and H_2O molecules and NH_4^+, CN^-, H^+, and OH^- ions (from water).

On the basis of the electroneutrality of solutions, we have:

$$[H^+] + [NH_4^+] = [OH^-] + [CN^-]$$

or, what is the same thing,

$$[H^+] - [OH^-] = [CN^-] - [NH_4^+]$$

This equation shows that only when $[H^+] = [OH^-]$, i.e., when the ionization constants of weak acids and bases are equal ($K_a = K_b$), the concentrations of the cation and anion are equal too and so are those of the base and acid: $[An^-] = [B^+]$ and $[HAn] = [BOH]$. In this case, $h_{An^-} = h_{B^+}$ (see the example in which the hydrolysis of the salt CH_3COONH_4 is considered). Since the preliminary calculation of $[H^+]$ by means of approximate formula (5.36) gave $[H^+] = 10^{-9.22}$, and $[OH^-] = 10^{-4.78}$, the absolute difference between these values is very small:

$$[OH^-] - [H^+] = 10^{-4.78} - 10^{-9.22} = [NH_4^+] - [CN^-]$$

It is for this reason that the absolute difference between the concentrations of the unhydrolysed cation and anion of the salt ($[NH_4^+] - [CN^-]$) must also be small. From this it follows that the concentrations of the hydrolysis products are approximately equal, i.e., $[NH_4OH] \approx [HCN]$ and $h_{NH_4^+} \approx h_{CN^-}$ (although $[NH_4^+]$ must in fact be greater than $[CN^-]$, and $[NH_4OH] < [HCN]$; therefore, NH_4^+ ions are somewhat less hydrolysed than CN^- ions, i.e., $h_{NH_4^+} < h_{CN^-}$).

And, finally, since as a first approximation the extent of hydrolysis of this salt reaches about 50 per cent, it follows that, according to the theory of buffer solutions, the solution of the salt NH_4CN has a maximum buffer capacity, and a slight change in the ratio $[NH_4OH]/[NH_4^+]$ or $[HCN]/[CN^-]$ will have practically no effect on the pH of the solution. Therefore, the pH of the solution calculated from the simplified equation is not far from its true value, and there is no need to make use of Eq. (5.58).

All that has been said justifies the use of simplified equations for calculation of the K_h, h, and pH for solutions of such salts.

Example 4. Calculate K_h, h, and pH for solutions of neutral salts resulting from the interaction of polybasic weak acids with monoacid strong bases.

(1) For a $0.1M$ solution of sodium sulphide, Na_2S:

$$K_{H_2S} = 8.9 \times 10^{-8} \ (pK_{H_2S} \approx 7.05)$$

$$K_{HS^-} = 1.3 \times 10^{-13} \ (pK_{HS^-} \approx 12.89)$$

The hydrolysis equation is:

$$S^{2-} + H_2O \rightleftharpoons HS^- + OH^-$$

Although HS^- ions are also capable of being hydrolysed, the degree of hydrolysis of these ions however is small, as we shall see later, when considering the hydrolysis of the acid salt NaHS. Neglecting this process, we may express the hydrolysis constant by Eq. (5.18):

$$K_h = \frac{[HS^-][OH^-]}{[S^{2-}]} = \frac{K_w}{K_{HS^-}} = \frac{10^{-14}}{1.3 \times 10^{-13}} \approx 7.7 \times 10^{-2} \approx 0.077$$

Taking into account the relatively large value of K_h, the degree of hydrolysis h of this salt should be calculated by means of Eq. (5.24):

$$h = \frac{-K_w}{2C_{Na_2S}K_{HS^-}} + \sqrt{\left(\frac{K_w}{2C_{Na_2S}K_{HS^-}}\right)^2 + \frac{K_w}{C_{Na_2S}K_{HS^-}}} =$$

$$= \frac{-10^{-14}}{2 \times 10^{-1} \times 1.3 \times 10^{-13}} +$$

$$+ \sqrt{\left(\frac{10^{-14}}{2 \times 10^{-1} \times 1.3 \times 10^{-13}}\right)^2 + \frac{10^{-14}}{10^{-1} \times 1.3 \times 10^{-13}}} \approx$$

$$\approx 0.573 \text{ or } 57.3 \text{ per cent}$$

Since 57.3 per cent of the $0.1M$ solution of the salt Na_2S is converted into HS^- and about the same amount of OH^- ions is formed, $[OH^-] = 0.1 \times 0.573 = 5.73 \times 10^{-2} \approx 10^{-1.24}$ and pOH = 1.24. From this, according to Eq. (5.26), $pH = pK_w - pOH = 14 - 1.24 = 12.76$.

(2) For a $0.1M$ solution of sodium orthophosphate, Na_3PO_4:

$$K_{H_3PO_4} = 7.6 \times 10^{-3} \quad (pK_{H_3PO_4} \approx 2.12)$$
$$K_{H_2PO_4^-} = 6.2 \times 10^{-8} \quad (pK_{H_2PO_4^-} \approx 7.21)$$
$$K_{HPO_4^{2-}} = 4.4 \times 10^{-13} \quad (pK_{HPO_4^{2-}} \approx 12.36)$$

The hydrolysis equation is:

$$PO_4^{3-} + H_2O \rightleftharpoons HPO_4^{2-} + OH^-$$

Although HPO_4^{2-} ions are hydrolysed further with the formation of $H_2PO_4^-$ and H_3PO_4, we can nevertheless calculate, as a first approximation, with the second and third hydrolysis steps being neglected, the values of K_h, h, and pH for the solution of this salt, using Eqs. (5.18), (5.24), (5.25), and (5.26):

$$K_h = \frac{[HPO_4^{2-}][OH^-]}{[PO_4^{3-}]} = \frac{K_w}{K_{HPO_4^{2-}}} = \frac{10^{-14}}{4.4 \times 10^{-13}} \approx 2.27 \times 10^{-2} \approx 0.023$$

$$h = \frac{-K_w}{2C_{Na_3PO_4}K_{HPO_4^{2-}}} + \sqrt{\left(\frac{K_w}{2C_{Na_3PO_4}K_{HPO_4^{2-}}}\right)^2 + \frac{K_w}{C_{Na_3PO_4}K_{HPO_4^{2-}}}} =$$
$$= \frac{-10^{-14}}{2 \times 10^{-1} \times 4.4 \times 10^{-13}} +$$
$$+ \sqrt{\left(\frac{10^{-14}}{2 \times 10^{-1} \times 4.4 \times 10^{-13}}\right)^2 + \frac{10^{-14}}{10^{-1} \times 4.4 \times 10^{-13}}} \approx$$
$$\approx 0.377 \text{ or } 37.7 \text{ per cent}$$

$$[OH^-] = 10^{-1} \times 0.377 \approx 3.77 \times 10^{-2} \approx 10^{-1.42}$$

Hence $pOH \approx 1.42$ and $pH \approx pK_w - pOH \approx 14.00 - 1.42 \approx 12.58$.

Example 5. Calculate K_h, h, and pH for solutions of neutral salts formed by weak polybasic acids and weak monoacid bases.

(1) For a $0.1M$ solution of ammonium sulphide, $(NH_4)_2S$:

$$K_{H_2S} = 8.9 \times 10^{-8} \quad (pK_{H_2S} \approx 7.05)$$
$$K_{HS^-} = 1.3 \times 10^{-13} \quad (pK_{HS^-} \approx 12.89)$$
$$K_{NH_4OH} = 1.76 \times 10^{-5} \quad (pK_{NH_4OH} \approx 4.75)$$

The first step of hydrolysis of the salt $(NH_4)_2S$ is expressed by the equation

$$NH_4^+ + S^{2-} + H_2O \rightleftharpoons HS^- + NH_4OH$$

As has already been said, the hydrolysis of HS^- is slight and, as a first approximation, this process may be disregarded. But in this case, taking account of the great difference between the values of K_a and K_b (1.3×10^{-13} and 1.76×10^{-5}) and also of the stepwise hydrolysis of S^{2-}, we cannot equate the concentrations of the HS^- and NH_4OH formed by hydrolysis. Here $[HS^-] \neq [NH_4OH]$ and therefore we should have done as indicated in Example 3 (2).

But, since the solution of a multigrade equation is very difficult, use may be made of simplified calculations for obtaining some information on the values of K_h, h, and pH for the solution of ammonium sulphide.

For calculation, use is made of Eqs. (5.31), (5.44), (5.45), (5.47) along with Eqs. (5.41), (5.42), and (5.43).

The hydrolysis constants of the salt under the given conditions are given by

$$K_h = \frac{[HS^-][NH_4OH]}{[S^{2-}][NH_4^+]} = \frac{K_w}{K_{HS^-}K_{NH_4OH}} =$$

$$= \frac{10^{-14}}{1.3 \times 10^{-13} \times 1.76 \times 10^{-5}} \approx 10^{3.64} \approx 4365$$

This value shows that the salt $(NH_4)_2S$ is practically completely hydrolysed in solution. Then, according to Eqs. (5.41) and (5.43), we have:

$$[S^{2-}] = \frac{C_{(NH_4)_2S}}{K_h} = \frac{10^{-1}}{10^{3.64}} \approx 10^{-4.64} \approx 2.3 \times 10^{-5}$$

$$[HS^-] \approx [NH_4OH] \approx C_{(NH_4)_2S} - [S^{2-}] \approx 10^{-1} - 10^{-4.64} \approx 0.099977$$

Substituting these values into Eqs. (5.44) and (5.45) gives:

$$h_{S^{2-}} = \frac{[HS^-]}{C_{(NH_4)_2S}} = \frac{0.099977}{0.1} \approx 0.99977 \text{ or } 99.98 \text{ per cent}$$

$$h_{NH_4^+} = \frac{[NH_4OH]}{2C_{(NH_4)_2S}} = \frac{0.099977}{0.2} \approx 0.49985 \text{ or } 49.985 \text{ per cent}$$

In other words, S^{2-} ions are practically completely converted into HS^-, while NH_4^+ ions are converted into NH_4OH by 50 per cent; hence, an ammonium buffer is formed in solution, with equal concentrations of the weak base and the salt of the weak base, for which pOH $\approx$ $\approx pK_{NH_4OH}$. Hence, according to Eq. (5.47), the pH of the $0.1M$ $(NH_4)_2S$ solution will be: $pH = pK_w - pK_{NH_4OH} \approx 14.00 - 4.75 \approx$ ≈ 9.25.

(2) For a 0.1*M* solution of ammonium carbonate, $(NH_4)_2CO_3$:

$$K_{H_2CO_3} = 4.5 \times 10^{-7} \ (pK_{H_2CO_3} \approx 6.35)$$

$$K_{HCO_3^-} = 4.8 \times 10^{-11} \ (pK_{HCO_3^-} \approx 10.32)$$

$$K_{NH_4OH} = 1.76 \times 10^{-5} \ (pK_{NH_4OH} \approx 4.75)$$

Since in this case the second step of hydrolysis of CO_3^{2-} is slight, the hydrolysis of the salt $(NH_4)_2CO_3$ may be represented, as a first approximation, as follows:

$$NH_4^+ + CO_3^{2-} + H_2O \rightleftarrows NH_4OH + HCO_3^-$$

The values of K_h, h, and pH for this solution can be calculated by using Eqs. (5.31), (5.44), (5.45), and (5.47) along with Eqs. (5.48), (5.42), and (5.43):

$$K_h = \frac{[NH_4OH][HCO_3^-]}{[NH_4^+][CO_3^{2-}]} = \frac{K_w}{K_{NH_4OH}K_{HCO_3^-}} \approx$$

$$\approx \frac{10^{-14}}{1.76 \times 10^{-5} \times 4.8 \times 10^{-11}} \approx 10^{1.07} \approx 11.84$$

According to Eqs. (5.48) and (5.43), we have:

$$[CO_3^{2-}] = \frac{-C_{(NH_4)_2CO_3}(2+K_h)}{2(K_h-1)} +$$

$$+\sqrt{\left(\frac{C_{(NH_4)_2CO_3}(2+K_h)}{2(K_h-1)}\right)^2 + \frac{C^2_{(NH_4)_2CO_3}}{K_h-1}} =$$

$$= \frac{-10^{-1}(2+11.84)}{2(11.84-1)} + \sqrt{\left(\frac{10^{-1}(2+11.84)}{2(11.84-1)}\right)^2 + \frac{(10^{-1})^2}{(11.84-1)}} \approx$$

$$\approx 6.86 \times 10^{-3} \approx 0.00686$$

$$[NH_4OH] \approx [HCO_3^-] \approx C_{(NH_4)_2CO_3} - [CO_3^{2-}] \approx 10^{-1} - 0.00686 \approx 0.09314$$

Hence,

$$h_{CO_3^{2-}} = \frac{[HCO_3^-]}{C_{(NH_4)_2CO_3}} \approx \frac{0.09314}{0.1} \approx 0.9314 \text{ or ca. } 93.14 \text{ per cent}$$

$$h_{NH_4^+} = \frac{[NH_4OH]}{2C_{(NH_4)_2CO_3}} \approx \frac{0.09314}{0.2} \approx 0.4657 \text{ or ca. } 46.57 \text{ per cent}$$

Just as in the preceding example, for the buffered solution of a mixture of NH_4OH and NH_4^+, we can calculate the pH on the basis of the NH_4^+ concentration [according to Eq. (5.42)]:

$$[NH_4^+] = C_{(NH_4)_2CO_3} + [CO_3^{2-}] \approx 0.1 + 0.00686 \approx 0.10686$$

or, which amounts to the same thing,

$$[NH_4^+] = 2C_{(NH_4)_2CO_3} - [NH_4OH] \approx 0.2 - 0.09314 \approx 0.10686$$

From this, according to Eq. (5.47), we obtain:

$$pH = pK_w - pK_{NH_4OH} - p[NH_4OH] + p[NH_4^+] \approx$$

$$\approx 14.00 - 4.75 + \bar{2}.969 - \bar{1}.029 \approx 9.19$$

Example 6. Calculate K_h, h, and pH for solutions of acid salts formed as a result of the reaction between weak polybasic acids and strong monoacid bases.

(1) For a $0.1M$ solution of sodium hydrosulphide, NaHS:

$$K_{H_2S} = 8.9 \times 10^{-8} \ (pK_{H_2S} \approx 7.05)$$

$$K_{HS^-} = 1.3 \times 10^{-13} \ (pK_{HS^-} \approx 12.89)$$

Hydrosulphide ions (HS^-) hydrolyse and ionize according to the following scheme:

$$HS^- + H_2O \rightleftharpoons H_2S + OH^-$$

$$HS^- \rightleftharpoons S^{2-} + H^+$$

$$2HS^- \rightleftharpoons H_2S + S^{2-}$$

Using Eqs. (5.52), (5.54), and (5.55), we find the values of K_h, h, and pH for the solution of this salt:

$$K_h = \frac{[H_2S][S^{2-}]}{[HS^-]^2} = \frac{K_{HS^-}}{K_{H_2S}} = \frac{1.3 \times 10^{-13}}{8.9 \times 10^{-8}} \approx 10^{-5.84} \approx 1.4 \times 10^{-6}$$

$$h = 2\sqrt{K_h} = 2\sqrt{1.4 \times 10^{-6}} \approx 2.4 \times 10^{-3} \approx 0.0024 \text{ or ca. } 0.24 \text{ per cent}$$

$$pH = \frac{1}{2}\, pK_{H_2S} + \frac{1}{2}\, pK_{HS^-} \approx 3.525 + 6.445 \approx 9.97$$

(2) For a $0.1M$ solution of sodium hydrophosphate, Na_2HPO_4:

$$K_{H_3PO_4} = 7.6 \times 10^{-3} \qquad (pK_{H_3PO_4} \approx 2.12)$$

$$K_{H_2PO_4^-} = 6.2 \times 10^{-8} \qquad (pK_{H_2PO_4^-} \approx 7.21)$$

$$K_{HPO_4^{2-}} = 4.4 \times 10^{-13} \qquad (pK_{HPO_4^{2-}} \approx 12.36)$$

The first step of hydrolysis of HPO_4^{2-} ions and their ionization may be represented by the following equations:

$$HPO_4^{2-} + H_2O \rightleftharpoons H_2PO_4^- + OH^-$$

$$HPO_4^{2-} \rightleftharpoons PO_4^{3-} + H^+$$

Although $H_2PO_4^-$ ions are capable of being hydrolysed, the process expressed by the equation

$$H_2PO_4^- + H_2O \rightleftharpoons H_3PO_4 + OH^-$$

may be neglected because of the slight degree of hydrolysis of HPO_4^{2-} present in a large amount in solution; the hydrolysis equation may be written, as a first approximation, in the following form:

$$2HPO_4^{2-} \rightleftharpoons H_2PO_4^- + PO_4^{3-}$$

Thus, the calculation of the K_h, h, and pH for a solution of this acid salt can be calculated approximately by using the same equations as in the preceding example, i.e., (5.52), (5.54) and (5.55):

$$K_h = \frac{[H_2PO_4^-][PO_4^{3-}]}{[HPO_4^{2-}]^2} \approx \frac{K_{HPO_4^{2-}}}{K_{H_2PO_4^-}} \approx \frac{4.4\times10^{-13}}{6.2\times10^{-8}} \approx 10^{-5.15} \approx 7.1\times10^{-6}$$

$$h = 2\sqrt{K_h} \approx 2\sqrt{7.1\times10^{-6}} \approx 5.3\times10^{-3} \approx 0.0053 \text{ or ca. } 0.53 \text{ per cent}$$

$$\text{pH} = \frac{1}{2}\,\text{p}K_{H_2PO_4^-} + \frac{1}{2}\,\text{p}K_{HPO_4^{2-}} \approx 3.61 + 6.18 \approx 9.79$$

(3) For a 0.1M solution of sodium dihydrogen phosphate, NaH_2PO_4:

$$K_{H_3PO_4} = 7.6\times10^{-3} \qquad (\text{p}K_{H_3PO_4} \approx 2.12)$$

$$K_{H_2PO_4^-} = 6.2\times10^{-8} \qquad (\text{p}K_{H_2PO_4^-} \approx 7.21)$$

$$K_{HPO_4^{2-}} = 4.4\times10^{-13} \qquad (\text{p}K_{HPO_4^{2-}} \approx 12.36)$$

Just as in Example 6 (2), in this case we have:

$$H_2PO_4^- + H_2O \rightleftharpoons H_3PO_4 + OH^-$$

$$H_2PO_4^- \rightleftharpoons HPO_4^{2-} + H^+$$

$$2H_2PO_4^- \rightleftharpoons H_3PO_4 + HPO_4^{2-}$$

$$K_h = \frac{[H_3PO_4][HPO_4^{2-}]}{[H_2PO_4^-]^2} = \frac{K_{H_2PO_4^-}}{K_{H_3PO_4}} \approx \frac{6.2\times10^{-8}}{7.6\times10^{-3}} \approx 10^{-5.09} \approx 8.1\times10^{-6}$$

$$h = 2\sqrt{K_h} = 2\sqrt{8.1\times10^{-6}} \approx 5.7\times10^{-3} \approx 0.0057 \text{ or ca. } 0.57 \text{ per cent}$$

$$\text{pH} = \frac{1}{2}\,\text{p}K_{H_3PO_4} + \frac{1}{2}\,\text{p}K_{H_2PO_4^-} \approx 1.06 + 3.61 \approx 4.67$$

The fact that the degree of hydrolysis of the salt NaH_2PO_4 (ca. 0.57 per cent) is found to be somewhat greater than that of the salt Na_2HPO_4 (ca. 0.53 per cent) might be thought of as contradicting to the conception that the salts of weaker acids are hydrolysed in water more strongly. This seeming disagreement is accounted for by two circumstances: first, in calculations we ignored the stepwise hydrolysis; second, $H_2PO_4^-$ is ionized more strongly than HPO_4^{2-}. Owing to the ionization of $H_2PO_4^-$, the amount of H^+ ions present in solution is greater than that of HPO_4^{2-}. These hydrogen ions bind a larger amount of hydroxyl ions and displace the hydrolysis of $H_2PO_4^-$ more strongly than that of HPO_4^{2-}.

The foregoing also applies to the calculation of the degree of hydrolysis of the salts Na_2SO_3 and $NaHSO_3$, the result of which is presented in Table 5.3. If, on the basis of the pH of solutions of these salts and the ionization constants of the corresponding acids, we calculate the concentrations of SO_3^{2-}, HSO_3^-, and H_2SO_3, and then calculate h from the data obtained, we shall see that the results are in satisfactory agreement with those given in Table 5.3.

TABLE 5.3. **Degree of Hydrolysis and pH for 0.1*M* Solutions of Selected Salts (at 25°C)***

Salt	Weak electrolyte formed by hydrolysis		Degree of hydrolysis, per cent	pH of solution
	Formula	Ionization constant		
CH_3COONa	CH_3COOH	1.74×10^{-5}	0.0076	8.88
NH_4Cl	NH_4OH	1.76×10^{-5}	0.0074	5.12
KCN	HCN	6.2×10^{-10}	1.3	11.10
NH_4CN	HCN and NH_4OH	see above	48.8	9.23
Na_2SO_3	HSO_3^-	6.2×10^{-8}	0.13	10.10
$NaHSO_3$	H_2SO_3	1.7×10^{-2}	0.38	4.49
Na_2CO_3	HCO_3^-	4.8×10^{-11}	4.46	11.65
$NaHCO_3$	H_2CO_3	4.5×10^{-7}	2.1	8.34
$(NH_4)_2CO_3$	HCO_3^- and NH_4OH	see above	93.14	9.19
CH_3COONH_4	CH_3COOH and NH_4OH	see above	0.57	7.00
Na_3PO_4	HPO_4^{2-}	4.4×10^{-13}	37.7	12.58
Na_2HPO_4	$H_2PO_4^-$	6.2×10^{-8}	0.53	9.79
NaH_2PO_4	H_3PO_4	7.6×10^{-3}	0.57	4.67
Na_2S	HS^-	1.3×10^{-13}	57.3	12.76
$NaHS$	H_2S	8.9×10^{-8}	0.24	9.97
$(NH_4)_2S$	HS^- and NH_4OH	see above	99.98	9.25
NH_4HS	H_2S and NH_4OH	see above	7.9	8.15
$NaClO$	$HClO$	5×10^{-8}	0.14	10.15

* The values of some of the ionization constants are rounded off. For Na_2CO_3 and Na_2SO_3 the data have been calculated in just the same way as for Na_2S [Example 4(1)]. The calculations for $NaHSO_3$ and $NaClO$ have been made in the same manner as for $NaHCO_3$ [Example 6(4)] and CH_3COONa [Example 1(1)], respectively.

For example, the pH of a 0.1*M* solution of $NaHSO_3$ is 4.49 and $[H^+] = 3.25 \times 10^{-5}$. Substituting these values into the expressions for both ionization constants

$$K_{H_2SO_3} = \frac{[H^+][HSO_3^-]}{[H_2SO_3]} \quad \text{and} \quad K_{HSO_3^-} = \frac{[H^+][SO_3^{2-}]}{[HSO_3^-]}$$

and assuming that $[HSO_3^-] \approx 0.1$, we find that

$$[H_2SO_3] = [SO_3^{2-}] = 1.9 \times 10^{-4}$$

Hence, owing to hydrolysis there is formed $[H_2SO_3] + [SO_3^{2-}] = 3.8 \times 10^{-4}\,M$, and the degree of hydrolysis of the 0.1*M* solution of $NaHSO_3$ will be equal to:

$$h = \frac{3.8 \times 10^{-4} \times 100}{0.1} = 0.38 \text{ per cent}$$

This agrees with the calculation performed in the same manner as in Example 6 (4).

Naturally, in all these cases, more accurate data for *h* and pH can be obtained if use is made of the equation for the electroneutrality of solutions. In particular, for phosphate solutions we have:

$$[H^+]+[Na^+]=[OH^-]+[H_2PO_4^-]+[HPO_4^{2-}]+[PO_4^{3-}]$$

(4) For a 0.1*M* solution of sodium hydrocarbonate, $NaHCO_3$:

$$K_{H_2CO_3}=4.5\times10^{-7}\ (pK_{H_2CO_3}\approx6.35)$$
$$K_{HCO_3^-}=4.8\times10^{-11}\ (pK_{HCO_3^-}\approx10.32)$$

Just as in the preceding examples, for this case we have:

$$HCO_3^-+H_2O \rightleftharpoons H_2CO_3+OH^-$$
$$HCO_3^- \rightleftharpoons CO_3^{2-}+H^+$$
$$\overline{2HCO_3^- \rightleftharpoons H_2CO_3+CO_3^{2-}}$$

$$K_h=\frac{[H_2CO_3][CO_3^{2-}]}{[HCO_3^-]^2}=\frac{K_{HCO_3^-}}{K_{H_2CO_3}}=\frac{4.8\times10^{-11}}{4.5\times10^{-7}}\approx10^{-3.97}\approx1.1\times10^{-4}$$

$$h=2\sqrt{K_h}=2\sqrt{1.1\times10^{-4}}\approx2.1\times10^{-2}\approx0.021 \text{ or ca. } 2.1 \text{ per cent}$$

$$pH=\frac{1}{2}pK_{H_2CO_3}+\frac{1}{2}pK_{HCO_3^-}\approx3.18+5.16\approx8.34$$

Example 7. Calculate K_h, h, and pH for solutions of acid salts formed from polybasic weak acids and monoacid weak bases.

(1) For a 0.1*M* solution of ammonium hydrosulphide, NH_4HS:

$$K_{H_2S}=8.9\times10^{-8}\qquad(pK_{H_2S}\approx7.05)$$
$$K_{HS^-}=1.3\times10^{-13}\qquad(pK_{HS^-}\approx12.89)$$
$$K_{NH_4OH}=1.76\times10^{-5}\qquad(pK_{NH_4OH}\approx4.75)$$

The HS^- ions undergo hydrolysis according to the equation

$$HS^-+H_2O \rightleftharpoons H_2S+OH^-$$

The OH^- ions formed react partly with NH_4^+ ions:

$$NH_4^++OH^- \rightleftharpoons NH_4OH$$

On the other hand, the HS^- ions ionize to give H^+ and S^{2-}:

$$HS^- \rightleftharpoons H^++S^{2-}$$

In their turn, hydrogen ions interact with hydroxyl ions with the formation of water.

This scheme of hydrolysis can be accounted for as follows. When equilibrium is established, the solution must contain all the ions (H^+, OH^-, NH_4^+, HS^-, S^{2-}) and molecules (NH_4OH, H_2S, H_2O), whose concentrations must satisfy the ionization constants of both the acid and the base—K_{H_2S}, K_{HS^-}, K_{NH_4OH}, and also the ion product constant of water—K_w.

To determine the values of K_h, h, and pH for solutions of the salt in question, the equation for the electroneutrality of solutions should have been used and the multigrade equation solved, but this is rather difficult.

Since the degree of hydrolysis of acid salts is in general quite small, in this particular case we may neglect the reaction of formation of S^{2-} and calculate K_h, h, and pH for the solution of the salt NH_4HS using Eqs. (5.31), (5.34), and (5.36), this salt being likened to a normal salt formed by the reaction of a monobasic weak acid with a monoacid weak base:

$$K_h = \frac{[NH_4OH][H_2S]}{[NH_4^+][HS^-]} = \frac{K_w}{K_{NH_4OH}K_{H_2S}} = \frac{10^{-14}}{1.76\times10^{-5}\times8.9\times10^{-8}} \approx$$

$$\approx 10^{-2.2} \approx 6.3\times10^{-3} \approx 0.0063$$

$$h \approx \sqrt{K_h} \approx \sqrt{10^{-2.2}} \approx 10^{-1.1} \approx 7.9\times10^{-2} \approx 0.079 \text{ or ca. } 7.9 \text{ per cent}$$

$$\text{pH} = \frac{1}{2}\,\text{p}K_w + \frac{1}{2}\,\text{p}K_{H_2S} - \frac{1}{2}\,\text{p}K_{NH_4OH} \approx 7.00 + 3.53 - 2.38 \approx 8.15$$

5.7. Amphoterism

The hydroxides of typical metals are known to be bases. The hydroxides of non-metals and certain, less typical metals, say, chromium and manganese (in their highest oxidation state) exhibit opposite chemical properties and belong to another class of compounds—acids. But there are hydroxides which are capable of functioning as both acids and bases. They are called **ampholytes**, or are said to be **amphiprotic** or **amphoteric**, and the phenomenon itself is known as **amphoterism.**

We shall encounter the phenomenon of amphoterism elsewhere in studying the cations of Group III-V; it is of great importance in analysis.

To illustrate, let us consider zinc hydroxide, $Zn(OH)_2$. Like other bases, this compound when reacted with acids dissolves to form the corresponding salts; for example,

$$Zn(OH)_2\downarrow + 2H^+ \rightleftharpoons Zn^{2+} + 2H_2O$$

But $Zn(OH)_2$ dissolves in bases too, forming zincates. In this reaction, zinc hydroxide behaves as an acid; this becomes especially clear if we write its formula in the same way as the formulas of acids are usually written:

$$H_2ZnO_2\downarrow + 2OH^- \rightleftharpoons ZnO_2^{2-} + 2H_2O$$

Thus, $Zn(OH)_2$ exhibits the properties of both acids and bases, i.e., is a typical amphoteric hydroxide. Like $Zn(OH)_2$, other compounds, such as $Al(OH)_3$, $Cr(OH)_3$, $Pb(OH)_2$, $Sn(OH)_2$, $As(OH)_3$, are also amphoteric.

This simplified interpretation of the composition of the hydroxides of elements, whose oxidation number exceeds 3, does not correspond to reality (see Sec. 5.5). These hydroxides are in fact hydrous oxides (e.g., $Al_2O_3 \cdot xH_2O$, $TiO_2 \cdot xH_2O$). The formation of an amphoteric hydroxide from a hydrated ion and the dissolution of this hydroxide in excess alkali may be illustrated by the example of aluminium hydroxide:

$$2[Al(OH_2)_6]^{3+} + 6OH^- \rightleftharpoons Al_2O_3 \cdot nH_2O\downarrow + (15-n)H_2O$$

$$Al_2O_3 \cdot nH_2O\downarrow + 2OH^- \rightleftharpoons 2[Al(OH)_4]^- + (n-3)H_2O$$

The dissolution of zinc hydroxide in an excess of alkali proceeds according to the equation

$$Zn(OH)_2\downarrow + 2OH^- \rightleftharpoons [Zn(OH)_4]^{2-}$$

This process may be regarded as a complex-formation reaction: when amphoteric hydroxides are dissolved in alkalis, their molecules combine with the OH^- ions of alkali to form the corresponding complex ions $[Al(OH)_4]^-$, $[Zn(OH)_4]^{2-}$, etc. There are actually no ZnO_2^{2-} and AlO_2^- ions in solutions.

It should be noted that the ions $[Zn(OH)_4]^{2-}$ and $[Al(OH)_4]^-$ differ from ZnO_2^{2-} and AlO_2^- in their water content only, and are in fact the hydrated ions ZnO_2^{2-} and AlO_2^-. For example,

$$ZnO_2^{2-} + 2H_2O \longrightarrow [Zn(OH)_4]^{2-}$$

and

$$AlO_2^- + 2H_2O \longrightarrow [Al(OH)_4]^-$$

Since here the hydration of ions is not reflected in their formulas, we will write them in the form of ZnO_2^{2-}, AlO_2^-, etc., in accordance with the classical theory of amphoterism.

In just the same way, in order to simplify the writing of reaction equations, we shall proceed from the existence of hydroxides of the compositions $Al(OH)_3$, $Cr(OH)_3$, etc.

In the case of $Al(OH)_3$ and $Cr(OH)_3$, the composition of salts (aluminates and chromites) formed under the action of alkalis corresponds not to these hydrates but to the compounds $HAlO_2$ and $HCrO_2$ formed from them when a molecule of water is lost. The equation for the reaction between $Al(OH)_3$ and NaOH has the following form:

$$H_3AlO_3\downarrow + OH^- \rightleftharpoons AlO_2^- + 2H_2O$$

Since amphoteric hydroxides exhibit both basic and acidic properties and, according to the theory of electrolytic dissociation, the former are ascribed to hydroxyl ions and the latter to hydrogen ions, it may be assumed that the ionization of amphoteric hydroxides results in the simultaneous formation of both ions. For example, in the case of $Zn(OH)_2$, part of the molecules ionize according to the equation*

$$Zn(OH)_2\downarrow \rightleftharpoons Zn(OH)_2 \rightleftharpoons Zn^{2+} + 2OH^-$$

and the other part according to the equation

$$Zn(OH)_2\downarrow \rightleftharpoons Zn(OH)_2 \rightleftharpoons 2H^+ + ZnO_2^{2-}$$

or

$$Zn(OH)_2\downarrow \rightleftharpoons H_2ZnO_2 \rightleftharpoons 2H^+ + ZnO_2^{2-}$$

Thus, here we have the following series of interrelated equilibria:

$$\begin{array}{ccccc} & & Zn(OH)_2\downarrow & & \\ & & \updownarrow & & \\ 2H^+ + ZnO_2^{2-} & \overset{I}{\rightleftharpoons} & Zn(OH)_2 & \overset{II}{\rightleftharpoons} & Zn^{2+} + 2OH^- \\ & & \text{(solution)} & & \end{array}$$

On the basis of this assumption, one can easily explain the ability of $Zn(OH)_2$ to dissolve in acids and alkalis. Indeed, when OH^- ions are introduced with alkali into the solution, the hydrogen ions formed by $Zn(OH)_2$ are removed. As a result, the solution becomes unsaturated with respect to $Zn(OH)_2$, which leads to the dissolution of the precipitate and build-up of ZnO_2^{2-} ions, i.e., to the formation of a zincate.

On the other hand, the action of acids must also result in the binding of OH^- ions and the dissolution of the precipitate with the formation of Zn^{2+} cations, i.e., a zinc salt of the corresponding acid.**

Thus, zinc ions are present mainly in the form of Zn^{2+} cations in acid solutions and as ZnO_2^{2-} anions (more precisely, as $[Zn(OH)_4]^{2-}$) in alkaline solutions. If, for example, ZnO_2^{2-} anions are to be converted into Zn^{2+} cations, the solution must be acidified. A precipitate of $Zn(OH)_2$ is formed, which then dissolves in excess acid:

$$ZnO_2^{2-} + 2H^+ = Zn(OH)_2\downarrow$$

$$Zn(OH)_2\downarrow + 2H^+ = Zn^{2+} + 2H_2O$$

The overall equation for the reaction is

$$ZnO_2^{2-} + 4H^+ \rightleftharpoons Zn^{2+} + 2H_2O \quad (5.85)$$

When $Zn(OH)_2$ is made to react with alkalis, which bind hydrogen ions, ZnO_2^{2-} anions will again be formed (in excess alkali); it means that reaction (5.85) is reversible and must lead to a chemical equilibrium. From this it follows that, under any conditions, solutions of zinc salts must contain both Zn^{2+} cations and ZnO_2^{2-} anions, the ratio of

* For simplicity, both types of equations for stepwise dissociation of $Zn(OH)_2$ are not given here, as should have been done in a more detailed treatment of the question.

** It is clear that addition of hydrogen ions also causes equilibrium (I) to be shifted to the right and addition of hydroxyl ions shifts equilibrium (II) to the left.

their concentrations being dependent on the hydrogen ion concentration in solution. Applying the law of mass action to this equilib rium, we may write:

$$\frac{[Zn^{2+}][H_2O]^2}{[ZnO_2^{2-}][H^+]^4}=K_1$$

Transposing the quantities $[H_2O]^2$ and $[H^+]^4$ to the right-hand side of the equation, we obtain:

$$\frac{[Zn^{2+}]}{[ZnO_2^{2-}]}=\frac{K_1}{[H_2O]^2}[H^+]^4$$

Since $[H_2O]$ is a practically constant quantity,

$$\frac{K_1}{[H_2O]^2}=K$$

And, finally, we have:

$$\frac{[Zn^{2+}]}{[ZnO_2^{2-}]}=K[H^+]^4 \qquad (5.86)$$

From Eq. (5.86) it is clear that with increasing concentration of H^+ the concentration of Zn^{2+} must increase very rapidly and that of ZnO_2^{2-} decrease. When the solution is made alkaline, the situation is reversed.

What has been said above about $Zn(OH)_2$ is equally applicable to other amphoteric hydroxides.

Since, according to the theory in question, all amphoteric hydroxides undergo two different types of ionization, a distinction should be made between their acid ionization constant (K_a) and their basic ionization constant (K_b). Comparing the values of these constants, we can find out which of the properties, acidic or basic, are more pronounced in a given hydroxide. For instance, if we compare the values of K_a and K_b for the primary ionization of lead hydroxide:

$$K_b=\frac{[Pb(OH)^+][OH^-]}{[Pb(OH)_2]}\approx 1\times 10^{-3}$$

and

$$K_a=\frac{[HPbO_2^-][H^+]}{[H_2PbO_2]}\approx 1\times 10^{-11}$$

we see that the basic properties of $Pb(OH)_2$ are much more strongly pronounced than its acidic properties. In the case of $As(OH)_3$, $K_a = 5.1\times 10^{-10}$ and $K_b \approx 1\times 10^{-14}$, i.e., it is the acidic properties that prevail over the basic properties.

An example of an amphoteric electrolyte, whose acidic and basic properties are equally well pronounced, is water; another example is gallium hydroxide, $Ga(OH)_3$, whose K_a and K_b are nearly equal (5×10^{-11} and 1×10^{-11}).

Slightly soluble hydroxides are precipitated at a definite concentration of OH^-, i.e., at a known pH of the solution, which depends on the value of the solubility product constant of the hydroxide and on the concentration of metal ions. For example, as shown in Sec. 4.3, the precipitation of $Mg(OH)_2$ at a concentration of Mg^{2+} equal to $1 \times \times 10^{-2}$ g-ion/litre begins at pH 10.4 and becomes practically complete at pH 12.42. If the pH is increased further, the precipitation of Mg^{2+} will be even more complete. No undesirable phenomena (except for the increase of the salt effect) are caused by addition of excess alkali. Conversely, in the case of amphoteric hydroxides, as we have already seen, excess alkali dissolves the precipitate. Therefore, the range of pH values within which an amphoteric hydroxide precipitates is limited.

For example, practically complete precipitation of $Al(OH)_3$ is attained at pH ≈ 5, and at pH ≈ 10 the precipitate begins to dissolve with the formation of AlO_2^-. It means that, when Al^{3+} is separated in the form of a hydroxide, the pH of the solution must be maintained within the limits of 5—10. Any reagent that creates the pH lying within these limits will evidently precipitate aluminium hydroxide. For example, $Al(OH)_3$ can be precipitated from acid solutions by the action of $(NH_4)_2S$ because the S^{2-} anions that combine with hydrogen ions provide the required value of pH (equal to ca. 9.2). The same is also observed when the precipitation is effected by the action of other salts of weak acids: Na_2CO_3, Na_2SO_3, $Na_2S_2O_3$, KCN, etc.

On the other hand, if aluminium hydroxide is to be prepared from a strongly alkaline solution (pH > 10), the value of pH should be depressed by removing OH^- ions. These ions can be removed by the action not only of H^+ but also of NH_4^+ or other cations of weak bases. Therefore, if to a solution of an aluminate there is added an excess of NH_4Cl, which forms with alkali an ammonium buffer (pH ≈ 9), $Al(OH)_3$ will be precipitated.

The reaction may be represented by the equations:

$$AlO_2^- + 2H_2O \rightarrow Al(OH)_3\downarrow + OH^-$$
$$OH^- + NH_4^+ \rightarrow NH_4OH$$
$$\overline{AlO_2^- + NH_4^+ + 2H_2O \rightarrow Al(OH)_3\downarrow + NH_4OH}$$

As can be seen from these equations, the reaction reduces essentially to the disturbance of the equilibrium of the aluminate hydrolysis due to the removal of OH^- ions by NH_4^+ ions.

From the standpoint of the theory of electrolytic dissociation, the basic, acidic, or amphoteric properties of hydroxides are determined by the type of their dissociation. If a given hydroxide dissociates to form OH^- ions, it is a base; but if the dissociation results in the formation of H^+ ions, the hydroxide is an acid. If both types of ions are formed upon dissociation, the hydroxide is said to be amphoteric.

How can the difference in the type of dissociation be explained for various hydroxides.

The answer to this question is given by the Kossel theory, according to which the molecules of hydroxides are regarded simply as consisting of ions whose mutual attraction is described by Coulomb's law. If the charges on the ions are denoted by e_1 and e_2, and their radii by r_1 and r_2 (Fig. 5.2), then the attractive force operating between them will be equal to (in vacuum):

$$f=\frac{e_1e_2}{(r_1+r_2)^2}$$

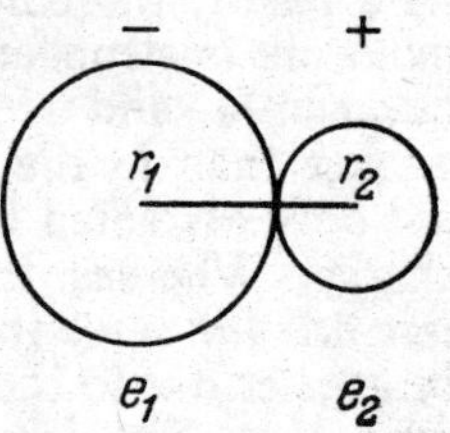

Fig. 5.2. Structure of the molecule formed by two ions (schematic).

The lower the magnitude of this force, the weaker must be the bond between the ions and the more easily will it be ruptured when the substance is dissolved in water. It is obvious from this that in the NaOH molecule (whose structure after Kossel is shown in Fig. 5.3) the bond between Na^+ and O^{2-} is weaker than that between H^+ and O^{2-} because the radius of Na^+ is much larger than that of H^+ (proton). It

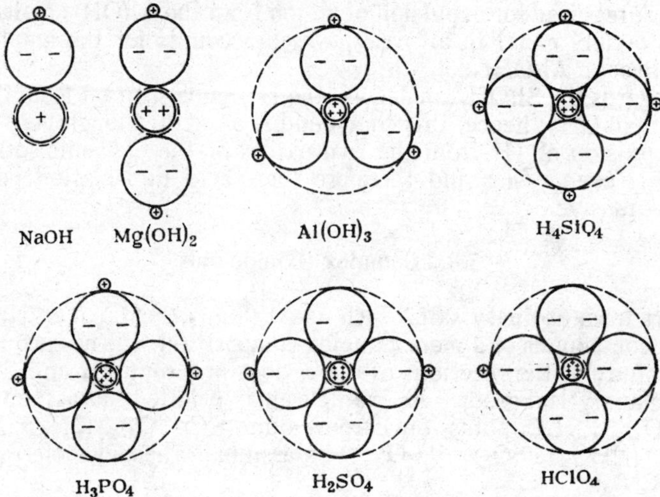

Fig. 5.3. Structure of the molecules of hydroxides and acids and types of their dissociation (after Kossel).

is precisely at the site of this weaker bond where the molecule breaks down (with the formation of Na^+ and OH^- ions), whioh takes place during the dissociation of NaOH in solution.

Turning to magnesium hydroxide, it is necessary first of all to note that Mg^{2+} has not one but two positive charges. Moreover, the radius of Mg^{2+} (0.78×10^{-8} cm) is smaller than that of Na^+ (0.98×10^{-8} cm). As a result, the bond between the oxygen and metal in the $Mg(OH)_2$ molecule is stronger than that in the NaOH molecule, and the bond between H^+ and O^{2-} in the molecule of magnesium hydroxide must be weaker than in the sodium hydroxide molecule. Indeed, H^+ ions are not only attracted by O^{2-} ions but also repelled by metal ions of like charge. Whereas the attractive force acting between H^+ and O^{2-} remains the same in both cases, the repulsion force for $Mg(OH)_2$ must be greater due to the high charge and small radius of Mg^{2+} as compared with Na^+. The ion OH^- in magnesium hydroxide however is more easily expelled than the H^+ ion, and therefore it is a base, though much weaker than NaOH.

Since on passing to the other elements of the same series in the Periodic System (Al, Si, P, S, and Cl) the number of electrons lost increases gradually (from 3 at aluminium up to 7 at chlorine) and the radii diminish, the direction of the changes remains the same: the splitting-off of OH^- by the hydroxide of each next element becomes increasingly difficult, and the expulsion of H^+ increasingly easier. Thus, the work required for expulsion of an ion from the $Al(OH)_3$ molecule is approximately equal in all cases, which accounts for the amphoteric properties of $Al(OH)_3$.

As regards the $Si(OH)_4$ molecule, the H^+ ion is split off from it more easily than OH^-; hence, this compound is an acid, though very weak. The expulsion of H^+ from the hydroxides of the elements following silicon is even easier and therefore their acid dissociation becomes more extensive.

5.8. Complex Compounds

Apart from ordinary salts, such as $Al_2(SO_4)_3$, $FeSO_4$, $Fe(CN)_2$, and others, compounds of a more complex composition will be encountered in the study of the reactions of the cations of Group III and the succeeding analytical groups. As examples may be cited: potassium alum, $KAl(SO_4)_2 \cdot 12H_2O$; potassium chrome alum, $KCr(SO_4)_2 \cdot 12H_2O$; Mohr's salt, $(NH_4)_2Fe(SO_4)_2 \cdot 6H_2O$; potassium hexacyanoferrate(II), $K_4[Fe(CN)_6]$.

The molecules of these compounds result from the combination of the electrically neutral molecules of salts of a simpler composition; for example,

$$K_2SO_4 + Al_2(SO_4)_3 \rightleftharpoons 2KAl(SO_4)_2$$

$$(NH_4)_2SO_4 + FeSO_4 \rightleftharpoons (NH_4)_2Fe(SO_4)_2$$

$$Fe(CN)_2 + 4KCN \rightleftharpoons K_4[Fe(CN)_6]$$

Such compounds are classified as compounds of higher order. How the ionization of these compounds actually occurs is important for analysis. Let us consider some examples.

A small amount of Mohr's salt, $(NH_4)_2Fe(SO_4)_2 \cdot 6H_2O$, is dissolved in water, and the resulting solution is treated with $(NH_4)_2S$; the formation of a black precipitate of FeS proves the presence of Fe^{2+} ions:

$$Fe^{2+} + S^{2-} \rightleftharpoons FeS\downarrow$$

This reaction shows that the solution of Mohr's salt contains Fe^{2+} ions. The reactions for NH_4^+ with NaOH (the evolution of NH_3 on heating) and for SO_4^{2-} with $BaCl_2$ (the precipitation of $BaSO_4$) are accomplished in the same way.

From this we may conclude that the dissociation of Mohr's salt proceeds according to the equation

$$(NH_4)_2Fe(SO_4)_2 \rightleftharpoons 2NH_4^+ + Fe^{2+} + 2SO_4^{2-}$$

The dissociation of potassium alum takes place in a similar way:

$$KAl(SO_4)_2 \rightleftharpoons K^+ + Al^{3+} + 2SO_4^{2-}$$

Thus, these salts dissociate in solution to give all the ions that were present in the corresponding simple salts, i.e., they behave as mixtures of simple salts.

Compounds of higher order do not always dissociate as described above. For example, if to a solution of potassium hexacyanoferrate(II), $K_4[Fe(CN)_6]$, there is added $NaHC_4H_4O_6$, a characteristic white precipitate of $KHC_4H_4O_6$ is formed, indicating the presence of K^+ in solution. When another portion of the solution is reacted with $(NH_4)_2S$, no precipitate of FeS is formed. Nor can the reactions for CN^- be carried out. At the same time, $K_4[Fe(CN)_6]$ gives a number of reactions which are typical neither of Fe^{2+} nor of CN^-, taken separately. For instance, $FeCl_3$ produces a beautiful dark-blue precipitate of ferric ferrocyanide (Prussian blue), $Fe_4[Fe(CN)_6]_3$:

$$4FeCl_3 + 3K_4[Fe(CN)_6] \rightarrow Fe_4[Fe(CN)_6]_3\downarrow + 12KCl$$

All these facts are evidence that the solution of $K_4[Fe(CN)_6]$ contains* neither Fe^{2+} nor CN^- but that it contains complex ions, $[Fe(CN)_6]^{4-}$, which give reactions of their own, different from the reactions of their parent ions, Fe^{2+} and CN^-. Hence, the dissociation of $K_4[Fe(CN)_6]$ proceeds according to the equation

$$K_4[Fe(CN)_6] \rightleftharpoons 4K^+ + [Fe(CN)_6]^{4-}$$

This is also verified by a study of the electrolysis of the salt in question, during which iron being in the form of the complex anion $[Fe(CN)_6]^{4-}$ moves towards the anode rather than the cathode.

---

* It would be more correct to say that there are *almost no* Fe^{2+} and CN^- ions because, as shown later, complex ions also break down into their constituent simple ions, but their degree of ionization is very small.

In order to reflect the properties indicated, the formula of the salt is written as $K_4[Fe(CN)_6]$, i.e., the group of atoms giving rise to a complex are enclosed in brackets.

Salts similar to $K_4[Fe(CN)_6]$ fall into the category of **complex compounds** (which are also called **coordination compounds**)*; ions or molecules that form complex ions cannot be identified by means of analytical reactions characteristic of these complex ions.

Compounds such as alums are often referred to as double salts, but double salts are in fact complex compounds having an extremely low strength.

The complex ion $[Fe(CN)_6]^{4-}$ is a compound made up of one positive ion (Fe^{2+}) and six negative ions (CN^-); in other cases, instead of negative ions, neutral molecules of NH_3, H_2O, and of certain other substances may enter into the composition of complexes.

For example, when ammonia is introduced into a solution of $AgNO_3$, the following reaction takes place:

$$Ag^+ + 2NH_4OH \rightarrow [Ag(NH_3)_2]^+ + 2H_2O$$

The formation of the complex ion $[Ag(NH_3)_2]^+$ is evidenced by the fact that some of the reactions for Ag^+ fail to fulfil their mission. For example, if a solution of $[Ag(NH_3)_2]NO_3$ is treated with a chloride, no precipitate of AgCl will form. Also, no precipitates of Ag_3PO_4 and $AgBrO_3$ will be formed by the action of Na_2HPO_4 or $KBrO_3$, etc.

5.9. Structure of Complex Compounds

According to Werner's coordination theory (1893), the central position in the molecules of complex compounds is usually occupied by a positive ion called the **central ion** or **complex-former.** Coordinated to the central atom are so-called **ligands** (coordinating groups) which are negative ions or neutral molecules residing in the immediate vicinity of the central atom. The central ion, or complex-former, and ligands constitute the **inner coordination sphere** of the compound. Apart from the inner sphere, in most cases there is also an **outer coordination sphere** which consists of negative or positive ions.

The ions of the outer sphere are attached to the complex ion ionogenically, i.e., when the substance dissolves in water, these ions are split off in the form of free ions. Conversely, the bond between the complex-former and ligands is non-ionogenic in character, and the entire inner coordination sphere is present in aqueous solutions in the form of a complex ion. For instance, in the case of $K_4[Fe(CN)_6]$, the Fe^{2+} ion is a complex-former and the CN^- ions are ligands. The outer coordination sphere contains K^+ ions. The group $[Fe(CN)_6]^{4-}$ is the inner sphere of the complex.

* The exact definition of the term "complex compound" is rather involved and is not therefore considered here.

The charge on a complex ion is the algebraic sum of the charges on the central ion and ligands. For the case under consideration, the charge on the ion $[Fe(CN)_6]^{4-}$ is equal to $(+2) + (-6) = -4$.

At this point it is rather appropriate to refer to such concepts as the ionic charge, oxidation state (or oxidation number), and valency of an element. It has been shown above that the ions present in solutions may be simple, i.e., consist only of a single atomic species (K^+, Na^+, Mg^{2+}, Cl^-, I^-, etc.), and complex, i.e., consist of more than one atomic species. The anions of oxygen-containing acids, SO_4^{2-}, CO_3^{2-}, MnO_4^-, CrO_4^{2-}, etc., are also classified as complex ions.

In simple or monoatomic ions, the charge on an ion is the same as the oxidation state and valency of the element. As regards complex ions, this is not necessarily true.

The charge on a simple ion shows the number of electrons lost or gained by the atom of the element that forms the given ion. For example, the ions K^+, Na^+, Li^+, Rb^+, and Cs^+ have a charge of +1 since the atoms of the corresponding elements have each lost one electron while forming these ions. The ions Cl^-, Br^-, and I^- have a charge of —1 because the atoms of the corresponding elements have each acquired one electron in forming these ions.

Complex ions have a definite charge too. For example, the ions SO_4^{2-}, SO_3^{2-}, and S^{2-}, all have in solution a charge of —2. But the oxidation state of sulphur which enters into the composition of these ions is different. The oxidation state of sulphur is +6 in the ions SO_4^{2-}, +4 in the ion SO_3^{2-}, and —2 in the ions S^{2-}, which coincides with the charge on the S^{2-} ion. Similarly, the oxidation state of iron is +2 both in the simple ion Fe^{2+} and in the complex ion $[Fe(CN)_6]^{4-}$. In the former case, the oxidation state of iron coincides with the charge on the simple ion Fe^{2+} and with the valency of the element. For the complex ion $[Fe(CN)_6]^{4-}$, the oxidation state of iron is +2; the charge on the complex ion is —4, and the valency of iron in the given complex compound is equal to 6 since the iron is coordinated to six CN^- groups, each having a charge of —1. Thus, the valency of the given element in a complex compound is defined as the sum of the bonds which the central atom (complex-former) possesses in the given compound.

If ligands are neutral molecules, their presence cannot evidently affect the charge on the complex. Its charge therefore is equal in such cases to the charge on the complex-former; for example, the charge on the ion $[Ag(NH_3)_2]^+$ is the same as that on the Ag^+ ion.

The charge on a complex can also be deduced from the composition of the outer coordination sphere. Having found the charge on the complex, one can easily determine the oxidation state of the complex-former if it is unknown. For example, in the outer coordination sphere of the complex compound $Na_3[Co(NO_2)_6]$, there are three unipositive Na^+ ions, and since the whole molecule is electrically neutral, the charge on the complex $[Co(NO_2)_6]^{3-}$ must equal —3. The NO_2^- ions, which are

ligands, bear in all 6 negative charges, which means that the complex-**former** must have a charge equal to +3.

Thus, the oxidation state of cobalt in the hexanitrocobaltate ion will be +3, while its valency in this ion is 6.

The total number of ligands which are coordinated to a given central ion (atom) to form a complex is called the **coordination number** of the given ion or atom. The coordination number of most complex-formers (Cr^{3+}, Co^{3+}, Fe^{2+}, Fe^{3+}, Zn^{2+}, Ni^{2+}, Pt^{4+}, etc.) is 6. The ions Cd^{2+}, Cu^{2+}, Hg^{2+}, Pt^{2+}, etc., are known to have a coordination number of 4. There also exist complexes, in which the central ion may have other coordination numbers, say, 2, 3, 8, etc.

The coordination number plays in the chemistry of complex compounds no less important role than the oxidation state of the element. A knowledge of the coordination number allows one to write the formulas of the various complex compounds. For example, knowing that the coordination number of Cd^{2+} is 4, we may conclude that the composition of its complex ammonia compounds must correspond to one of the following formulas: $[Cd(NH_3)_4]Cl_2$, $[Cd(NH_3)_4]SO_4$, $[Cd(NH_3)_4](OH)_2$, etc.

Analogously, the cyanide complexes of cadmium have the formulas $K_2[Cd(CN)_4]$, $Na_2[Cd(CN)_4]$, etc., since the complex ion, which has one dipositive ions (Cd^{2+}) and four uninegative ions (CN^-), must be an anion with two negative charges.

Under known conditions, the maximum coordination number characteristic of a given complex-former may not be attained, in which case so-called **coordination unsaturated compounds** are formed.

As shown by investigations*, of great importance are concentration conditions. For example, the complex ion $[Fe(SCN)_6]^{3-}$ resulting from the reaction between iron(III) salts and thiocyanates (salts of HSCN) and having the maximum coordination number of Fe^{3+}, is formed only in the presence of a large excess of SCN^-. At smaller concentrations of SCN^-, other complexes of iron are formed, including $[Fe(SCN)]^{2+}$, which have smaller coordination numbers.

Ligands are in most cases uninegative ions and also molecules such as NH_3, H_2O, C_2H_5OH (ethyl alcohol), and others, which are capable of occupying only one coordination position in a complex. There are ligands, however, which can bind at more than one position. Such ligands are called bidentates (literally, "two-toothed" or "two-clawed"), tridentates, quadridentates, etc., depending on the number of spatial positions which may be occupied by ligands in the inner sphere of a complex compound.

The tendency towards complex-formation is most strongly displayed by elements in the middle of the long periods of the Periodic System, especially by the elements of analytical Groups III, IV, and V. It is

* Babko, A. K., *Zhurn. Obshchei Khimii*, **18**, 816 (1948); *Doklady Akad. Nauk SSSR*, **52**, 37 (1946).

precisely the cations of the metals of these groups that frequently play the role of complexing agents and give rise to a large variety of complex compounds.

From the viewpoint of the electronic theory of the structure of matter, the formation of complex compounds, which does not fit into the framework of ordinary valency concepts, may be interpreted as being the result of the manifestation of electrostatic attractive forces acting between ligands and complex-forming ions.

Here it should be borne in mind that the forces of electrostatic attraction between opposite charges operate in all cases, regardless of whether these charges are equal or not in magnitude. For example, the Fe^{2+} ion attracts two CN^- ions, forming an electrically neutral molecule, $Fe(CN)_2$; but even after the molecule is formed, the Fe^{2+} ion in its composition is still capable of attracting negatively charged ions. Thus, if a solution contains an excess of CN^-, the Fe^{2+} ion will attract more than two anions. It is clear, however, that it will be able to hold only a limited number of anions because the number of positions that can be occupied by CN^- ions in the immediate vicinity of the Fe^{2+} ion is limited.

On the other hand, CN^- ions, when being attracted by the Fe^{2+} ion, repel each other because they have like charges. The larger the number of CN^- ions attached to the Fe^{2+} ion, the greater the magnitude of these repulsion forces. In the long run, they will overcome the attraction forces acting between the Fe^{2+} and CN^- ions, and the addition of further CN^- ions will become impossible. In this case, it will happen only after six CN^- ions have been linked to the Fe^{2+} ion, i.e., after the complex $[Fe(CN)_6]^{4-}$ has been formed.

A theoretical explanation of the valency of and affinity between ions during the formation of complex compounds was given in 1915-1916, on the basis of the electrostatic conceptions, by the German scientist Kossel. In the same period, the American researcher Lewis developed the theory of covalent bonds that is especially important for interpretation of the formation of complex compounds, in which neutral molecules are ligands. If ligands are neutral molecules such as NH_3 or H_2O, their combination with complexing agents may be accounted for by the fact that these molecules are dipoles, which means that they must be oriented towards the central ion and attracted by it.

The formation of such complexes is also explained by a special type of **covalent bonds** between the ions of a complex-former and the ligands. The *covalent bond results when two electrons (an electron pair), one from each atom, are shared equally between two atoms* and do not belong exclusively to one alone. Examples of the formation of a covalent bond are

```
         H                    H        H
     ·  ·×  ·             ·  ·×       ·×
  H×     C  ×H         H×     N:      :O:
        ·×                   ·×       ·×
         H                    H        H
```

In the above schemes, the electrons donated by hydrogen atoms for the formation of electron pairs are designated by crosses, and those donated by the other atoms involved in the bond by dots. From these schemes it is seen that each electron pair is formed by the two atoms being combined.

The formation of shared electron pairs may however occur in a different way too: both electrons are donated by only one of the atoms involved in the bond. To illustrate, the ammonia molecule, which has a pair of "free" (i.e., belonging only to the nitrogen atom) electrons, may combine with an H^+ ion by way of sharing this free pair of electrons. Like other ammines, the resulting ammonium ion may be looked upon as a complex ion. The formation of the ammonium ion may be depicted by means of the following scheme:

$$\begin{array}{c} \mathrm{H} \\ \cdot\times \\ \mathrm{H}\times\ \mathrm{N}: \\ \cdot\ \ \cdot\times \\ \mathrm{H} \end{array} + \mathrm{H}^+ \rightarrow \left[\begin{array}{c} \mathrm{H} \\ \cdot\times \\ \mathrm{H}\times\ \mathrm{N}:\ \mathrm{H} \\ \cdot\ \ \cdot\times \\ \mathrm{H} \end{array} \right]^+$$

or

$$\begin{array}{c} \mathrm{H} \\ | \\ \mathrm{H}-\mathrm{N} \\ | \\ \mathrm{H} \end{array} + \mathrm{H}^+ \rightarrow \left[\begin{array}{c} \mathrm{H} \\ | \\ \mathrm{H}-\mathrm{N}\rightarrow\mathrm{H} \\ | \\ \mathrm{H} \end{array} \right]^+$$

This type of bond in which both electrons are donated by only one of the atoms involved in bond formation is called the ***dative*** *or* ***coordinate-covalent bond*** (also called the ***donor-acceptor bond***). The atom that donates the electrons is one which has an unshared electron pair in its outer valence shell (it is called the **donor atom**). The electron deficient atom that gains the electrons is known as the **acceptor atom**. The coordinate-covalent bond is schematically depicted in the form of an arrow (→) directed from the donor atom to the acceptor atom or else by a dashed line.

Other complexes may also owe their origin to the formation of coordinate-covalent bonds, for example,

$$4NH_3 + Cd^{2+} \rightarrow \left[\begin{array}{c} NH_3 \\ \downarrow \\ H_3N \rightarrow Cd \leftarrow NH_3 \\ \uparrow \\ NH_3 \end{array} \right]^{2+}$$

In the same way one can account for the formation of various crystal hydrates which should be assigned, according to their properties, to typical complex compounds. Examples* are $[Cr(OH_2)_6]Cl_3$, $[Co(OH_2)_6](NO_3)_2$, $[Cu(OH_2)_4]SO_4 \cdot H_2O$.

* Since the central ion and water molecules in crystal hydrates are bound through oxygen atoms, water is represented by the formula OH_2.

A large number of complex compounds are known; study of these compounds has a great scientific and practical importance, which was first stressed by D. Mendeleyev.

Especially important for investigation of complex compounds were the works of L. Chugaev (see Sec. 1.6) who not only carried out a number of classical investigations but also created a school of chemists.

The two theories described above (those of Kossel and Lewis) fail to provide an exhaustive explanation for the existance and structure of complex compounds. The next stage in the development of the theory of coordination compounds is the **molecular orbital theory**, which in fact is a further extension of the covalent-bond theory, and the **crystal field theory** based on the electrochemical conceptions. And, finally, a theory has recently come into being, which combines both the theories mentioned. This is known as the **ligand field theory**.

Problems relating to the structure of complex compounds and the nature of their bonding constitute the subject matter of inorganic and general chemistry and are beyond the scope of this book.*

5.10. Stability of Complex Compounds

A sharp borderline between complex compounds and double salts cannot be drawn (see Sec. 5.8). It has been verified experimentally that, along with simple ions, more concentrated solutions of double salts also contain complex ions, for example, the $[Fe(SO_4)_2]^{2-}$ ions in a solution of Mohr's salt. On the other hand, complex ions too are capable of dissociating into the corresponding simpler ions (or into ions and molecules), which can be seen from the following examples.

To a solution of $AgNO_3$ is added a solution of NH_4OH until an odour evolves. This gives rise to the complex ammonia salt of silver $[Ag(NH_3)_2]NO_3$, and no precipitates of $AgBrO_3$ and AgCl are formed in the resulting solution by the action of $KBrO_3$ and KCl, respectively. If however separate portions of the $[Ag(NH_3)_2]NO_3$ solution are treated with reagents for the silver ion, such as KBr, KI, and H_2S, which form with the Ag^+ ions precipitates of AgBr (yellowish in colour), AgI (yellow), and Ag_2S (black), it will turn out that these reactions can also be accomplished with a solution of $[Ag(NH_3)_2]NO_3$.

Thus, in the first two reactions the complex $[Ag(NH_3)_2]NO_3$ exhibits the typical properties of complex compounds, and in the last three reactions it behaves as a double salt.

---

* For a more detailed study of the chemistry of complex compounds, see the following books: Grinberg, A. A., *An Introduction to the Chemistry of Complex Compounds*, English translation available from Pergamon Press, New York, 1962; Gray, H. B., *Electrons and Chemical Bonding*, W. A. Benjamin, Inc., New York, 1964; Cotton, F. A. and G. Wilkinson, *Advanced Inorganic Chemistry*, 2nd ed., Wiley Interscience, 1966; Basolo, F. and R. Johnson, *Coordination Chemistry*, Benjamin, New York, 1964; Day C. and D. D. Selbin, *Theoretical Inorganic Chemistry*.

These facts evidently indicate that, along with complex $[Ag(NH_3)_2]^+$ ions, the $[Ag(NH_3)_2]NO_3$ solution also contains Ag^+ cations. The concentration of Ag^+ cations however is so small that only the solubility product constants of the least soluble salts (AgBr, AgI, and Ag_2S) can be attained by the action of the appropriate reagents, but those of the more soluble salts ($AgBrO_3$ and AgCl) cannot be reached. Thus, the salt under consideration first undergoes ionization as a strong electrolyte, i.e., practically completely:

$$[Ag(NH_3)_2]NO_3 \rightarrow [Ag(NH_3)_2]^+ + NO_3^- \quad (5.87)$$

Then the complex ion ionizes by steps:
(1) first step

$$Ag(NH_3)_2^+ \rightleftharpoons AgNH_3^+ + NH_3 \quad (5.88)$$

$$K_{[Ag(NH_3)_2]^+} = \frac{[AgNH_3^+][NH_3]}{[Ag(NH_3)_2^+]}$$

(2) second step

$$AgNH_3^+ \rightleftharpoons Ag^+ + NH_3 \quad (5.89)$$

$$K_{[AgNH_3]^+} = \frac{[Ag^+][NH_3]}{[AgNH_3^+]}$$

The dissociation of the complex proceeds but to a small degree and may be characterized by the dissociation or ionization constant of the given complex, which is the product of the successive ionization constants (for the individual steps). In this particular case

$$K_{ins} = K_{[Ag(NH_3)_2]^+} K_{[AgNH_3]^+} = \frac{[Ag^+][NH_3]^2}{[Ag(NH_3)_2^+]} = 5.8 \times 10^{-8}$$

The greater the value of this constant, the more extensive is the dissociation of the complex and the less stable it is. This constant is called the **instability constant** of the complex.

The reverse of the instability constant is known as the **formation constant** or **stability constant** of the complex. They are interrelated by the equation

$$K_{st} = \frac{1}{K_{ins}}$$

Complexes differ in their instability constants.*

For example, the instability constant of the complex $[AgS_2O_3]^-$ is equal to 1.5×10^{-9} and that of the complex $[Ag(CN)_2]^-$ is $1.4 \times \times 10^{-20}$. Consequently, the complex $[Ag(NH_3)_2]^+$ is the least and the complex $[Ag(CN)_2]^-$ the most stable among the three complex silver ions indicated.

---

* Yatsimirskii, K. B. and V. P. Vasil'ev, *Instability Constants in Complex Compounds*, English translation available from Pergamon Press, New York, 1960; Lurie, Ju.. *Handbook of Analytical Chemistry*, Mir Publishers, Moscow, 1974.

If the instability constants of various complexes are known, one can calculate theoretically the concentrations of the corresponding simple ions in solutions of complex salts.

Example. Find the concentration of Ag^+ in $0.01M$ solutions of: (a) $[Ag(NH_3)_2]NO_3$; (b) $K[Ag(CN)_2]$.

Solution. Denoting the sought-for concentration of Ag^+ by x, we find from the ionization equation for $[Ag(NH_3)_2]^+$:

$$[NH_3]=2x \quad \text{and} \quad [Ag(NH_3)_2^+]=0.01-x\approx 0.01$$

since the degree of ionization of the complex is small. Hence, we may write:

$$K_{ins}=\frac{x(2x)^2}{0.01}=5.8\times 10^{-8}$$

which solves to

$$x=[Ag^+]=\sqrt[3]{\frac{5.8\times 10^{-10}}{4}}=5.25\times 10^{-4} \text{ g-ion/litre}$$

The instability constant of the complex $[Ag(CN)_2]^-$ is calculated in the same manner:

$$K_{ins}=\frac{x(2x)^2}{0.01}=1.4\times 10^{-20}$$

$$x=[Ag^+]=\sqrt[3]{\frac{1.4\times 10^{-22}}{4}}=3.3\times 10^{-8} \text{ g-ion/litre}$$

The example of calculation given shows that, at equal molar concentrations, the greater the instability constant of the complex, the higher the concentrations of the simple ions present in solutions of complex salts. For example, the concentration of Ag^+ in a $0.01M$ solution of $[Ag(NH_3)_2]^+$ is approximately 16,600 times as great as that in a $0.01M$ solution of $[Ag(CN)_2]^-$ since the instability constant of the former complex (5.8×10^{-8}) is much higher than that of the latter ($1.4\times\times 10^{-20}$). This difference in the concentration of Ag^+ will of course manifest itself in reactions. It should be expected that the number of precipitation reactions of the Ag^+ ion would decrease with decreasing Ag^+ concentration, because the number of cases, where the solubility product constants of slightly soluble silver salts become unattainable, must increase. These theoretical considerations are fully confirmed by experiment.

Excess of complexing agent plays an important role in the dissolution of slightly soluble compounds. To illustrate, let us consider an example. Depending on the concentration of $S_2O_3^{2-}$ in the solution, there may be formed complex ions of different composition and different stability: $[Ag(S_2O_3)]^-$, $K_{ins}=1.5\times 10^{-9}$; $[Ag(S_2O_3)_2]^{3-}$, $K_{ins}= 3.5\times 10^{-14}$; $[Ag(S_2O_3)_3]^{5-}$, $K_{ins}=9.3\times 10^{-15}$.

Let us perform calculations for comparison purposes. What are the concentrations of BrO_3^-, Cl^-, Br^-, I^-, and S^{2-} required for precipitation of $AgBrO_3$, AgCl, AgBr, AgI, and Ag_2S from a pure solution of a silver salt and from solutions of certain complex silver ions, which contain an excess of the corresponding ligand?

Suppose that the concentration of Ag^+ in a pure solution of a silver salt is 10^{-2} g-ion/litre.

Using the solubility product constants of the corresponding compounds, we obtain:

$$[BrO_3]^- > \frac{5.5 \times 10^{-5}}{10^{-2}} \approx 5.5 \times 10^{-3} \text{ g-ion/litre}$$

$$[Cl^-] > \frac{1.78 \times 10^{-10}}{10^{-2}} \approx 1.8 \times 10^{-8} \text{ g-ion/litre}$$

$$[Br^-] > \frac{5.3 \times 10^{-13}}{10^{-2}} \approx 5.3 \times 10^{-11} \text{ g-ion/litre}$$

$$[I^-] > \frac{8.3 \times 10^{-17}}{10^{-2}} \approx 8.3 \times 10^{-15} \text{ g-ion/litre}$$

$$[S^{2-}] > \frac{6.3 \times 10^{-50}}{(10^{-2})^2} \approx 6.3 \times 10^{-46} \text{ g-ion/litre}$$

Thus, for a precipitate of Ag_2S to be formed in the solution of a silver salt, in which $[Ag^+] = 10^{-2}$ g-ion/litre, it will suffice if the concentration of S^{2-} is greater than 6.3×10^{-46} g-ion/litre.

If a similar calculation is to be made for a solution of a complex silver compound which contains a definite amount of the corresponding ligand, one must first calculate the concentration of Ag^+ from the instability constant of the complex ion, and then the minimum concentration of the precipitating ion in solution from the solubility product constant of the corresponding slightly soluble compound. For example, the instability constant of the complex ion $[Ag(NH_3)_2]^+$ is:

$$K_{ins} = \frac{[Ag^+][NH_3]^2}{[Ag(NH_3)_2^+]} = 5.8 \times 10^{-8}$$

Suppose that the equilibrium concentrations of $[Ag(NH_3)_2]^+$ and NH_3 are equal to 10^{-2} g-ion/litre; then $[Ag^+] = 5.8 \times 10^{-6}$ g-ion/litre.

On this basis we find that, for the precipitates to be formed, the concentrations of the precipitating ions must be equal to:

$$[BrO_3^-] > \frac{5.5 \times 10^{-5}}{5.8 \times 10^{-6}} \approx 9.5 \text{ g-ion/litre}$$

$$[Cl^-] > \frac{1.78 \times 10^{-10}}{5.8 \times 10^{-6}} \approx 3.1 \times 10^{-5} \text{ g-ion/litre}$$

$$[Br^-] > \frac{5.3 \times 10^{-13}}{5.8 \times 10^{-6}} \approx 9.1 \times 10^{-8} \text{ g-ion/litre}$$

$$[I^-] > \frac{8.3 \times 10^{-17}}{5.8 \times 10^{-6}} \approx 1.4 \times 10^{-11} \text{ g-ion/litre}$$

$$[S^{2-}] > \frac{6.3 \times 10^{-50}}{(5.8 \times 10^{-6})^2} \approx 1.8 \times 10^{-39} \text{ g-ion/litre}$$

The instability constant of the complex ion $[Ag(CN)_2]^-$ is:

$$K_{ins} = \frac{[Ag^+][CN^-]^2}{[Ag(CN)_2^-]} = 1.4 \times 10^{-20}$$

If the concentrations of this complex compound and the ligand in solution are assumed to be equal to 10^{-2} g-ion/litre, we obtain:

$$[Ag^+] = 1.4 \times 10^{-18} \text{ g-ion/litre}$$

Then, for the corresponding slightly soluble compounds to be precipitated, the concentrations of the precipitating ions must be equal to:

$$[BrO_3^-] > \frac{5.5 \times 10^{-5}}{1.4 \times 10^{-18}} \approx 3.9 \times 10^{13} \text{ g-ion/litre}$$

$$[Cl^-] > \frac{1.78 \times 10^{-10}}{1.4 \times 10^{-18}} \approx 1.3 \times 10^{8} \text{ g-ion/litre}$$

$$[Br^-] > \frac{5.3 \times 10^{-13}}{1.4 \times 10^{-18}} \approx 3.8 \times 10^{5} \text{ g-ion/litre}$$

$$[I^-] > \frac{8.3 \times 10^{-17}}{1.4 \times 10^{-18}} \approx 5.9 \times 10^{1} \text{ g-ion/litre}$$

$$[S^{2-}] > \frac{6.3 \times 10^{-50}}{(1.4 \times 10^{-18})^2} \approx 3.2 \times 10^{-14} \text{ g-ion/litre}$$

For thiosulphate complexes of silver, the required concentration of the precipitating ion will change, depending on the composition of the complex ion present in solution.

For example, if the solution contains the complex ion $[Ag(S_2O_3)]^-$ and the ligand $S_2O_3^{2-}$ in a concentration of 10^{-2} g-ion/litre, from the value of the instability constant of this complex ion:

$$K_{ins} = \frac{[Ag^+][S_2O_3^{2-}]}{[Ag(S_2O_3)^-]} = 1.5 \times 10^{-9}$$

we find

$$[Ag^+] = K_{ins} = 1.5 \times 10^{-9} \text{ g-ion/litre}$$

The concentrations of the precipitating ions must then be, respectively, equal to:

$$[BrO_3^-] > \frac{5.5 \times 10^{-5}}{1.5 \times 10^{-9}} \approx 3.7 \times 10^{4} \text{ g-ion/litre}$$

$$[Cl^-] > \frac{1.78 \times 10^{-10}}{1.5 \times 10^{-9}} \approx 1.2 \times 10^{-1} \text{ g-ion/litre}$$

$$[Br^-] > \frac{5.3 \times 10^{-13}}{1.5 \times 10^{-9}} \approx 3.5 \times 10^{-4} \text{ g-ion/litre}$$

$$[I^-] > \frac{8.3 \times 10^{-17}}{1.5 \times 10^{-9}} \approx 5.5 \times 10^{-8} \text{ g-ion/litre}$$

$$[S^{2-}] > \frac{6.3 \times 10^{-50}}{(1.5 \times 10^{-9})^2} \approx 2.8 \times 10^{-32} \text{ g-ion/litre}$$

For a solution of the complex ion $[Ag(S_2O_3)_3]^{5-}$ under the same conditions, we use the instability constant of this ion ($K_{ins} = 9.3 \times 10^{-15}$), substitute the corresponding numerical values, and find that

$$[Ag^+] = K_{ins} \frac{[Ag(S_2O_3)_3^{5-}]}{[S_2O_3^{2-}]^3} = 9.3 \times 10^{-11} \text{ g-ion/litre}$$

The concentrations of the precipitating ions must thus be equal to:

$$[BrO_3^-] > \frac{5.5 \times 10^{-5}}{9.3 \times 10^{-11}} \approx 5.9 \times 10^5 \text{ g-ion/litre}$$

$$[Cl^-] > \frac{1.78 \times 10^{-10}}{9.3 \times 10^{-11}} \approx 1.9 \text{ g-ion/litre}$$

$$[Br^-] > \frac{5.3 \times 10^{-13}}{9.3 \times 10^{-11}} \approx 5.7 \times 10^{-3} \text{ g-ion/litre}$$

$$[I^-] > \frac{8.3 \times 10^{-17}}{9.3 \times 10^{-11}} \approx 8.9 \times 10^{-7} \text{ g-ion/litre}$$

$$[S^{2-}] > \frac{6.3 \times 10^{-50}}{(9.3 \times 10^{-11})^2} \approx 7.3 \times 10^{-30} \text{ g-ion/litre}$$

The results of these calculations compared in Table 5.4 show that bromate ions cannot precipitate $AgBrO_3$ from solutions of the indicated complex compounds in the presence of even a small excess of ligand (10^{-2} g-mole/litre). This would require a rather high concentration of BrO_3^-.

For the ions Cl^-, Br^-, and I^-, the minimum concentration required for precipitation of AgCl, AgBr, or AgI from a solution of the given complex decreases according to the decrease of the solubility product constants of these compounds.

The concentration of S^{2-} required for a precipitate of Ag_2S to be formed is so small that this sulphide will precipitate from solutions of all the complex compounds of silver considered above. In other words, silver sulphide, Ag_2S, cannot be dissolved whatever the excess of ligand used. Precipitates of AgI, AgBr, and the more so of AgCl, however, dissolve relatively readily under the action of sodium thiosulphate present in an appropriate concentration.

From all that has been said above it follows that *the behaviour of complex salts in precipitation reaction is primarily determined by the ratio of the instability constants of the complex to the solubility product of the compound being precipitated*. The greater the value of K_{ins} and the lower the solubility product constant, the greater the probability of precipitation, and vice versa.*

The course of a precipitation reaction is strongly influenced by excess ions (or un-ionized molecules) in solution, which enter as ligands into

* Of course, this is valid only when the concentrations of the reagent and complex compound are constant, i.e., only when the conditions of the experiment are constant.

TABLE 5.4. **The Minimum Concentration of Precipitating Ion Required for Formation of Precipitates from 0.01*M* Solutions of Complex Silver Compounds in the Presence of 0.01 Gram-Ion of Excess Ligand**

Reagent (ion forming a precipitate with silver ion)	Silver ion				Precipitate formed with reagent ion
	$[Ag(CN)_2]^-$ $K_{ins} = 1.4\times10^{-20}$	$[Ag(S_2O_3)_3]^{5-}$ $K_{ins} = 9.3\times10^{-15}$	$[Ag(NH_3)_2]^+$ $K_{ins} = 5.8\times10^{-8}$	Ag^+ 0.01 *M* solution of simple salt	
BrO_3^-	$\geqslant 3.9\times10^{13}$	$\geqslant 5.9\times10^{5}$	$\geqslant 9.5$	$\geqslant 5.5\times10^{-3}$	$AgBrO_3$, $SP = 5.5\times10^{-5}$
Cl^-	$\geqslant 1.3\times10^{8}$	$\geqslant 1.9$	$\geqslant 3.1\times10^{-5}$	$\geqslant 1.8\times10^{-8}$	$AgCl$, $SP = 1.78\times10^{-10}$
Br^-	$\geqslant 3.8\times10^{5}$	$\geqslant 5.7\times10^{-3}$	$\geqslant 9.1\times10^{-8}$	$\geqslant 5.3\times10^{-11}$	$AgBr$, $SP = 5.3\times10^{-13}$
I^-	$\geqslant 5.9\times10^{1}$	$\geqslant 8.9\times10^{-7}$	$\geqslant 1.4\times10^{-11}$	$\geqslant 8.3\times10^{-15}$	AgI, $SP = 8.3\times10^{-17}$
S^{2-}	$\geqslant 3.2\times10^{-14}$	$\geqslant 7.3\times10^{-30}$	$\geqslant 1.9\times10^{-39}$	$\geqslant 6.3\times10^{-46}$	Ag_2S, $SP = 6.3\times10^{-50}$

the complex. For example, if to a solution of the complex salt $Na[AgS_2O_3]$ is added $Na_2S_2O_3$, the common ions $S_2O_3^{2-}$ introduced with it will suppress the degree of ionization of the complex and hence decrease the concentration of Ag^+ in solution. Calculations show that it is precisely due to the presence of a small excess* of $Na_2S_2O_3$ that the solution gives no precipitate when treated with KBr. Conversely, had no such excess been present, a precipitate of AgBr should have been formed.

In the precipitation reactions considered above, the complexes were destroyed as a result of the binding of the complexing agent and the formation of slightly soluble compounds. Such breakdown of complexes may also result due to the formation of even more stable complexes. For example, when to a solution of $[Ag(NH_3)_2]Cl$ there is added KCN, the following reaction takes place:

$$[Ag(NH_3)_2]^+ + 2CN^- \rightarrow [Ag(CN)_2]^- + 2NH_3$$

Complexes may break down as a result of the binding of the corresponding ligands. As an example, if to a solution of a complex salt of copper, $[Cu(NH_3)_4]SO_4$, which has a dark-blue colour, there is added an acid, say, H_2SO_4, the colour will change to pale-blue. This change of colour is caused by the breakdown of the $[Cu(NH_3)_4]^{2+}$ ions (which

* An excess of $Na_2S_2O_3$ is introduced for preparation of $Na[AgS_2O_3]$.

are responsible for the dark-blue colour) due to the interaction between the NH_3 molecules and the hydrogen ions. This leads to the formations of more stable complex ions, NH_4^+, the pale-blue colour of the solution being due to the presence of Cu^{2+} ions:

$$[Cu(NH_3)_4]^{2+} + 4H^+ \rightarrow Cu^{2+} + 4NH_4^+$$

Acidification of the solution may lead to the breakdown of complexes, whose ligands are the anions of very weak acids, because molecules of the corresponding slightly dissociated acid are formed. This is observed, for example, in the case of complexes formed by the cations of heavy metals with certain organic compounds of acidic character (e.g., phenols, dimethylglyoxime, dyes such as alizarin, etc.).

The foregoing can be summarized as follows: there is no fundamental difference between double and complex salts. Both types of compounds are but the extreme cases of higher-order compounds, which are characterized by a range of transitions, from practically complete ionization up to its practically complete absence.

5.11. Inner-Complex Compounds. Polyacids and Polysulphides

Complex compounds formed by various cations with organic reagents fall into the category of so-called inner-complex salts which are of great importance to analytical chemistry.

Organic compounds were first used in inorganic analysis by M. A. Il'inski who proposed in 1884 the organic compound α-nitroso-β-naphthol as a reagent for detection of cobalt. But it was only after the investigations carried out by P. A. Chugaev that organic compounds found wide application as reagents. In 1905 Chugaev proposed the organic reagent dimethylglyoxime which is still the best reagent for Ni^{2+}.

The salt-forming properties of organic compounds depend on the presence in their molecules of certain atomic groups which possess acidic properties. These groups include —COOH, —OH, =NOH, =NH, $—NH_2$, $—SO_3H$. The hydrogen atoms of such groups are capable, under definite conditions, of being replaced by a metal ion. But if, apart from such a group, the molecule of an organic compound contains another group, which may function as a ligand for a given cation, the cation displaces the hydrogen atom in the acid group and is capable of forming a coordinate covalent bond with the ligand group. The salts formed in this way are called **inner-complex salts (chelates)**. The simplest example of an inner-complex salt is the compound copper glycocollate, i.e., the copper salt of aminoacetic acid (glycine or glycocoll, $NH_2—CH_2—COOH$), which, in addition to the acid group —COOH, also contains a $—NH_2$ group capable of being coordinated to the Cu^{2+}

ion. The structure of copper glycocollate corresponds to the formula

```
OC—O     O—CO
|    \   /    |
|     Cu      |
|    ↗  ↖     |
H₂C—NH₂ H₂N—CH₂
```

The Cu^{2+} ion in this compound has all the properties of an ion with all its coordination sites being occupied. A similar inner-complex salt is the compound formed by Ni^{2+} with dimethylglyoxime. The reaction of formation of nickel dimethylglyoximate may be represented by the equation

```
                                         H
                                        / \
                                       O   O
                                       ↑   |
   H₃C—C=NOH                     H₃C—C=N   N=C—CH₃
       |                             |  \ ↙   |
2      |          + Ni²⁺ =           |   Ni   |        + 2H⁺
       |                             |  ↗ \   |
   H₃C—C=NOH                     H₃C—C=N   N=C—CH₃
                                       |   ↓
                                       O   O
                                        \ /
                                         H
```

The equation shows that, in addition to replacing two hydrogen atoms in the acidic =NOH groups of two dimethylglyoxime molecules, the Ni^{2+} ion also forms two coordinate-covalent bonds with the nitrogen atoms of two more =NOH groups.

From these examples it is seen that the molecules of inner-complex salts have a cyclic (ring-like) structure.* Inner-complex compounds are important for analysis because they are usually slightly soluble in water, are brightly coloured, and dissociate very slightly into ions.

It is essential that the formation of inner-complex compounds often reveals very distinctly the individual properties of the cations involved, which remain unnoticed in ordinary salt-formation processes. This circumstance makes the action of organic reagents more specific.

It has been found experimentally that the Ni^{2+} ion produces characteristic, mostly red, precipitates not only with dimethylglyoxime but also with other organic compounds (dioximes) containing the group

```
—C=NOH
 |
—C=NOH
```

* Complex compounds having a cyclic (ring-like) structure are called chelates (the word is derived from the Greek, meaning a claw, as of a lobster). Inner-complex compounds constitute a special type of chelates. The most stable inner-complex compounds are those with five- and six-membered rings.

in their molecules. This group has been found to be specific for Ni^{2+}. There are similar specific groups for many other ions.

The analytical properties of reagents containing a specific group are also influenced by the radicals to which this group is linked. Therefore, by substituting these radicals, one can sometimes improve considerably the properties of reagents in a direction that will increase their sensitivity or specificity.

As the number of organic compounds is very great and the structure and composition of their molecules can be altered by a large variety of methods through addition of various groups, it appears possible to develop new specific and highly sensitive reagents for ions.

Let us illustrate this by the example borrowed from the works of V. I. Kuznetsov and I. P. Alimarin. Arsenic acid precipitates most of the metals in the oxidation states +3 and +4. If one of the hydroxyl groups of this acid is replaced by a benzene residue (i.e., the phenyl radical $—C_6H_5$), phenylarsonic acid will result, which is highly selective. In the presence of mineral acids, it precipitates only the ions of zirconium, hafnium, tin(IV), niobium, tantalum, and (incompletely) titanium. If in the molecule of phenylarsonic acid we substitute a nitro group ($—NO_2$) for the hydrogen atom of the benzene ring in the ortho position*, we shall obtain *o*-nitrophenylarsonic acid which is a highly specific reagent for Sn(IV) (this element alone is precipitated in the presence of hydrochloric and tartaric acids):

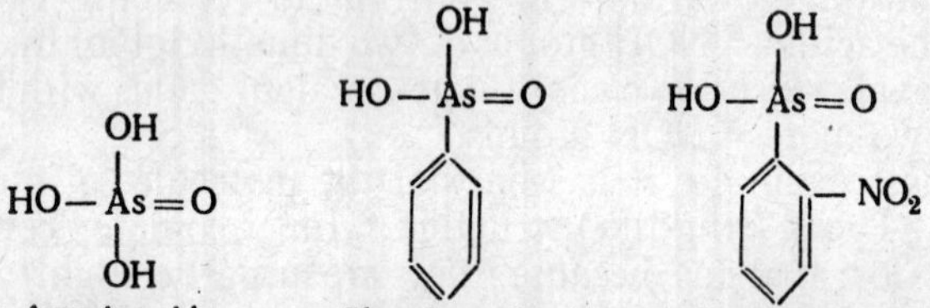

Arsenic acid   Phenylarsonic acid   *o*-Nitrophenylarsonic acid

In analytical chemistry, wide use is made of reagents which enable fractional detection of numerous cations, and also of reagents capable of forming complex compounds with most of the cations, i.e., which are, as it were, universal. One such universal precipitant capable of bringing down, at various pH values, all the cations, except those of alkaline metals, is 8-hydroxyquinoline (oxine), C_9H_6NOH.

Organic substances that form stable soluble complexes with most of the cations have also found wide application. These substances are collectively known as **complexones.** Complexones are usually polybasic organic acids (or their salts) containing amino groups (amino acids).

* That is, at the carbon atom adjacent to the one linked to the arsenic acid residue.

The following are examples of complexones:

$$HOOC-CH_2-N(CH_2COOH)_2$$

nitrilotriacetic acid (NTA)

$$(HOOC-CH_2)_2N-CH_2-CH_2-N(CH_2-COOH)_2$$

ethylenediaminetetraacetic acid (EDTA)

The disodium salt of ethylenediaminetetraacetic acid or complexone III (Trilon B) forms soluble complexes with all doubly, triply, and quadruply charged cations at various pH values. This property can be used to advantage for separation of cations and for analytical masking.

The use of complexones is particularly widespread at present in quantitative analysis.*

A special class of complex compounds is called **polyacids**. These compounds are oxyacids in which some or all the oxygen atoms are replaced by the anions of acids.** For example, the oxygen atoms in the CrO_4^{2-} ion (which itself is a complex ion) can be replaced by CrO_4^{2-} ions, which results in the formation of various polychromic acids:

$$CrO_4^{2-} \qquad Cr_2O_7^{2-} \qquad Cr_3O_{10}^{2-}$$

* See: Přibil, R., *Complexones in Chemical Analysis*, Dyatlova, N. M., V. Ya. Temkina, and I. D. Kolpakova, *Complexones*, Khimiya, Moscow, 1970 (in Russian).

** Polyacids may also be described as compounds consisting of several molecules of an acid anhydride and water, for example,

$$2CrO_3 + H_2O \rightleftharpoons H_2Cr_2O_7$$

Such acids in which the oxygen is replaced by the anions of the same acid are called **isopolyacids.**

In some cases, the oxygen of an anion is replaced by the anions of other acids. As an example, orthophosphoric acid, H_3PO_4, is capable of forming an ammonium salt of the composition $(NH_4)_3[PMo_{12}O_{40}]\cdot xH_2O$. As seen, the oxygen of orthophosphoric acid is replaced by the complex anions of molybdic acid. An analogous compound is formed by arsenic acid: $(NH_4)_3[AsMo_{12}O_{40}]\cdot yH_2O$.

In contrast to isopolyacids, these compounds are referred to as heteropolyacids.* Heteropolyacids often contain Si and B as complexing agents (complex-formers) and also the anions of tungstic, vanadic, and other acids as ligands.

Polysulphides, such as $(NH_4)_2S_2$, are the analogues of peroxides and are also oxidizing agents. As distinct from oxygen chains, the chains formed by sulphur are somewhat stabler. The existence of polysulphides up to $Me^{I}_2S_9$ is possible. They may be visualized as complex compounds of the general formula $Me_2[S \ldots S_n]$:

$$Me—S—S\ldots—S—S—Me$$

5.12. Importance of Complex Compounds in Analysis

As a result of complex formation, many ions of complex compounds fail to react. This circumstance has to be given due consideration in analysis since otherwise wrong results may be obtained.

For example, organic compounds, whose molecules contain the group =CHOH, say, various sugars, tartaric and citric acids, glycerine (glycerol), form complexes with many cations which are therefore found to have no chance to be precipitated by OH^- ions and by some other reagents used in analysis. For this reason, before starting an analysis, it is necessary to remove organic substances from the sample to be analysed, the more so that they may turn out to be protective colloids tending to prevent the coagulation of some precipitates.

Complex formation however may be not only an obstacle but, fortunately, also a boon to the analyst. Let us cite some examples.

1. The presence in solution of Fe^{3+} not infrequently interferes with detection of a number of other ions (Co^{2+}, Ti^{4+}, etc.). In such cases, instead of removing Fe^{3+} in the form of any slightly soluble compound, say, $Fe(OH)_3$, it is more convenient, in the identification of the corresponding ion, to bind Fe^{3+} into a stable complex by addition of tartaric or citric acid. In this way one can lower the concentration of Fe^{3+} in solution, and it will no longer interfere with the detection of ions of interest.

---

* The prefix "iso-" is derived from the Greek (isos), meaning the same; the prefix "hetero-" means unequal, different.

Instead of tartaric or citric acid, orthophosphoric acid or ammonium fluoride may also be added, with which the Fe^{3+} ion forms the complexes $[Fe(PO_4)_2]^{3+}$ and $[FeF_6]^{3-}$.

2. In the systematic analysis of the cations of Group IV, Cu^{2+} and Cd^{2+} are found to be present in solution after all the other cations have been separated. The most characteristic reaction of the Cd^{2+} ion, which consists of the formation of a bright-yellow precipitate of CdS by the action of H_2S, is interfered with by Cu^{2+} forming with hydrogen sulphide a black precipitate of CuS. Therefore, Cu^{2+} must be removed from the solution. This can be achieved, for example, by reducing Cu^{2+} to metallic copper by the metals Al, Zn, or Fe, which do not reduce Cd^{2+} in acid solution. It is simpler however to bind the ions Cu^{2+} and Cd^{2+} into the soluble cyanide complexes* $[Cu(CN)_4]^{3-}$ and $[Cd(CN)_4]^{2-}$ by addition of KCN to the solution. The concentration of cuprous ions in solution decreases to such an extent that the solubility product constant of cupric sulphide cannot be reached. Conversely, the less stable ion $[Cd(CN)_4]^{2-}$ gives a sufficient amount of Cd^{2+} ions for a precipitate of CdS to be formed by the action of H_2S.

Thus, the non-specific reaction for Cd^{2+} with hydrogen sulphide becomes sufficiently specific if it is carried out in the presence of KCN.

3. The analysis of Group IV cations can be simplified even more by adding glycerol to a solution containing the cations Cu^{2+}, Cd^{2+}, Bi^{3+}, and Pb^{2+}. All these cations, except Cd^{2+}, form with glycerol, $CH_2OH—CHOH—CH_2OH$, complexes non-decomposable by alkalis. If, after glycerine is added, the solution is made to react with sodium hydroxide, the Cd^{2+} ion will be precipitated as $Cd(OH)_2$, and the other cations can be identified by means of characteristic reactions in separate portions of the solution since the glycerol complexes present in solution give sufficient concentrations of the corresponding simple ions for the reaction to proceed (or, if necessary, they can be destroyed by addition of acid).

The binding of interfering ions (Fe^{3+}, Cu^{2+}, etc.) into sufficiently stable complexes, the process known as **masking**, is often used in analysis and is of great practical value since it allows many ions to be fractionally detected in those cases where other methods would be useless.

The importance of complex compounds is due, to an even greater degree, to the fact that complex-formation reactions are often very sensitive and specific for individual ions. Often complex ions themselves are sensitive and specific reagents.

For example, the Fe^{3+} ion can be identified by the formation of a blue precipitate of ferric ferrocyanide, $Fe_4[Fe(CN)_6]_3$, under the action of the reagent $K_4[Fe(CN)_6]$, or else by the formation of red soluble complexes of various compositions by the action of thiocyanate ion.

---

* When the complex ion $[Cu(CN)_4]^{3-}$ is being formed, the Cu^{2+} ion is reduced to Cu^+. The equations are given in Sec. 6.6, Reaction 7.

The Cu^{2+} ion can be identified by the formation of a brown precipitate of $Cu_2[Fe(CN)_6]$ or of a blue soluble ammonia complex (ammine), $[Cu(NH_3)_4]^{2+}$.

The Co^{2+} ion can easily be detected by the formation of the soluble thiocyanate complex or of the insoluble complex $Co[Hg(SCN)_4]$ having a blue colour.

As indicated above, the Ni^{2+} ion gives a characteristic pink compound with dimethylglyoxime.

Many other examples could be quoted. There are sensitive reagents for almost all the cations. These reagents give complex compounds having characteristic properties.

As far as the importance of complex compounds is concerned, mention must be made of the effect of complex formation on the acidic-basic and oxidizing-reducing properties of compounds a given complex is composed of. If the anion of a weak acid is bound into a complex, its bonding with H^+ is weakened and an acid stronger than the original one is formed. For example, the acidic properties of oxalic acid are considerably enhanced in the presence of Mg^{2+}, which binds the $C_2O_4^{2-}$ anions into a complex. The acidic properties of boric acid, H_3BO_3, also become more pronounced owing to the presence of glycerol, mannite (mannitol), glucose, etc.

If to a solution of an aluminium salt there is added a fluoride, the resulting complex $[AlF_6]^{3-}$ is so strong that aluminium hydroxide cannot be formed. Accordingly, the degree of hydrolysis of the aluminium salt is decreased. Many other examples could be cited.

The oxidizing and reducing properties of compounds are also changed as a result of complex formation. Usually, the chemical activity of oxidizing (or reducing) agents is diminished in such cases due to the decrease of their concentration in solution. For instance, Fe^{3+} loses its ability to oxidize I^- to I_2 upon addition of any fluoride because of the formation of the complex $[FeF_6]^{3-}$. Likewise, the reducing ability of oxalic acid decreases to such an extent in the presence of an excess of molybdic acid, which forms a complex with oxalic acid, that potassium permanganate no longer oxidizes it. Cases however are known where the oxidizing power of compounds increases due to complex formation. Advantage is taken of this fact in analysis. For example, Mo(VI), which enters into the composition of molybdic acid, $MoO_3 \cdot H_2O$, and normal molybdates, does not oxidize the organic reagent benzidine, $NH_2—C_6H_4—C_6H_4—NH_2$, but once the complex phosphomolybdic acid is formed, benzidine is readily oxidized.

It should also be noted that complex formation can be used for dissolution of certain slightly soluble compounds as well (see Sec. 4.5)*.

* For more detailed information on the importance and use of complex compounds in analysis, see: Lewis, J. and R. G. Wilkins, *Modern Co-ordination Chemistry*, Interscience, New York, 1960; Ringbom, A., *Complexation in Analytical Chemistry*, Interscience, New York, 1963.

5.13. Oxidation-Reduction Reactions

As known, the process of oxidation was originally thought of as the addition of oxygen atoms to or removal of hydrogen atoms from a substance, and reduction as the converse process. There are however many reactions in which neither oxygen nor hydrogen takes part and which must nevertheless be classified as the oxidation-reduction processes. The concepts "oxidation" and "reduction" have been quite clearly defined owing to the modern electronic theory of the structure of matter.

In the light of the electronic theory, *oxidation-reduction (redox) reactions include all chemical processes involving a transfer of electrons from some atoms or ions to others.*

Such is, for example, the reaction

$$Sn^{2+} + 2Fe^{3+} \rightleftarrows Sn^{4+} + 2Fe^{2+} \quad (5.90)$$

The changes taking place in this reaction consist in the following: the Sn^{2+} ion loses two electrons and is converted into an Sn^{4+} ion. This process may be represented by the equation

$$Sn^{2+} - 2e = Sn^{4+} \quad (5.91)$$

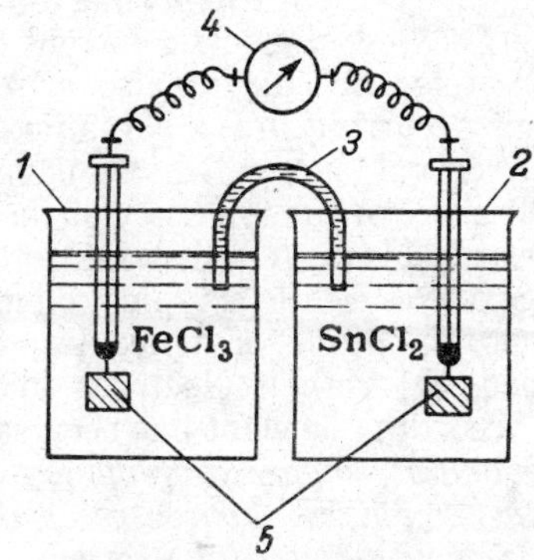

Fig. 5.4. Galvanic cell:
1, *2* — beakers; *3* — salt bridge; *4* — voltmeter; *5* — platinum electrodes.

The electrons lost by the Sn^{2+} ion are transferred to two Fe^{3+} ions, thereby lowering their charge (and the oxidation state of iron):

$$2Fe^{3+} + 2e \rightleftarrows 2Fe^{2+} \quad (5.92)$$

It can be proved by experiment that the reaction in question involves transfer of electrons. This is illustrated by the following example. We place 0.1*M* solutions of $FeCl_3$ and $SnCl_2$, acidified with HCl to increase the electrical conductivity and to suppress the hydrolysis of these salts, into beakers *1* and *2* (Fig. 5.4) and connect them with the aid of a so-called salt bridge, i.e., an inverted U-tube *3* filled with a solution of KCl, through which the ions could diffuse from one beaker into the other. Into either of the solutions are dipped platinum electrodes *5* which are connected with a copper wire (a conductor of the first kind), sensitive voltmeter *4* being inserted into the external circuit. The deflection of the pointer of the voltmeter shows that electric current is flowing in the external circuit of the galvanic cell thus formed and permits its direction to be established. In this case, the electrons move from beaker *2* with the $SnCl_2$ solution into beaker *1* with the $FeCl_3$ solution. If after some time these solutions are tested with the corresponding reagents, the $SnCl_2$ solution will be found to contain Sn^{4+} ions and the $FeCl_3$ solution, Fe^{2+} ions.

This is evidence that the processes expressed by Eqs. (5.91) and (5.92) have actually taken place in beakers *2* and *1*, respectively, and that the electrons have been transferred from Sn^{2+} to Fe^{3+}. Adding of these equations gives the net equation for reaction (5.90) taking place in the galvanic cell.

When the galvanic cell is operating, the Cl^- ions undergo no changes and only move through tube *3* from beaker *1* to beaker *2*, thereby inducing the flow of electric current in the internal circuit.

An electric current also arises in other oxidation-reduction reactions occurring in the galvanic cell.

Each galvanic cell has a definite electromotive force (emf) which is a measure of the tendency of the substances taking part in the reaction towards electron redistribution.

Since oxygen is a typical element, whose atoms tend to add up electrons, the atoms or molecules, which react with oxygen (i.e., are oxidized), give their electrons to oxygen.

According to what has been said, *an oxidation process is any chemical reaction accompanied by the loss of one or more electrons by atoms, molecules, or ions. Reduction being the reverse process of oxidation is characterized by the gain of electrons.*

In the reaction considered above, the Sn^{2+} ions are oxidized to Sn^{4+}, whereas the Fe^{3+} ions are reduced to Fe^{2+}

Let us cite some more examples of oxidation-reduction processes*:

$$2Cu + O_2 \rightleftarrows 2CuO \;(4e) \qquad Cu + Cl_2 \rightleftarrows CuCl_2 \;(2e) \qquad 2I^- + Cl_2 \rightleftarrows I_2 + 2Cl^- \;(2e)$$

The loss of electrons by the atoms, which enter into the composition of ions or molecules and which are oxidized, leads to an increase of their positive charges. In the above examples, the copper atoms when being oxidized are converted into copper(II) ions; the tin(II) ions lose two electrons more and change into tin(IV) ions; the I^- ions lose one electron and convert into an electrically neutral iodine, I_2, etc.

In contrast to oxidation, a reduction process involves a gain of electrons, which leads to a decrease in the oxidation state of an element in atoms, ions, or molecules. For example, in the successive reduction of sulphur in SO_4^{2-} ions, there are formed SO_3^{2-} ions, free sulphur S, and S^{2-} ions, as has already been pointed out (see Sec. 5.9).

In solutions, there exist no simple positive ions with a charge higher than +3. These ions are either hydrated, forming aquo-complexes, or give complexes with the anions of those acids which are present in solution (e.g., $[SnCl_5]^-$, $[SnCl_6]^{2-}$). To simplify the writing of oxidation-reduction reactions, these ions are represented as simple ions

* The arrows indicate the electron transfer involved in the reaction.

(e.g., Sn^{4+}). One must however remember that this is not the actual state of the ion in solution.

Since in reactions the electrons do not remain free but are only transferred from some atoms, molecules, or ions to others, the oxidation of some substances is always accompanied by the reduction of others. These processes cannot take place independently, one without the other.

A substance that causes the oxidation of any other substance, being itself reduced, is called an **oxidizing agent (oxidant)**. A substance that is oxidized in the reaction, thereby reducing another substance, is called a **reducing agent (reductant)**.

It becomes clear from the foregoing that in redox reactions the role of an oxidizing agent consists in accepting the electrons given up by the substance being oxidized. Conversely, the reducing agent, i.e., the substance being oxidized, donates these electrons.

Since electrons do not remain free, the number of electrons donated by the reducing agent must equal the number of electrons accepted by the oxidizing agent. This circumstance may be used to find the coefficients in equations for redox reactions. The number of electrons lost or gained can be deduced from the change of the oxidation state (oxidation number) of the elements taking part in the reaction. The oxidation states of the elements can be established on the basis of the following rules:

1. Hydrogen and metals nearly always lose electrons in redox reactions. The oxidation number of hydrogen is generally +1 and its ion has one positive charge.

2. Oxygen usually accepts two electrons, i.e., it is generally assigned the oxidation number —2 (exceptions to this rule are extremely rare).

3. The other elements may have either positive or negative charges, depending on what ions are formed by a given element.

4. The sum of the charges of the elementary particles making up a molecule is equal to zero, i.e., the molecule is electrically neutral. The charge on the atoms is also equal to zero.

As an example, let us find the oxidation state (or number) of the nitrogen in the HNO_3 molecule. The oxidation state of one hydrogen atom is +1, i.e., it has lost one electron; the oxidation state of each oxygen atom is —2; it means that three atoms of oxygen have acquired six electrons.

Since molecules are electrically neutral, one nitrogen atom must donate five electrons.

5. The difference between the number of electrons lost and the number of electrons gained by the different atomic species in a complex ion determines its valency and the number of charges.

For example, the sulphur atom in the complex ion SO_4^{2-} has lost 6 electrons and each oxygen atom has gained 2 electrons, which means that four oxygen atoms have in all acquired 8 electrons. The difference is 2 electrons, and therefore the sulphate ion is divalent and have two negative charges. In the same way, one can figure out the valency and charge for $Cr_2O_7^{2-}$, etc.

In setting up equations for oxidation-reduction (redox) reactions, one should make use of detailed ion-electron schemes. The advantage of such schemes is that not only the change in the oxidation state of the elements taking part in the reaction is shown but also the change in the composition of the ions or molecules involved. It has already been pointed out that any oxidation-reduction process may serve as a source of electric current if it takes place in a galvanic cell. To achieve this purpose, the oxidizing agent must be separated spatially from the reducing agent, i.e., they must be placed in different vessels and be able to exchange their electrons. The oxidation of the reducing agent occurs in one vessel, and the reduction of the oxidizing agent (which accepts the electrons donated by the reducing agent through the electrode) in the other. If the equations for the reactions taking place in both vessels are written separately (half-reactions) and added up termwise, after the loss and gain of the electrons are balanced, the net equation for the redox reaction occurring in the galvanic cell is obtained.

This method of balancing equations for oxidation-reduction reactions is known as the **ion-electron (half-reaction) method.**

Let us consider some examples.

1. We are to set up an equation for the oxidation of iron(II) by nitric acid, i.e., the nitrate ion.

The Fe^{2+} ion is oxidized to Fe^{3+} giving up one electron:

$$Fe^{2+} - e \rightarrow Fe^{3+} \tag{5.93}$$

The NO_3^- ion, which is an oxidizing agent, is reduced in a moderately acid solution to NO, i.e., accepts three electrons:

$$NO_3^- + 3e \rightarrow NO \tag{5.94}$$

This equation however is written incorrectly (is unbalanced) because there are two oxygen atoms more on the left than on the right. To balance the oxygen atoms, we would need four hydrogen atoms:

$$NO_3^- + 4H^+ + 3e \rightarrow NO + 2H_2O$$

In order to balance the electrons donated by the reducing agent and accepted by the oxidizing agent, we multiply Eq. (5.93) by 3. Adding up the two equations term by term, we obtain:

$$\begin{array}{c} 3Fe^{2+} - 3e \rightarrow 3Fe^{3+} \\ NO_3^- + 4H^+ + 3e \rightarrow NO\uparrow + 2H_2O \\ \hline 3Fe^{2+} + NO_3^- + 4H^+ \rightarrow 3Fe^{3+} + NO\uparrow + 2H_2O \end{array}$$

We can note that the number of oxygen atoms on both sides of the equation is the same, which means that this equation is written correctly. Moreover, it is given in an ionic form. This is an advantage of the ion-electron method since ionic equations reflect more exactly the processes that actually occur in solutions.

This method is especially convenient in those cases where the oxidation number method would lead to wrong results. Such cases are encountered in reactions involving hydrogen peroxide or other peroxide compounds, say, ammonium persulphate, $(NH_4)_2S_2O_8$, and also in reactions with organic substances in which there are carbon atoms linked to one another (for example, in oxalic acid, $H_2C_2O_4$).

2. Consider now the reaction of oxidation of the oxalate ions, $C_2O_4^{2-}$, by MnO_4^- ions in acid solution. The structural formula of oxalic acid is

```
O=C—OH
  |
O=C—OH
```

The carbon in this acid has an oxidation number of 4. Oxalic acid is oxidized by MnO_4^- to CO_2, i.e., the oxidation state of the carbon remains the same in this reaction. The donation of electrons occurs due to the rupture of the covalent bond between the carbon atoms. These electrons are supplied by the $C_2O_4^{2-}$ ion:

$$C_2O_4^{2-} - 2e \rightarrow 2CO_2\uparrow \qquad (5.95)$$

The MnO_4^- ion is reduced in acid solution to Mn^{2+}, i.e., it accepts 5 electrons.

To bind the oxygen from the permanganate ion, 8 hydrogen ions are required. Hence,

$$MnO_4^- + 8H^+ + 5e \rightarrow Mn^{2+} + 4H_2O \qquad (5.96)$$

To balance the numbers of electrons lost by the $C_2O_4^{2-}$ ions and gained by the MnO_4^- ions, we multiply Eq. (5.95) by 5 and Eq. (5.96) by 2, and add them up:

$$\begin{array}{c} 5C_2O_4^{2-} - 10e \rightarrow 10CO_2 \\ 2MnO_4^- + 16H^+ + 10e \rightarrow 2Mn^{2+} + 8H_2O \\ \hline 5C_2O_4^{2-} + 2MnO_4^- + 16H^+ \rightarrow 2Mn^{2+} + 10CO_2\uparrow + 8H_2O \end{array}$$

To check, we count up the number of oxygen atoms that appear on the left and on the right side of the ionic equation obtained: $20 + + 8 = 20 + 8$.

The equation is thus balanced.

3. Write an equation for the oxidation of Cr^{3+} to $Cr_2O_7^{2-}$ by the persulphate ion, $S_2O_8^{2-}$.

When an oxidizing agent contains an oxygen atom (as in NO_3^- and MnO_4^-), hydrogen ions are required to bind this oxygen. For this reason, reactions involving such oxidizing agents are carried out in acid medium.

In this example, we deal with the reverse case: the oxidation of the simple ion Cr^{3+} results in the formation of an oxygen-containing ion,

$Cr_2O_7^{2-}$. This oxygen can be obtained from water molecules. Therefore, the number of water molecules on the left-hand side of the equation must equal the number of oxygen atoms required for the formation of the oxygen-containing ion.

The process of oxidation of Cr^{3+} ions may be written thus:

$$2Cr^{3+} + 7H_2O - 6e \rightarrow Cr_2O_7^{2-} + 14H^+$$

The oxidizing agent here is the anion of persulphuric acid.* The sulphur atoms in this compound have an oxidation number of 6.

The product of the reduction of persulphuric acid is SO_4^{2-} in which the sulphur atoms are also in the oxidation state 6. Therefore, the electrons responsible for the reduction are released as a result of the rupture of the coordinate bond between the oxygen atoms of the peroxide group:

$$S_2O_8^{2-} + 2e \rightarrow 2SO_4^{2-}$$

Thus, each Cr^{3+} ion gives 3 electrons and each $S_2O_8^{2-}$ ion accepts 2 electrons. Hence,

$$2Cr^{3+} + 7H_2O - 6e \rightarrow Cr_2O_7^{2-} + 14H^+$$
$$3S_2O_8^{2-} + 6e \rightarrow 6SO_4^{2-}$$
$$\overline{2Cr^{3+} + 3S_2O_8^{2-} + 7H_2O \rightarrow Cr_2O_7^{2-} + 6SO_4^{2-} + 14H^+}$$

4. Let us now write an equation for the oxidation of CrO_2^- to CrO_4^{2-} by the action of H_2O_2 in alkaline medium. For the reducing agent, i.e., for CrO_2^-, we have:

$$CrO_2^- + 4OH^- - 3e \rightarrow CrO_4^{2-} + 2H_2O$$

The oxidizing agent is hydrogen peroxide, H—O—O—H. When it is reduced, the peroxide bond is ruptured. As a result, the H_2O_2 molecule acquires 2 electrons:

$$H_2O_2 + 2e \rightarrow 2OH^-$$

To balance the number of electrons lost by the reducing agent and gained by the oxidizing agent, we multiply the first equation by 2 and the second by 3:

$$2CrO_2^- + 8OH^- - 6e \rightarrow 2CrO_4^{2-} + 4H_2O$$
$$3H_2O + 6e \rightarrow 6OH^-$$
$$\overline{2CrO_2^- + 3H_2O_2 + 2OH^- \rightarrow 2CrO_4^{2-} + 4H_2O}$$

5. In the example just considered, the oxidizing agent is H_2O_2, this being generally typical of this compound in alkaline solution.

But with respect to some very strong oxidants, such as $KMnO_4$, $K_2Cr_2O_7$, and $MnO(OH)_2$, in acid medium hydrogen peroxide behaves as a reducing agent, being oxidized to oxygen:

$$H_2O_2 - 2e \rightarrow O_2 \uparrow + 2H^+$$

* All peracids are peroxide compounds and contain the peroxide group —O—O—.

As an example, let us write an equation for the reduction of MnO_4^- to Mn^{2+} by the action of H_2O_2 in acid solution:

$$5H_2O_2 - 10e \rightarrow 5O_2\uparrow + 10H^+$$
$$2MnO_4^- + 16H^+ + 10e \rightarrow 2Mn^{2+} + 8H_2O$$
$$\overline{2MnO_4^- + 5H_2O_2 + 6H^+ \rightarrow 2Mn^{2+} + 5O_2\uparrow + 8H_2O}$$

6. The oxidation of I^- ions by $Cr_2O_7^{2-}$ ions in acid solution:

$$6I^- - 6e \rightarrow 3I_2$$
$$Cr_2O_7^{2-} + 14H^+ + 6e \rightarrow 2Cr^{3+} + 7H_2O$$
$$\overline{6I^- + Cr_2O_7^{2-} + 14H^+ \rightarrow 3I_2 + 2Cr^{3+} + 7H_2O}$$

7. The dissolution of FeS in a dilute solution of HNO_3:

$$FeS\downarrow - 3e \rightarrow Fe^{3+} + S\downarrow$$
$$NO_3^- + 4H^+ + 3e \rightarrow NO\uparrow + 2H_2O$$
$$\overline{FeS\downarrow + NO_3^- + 4H^+ \rightarrow Fe^{3+} + S\downarrow + NO\uparrow + 2H_2O}$$

8. The oxidation of Mn^{2+} to MnO_4^- by $NaBiO_3$:

$$2Mn^{2+} + 8H_2O - 10e \rightarrow 2MnO_4^- + 16H^+$$
$$5NaBiO_3\downarrow + 30H^+ + 10e \rightarrow 5Na^+ + 5Bi^{3+} + 15H_2O$$
$$\overline{2Mn^{2+} + 5NaBiO_3\downarrow + 14H^+ \rightarrow 2MnO_4^- + 5Na^+ + 5Bi^{3+} + 7H_2O}$$

9. The oxidation of Mn^{2+} to MnO_4^- by $S_2O_8^{2-}$:

$$2Mn^{2+} + 8H_2O - 10e \rightarrow 2MnO_4^- + 16H^+$$
$$5S_2O_8^{2-} + 10e \rightarrow 10SO_4^{2-}$$
$$\overline{2Mn^{2+} + 5S_2O_8^{2-} + 8H_2O \rightarrow 2MnO_4^- + 10SO_4^{2-} + 16H^+}$$

10. The oxidation of Mn^{IV} to Mn^{VI} by fusing with sodium carbonate and potassium chlorate.

Since it is solid substances that are fused (the reaction proceeds in a solid phase), the reaction equation cannot be written in an ionic form. To calculate the number of electrons transferred from the reducing agent to the oxidizing agent, we set up a scheme:

$$2KClO_3 + 6Na_2CO_3 + 12e \rightarrow 2KCl + 3O_2\uparrow + 6CO_2\uparrow + 6Na_2O$$
$$6MnO(OH)_2 + 3O_2\uparrow + 6Na_2O - 12e \rightarrow 6Na_2MnO_4 + 6H_2O$$
$$\overline{6MnO(OH)_2 + 2KClO_3 + 6Na_2CO_3 \rightarrow 6Na_2MnO_4 + 2KCl + 6H_2O + 6CO_2\uparrow}$$

Dividing all the coefficients by 2, we obtain the final balanced equation:

$$3MnO(OH)_2 + KClO_3 + 3Na_2CO_3 \rightarrow 3Na_2MnO_4 + KCl + 3H_2O + 3CO_2\uparrow$$

11. The oxidation of As_2S_3 by concentrated nitric acid.

This is a more complicated case of an oxidation-reduction reaction since both the elements in the As_2S_3 molecule undergo oxidation.

Arsenic is here in the oxidation state +3 and, when As_2S_3 is oxidized to arsenic acid, it acquires the oxidation state +5. Thus, two arsenic atoms lose 4 electrons.

The sulphur in As_2S_3 is in the oxidation state —2. As a result of the reaction, there is obtained sulphuric acid in which the oxidation state of sulphur is +6. Each sulphur atom loses 8 electrons, which makes 24 electrons for three atoms. Thus, when As_2S_3 is oxidized by concentrated nitric acid, 28 electrons are lost by each As_2S_3 molecule. It means that the same number of electrons must be accepted by the oxidizing agent, i.e., by nitric acid. Concentrated nitric acid is reduced to NO_2, in contrast to dilute nitric acid. With all this taken into account, the following equations may be written:

$$As_2S_3\downarrow + 20H_2O - 28e \rightarrow 2H_3AsO_4 + 3H_2SO_4 + 28H^+$$
$$28NO_3^- + 56H^+ + 28e \rightarrow 28NO_2\uparrow + 28H_2O$$

Adding up the two equations, we finally obtain the complete balanced equation

$$As_2S_3\downarrow + 28HNO_3 \rightarrow 2H_3AsO_4 + 3H_2SO_4 + 28NO_2\uparrow + 8H_2O$$

Questions and Problems

General Directions. In all the problems, the precipitation should be assumed to be practically complete when the concentration of the ions being precipitated falls off to 10^{-6} g-ion/litre. When a given slightly soluble compound is being precipitated, a more soluble form is first formed. Therefore, the solubility product constant of the more soluble form must be used for calculations. After aging (ripening), the slightly soluble compound changes to its less soluble modification. It is for this reason that the solubility product constant to be used in solving solubility problems must be the one corresponding to the less soluble form a given compound.

The values of *SP* for various forms of slightly soluble compounds are listed in Table 5.1 (Sec. 5.2).

It should also be remembered that, in saturated aqueous solutions at room temperature, the concentration of hydrogen sulphide may be assumed to be equal to its solubility, i.e., to $10^{-1}M$.

1. Will a precipitate of CdS be formed if H_2S is passed into a $0.001M$ solution of $CdCl_2$ acidified with HCl to a concentration of $0.5N$?

Answer: yes, it will.

2. At what values of the solution pH could iron(II) sulphide be completely precipitated by hydrogen sulphide?

Answer: at pH $\geqslant$ 4.81.

3. At what values of pH do Fe^{2+} ions fail to precipitate from $0.1M$ solutions of the corresponding salts under the action of hydrogen sulphide?

Answer: at pH $\leqslant$ 2.3.

4. At what values of pH could you prevent the precipitation of ZnS by hydrogen sulphide from a $0.01M$ solution of a zinc salt?

Answer: at pH $\leqslant$ 1.46.

5. What is the amount of zinc (in per cent) that will be precipitated if H_2S is passed into a $0.1M$ solution of $ZnCl_2$ in the presence of formate buffer which maintains a pH of 2.0 in solution?

Answer: 99.98 per cent.

6. A $0.1M$ solution of $CuCl_2$, which also contains HCl at a concentration of $0.5N$, is saturated with hydrogen sulphide at room temperature. Taking into account the formation of H^+ in the reaction, calculate the concentration of Cu^{2+} ions left unprecipitated in solution.

Hint. According to the reaction equation

$$Cu^{2+} + H_2S \rightarrow CuS\downarrow + 2H^+$$

the precipitation of 0.1 g-ion of Cu^{2+} is accompanied by the formation of 0.2 g-ion of H^+. Hence, at the end of precipitation, the concentration of H^+ will be equal to $0.5 + 0.2 = 0.7$ g-ion/litre. On the basis of this value of $[H^+]$, calculate the value of $[S^{2-}]$ at equilibrium and the concentration of Cu^{2+} corresponding to it.

Answer: $[Cu^{2+}] = 2.6 \times 10^{-15}$ g-ion/litre.

7. What will be the concentration of Fe^{2+} in a solution after the precipitation by ammonium sulphide is completed if the excess of the reagent added is $0.01M$ and the pH of the solution is kept at 9.25? The calculation is to be done: (a) with the second-stage hydrolysis of $(NH_4)_2S$ being neglected; and (b) with the second-stage hydrolysis of $(NH_4)_2S$ being taken into account. Compare the results obtained.

Hint. The calculation taking account of the second-stage hydrolysis can be done as follows. First, from the value of pH find the value of $[H^+]$, then substitute it into the equations

$$\frac{[H^+][S^{2-}]}{[HS^-]} = K_2 = 1.3 \times 10^{-13} \quad \text{and} \quad \frac{[H^+]^2[S^{2-}]}{[H_2S]} = K_1K_2 = 1.2 \times 10^{-20}$$

and express the quantities $[HS^-]$ and $[H_2S]$ in terms of $[S^{2-}]$.

Since according to the law of conservation of mass

$$[S^{2-}] + [HS^-] + [H_2S] = 10^{-2}$$

it is easy to calculate the concentration of S^{2-} by substituting the expressions for $[HS^-]$ and $[H_2S]$ into this equation. Then the sought-for concentration of Fe^{2+} is calculated in the usual manner.

Answer: the value obtained in both cases is approximately equal to 2.21×10^{-12} g-ion/litre.

8. Solve the same problem for the concentration of Zn^{2+} precipitated by ammonium sulphide at pH = 9.25.

Answer: $[Zn^{2+}] \approx 1 \times 10^{-16}$ g-ion/litre.

9. What will be the concentrations of Mn^{2+} and Zn^{2+} after they are precipitated by ammonium sulphide (added in an excess of $0.1M$)* if the precipitation is effected at pH 5?

---

* $SP_{MnS} = 2.5 \times 10^{-10}$; $SP_{\beta\text{-}ZnS} = 2.5 \times 10^{-22}$.

Answer: $[Mn^{2+}] \approx 20$ g-ion/litre; $[Zn^{2+}] = 2 \times 10^{-11}$ g-ion/litre.

10. What is the concentration of S^{2-} in a saturated aqueous solution of H_2S (at room temperature)? By how many times will it change if the solution is acidified to pH 0.5?

Hint. In answering the first question use should be made of the equation for the K_2 of hydrogen sulphide; since the secondary dissociation is negligibly slight, the concentrations of H^+ and HS^- may be assumed to be practically equal.

Answer: $[S^{2-}] = 1.3 \times 10^{-13}$ g-ion/litre; the concentration sought for will decrease by a factor of about 10 million times.

11. Carbon dioxide is used to saturate (at 20°C): (a) pure water; (b) a $0.1N$ solution of HCl. Assuming the concentration of the saturated solution of CO_2 at the stated temperature to be equal to 3.8×10^{-2} M, calculate the concentration of CO_3^{2-} in solution for both cases. (c) How is the concentration influenced by the presence of HCl?

Answer: (a) 4.8×10^{-11} M; (b) 8.2×10^{-17} M; (c) decreases by about 600,000 times.

12. Figure out whether $CaCO_3$ ($SP = 4.8 \times 10^{-9}$) and $PbCO_3$ ($SP = 1.0 \times 10^{-13}$) are precipitated if gaseous CO_2 is passed into $0.2N$ solutions of the acetates of the corresponding metals (see the previous problem).

Answer: $CaCO_3$ is not precipitated, while $PbCO_3$ is precipitated (partially).

13. On the basis of the solubility product principle explain: (a) why $KHC_4H_4O_6$ is not precipitated by the action of $H_2C_4H_4O_6$ in the presence of HCl, and why it is precipitated incompletely in the absence of HCl; (b) why the addition of CH_3COONa affects the completeness of precipitation of the salt indicated.

14. Explain: (a) why ZnS dissolves in H_2SO_4 and does not dissolve in CH_3COOH; (b) why, in contrast to zinc sulphide, the sulphides of manganese and iron(II) are soluble in both acids; (c) why bismuth sulphide is insoluble not only in acetic but also in dilute sulphuric or hydrochloric acid; (d) why Bi_2S_3, being insoluble in the acids indicated, dissolves on heating with HNO_3; (e) why HgS is insoluble also in nitric acid.

15. Explain: (a) why aluminium and iron(III) phosphates are insoluble in CH_3COOH, whereas most of other phosphates slightly soluble in water dissolve in it; (b) why iron(II) hydroxide is soluble in acids and in ammonium salts, whereas iron(III) hydroxide is soluble only in acids but insoluble in ammonium salts.

16. The solubility product constant of the least soluble modification of nickel sulphide is 2×10^{-26}. Calculate the solubility of the indicated modification of NiS in a $0.1N$ HCl solution.

17. In what cases and why does a solution of a salt show: (a) a neutral reaction; (b) an acid reaction; (c) an alkaline reaction?

18. Name the products of hydrolysis of salts on the basis of the oxidation states of their cation and anion. Consider, in particular, the hydrolysis of salts of the following types: $Me^{1+}An^{1-}$, $Me_2^{1+}An^{2-}$, $Me_3^{1+}An^{3-}$, $Me^{2+}An_2^{1-}$, and $Me^{3+}An_3^{1-}$ and illustrate the hydrolysis by the corresponding reaction equations.

19. Why cannot the hydrolysis of Na_2SO_3 be represented by the net equation given below?

$$Na_2SO_3 + 2H_2O \rightleftarrows 2NaOH + H_2SO_3$$

20. Why does a precipitate of $Al(OH)_3$ form by the action of sodium carbonate on a solution of $AlCl_3$? Give the equation for the reaction. Why is the Mn^{2+} ion precipitated by Na_2CO_3 as carbonate and not as hydroxide, in contrast to Al^{3+}?

21. How can hydrolysis be enhanced and suppressed?

22. Calculate the values of the hydrolysis constants of Na_2SO_3 and $NaHSO_3$.

Answer: $K_{Na_2SO_3} \approx 1.6 \times 10^{-7}$; $K_{NaHSO_3} \approx 3.7 \times 10^{-6}$.

23. Calculate the degree of hydrolysis (h) of Na_2SO_3 and $NaHSO_3$ in a $1M$ solution.

Answer: $h_{Na_2SO_3} \approx 0.13$ per cent; $h_{NaHSO_3} \approx 0.38$ per cent.

24. One of the characteristic reactions for Fe^{3+} is the action of ammonium thiocyanate, NH_4SCN, resulting in the formation of ferric thiocyanate*, $Fe(SCN)_3$, which imparts a blood-red colour to the solution. Would this colour appear if ammonium thiocyanate is made to react: (a) with a solution of the salt $(NH_4)_2FeSO_4 \cdot 12H_2O$; (b) with a solution of the complex iron salt $K_3[Fe(CN)_6]$?

25. A solution of $K_3[Fe(CN)_6]$ reacts with iron(II) salts according to the equation

$$2K_3[Fe(CN)_6] + 3FeSO_4 \rightarrow Fe_3[Fe(CN)_6]_2 \downarrow + 3K_2SO_4$$

Write an ionic equation for the reaction.

26. What are: (a) the valency of the complex ions; and (b) the coordination numbers and charges of the central ions (complexing agents) in the following compounds: $K_2[PtCl_6]$, $H[AuCl_4]$, $[Co(NH_3)_6]SO_4$?

27. Identify the changes: (a) in the valency of the complexes; and (b) in the number of ions in the outer coordination sphere in the following series of compounds: $[Co(NH_3)_6]Cl_3$; $[Co(NH_3)_5NO_2]Cl_2$; $[Co(NH_3)_3(NO_2)_3]$; $K_3[Co(NO_2)_6]$.

How must the electrical conductivity of solutions of the same molar concentration change in the series indicated?

28. Two compounds of the same empirical formula $PtCl_2Br_2 \cdot 4NH_3$ react differently with a solution of $AgNO_3$. One of them produces with $AgNO_3$ a precipitate of AgCl, and the other, a precipitate of AgBr. Account for this difference and write a coordination formula for each of these compounds.

* Other ferrithiocyanate complexes may also be formed (see Sec. 5.19).

29. What reagents will reveal the difference between two isomers of the composition corresponding to the empirical formula $CoBrSO_4 \cdot$ $\cdot 5NH_3$?

30. Calculate the concentration of Ag^+ (in g-ion/litre) in $0.1M$ solutions of the following complex salts: $[Ag(NH_3)_2]NO_3$; $Na[AgS_2O_3]$; $K[Ag(CN)_2]$.

Answer: 1.1×10^{-3} g-ion/litre; 1.2×10^{-5} g-ion/litre; $7.05 \times$ $\times 10^{-8}$ g-ion/litre.

31. Taking into account the results obtained in the previous problem, figure out whether $AgBrO_3$ ($SP = 5.5 \times 10^{-5}$) and AgI ($SP =$ $= 8.3 \times 10^{-17}$) are precipitated by the action of equal volumes of $0.2M$ solutions of $KBrO_3$ and KI on a $0.2M$ solution of $Na[AgS_2O_3]$.

32. Will a precipitate of AgI be formed if a $0.2M$ solution of $K[Ag(CN)_2]$ is mixed with an equal volume of a $0.2M$ solution of KI? Does the conclusion made agree with experimental data (see Table 5.4 and also Problem 33)?

33. To prepare $K[Ag(CN)_2]$, a solution of $AgNO_3$ is treated with an excess of KCN. The following reactions take place:

$$AgNO_3 + KCN \rightarrow AgCN\downarrow + KNO_3$$
$$AgCN\downarrow + KCN \rightarrow K[Ag(CN)_2]$$

Thus, a solution of $K[Ag(CN)_2]$ always contains a certain excess of KCN. Calculate the concentration of KCN in the solution indicated at which no precipitate of AgI can be formed under the conditions stated in the previous problem.

Answer: $1.3 \times 10^{-3}M$ or higher.

34. If a solution of the cadmium ammine complex, $[Cd(NH_3)_4]SO_4$, is treated with KCN, it will change into a cadmium cyanide complex. Write the equation for the reaction in molecular and ionic forms and account for the course of this reaction.

35. The Cd^{2+} ion forms a white precipitate of $Cd(OH)_2$ when acted on by alkalis, and a yellow precipitate of CdS when reacted with H_2S. But if to a solution of a cadmium salt we first add excess KCN and make it react with the reagents specified, a precipitate is formed only under the action of H_2S and not of alkalis. Explain this.

36. Why, in spite of a very small concentration of Ag^+ in a solution of $K[Ag(CN)_2]$, is a large amount of a precipitate of Ag_2S formed under the action of H_2S?

37. How can the interfering effect of Fe^{3+} be eliminated in detecting other cations without resorting to precipitation?

38. In a quantitative determination of nickel in steels, a sample of steel is dissolved in HCl, and after the oxidation of Fe^{2+} to Fe^{3+} by nitric acid, tartaric or citric acid is added to the solution. Then the solution is made slightly alkaline by ammonia, and Ni^{2+} is precipitated by dimethylglyoxime. What part is played by the addition of tartaric (or citric) acid in the precipitation of Ni^{2+}?

39. Write ionic equations for the following reactions and indicate which ions or atoms are oxidized and which are reduced:

$$CuSO_4 + Fe \rightarrow FeSO_4 + Cu\downarrow$$
$$Ag_2SO_4 + 2FeSO_4 \rightarrow 2Ag\downarrow + Fe_2(SO_4)_3$$
$$2KI + Br_2 \rightarrow 2KBr + I_2$$
$$SnCl_2 + 2HgCl_2 \rightarrow Hg_2Cl_2\downarrow + SnCl_4$$

What changes does the charge on the atoms in the molecules of the compounds indicated undergo?

40. Balance the following oxidation-reduction reactions and indicate which of the substances are oxidizing agents and which are reducing agents:

$$FeSO_4 + KMnO_4 + H_2SO_4 \rightarrow Fe_2(SO_4)_3 + K_2SO_4 + MnSO_4 + H_2O$$
$$KMnO_4 + HNO_2 + H_2SO_4 \rightarrow HNO_3 + K_2SO_4 + MnSO_4 + H_2O$$
$$H_2O_2 + KI + HCl \rightarrow I_2 + KCl + H_2O$$
$$MnO(OH)_2\downarrow + H_2O_2 + H_2SO_4 \rightarrow MnSO_4 + H_2O + O_2\uparrow$$
$$KClO_3 + FeCl_2 + HCl \rightarrow KCl + FeCl_3 + H_2O$$
$$Na_2SO_3 + I_2 + H_2O \rightarrow Na_2SO_4 + HI$$
$$Na_2SO_3 + Na_2Cr_2O_7 + H_2SO_4 \rightarrow Na_2SO_4 + Cr_2(SO_4)_3 + H_2O$$
$$CuS\downarrow + HNO_3 \rightarrow Cu(NO_3)_2 + NO\uparrow + H_2O + S\downarrow$$
$$As_2S_5\downarrow + HNO_3 \rightarrow H_3AsO_4 + H_2SO_4 + NO\uparrow + H_2O$$
$$K_2Cr_2O_7 + H_2SO_4 + H_2S \rightarrow K_2SO_4 + Cr_2(SO_4)_3 + S\downarrow + H_2O$$
$$KMnO_4 + H_2SO_4 + H_2S \rightarrow K_2SO_4 + MnSO_4 + S\downarrow + H_2O$$
$$KMnO_4 + KCrO_2 + H_2O \rightarrow K_2CrO_4 + MnO(OH)_2\downarrow$$

41. Using the ion-electron method, write the following schemes in the form of ionic equations:

$$Sn^{2+} + Fe^{3+} \rightarrow Sn^{4+} + Fe^{2+}$$
$$Mn^{2+} + PbO_2 + H^+ \rightarrow MnO_4^- + Pb^{2+} + H_2O$$
$$ClO_3^- + SO_3^{2-} + H^+ \rightarrow Cl^- + SO_4^{2-} + H_2O$$
$$I^- + NO_2^- + H^+ \rightarrow I_2 + NO\uparrow + H_2O$$
$$Cr_2O_7^{2-} + SO_3^{2-} + H^+ \rightarrow SO_4^{2-} + Cr^{3+} + H_2O$$
$$Cr_2O_7^{2-} + H_2S + H^+ \rightarrow Cr^{3+} + S^{2-}\downarrow + H_2O$$
$$Fe^{2+} + H_2O_2 + OH^- \rightarrow Fe(OH)_3\downarrow$$
$$Co(OH)_3\downarrow + NO_2^- + H^+ \rightarrow Co^{2+} + NO_3^- + H_2O$$
$$IO_3^- + I^- + H^+ \rightarrow I_2 + H_2O$$
$$As_2S_3\downarrow + OH^- + H_2O_2 \rightarrow AsO_4^{3-} + SO_4^{2-} + H_2O$$
$$MnO(OH)_2\downarrow + H_2O_2 + H^+ \rightarrow Mn^{2+} + H_2O + O_2\uparrow$$
$$As_2S_5\downarrow + NO_3^- + H^+ \rightarrow H_3AsO_4 + SO_4^{2-} + H_2O + NO\uparrow$$

B. REACTIONS AND ANALYTICAL PROCEDURE FOR GROUP III CATIONS

5.14. General Characteristics of Group III

Analytical Group III comprises the ions of the metals aluminium (Al), chromium (Cr), titanium (Ti), iron (Fe), manganese (Mn), zinc (Zn), nickel (Ni), cobalt (Co), and certain other cations of the less widespread elements.

The cations of this group differ from those of Groups I and II by forming water-insoluble sulphides, which however dissolve readily in dilute acids, a feature distinguishing Group III from Group IV and V.

Being soluble in acids, Group III sulphides cannot be precipitated by hydrogen sulphide from acidified solutions of the corresponding salts. In un-acidified solutions, hydrogen sulphide either does not precipitate Group III cations at all or precipitates them incompletely (ZnS). Separation of the cations of Group III is effected by the action of H_2S at pH $\geqslant$ 8.7 (for example, in the presence of ammonium buffer with pH 8-9) or by the action of ammonium sulphide, $(NH_4)_2S$, which is the group reagent for Group III.*

Under suitable conditions (see Sec. 5.15), ammonium sulphide precipitates most of Group III cations in the form of sulphides: Fe_2S_3, FeS, MnS, ZnS, NiS, and CoS. The ions Al^{3+}, Cr^{3+}, and Ti^{IV} whose hydroxides** are less soluble than the sulphides, are precipitated by $(NH_4)_2S$ in the form of hydroxides. Apart from Al^{3+}, Cr^{3+}, and Ti^{IV}, many other Group III cations not considered in this book are precipitated as the hydroxides (for example, Be^{2+}, Sc^{3+}, La^{3+}, Zr^{IV}).

The cations of Group III precipitated as sulphides are in the middle of the fourth long period of the Periodic System, i.e., these cations correspond to elements whose third energy levels are built up of from 8 to 18 electrons. Because of this, the cations in question have, as a rule, incomplete outer electron levels (see Table 1.3, Sec. 1.5). An exception is Zn^{2+}, which has a complete 18-electron shell, but in its properties (by the ability to be precipitated by hydrogen sulphide) this cation is intermediate between the cations of Group III and those of Group IV.

The cations of Group III precipitated by the group reagent in the form of hydroxides are in Groups III-V of the Periodic System, in the short periods or in the first half of long periods, with the exception of Cr^{3+} and Be^{2+}. Like the cations of the first two groups, they have complete outer 8-electron shells (except Cr^{3+}).

---

* A theoretical interpretation of all these analytically important facts is presented in Sections 5.1 and 5.2.

** As has already been pointed out, the hydroxides of these metals should be referred to more exactly as the hydrous oxides (hydrates) whose compositions are expressed by the formulas $Al_2O_3 \cdot xH_2O$, $Cr_2O_3 \cdot xH_2O$, $TiO_2 \cdot xH_2O$, etc. In a number of cases, it is precisely the formulas of hydrous oxides that are given in the book (for example, in Tables 5.5, 5.6, 5.7).

Associated with the positions in the Periodic System of the elements forming Group III cations are a number of specific features which are very important for analysis and which distinguish them from Groups I and II. Let us dwell on some of these characteristics.

1. Ionic Charge. In the analytical groups studied so far, we were concerned with the cations of the elements in Groups I and II of the Periodic System. Each of these elements forms a single salt-forming oxide and is therefore present in solution always in the form of an ion having a definite charge. The cations of analytical Group III are formed by the elements belonging to the various groups of the Periodic Systems (II, III, IV, VI, VII, and VIII).* Of these, only the elements of the lowest groups (II and III), zinc with its complete 18-electron shell and aluminium with a complete 8-electron shell, have constant charges. They form the cations Zn^{2+} and Al^{3+}. The elements of the higher groups of the Periodic System, namely titanium (Group IV), chromium (Group VI), manganese (Group VII), iron, cobalt, and nickel (Group VIII), which all have incomplete 18-electron shells, exist in several oxidation states and may therefore be present in solution in the form of ions of different charge.

Ions of different charge are known to give different reactions. But not all these ions are sufficiently stable. For example, Mn^{3+} and Cr^{2+} are very unstable and change readily under analytical conditions into Mn^{2+} and Cr^{3+}. Conversely, both iron cations, Fe^{2+} and Fe^{3+}, are fairly stable.** We shall study the cations Al^{3+}, Cr^{3+}, Ti^{IV}, Fe^{3+}, Fe^{2+}, Mn^{2+}, Zn^{2+}, Co^{2+}, and Ni^{2+}.

2. Chemical Properties of Oxides and Their Hydrates. Analytical Groups I and II contain the cations of the alkali and alkaline-earth metals, i.e., elements with the most strongly pronounced metallic properties. For example, the hydroxides of Group I and II cations belong, with the exception of magnesium hydroxide, to the strongest and most water-soluble bases.

But as we pass further to the right in the Periodic System, the metallic properties of the elements become gradually weaker and the non-metallic properties are enhanced and become the strongest at elements of Group VII in the second half of long periods. Accordingly, the hydroxides of the elements of analytical Group III are either weak and slightly water-soluble bases or have amphoteric properties, or else may even be typical acids. The elements of Groups VI and VII in the Periodic System have the strongest acidic properties in their highest oxidation states.

---

* If account is also taken of those cations of Group III which are not discussed in this book, it will turn out that this group also includes the ions of the elements of all the groups of the Periodic System, with the exception of the first and zero groups.

** Cobalt(III) may exist in solution only in the form of complex ions. Nickel(III) forms only one oxide, Ni_2O_3, and the corresponding hydrate, $Ni(OH)_3$.

Chromium, an element of Group VI, forms the oxides CrO, Cr_2O_3, and CrO_3. The first, CrO, chromous oxide, is a basic oxide, to which there corresponds $Cr(OH)_2$, a weak base, that forms salts with acids only. Chromic oxide, Cr_2O_3, is amphoteric since it is capable of forming salts not only with acids but also with bases (e.g., $CrCl_3$, $KCrO_2$). As regards the oxide CrO_3, it is an acid oxide (chromic anhydride), to which there correspond chromic acid, H_2CrO_4, and dichromic acid, $H_2Cr_2O_7$. From this it can be concluded that chromium(III) may be present in solution either as Cr^{3+} cation or as CrO_2^- anion, the former predominating in acid solution and the latter in alkaline solution (see Sec. 5.7). Chromium(VI) is present in solution in the form of CrO_4^{2-} anions (alkaline solution) or $Cr_2O_7^{2-}$ (acid solution).

Titanium, an element of Group IV, forms the oxides Ti_2O_3 and TiO_2. The former is a basic oxide and gives salts with acids, for example, $TiCl_3$, $Ti_2(SO_4)_3$; these salts have a violet colour. They are strong reducing agents. Titanium dioxide, TiO_2, is an amphoteric oxide to which there correspond salts such as $Ti(SO_4)_2$ and $TiCl_4$ in whose solutions titanium is present in the form of hydrated complex ions: $[TiCl_5(H_2O)]^-$, $[TiCl_6]^{2-}$, $[Ti(OH)_2(H_2O)_4]^{2+}$, $[Ti(OH)_2HSO_4]^+$, etc.

In writing characteristic reactions for titanium further in the text, we will represent its ion as Ti^{4+}. But it should be kept in mind that, depending on the conditions, the mechanism of these reactions is highly complicated. There also exist salts, such as Na_2TiO_3, K_2TiO_3, in which titanium forms TiO_3^{2-} anions. The corresponding acid, H_2TiO_3, is known as metatitanic.

The element manganese (Group VII) is known to form the following oxides: MnO, Mn_2O_3, MnO_2, and Mn_2O_7. The first two are basic and the last three are acidic oxides. To the acidic oxides there correspond the acids $MnO(OH)_2$ (H_2MnO_3 or, more exactly, $MnO_2 \cdot xH_2O$)—manganous acid, H_2MnO_4—manganic, and $HMnO_4$—permanganic. Of the ions formed by manganese, we shall consider the cation Mn^{2+} and the anion MnO_4^- (permanganate ion). Although MnO_4^- as well as the anions CrO_4^{2-} and $Cr_2O_7^{2-}$ are detected in the analysis of solutions for anions, their most important reactions will be discussed together with the reactions for the cations Mn^{2+} and Cr^{3+} with which they are closely associated.

The hydroxides of iron, cobalt, and nickel, which are the elements of Group VIII in the Periodic System, are weak bases.

The hydroxides of aluminium and zinc are amphoteric; the fact that they are amphoteric is used in analysis for separation from most of the other cations of Group III.

3. Hydrolysis of Salts. The basic properties of the hydroxides formed by the cations of Group III are feeble, and therefore the corresponding salts must undergo hydrolysis to a more or less considerable extent. The effect of hydrolysis of salts of strong acids is that solutions of such salts give a more or less strongly acid reaction (pH < 7). The order of pH values for solutions of various salts can be approximately deduced from the data of Table 4.1 (see Sec. 4.3). For example, the precipitation of $Fe(OH)_3$ from a $0.01M$ solution of $FeCl_3$ begins at pH 2.2, and therefore we may conclude that the pH of this solution must be less than 2.2.

At large concentrations, the acidity of solutions of iron(III) salts must be even greater. For example, the pH of a $1M$ solution of $FeCl_3$ or $Fe_2(SO_4)_3$ cannot exceed 1.5 at which, in this particular case, $Fe(OH)_3$ begins to precipitate. Like iron(III) salts, a fairly strongly acid reaction must also be shown by the salts of aluminium and chromium(III) since the precipitation of $Al(OH)_3$ from $0.01M$ solutions of the salts begins at pH 4.0 and that of $Cr(OH)_3$ at pH 4.25.

An even stronger acid reaction is given by solutions of titanium(IV) salts since the precipitation of hydrous titanium oxide from a $0.01N$ solution begins at pH ca. 1.4. Solutions of the salts of cobalt, nickel, manganese, and zinc must show but a very weakly acid reaction, in distinction to solutions of the salts of the cations indicated above. Although the calculations of pH for the start of precipitation of hydroxides, on which these conclusions are based, are approximate, they are, with rare exceptions, in good agreement with experimental results.

Sometimes, in order to suppress the hydrolysis which may lead to the precipitation of basic salts or hydroxides when solutions are stored, certain acids are added to them. For example, solutions of the salts of Ti^{IV}, Fe^{3+}, and Fe^{2+}, and also a solution of $AlCl_3$, are always acidified. Therefore, the pH values of these solutions are usually found to be less than or equal to unity.

Because of the sharp differences in pH values for solutions of various salts, the pH of the test solution when determined may serve as an important sign indicating the possible presence (or absence) of ions in solution.

The hydrolysis of the salts of very weak acids, for example, H_2S, H_2CO_3, HCN, and others, sometimes (in the case of the salts of Al^{3+}, Cr^{3+}, and Ti^{IV}) proceeds so far that the corresponding salt cannot exist in aqueous solution at all. Therefore, if the action of a reagent leads to the appearance in solution of ions which form such a completely hydrolysable salt, a corresponding hydroxide is obtained.

4. Oxidation-Reduction Properties. In studying the cations of analytical Groups I and II, we dealt almost exclusively with exchange reactions. Oxidation-reduction reactions are not employed for detection of the cations of these groups because their oxidation state is constant. In analytical Group III, we are concerned with the elements of the higher groups of the Periodic System, which have variable oxidation states. Wide use is therefore made of oxidation-reduction (redox) reactions for analysis of Group III cations. These reactions are especially important for detection and separation of manganese, chromium, and iron.

It is obvious that oxidizing agents are those ions (or compounds) in which the corresponding element is in its highest oxidation state. Examples are MnO_4^-, CrO_4^{2-} (or $Cr_2O_7^{2-}$), Fe^{3+}, since they are capable of gaining electrons and being reduced to ions (or compounds) in which the element is in the lowest oxidation state, i.e., Mn^{2+}, Cr^{3+}, Fe^{2+}. These are reducing agents capable of being oxidized. Substances or ions, in which a given element is in an intermediate oxidation state, serve as oxidizing agents in some reactions and as reducing agents in others. For example, the compounds of manganese(IV)—$MnO_2 \cdot xH_2O$ or $MnO(OH)_2$—can either oxidize such substances as nitrites or H_2O_2, being thus reduced to Mn^{2+}, or be oxidized by the action of PbO_2 and some other oxidants to the ion MnO_4^-, corresponding to manganese(VII).

5. Complex Formation. The ability to form complexes is exhibited mainly by those elements which are in the middle of the long periods in the Periodic System. From this it is clear that complex-formation processes play an important part in the analysis of Group III cations. In the systematic analysis of these cations, use is made of the ability of the cations Co^{2+}, Ni^{2+}, and Zn^{2+} to form the ammine complexes $[Co(NH_3)_6]^{2+}$, $[Zn(NH_3)_6]^{2+}$, and $[Ni(NH_3)_6]^{2+}$ with an excess of ammonia. The reaction is carried out in the presence of NH_4Cl which is added for the solubility product constant of manganese hydroxide not to be exceeded and for the Mn^{2+} ion to be left in solution together with the complexes indicated. Under these conditions, Fe^{3+}, Al^{3+}, Cr^{3+}, and Ti^{IV} are precipitated as the hydroxides since the values of their solubility product constants are found to be exceeded at the pH value (ca. 9) developed in a mixture of NH_4OH and NH_4Cl (see Table 4.1) and since under the conditions of the experiment these cations do not form ammine complexes.

Thus, by using a mixture of ammonia and an ammonium salt, it is possible to separate the cations of Group III into two subgroups. Instead of ammonia, use may be made of an organic base—pyridine, C_5H_5N, which forms complexes of the composition $[Me(C_5H_5N)_2]^{2+}$ with Mn^{2+}, Zn^{2+}, Ni^{2+}, and Co^{2+}. These complexes remain in solution. The presence in solution of pyridine maintains a pH value of 6.5, at which (as seen from Table 4.1) the ions Fe^{3+}, Al^{3+}, Cr^{3+}, and Ti^{IV} are completely precipitated as the hydroxides.

Of great importance is the formation of complex cyanides, thiocyanates, and mercury-thiocyanates* compounds used for detection of the ions of iron, zinc, and cobalt, and also of complexes formed by certain cations of Group III and organic reagents—dimethylglyoxime, α-nitroso-β-naphthol, etc.

---

* Such are, for example, complex salts of the compositions $Co[Hg(SCN)_4]$ and $Zn[Hg(SCN)_4]$.

In studying the reactions of Group III cations, we shall also encounter cases where complex-formation is used for masking ions which interfere with the reactions of interest.

5.15. Action of the Group Reagent

The precipitation of Group III cations (see Sec. 5.1) can be carried out either with hydrogen sulphide, H_2S, in a weakly alkaline solution or with ammonium sulphide, $(NH_4)_2S$, which is the group reagent. As is always the case in the precipitation of salts of weak acids, one of the most important conditions is the pH of the solution. Calculations show that for sufficiently complete precipitation of Group III cations by hydrogen sulphide it is necessary that the pH of the solution be equal to or greater than 8.7 (see Sec. 5.1). Solutions of the salts of Group III cations have $pH < 7$ and, besides, hydrogen ions build up as the result of the reaction, for example,

$$Zn^{2+} + H_2S \rightarrow ZnS\downarrow + 2H^+$$

and therefore a buffer mixture must be introduced into the solution, so that the pH is maintained practically constant (not less than 8.7). Of the buffers usually employed in analysis, the most suitable for this particular case is the ammonium buffer with pH 8-9 (depending on the concentrations of NH_4OH and NH_4Cl).

It might seem at first glance that, when the cations of Group III are precipitated by ammonium sulphide, it is useless to add a buffer since because of the 100-percent hydrolysis at the first stage:

$$2NH_4^+ + S^{2-} + H_2O \rightarrow NH_4^+ + HS^- + NH_4OH$$

the solution of $(NH_4)_2S$ is itself an ammonium buffer with pH 9.25 (see Table 4.1). But this is not actually the case. Just as in the precipitation of Group II cations (see Sec. 4.9), here it is necessary not only to introduce into the solution a buffer of the appropriate pH but also to maintain the pH value practically constant, i.e., the capacity of the buffer mixture must be sufficient. Indeed, mere addition of $(NH_4)_2S$ to the solution will not suffice in this case, which is verified by the following considerations.

1. Solutions of the salts of Group III cations show acid reactions due to hydrolysis; besides, in the systematic analysis, HCl is introduced into the solution in the precipitation of the cations of Groups IV and V by hydrogen sulphide.

For these acids to be neutralized, there would be required too large amounts of $(NH_4)_2S$, and poisonous hydrogen sulphide would be evolved into the air:

$$(NH_4)_2S + 2HCl \rightarrow 2NH_4Cl + H_2S\uparrow$$

2. The equation written above does not take into account the hydrolysis of $(NH_4)_2S$, as a result of which its solutions become in

fact mixtures of NH_4OH and NH_4HS. With this in mind, the reaction may be written in the form of an ionic equation as follows:

$$NH_4OH + HS^- + 2H^+ \rightarrow H_2S\uparrow + H_2O + NH_4^+$$

From this equation it is seen that the interaction of ammonium sulphide with acids is accompanied by the consumption of NH_4OH and build-up of NH_4^+.

A similar reaction takes place in the precipitation of Group III cations:

$$Me^{2+} + NH_4OH + HS^- \rightarrow MeS\downarrow + H_2O + NH_4^+$$

Thus, although the $(NH_4)_2S$ solution contains equal numbers of NH_4OH molecules and NH_4^+ ions, the equality of their concentrations will be disturbed as a result of the reactions indicated. The concentration ratio will become less than unity and the pH of the solution will accordingly fall off to below 9.25, as it follows from the equation (see Sec. 3.9):

$$pH = 14.00 - 4.76 + \log\frac{[NH_4OH]}{[NH_4^+]}$$

If this decrease of pH is considerable, i.e., if the capacity of the buffer is insufficient, the precipitation of Group III cations may be found to be incomplete.

3. It must also be noted that solutions of $(NH_4)_2S$ often contain an impurity of polysulphides, say, $(NH_4)_2S_2$, formed as a result of the oxidation of ammonium sulphide by the oxygen of the air. Polysulphides when acted on by acids give a precipitate of sulphur:

$$(NH_4)_2S_2 + 2HCl \rightarrow 2NH_4Cl + H_2S\uparrow + S\downarrow$$

Contamination of precipitated sulphides with sulphur is also undesirable.

All that has been said above show that it is not expedient that the acid solution be directly acted on by ammonium sulphide, although it has the required pH value. It is necessary first of all to neutralize the acid by ammonia. And since the addition of ammonia may cause precipitation of Mg^{2+}, a sufficient amount of NH_4Cl is also added so that the pH of the solution does not exceed 9.3.* In other words, before $(NH_4)_2S$ is added, one has to introduce an ammonium buffer into the solution in order to increase its buffer capacity to such an extent that the build-up of NH_4^+ and consumption of NH_4OH during the precipitation of Group III cations can no longer be able to change its pH to any appreciable degree.

The addition of NH_4Cl is useful for another reason too: it is a coagulant-electrolyte which favours the coagulation of colloidal solutions of

* As seen from Table 4.1 (see Sec. 4.3), at pH 9.4 magnesium hydroxide, $Mg(OH)_2$, begins to precipitate from solution if $[Mg^{2+}]$ = 1g-ion/litre.

sulphides; many of the sulphides, particularly NiS, are capable of producing colloidal solutions. Coagulation is also favoured by heating the solution during the precipitation.

When precipitation is carried out under such conditions (see Sec. 5.1.), all the cations of Group III, except Al^{3+}, Cr^{3+}, and Ti^{IV}, precipitate completely as sulphides. The ions Al^{3+}, Cr^{3+}, and Ti^{IV} are precipitated as the hydroxides which are less soluble than the corresponding sulphides. It is because of this that the solubility product constants of the hydroxides are attained before those of the sulphides, which means that the hydroxides (and not the sulphides) are precipitated.

It has already been pointed out that the precipitation of Group III cations is carried out in a weakly alkaline solution at pH ca. 9. It should be noted that a strongly alkaline solution must not be produced in this case since otherwise: (a) the Mg^{2+} ion will precipitate completely together with Group III cations; (b) the amphoteric hydroxides of aluminium and chromium(III) formed in the reaction will dissolve with the formation of AlO_2^- and CrO_2^-, which remain in solution together with the cations of Groups II and I.

Thus, the precipitation of Group III cations by the group reagent, $(NH_4)_2S$, is accomplished in the presence of NH_4OH and NH_4Cl at pH ≈ 9. The sulphides of all the cations of Group III are precipitated, with the exception of Al^{3+}, Cr^{3+}, and Ti^{IV}, which precipitate as hydroxides. The cations of Groups II and I remain in solution together with excess of $(NH_4)_2S$ and other compounds of ammonia.

One must remember that not infrequently the solution of $(NH_4)_2S$ contains SO_4^{2-} as an impurity due to the oxidation of S^{2-} by the oxygen of the air. On the other hand, the $(NH_4)_2S$ solution and also the ammonia used for neutralization of acids contain CO_3^{2-} as an impurity (due to the absorption of CO_2 from the air). Both these anions are known to precipitate Group II cations. Moreover, one has also to consider the coprecipitation of these anions with the hydroxides of triply charged cations of Group III. The separation of the cations of Group III from those of Group II therefore involves considerable losses of Group II cations which are partially precipitated with Group III cations. In especially unfavoruble cases, Group II cations may even be lost completely. What has to be done in such cases is indicated in the description of the analysis of Group II cations.

In conclusion, let us consider the action of acids on the sulphides of Group III cations. Since the values of their solubility product constants are sufficiently high, they (in contrast to Group IV sulphides with smaller solubility product constants) dissolve readily in dilute acids. The only exceptions are the sulphides of nickel and cobalt which are practically insoluble (in the cold) in dilute HCl and H_2SO_4. Despite this, Co^{2+} and Ni^{2+} are not precipitated by H_2S in acid solution (for example, at pH $\leqslant 2$). The cause of this seeming contradiction lies in the existence of several allotropic modifications of NiS and CoS which differ in solubility. At the beginning of the precipitation the most soluble α-NiS and α-CoS are always formed, whose solubility product constants are, respectively, 3.2×10^{-19} and 4.0×10^{-21}. According to the values of the solubility products, these forms of sulphides are well soluble in acids and are not precipitated by H_2S in acid solution. On standing, however, being in contact with the solution, they gradually change into other modifications, γ-NiS ($SP = 2 \times 10^{-26}$) and β-CoS ($SP = 2 \times 10^{-25}$), which are insoluble in HCl (see Table 5.1). This fact is occasionally used for separation of Ni^{2+} and Co^{2+} from other cations fo Group III.

Let us now consider the action of the group reagent on the individual cations of Group III.

1. Place 2 drops each of solutions of the salts of iron(II), iron(III), manganese, zinc, nickel, and cobalt into conical test tubes and add to them 2 or 3 drops each of solutions of NH_4OH, NH_4Cl, and $(NH_4)_2S$. This causes the corresponding sulphides to be precipitated:

$$Fe^{2+} + S^{2-} \rightarrow FeS \downarrow$$
$$2Fe^{3+} + 3S^{2-} \rightarrow Fe_2S_3 \downarrow$$
$$Mn^{2+} + S^{2-} \rightarrow MnS \downarrow$$

The precipitates of FeS, Fe_2S_3, CoS, and NiS are black in colour; those of MnS and ZnS are pink* and white, respectively. All the sulphides of Group III, except NiS and CoS, dissolve in dilute HCl and H_2SO_4 with the evolution of gaseous H_2S; for example,

$$FeS \downarrow + 2H^+ \rightarrow Fe^{2+} + H_2S \uparrow$$
$$MnS \downarrow + 2H^+ \rightarrow Mn^{2+} + H_2S \uparrow$$

When Fe_2S_3 dissolves, the Fe^{3+} ion is reduced to Fe^{2+} by the hydrogen sulphide evolved, a white turbidity of sulphur being formed:

$$Fe_2S_3 \downarrow + 6H^+ \rightarrow 2Fe^{3+} + 3H_2S \uparrow$$
$$2Fe^{3+} + H_2S \rightarrow 2Fe^{2+} + 2H^+ + S \downarrow$$

Adding up both these equations gives the net equation for the reaction:

$$Fe_2S_3 \downarrow + 4H^+ \rightarrow 2Fe^{2+} + S \downarrow + 2H_2S \uparrow$$

The dissolution of NiS and CoS (see above) is achieved if they are acted on not only by an acid but also by an oxidizing agent, say, by hydrogen peroxide. The reaction proceeds on heating:

$$NiS \downarrow + 2H^+ + H_2O_2 \rightarrow Ni^{2+} + 2H_2O + S \downarrow$$
$$CoS \downarrow + 2H^+ + H_2O_2 \rightarrow CO^{2+} + 2H_2O + S \downarrow$$

* Depending on the precipitation conditions, MnS may also separate out as a greenish precipitate.

The same effect is shown by $6N$ nitric acid on heating. Addition of several small crystals of KNO_2 or $NaNO_2$ accelerates catalytically the oxidation reaction:

$$3CoS\downarrow + 8H^+ + 2NO_3^- \rightarrow 3Co^{2+} + 3S\downarrow + 2NO\uparrow + 4H_2O$$

The sulphides CoS and NiS can also be dissolved by heating in aqua regia (a mixture of 1 volume of concentrated HNO_3 and 3 volumes of concentrated HCl). The action of nitric acid and aqua regia will of course dissolve all the other sulphides (and hydroxides) of Group III cations as well.

2. When solutions of salts of aluminium, chromium, and titanium are acted on by a solution of $(NH_4)_2S$, instead of the sulphides Al_2S_3, Cr_2S_3, and TiS_2, the following hydroxides are precipitated: $Al(OH)_3$ (white), $Cr(OH)_3$ (grey-violet or grey-green), and $Ti(OH)_4$ (white). For Cr^{3+} the reaction may be written as follows:

$$Cr^{3+} + 3S^{2-} + 3H_2O \rightarrow Cr(OH)_3\downarrow + 3HS^-$$

Since part of the HS^- ions formed in this reaction are transformed into H_2S due to further hydrolysis, the following reaction partially takes place simultaneously:

$$Cr^{3+} + 3HS^- + 3H_2O \rightarrow Cr(OH)_3\downarrow + 3H_2S\uparrow$$

Analogous equations can also be written for Al^{3+} and Ti^{IV}.

Like Group III sulphides, the precipitates of $Al(OH)_3$, $Cr(OH)_3$, and $Ti(OH)_4$ are soluble in dilute acids.

5.16. Reactions of the Aluminium Ion

In aqueous solutions the Al^{3+} ion is colourless.

1. Alkali Hydroxides, NaOH and KOH. When these reagents are added carefully (drop by drop), the Al^{3+} ion separates in the form of a white precipitate of $Al(OH)_3$:

$$Al^{3+} + 3OH^- \rightarrow Al(OH)_3\downarrow$$

Place 5 or 6 drops of a solution of an aluminium salt in a test tube and add 1 drop of $2N$ NaOH solution. Transfer part of the liquid together with the suspended precipitate to another test tube, add several drops of an acid, and observe the dissolution of the precipitate:

$$Al(OH)_3\downarrow + 3H^+ \rightarrow Al^{3+} + 3H_2O \tag{5.97}$$

Add a few drops of a solution of NaOH or KOH to the remaining liquid in the first tube and mix. The precipitate dissolves with the formation of aluminate:

$$Al(OH)_3\downarrow + OH^- \rightarrow AlO_2^- + 2H_2O \tag{5.98}$$

Thus, $Al(OH)_3$ is a typical amphoteric hydroxide. Its basic properties are exhibited in reaction (5.97) and its acidic properties in reaction (5.98).

Complete precipitation of $Al(OH)_3$ is achieved at pH ≈ 5 (see Sec. 5.6); further addition of alkali (pH > 10) results in the dissolution of $Al(OH)_3$ with the formation of AlO_2^-. From this it follows that, in order to precipitate $Al(OH)_3$ from an aluminate solution and then to obtain the Al^{3+} cations, it is necessary to lower the pH of the solution by using an acid. To make certain, to the aluminate solution obtained in the preceding experiments add, drop by drop, a dilute solution of HCl. The first to precipitate is $Al(OH)_3$. If the addition of the acid is continued (i.e., with decreasing pH), the precipitate redissolves:

$$AlO_2^- + H^+ + H_2O \rightarrow Al(OH)_3 \downarrow$$
$$Al(OH)_3 \downarrow + 3H^+ \rightarrow Al^{3+} + 3H_2O$$

Adding up these equations, we obtain the net equation for the reaction involved in the conversion of AlO_2^- anions into Al^{3+} cations:

$$AlO_2^- + 4H^+ \rightleftharpoons Al^{3+} + 2H_2O$$

If the reaction is to be stopped at the stage of formation of $Al(OH)_3$, use must be made of such a reagent which would decrease the pH of the solution to 5 (but not below). A suitable reagent is NH_4Cl. Indeed, the NH_4^+ ions combine with OH^- ions, thereby decreasing the solution pH. The NH_4OH formed and excess NH_4Cl constitute an ammonium buffer with pH ≈ 9 (at equal concentrations of both components). Hence, the precipitate will not dissolve and the reaction will stop once $Al(OH)_3$ has been formed:

$$AlO_2^- + 2H_2O + NH_4^+ \rightarrow Al(OH)_3 \downarrow + NH_4OH$$

Not infrequently this reaction is employed for detection of Al^{3+} in the systematic analysis. To carry out this reaction, to a few drops of the aluminate solution (see above for its preparation) add a saturated solution of NH_4Cl (or a little of the solid salt) and heat the mixture for several minutes. A precipitate of $Al(OH)_3$ separates out in the form of white gelatinous flocs, often floating to the surface.

To make this reaction sufficiently reliable, it must be conducted on prolonged heating (boiling) of the solution until the ammonia is nearly completely removed. If excess ammonia, which may result from the interaction of the alkali present in solution with NH_4Cl, proves to be so large that the solution pH becomes greater than 10, no precipitate of $Al(OH)_3$ will be formed. It is therefore more reliable to precipitate aluminium hydroxide by the action of NH_4OH after the AlO_2^- ions have been preliminarily converted into Al^{3+} cations by adding HCl until the reaction is acid (see Reaction *2*).

2. Ammonium Hydroxide, NH_4OH. This reagent also precipitates Al^{3+} as $Al(OH)_3$ which is partially soluble in excess of reagent* (due

* Complete dissolution takes place only with a very large excess of concentrated NH_4OH.

to the amphoteric properties of ammonium hydroxide). Since the solubility product constant of $Al(OH)_3$ is very low and Al^{3+} does not form complexes with ammonia, the precipitated $Al(OH)_3$ is insoluble in ammonium salts. The Al^{3+} ion therefore is also precipitated from the solution by the action of a mixture of ammonia and an ammonium salt.

This also follows from the fact that the complete precipitation of $Al(OH)_3$ is achieved even at pH 5, while mixtures of NH_4OH and NH_4Cl have pH values between 8 and 10, depending on the ratio of the concentrations of both components.

3. Salts of Weak Acids and Strong Bases. For example, Na_2CO_3, Na_2SO_3, and KNO_2 precipitate Al^{3+} in the form of $Al(OH)_3$, since solutions of these salts contain OH^- in a sufficient concentration owing to hydrolysis. For example,

$$3CO_3^{2-} + 2Al^{3+} + 3H_2O \rightarrow 2Al(OH)_3\downarrow + 3CO_2\uparrow$$

It will be recalled that the group reagent $(NH_4)_2S$ has the same effect on Al^{3+}.

4. Sodium acetate CH_3COONa. This salt produces in solutions of aluminium salts a white flocculent precipitate. The exact composition of this compound has not yet been established.

To enhance the hydrolysis, the solution should be strongly diluted with water and boiled.*

5. Disodium Hydrogen Phosphate Na_2HPO_4. It brings down Al^{3+} in the form of a white precipitate of $AlPO_4$ which is soluble in strong acids but practically insoluble in acetic acid:

$$Al^{3+} + HPO_4^{2-} \rightarrow AlPO_4\downarrow + H^+$$

The hydrogen ions are bound by an excess of the HPO_4^{2-} ions added with the formation of $H_2PO_4^-$:

$$H^+ + HPO_4^{2-} \rightleftarrows H_2PO_4^-$$

Thus, a phosphate buffer ($NaH_2PO_4 + Na_2HPO_4$) is formed in solution, which maintains the pH at about 7, this being quite enough for complete precipitation of $AlPO_4$.

Adding up the above equations, we arrive at the net equation of the reaction

$$Al^{3+} + 2HPO_4^{2-} \rightarrow AlPO_4\downarrow + H_2PO_4^-$$

Attention should be called to the fact that Al^{3+} is precipitated by Na_2HPO_4 not as the hydrophosphate $Al_2(HPO_4)_3$ but in the form of the phosphate $AlPO_4$.

* The addition of solid CH_3COONa is more effective.

This may be accounted for as follows. Apart from a large amount of HPO_4^{2-}, the reagent used (Na_2HPO_4) also contains PO_4^{3-} which is formed according to the equation

$$HPO_4^{2-} \rightleftharpoons H^+ + PO_4^{3-} \qquad (K_3 = 4.4 \times 10^{-13})$$

In spite of the fact that the concentration of PO_4^{3-} in solution is very low, because of the considerably low solubility of $AlPO_4$ as compared with the solubilities of other aluminium phosphates, the solubility product constant of the given salt is reached first in the reaction and that is why it is precipitated. But once the precipitation has started, the PO_4^{3-} ions removed from the solution are immediately compensated for by those formed due to further dissociation of HPO_4^{2-}, so that practically all the Al^{3+} ions present in solution will be precipitated as $AlPO_4$.

In contrast to Al^{3+}, the cations of Group II are precipitated by the same reagent in the form of the hydrophosphates $BaHPO_4$, $CaHPO_4$, etc., since it is precisely the solubility product constants of these salts that are reached first. Neutral salts can also be produced, e.g., $Ba_3(PO_4)_2$, if the reaction is carried out in the presence of alkali whose OH^- ions react with HPO_4^{2-}:

$$HPO_4^{2-} + OH^- \rightleftharpoons PO_4^{3-} + H_2O$$

As a result, the concentration of PO_4^{3-} in solution increases and that of HPO_4^{2-} decreases to such an extent that the solubility product constants of the phosphates, $Ba_3(PO_4)_2$, etc., are attained first.

All that has been said serves as another good illustration of the importance of the principle of fractional precipitation (see Sec. 4.4): every time when the reaction may result in the formation of various slightly soluble compounds, the compound actually formed is one whose solubility product constant is reached first. This also happens when salts of weak acids act on Al^{3+} (see above), a solution of $K_2Cr_2O_7$ is reacted with Ba^{2+}, etc.

6. Alizarin (1, 2-Dihydroxyanthraquinone). This reagent produces with aluminium hydroxide a slightly soluble compound of a bright-red colour, called an aluminium lake.

The formula of alizarin is

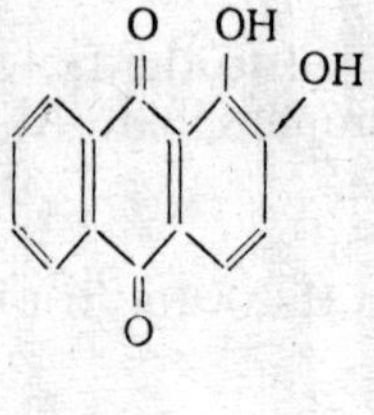

Probably, the composition of the compound formed corresponds to the formula

HO—Al—OH

O O

OH

O

Another point of view existed formerly, according to which this reaction involved the formation of an adsorption compound of variable composition. We have already encountered the formation of similar compounds in studying the reactions for Mg^{2+} (see Sec. 3.15, Reaction *5*).

Similar coloured lakes are formed by alizarin with other cations too. Therefore, they must be separated from Al^{3+}. When the reaction is carried out by the spot-test method, this is achieved by the use of a substrate (see Sec. 2.4) of $K_4[Fe(CN)_6]$, with which the interfering cations form slightly soluble hexacyanoferrates(II), and so remain at the centre of the spot. The Al^{3+} ions, which are not precipitated by $K_4[Fe(CN)_6]$, diffuse to the periphery of the spot where they can be detected by the action of alizarin in the presence of NH_4OH.

The technique of carrying out this reaction is as follows.* Place a drop of a solution of $K_4[Fe(CN)_6]$ on a strip of filter paper. Then apply a drop of the test solution to the centre of the moistened spot. To do this, touch the paper with the tip of the capillary, in which a small amount of the test solution is held by capillarity, and hold it in this position for several seconds.** If, apart from Al^{3+}, the solution also contains interfering cations (e.g., Fe^{3+}), a precipitate of slightly soluble hexacyanoferrates(II) appears at the centre of the spot. To wash off the adsorbed Al^{3+} ions, it is useful to apply in just the same way a drop of water to the precipitate.

The spot is treated with gaseous ammonia, for which purpose the paper is held over the mouth of the bottle containing a concentrated solution of ammonia.

Move the capillary with the alizarin solution along the spot periphery and again treat the spot with gaseous ammonia. No yellow colouration typical of alizarin in acid medium must persist in solution.

---

* To study this reaction, use should be made of a mixture of solutions of ammonium and ferric salts.

** The success of the test depends on how correctly the drops of the test solution, reagent, and water are placed on the paper. The spot-test technique is described in Sec. 2.4.

If Al^{3+} is present, a pink ring appears against the violet background. This background is the colour of alizarin, which it assumes in alkaline solution. Since the background complicates somewhat the observation of the pink colour caused by Al^{3+}, the paper is carefully dried by holding it high over the flame of a burner. The ammonia volatilizes and the violet colour of alizarin changes to yellow colouration, which does not interfere with the observation of the colour of the aluminium lake. The reaction is used for fractional detection of Al^{3+} in the presence of other cations.

This reaction can also be carried out in a test tube in a weak solution of acetic acid. For better coagulation of the pink precipitate, the solution must be heated.

7. **Aluminon.** This reagent, which is the ammonium salt of aurintricarboxylic acid, $C_{22}H_{11}O_9(NH_4)_3$, also forms with $Al(OH)_3$ a red-coloured lake to which is assigned the following structure:

$$\left(HO-C_6H_3(COONH_4)-\right)_2=C=C_6H_3(=O)(-C(=O)-O-)\rightarrow Al(OH)_2$$

To carry out the test, to 2 drops of a solution of an aluminium salt containing a little acetic acid add 1 or 2 drops of a 0.01-percent solution of aluminon and heat in a water bath. Then add a solution of NH_4OH (until it smells of ammonia) and 2 or 3 drops of $(NH_4)_2CO_3$ solution. The red flocs of the aluminium lake prove the presence of Al^{3+}. The reaction is very sensitive. It should be borne in mind that, apart from Al^{3+}, red-coloured lakes are also formed by aluminon with Fe^{3+}, Cr^{3+}, and Ca^{2+}. As distinct from the aluminium lake, the lake produced by chromium is destroyed by the action of NH_4OH, and that formed by calcium, by the action of $(NH_4)_2CO_3$. The Fe^{3+} ion must preliminarily be removed completely (by the action of excess alkali, following which the solution containing AlO_2^- ions is acidified with CH_3COOH).

8. **Morin***. This compound, when added to neutral or acetic acid solutions of aluminium salts, produces a beautiful green fluorescence.

The test is carried out as follows. To 1 or 2 drops of a solution of an aluminium salt add a few drops of CH_3COOH and about 1 ml of wate , after which the solution is treated with 1 drop of morin solution. The reaction is very sensitive. The fluorescence however is difficult to see by electric light.

9. **8-Hydroxyquinoline (Oxine).** This reagent produces in an acetate buffer solution of an aluminium salt at pH ≈ 5 (distinction from

* Morin is a flavone (pentahydroxyflavone), yellow colouring matter of the wood of *Morus tinctoria*. Its composition is represented by the formula $C_{15}H_{10}O_7 \cdot 2H_2O$, and the compound formed by morin with aluminium has the formula $(C_{15}H_9O_7)_3Al$.

magnesium) a greenish-yellow crystalline precipitate of aluminium hydroxyquinolate which is soluble in mineral acids.

The precipitate has the composition corresponding to the formula*·

O—Al3+/3 ← N (8-hydroxyquinoline ring) or $(C_9H_6NO)_3Al$

10. Dry Test. When solid compounds of aluminium are treated with a very dilute solution of $Co(NO_3)_2$ and ignited, they assume a blue colour due to the formation of cobalt metaaluminate, $Co(AlO_2)_2$, which is known as Thenard's blue; for example,

$$2Al_2(SO_4)_3 + 2Co(NO_3)_2 \rightarrow 2Co(AlO_2)_2 + 6SO_3\uparrow + 4NO_2\uparrow + O_2\uparrow$$

The test is made as follows. A little of a solid aluminium salt is put in the loop of a nichrome wire, moistened with 1 or 2 drops of a solution of a cobalt salt, and ignited in the flame of a gas burner.

5.17. Reactions of Chromium Ions

Chromium forms two series of stable salts: chromium(III) salts and chromium(VI) salts.

Solutions of salts of chromic oxide contain chromium as tripositive Cr^{3+} ions.

Under the action of various oxidizing agents, chromium(III) changes to Cr^{VI}, forming the anions of chromic acid, CrO_4^{2-} (chromate ion), or the anions of dichromic acid, $Cr_2O_7^{2-}$ (dichromate ion).

Solutions of chromium(III) salts have a green or violet colour, solutions of chromates are yellow, and those of dichromates are orange.**

(a) Reactions of the Chromic Ion

1. Alkali Hydroxide, NaOH and KOH. These reagents produce with Cr^{3+} a precipitate of $Cr(OH)_3$ which is gray-violet or gray-green in colour and has amphoteric properties. The equations of the reactions are analogous to those of the corresponding reactions of the Al^{3+} ion.

The chromites $NaCrO_2$ and $KCrO_2$ formed by the action of alkalis on $Cr(OH)_3$ are coloured bright-green.

---

* The method of writing the formula shows that there are three molecules of 8-hydroxyquinoline for one tripositive Al^{3+} ion.

** The fact that solutions of chromium(III) are coloured differently depends on the presence in these solutions of complex ions of different composition: $[Cr(OH_2)_6]^{3+}$ (violet); $[Cr(OH_2)_5Cl]^{2+}$ (light-green); $[Cr(OH)_2)_4Cl_2]^+$ (dark-green), etc.

In contrast to aluminates, they decompose irreversibly on boiling (hydrolysis) to give $Cr(OH)_3$:

$$NaCrO_2 + 2H_2O \rightarrow Cr(OH)_3\downarrow + NaOH$$

2. Ammonium Hydroxide, NH_4OH, and Salts of Weak Acids [Na_2CO_3, $(NH_4)_2S$, etc.]. These also precipitate Cr^{3+} as $Cr(OH)_3$.

Since complete precipitation of $Cr(OH)_3$ takes place at pH ≈ 6 (see Table 4.1), it can also be accomplished through the action of ammonium buffer (NH_4OH and NH_4Cl) which produces pH $\approx$ 8-10 (compare the reactions of the Al^{3+} ion).

3. Sodium Acetate, CH_3COONa. This reagent does not precipitate (on heating) Cr^{3+} (distinction from Al^{3+} and Fe^{3+}) because of the formation of a complex cation*. In the presence of Al^{3+} and Fe^{3+}, however, the Cr^{3+} ion is also precipitated partially.

4. Disodium Hydrogen Phosphate, Na_2HPO_4. This compound produces with Cr^{3+} a greenish precipitate of $CrPO_4$. The reaction proceeds in the way as with Al^{3+} (see Sec. 5.16). The precipitate is soluble in mineral acids and bases.

5. Oxidation of Chromium (III) to Chromium(VI). This process can be accomplished by the action of various oxidizing agents, say, H_2O_2, Na_2O_2, Cl_2, Br_2, $KMnO_4$, $(NH_4)_2S_2O_8$, etc. In alkaline solution, there are formed CrO_4^{2-} (chromate) ions. The oxidation in acid solution leads to the formation of $Cr_2O_7^{2-}$ (dichromate) ions.**

With all that has just been said taken into account, let us consider the two cases of oxidation separately.

(1) *The oxidation in alkaline solution* is most often carried out by using hydrogen peroxide or sodium peroxide. We shall discuss the oxidation by hydrogen peroxide. In alkaline solution chromium(III) exists in the form of chromites, say, $KCrO_2$, and the reaction may therefore be represented by the equation***

$$2CrO_2^- + 3H_2O_2 + 2OH^- \rightarrow 2CrO_4^{2-} + 4H_2O$$

To 2 or 3 drops of a solution of a chromium(III) salt add 4 or 5 drops of 2*N* NaOH, 2 or 3 drops of a 3-percent solution of hydrogen peroxide, and heat for several minutes until the green colour of the solution changes to yellow.

The solution is stored for further tests (for detection of CrO_4^{2-}).

In the experiment just described, hydrogen peroxide may be replaced by sodium peroxide (which reacts with water to form H_2O_2 and NaOH) or bromine (or chlorine) water.

With bromine water the reaction takes place according to the equation

$$2CrO_2^- + 3Br_2 + 8OH^- \rightarrow 2CrO_4^{2-} + 6Br^- + 4H_2O$$

---

* Its composition is $[Cr_3(CH_3COO)_6O]^+$.

** For the conditions of existence of CrO_4^{2-} and $Cr_2O_7^{2-}$ ions, see Sec. 4.10.

*** The method of writing oxidation-reduction reactions is described in Sec. 5.13.

(2) *The oxidation in acid solution* can be carried out by using $KMnO_4$, $(NH_4)_2S_2O_8$, and certain other strong oxidizing agents. As an example, we consider the action of ammonium persulphate, $(NH_4)_2S_2O_8$, on $Cr_2(SO_4)_3$.

The reaction proceeds according to the equation

$$2Cr^{3+} + 3S_2O_8^{2-} + 7H_2O \rightarrow Cr_2O_7^{2-} + 6SO_4^{2-} + 14H^+$$

To carry out this reaction, place 5-6 drops of a solution of $(NH_4)_2S_2O_8$ in a test tube, and add a drop each of $2N$ solutions of H_2SO_4 and a silver salt (as a catalyst; see Sec. 5.20). To this oxidizing mixture add 2-3 drops of a solution of $Cr_2(SO_4)_3$ or $Cr(NO_3)_3$ (not $CrCl_3$!)* and heat. Note the change of the colour. The resulting solution is to be preserved for further tests.

The action of potassium permanganate on Cr^{3+} is analogous to that of persulphate:

$$2Cr^{3+} + 2MnO_4^- + 5H_2O \rightarrow 2MnO(OH)_2 \downarrow + Cr_2O_7^{2-} + 6H^+$$

The reaction is accompanied by the formation of a brown precipitate of manganic acid, $MnO(OH)_2$, and proceeds on heating.** The use of this acid for detection of Cr^{3+} in the presence of other cations of Group III is discussed below.

6. Dry Test (Bead Formation). Prepare a precipitate of $Cr(OH)_3$ by one of the methods described above, and centrifuge. Discard the centrifugate and transfer the precipitate with a dropper to a strip of filter paper. Heat the loop of a platinum (or nichrome) wire loop until it is red-hot and dip it into powdered sodium tetraborate (borax), $Na_2B_4O_7 \cdot 10H_2O$, or (preferably) sodium ammonium hydrophosphate, $NaNH_4HPO_4 \cdot 4H_2O$. Heat the adhering small crystals in the Bunsen flame until the mass ceases to swell up. Withdraw the wire from the flame and cool the bead obtained. Now put a little of the $Cr(OH)_3$ precipitate on the bead and heat again as before. After the bead is cooled, examine it in transmitted light. In the presence of chromium, the bead assumes an emerald-green colour.

The colour of the bead is due to the formation of chromium phosphate, $CrPO_4$, or chromium metaborate, $Cr(BO_2)_3$:

$$3NaNH_4HPO_4 + 2Cr(OH)_3 \rightarrow 2CrPO_4 + Na_3PO_4 + 3NH_3 \uparrow + 6H_2O \uparrow$$

$$3Na_2B_4O_7 + 2Cr(OH)_3 \rightarrow 2Cr(BO_2)_3 + 6NaBO_2 + 3H_2O \uparrow$$

To clean the platinum wire, heat it to redness and shake the hot drop off by a sudden jerk.

---

* Strong oxidizing agents oxidize the Cl^- ion to free chlorine.

** The formation of $MnO(OH)_2$ in this particular case, instead of a manganous salt, as is usually the case in reactions involving the oxidation of permanganate in acid solution, is accounted for by the fact that the reaction is conducted in a weakly acid solution.

(b) Reactions of Chromate and Dichromate Ions

1. Formation of Slightly Soluble Salts. The CrO_4^{2-} ion (chromate ion) gives slightly soluble salts with Pb^{2+}, Ag^+, and Ba^{2+}, a fact that can be used to advantage for its detection and separation. The reactions proceed according to the equations

$$CrO_4^{2-} + Pb^{2+} \rightarrow \underset{\text{yellow}}{PbCrO_4} \downarrow$$

$$CrO_4^{2-} + Ba^{2+} \rightarrow \underset{\text{yellow}}{BaCrO_4} \downarrow$$

$$CrO_4^{2-} + 2Ag^+ \rightarrow \underset{\text{brick-red}}{Ag_2CrO_4} \downarrow$$

The reactions should be conducted in the presence of acetate buffer.

2. Benzidine, $C_{12}H_8(NH_2)_2$*. This reagent is oxidized by chromium(VI) ions in the presence of acetic acid to give a blue-coloured product. This reaction used for detection of Cr^{3+} is carried out by the spot-test technique after Cr^{3+} is oxidized to CrO_4^{2-} by the action of H_2O_2 (or Na_2O_2). Place a drop of the solution, resulting from the reaction with H_2O_2, on a strip of filter paper and treat it with a drop of benzidine acetate solution. The sport turns blue.** The reaction enables detection of 0.25μg of chromium. The dilution limit is 1 in 200,000.

The reaction in question can be used for fractional detection of Cr^{3+} in the presence of all other cations.

To do this, the reaction must be carried out on a watch glass (or on a crucible lid). Put 3-5 drops of the test solution on it, add a little of solid sodium peroxide, Na_2O_2, and mix thoroughly. The Cr^{3+} ion is thus oxidized to CrO_4^{2-}.

Then dilute the reaction mixture with 2-3 drops of water and introduce into it the capillary tip of a glass tubing which will draw in a small amount of the liquid with the precipitate due to capillarity. Touch a sheet of filter paper with the capillary tip and hold it in this position until a watery zone several millimetres wide is formed around the precipitate which remains in the centre. This zone contains

---

* The formula of benzidine is

$$H_2N-C_6H_4-C_6H_4-NH_2$$

** The product of the oxidation of benzidine, known as benzidine blue, has the following structure:

$$\left[\begin{array}{c} H_2N-C_6H_4-C_6H_4-NH_2 \\ HN=C_6H_4=C_6H_4=NH \end{array}\right]\cdot 2HAn$$

(where An is the singly charged anion of the acid).

CrO_4^{2-} ions and the precipitate consists of various hydroxides, including $Co(OH)_3$ and $MnO(OH)_2$. Draw the capillary tip containing benzidine solution around the moist spot on the paper. If a blue ring appears around the spot, this proves the presence of chromium in the test solution.

In carrying out this reaction, care must be exercised not to allow the benzidine solution to reach the centre of the spot, otherwise it may turn blue even in the absence of CrO_4^{2-}, due to the presence in the precipitate of $Co(OH)_3$ and $MnO(OH)_2$ which also oxidize benzidine. Hence, the test is conclusive provided there is a colourless zone between the blue ring and the precipitate.

3. Formation of Perchromic Acid, H_2CrO_6. The action of H_2O_2 on an acidified solution of chromate or dichromate results in the formation of perchromic acid*, H_2CrO_6, as a result of which the solution acquires a deep-blue colour:

$$Cr_2O_7^{2-} + 4H_2O_2 + 2H^+ \rightarrow 2H_2CrO_6 + 3H_2O$$

In aqueous solution, perchromic acid is highly unstable. It decomposes to give Cr^{3+} and the blue colour immediately turns green. If however we add amyl alcohol, $C_5H_{11}OH$, or diethyl ether, $(C_2H_5)_2O$, to the solution, perchromic acid passes on shaking into the layer of these solvents, which considerably increases its stability.

If the reaction described is used in conjunction with any of the reactions of oxidation of Cr^{3+} to $Cr_2O_7^{2-}$, it can be employed for fractional detection of chromium in the presence of all other Group III cations. The reaction with $KMnO_4$ in acid solution is most reliable.

Acidify a few drops of the solution to be analysed with nitric acid, add an excess of $KMnO_4$, and heat for several minutes in a water bath. Then observe the colouration of the solution. It must be crimson-coloured. This colouration is evidence that the amount of $KMnO_4$ taken was enough for all the reducing agents present in solution, including Cr^{3+}, to be oxidized. If no crimson colouration appears, it is necessary to add more $KMnO_4$ and heat again. When the crimson colour is obtained, centrifuge the $MnO(OH)_2$ precipitate and discard. Cool the centrifugate thoroughly (otherwise, perchromic acid will rapidly

---

* The structural formula of H_2CrO_6 is

$$(O{=})_2Cr(OOH)_2$$

Some authors assign the formulas H_3CrO_8 and H_7CrO_{10} to the reaction product, or else consider it to be chromium peroxide of the composition CrO_5, i.e.,

$$O{=}Cr(O_2)_2$$

decompose), add 8-10 drops of amyl alcohol (or diethyl ether) and a few drops of H_2O_2. The excess permanganate is reduced to Mn^{2+} and the $Cr_2O_7^{2-}$ ions form perchromic acid which passes on shaking into the amyl-alcohol layer, imparting a blue colour to it. This proves the presence of chromium in the test solution.

In carrying out this reaction, one has to remember that the oxidation of Cr^{3+} takes place only after the Fe^{2+} ions are oxidized by permanganate. Hence, if the amount of $KMnO_4$ taken is small, perchromic acid may fail to form even in the presence of chromium. The solution must not be acidified too strongly since otherwise permanganate may decompose to give $MnO(OH)_2$ and oxygen.

4. Reduction of Cr^{VI} to Cr^{III}. This process can be effected through the use of various reducing agents, say, Na_2SO_3, H_2S, ethyl alcohol (C_2H_5OH), iron(II) salts, and other substances in acid solution.

To carry out this reaction, add a few drops of Na_2SO_3 solution to a solution of $K_2Cr_2O_7$ made acid by H_2SO_4:

$$Cr_2O_7^{2-} + 3SO_3^{2-} + 8H^+ \rightarrow 2Cr^{3+} + 3SO_4^{2-} + 4H_2O$$

A mixture of $K_2Cr_2O_7$ and concentrated H_2SO_4 (chromic acid mixture) exhibits oxidizing properties and has found wide application in chemical laboratories, particularly for washing glassware to remove grease from the glass surface.

5.18. Reactions of the Titanium(IV) Ion

Solutions of titanium(IV) ions* are colourless.

1. Ammonium Sulphide. Ammonium sulphide produces with the Ti^{IV} ion a white precipitate of orthotitanic acid, H_4TiO_4:

$$Ti^{IV} + 4S^{2-} + 4H_2O \rightarrow H_4TiO_4\downarrow + 4HS^-$$

The precipitate is soluble in acids. On prolonged boiling it converts into metatitanic acid, H_2TiO_3, which is slightly soluble in dilute acids:

$$H_4TiO_4\downarrow \rightarrow H_2TiO_3\downarrow + H_2O$$

2. Alkali Hydroxides and Ammonia. These reagents also form orthotitanic acid, H_4TiO_4.

3. Hydrolysis of Titanium(IV) Salts. The compound H_4TiO_4 is not only an acid but it has basic properties as well, i.e., is capable of dissolving in acids. The basic properties of this compound however are very faint. Therefore, titanium(IV) salts formed by strong acids are hydrolysed to a considerable extent and show an acid reaction:

$$Ti(SO_4)_2 + 2H_2O \rightleftharpoons Ti(OH)_2(HSO_4)_2$$

---

* As has already been stated (see Sec. 5.14), no Ti^{IV} ions exist in solution; depending on the medium, Ti^{IV} forms various complex ions. To simplify, however, we shall use this symbol to designate the titanium ion.

The salts of weak acids are absolutely incapable of existence in aqueous solution.

If a solution containing Ti^{IV} cations is treated with a salt of a weak acid, a white precipitate of $TiO_2 \cdot nH_2O$ is formed.

As pointed out above, ammonium sulphide exerts the same effect.

4. Reduction of Ti^{IV} to Ti^{3+}. A solution of a titanium(IV) salt is made acid by adding a drop or two of concentrated HCl and allowed to stand after a piece of metallic zinc is added; the following reaction takes place:

$$2Ti^{IV} + Zn \rightarrow 2Ti^{3+} + Zn^{2+}$$

The reaction is accompanied by the appearance of a pale-violet colouration typical of Ti^{3+}. This colour becomes noticeable only after the prolonged interaction of the solution with zinc and at a sufficient concentration of the titanium salt.

5. Hydrogen Peroxide, H_2O_2. This compound causes an orange-yellow colour in acid solution due to the formation of a peroxide compound of titanium:

$$Ti(SO_4)_2 + H_2O_2 \rightarrow [TiO_2(SO_4)_2]^{2-} + 2H^+$$

or

$$Ti(OH)_2^{2+} + H_2O_2 + (x-2)H_2O \rightarrow [TiO_2(H_2O)_x]^{2+}$$

This is one of the most characteristic reactions of Ti^{IV}

6. Chromotropic Acid. Mix a drop of the test solution with a drop of a solution of chromotropic acid on filter paper or on a spot plate. This results in the formation of a characteristic reddish-brown colour.

The formula of chromotropic acid is

SO₃H
HO—
HO—
SO₃H

The composition of the compound formed by chromotropic acid and titanium is believed to be

SO₃H
O—
OTi
O—
SO₃H

Since Fe^{3+} produces with this reagent a green colour, it must be preliminarily reduced to Fe^{2+} by $SnCl_2$.

Place a drop of $SnCl_2$ on filter paper, followed by a drop of the test solution. Add another drop of $SnCl_2$, after which moisten the spot with a solution of chromotropic acid. In the presence of Ti^{IV}, a brown sport or ring appears. The reaction enables detection of 2 μg of Ti^{IV}.

According to Alimarin, the sensitivity of the reaction is increased if it is carried out in a neutral solution, for example, in the presence of acetate buffer (a mixture of CH_3COOH and CH_3COONa).

Because of the instability of the solution of chromotropic acid, it is preferable to make use of filter paper impregnated with this solution and dried.* Place a drop of the test solution on a sheet of such filter paper. If Ti^{IV} ion is present, a characteristic reddish-brown spot is formed. The Fe^{3+} ion gives with chromotropic acid a green colouration. Therefore, in the presence of Fe^{3+} in solution, the test should be repeated with a drop of a solution of $SnCl_2$ (to reduce Fe^{3+} to Fe^{2+}) being preliminarily placed on a sheet of paper impregnated with chromotropic acid. Other cations of Group III do not interfere with the reaction.

5.19. Reactions of Iron Ions

Iron forms two series of salts: the ferrous salts in which the metal is divalent and the ferric salts containing trivalent iron. Solutions of ferric salts contain tripositive Fe^{3+} ions, while those of ferrous salts contain dipositive Fe^{2+} ions. Since these ions give quite different reactions, they are considered separately.

(a) Reactions of the Ferric Ion

Solutions of iron(III) salts are yellow or red-brown in colour.**

1. Alkali Hydroxides, NaOH and KOH, and Ammonium Hydroxide, NH_4OH. These reagents produce with Fe^{3+} a red-brown precipitate of $Fe(OH)_3$. In contrast to $Al(OH)_3$ and $Cr(OH)_3$, ferric hydroxide has practically no amphoteric properties and is therefore insoluble in excess alkali (*Test!*). Since the complete precipitation of $Fe(OH)_3$ occurs at pH = 3.5 (see Table 4.1), it can also be precipitated, like $Al(OH)_3$ and $Cr(OH)_3$, by the action of ammonia and a mixture of ammonia with ammonium salts.

2. Sodium Acetate, CH_3COONa. This salt gives with Fe^{3+} in the cold a water-soluble red complex compound, $[Fe_3(CH_3COO)_6O]^+$ (just as in the case of Cr^{3+}). On heating, this compound separates out as a precipitate (distinction from Cr^{3+}).*** Under the same conditions, aluminium gives a precipitate of a white basic salt. The other cations of Group III form no precipitates with CH_3COONa.

---

* Filter paper is impregnated with a 10-percent aqueous solution of chromotropic acid, dried at room temperature, and stored in a bottle fitted with a ground stopper.

** The fact that solutions of the salts of iron, chromium, nickel, and cobalt are variously coloured depends on what complexes exist under given conditions in solution.

*** The behaviour of Cr^{3+} is discussed in Sec. 5.17, Reaction *3*.

3. Alkali and Ammonium Carbonates. These compounds produce with Fe^{3+} a brown precipitate of basic salts which are converted into hydroxides on heating.

4. Ammonium Thiocyanate, NH_4SCN, and Potassium Thiocyanate*, KSCN. These substances produce with Fe^{3+} ferric thiocyanates which impart a blood-red colour to the solution.

Depending on the concentration of the thiocyanate ion, complexes of different composition may be formed:

$$Fe^{3+} + SCN^- \rightarrow [FeSCN]^{2+}$$
$$Fe^{3+} + 2SCN^- \rightarrow [Fe(SCN)_2]^+$$
$$Fe^{3+} + 3SCN^- \rightarrow [Fe(SCN)_3]$$

and so on until the ferric thiocyanate complex is formed:

$$Fe^{3+} + 6SCN^- \rightarrow [Fe(SCN)_6]^{3-}$$

Since the reaction is reversible, addition of an excess of the reagent intensifies the colouration. This is one of the most characteristic and most sensitive reactions of the Fe^{3+} ion. But it is not always reliable because a number of subtances, which produce complexes with Fe^{3+}, prevent the colour formation. Such substances include, for example, fluorides since F^- forms with Fe^{3+} the slightly ionizing complex $[FeF_6]^{3-}$.

When 1-2 drops of a solution of NH_4SCN are added to a solution of an iron(III) salt, a red colouration is produced; if the solution is then treated with NH_4F or NaF, the colour disappears:

$$Fe(SCN)_3 + 6F^- \rightarrow [FeF_6]^{3-} + 3SCN^-$$

The same effect is produced by addition of phosphoric acid, oxalates, tartrates, citrates, and other salts of organic acids.

5. Potassium Hexacyanoferrate(II), $K_4[Fe(CN)_6]$. This complex produces with Fe^{3+} an intense-blue precipitate of Prussian blue (ferric ferrocyanide):

$$4Fe^{3+} + 3[Fe(CN)_6]^{4-} \rightarrow Fe_4[Fe(CN)_6]_3 \downarrow$$

The occurrence of this most important reaction is favoured by making the solution moderately acid. Too strong an acidification as well as a large excess of the reagent is not recommended since the precipitate dissolves. Alkalis decompose Prussian blue to give ferric hydroxide, $Fe(OH)_3$.

6. Disodium Hydrogen Phosphate, Na_2HPO_4. The salt forms with Fe^{3+} a pale-yellow precipitate of $FePO_4$:

$$Fe^{3+} + 2HPO_4^{2-} \rightarrow FePO_4 \downarrow + H_2PO_4^-$$

* The salts of thiocyanic acid, HSCN.

The precipitate is soluble in strong acids but insoluble in acetic acid (*Test!*).

7. Oxidation-Reduction Reaction. Since Fe^{3+} is capable of being reduced to Fe^{2+}, it is an oxidizing agent. For example, if hydrogen sulphide is passed into an acid solution of an iron(III) salt, it is oxidized to free sulphur which precipitates as a white turbidity:

$$2Fe^{3+} + H_2S \rightarrow 2Fe^{2+} + 2H^+ + S\downarrow$$

It will be recalled that this reaction also proceeds on dissolution of precipitated Fe_2S_3 in acids (see Sec. 5.15).

The Fe^{3+} ion also oxidizes Sn^{2+} to Sn^{IV} and I^- to I_2:

$$Sn^{2+} + 2Fe^{3+} \rightarrow Sn^{IV} + 2Fe^{2+}$$

$$2I^- + 2Fe^{3+} \rightarrow I_2 + 2Fe^{2+}$$

The occurrence of the second reaction can be recognized by the brown colouration which is caused by addition of a ferric salt to the acidified KI solution. If we add to the same test tube a few drops of benzene, C_6H_6, (or chloroform, $CHCl_3$), in which I_2 is more soluble than in water, and shake, most of the iodine passes into the solvent layer and imparts to it a violet colour characteristic of iodine (solvent extraction).

The Fe^{3+} ion however is a weak oxidizing agent. Therefore, it can oxidize only relatively strong reducing agents. Weaker reducing agents, such as Br^- and Cl^-, are not oxidized by Fe^{3+}.*

8. Detection of Fe^{III} by the Grinding Method (F. M. Flavitsky's Method). Put a few small crystals of solid $Fe_2(SO_4)_3$ in a small evaporating dish (or on a porcelain plate) and grind using the narrow end of a porcelain pestle with an approximately equal amount of solid ammonium thiocyanate, NH_4SCN. The reaction mixture assumes then a red-brown colour due to the formation of $Fe(SCN)_3$ or complexes of other compositions.

If a solution (not the solid salt) is to be analysed, the Fe^{3+} ion can be detected as follows. To a few drops of the solution in a small evaporating dish add one drop or two of $2N$ H_2SO_4 and evaporate to dryness. After the dry residue is cooled, proceed as described above.

(b) Reactions of the Ferrous Ion

Solutions of ferrous salts are pale-green in colour. Dilute solutions are colourless.

1. Alkali Hydroxides, NaOH and KOH. These alkalis precipitate Fe^{2+} as $Fe(OH)_2$:

$$Fe^{2+} + 2OH^- \rightarrow Fe(OH)_2\downarrow$$

---

* The power of various oxidizing and reducing agents will be discussed in greater detail in Sections 6.1 through 6.4.

If the precipitation is carried out in the absence of air, a white precipitate of $Fe(OH)_2$ is formed. Under ordinary conditions, the precipitate is partially oxidized, acquiring a dingy-green colour. The rapid change of the colour of the precipitate in air can conveniently be observed by filtering it off. The end oxidation product is $Fe(OH)_3$. The precipitated $Fe(OH)_2$ and $Fe(OH)_3$ are soluble in acids but insoluble in bases.

2. Ammonium Hydroxide, NH_4OH. It also precipitates $Fe(OH)_2$, but the precipitation is incomplete. In the presence of ammonium salts, $Fe(OH)_2$ is not precipitated at all. Just as in the case of $Mg(OH)_2$, this is accounted for by the relatively high solubility product constant of $Fe(OH)_2$ ($SP = 1 \times 10^{-15}$), which is not attained at the OH^- concentration produced by the ammonium buffer. On standing however the products of oxidation of $Fe(OH)_2$ by atmospheric oxygen are gradually precipitated.

3. Alkali and Ammonium Carbonates. These salts produce with Fe^{2+} a white precipitate of $FeCO_3$ which turns brown in the air, being oxidized to $Fe(OH)_3$:

$$4FeCO_3\downarrow + 6H_2O + O_2 \rightarrow 4CO_2\uparrow + 4Fe(OH)_3\downarrow$$

4. Potassium Hexacyanoferrate(III)*, $K_3[Fe(CN)_6]$. The complex forms with Fe^{2+} a dark-blue precipitate of ferrous ferricyanide (Turnbull's blue).

The composition of this precipitate is now considered to be identical with Prussian blue (ferric ferrocyanide). The reactions proceed according to the equations

$$Fe^{2+} + [Fe(CN)_6]^{3-} \rightarrow Fe^{3+} + [Fe(CN)_6]^{4-}$$

$$4Fe^{3+} + 3[Fe(CN)_6]^{4-} \rightarrow Fe_4[Fe(CN)_6]_3\downarrow$$

The difference in colour between Prussian blue and Turnbull's blue is accounted for by the fact that the formula written above does not fully correspond to the composition of the precipitate. In particular, K^+ ions enter into its composition which varies, depending on the conditions under which precipitation takes place. The precipitate is insoluble in acids but decomposed by alkalis.

The reaction considered above** is the most characteristic of Fe^{2+} ion.

5. Dimethylglyoxime (Chygaev's Reagent).*** This compound produced with Fe^{2+} water-soluble inner-complex salts having a red colour. The reaction is conducted in ammonical solution in the presence of

* It is a salt of the acid $H_3[Fe(CN)_6]$ in which iron is in the oxidation state +3. This reagent must not be confused with potassium hexacyanoferrate (II), $K_4[Fe(CN)_6]$ which is used for detection of Fe^{3+}.

** $K_3[Fe(CN)_6]$ does not give a precipitate with Fe^{3+}, but the solution acquires a dark-brown colour. The Fe^{2+} ions form with $K_4[Fe(CN)_6]$ a white precipitate of $Fe_2[Fe(CN)_6]$ which turns blue in air because of the oxidation of iron and conversion of the precipitate into Prussian blue.

*** The structural formula of the reagent is given in Sec. 5.11.

tartaric acid added in order to bind the Fe^{3+} ions into a complex because these ions are capable of producing a precipitate of $Fe(OH)_3$ with ammonia.

Since dimethylglyoxime is the most important reagent for Ni^{2+}, with which it forms a pink precipitate, the presence of Fe^{2+} interferes with detection of Ni^{2+}. The reaction in question may be used for detection of Fe^{2+} only if the absence of Ni^{2+} has been proved, or after Ni^{2+} has been bound by KCN into the stable complex $[Ni(CN)_4]^{2-}$.

6. Oxidation of Fe^{2+} to Fe^{3+}. The Fe^{3+} ion possesses the properties of a weak oxidizing agent, whereas the Fe^{2+} ion is a fairly strong reducing agent and is capable of being oxidized by the action of a number of oxidizing agents, such as H_2O_2, $KMnO_4$, $K_2Cr_2O_7$ in acid solution, HNO_3, etc. Let us now consider in greater detail the oxidation of Fe^{2+} by nitric acid.

The reaction proceeds according to the equation

$$3Fe^{2+} + 4H^+ + NO_3^- \rightarrow 3Fe^{3+} + 2H_2O + NO\uparrow$$

This reaction is carried out as follows. To 2-3 drops of a solution of a ferrous salt add 2 drops of $2N$ H_2SO_4 and one drop or two of $6N$ HNO_3. Heat the resulting solution until the brown colouration disappears, this colour being due to the formation of the unstable complex compound $[Fe(NO)]SO_4$ which decomposes on heating.

The same reaction occurs when ferrous sulphide is dissolved by nitric acid:

$$FeS\downarrow + 4H^+ + NO_3^- \rightarrow Fe^{3+} + S\downarrow + 2H_2O + NO\uparrow$$

Thus, after the cations of Group III are separated from those of Groups II and I, iron will be present in solution in the form of Fe^{3+}.

This is convenient for analysis since, in separation reactions, Fe^{3+} is precipitated from solution more completely than Fe^{2+}.

The oxidation of Fe^{2+} in alkaline solution can be accomplished, for example, by hydrogen peroxide:

$$2Fe^{2+} + 4OH^- + H_2O_2 \rightarrow 2Fe(OH)_3\downarrow$$

5.20. Reactions of the Manganous Ion

Solutions of manganese salts are pale-pink in colour. Dilute solutions are colourless.

1. Alkali Hydroxides, NaOH and KOH. These form with Mn^{2+} a white precipitate of $Mn(OH)_2$ which is soluble in acids but insoluble in bases:

$$Mn^{2+} + 2OH^- \rightarrow Mn(OH)_2\downarrow$$

The precipitate quickly turns brown as a result of the oxidation of Mn^{2+} to manganese(IV), i.e., due to the formation of manganous acid, $MnO(OH)_2$ (or H_2MnO_3). If, along with alkali, an oxidizing agent (hydrogen peroxide or bromine water) is allowed to react with a manganese(II) salt, a dark-brown precipitate of $MnO(OH)_2$ is immediately brought down:

$$Mn^{2+} + 2OH^- + H_2O_2 \rightarrow MnO(OH)_2 \downarrow + H_2O$$

In contrast to $Mn(OH)_2$, the $MnO(OH)_2$ precipitate is insoluble in dilute H_2SO_4. If, however, along with H_2SO_4, we add 1 or 2 drops of H_2O_2 solution, the precipitate will dissolve (*Test!*). Under these conditions (i.e., in acid solution) H_2O_2 acts on Mn^{IV} as a reducing agent, being itself oxidized to free oxygen:

$$MnO(OH)_2 \downarrow + 2H^+ + H_2O_2 \rightarrow Mn^{2+} + 3H_2O + O_2 \uparrow$$

The reduction of manganese(IV) to the divalent state on dissolution of $MnO(OH)_2$ in H_2SO_4 can also be effected by addition of a solution of a nitrite ($NaNO_2$ or KNO_2) which is oxidized to the corresponding nitrate ($NaNO_3$ or KNO_3). In distinction to sulphuric acid, hydrochloric acid dissolves $MnO(OH)_2$ (on heating), the role of a reducing agent being played by Cl^- ions which are oxidized to Cl_2:

$$MnO(OH)_2 \downarrow + 4H^+ + 2Cl^- \rightarrow Mn^{2+} + 3H_2O + Cl_2 \uparrow$$

2. **Ammonium Hydroxide, NH_4OH.** This reagent precipitates Mn^{2+}, like alkali hydroxides (caustic alkalis). The precipitation however is incomplete because of the relatively high solubility product constant of $Mn(OH)_2$ and the build-up of NH_4^+ ions in the course of the reaction:

$$Mn^{2+} + 2NH_4OH \rightarrow Mn(OH)_2 \downarrow + 2NH_4^+$$

As a result of this, the pH of the solution falls below 10.83 (the value required for complete precipitation of manganous hydroxide). But if sufficient NH_4Cl is added to the solution, the pH of the solution will be decreased to such an extent that a precipitate of $Mn(OH)_2$ will not form at all. It will be recalled that the same phenomenon was also observed for $Mg(OH)_2$ and $Fe(OH)_2$. The action of a mixture of ammonium hydroxide and an ammonium salt on Mn^{2+} is used as the basis for its separation from Al^{3+}, Cr^{3+}, Fe^{3+}, Ti^{IV}, which are precipitated by this mixture practically completely in the form of $Al(OH)_3$, $Cr(OH)_3$, $Fe(OH)_3$, and $Ti(OH)_4$, respectively. This separation however is incomplete because Mn^{2+} is gradually oxidized in ammoniacal solution by atmospheric oxygen to manganese(IV) which precipitated as $MnO(OH)_2$.

3. **Alkali and Ammonium Carbonates.** These salts form a white precipitate of $MnCO_3$ which is soluble in acids.

4. **Disodium Hydrogen Phosphate, Na_2HPO_4.** This reagent brings down a white precipitate of $Mn_3(PO_4)_2$ which is soluble in mineral acids and in acetic acid (distinction from Al^{3+}, Cr^{3+}, Fe^{3+}, and Ti^{IV}).

5. **Oxidation of Mn^{2+} to MnO_4^-.** This reaction, which is very important for detection of Mn^{2+}, can be accomplished by using various oxidizing agents.

Let us consider three methods.

(1) *Oxidation by Lead Dioxide, PbO_2.* Take a little powdered PbO_2 (or Pb_3O_4) with a glass spatula and place it in a test tube. Add 4-5 drops of 6*N* HNO_3 and heat. After 1-2 minutes centrifuge and, without separating the precipitate, note the colour of the solution. If it is crimson-violet (the colour of MnO_4^-), this means that the reagents contain manganese and cannot be used for the test. If no colour is observed, to the same test tube with a mixture of HNO_3 and PbO_2 add a very small amount (1 drop) of the solution to be tested for Mn^{2+}, which contains no Cl^- ions [$MnSO_4$ or $Mn(NO_3)_2$], mix, and heat again. The Mn^{2+} ion is thus oxidized to permanganic acid, $HMnO_4$, which has a characteristic crimson-violet colour:

$$2Mn^{2+} + 5PbO_2\downarrow + 4H^+ \rightarrow 2MnO_4^- + 5Pb^{2+} + 2H_2O$$

As can be seen from this equation, the oxidizing agent is PbO_2, more precisely, lead(IV) entering into its composition, which is reduced to lead(II).

In carrying out this reaction and the succeeding tests, it is necessary to add a very small amount of manganese(II) salt since an excess of the salt reduces the resultant MnO_4^- to $MnO(OH)_2$:

$$2MnO_4^- + 3Mn^{2+} + 7H_2O \rightarrow 5MnO(OH)_2\downarrow + 4H^+$$

If to the crimson-violet solution obtained we add 3-4 drops more of the solution of the manganese(II) salt and heat again, the colour will disappear. For the same reason the solution must not contain any other reducing agents, in particular Cl^- ions, because they also reduce MnO_4^- to $MnO(OH)_2$ or even to Mn^{2+}:

$$2MnO_4^- + 16H^+ + 10Cl^- \rightarrow 2Mn^{2+} + 8H_2O + 5Cl_2\uparrow$$

(2) *Oxidation by Ammonium Persulphate, $(NH_4)_2S_2O_8$.* The reaction is carried out in the presence of Ag^+ as a catalyst.

The silver ions Ag^+ are evidently oxidized by ammonium persulphate to Ag^{2+} ions:

$$2Ag^+ + S_2O_8^{2-} \rightarrow 2Ag^{2+} + 2SO_4^{2-}$$

The Ag^{2+} ions then oxidize Mn^{2+} to MnO_4^-, being themselves reduced to Ag^+:

$$Mn^{2+} + 5Ag^{2+} + 4H_2O \rightarrow MnO_4^- + 5Ag^+ + 8H^+$$

Hence, the catalyst is recovered and again enters into the reaction. The net reaction may be represented by the equation*

$$2Mn^{2+} + 5S_2O_8^{2-} + 8H_2O \rightarrow 2MnO_4^- + 10SO_4^{2-} + 16H^+$$

In the absence of Ag^+ and also with an excess of Mn^{2+}, instead of MnO_4^-, there is formed a brown precipitate of $MnO(OH)_2$.

* See Sec. 5.13, Reaction *9*.

To perform the test, to 5-6 drops of a solution of $(NH_4)_2S_2O_8$ add 1 drop of $2N$ H_2SO_4 (or HNO_3), 1 or 2 drops of concentrated H_3PO_4*, and 2 drops of a solution of a silver salt, and heat. Into the heated oxidizing mixture introduce, using a glass rod, a minimum amount of a solution to be analysed for Mn^{2+} and containing no Cl^- ions, and mix.

If the resulting crimson-violet colouration, indicating the formation of MnO^-, is too faint, a little more of the solution being analysed may be added.

If, conversely, a brown precipitate of $MnO(OH)_2$ is formed, indicating that the concentration of Mn^{2+} is too high**, the test is repeated, a drop of the test solution being diluted preliminarily with a few drops of water.

Since the Cr^{3+} ion is also oxidized by $(NH_4)_2S_2O_8$ to $Cr_2O_7^{2-}$, the Mn^{2+} ion may be detected by this reaction simultaneously with the detection of Cr^{3+}. Having carried out the test for Mn^{2+} as described above, to the same test tube add more (2-3 drops) of the test solution and heat again. Centrifuge the precipitated $MnO(OH)_2$ and detect in the centrifugate (after it is cooled) $Cr_2O_7^{2-}$ by converting it to perchromic acid, H_2CrO_6.

(3) *Oxidation by Sodium Bismuthate****, $NaBiO_3$. The reaction proceeds according to the equation

$$2Mn^{2+} + 5NaBiO_3\downarrow + 14H^+ \rightarrow 2MnO_4^- + 5Bi^{3+} + 5Na^+ + 7H_2O$$

Unlike the reactions so far studied, this reaction takes place in the cold and therefore a small excess of a manganous salt does not interfere.

In the cold, Cl^- ions react with MnO_4^- relatively slowly; it is for this reason that even in their presence the appearance of a crimson colouration of the solution can be observed at the very beginning. But the presence of large amounts of chlorides in solution must of course be avoided.

To carry out the reaction, to 1 or 2 drops of a solution of a manganese(II) salt add 3-4 drops of $6N$ HNO_3 and 5-6 drops of water, and then introduce into the solution a little powdered $NaBiO_3$ with a glass spatula. Mix, centrifuge the excess reagent, and observe the colouration of the solution. This reaction is extremely sensitive and is very convenient.

Permanganic acid, $HMnO_4$, resulting from the reactions considered, is one of the most powerful oxidizing agents capable of oxidizing a

---

* Orthophosphoric acid stabilizes MnO_4^- ions and prevents their decomposition.

** If, of course, there are no other reducing agents in solution.

*** The reagent is prepared by fusing a bismuth salt with sodium peroxide. Just as in the case of the reaction with PbO_2, it is necessary to make certain that the reagent is pure, for which purpose a corresponding control test is carried out. For derivation of the equation for this reaction, see Sec. 5.13, Reaction 8.

large variety of reducing agents. If such an oxidation reaction takes place in acid solution, Mn^{VII} is reduced to Mn^{II}, i.e., to the Mn^{2+} ion. In oxidation reactions occurring in alkaline or neutral solution, Mn^{VII} is reduced only to Mn^{IV} which separates as a brown precipitate of $MnO(OH)_2$.

To a solution of an iron(II) salt acidified with H_2SO_4 add, drop by drop, a dilute solution of potassium permanganate and observe the decolourization.

The reaction takes place according to the equation

$$5Fe^{2+} + MnO_4^- + 8H^+ \rightarrow 5Fe^{2+} + Mn^{2+} + 4H_2O$$

Potassium permanganate becomes also colourless when it is added to a solution of oxalic acid, or of one of its salts, which has been acidified with H_2SO_4 and heated to near boiling (for balancing the equation for this reaction, see Sec. 5.13, Reaction *2*):

$$5C_2O_4^{2-} + 2MnO_4^- + 16H^+ \rightarrow 2Mn^{2+} + 10CO_2\uparrow + 8H_2O$$

Both these reactions are of great practical value in quantitative analysis.

The reduction of MnO_4^- to Mn^{2+} is achieved also by the action of hydrogen peroxide, H_2O_2, which functions here as a reducing agent (for derivation of the equation for this reaction, see Sec. 5.13, Reaction *5*):

$$2MnO_4^- + 5H_2O_2 + 6H^+ \rightarrow 2Mn^{2+} + 5O_2\uparrow + 8H_2O$$

6. Benzidine, $C_{12}H_8(NH_2)_2$. Benzidine is oxidized by manganese(IV) compounds with the formation of a blue-coloured product. The reaction can conveniently be carried out by the spot-test technique.

Place a drop of a solution of a manganous salt on a strip of filter paper and treat with ammonia vapour. The resulting manganous hydroxide, $Mn(OH)_2$, rapidly oxidizes to $MnO(OH)_2$ on exposure to air and the spot turns brown. If the spot is then treated with a drop of a solution of benzidine acetate, the latter will be oxidized by $MnO(OH)_2$ and the spot will turn blue* (compare the corresponding

* The reaction between $MnO(OH)_2$ and benzidine may be represented by the following equations:

$$C_{12}H_8(NH_2)_2 + MnO(OH)_2\downarrow + 2CH_3COOH \rightarrow$$
$$\rightarrow C_{12}H_8(NH)_2 + Mn^{2+} + 2CH_3COO^- + 3H_2O$$
$$C_{12}H_8(NH)_2 + C_{12}H_8(NH_2)_2 \rightarrow C_{12}H_8(NH)_2 \cdot C_{12}H_8(NH_2)_2$$

At the first stage of the reaction, the amine groups $-NH_2$ of benzidine are oxidized to the imide group $=NH$ under the action of $MnO(OH)_2$. At the second stage, the molecule of the compound obtained, $C_{12}H_8(NH)_2$, combines with the second benzidine molecule to form a dark-blue compound.

reaction for CrO_4^{2-}). The reaction enables detection of 0.15 μg of Mn^{2+} The dilution limit is 1 in 330,000.

When large quantities of Co^{2+} are present, a pale-blue colouration may appear even in the absence of Mn^{2+}. To avoid this, the drop should be treated with a drop of a solution of ammonium chloride prior to addition of ammonia.

7. **Silver Chloride, AgCl.** This salt is reduced by Mn^{2+} to metallic silver. The reaction is also carried out by the spot-test technique.

Place a drop of a solution of HCl on a strip of filter paper. Apply a capillary with a solution of a silver salt to the centre of the moist spot formed. A precipitate of AgCl is formed on the paper. Wash the precipitate thoroughly by placing a capillary, with as much water as can be held by capillarity, in the centre of the spot. Hold the capillary until the diameter of the spot increases by 2 or 3 times.* Then moisten the washed precipitate with a drop of the solution to be tested for Mn^{2+} and add a small drop of a concentrated solution of NaOH. In the presence of Mn^{2+}, which reduces AgCl to metallic silver, the spot becomes black**:

$$2AgCl + Mn^{2+} + 4OH^- \rightarrow 2Ag\downarrow + MnO(OH)_2\downarrow + 2Cl^- + H_2O.$$

8. **Dry Test.** Prepare a bead of Na_2CO_3 as described in the reactions for chromium. Pick up a little of the solid substance to be tested for manganese, such as $MnSO_4$ or precipitated $MnO(OH)_2$, on the bead and ignite. Now put a few crystals of $KClO_3$ on the bead and again ignite cautiously. A green (or light-blue) bead confirms the presence of manganese in the substance being tested. In the case of $MnO(OH)_2$, the reaction proceeds according to the equation

$$3MnO(OH)_2 + 3Na_2CO_3 + KClO_3 \rightarrow 3Na_2MnO_4 + KCl + 3H_2O\uparrow + 3CO_2\uparrow$$

The resulting manganese compound, Na_2MnO_4, is the salt of manganic acid, H_2MnO_4, in which manganese is in the oxidation state VI. The green colour of the bead is due to the presence of the salt Na_2MnO_4.

5.21. Reactions of the Zinc Ion

Solutions of zinc salts are colourless.

1. **Alkali Hydroxides, NaOH and KOH.** These alkalis produce in solutions of zinc salts (Zn^{2+}) a white precipitate of zinc hydroxide, $Zn(OH)_2$ (or H_2ZnO_2), which is soluble in acids and bases (it is ampho-

---

* If necessary, this operation may be repeated.

** This reaction was suggested by N. A. Tananaev and I. V. Tananaev. It is more convenient to use an ammoniacal solution of $AgNO_3$ as the reagent. To prepare it, add to a saturated solution of $AgNO_3$ as much concentrated ammonia as required for the dissolution of the initially separated precipitate. Then add as much ammonia as has already been added. To detect Mn^{2+} in the presence of this reagent, place successively a drop of the test solution and a drop of the reagent on a strip of filter paper.

teric). For example,

$$Zn(OH)_2\downarrow + 2H^+ \rightarrow Zn^{2+} + 2H_2O$$

$$H_2ZnO_2\downarrow + 2OH^- \rightarrow ZnO_2^{2-} + 2H_2O$$

In contrast to aluminates, the zincate ion, ZnO_2^{2-}, does not form a precipitate of $Zn(OH)_2$ on being treated with NH_4Cl because it is soluble in ammonium salts (see below).

2. Ammonium Hydroxide, NH_4OH. Ammonium hydroxide forms a precipitate of $Zn(OH)_2$ which is soluble in excess NH_4OH due to the formation of $[Zn(NH_3)_6]^{2+}$ or $[Zn(NH_3)_4]^{2+}$; for example,

$$Zn(OH)_2\downarrow + 6NH_4OH \rightarrow [Zn(NH_3)_6]^{2+} + 2OH^- + 6H_2O$$

Owing to the relatively high value of the solubility product constant ($SP = 7.1 \times 10^{-18}$), zinc hydroxide dissolves also in ammonium chloride.

Therefore, in the presence of an ammonium salt, Zn^{2+} is not precipitated as the hydroxide.

3. Carbonates of the Alkali Metals and Ammonium Carbonate. These form a white precipitate of variable composition, which is soluble in acids.

4. Disodium Hydrogen Phosphate, Na_2HPO_4. The phosphate produces a white precipitate of $Zn_3(PO_4)_2$ which is soluble in acids, bases, and ammonia. In the presence of ammonium salts, the less soluble double salt $ZnNH_4PO_4$ is precipitated.

5. Hydrogen Sulphide, H_2S. This reagent precipitates Zn^{2+} in the form of a white precipitate of ZnS:

$$Zn^{2+} + H_2S \rightarrow ZnS\downarrow + 3H^+$$

The behaviour of Zn^{2+} toward the action of H_2S distinguishes it from all other cations of Group III, which are not precipitated by hydrogen sulphide, and makes it comparable with the cations of Group IV. The precipitation of Zn^{2+} is accounted for by the fact that the value of the solubility product constant of ZnS is considerably lower than those of the sulphides of other cations of Group III. Calculation and experiment show that the near-complete precipitation of Zn^{2+} is achieved even in acid solutions if the pH of the solution is not less than 2 (see Sec. 5.1).

When the solution is treated with hydrogen sulphide at pH ≈ 2, the Zn^{2+} ion can be detected in the presence of all other cations of Group III (and also of Groups II and I).

The reaction is interfered with by Fe^{3+} because it oxidizes hydrogen sulphide to free sulphur which separates out as a white precipitate resembling that of ZnS. Of course, the solution must not contain other oxidizing agents, say, MnO_4^-, $Cr_2O_7^{2-}$, or H_2O_2.

In the systematic analysis of the cations, the analysis of Group III is started after the cations of Group IV have been precipitated by hydrogen sulphide in acid medium (at pH ≈ 0.5). The Fe^{3+} ion is reduced in this case to Fe^{2+} which does not interfere with the detection of Zn^{2+}. If the pH of the solution, obtained after the separation of Group IV cations, is increased from 0.5 to 2 (i.e., if the concentration of H^+ is decreased from 0.3 to 0.01 g-ion/litre) and the solution is treated with hydrogen sulphide, the presence of Zn^{2+} can be deduced from the formation of a white precipitate of ZnS soluble in acids. The pH can most simply be brought to a value of 2 by addition of formate buffer (HCOOH and $HCOONH_4$), i.e., a mixture of formic acid and its salt, taken in appropriate concentrations.

If Groups IV and V are absent in solution, the interfering Fe^{3+} ions can more conveniently be removed by shaking the solution with powdered $CaCO_3$ in the cold. This neutralizes the acids present in solution and increases the pH to about 5. Since the complete precipitation of $Fe(OH)_3$ is achieved even at pH 3.5 (see Table 4.1), the Fe^{3+} ions are thus found to have been removed. The hydroxides $Al(OH)_3$, $Ti(OH)_4$, and (partially) $Cr(OH)_3$ coprecipitate with $Fe(OH)_3$. Conversely, the Zn^{2+} ions remain completely in solution because the precipitation of $Zn(OH)_2$ begins only at pH 6.43 (at a Zn^{2+} concentration of 10^{-2} g-ion/litre). It must however be remembered that the reaction with $CaCO_3$ must be conducted in the cold since on heating the hydrolysis of this salt is enhanced to such an extent that $Zn(OH)_2$ can also be precipitated.

When the excess of reagent and the percipitated hydroxides formed under the action of $CaCO_3$ are removed, the Zn^{2+} ion is detected in the centrifugate by means of hydrogen sulphide at pH ≈ 2. The pH of the solution can be brought to 2 by addition of formate buffer to the almost neutral solution prepared by the method just described.

The technique of fractional detection of Zn^{2+} by means of the reaction in question is discussed in the description of the analysis of Group III. If this technique is strictly followed, the results obtained are quite reliable.

6. **Potassium Hexacyanoferrate(II), $K_4[Fe(CN)_6]$.** This reagent produces with Zn^{2+} a white precipitate of a double salt—zinc potassium hexacyanoferrate(II) soluble in bases:

$$3Zn^{2+} + 2K^+ + 2[Fe(CN)_6]^{4-} \rightarrow K_2Zn_3[Fe(CN)_6]_2 \downarrow$$

This reaction distinguishes Zn^{2+} from Al^{3+} which does not form a precipitate with $K_4[Fe(CN)_6]$.

7. **Potassium Hexacyanoferrate(III), $K_3[Fe(CN)_6]$.** The reagent gives with Zn^{2+} a brownish-yellow precipitate of $Zn_3[Fe(CN)_6]_2$ which is soluble in HCl and NH_4OH.

8. Dithizone*. This reagent, which is used in the form of a solution in chloroform, $CHCl_3$, or carbon tetrachloride, CCl_4, forms with Zn^{2+} an inner-complex salt having a crimson-red colour; in distinction to the analogous compounds (dithizonates) of other elements this salt colours in alkaline solution not only the organic layer ($CHCl_3$ or CCl_4) but also the water layer. The reaction is used for fractional detection of Zn^{2+} in the presence of other cations, in particular, all the cations of Group III. The test is most conveniently performed as follows.

Treat 3 drops of the test solution in a micro crucible (or on a watch glass) with a small amount of solid sodium peroxide, Na_2O_2. After the vigorous reaction has stopped, place a drop of the solution, which is likely to contain ZnO_2^{2-}, AlO_2^-, CrO_4^{2-}, and also excess of NaOH, with the precipitated hydroxides on a strip of filter paper using a capillary tubing. Do not take the tip of the capillary off the paper until a watery zone several millimetres wide is formed around the precipitate. Now draw a ring around the moist spot with a capillary containing dithizone solution. The appearance of a characteristic crimson-red ring indicates the presence of Zn^{2+}. In case Zn^{2+} is absent, an orange-coloured ring is formed, which is characteristic of dithizone in alkaline solution. To distinguish the colours with certainty, place a drop of 2*N* NaOH on the paper next to the spot and treat it with a drop of dithizone solution. This test may conveniently be used in combination with the benzidine spot test for chromium described in Sec. 5.17.

9. Ammonium Mercuri-Thiocyanate, $(NH_4)_2[Hg(SCN)_4]$, in the Presence of $CoCl_2$. Place 2-3 drops of a very dilute solution of $CoCl_2$ (not stronger than 0.02-percent) into a test tube and add an equal volume of $(NH_4)_2[Hg(SCN)_4]$. Scratch the sides of the test tube with a glass rod for half a minute; no blue precipitate must be formed. Then add to the same test tube 1 drop of a solution of a zinc salt and again scraten the tube sides with a glass rod for not more than half a minute. A blue or azure precipitate, consisting of mixed crystals of $Co[Hg(SCN)_4]$ and $Zn[Hg(SCN)_4]$, appears rapidly:

$$Co^{2+} + [Hg(SCN)_4]^{2-} \rightarrow Co[Hg(SCN)_4] \downarrow$$
$$Zn^{2+} + [Hg(SCN)_4]^{2-} \rightarrow Zn[Hg(SCN)_4] \downarrow$$

---

* Dithizone is the shortened name of diphenylthiocarbazone

$$SC\begin{matrix} \diagup NH-NH-C_6H_5 \\ \diagdown N=N-C_6H_5 \end{matrix}$$

This reagent forms coloured inner-complex salts with many heavy metals; the following structures have been ascribed to these salts:

$$\begin{matrix} NH-N-C_6H_5 \\ | \quad\;\; | \quad\quad\; \\ CS \rightarrow Me \quad\; \\ | \quad\quad\quad\quad\;\; \\ N=N-C_6H_5 \end{matrix} \quad \text{or} \quad \begin{matrix} NH-N-C_6H_5 \\ | \quad\;\; | \quad\quad\; \\ CS \quad Me \quad\; \\ | \quad\;\; \uparrow \quad\quad\; \\ N=N-C_6H_5 \end{matrix}$$

The role of Zn^{2+} consists in that the white precipitate of $Zn[Hg(SCN)_4]$ formed by the ion serves to speed up (as a seed) the formation of a blue precipitate of $Co[Hg(SCN)_4]$ which may fail to settle for hours in the absence of $Zn[Hg(SCN)_4]$ (formation of a supersaturated solution). This gives mixed crystals of the two compounds coloured pale-blue or dark-blue, depending on the amount of Co^{2+} present. It should be kept in mind that a precipitate of $Co[Hg(SCN)_4]$ separates from more concentrated solutions of cobalt salts even in the absence of Zn^{2+}. Therefore, solutions of $CoCl_2$ having a concentration higher than 0.02 percent must not be used.

The reaction is interfered with by large amounts of Co^{2+}, and also by Ni^{2+} and Fe^{2+}, which form greenish precipitates with $(NH_4)_2[Hg(SCN)_4]$. In the presence of Fe^{3+}, there may be formed a violet-coloured precipitate. Moreover, the colour of the precipitate is covered up by the formation of intensely coloured ferric thiocyanate, $Fe(SCN)_3$. These complications can be avoided by binding the Fe^{3+} ions into the complex $[FeF_6]^{3-}$ through the addition of NaF or NH_4F.

In carrying out the reaction, one should also remember that it must not be conducted in a strongly acid solution since the precipitate of $Zn[Hg(SCN)_4]$ is soluble in concentrated acids. Under the action of alkalis, this precipitate, like the reagent $(NH_4)_2[Hg(SCN)_4]$ used, decomposes with the formation of a yellow precipitate of HgO. Consequently, the test must be carried out in a weakly or moderately acid solution (the concentration of the acid must not exceed $0.5N$).

10. Methyl Violet*. If a neutral or a very faintly acid solution of a zinc salt is acted on by the dye methyl violet (also called crystal violet) in the presence of NH_4SCN, a suspension of tiny crystals of a salt results, which has a violet colour in transmitted light. The salt is formed by the organic cation of methyl violet and the complex anion $[Zn(SCN)_4]^{2-}$.

The reaction is best carried out as follows. To 2 ml of water add 5 drops of a saturated solution of tartaric acid**, 1 drop of HCl solution (density 1.12 g/cm³), 2 drops of 20-percent solution of NH_4SCN, and 1 drop of 0.06-percent solution of methyl violet. The solution acquires a bluish-green colour due to the presence of H^+. Divide it into two portions; leave one for comparison and to the other add 1-2 drops of the solution to be tested for Zn^{2+} and shake. A change of the colour to violet (or blue at a small concentration of zinc) indicates

* Methyl violet is the chloride of an organic base; its structural formula is

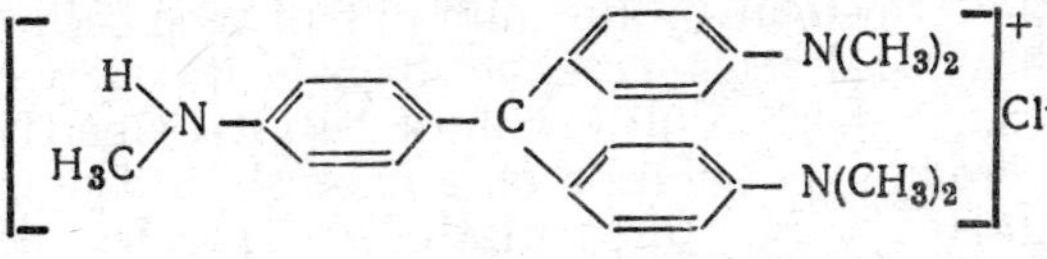

$$\left[\begin{matrix} H \\ H_3C \end{matrix} \!>\! N-C_6H_4-C \begin{matrix} \diagup C_6H_4 - N(CH_3)_2 \\ \diagdown C_6H_4 - N(CH_3)_2 \end{matrix} \right]^+ Cl^-$$

** For the purpose of binding certain interfering cations into complexes.

the presence of Zn^{2+}. An excess of acid is detrimental to the test. Therefore, acid solutions must be neutralized as exactly as possible before the test is started.

All the directions regarding the amounts and concentrations of the reagents used must be strictly adhered to, otherwise the result of the test may prove to be wrong.

This reaction is interfered with by Fe^{3+} and Cu^{2+}. In their presence the solution must be heated with metallic lead which reduces Fe^{3+} to Fe^{2+}, and Cu^{2+} to metallic copper.

The use of this reaction however is more reliable in the systematic analysis after the Group III cations are separated by the action of

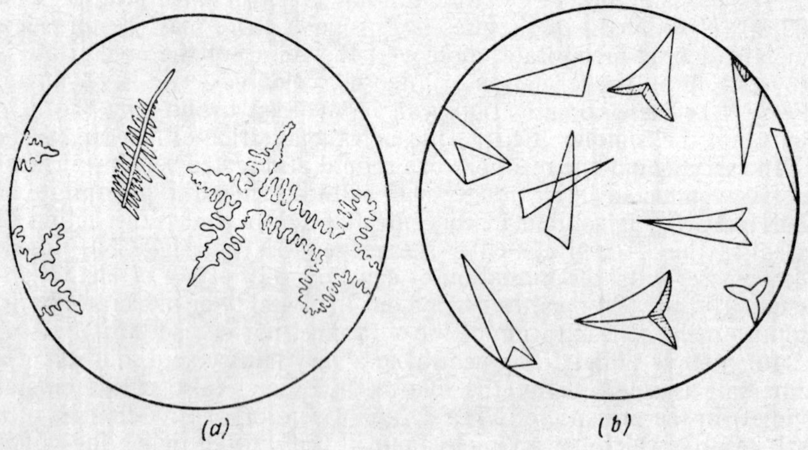

(a) (b)

Fig. 5.5. Crystals of $Zn[Hg(SCN)_4]$:
(a) in ordinary conditions; (b) formed in the presence of mineral acids (or from very dilute solutions).

ammonia in the presence of an ammonium salt. The ammoniacal solution obtained, which contains Mn^{2+} and the ammine complexes of cobalt, nickel, and zinc, must be neutralized as exactly as possible with HCl and tested for Zn^{2+} as described above.

11. Microcrystalloscopic Test. To a drop of a solution of a zinc salt made acid with CH_3COOH and heated add a drop of $(NH_4)_2[Hg(SCN)_4]$ solution on a glass slide and examine the crystals of $Zn[Hg(SCN)_4]$ formed under the microscope. They appear as characteristic crosses and dendrites (Fig. 5.5*a*).

The crystals formed in dilute solutions of zinc salts or solutions acidified with mineral acids have the shape of irregular triangles and wedges (Fig. 5.5*b*). Such solutions must be preliminarily evaporated by placing a drop of the solution on a glass slide. The dry residue is cooled and treated with a drop of the reagent. The dilution limit (concentration limit) is 1 in 10,000.

12. Dry Test. Moisten a small strip of filter paper with a solution of a zinc salt and a dilute solution of $Co(NO_3)_2$, dry, and burn the paper. The ash is found to be green in colour due to the formation of cobalt zincate (Rinmann's green):

$$Zn(NO_3)_2 + Co(NO_3)_2 \rightarrow CoZnO_2 + 4NO_2 \uparrow + O_2 \uparrow$$

5.22. Reactions of the Cobaltous Ion

The crystal hydrates and dilute solutions of cobaltous salts are pink. This colour is due to the hydrated $[Co(OH_2)_6]^{2+}$ ions. Evaporation of the solutions or the action of dehydrating agents, such as alcohol and concentrated HCl, changes the colour from pink to blue as a result of the dehydration of the complex ions and the replacement (substitution) of the H_2O molecules by other ligands.

1. Alkali Hydroxides, NaOH and KOH. These alkalis produce with Co^{2+} a blue precipitate of a basic salt:

$$Co^{2+} + Cl^- + OH^- \rightarrow CoOHCl \downarrow$$

Upon warming with excess alkali, the basic salt is converted into pink cobaltous hydroxide:

$$CoOHCl \downarrow + OH^- \rightarrow Co(OH)_2 \downarrow + Cl^-$$

The $Co(OH)_2$ precipitate is gradually transformed into brown cobaltic hydroxide on exposure to air:

$$4Co(OH)_2 \downarrow + O_2 + 2H_2O \rightarrow 4Co(OH)_3 \downarrow$$

If the cobalt salt solution is acted on by alkali and hydrogen peroxide, a dark-brown precipitate of $Co(OH)_3$ is immediately formed. For example,

$$2Co^{2+} + 4OH^- + H_2O_2 \rightarrow 2Co(OH)_3 \downarrow,$$

The precipitate is insoluble in H_2SO_4 but dissolves in mixtures of sulphuric acid with H_2O_2 or $NaNO_2$, which reduce Co^{III} to Co^{II} (compare Mn).

2. Ammonium Hydroxide, NH_4OH. The reagent produces with Co^{2+} a blue precipitate of a basic salt as in Reaction *1*, which dissolves in the presence of a large excess of NH_4OH or ammonium salts to give a compound of a dingy yellow colour*:

$$CoOHCl \downarrow + 7NH_4OH \rightarrow [Co(NH_3)_6]^{2+} + 2OH^- + NH_4^+ + Cl^- + 6H_2O$$

It is for this reason that ammonia does not precipitate Co^{2+} in the presence of ammonium salts. If however the solution contains Fe^{3+}, Al^{3+}, or Cr^{3+}, the Co^{2+} ion coprecipitates partially with them.

---

* When exposed to air and allowed to stand, the solution gradually turns red because of the formation of the cobalt(III) complex $[Co(NH_3)_6]^{3+}$, whose instability constant is much lower than that of the cobalt(II) ammine complex.

The cobalt (III) ammine complex is also obtained by the action of a mixture of NH_4OH and H_2O_2 on a solution of a cobalt(II) salt.

3. Carbonates of the Alkali Metals. These form with Co^{2+} a reddish precipitate of basic salts of variable composition. The precipitate is soluble in dilute acids and in solutions of ammonia and ammonium carbonate.

4. Disodium Hydrogen Phosphate, Na_2HPO_4. This reagent forms a violet precipitate of $Co_3(PO_4)_2$ which is soluble in acids, including acetic acid. The action of alkalis converts it to $Co(OH)_2$. When acted on by ammonia, the precipitate dissolves to give a complex salt having a dingy-yellow colour.

5. Ammonium Thiocyanate, NH_4SCN. Ammonium thiocyanate produces with Co^{2+} a complex salt of the composition $(NH_4)_2[Co(SCN)_4]$:

$$Co^{2+} + 4SCN^- \rightarrow [Co(SCN]_4]^{2-}$$

If amyl alcohol (or a mixture of amyl alcohol and ether) is added and the solution shaken, the blue complex compound passes into the solvent layer (in which it is more soluble than in water), rendering it blue. In carrying out this reaction, which is most characteristic of Co^{2+}, one should bear in mind that the complex $[Co(SCN)_4]^{2-}$ is highly unstable ($K_{ins} = 1 \times 10^{-3}$).

To make the reaction work, it is necessary to repress the dissociation of the complex into the ions Co^{2+} and SCH^- by introducing a large excess of SCN^-. For 2 drops of a cobalt(II) salt solution, there are taken 8 drops of a saturated solution of NH_4SCN (it is even better to add a small amount of the solid salt) and 5-6 drops of amyl alcohol. The reaction is interfered with by Fe^{3+} which produces with ammonium thiocyanate a blood-red colour; this colour covers up the colour formed by Co^{2+}. Therefore, the Fe^{3+} ion is bound into the complex ferrifluoride ion $[FeF_6]^{3-}$ by addition of NH_4F or NaF in the form of dry salt until the red colouration of the solution disappears.

The sensitivity of the reaction can be increased as follows. Treat 1-2 drops of the weakly acid test solution (a few milligrams of a solid fluoride is added if Fe^{3+} is present) on stirring with 5 drops of a saturated solution of NH_4SCN in acetone. A blue to green colouration appears, depending on the amount of Co^{2+} present. If the concentration of Co^{2+} is low, ammonium thiocyanate is to be taken in the form of dry salt.

And, finally, the test can very conveniently be carried out as a spot test* on a strip of newspaper. This type of paper contains lignin, in the presence of which the cobalt thiocyanate complex becomes more stable and is not decomposed by water. The reaction can even better be carried out if the paper is preliminarily impregnated with solutions of NH_4SCN and NH_4F and dried. When the test solution is applied to such a paper a blue-coloured spot is produced. If the solution also contains Fe^{3+}, a brown colour appears first, due to the for-

* Isakov, P. M., "The Colour Reactions of Lignin and Their Use in Analytical Chemistry", *Zhurn. Priklad. Khim.*, **16**, 234 (1943).

mation of $Fe(SCN)_3$. This colouration however gradually fades away as the result of the Fe^{3+} ions being bound by F^- ions, and is replaced by the blue colour of the cobalt thiocyanate complex.

6. Alpha-Nitroso-Beta-Naphthol, $C_{10}N_6(NO)(OH)$ (Il'inski's Reagent). The reagent oxidizes Co^{2+} to Co^{3+} (the cobaltic ion) with the formation of a voluminous reddish-brown precipitate of cobalt nitroso-β-naphthol (an inner-complex salt), $[C_{10}H_6(NO)O]_3Co$:

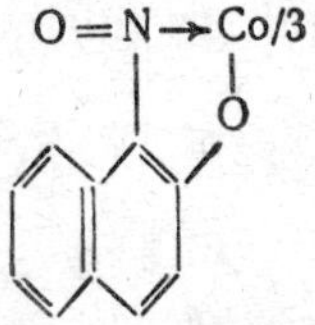

$O{=}N \rightarrow Co/3$

O

To carry out the reaction, dilute 2-3 drops of a neutral or faintly acidified solution of a cobalt(II) salt (to strongly acid solutions there should be added CH_3COONa) with several drops of water, add 2-3 drops of the reagent, and heat.

The presence of Fe^{3+}, which forms a brownish-black precipitate with the reagent, interferes with the reaction.

7. Ammonium Mercuri-Thiocyanate, $(NH_4)_2[Hg(SCN)_4]$. The action of this reagent on Co^{2+} produces a blue precipitate of $Co[Hg(SCN)_4]$ (see Sec. 5.21, Reaction *9*). The addition of a small amount of a zinc salt speeds up its precipitation, the colour of the precipitate being changed from blue to pale blue.

The precipitate dissolves in concentrated HCl with the formation of a blue solution which turns pink on dilution with water. Alkali hydroxides and ammonia decompose the reagent, giving a yellow precipitate of mercuric oxide. The precipitate is almost insoluble in dilute HCl.

8. Potassium Nitrite, KNO_2. An excess of this reagent added to not too dilute a solution of a cobalt(II) salt, acidified with acetic acid, produces a yellow crystalline precipitate of the complex salt $K_3[Co(NO_2)_6]$:

$$Co^{2+} + 7NO_2^- + 3K^+ + 2CH_3COOH \rightarrow$$

$$\rightarrow K_3[Co(NO_2)_6]\downarrow + NO\uparrow + 2CH_3COO^- + H_2O$$

This reaction allows Co^{2+} to be distinguished from Ni^{2+} and can be used for their separation. In dilute solutions, the precipitate separates out if the solution is heated and allowed to stand.

9. Microcrystalloscopic Test. Under the microscope the Co^{2+} ion is detected with the aid of the above reaction of formation of bright-blue crystals of $Co[Hg(SCN)_4]$ (Fig. 5.6).

The reaction is carried out as follows. Evaporate a drop of a neutral or faintly acid solution of a cobalt(II) salt to dryness on a glass slide and after it is cooled treat with a drop of the reagent. If required, allow it to stand. If Fe^{3+} is present, a small grain of NaF is introduced into the drop before the reagent is added.

10. Dry Test. (1) *Borax Bead Test.* Cobalt compounds give blue beads with borax, $Na_2B_4O_7 \cdot 10H_2O$ (see Sec. 5.17). The presence of large amounts of Ni^{2+}, which forms brown beads, interferes with the reaction.

(2) *Grinding Method* (*Flavitsky's Method*). Grind a few crystals of the test mixture with several crystals of NH_4SCN in an evaporating

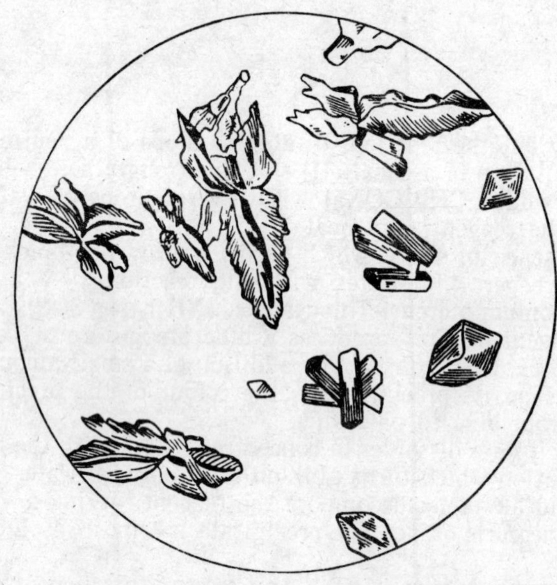

Fig. 5.6. Crystals of $Co[Hg(SCN)_4]$.

dish (or on a porcelain plate) with the aid of the narrow end of the pestle. If the mixture turns brown (this is an indication of the presence of Fe^{3+}), grind it with several crystals of sodium thiosulphate, $Na_2S_2O_3$. When this is done, iron(III) is reduced to iron(II) and the red-brown colouration of $Fe(SCN)_3$ disappears. In the presence of cobalt, this colour of the mixture changes to the blue (or light-blue) colour of $(NH_4)_2[Co(SCN)_4]$. If the resulting blue reaction mixture is treated with a drop of water, the cobalt thiocyanate complex is transformed to hydrated $[Co(OH)_2)_6]^{2+}$ ions and the blue colour changes to pink. Hence, the presence of water interferes with the reaction. Therefore, in carrying out the reaction by the wet method (see Sec. 5.22, Reaction 5), the reaction product has to be extracted by an organic water-immiscible solvent (amyl alcohol).

It is interesting to note that a number of other reactions that fail to proceed in the presence of water can be accomplished by the grinding method. For example, it is possible to detect Co^{2+} by grinding the test mixture with $Na_2S_2O_3$ and in this case tco the mixture becomes blue due to the formation of the cobalt thiosulphate complex. In aqueous solution, this reaction does not proceed.

5.23. Reactions of the Nickel Ion

Aqueous solutions containing nickel(II) ions are green due to the formation of hydrated ions.

1. Alkali Hydroxides, NaOH and KOH. These reagents produce with Ni^{2+} a green precipitate of $Ni(OH)_2$:

$$Ni^{2+} + 2OH^- \rightarrow Ni(OH)_2 \downarrow$$

The precipitate dissolves in acids and in solutions of ammonia and ammonium salts.

2. Ammonium Hydroxide, NH_4OH. Ammonium hydroxide gives first a green precipitate of the basic salt $(NiOH)_2SO_4$, which then dissolves in an excess of NH_4OH to form a complex salt having a blue colour:

$$Ni^{2+} + 6NH_4OH \rightarrow [Ni(NH_3)_6]^{2+} + 6H_2O$$

No precipitation takes place with ammonia solution because of the formation of the complex cation $[Ni(NH_3)_6]^{2+}$.

3. Alkali Carbonates. These produce a green precipitate of nickel carbonate, $NiCO_3$. The precipitate is soluble in acids and in solutions of ammonia and ammonium carbonate.

4. Disodium Hydrogen Phosphate, Na_2HPO_4. This reagent gives a green precipitate of $Ni_3(PO_4)_2$ which is soluble in acids, including acetic acid. The precipitate also dissolves in ammonium hydroxide but is insoluble in caustic alkalis.

5. Dimethylglyoxime (Chugaev's Reagent). This reagent produces with Ni^{2+} in ammoniacal solution a characteristic red precipitate of an inner-complex salt. The equation for the reaction is given in Sec. 5.11. To carry out the reaction, treat a drop of a nickel salt solution with a drop of dimethylglyoxime solution and a drop of a dilute ammoniacal solution.

As regards the conditions of this most important reaction for Ni^{2+}, mention must first of all be made of the influence of the pH of the solution.

With the solution having a considerable acidity, the reaction fails to proceed since dimethylglyoxime is a weak acid and the complex formed by it is broken down by hydrogen ions. Excess of ammonia should also be avoided because this favours the formation of a complex nickel ammine compound. In general, a sufficiently complete precipitation of Ni^{2+} by dimethylglyoxime occurs only at pH values ranging from 5 to 10. The reaction may be carried out in the presence of acetate buffer (a mixture of CH_3COOH and an excess of CH_3COONa) with pH ≈ 5.

The reaction under consideration is interfered with by Fe^{2+} which develops a red colouration with dimethylglyoxime (see Sec. 5.19, Reaction *5*) due to the formation of a soluble inner-complex salt. Cations that form with NH_4OH coloured precipitates of hydroxides (for example, Fe^{3+}) must also be absent: they may mask the reaction.

With the aid of the reaction carried out with Na_2HPO_4 on paper, it is possible to detect Ni^{2+} even in the presence of interfering ions. The spot-test technique is as follows. Place a drop of Na_2HPO_4 solution (substrate) on a strip of paper, following which touch the centre of the resulting moistened spot with a capillary containing the test solution and add another drop of Na_2HPO_4 solution. Now draw a capillary containing dimethylglyoxime solution around the spot and hold the paper over ammonia vapour. If Ni^{2+} is present, a pink ring is formed, and in the presence of a large amount of Ni^{2+} the whole spot becomes pink.

The role of Na_2HPO_4 is to form with interfering cations slightly soluble phosphates (for example, $FePO_4$) which remain in the centre of the spot. Conversely, the more soluble nickel phosphate is found to be situated along the periphery of the spot.

5.24. Analysis of Mixtures of Group III, II and I Cations

The most important reactions of Group III cations are listed in Table 5.5.

The chemical transformations we have encountered in our study of Group III are more diverse and complicated than those involved in Groups II and I. On account of this, a number of various methods can be planned for analysis of Group III. These methods differ primarily in the way the Group III cations are initially divided into subgroups.

In the *hydrogen peroxide method*, the separation of cations into subgroups is based on the action of excess alkali in the presence of hydrogen peroxide (or sodium peroxide). The cations Al^{3+} and Zn^{2+} are transformed into the anions AlO_2^- and ZnO_2^{2-} and remain in solution together with CrO_4^{2-} ions which are formed on oxidation of chromium(III) by hydrogen peroxide in alkaline solution. The other cations are precipitated as $Ti(OH)_4$, $Fe(OH)_3$, $MnO(OH)_2$, $Co(OH)_3$, and $Ni(OH)_2$.

In the *ammonia method*, the separation into subgroups is based on the action of NH_4OH in the presence of ammonium salts, the cations Al^{3+}, Cr^{3+}, Ti^{IV}, and Fe^{3+} being precipitated as the hydroxides, and the cations Fe^{2+}, Mn^{2+}, $[Zn(NH_3)_6]^{2+}$, $[Co(NH_3)_6]^{2+}$, and $[Ni(NH_3)_6]^{2+}$ remaining in solution. Instead of ammonia, use may be made of pyridine, C_5H_5N, in the presence of an ammonium salt, developing a pH of 6.5 at which the hydroxides of aluminium, iron(III), chromium(III), and titanium(IV) are precipitated and the other cations of Group III remain in solution in the form of soluble complex salts with pyridine (see Sec. 5.14).

In the *alkali method*, the separation into subgroups is effected through the action of excess NaOH on heating. The Al^{3+} and Zn^{2+} ions, which form the aluminate ion AlO_2^- and the zincate ion ZnO_4^{2-}, remain in solution, the other cations of Group III being precipitated in the form of hydroxides.*

In the *acetate method* (suitable mainly in the absence of chromium), the ions Al^{3+} and Fe^{3+} are precipitated by heating with CH_3COONa in the form of basic acetates, titanium** settles down as $Ti(OH)_4$, and other cations remain in solution.

When the above methods are used, Co^{2+} and Ni^{2+} are usually separated first, the separation being based on the insolubility of CoS and NiS and the solubility of the other sulphides of Group III in dilute HCl. The separation however is far from being complete and is associated with considerable loss of time, and is thus inconvenient under the semimicroanalysis conditions.

It should however be noted that all other methods of separation of Group III cations are greatly complicated by coprecipitation (see Sec. 5.4) and are therefore far from perfect. Thus, in the ammonia method, Co^{2+}, Ni^{2+}, and, partly, Mn^{2+} coprecipitate appreciably with the hydroxides of aluminium, chromium, titanium, and iron. Therefore, they should be detected prior to the separation into subgroups, before they are distributed between the precipitate and the solution. In the method using hydrogen peroxide (and especially in the alkali method), Zn^{2+} coprecipitates strongly with the ions being precipitated and is often "missed" in the systematic analysis.

Below is described the analysis of mixtures of the cations of Groups III, II, and I by the hydrogen-peroxide, ammonia, and pyridine methods. The first two methods are illustrated by schemes (Tables 5.6 and 5.7) which are not meant to serve as analytical instructions but merely show the sequence of the individual operations to be carried out and the separations involved.

It must be kept in mind that all the methods of analysis under consideration are applicable only in the absence of PO_4^{3-}. If this anion is present in the test solution, follow the procedure described in Sec. 6.21.

---

* In the cold, chromium too passes into solution in the form of chromite. But on heating, chromites undergo hydrolysis and $Cr(OH)_3$ is precipitated (see Sec. 5.17, Reaction *1*).

** On prolonged heating, titanium is precipitated as H_2TiO_3.

TABLE 5.5. Reactions

Reagent	Cation			
	Al^{3+}	Cr^{3+}	Ti^{IV}	Fe^{3+}
$(NH_4)_2S$ in the presence of NH_4OH and ammonium salts	A white ppt.: $Al_2O_3 \cdot xH_2O$	A gray-green or gray-violet ppt.: $Cr_2O_3 \cdot xH_2O$	A white ppt.: $TiO_2 \cdot xH_2O$	A black ppt.: Fe_2S_3
H_2S in the presence of HCOOH and $HCOONH_4$	—*	—	—	Reduction to Fe^{2+}, a ppt.: S
NaOH or KOH not in excess	A white ppt.: $Al_2O_3 \cdot xH_2O$	A gray-green or gray-violet ppt.: $Cr_2O_3 \cdot xH_2O$	A white ppt.: $TiO_2 \cdot xH_2O$	A brown ppt.: $Fe_2O_3 \cdot xH_2O$
NaOH or KOH in excess on heating	A colourless solution: $[Al(OH)_4]^-$	Ditto	Ditto	Ditto
H_2O_2 in alkaline solution on heating	Ditto	A yellow solution: CrO_4^{2-}	Ditto	Ditto
NH_4OH not in excess	A white ppt.: $Al_2O_3 \cdot xH_2O$	A gray-green or gray-violet ppt.: $Cr_2O_3 \cdot xH_2O$	Ditto	Ditto
NH_4OH (excess) in the presence of ammonium salts	Ditto	Ditto	Ditto	Ditto
CH_3COONa on heating	A white ppt. of basic salts. The exact composition has not been established	In solution***: $[Cr_3O(CH_3COO)_6]^+$	Ditto	A brown ppt.: $[Fe_3O(CH_3COO)_6]OH$

* The dash signifies the absence of a precipitate or characteristic colouration.
** The oxidation products $Fe_2O_3 \cdot xH_2O$ and $MnO_2 \cdot xH_2O$ are gradually precipitated from
*** If the solution contains Fe^{3+} and Al^{3+}, the Cr^{3+} ion coprecipitates with them in the

of Group III Cations

Fe^{2+}	Mn^{2+}	Zn^{2+}	Co^{2+}	Ni^{2+}
A black ppt.: FeS	A pale-pink ppt.: MnS	A white ppt.: ZnS	A black ppt.: CoS	A black ppt.: NiS
—	—	Ditto	—	—
A greenish ppt. turning brown: $Fe(OH)_2$	A white ppt. turning brown: $Mn(OH)_2$	A white ppt.: $Zn(OH)_2$	A blue ppt. of basic salts of variable composition	A greenish ppt.: $Ni(OH)_2$
Ditto	Ditto	A colourless solution: $[Zn(OH)_4]^{2-}$	A pink ppt.: $Co(OH)_2$	Ditto
A brown ppt.: $Fe_2O_3 \cdot xH_2O$	A brown ppt.: $MnO_2 \cdot xH_2O$	Ditto	A dark-brown ppt.: $Co_2O_3 \cdot xH_2O$	Ditto
A greenish ppt. turning brown: $Fe(OH)_2$	A white ppt. turning brown: $Mn(OH)_2$	A white ppt.: $Zn(OH_2)$	A blue ppt. of basic salts of variable composition	A greenish ppt. of basic salts of variable composition
—**	—**	A colourless solution: $[Zn(NH_3)_6]^{2+}$	An orange-brown solution: $[Co(NH_3)_6]^{2+}$	A lilac-coloured solution: $[Ni(NH_3)_6]^{2+}$
—	—	—	—	—

solution.
form of a basic acetate.

Reagent	Cation			
	Al^{3+}	Cr^{3+}	Ti^{IV}	Fe^{3+}
Na_2CO_3, K_2CO_3, $(NH_4)_2CO_3$	A white ppt.: $Al_2O_3 \cdot xH_2O$	A gray-green or gray-violet ppt.: $Cr_2O_3 \cdot xH_2O$	A white ppt.: $TiO_2 \cdot xH_2O$	A brown ppt.: $Fe_2O_3 \cdot xH_2O$
Na_2HPO_4	A white ppt: $AlPO_4$	A green ppt.: $CrPO_4$	A white precipitate: $Ti(HPO_4)_2$	A yellowish ppt.: $FePO_4$
$K_4[Fe(CN)_6]$	—	—	A brown ppt.: $Ti[Fe(CN)_6]$	A blue ppt.: $Fe_4[Fe(CN)_6]_3$
$K_3[Fe(CN)_6]$	—	—	A yellow ppt.: $Ti_3[Fe(CN)_6]_4$	—
Oxidizing agents in alkaline solution	A colourless solution: $[Al(OH)_4]^-$	A yellow solution: CrO_4^{2-}	A white ppt.: $TiO_2 \cdot xH_2O$	A brown ppt.: $Fe_2O_3 \cdot xH_2O$
Oxidizing agents in acid solution	—	An orange solution: $Cr_2O_7^{2-}$	Produces with H_2O_2 a yellow solution: $[TiO_2(H_2O)x]^{2+}$	—
Bead with $Na_2B_4O_7$ or $NaNH_4HPO_4$	—	Emerald green	—	—

Table 5.5, continued

Fe^{2+}	Mn^{2+}	Zn^{2+}	Co^{2+}	Ni^{2+}
A white ppt. of carbonate turning brown: $FeCO_3$	A white ppt.: $MnCO_3$	A white ppt. of basic carbonate of variable composition	A blue ppt. of basic carbonate of variable composition	A green ppt.: $NiCO_3$
A white ppt.: $Fe(PO_4)_2$	A white ppt.: $Mn_3(PO_4)_2$	A white ppt.: $Zn_3(PO_4)_2$	A violet ppt.: $Co_3(PO_4)_2$	A green ppt.: $Ni_3(PO_4)_2$
A white ppt., $Fe_2[Fe(CN)_6]$	A white ppt. of hexacyanoferrates(II) of variable composition	A white ppt.: $K_2Zn_3[Fe(CN)_6]_2$	A green ppt. of hexacyanoferrates(II) of variable composition	A pale-green ppt. of hexacyanoferrates (II) of variable composition
A blue ppt.: $Fe_4[Fe(CN)_6]_3$	A brown ppt. of hexacyanoferrates(III) of variable composition	A yellow ppt.: $Zn_3[Fe(CN)_6]_2$	A dark-red ppt. of hexacyanoferrates(III) of variable composition	A yellow-brown ppt. of hexacyanoferrates (III) of variable composition
A brown ppt.: $Fe_2O_3 \cdot xH_2O$	A brown ppt.: $MnO_2 \cdot xH_2O$	A colourless solution: $[Zn(OH)_4]^{2-}$	A dark-brown ppt.: $Co_2O_3 \cdot xH_2O$	A black ppt.: $Ni_2O_3 \cdot xH_2O$
Oxidation to Fe^{3+} (solution)	A crimson-violet solution: MnO_4^-	—	—	—
Brown or greenish (in the reducing flame)	Violet-amethyst	—	Blue	Brown

The cations of Groups II and I are colourless and give, in most cases, white precipitates, while some of the cations of Group III are coloured (Cr^{3+}, Ni^{2+}, Co^{3+}, Fe^{3+}) and form coloured precipitates. The *colouration* of all these cations is a very important analytical sign which should be given considerable attention. Sometimes this facilitates the analysis and makes the analyst more certain as to the correctness of the results obtained. For example, if the solution is colourless, the above coloured cations cannot be present in it in any appreciable amounts.

Likewise, if a light-coloured precipitate is obtained by the action of $(NH_4)_2S$, one may draw a conclusion as to the absence in the test solution of Fe^{3+}, Fe^{2+}, Ni^{2+}, and Co^{2+}, whose sulphides are black. If the precipitate is pure white, Al^{3+}, Zn^{2+}, or Ti^{IV} may be expected to be present; the Cr^{3+} ion is likely to be present if the precipitate is gray-green in colour; the pale-pink precipitate signifies the presence of Mn^{2+}. Certainly, if different ions are jointly present, their colours are mutually masked; but not infrequently such masking does not occur. Observation of colours makes the analysis much easier. On the other hand, errors are often revealed if the colours of the solutions or precipitates are taken into account. For example, if Fe^{3+}, Fe^{2+}, Co^{2+}, and Ni^{2+} are not found during analysis of Group III but the sulphide precipitate is black, it is obvious that an error has been made. From this example we see how important it is to note the colouration both of the test solution and of the precipitates obtained during analysis.

Valuable conclusions as to the probable presence or absence of certain cations in solution can also sometimes be made on the basis of *pH determinations*. Thus, if the solution shows a strongly alkaline reaction ($pH > 10$) and does not smell of ammonia, it may not be expected to contain any cations whose hydroxides possess no amphoteric properties. Hence, in such cases, of the cations of Group III only aluminium, chromium, and zinc should be sought for since it is these elements that are present in alkaline solution as the anions AlO_2^-, CrO_2^- (or CrO_4^{2-}), and ZnO_2^{2-}. If the alkalinity of the solution is due to ammonia or to a mixture of ammonia and an ammonium salt, the solution may be expected to contain the Fe^{2+}, Mn^{2+}, $[Zn(NH_3)_6]^{2+}$, $[Ni(NH_3)_6]^{2+}$, and $[Co(NH_3)_6]^{2+}$ ions*, but the presence of Fe^{3+}, Ti^{IV}, and also of Al^{3+} and Cr^{3+} (if the pH does not exceed 10), is impossible.

If the solution is acid, it may contain any of the Group III cations; but even in this case the value of pH may count. Thus, if the solution pH is equal to or greater than 5, we may state, with a high degree of certainty, that the solution does not contain Ti^{IV} and Fe^{3+} because iron is completely precipitated as $Fe(OH)_3$ even at pH 3.5, and titanium in an even more strongly acid solution (see Table 4.1). Likewise,

* The ions Fe^{2+}, Mn^{2+}, and $[Co(NH_3)_6]^{2+}$ are gradually oxidized by atmospheric oxygen to $Fe(OH)_3$, $MnO(OH)_2$, and $[Co(NH_3)_6]^{3+}$, respectively.

at pH 5-6 the presence of Al^{3+} is unlikely since it is completely precipitated* at pH 5.32.

Such conclusions however must not be regarded as being final and must be confirmed by subsequent tests. One must remember that pH values are determined only to the nearest unit and that the situation is considerably complicated by the joint presence of several cations, which makes conclusions more difficult to draw. Nevertheless, determinations of the solution pH and observations of colours are very helpful for they serve as a guide for the analyst. Combined with the results of the analysis, they increase the validity of the analysis, especially in the case of chemically individual substances.

It must not be forgotten that the acid reaction of the solution may depend not only on the presence of any of the Group III cations but also on the presence of free acids which are often introduced during analysis, for example, in the precipitation of Group IV cations by hydrogen peroxide, in the separation of Group II sulphates, etc. Of course, direct determinations of pH values in such cases are pointless. To obtain a pH value that might serve as an analytical indication of the presence of cations in solution, the free acids must be neutralized until a turbidity appears, which persists on stirring; only after this is done, may the pH be determined. The pH value is best determined after the solution has been neutralized, before the Group III cations are precipitated by ammonium sulphide (see below).

(a) The Hydrogen Peroxide Method

1. Preliminary Observations and Tests. *Colouration.* (1) Note the colour of the solution and of the precipitate; draw conclusions as to the presence or absence of coloured cations (see above).

(2) *Dissolution of the Precipitate and Separation of Group II Sulphates.* If the test solution contains a precipitate, stir it with a glass rod, transfer 2 drops of the turbid liquid to another test tube, and add 3-4 drops of $2N$ HCl solution. Stir and, if required, heat. If the precipitate can be completely dissolved by HCl, dissolve it (in the entire volume of the test solution) in as little $2N$ HCl solution as possible. Analyse the resulting solution as directed below (see Procedures *3*, *4*, etc.).

If the precipitate does not dissolve completely, this proves the presence of the sulphates of Ba^{2+} and Sr^{2+}. In such a case, make the solution acid with a few drops of $2N$ H_2SO_4 solution. This is done to achieve a double purpose: on the one hand, all the acid-soluble precipitates (carbonates, hydroxides, phosphates, etc.) are dissolved, and, on the other hand, if the concentration of SO_4^{2-} in solution is not suf-

---

* To suppress hydrolysis, solutions of the salts of titanium(IV), iron(III), iron(II), chromium(III), and of certain salts of aluminium (e.g., $AlCl_3$) are acidified, so that they show a strongly acid reaction.

ficient for complete precipitation of all the cations of Group II, they can be postprecipitated by adding H_2SO_4. Then test for completeness of precipitation. To do this, centrifuge a small portion of the solution in a test tube after addition of H_2SO_4 and treat the clear centrifugate obtained with a few drops of H_2SO_4. The absence of a precipitate indicates that the sulphates of Ba^{2+} and Sr^{2+} have been precipitated completely.

Heat the solution with the precipitated sulphates for 2-3 minutes in a boiling water bath. Centrifuge the undissolved $BaSO_4$ and $SrSO_4$ precipitate, wash, and analyse as directed in Procedure *3* (see Sec. 4.13). Investigate the solution as described below.

(3) *Determination of pH*. If the solution was acidified at the previous stages of analysis, the pH is determined after the solution is neutralized until a slight turbidity appears, which persists on stirring, before the cations of Group III are precipitated by ammonium sulphide (see Procedure *2*). This should also be done when the pH of the test solution is less than 2, i.e., if the solution is strongly acid. This may serve as proof of the presence of free acids.

(4) *Detection of* NH_4^+. Ammonium ions are introduced into the solution in the precipitation of Group III cations by the group reagent and therefore ammonium must be detected by the fractional method before the precipitation. The test is performed in the usual way. To 1-2 drops of the test solution add excess NaOH in a gas chamber, or treat with Nessler's reagent. In the latter case, first precipitate the hydroxides of Group III cations in a separate portion of the solution (1-2 drops) by the action of 3-4 drops of alkali. Then centrifuge the precipitate and test the centrifugate for NH_4^+ (see Sec. 4.13).

(5) *Detection of* Fe^{3+} *and* Fe^{2+}. This must be carried out prior to precipitation of Group III cations because, after the precipitation and dissolution of the precipitate in HNO_3, all the iron will be present as Fe^{3+}.

If the precipitated sulphides and hydroxides of Group III cations are treated with hydrochloric acid (in order to separate NiS and CoS), all the iron will be reduced by hydrogen sulphide and will be present as Fe^{2+}.

To detect Fe^{3+}, mix 1 drop each of the test solution, 2*N* HCl, and a solution of $K_4[Fe(CN)_6]$ on a spot plate or in a test tube. A blue precipitate of ferric ferrocyanide (Prussian blue), $Fe_4[Fe(CN)_6]_3$, confirms the presence of Fe^{3+}. Another portion of the acidified solution may be treated with a drop of NH_4SCN solution; a red colouration is produced due to the formation of $Fe(SCH)_3$.

The Fe^{2+} ion is detected by the action of $K_3[Fe(CN)_6]$ on a separate portion of the acidified solution. A blue precipitate shows the presence of Fe^{2+}.

(6) *Detection of* Zn^{2+}. Here it is expedient to use the fractional method since this ion cannot sometimes be identified in the systematic ana-

lysis because of the coprecipitation. The ion is detected by using the reaction either with H_2S or with dithizone.

In the first case, after the sulphates of Group II cations are separated*, 5 drops of the solution are neutralized by Na_2CO_3 until a turbidity appears, which persists on stirring. Dissolve the turbidity in 1-2 drops of 2*N* HCl. To the weakly acid solution add about twice the volume of water and such an amount of powdered $CaCO_3$ that part of it remains undissolved after vigorous agitation for 1 minute**. The pH of the solution increases approximately up to 5 and the interfering Fe^{3+} ion is precipitated as $Fe(OH)_3$, the Zn^{2+} ion remaining in solution. Centrifuge and discard the precipitate. To the centrifugate add 10 drops of formate buffer, check the pH of the solution (it must equal 2), and pass gaseous hydrogen sulphide into the solution for 1-2 min. The formation of a white precipitate (turbidity) of ZnS, which is dissolved on addition of HCl, shows the presence of Zn^{2+}. Instead of formate buffer, use may be made of one-fourth of the volume (count up drops) of 0.05*N* HCl solution. Then, having checked the pH, treat with hydrogen sulphide.

The ions Zn^{2+} and Al^{3+} can also be separated from the other cations of Group III by the action of a 30-percent solution of NaOH or KOH. The solution will be found to contain only ZnO_2^{2-} and AlO_2^-. The solution is acted on by hydrogen sulphide: the appearance of a white precipitate is evidence of the presence of Zn^{2+}. Separate ZnS, acidify the filtrate with CH_3COOH, and test for Al^{3+} after the hydrogen sulphide is boiled off.

In certain cases, for example, in the presence of other oxidizing agents (apart from Fe^{3+}), free sulphur may precipitate. Sometimes, the precipitate may be found to contain traces of the sulphides of nickel and cobalt, in which case it is coloured gray or even black. Titanic acid, H_4TiO_4, may also be partially precipitated. Therefore, when a precipitate is formed by the action of hydrogen sulphide, it is necessary to conduct a confirmatory test for Zn^{2+}. To do this, treat the precipitate in the cold with a few drops of 2*N* HCl, centrifuge, and test the centrifugate for the presence of Zn^{2+} by carrying out a microcrystalloscopic reaction with $(NH_4)_2[Hg(SCN)_4]$ (see Sec. 5.21).

The Zn^{2+} ion can be detected by the drop reaction with dithizone described in Sec. 5.21. The formation of a characteristic crimson-red ring of dithizonate indicates the presence of Zn^{2+}. If it is absent, an orange ring is formed due to the formation of sodium dithizonate. To avoid an error, it is necessary to place a drop of 2*N* NaOH solution next to the spot of the test solution and to add dithizone. Compare the colours formed.

---

* Of course, the cations of Groups IV and V must not be present in the solution.

** The $CaCO_3$ solution must be treated in the cold, otherwise Zn^{2+} will also be precipitated (see Sec 5.21).

(7) *Detection of Ti^{IV}*. The reaction is better to be performed by the fractional method since in the subsequent analysis, especially if the precipitation of Group III has been carried out with ammonium sulphide on heating, titanium may pass into the acid-insoluble metatitanic acid, H_2TiO_3, and it will then be impossible to detect it in the course of analysis.

It is best to detect titanium by the spot-test technique using chromotropic acid. It can also be detected by the action of H_2O_2 in acid solution, in which case the Fe^{3+} ion has to be preliminarily bound into a complex with orthophosphoric acid, H_3PO_4. This is necessary because Fe^{3+} has a yellow colour, just like the peroxide compound of titanium formed.

To 2 drops of the test solution add 1 drop of a solution of orthophosphoric acid and 2-3 drops of a 3-percent H_2O_2. If the test solution is found to be coloured (green or pink) and the yellow colouration is poorly visible, a solution should be prepared for comparison (a so-called blank experiment should be carried out). To this end, place 2 drops of the test solution into another test tube and add 1 drop of H_3PO_4 and 2-3 drops of water. If the solution in the first test tube is more intensely yellow-coloured than in the second, it means that titanium is present.*

2. Separation of Group III Cations from Cations of Groups II and I. The cations of Group III form precipitates with Na_2HPO_4, $(NH_4)_2CO_3$, $(NH_4)_2C_2O_4$, and other reagents used in analysis of Groups II and I; therefore, Group III must first be separated from the other two groups**.

Place 20-25 drops (not more) of the test solution into a conical test tube, add 6-8 drops of NH_4Cl solution, and neutralize the acid with ammonia. Since excess ammonia favours the formation of colloidal solutions of NiS and other sulphides, care must be exercised in carrying out the neutralization: add, a drop at a time, 2*N* ammonia solution and stir the contents of the test tube after each addition.*** The addition of ammonia is continued until a permanent turbidity appears or the solution becomes weakly alkaline (pH $\approx$ 9) if no turbidity is formed due to NH_4OH.**** Having thus prepared the test solution, heat it

---

* For the colour of the peroxide compound of titanium to be more intense, the drops of H_2O_2 should be poured along the side of the test tube so that they do not mix with the solution. If Ti^{IV} is present, a yellow or orange ring is formed at the interface between the solution and hydrogen peroxide.

** The conditions for separation of the cations of Group III from those of Groups II and I are discussed in detail in Sec. 5.15.

*** If too much ammonia has been added and a precipitate is formed instead of a turbidity, dissolve it in as little 2*N* HCl as possible, and then add one drop or two more of NH_4OH.

**** No turbidity is formed if the solution does not contain Fe^{3+}, Cr^{3+}, Al^{3+}, and Ti^{IV}. As pointed out in Procedure *1*, in certain cases it may be useful to carry out the neutralization simultaneously with the determination of the solution pH. If no turbidity results from the action of ammonium hydroxide,

nearly to boiling and, with stirring, precipitate the cations of Group III, using 12-15 drops of $(NH_4)_2S$ solution. For a better coagulation of the particles to be achieved, heat again the contents of the test tube in a water bath for several minutes, centrifuge off the precipitate, and test for complete precipitation by addition of 1 drop of $(NH_4)_2S$ solution. If the precipitation is incomplete, add another 5-6 drops of $(NH_4)_2S$ solution, heat, centrifuge, and test again for complete precipitation. Having achieved complete precipitation, transfer the centrifugate to another test tube and wash the precipitate two times with hot water containing NH_4NO_3; add ammonium nitrate to avoid peptization. The washing is aimed at freeing the precipitate from the major part of adsorbed Cl^- ions which interfere with the detection of Mn^{2+} by oxidizing it to MnO_4^-.

Having obtained the precipitate of the sulphides and hydroxides of Group III cations, note the colour of the precipitate and draw a preliminary conclusion as to the probable composition. Further analysis is to be carried out according to Procedure *4*; the centrifugate containing the cations of Groups II and I with excess $(NH_4)_2S$ is to be investigated as directed in Procedure *3*. If the precipitate is difficult to separate from the solution by centrifuging because of the errors made, it is necessary to induce coagulation of the colloidal solution by addition of a few drops of a saturated solution of NH_4NO_3 and by heating.

3. Preparing the Solution for Detection of Group II and I Cations. To avoid the oxidation of S^{2-} to SO_4^{2-} by the oxygen of the air and the associated precipitation of Group II cations in the form of sulphates, it is necessary to destroy the excess $(NH_4)_2S$ in the centrifugate, immediately after centrifuging, by heating with acid.

Transfer the solution to an evaporating dish, acidify with CH_3COOH, and evaporate to about half its volume. Centrifuge the precipitated sulphur*, evaporate the centrifugate to dryness, and ignite to remove ammonium salts (a large excess of these salts would interfere with complete precipitation of Group II cations by ammonium carbonate**).

Dissolve the dry residue, after it is cooled, in 10-12 drops of $2N$ HCl and test for Groups II and I as directed in Sec. 4.13.

4. Dissolving the Precipitated Sulphides and Hydroxides of Group III Cations and Oxidation of Fe^{2+} to Fe^{3+}. Treat, on heating, the thoro-

Note. If the dry residue does not dissolve completely in HCl, test the undissolved part for the sulphates of Group II cations (see Sec. 4.10, Reaction *2*).

---

a fresh portion of the solution has to be taken for determination of the pH; it is to be neutralized with a solution of Na_2CO_3 or NaOH until a turbidity appears.

* The sulphur precipitate formed in this case (it will also be encountered further in the dissolution of sulphides) may be found to be dark-coloured due to traces of Group III sulphides.

** The carbonates of Group II cations are appreciably soluble in solutions of ammonium salts because these solutions show an acid reaction due to hydrolysis.

ughly washed precipitate of the sulphides and hydroxides of Group III cations (see Procedure *3*) with 4-6 drops of 6*N* NHO_3, 1-2 drops of KNO_2 solution being added to speed up the reaction. If a black precipitate is left after heating for 2-3 min, add 3-4 drops more of HNO_3 and heat again. Centrifuge off the sulphur precipitate and discard. Dissolve the centrifugate in 5-6 drops of water.

Nitric acid is used because the precipitates of NiS and CoS do not dissolve in hydrochloric and sulphuric acids. Moreover, NHO_3 oxidizes Fe^{2+} to Fe^{3+}, which is also convenient for analysis. Test the resulting solution as directed in Procedure *5*.

There is another method for dissolving the precipitated sulphides and hydroxides of Group III cations, which leads to separation of NiS and CoS from other cations of this group.

The precipitated sulphides and hydroxides should be left below the mother liquor for 15-20 min. The precipitate is thus allowed to age: the most soluble modifications α-NiS and α-CoS formed at the first moment convert into the less soluble modifications γ-NiS and β-CoS which are insoluble in 2*N* HCl, while the other sulphides and hydroxides of Group III dissolve in it.

Therefore, with the sulphide and hydroxide precipitate having been washed and treated with 2*N* HCl, dissolve all the cations of Group III, except Ni^{2+} and Co^{2+} which will remain in the precipitate in the form of γ-NiS and β-CoS. Separate the undissolved sulphides of Ni^{2+} and Co^{2+} by centrifuging, wash, and treat with 3-percent H_2O_2 or sodium nitrite in acetic acid solution (2*N* CH_3COOH). The precipitate is dissolved (the equations of the reactions are given in Sec. 5.15).

After the solution has been heated to completely remove hydrogen sulphide, divide it into two portions and try to detect Co^{2+} and Ni^{2+} with the aid of the corresponding characteristic reactions under suitable conditions (see Sections 5.22 and 5.23).

5. Separation of Fe^{3+}, Mn^{2+}, Co^{2+}, Ni^{2+}, and Ti^{IV} from Al^{3+}, Cr^{3+}, and Zn^{2+}. To the centrifugate (see Procedure *4*) containing the cations Ti^{IV}, Fe^{3+}, Al^{3+}, Cr^{3+}, Mn^{2+}, Ni^{2+}, Co^{2+}, and Zn^{2+}, along with excess HNO_3, add a concentrated (30-percent) solution of NaOH until alkaline. This leads to the neutralization of HNO_3, and all the cations mentioned above are precipitated as hydroxides. Then add 5-8 drops more of the same NaOH solution and 4-5 drops of a 3-percent solution of H_2O_2. After the vigorous reaction is over, heat the liquid in a water bath for several minutes.

In the presence of excess alkali, the amphoteric hydroxides $Al(OH)_3$ and $Zn(OH)_2$ form soluble aluminate and zincate ions. Chromium hydroxide, $Cr(OH)_3$, when acted on by excess NaOH, forms a chromite, CrO_2^-, which is then oxidized to a chromate, CrO_4^{2-}, on heating under the influence of H_2O_2. At the same time, the other cations of Group III, whose hydroxides possess no amphoteric properties, remain in the precipitate, the manganese being oxidized to Mn^{IV} by hydrogen peroxide, and cobalt to Co^{III}. Consequently, the precipitate will contain $Fe(OH)_3$, $Ni(OH)_2$, $Co(OH)_3$, $MnO(OH)_2$, and $Ti(OH)_4$. When the precipitate is obtained, note its colour*, and then wash and test it (see Procedure *6*). Test the centrifugate as directed in Procedure *10*.

6. Dissolving the Precipitate and Detection of Co^{2+}. Dissolve the washed hydroxide precipitate on heating in a few drops of 2*N* H_2SO_4. Since $Co(OH)_3$ and $MnO(OH)_2$ do not dissolve in sulphuric acid, a suitable reducing agent capable of reducing Co^{III} and Mn^{IV} to Co^{II} and Mn^{II} must be introduced together with H_2SO_4 in order to dissolve these hydroxides. Add, as a reducing agent, 2-3 drops of a solution of KNO_2 to the solution. Decomposition of excess KNO_2 can be brought about by heating the solution in the water bath for 2-3 min. The Co^{2+} ion is detected in a separate portion of the solution. To do this, to 2 drops of the solution add 8-10 drops of a saturated solution of NH_4SCN (it is better to add a little solid salt).** If the Fe^{3+} ion, which develops a blood-red colouration due to the formation of $Fe(SCN)_3$, is present, it is bound into the complex $[FeF_6]^{3-}$ by adding solid NaF or NH_4F until the red colour disappears. Then add 5-7 drops of amyl alcohol, $C_5H_{11}OH$, or a mixture of amyl alcohol and ether, and shake. The blue colour of the organic layer indicates the formation of $(NH_4)_2[Co(SCN)_4]$.

The Co^{2+} ion can be detected even more conveniently by a drop reaction on a newspaper impregnated with solutions of NH_4SCN and NH_4F and dried. A blue stain appears on the paper if Co^{2+} is present. If Fe^{3+} is also present, the brown colour of $Fe(SCN)_3$ is formed first, which however disappears rapidly under the action of F^- ions, being replaced by the blue colour of $[Co(SCN)_4]^{2-}$. The Co^{2+} ion can also be detected by a microcrystalloscopic reaction or Il'inski's reaction (in the absence of Fe^{3+}) (see Sec. 5.22). Test the remaining portion of the solution as described in Procedure *7*.

7. Separation of Fe^{3+}, Mn^{2+}, and Ti^{IV} from Ni^{2+} and Co^{2+}. To the solution add a concentrated (25-percent) solution of NH_4OH until alkaline, and then add 5-6 drops more of it. Then introduce 4-5 drops of 3-percent H_2O_2 and heat the contents of the test tube until the evolution of oxygen bubbles stops (after 3-4 min). Centrifuge the precipitate consisting of $Fe(OH)_3$, $MnO(OH)_2$, and $Ti(OH)_4$ (note the colour of the precipitate), wash thoroughly (to remove Cl^- ions which interfere with detection of Mn), and test as directed in Procedure *8*.

The centrifugate, which is likely to contain the complex ions $[Ni(NH_3)_6]^{2+}$ (bluish) and $[Co(NH_3)_6]^{3+}$ (pink), is to be investigated according to Procedure *9*.

---

* Let us recall that $Fe(OH)_3$ is red-brown, $MnO(OH)_2$ and $Co(OH)_3$ are dark-brown, and $Ti(OH)_4$ is white.

** It should be borne in mind that the complex $[Co(SCN)_4]^{2-}$ is unstable and may form only at a high concentration of SCN^- which depresses the dissociation of the complex.

8. Detection of Mn^{2+} and Fe^{3+}. (1) Stir the precipitate in 15-20 drops of water, introduce a drop of the turbid liquid into a mixture of 5 drops of 6*N* HNO_3 and 10 drops of water, and add a little powdered $NaBiO_3$. Mix, let stand 1-2 min, and then separate the excess $NaBiO_3$ by centrifuging. The crimson-violet colour of the solution (MnO_4^-) indicates the presence of manganese.

The reactions for Mn^{2+} with PbO_2 or with $(NH_4)_2S_2O_8$ on heating may also be used (see Sec. 5.20, Reaction *5*). Let us recall that the presence in the solution of Cl^- ions, which reduce MnO_4^- to $MnO(OH)_2$ or even to Mn^{2+}, strongly interferes. Less sensitive to small amounts of Cl^- is the reaction with $NaBiO_3$.

(2) If, after the preliminary tests, there is any doubt as to the presence of a ferric ion, confirmatory tests may be carried out. To do this, dissolve part of the precipitate in HCl and treat the resulting solution with $K_4[Fe(CN)_6]$ or NH_4SCN. But it should be remembered that the initial oxidation state of iron in the sample being analysed can no longer be established in this test.

(3) A confirmatory test for titanium with chromotropic acid or with H_2O_2 can be carried out in hydrochloric acid solution (see Procedure *2*) by binding Fe^{3+} with orthophosphoric acid into a complex (see Sec. 5.18, Reactions *5* and *6*). If however the previous separations have been carried out on heating, titanium cannot usually be detected here.

9. Detection of Ni^{2+}. The Ni^{2+} ion is detected in a separate portion of the solution which may contain complex nickel and cobalt ammine compounds. Treat 1-2 drops of the solution on a spot plate with 3-4 drops of an ammoniacal solution of dimethylglyoxime. The red colour shows the presence of Ni^{2+}.

The same reaction can be carried out by the spot-test technique on a substrate of Na_2HPO_4 (see Sec. 5.23).

If the characteristic colouration of Ni^{2+} is not produced, check the pH of the solution; if it is strongly alkaline (pH $\geqslant$ 10), adjust the pH by adding acetic acid (25-percent) so that its value remains within the limits of 5-9, and then repeat the reaction again.

Since the question as to the presence of Co^{2+} has already been solved, there is no need to carry out the test for Co^{2+} again.

10. Detection and Separation of CrO_4^{2-}. The yellow colour of the centrifugate is proof of the presence of CrO_4^{2-} and hence of Cr^{3+} in the original solution. Test this by the following reactions for CrO_4^{2-}.

(1) Touch a strip of filter paper with the tip of a capillary containing the centrifugate being tested and hold it in this position for a few seconds. Draw a capillary with a solution of benzidine around the moistened spot. The formation of a blue ring confirms the presence of chromium.

(2) To 2-3 drops of the centrifugate being tested add a drop of phenolphthalein solution and, drop by drop, 2*N* H_2SO_4 solution until the red colour disappears. To the colourless solution add 1-2 drops

more of acid, cool the solution (if it is warm), and try to detect CrO_4^{2-} by the action of H_2O_2 in the presence of amyl alcohol. The blue colouration of the alcohol layer due to perchromic acid, H_2CrO_6, confirms the presence* of CrO_4^{2-}.

If CrO_4^{2-} is present, it is to be removed by precipitation in the form of $BaCrO_4$ in acetic acid solution. Adjust the pH of the solution to 4-5 by adding acetic acid (30-percent), heat, and precipitate the chromate by a solution of $BaCl_2$. When precipitation is complete, centrifuge the $BaCrO_4$ precipitate (and, possibly, $BaSO_4$**) and discard. Examine the centrifugate which may possibly contain Al^{3+} and Zn^{2+} with excess Ba^{2+} as directed in Procedure *11*.

11. Detection of Al^{3+} and Zn^{2+}. Aluminium and zinc are detected in separate portions of the centrifugate obtained in Procedure *10* or *5* if the CrO_4^{2-} ion is absent. In the latter case, the solution must be acidified with 30-percent acetic acid up to pH 5. The solution is divided into two portions.

(1) *Detection of Al^{3+}*. To one portion of acetic acid solution add a few drops of a saturated solution of NH_4Cl (to prevent precipitation of zinc hydroxide which is soluble in solutions of ammonium salts) and as much of NH_4OH as required for the solution to become slightly alkaline (pH $\approx$ 9), and heat in a water bath for a few minutes. The formation of a white flocculent precipitate of $Al(OH)_3$ indicates the presence of Al^{3+}. When its amount is small, the flakes of $Al(OH)_3$ can be noticed only if the contents of the test tube is observed carefully; the flakes often float to the top.

When the precipitate is obtained, confirm the presence of Al^{3+} by means of tests with aluminon, alizarin, or morin. To do this, it is necessary to centrifuge off the precipitate, wash it, dissolve on heating in a few drops of CH_3COOH, and then proceed as directed in Sec. 5.16.

The success in the detection of Al^{3+} depends on how correctly the solution has been acidified with acetic acid and the precipitation effected by ammonia. If the amount of CH_3COOH added is not enough and the solution shows an alkaline reaction, aluminium will be present in the form of AlO_2^- ions which give no precipitate upon addition of ammonia.

On the other hand, if the amount of ammonia added is too small, the solution will give an acid reaction and no precipitate will be formed either. Too large an excess of ammonia should also be avoided since this favours the dissolution of the precipitate and formation of the aluminate. This means that, in order to avoid an error, it is necessary to

* If the reaction is not sufficiently distinct and the presence of chromium(IV) is doubtful, it can be carried out by the fractional method: a few drops of the original solution are taken and the Cr^{3+} ion is oxidized to $Cr_2O_7^{2-}$ by the action of $KMnO_4$ in a faintly acid solution (see Sec. 5.17).

** $BaSO_4$ may form at the expense of the SO_4^{2-} ions adsorbed from the original solution by the precipitated sulphides of Group III cations (if this precipitate has not been thoroughly washed).

TABLE 5.6. **Scheme of Analysis of Mixtures of Group III-I Cations (Hydrogen Peroxide Method)**

<table>
<tr><td colspan="5">1. Preliminary tests: (a) test for colour of the solution, odour, formation of a ppt., its colour; test the ppt. for solubility in HCl; determine solution pH; (b) look for NH_4^+; Fe^{3+} Fe^{2+}, Zn^{2+}, Ti^{IV} (in separate portions of solution).</td></tr>
<tr><td colspan="5">2. Treat the solution with the ppt. with HCl; if required (see page 359, Procedure *1.2*), postprecipitate Group II sulphates, separate $BaSO_4$, $SrSO_4$ ($CaSO_4$). Test the pptd. sulphates according to Procedure *3*.</td></tr>
<tr><td colspan="5">3. Precipitation of Group III cations by $(NH_4)_2S$ in the presence of NH_4OH and ammonium salts. Wash the ppt. with water containing NH_4NO_3.</td></tr>
<tr><td colspan="4">Precipitate: $Al_2O_3 \cdot xH_2O$, $Cr_2O_3 \cdot xH_2O$, $TiO_2 \cdot xH_2O$, Fe_2S_3, FeS, MnS, ZnS, CoS and NiS.</td><td>Centrifugate: Ba^{2+}, Sr^{2+}, Ca^{2+}, Mg^{2+}, K^+, Na^+(NH_4^+)</td></tr>
<tr><td colspan="4">5. Dissolve the ppt. in HNO_3 with heating. Remove the sulphur ppt.*.</td><td rowspan="7">4. Decompose excess $(NH_4)_2S$ by heating with CH_3COOH (Hood!). Remove the pptd. sulphur. Remove the major portion of ammonium salts by evaporation with concentrated HNO_3 and ignite the dry residue. Dissolve this residue in water acidified with HCl. Look for the cations of Groups II and I according to Table 5.1.</td></tr>
<tr><td colspan="4">Solution: Al^{3+}, Cr^{3+}, Ti^{IV}, Fe^{3+}, Mn^{2+}, Zn^{2+}, Co^{2+}, Ni^{2+}.</td></tr>
<tr><td colspan="4">6. Separate Al^{3+}, Cr^{3+} and Zn^{2+} from Fe^{3+}, Mn^{2+}, Co^{2+}, Ni^{2+} and Ti^{IV} by excess of 30-percent NaOH, 3-percent H_2O_2 with heating.</td></tr>
<tr><td colspan="2">Precipitate: $Fe_2O_3 \cdot xH_2O$, $MnO_2 \cdot xH_2O$, $Co_2O_3 \cdot xH_2O$, $Ni(OH)_2$ and $TiO_2 \cdot xH_2O$.</td><td colspan="2">Solution: CrO_4^{2-}, $[Al(OH)_4]^-$, $[Zn(OH)_4]^{2-}$.</td></tr>
<tr><td colspan="2">7. Dissolve the ppt. in a mixture of H_2SO_4 and KNO_2. Solution: Fe^{3+}, Mn^{2+}, Co^{2+}, Ni^{2+}, Ti^{IV}. Detection of Co^{2+} in a portion of the solution.</td><td colspan="2" rowspan="2">9. Look CrO_4^{2-} in a portion of the solution. If present, remove it from the whole solution by $BaCl_2$ in acetate buffered solution in the presence of Na_2SO_4.</td></tr>
<tr><td colspan="2">8. Separate Fe^{3+}, Mn^{2+} and Ti^{IV} from Co^{2+} and Ni^{2+} by the action of excess NH_4OH and 3-percent H_2O_2.</td></tr>
<tr><td>Precipitate: $Fe_2O_3 \cdot xH_2O$, $MnO_2 \cdot xH_2O$, $TiO_2 \cdot xH_2O$.
Look for Mn^{2+}; carry out confirmatory tests for iron and titanium.</td><td>Solution: $[Co(NH_3)_6]^{3+}$, $[Ni(NH_3)_6]^{2+}$.
Look for Ni^{2+}.</td><td>Precipitate: $BaCrO_4$ ($BaSO_4$).
Not to be examined.</td><td>Solution: Al^{3+}, Zn^{2+}
Look for Al^{3+} and Zn^{2+} in separate portions of the solution.</td></tr>
</table>

* The precipitate may be aged for 20-30 min under the mother liquor, then centrifuged off and treated with 2*N* HCl. All the precipitates, except NiS and CoS, dissolve. Nickel and cobalt sulphides are dissolved by the action of 3-percent H_2O_2 in the presence of CH_3COOH.

adhere strictly to all the directions concerning the value of pH for both the operations considered.

(2) *Detection of Zn^{2+}*. Heat the other portion of the acetic acid solution and pass H_2S into it. A white precipitate of ZnS soluble in HCl is proof of the presence of Zn^{2+}.

To check, centrifuge off the precipitate formed, dissolve in 3-4 drops of 2*N* HCl, and test the resulting solution after heating (to expel H_2S) for Zn^{2+} microcrystalloscopically with $(NH_4)_2[Hg(SCN)_4]$ or by the reaction of formation of mixed crystals of $Zn[Hg(SCN)_4]$ and $Co[Hg(SCN)_4]$ (see Sec. 5.21).

Since in the method in question the Zn^{2+} ion is considerably coprecipitated with $Fe(OH)_3$, $MnO(OH)_2$, $Co(OH)_3$, and $Ni(OH)_2$, the possibility is not excluded that, with Zn^{2+} detected in the preliminary tests, we will not be able to establish its presence in subsequent analysis. It is for this purpose that a preliminary test for Zn^{2+} is recommended (see Procedure *1.6*).

The scheme for analysis of mixtures of the cations of Groups III, II, and I using the hydrogen peroxide method is given in Table 5.6.

(b) The Ammonia Method

Analysis using the ammonia method is started in just the same way as the method using H_2O_2, i.e., in accordance with Procedures *1-4* in the description of the analysis. Further analysis is to be done as follows.

5. Detection of Co^{2+} and Mn^{2+}. In subsequent precipitation of the hydroxides of aluminium, chromium, iron(III), and titanium through the action of ammonia in the presence of NH_4Cl, the Co^{2+} and Mn^{2+} ions are rather heavily coprecipitated (Ni^{2+} also coprecipitates partially) and hence are distributed between the precipitate and the solution. It is reasonable therefore to detect them before the precipitation indicated. Identification tests are carried out in separate portions of the solution obtained after the precipitated sulphides and hydroxides of Group III cations have been dissolved in HNO_3 (see Procedure *4*, in the hydrogen peroxide method).

(1) The Co^{2+} ion is detected by the reactions employed in the previous method (see Procedure *6*), using 1-2 drops of the nitric acid solution for the purpose.

(2) To detect Mn^{2+}, dilute 1 drop of the solution with 10-15 drops of water, add 3-5 drops of 6*N* HNO_3, and introduce a little powdered $NaBiO_3$. Mix with a glass rod and separate excess $NaBiO_3$ by centrifuging. The crimson-violet or pink colouration of the solution (MnO_4^-) indicates the presence of Mn^{2+}.

Other reactions for Mn^{2+} can also be used, for example, the reactions with PbO_2 or with $(NH_4)_2S_2O_8$ (see Sec. 5.20, Procedure *5*). It should however be kept in mind that these reactions are even interfered with by very small amounts of Cl^- ions which reduce MnO_4^- to $MnO(OH)_2$ (a brown-coloured precipitate) or even to Mn^{2+}. Incidentally, in the precipitation of Group III cations, the Cl^- ions are adsorbed by the precipitate, and appreciable amounts of them (especially if the precipitate was not washed well) find their way into solution on dissolution of the precipitate. This interferes but very little with the detection of Mn^{2+} by means of the reaction with $NaBiO_3$, which is carried out in the cold. If it is desired to use the above reactions, one has to separate the Cl^- ion. For this purpose, carry out the precipitation in 1-2 drops of the solution by 2*N* NaOH on heating, centrifuge the precipitate, wash it two times, dissolve in 2-3 drops of 2*N* H_2SO_4, dilute with a few drops of water, and identify Mn^{2+} in the resulting dilute solution. Let us recall that the reactions with PbO_2 and $(NH_4)_2S_2O_8$ are effective only with very small amounts of Mn^{2+}.

The Cl^- ions can also be removed in the form of gaseous HCl by evaporating to dryness 1-2 drops of the test solution in a micro crucible with 3-4 drops of 2*N* H_2SO_4 or concentrated HNO_3. Dissolve the dry residue, after cooling, in a few drops of water and carry out the above reactions for Mn^{2+} with a drop of the solution.

6. Precipitation of Al^{3+}, Cr^{3+}, Fe^{3+}, and Ti^{IV}. To the nitric acid solution containing Group III cations (obtained in Procedure *4*, by the hydrogen peroxide method) add 8-10 drops of a saturated solution of NH_4Cl, heat, and precipitate Al^{3+}, Cr^{3+}, Fe^{3+}, and Ti^{IV} by concentrated (25-percent) solution of NH_4OH. Add ammonium hydroxide until the solution is alkaline, and then add 5 more drops of NH_4OH. Note the colour of the precipitate formed and draw conclusions. Centrifuge the precipitate* and wash once with water containing NH_4Cl.

In this operation, the hydroxides of Al^{3+}, Cr^{3+}, Fe^{3+}, and Ti^{IV} are precipitated. The other cations of Group III (whose hydroxides are soluble in excess ammonia or in ammonium salts) will be present in solution mainly in the form of Mn^{2+}, $[Zn(NH_3)_6]^{2+}$, $[Ni(NH_3)_6]^{2+}$, and $[Co(NH_3)_6]^{2+}$. But part of Mn^{2+} and Co^{2+} being in the form of the corresponding hydroxides may also be found to be present in the precipitate.** Test the precipitate as directed in Procedure *7* and the centrifugate as indicated in Procedure *9*.

7. Separation of Fe^{3+} and Ti^{IV} from Cr^{3+} and Al^{3+}. Treat the precipitate (see Procedure *6*) with 5-6 drops of 6*N* NaOH solution and heat. While heating, add (drop by drop) 5-6 drops of a 3-percent solu-

* If the centrifugate obtained is turbid, it is to be heated for a few minutes in the water bath and centrifuged again. After the centrifugate is separated, the precipitates are combined and washed. Alternatively, the turbidity is removed from the centrifugate by filtering the latter through a small filter paper in the apparatus shown in Fig. 2.17. Since the filter retains part of the solution, it should be washed with 5-6 drops of water; only after this is done may the test tube with the filtrate be withdrawn from the flask.

** It is for this reason that they have to be detected before the separation of ions into subgroups.

tion of H_2O_2, the contents of test tube being stirred for some time after each drop. Heat the precipitate for 2-3 min, centrifuge, and separate from the solution.

Aluminium hydroxide, $Al(OH)_3$, dissolves in alkali with the formation of the aluminate, AlO_2^-; $Cr(OH)_3$ forms CrO_4^{2-} ions with alkali in the presence of H_2O_2. Thus, the precipitate is found to contain $Fe(OH)_3$, $Ti(OH)_4$, and also the previously coprecipitated hydroxides of manganese(II) and cobalt(II), which are oxidized by H_2O_2 to $MnO(OH)_2$ and $Co(OH)_3$. Since all the ions present in the precipitate have already been identified, further tests may be omitted.* The centrifugate containing CrO_4^{2-}, AlO_2^-, and excess NaOH is to be analysed as directed in Procedure *8*.

8. Detection of Chromium and Aluminium Ions. These tests are carried out in separate portions of the solution obtained in Procedure 7.

(1) Detect chromium(VI) as indicated in Procedure *10* of the previously discussed hydrogen peroxide method.

(2) To detect aluminium, you need not separate CrO_4^{2-}; acidify the strongly alkaline test solution by $2N$ H_2SO_4 to pH $\geqslant 4$. Then add ammonia until the solution is weakly alkaline (pH ≈ 9) and heat. A white flocculent precipitate of $Al(OH)_3$ indicates the presence of Al^{3+}. If Al^{3+} is present, centrifuge the precipitate, wash, dissolve in acetic acid, and perform confirmatory tests for Al^{3+} with alizarin, aluminon, or morin (see Sec. 5.16).

9. Detection of Ni^{2+}. The Ni^{2+} ion is detected by the action of dimethylglyoxime in a separate portion of the ammoniacal solution which may contain nickel, cobalt, and zinc ammine complexes and also Mn^{2+}. The reaction is to be carried out as indicated in the description of the hydrogen peroxide method (Procedure *9*).

10. Confirmatory Tests for Zn^{2+}. If no Zn^{2+} has been detected by preliminary tests, the reactions for Zn^{2+} may be repeated, in which case the solution is to be acidified slightly with HCl. Then shake the resulting faintly acid solution in the cold with powdered $CaCO_3$ and proceed as directed in Procedure *1.6* of the hydrogen peroxide method.

The scheme of analysis of mixtures of the cations of Groups III, II, and I by the ammonia method is presented in Table 5.7.

(c) The Pyridine Method

Pyridine, C_5H_5N, is a weak organic base ($K_{base} = 1.5 \times 10^{-9}$). In a weakly acid solution, it protonates with the formation of $(C_5H_5NH)^+$, for example, $(C_5H_5NH)NO_3$ which when mixed with excess pyridine forms a buffered solution with pH ≈ 6.5. From Table 4.1 (see Sec. 4.3) it is seen that at this pH the ions Fe^{3+}, Al^{3+},

* The precipitate may be washed and dissolved in H_2SO_4 containing a drop or two of KNO_2 or H_2O_2 solution (to reduce Mn^{IV}), and the ions Fe^{3+}, Mn^{2+}, Co^{2+}, and Ti^{IV} can again be detected by the fractional method.

TABLE 5.7. **Scheme of Analysis of Mixtures of Group III-I Cations (Ammonia Method)**

<table>
<tr><td colspan="5">1. Preliminary tests: ( ) test for colour of the solution, odour, the presence of a ppt., its colour; test for solubility of the ppt. in HCl, determine solution pH; (b) look for NH_4^+, Fe^{3+}, Fe^{2+}, Zn^{2+}, Ti^{IV} (in separate portions of the solution).</td></tr>
<tr><td colspan="5">2. Treat the solution with the ppt. with HCl; if necessary (see page 359, Procedure *1.2*), postprecipitate Group II sulphates, separate $BaSO_4$ $SrSO_4$ ($CaSO_4$). Analyse the pptd. sulphates according to Procedure *3* in Table 4.3.</td></tr>
<tr><td colspan="5">3. Precipitation of Group III cations by $(NH_4)_2S$ in the presence of NH_4OH and ammonium salts. Wash the ppt. with water containing NH_4NO_3.</td></tr>
<tr><td colspan="4">Precipitate: $Al_2O_3 \cdot xH_2O$, $Cr_2O_3 \cdot xH_2O$, $TiO_2 \cdot xH_2O$, Fe_2S_3, FeS, MnS, ZnS, CoS and NiS.</td><td>Centrifugate: Ba^{2+}, Sr^{2+}, Ca^{2+}, Mg^{2+}, K^+, $Na^+(NH_4^+)$.</td></tr>
<tr><td colspan="4">5. Dissolve the ppt. in HNO_3 with heating. Remove the sulphur ppt.</td><td rowspan="6">4. Decomposition of excess $(NH_4)_2S$ by heating with CH_3COOH (Hood!). Remove the sulphur ppt. Remove the major part of ammonium salts by evaporation with concentrated HNO_3 and ignition of the dry residue. Dissolve this residue in water. Look for the cations of Groups II and I according to Table 4.3.</td></tr>
<tr><td colspan="4">Solution: Al^{3+}, Cr^{3+}, Ti^{IV}, Fe^{3+}, Mn^{2+}, Zn^{2+}, Co^{2+}, Ni^{2+}. Look for Co^{2+} and Mn^{2+} in separate portions of the solution.</td></tr>
<tr><td colspan="4">6. Precipitation of Al^{3+}, Cr^{3+}, Fe^{3+} and Ti^{IV} by excess NH_4OH in the presence of ammonium salts.</td></tr>
<tr><td colspan="2">Precipitate: $Fe_2O_3 \cdot xH_2O$, $Cr_2O_3 \cdot xH_2O$, $Al_2O_3 \cdot xH_2O$, $TiO_2 \cdot xH_2O$ and, partially, the hydroxides of manganese, cobalt and nickel.</td><td colspan="2">Centrifugate: Mn^{2+}, $[Zn(NH_3)_6]^{2+}$, $[Ni(NH_3)_6]^{2+}$, $[Co(NH_3)_6]^{2+}$. Look for Ni^{2+} in a portion of the solution.</td></tr>
<tr><td colspan="2">7. Separate Al^{3+} and Cr^{3+} from Fe^{3+} and Ti^{IV} (and also from small amounts of Mn^{2+}, Co^{2+} and Ni^{2+}) by excess of 6N NaOH and 3-percent H_2O_2.</td><td colspan="2">8. Precipitation of Zn^{2+} by H_2S in the presence of formate buffer.</td></tr>
<tr><td>Precipitate: $Fe_2O_3 \cdot xH_2O$, $TiO_2 \cdot xH_2O$, $[MnO_2 \cdot xH_2O$, $Co_2O_3 \cdot xH_2O$, $Ni(ON)_2]$. Not to be examined.</td><td>Centrifugate: $[Al(OH)_4]^-$ and CrO_4^{2-}.
(a) detection and separation of CrO_4^{2-};
(b) detection of Al^{3+}.</td><td>Precipitate: ZnS (CoS and NiS may partially precipitate). Dissolution in HCl. Confirmatory test for Zn^{2+}.</td><td>Centrifugate: Mn^{2+}, Co^{2+}, Ni^{2+}.
Not to be examined.</td></tr>
</table>

Cr^{3+}, and Ti^{IV} will be completely precipitated in the form of hydroxides. Conversely, Mn^{2+}, Zn^{2+}, Ni^{2+}, and Co^{2+} are not precipitated at that pH and remain in solution in the form of pyridine complexes of the composition $[Me(C_5H_5N)_2]^{2+}$, which can be detected by means of ordinary reactions.

The analysis is started just as in the hydrogen peroxide method (see Procedures *1-4*). Then proceed as follows.

5. Separation of Al^{3+}, Cr^{3+}, Fe^{3+}, and Ti^{IV} from Other Group III Cations. Dilute the centrifugate, obtained after the dissolution of the sulphides and hydroxides of Group III cations in HNO_3 (see Procedure *4*), with 15-20 drops of water and neutralize with a solution of Na_2CO_3 until a permanent turbidity appears. Dissolve this turbidity in 1-2 drops of 6*N* HNO_3. Heat the weakly acid solution obtained, add 8-10 drops of a saturated solution of NH_4NO_3, and precipitate Al^{3+}, Cr^{3+}, Fe^{3+}, and Ti^{IV} by adding 10-15 drops of a 20-percent solution of pyridine, C_5H_5N. Add pyridine, drop by drop, with constant stirring using a glass rod. If the precipitation is complete (make certain), separate the precipitate containing $Al(OH)_3$, $Cr(OH)_3$, $Fe(OH)_3$, and $Ti(OH)_4$ by centrifuging, without allowing the solution to cool. Note the colour of the precipitate and draw a conclusion as to the possible presence of individual ions in it. Wash the precipitate with a hot 3-percent solution of NH_4NO_3, adding 1-2 drops of pyridine, and test just as in the analysis using the ammonia method (see Procedures *7* and *8*).

Test the centrifugate which may contain $[Mn(C_5H_5N)_2]^{2+}$, $[Zn(C_5H_5N)_2]^{2+}$, $[Ni(C_5H_5N)_2]^{2+}$, and $[Co(C_5H_5N)_2]^{2+}$ as directed in Procedure *6*.

6. Detection of Zn^{2+}, Mn^{2+}, Co^{2+}, and Ni^{2+}. These ions are detected by the fractional method in separate portions of the centrifugate (see Procedure *5*).

(1) To detect Zn^{2+}, to 10 drops of the solution add 8-10 drops of formate buffer. Check the pH of the solution (it must be equal to about 2) and pass H_2S into the solution for 1-2 min. A white precipitate of ZnS indicates the presence of Zn^{2+}. If it is desired, centrifuge the precipitate, dissolve in HCl, remove H_2S by heating, and perform confirmatory tests (see Sec. 5.21, Reactions *9* and *11*).

(2) The Mn^{2+} ion is detected in just the same way as in the ammonia method (see Procedure *5.2*).

(3) The Co^{2+} ion is detected as follows. To 2 drops of the solution add 8 drops of a saturated solution of NH_4SCN and a few drops of amyl alcohol. The blue colouration of the alcohol layer on shaking and separation into layers indicates the presence of Co^{2+}.

(4) To detect Ni^{2+}, to 1-2 drops of the solution add 3-4 drops of the ammoniacal solution of dimethylglyoxime. A red colour proves the presence of Ni^{2+}. The reaction can also be carried out on a spot plate using Na_2HPO_4 as a substrate (see Sec. 5.23).

Chapter 6

Cations of Groups IV and V

As has already been shown (see Sec. 1.5), all cations of analytical Groups IV and V are precipitated by hydrogen sulphide from acid solution in the form of slightly soluble sulphides. They have many properties in common: the tendency to complex-formation, the ability to undergo oxidation-reduction reactions, etc. The cations of Group V, silver(I), lead(II), and mercury(I), are characterized by the low solubility of chlorides, in contrast to the other cations of both groups. Since the precipitation by hydrogen sulphide is effected from hydrochloric acid solution, the cations indicated are separated from the other cations in the form of slightly soluble chlorides, AgCl, Hg_2Cl_2, and $PbCl_2$, before hydrogen sulphide is passed into the solution.

Since the properties common to the cations of Groups IV and V are rather similar, the theoretical questions involved in the study of these properties are discussed jointly for both groups.

This chapter concludes the study of the theory of redox reactions. We shall consider the oxidation-reduction potentials and directions of redox reactions, the effect of the pH of the medium and of the reactant concentrations on the course of redox reactions, and also the equilibrium constants for redox reactions.

A. THEORETICAL PRINCIPLES

6.1. Oxidation-Reduction Potentials

In studying redox reactions, we constantly find that, while some oxidizing agents oxidize readily a given substance or ion, others cannot effect this reaction. For example, Cl_2 oxidizes Br^-:

$$Cl_2 + 2Br^- \rightarrow Br_2 + 2Cl^-$$

Iron(III) is incapable of oxidizing Br^- to Br_2 but it easily oxidizes I^-:

$$2I^- + 2Fe^{3+} \rightarrow I_2 + 2Fe^{2+}$$

These facts evidently show that oxidizing and reducing agents differ in "strength", i.e., in chemical activity. In the above examples, Cl_2 is a stronger oxidizing agent than Fe^{3+}; iodide ions capable of reducing

Fe^{3+} to Fe^{2+} act as a stronger reducing agent than bromide ions which cannot reduce Fe^{3+}, etc.

If it is taken into account that an oxidizing agent gains electrons, while a reducing agent loses them, it will be clear that the difference in the strength of various oxidizing or reducing agents is associated with unequal tendencies of their atoms or ions to gain or lose electrons. *The greater the tendency of an atom, ion, or molecule to lose electrons, the more powerful its action as a reducing agent; the greater the tendency of an atom, ion, or molecule to gain electrons, the more active it is as an oxidizing agent.*

A quantitative measure of the tendency of various atoms, ions, or molecules to lose or gain electrons is furnished by a study of the corresponding reactions taking place in a galvanic (or voltaic) cell (see Sec. 5.13). Indeed, the measured electromotive force (emf) of a cell, which is the difference in equilibrium potentials between its two constituent half-cells, serves as a measure of the ability of electrons to be transferred from atoms (ions or molecules) of the reducing agent to atoms (ions or molecules) of the oxidizing agent.

Half-cells that form a galvanic cell invariably contain not only an oxidizing agent (or a reducing agent) but also those substances that are formed from them. For example, in the galvanic cell (considered in Sec. 5.13) which produces an electric current as a result of the reaction

$$Sn^{2+} + 2Fe^{3+} \rightarrow Sn^{4+} + 2Fe^{2+}$$

one of the compartments contains not only Sn^{2+} ions but also the Sn^{4+} ions formed from them. These both constitute a so-called **oxidation-reduction (redox) couple** designated as Sn^{4+}/Sn^{2+}. Each half-cell of the given cell contains a corresponding redox couple: Sn^{4+}/Sn^{2+} and Fe^{3+}/Fe^{2+}.

In each redox couple, a distinction is made between the oxidized form produced by an element in a higher oxidation state (Sn^{4+} and Fe^{3+}) and the reduced form produced by an element in a lower oxidation state (Sn^{2+} and Fe^{2+}). The oxidized form of a redox couple is an oxidizing agent and the reduced form is a reducing agent.

By combining various redox couples, one can construct galvanic cells in which various oxidation-reduction reactions may take place. In order to obtain a quantitative characteristic of the strength of various oxidizing and reducing agents by measuring the emf's of these cells, it is necessary to combine individual half-cells with the same **standard** half-cell. The standard system used for this purpose is **the couple** $2H^+/H_2$ at a concentration (more exactly, activity) of H^+ ions equal to 1 g-ion/litre and a gaseous hydrogen pressure of 1 atm. It is called the **standard (normal) hydrogen electrode.** As shown in Fig. 6.1, it consists of a piece of platinum, coated electrolytically with finely divided platinum called platinum black, dipping into a solution of H_2SO_4 with a concentration of 1 mole/litre, placed in a beaker *1*.

Chemically pure hydrogen gas is continuously bubbled through a tube *4* into the solution over the platinum electrode and is adsorbed by the platinum black. Therefore, the electrode behaves so as if it were made of hydrogen.

The potential of the standard hydrogen electrode is arbitrarily assumed to be equal to zero (at any temperature), just as the temperature of melting ice is taken as zero for temperature measurements.

If we connect a standard hydrogen electrode to another couple, say, Fe^{3+}/Fe^{2+}, placed in a beaker *2* into which is dipped an indicator electrode—a smooth platinum plate, a galvanic cell is obtained (Fig. 6.1).

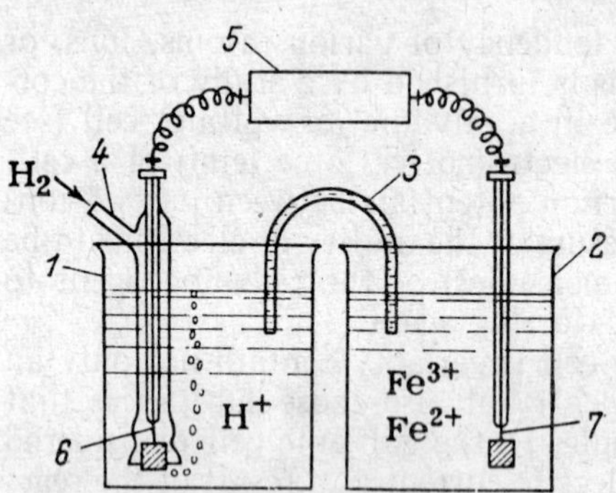

Fig. 6.1. Schematic diagram of apparatus for determination of the standard oxidation potential of the Fe^{3+}/Fe^{2+} couple:

1, 2 — beakers; *3* — salt bridge; *4* — tubing for passage of hydrogen; *5* — potentiometer; *6* — platinized platinum electrode; *7* — platinum electrode.

Beakers *1* and *2* are connected with the aid of an inverted U-shaped tube *3* (a salt bridge) filled with a solution of KCl. The electromotive force (emf) of the resulting cell can be measured with the aid of an instrument called a potentiometer. This emf characterizes the potential of the given couple in accordance with the arbitrary scale. It has been established experimentally that, at concentrations (activities) of Fe^{III} and Fe^{II} equal to unity, the potential E^0 is $+0.77$ volt.

This potential is known as the **standard oxidation-reduction potential** (or **oxidation potential**) of the given couple. These potentials are sometimes referred to as redox potentials.

In the arrangement shown in Fig. 6.1, the role of an indifferent electrode at which electrons are accepted (given up) from the oxidizing agent (by the reducing agent) is played by the platinum electrode, and the tube filled with a solution of KCl is a bridge permitting electrical contact between the two solutions.

The sign of the potential shows that the couple in the cell serves as the positive electrode (cathode) at which reduction takes place. The standard hydrogen electrode is the negative electrode (anode) at which oxidation occurs.

In conformity with this, since the electrons are delivered from the negative electrode and accepted at the positive electrode, the following electrochemical processes take place in the cell:

(1) in beaker *1*

$$H_2\uparrow - 2e \rightarrow 2H^+$$

(2) in beaker 2

$$2Fe^{3+} + 2e \rightarrow 2Fe^{2+}$$

The net equation for the reaction occurring in the complete cell is

$$2Fe^{3+} + 2H_2\uparrow \rightarrow 2H^+ + 2Fe^{2+}$$

If, instead of the couple Fe^{3+}/Fe^{2+}, we combine another couple*, say, $Cl_2\uparrow/2Cl^-$, with a standard hydrogen electrode (also at activities of Cl_2 and Cl^- equal to unity), we shall find that the standard oxidation potential is

$$E^0_{Cl_2\uparrow/2Cl^-} = +1.36 \text{ volt}$$

This means that the reaction taking place in the cell is

$$\overset{2e}{H_2\uparrow + Cl_2\uparrow} \rightarrow 2H^+ + 2Cl^-$$

which, just as in the case considered above, consists in removing electrons from H_2, the only difference being that the oxidizing agent, which takes these electrons, is gaseous Cl_2 and not Fe^{3+} ions. But since in this case the cell emf (1.36 volts) is considerably higher than 0.77 volt we may conclude that Cl_2 has a greater tendency to gain electrons, i.e., is a more powerful oxidizing agent than Fe^{3+}.

But these figures, however, show not only the power of oxidizing agents making up the given couples. They also provide information concerning the relative power of the corresponding reducing agents, i.e., Fe^{2+} and Cl^-.

Indeed, if we say, for example, that Cl_2 is a powerful oxidizing agent, it means that its atoms possess a strongly pronounced tendency to gain electrons, i.e., to be converted into Cl^- ions. But in this case Cl^- ions must evidently hold electrons firmly, i.e., be a very weak reducing agent. Conversely, the Fe^{2+} ions, resulting from the acception of electrons by Fe^{3+} ions, must hold electrons less firmly, i.e., be a more powerful reducing agent than Cl^-, since the standard oxidation potential of the Fe^{3+}/Fe^{2+} couple is lower than that of the $Cl_2\uparrow/2Cl^-$ couple.

Thus, the oxidizing and reducing activity of the oxidants and reductants, of which the couples are composed, can be judged from the values of the standard oxidation potentials of these couples measured relative to the standard hydrogen electrode.

* This half-cell is made up in the same way as the standard hydrogen electrode, except that instead of H_2SO_4 use is made of a solution of HCl in which the Cl^- concentration (more precisely, activity) is equal to 1 g-ion/litre, and gaseous Cl_2 is passed instead of H_2. To prevent the chlorine from interacting with the platinum electrode, the gaseous chlorine pressure must be much lower than 1 atm. The emf found in this way is recalculated for a pressure of 1 atm with the aid of a special formula.

The greater the standard oxidation potential of a given couple, the more powerful an oxidizing agent is the corresponding oxidized form and the weaker a reducing agent is the reduced form.

Obviously, the oxidized form of the $2H^+/H_2\uparrow$ couple, i.e., H^+ ions, is an even weaker oxidizing agent than Fe^{3+} ions, and the molecular H_2 is a more active reducing agent than Fe^{2+} ions. Since for the $Zn^{2+}/Zn\downarrow$ couple the standard oxidation potential is equal to*

$$E^0_{Zn^{2+}/Zn\downarrow} = -0.76 \text{ volt}$$

it follows that Zn^{2+} ions are an even weaker oxidizing agent and metallic zinc an even stronger reducing agent than the corresponding components of the $2H^+/H_2\uparrow$ couple, etc.

The equilibrium potentials of redox couples depend not only on the nature of the oxidizing and reducing agents involved, but also on their concentration. The effect of the concentration of an oxidizing and a reducing agent is not difficult to predict. It is clear, for example, that the oxidizing activity of Fe^{3+} ions must be the higher the greater their concentration in solution and the lower the concentration of Fe^{2+} ions which are the product of the corresponding reaction and which therefore favour the reverse process. Hence, the oxidation-reduction potential of a given couple must also increase with increasing ratio $[Fe^{3+}]/[Fe^{2+}]$ or, in a general case, with increasing ratio of the concentrations of the oxidized and reduced forms.

The quantitative relationship** between the value of the equilibrium oxidation-reduction potential (E) in volts and the indicated ratio of the concentrations of the components of a redox couple is defined by the Nernst equation

$$E = E^0 + \frac{RT}{nF} \ln \frac{[Ox]}{[Red]} \tag{6.1}$$

where E^0 = standard oxidation potential of a given couple, volts
[Ox] = concentration (more precisely, activity) of oxidized form
[Red] = concentration (activity) of reduced form
R = gas constant, 8.314 joule/deg·mole
T = absolute temperature
F = Faraday's number equal to 96,500 coulombs
n = number of electrons gained (or lost) on conversion of oxidized form (or reduced form).

* The negative sign of this potential signifies that, when the given couple is combined with a standard hydrogen electrode, it plays the part of the negative pole of the cell, while the couple $2H^+/H_2\uparrow$ is the positive pole. Therefore the following reaction takes place in the cell:

$$\overset{2e}{\overbrace{Zn\downarrow + 2H^+}} \rightarrow Zn^{2+} + H_2\uparrow$$

** The theory of oxidation-reduction processes is discussed in greater detail in· Alexeyev, V. N., *Quantitative Analysis*, Mir Publishers, Moscow, 1972.

As has already been said, the standard oxidation potential for a half-cell is the potential which the couple has on condition that [Ox] = = [Red] = 1. Indeed, in this case, ln ([Ox]/[Red]) = 0 and in Eq. (6.1) $E = E^0$.

If we substitute the numerical values of the constants and pass from natural to common logarithms, we shall have at 25°C:

$$E = E^0 + \frac{0.059}{n} \log \frac{[Ox]}{[Red]} \tag{6.2}$$

For the Fe^{3+}/Fe^{2+} couple, the equilibrium oxidation potential can be calculated from the formula

$$E_{Fe^{3+}/Fe^{2+}} = 0.77 + \frac{0.059}{1} \log \frac{[Ox]}{[Red]}$$

If, for example, $[Fe^{3+}] = 1$ g-ion/litre and $[Fe^{2+}] = 0.0001$ g-ion/litre, then

$$E_{Fe^{3+}/Fe^{2+}} = 0.77 + \frac{0.059}{1} \log 10^4 = 1.006 \text{ volt}$$

If the equation for the reaction involved in the conversion of the oxidized into the reduced form contains coefficients not equal to unity, they must enter into the Nernst equation as the exponents for the corresponding concentrations. For example, for the $Br_2/2Br^-$ half-cell,

$$E = E^0 + \frac{0.059}{2} \log \frac{[Br_2]}{[Br^-]^2}$$

In the case of half-cells such as $Zn^{2+}/Zn\downarrow$, where the reduced form is a metal, its concentration is a constant value and therefore, is not included in the expression for E. Hence,

$$E_{Zn^{2+}/Zn\downarrow} = E^0 + \frac{0.059}{2} \log [Zn^{2+}]$$

Obviously, the value of E^0 (—0.76 volt) is equal to the value of the equilibrium potential if $[Zn^{2+}] = 1$ g-ion/litre since it is only in this case that $\log [Zn^{2+}] = 0$ and, hence, $E = E^0$.

So far we have considered the application of the Nernst equation to the oxidation-reduction reactions in which the oxidizing or reducing agents are simple ions. Such ions change only their charge, their composition remaining unaltered. Not infrequently the reactions involve complex ions acting as the oxidizing or reducing agents. As an example, $[Fe(CN)_6]^{3-}$ ions are reduced to $[Fe(CN)_6]^{4-}$ ions, and MnO_4^- ions can be reduced to MnO_4^{2-} ions. Complex ions very often undergo radical changes in their composition during the oxidation-reduction reaction. The following changes taking place upon oxidation in acid solution serve as examples: $MnO_4^- \rightarrow Mn^{2+}$; $Cr_2O_7^{2-} \rightarrow Cr^{3+}$; $AsO_4^{3-} \rightarrow AsO_3^{3-}$; $SO_4^{2-} \rightarrow SO_3^{2-}$, free sulphur, or even H_2S. These transformations may be represented by the equations

$$MnO_4^- + 8H^+ + 5e \rightarrow Mn^{2+} + 4H_2O$$
$$Cr_2O_7^{2-} + 14H^+ + 6e \rightarrow 2Cr^{3+} + 7H_2O$$
$$AsO_4^{3-} + 2H^+ + 2e \rightarrow AsO_3^{3-} + H_2O$$
$$SO_4^{2-} + 2H^+ + 2e \rightarrow SO_3^{2-} + H_2O$$
$$SO_4^{2-} + 8H^+ + 6e \rightarrow S + 4H_2O$$
$$SO_4^{2-} + 10H^+ + 8e \rightarrow H_2S + 4H_2O, \text{ etc.}$$

From the equations given we can see that the deciding factor in the rebuilding of complex ions in the course of their reduction or oxidation is hydrogen ions which bind the oxygen atoms of a complex ion into water molecules. It is clear that in such cases the occurrence of oxidation-reduction processes and, hence, the value of the oxidation potential for the corresponding couple, must depend on the concentration of hydrogen ions in solution.

The hydrogen ion concentration is included in the numerator of the fraction after the logarithm sign in the Nernst equation; it is raised to a power equal to the coefficient for H^+ in the reaction equation. Thus, in the case of the MnO_4^-/Mn^{2+} couple, we have

$$E_{MnO_4^-/Mn^{2+}} = E^0 + \frac{0.059}{5} \text{ og } \frac{[MnO_4^-][H^+]^8}{[Mn^{2+}]}$$

Thus, the change of the hydrogen ion concentration has a great effect on the value of the oxidation potential for the given couple and hence on the oxidizing power of MnO_4^-; this power increases rapidly with increasing concentration of hydrogen ions, i.e., with decreasing pH of the solution.

The equation given above shows that the given couple has a standard oxidation potential at $[MnO_4^-] = [Mn^{2+}] = 1$ g-ion/litre and $[H^+] = 1$ g-ion/litre since it is only under these conditions that $E = E^0$.

6.2. Direction of Oxidation-Reduction Reactions

The values of the standard oxidation potentials, E^0, which are included in the Nernst equation, have been measured for a large number of various redox couples; the values for the most commonly used redox couples are listed in Table 6.1. The first and third columns of this table give the formulas of the individual components of various couples, which are arranged in the order of increasing standard oxidation potentials (the fourth column). The second column indicates the number of electrons, n, gained during the conversion of oxidizing agents (the first column) into the corresponding reducing agents (the third column).

Since the power of oxidizing agents increases and that of reducing agents decreases with increasing standard oxidation potentials, the

TABLE 6.1. **Standard Oxidation-Reduction Potentials for Some Redox Couples (at 25°C)**

Symbols: (*g*)—gas; (*l*)—liquid; (*s*)—solid

Oxidized form (oxidant)	Number of electrons n	Reduced form (reductant)	Standard oxidation potential E^0, volts
Zn^{2+}	2	Zn (*s*)	−0.7628
Fe^{2+}	2	Fe (*s*)	−0.440
Sn^{2+}	2	Sn (*s*)	−0.140
Pb^{2+}	2	Pb (*s*)	−0.126
Fe^{3+}	3	Fe (*s*)	−0.04
$2H^+$	2	H_2 (*g*)	0.00
S (*s*) + $2H^+$	2	H_2S (*g*)	+0.14
Sn^{4+}	2	Sn^{2+}	+0.15
$SO_4^{2-} + 4H^+$	2	$H_2SO_3 + H_2O$	+0.17
Cu^{2+}	2	Cu (*s*)	+0.337
I_2 (*s*)	2	$2I^-$	+0.536
Fe^{3+}	1	Fe^{2+}	+0.771
$NO_3^- + 4H^+$	3	NO (*g*) + $2H_2O$	+0.96
$HNO_2 + H^+$	1	NO (*g*) + H_2O	+0.99
Br_2 (*l*)	2	$2Br^-$	+1.087
Cl_2 (*g*)	2	$2Cl^-$	+1.359
$MnO_4^- + 8H^+$	5	$Mn^{2+} + 4H_2O$	+1.51
F_2 (*g*)	2	$2F^-$	+2.87

most powerful oxidizing agents are given at the bottom of the first column, and the most powerful reducing agents are given at the beginning of the third column.

For example, the strongest oxidizing agent is F_2 to which there corresponds the highest value of E^0 (+2.87 volts). Other very powerful oxidizing agents include ozone, O_3 (+2.07 volts), ammonium persulphate, $(NH_4)_2S_2O_8$ (+2.0 volts), sodium bismuthate, $NaBiO_3$ (> +1.8 volts), hydrogen peroxide, H_2O_2 (+1.77 volts), potassium permanganate, $KMnO_4$, in acid solution (+1.51 volts), chlorine, Cl_2 (+1.36 volts), the ions Au^+, Au^{3+}, Ag^+, etc.

The most powerful reducing agents are the alkali and alkaline-earth metals and also Mg, Al, Zn, and others.

Among the ions, the S^{2-} ion is an active reducing agent; the ions SO_3^{2-}, I^-, Fe^{2+}, etc., are less active. No reducing properties are exhibited by F^- ions. As known, electrons can be removed from the F^- ion only by electrolysis.

Knowing the relative strength of various oxidizing and reducing agents, it is possible to predict the direction of the reactions involved. Suppose we are to predict the direction of the reaction

$$2Fe^{3+} + Sn^{2+} \rightleftharpoons Sn^{4+} + 2Fe^{2+}$$

that is, to establish whether it will result in the oxidation of Sn^{2+} ions by Fe^{3+} to Sn^{4+} or will proceed in the reverse direction.

The standard oxidation potential of the Fe^{3+}/Fe^{2+} couple (+0.77 volt) is greater than that of the couple Sn^{4+}/Sn^{2+} (+0.15 volt). Therefore, if a galvanic cell is built up of these couples, the latter will be found to be the negative pole and the former, the positive pole. This means that the following processes will take place in the cell:

$$Sn^{2+} - 2e \rightarrow Sn^{4+}$$

$$2Fe^{3+} + 2e \rightarrow 2Fe^{2+}$$

and the reaction will proceed in the direction of oxidation of Sn^{2+} to Sn^{4+}:

$$Sn^{2+} + 2Fe^{3+} \rightarrow Sn^{4+} + 2Fe^{2+}$$

This reaction will also proceed in the same direction outside the cell on mixing the solutions of the corresponding substances.

It is not difficult to see that any oxidation-reduction process involves the formation of a new reducing and a new oxidizing agent from the oxidant and reductant taken for the reaction:

$$Ox_1 + Red_2 \rightarrow Red_1 + Ox_2$$

the products being weaker than the reactants. In the above example, according to the values of the standard oxidation potentials of the couples Sn^{4+}/Sn^{2+} and Fe^{3+}/Fe^{2+}; the oxidant (Sn^{4+}) and the reductant (Fe^{2+}) are weaker than the reactants (Fe^{3+} and Sn^{2+}). The same also applies to all other oxidation-reduction reactions.

It should also be noted that each oxidizing agent can oxidize only substances above it in the series (in the third column of Table 6.1), those with lower standard potentials, and, conversely, each reducing agent can reduce only substances that lie below it in the series (in the first column), those with higher standard potentials.

Attention should be called to the obvious analogy between oxidation-reduction processes and protolytic reactions (Sec. 3.7).

(1) Oxidation-reduction reactions are accompanied by a transfer of electrons, and protolytic reactions involve a transfer of protons.

(2) Redox reactions give rise to new reducing and oxidizing agents formed from those taken for the reaction; in protolytic reactions, new acids and bases are obtained from the acids and bases used as reactants. This is clearly seen from the following schemes:

$$Ox_1 + Red_2 \rightarrow Ox_2 + Red_1$$

$$Acid_1 + Base_2 \rightarrow Acid_2 + Base_1$$

(3) The reducing agent is the stronger the more readily it donates electrons; the oxidizing agent is the stronger the greater its tendency to accept electrons. Likewise, according to the protolytic theory, the acid is the stronger the more easily it gives up protons, and the base is the stronger the greater its tendency to accept protons.

(4) A strong oxidizing agent gains electrons and is converted into a weak reducing agent, and vice versa. Likewise, a strong base accepts protons and is thus converted into a weak acid, and a weak acid gives up protons, being converted into a strong base.

Let us consider some examples to illustrate the above rule.

Example 1. From Table 6.1 it is seen that metallic zinc ($E^0 \approx -0.76$ volt) is a stronger reducing agent than Pb ($E^0 \approx -0.13$ volt) or Cu ($E^0 \approx +0.34$ volt). Hence, when metallic zinc is placed into solutions of lead or copper salts, the following reactions take place:

$$Pb^{2+} + Zn \rightarrow Pb\downarrow + Zn^{2+}$$

$$Cu^{2+} + Zn \rightarrow Cu\downarrow + Zn^{2+}$$

since in these reactions the action of a stronger reducing agent (Zn) gives rise to a weaker reducing agent (Pb or Cu). Conversely, Pb and Cu cannot reduce Zn^{2+} to Zn.

If all the metals are arranged in the order of increasing standard oxidation potentials of the corresponding couples, we will obtain what is known as the electromotive or electrochemical series of metals: Li, K, Ba, Ca, Na, Sr, Mg, Al, Mn, Zn, Fe Cd, Co, Ni, Sn, Pb, H, Cu, Sb, Bi, Hg, Ag, Pt, Au.

The farther is the metal to the left in this series the more stronger it is as a reducing agent and the greater is its chemical activity. As a rule, none of the metals standing on the right are capable of displacing (i.e., reducing) the ion of a metal on the left from its salts in solution, and the metals standing on the left are capable of displacing the ions of the succeeding metals from their salts in solution. All the metals preceding hydrogen in the series are capable of liberating hydrogen from acid solutions and therefore, as a rule, they are soluble in acids. The metals standing to the right from hydrogen do not dissolve in dilute acids (with the exception of nitric acid which dissolves almost all the metals because the NO_3^- ion has a stronger oxidizing power than H^+).

Example 2. Let us consider the action of various oxidizing agents on I^-, Br^-, and Cl^-, this being very important for their identification in analysis. It has already been pointed out that Br^- can be oxidized to Br_2 by free Cl_2 and not by Fe^{3+}. This is easily understood if we consider the positions of Cl_2, Br_2, and Fe^{3+} in the table of standard oxidation potentials. Indeed, Br_2 is weaker as an oxidizing agent than Cl_2 but stronger than Fe^{3+}. Hence, the reaction

$$Cl_2 + 2Br^- \rightarrow 2Cl^- + Br_2$$

satisfies the above rule. Conversely, the reaction

$$2Fe^{3+} + 2Br^- \rightarrow Br_2 + 2Fe^{2+}$$

would not obey the rule because it would produce a stronger oxidizing agent (Br_2) and a stronger reducing agent (Fe^{2+}) from weaker ones

(Fe^{3+} and Br^-). The reverse reaction

$$2Fe^{2+} + Br_2 \rightarrow 2Fe^{3+} + 2Br^-$$

should proceed easily, which is found to be the case.

Using the table of standard oxidation potentials, one may conclude that nitrous acid is capable of oxidizing I^- but cannot oxidize Br^-.

In contrast to nitrous acid, the permanganate ion in acid solution ($E^0 = +1.51$ volts) must evidently oxidize not only I^- but also Br^- and Cl^-. It is clear that Br_2 can oxidize I^- and not Cl^-, and Cl_2 is capable of oxidizing both I^- and Br^-.

Example 3. In analysis it often happens that the test solution contains not one but two or more reducing agents (for example, Br^- and I^-) which can react with the oxidizing agent added (e.g., a solution of Cl_2). There arises a question: In what sequence will their oxidation proceed?

Referring to Table 6.1, one can easily see that Br^- ions can start to be oxidized only after the I^- ions have been oxidized.

Indeed, if Br^- ions were oxidized to any degree by chlorine, Br_2 molecules would appear in solution. According to the position of the $Br_2/2Br^-$ couple in the table of standard oxidation potentials, bromine molecules would immediately oxidize I^- to I_2, being themselves reduced back to Br^-. This process will continue until all the I^- ions are oxidized to I_2 (irrespective of whether the oxidation is effected by the direct action of chlorine or by the action of the free bromine formed).

Similarly, we find that, in those cases where, in addition to Br^- and I^-, the solution also contains other, stronger reducing agents (e.g., S^{2-}, SO_3^{2-}, $S_2O_3^{2-}$, etc.), the oxidation of I^- cannot start until all the reducing agents involved are practically completely oxidized. This circumstance has to be taken into account when I^- and Br^- are to be detected.

From all the foregoing it follows that any oxidizing agent first attacks the strongest reducing agent present in solution. The action of any reducing agent is primarily directed to the most powerful oxidizing agent.

In other words, *of all the oxidation-reduction processes possible under given conditions the first to occur is the one corresponding to the largest difference of standard oxidation potentials.**

If, for example, a bar of Fe is placed into a solution of $SnCl_4$, the following reaction takes place:

$$Sn^{4+} + Fe \rightarrow Sn^{2+} + Fe^{2+}$$

rather than the reduction of Sn^{4+} to metallic tin since to this reaction there corresponds a smaller potential difference

$$+0.01 - (-0.44) = +0.45 \text{ volt}$$

* The effect of the values of [Ox] and [Red], the hydrogen ion concentration, and also of the kinetic factor, on the direction of an oxidation-reduction reaction will be considered in Sections 6.3 and 6.4.

than to the reduction to Sn^{2+}:

$$+0.15-(-0.44)=+0.59 \text{ volt}$$

Example 4. And, finally, suppose metallic iron is to be dissolved in dilute HNO_3 and we are to predict the equation for the reaction. How could this be done with the aid of the table of standard oxidation potentials?

First we note that there are two oxidizing agents, H^+ and NO_3^-. But NO_3^- stands well below H^+ in the series and has therefore a higher oxidizing activity. Hence, in accordance with the rule given above, NO_3^- (and not H^+) will function as an oxidizing agent in the reaction.

Further, there is another question, namely, to what oxidation state of iron the reaction will proceed. It is not difficult to see that, although to the formation of Fe^{2+} there corresponds a larger potential difference (1.40 volts) than to the formation of Fe^{3+} ions (1.00 volt)*, there will nevertheless be obtained Fe^{3+} in excess HNO_3. Indeed, if Fe^{2+} ions were obtained, then, according to the position in the table, they would immediately be oxidized to Fe^{3+} by the action of excess nitric acid. Thus, the reaction in question may be described by the equation

$$Fe\downarrow+4H^++NO_3^-\rightarrow Fe^{3+}+NO\uparrow+2H_2O$$

6.3. Effect of Concentrations of Redox-Couple Components and Solution pH on Oxidation Potential

The examples considered in the preceding section show that the use of a table of standard oxidation potentials allows one to determine the direction of oxidation-reduction reactions, to select the most suitable oxidizing and reducing agents, and also to solve a number of other important problems.

It should however be remembered that the potentials given in the table are standard potentials, i.e., they refer to definite concentrations ($[Ox]=[Red]=1$) and change with change of the concentrations. As a result, the direction of the reaction may be completely reversed.

Let us consider an example. Judging by the values of E^0 for the couples $Sn^{2+}/Sn\downarrow$ (—0.14 volt) and $Pb^{2+}/Pb\downarrow$ (—0.13 volt), metallic tin must reduce Pb^{2+}:

$$Pb^{2+}+Sn\downarrow\rightarrow Pb\downarrow+Sn^{2+}$$

If the concentration of Pb^{2+} ions is reduced to 0.1 g-ion/litre, then it can easily be calculated from the Nernst equation that the equilibrium oxidation potential of the $Pb^{2+}/Pb\downarrow$ couple will fall to

---

* $E^0_{NO_3^-/NO\uparrow}=+0.96$ volt $E^0_{Fe^{3+}/Fe\downarrow}=-0.04$ volt

$E^0_{Fe^{2+}/Fe\downarrow}=-0.44$ volt $E=+0.96-(-0.04)=1.00$ volt.

$E=+0.96-(-0.44)=1.40$ volts

—0.159 volt which is lower than the value for the $Sn^{2+}/Sn\downarrow$ couple (—0.14 volt). Therefore, at the stated concentrations of Sn^{2+} (1 g-ion/litre) and Pb^{2+} (0.1 g-ion/litre), the reaction is reversed:

$$Pb\downarrow + Sn^{2+} \rightarrow Sn\downarrow + Pb^{2+}$$

Evidently, the reason why the direction of the reaction is changed by such a relatively small change in the concentration of one of the ions is that the difference between the standard oxidation potentials of the two couples is small (0.01 volt). In the reaction

$$2Ag^{+} + H_2\uparrow \rightarrow 2H^{+} + 2Ag\downarrow$$

the potential difference is 0.80 volt, and therefore the direction of the reaction will be reversed only when the concentration of Ag^{+} is very strongly decreased. Such a decrease may take place if the reaction results in the formation of a very slightly soluble salt or a slightly dissociated complex of silver. Suppose, for example, that metallic silver is placed into a 1*M* solution of hydriodic acid. Since the solubility product constant of AgI is equal to about 10^{-16} and the concentration of I^{-} in solution is 1 g-ion/litre, the concentration of Ag^{+} does not exceed 10^{-16} g-ion/litre.

At this value of $[Ag^{+}]$, the real potential of the $Ag^{+}/Ag\downarrow$ couple is

$$E_{Ag^{+}/Ag\downarrow} = 0.80 + 0.058 \log 10^{-16} = -0.13 \text{ volt}$$

The negative value of the potential of the $Ag^{+}/Ag\downarrow$ couple indicates that under the given conditions the above reaction must proceed from the right to the left. Thus, metallic silver is capable of displacing hydrogen from solutions of hydriodic (and also of hydrosulphuric) acid (i.e., capable of dissolving with the evolution of gaseous hydrogen) although it stands on the right of hydrogen in the electromotive series and cannot therefore displace it from solutions of other acids. There are many examples which illustrate the effect of the formation of precipitates (or complexes) on the direction of oxidation-reduction reactions.

It has been stated above that, if hydrogen ions take part in the reaction, their concentration has also a strong effect on the oxidation potential and a change in the H^{+} concentration may bring about a change in the direction of the reaction.

For example, from the values of E^0 for the couples $H_3AsO_4/HAsO_2$ (+0.56 volt) and $I_2\downarrow/2I^{-}$ (or $I_3^{-}/3I^{-}$) (+0.536 volt), we may conclude that AsO_4^{3+} ions must oxidize I^{-} ions to I_2:

$$H_3AsO_4 + 2I^{-} + 2H^{+} \rightarrow HAsO_2 + I_2 + 2H_2O \quad (6.3)$$

At equal concentrations of H_3AsO_4 and $HAsO_2$ and at $[H^{+}] = 1$ (or higher), this is what actually takes place.

But if the reaction is conducted in the presence of excess $NaHCO_3$, whose solutions have pH ≈ 8, the decrease of the H^{+} concentration to

10^{-8} g-ion/litre caused by the addition of $NaHCO_3$ does not affect the potential of the $I_2/2I^-$ couple. Conversely, in the case of the couple $H_3AsO_4/HAsO_2$, where hydrogen ions take part in the conversion of the oxidized to the reduced form, the equilibrium potential will fall, according to Eq. (6.3), to the value

$$E_{H_3AsO_4/HAsO_2} = 0.56 + \frac{0.059}{2} \log \frac{[H_3AsO_4][H^+]^2}{[HAsO_2]}$$

At $[H_3AsO_4] = [HAsO_2]$ and $[H^+] = 10^{-8}$, we have

$$E_{H_3AsO_4/HAsO_2} = 0.56 + \frac{0.059}{2} \log (10^{-8})^2 = 0.068 \text{ volt}$$

Since $0.088 < 0.54$, reaction (6.3) will proceed in the reverse direction.

From the example just considered it follows that if hydrogen ions are consumed in the reaction, it must be carried out in acid solution; but if H^+ ions are formed as a result of the reaction, they must be bound by adding alkali or substances, such as $NaHCO_3$, to the solution.

Since the concentrations of H^+ and OH^- are interrelated by the equation

$$[H^+][OH^-] = K_w$$

the consumption of OH^- in the reaction is equivalent to the formation of H^+, and vice versa.

All these conclusions can be obtained on the basis of the law of mass action.

As has been pointed out above, the consumption of H^+ is usually observed in oxidation reactions brought about by the anions of oxygen-containing acids, whose oxygen atoms are bound by hydrogen ions into water molecules. Therefore, the acidity of the solution plays a very important part in these reactions.

Since the values of E^0 for redox couples, in which the oxidized components are reduced with the aid of H^+, are valid only at $[H^+] = 1$, the direction of an oxidation-reduction reaction can be determined by comparing the standard oxidation potentials of the two couples participating in the reaction only if $[H^+] = 1$. With other values of the hydrogen ion concentration, the potential of individual half-cells may be changed so strongly that the reaction will be reversed.

This can be proved by the following experiments.

1. Mix solutions of KI and KNO_2 in a test tube. Although the standard oxidation potential of the half-cell $HNO_2/NO\uparrow$ (0.99 volt) is large as compared with that of the half-cell $I_2\downarrow/2I^-$ (0.54 volt), no oxidation of I^- to I_2 occurs. But as soon as a little H_2SO_4, HCl, or even CH_3COOH is added to the solution, a reaction begins, accompanied by the evolution of NO and the formation of a dark-brown precipitate of iodine:

$$2I^- + 2NO_2^- + 4H^+ \rightarrow I_2\downarrow + 2NO\uparrow + 2H_2O$$

Similarly, in alkaline or neutral solution, I^- ions do not react with most of the other oxidizing anions, such as CrO_4^{2-}, IO_3^-, BrO_3^-, ClO_3^-, AsO_4^{3-}, and when the solutions are acidified, I^- ions are oxidized to I_2.

2. When solutions of K_2CrO_4 and Na_2S are mixed, no changes take place. If the resulting mixture is acidified, the following reaction occurs:

$$3S^{2-} + 2CrO_4^{2-} + 16H^+ \rightarrow 3S\downarrow + 2Cr^{3+} + 8H_2O$$

In the same way, other reducing anions (Cl^-, Br^-, I^-, SO_3^{2-}, $S_2O_3^{2-}$, SCN^-, $C_2O_4^{2-}$, AsO_3^{3-}, etc.) are oxidized by CrO_4^{2-} ions only in acid solution.

In contrast to NO_3^-, CrO_4^{2-} and many other oxidizing agents*, MnO_4^- can oxidize not only in acid but also in alkaline solution. The reduction products however are different in the two cases. In the oxidation reaction in acid solution, the MnO_4^- ion gains 5 electrons and is reduced to Mn^{2+}. In alkaline (or neutral) solution, this anion gains only 3 electrons and is reduced to MnO_2, or more precisely, to manganous acid, $MnO(OH)_2$, which separates out as a brown precipitate. Accordingly, there is a large difference between the oxidizing activities of permanganate in acid and alkaline solutions, which can be seen from a comparison of the standard potentials:

$$MnO_4^-/Mn^{2+}\ (E^0 = +1.51 \text{ volts}) \text{ and } MnO_4^-/MnO_2\ (E^0 = +0.60 \text{ volt})$$

The difference in the reduction of MnO_4^- in acid and alkaline solutions becomes clear if we consider the equation for the reduction of MnO_2 to Mn^{2+}:

$$MnO_2\downarrow + 4H^+ + 2e \rightarrow Mn^{2+} + 2H_2O$$

The equation shows that in acid solution the more stable form is Mn^{2+} (the lowest oxidation state of manganese). If MnO_2 was first formed in the reaction, then because of the higher concentration of hydrogen ions it would be reduced to Mn^{2+}. Conversely, in alkaline or neutral solution the equilibrium under consideration is found to be shifted towards the formation of MnO_2, or more precisely, of $MnO(OH)_2$, i.e., of the hydrated compound $MnO_2 \cdot H_2O$. And it is precisely this compound which is the final product of the reduction of MnO_4^- under the given conditions.

6.4. Equilibrium in Oxidation-Reduction Reactions

The fact that oxidation-reduction reactions may proceed in either direction is evidently the consequence of the reversibility of such reactions. We know that reversible reactions lead to the establishment of chemical equilibrium. The equilibrium constant can be calculated if the standard oxidation potentials of both half-cells are known.

* An exception is the iodate ion, IO_3^-, which oxidizes S^{2-} even in alkaline solution with the formation of free iodine and sulphur. These ions cannot coexist in neutral and alkaline solutions.

As an example, consider the reaction

$$Sn^{2+} + 2Fe^{3+} \rightleftharpoons Sn^{4+} + 2Fe^{2+}$$

The equilibrium constant of this reaction is expressed by the equation

$$K = \frac{[Sn^{4+}][Fe^{2+}]^2}{[Sn^{2+}][Fe^{3+}]^2} \qquad (6.4)$$

Let us write the expressions for the oxidation potentials of the half-cells Sn^{4+}/Sn^{2+} and Fe^{3+}/Fe^{2+}:

$$E_{Sn^{4+}/Sn^{2+}} = E^0_{Sn^{4+}/Sn^{2+}} + \frac{0.059}{2} \log \frac{[Sn^{4+}]}{[Sn^{2+}]} \qquad (6.5)$$

$$E_{Fe^{3+}/Fe^{2+}} = E^0_{Fe^{3+}/Fe^{2+}} + 0.059 \log \frac{[Fe^{3+}]}{[Fe^{2+}]} \qquad (6.6)$$

From these equations we can see that, as the concentrations of Sn^{4+} and Fe^{2+} increase and those of Sn^{2+} and Fe^{3+} decrease during the reaction, the equilibrium potential of the first half-cell, which is smaller at first, must gradually increase, while the potential of the second one must decrease. In the long run these potentials will become equal.

It is known, however, that electron transfer is possible only when there is a difference between the potentials and must cease when there is no such a difference. Hence, at $E_{Sn^{4+}/Sn^{2+}} = E_{Fe^{3+}/Fe^{2+}}$, equilibrium will be established. Inserting into this equation the values of $E_{Sn^{4+}/Sn^{2+}}$ and $E_{Fe^{3+}/Fe^{2+}}$ from Eqs. (6.5) and (6.6), we obtain:

$$E^0_{Sn^{4+}/Sn^{2+}} + \frac{0.059}{2} \log \frac{[Sn^{4+}]}{[Sn^{2+}]} = E^0_{Fe^{3+}/Fe^{2+}} + 0.059 \log \frac{[Fe^{3+}]}{[Fe^{2+}]}$$

which gives

$$\frac{0.059}{2} \log \frac{[Sn^{4+}]}{[Sn^{2+}]} - 0.059 \log \frac{[Fe^{3+}]}{[Fe^{2+}]} = E^0_{Fe^{3+}/Fe^{2+}} - E^0_{Sn^{4+}/Sn^{2+}}$$

We transform the second term in the left-hand side of this equation as follows:

$$0.059 \log \frac{[Fe^{3+}]}{[Fe^{2+}]} = \frac{0.059}{2} \log \frac{[Fe^{3+}]^2}{[Fe^{2+}]^2}$$

Now the coefficient 0.059/2 can be taken outside the parentheses:

$$\frac{0.059}{2} \left(\log \frac{[Sn^{4+}]}{[Sn^{2+}]} - \log \frac{[Fe^{3+}]^2}{[Fe^{2+}]^2} \right) = E^0_{Fe^{3+}/Fe^{2+}} - E^0_{Sn^{4+}/Sn^{2+}}$$

from which we have

$$\log \frac{[Sn^{4+}][Fe^{2+}]^2}{[Sn^{2+}][Fe^{3+}]^2} = \frac{2\,(E^0_{Fe^{3+}/Fe^{2+}} - E^0_{Sn^{4+}/Sn^{2+}})}{0.059}$$

Since the expression after the logarithm sign is the equilibrium constant of the reaction in question, we can write:

$$\log K = \frac{2\,(E^0_{Fe^{3+}/Fe^{2+}} - E^0_{Sn^{4+}/Sn^{2+}})}{0.059} = \frac{2\,(0.77-0.15)}{0.059} \approx 21$$

whence

$$K \approx 10^{21}$$

The result obtained shows that the product of the concentrations of the reaction products at equilibrium is about 10^{21} times the product of the concentrations of the substances that have not reacted.

This evidently means that the reaction actually proceeds in the direction of the formation of Sn^{4+} and Fe^{2+} from Sn^{2+} and Fe^{3+}. In a general form, the relation between the equilibrium constant of the reaction and the values of the corresponding standard oxidation potentials may be represented (at 25°C) by the equation

$$\log K = \frac{(E^0_{I} - E^0_{II})\,n}{0.059} \tag{6.7}$$

Here E^0_I and E^0_{II} are the standard oxidation potentials of the half-cells, corresponding to the oxidizing agent (E^0_I) and the reducing agent (E^0_{II}) taken.

If $E^0_I > E^0_{II}$, then $\log K > 0$ and $K > 1$. This means that at equilibrium the product of the concentrations of the substances formed is greater than that of the unreacted substances introduced into the reaction. In other words, the reaction in fact proceeds in the direction indicated, and if the difference $E^0_I - E^0_{II}$ is large enough, the reaction will go practically to completion. Conversely, if $E^0_I < E^0_{II}$ (i.e., if the oxidizing agent introduced is weaker than that formed in the reaction), then K will be very small provided the potential difference is sufficiently great. From this we can conclude that the equilibrium is strongly shifted to the left, i.e., in this particular case the reaction must proceed in the opposite direction.

If the difference between the standard oxidation potentials of both half-cells is small, then the equilibrium constant is close to unity; hence, the corresponding reaction cannot go to completion unless the conditions are adjusted so as to shift the equilibrium in the desired direction. For example, in the case of the reaction

$$H_3AsO_4 + 2I^- + 2H^+ \rightarrow HAsO_2 + I_2 + 2H_2O$$

whose equilibrium constant (according to a difference of about 0.02 volt between the standard potentials) is only about 10, a practically complete conversion of the substances taken can be achieved only by increasing the hydrogen ion concentration through the addition of HCl to the solution. The equilibrium can also be shifted if we increase the concentration of I^- in solution, which becomes evident from the equation

$$\frac{[HAsO_2]\,[I_2]}{[H_3AsO_4]\,[I^-]^2\,[H^+]^2}=K\approx 10$$

It should be noted that the values of the standard oxidation potentials provide no information on the rate at which the equilibrium is attained. Sometimes a larger difference of potentials corresponds to a lower rather than to a higher reaction rate. For example, the oxidation of hydrogen by oxygen, which is characterized by a standard potential difference of about 1.23 volts, proceeds at room temperature at an immeasurably low rate. Conversely, the oxidation of Fe^{2+} to Fe^{3+} by oxygen proceeds much faster in spite of the fact that the standard potential difference in this case is smaller (about 0.46 volt).

Slow reactions can be made to proceed faster in various ways. The first to be noted is increase of temperature; the reaction is accelerated on the average by 2 or 3 times by a temperature rise of 10°. Hence, the reaction rate increases in geometrical progression as the temperature increases in arithmetical progression.

Another factor influencing the reaction rate is the concentration of the reacting substances. As pointed out in Sec. 3.1, according to the law of mass action, the rate of reactions in a homogeneous medium is directly proportional to the product of the concentrations of the reactants if these concentrations are raised to powers equal to their stoichiometric coefficients. For example, according to the law of mass action, the rate of the reaction

$$2I^-+H_2O_2+2H^+\rightarrow I_2\downarrow+2H_2O$$

should be equal to

$$v=K\,[I^-]^2\,[H_2O_2]\,[H^+]^2$$

If it is assumed that

$$[I^-]=[H_2O_2]=[H^+]=C$$

we obtain

$$v=KC^5$$

Experiment shows, however, that the rate of the reaction under consideration is proportional not to the fifth power of the concentration, as it should be expected in accordance with the law of mass action, but rather to the square of the concentration, i.e.,

$$v=KC^2$$

Many other oxidation-reduction reactions show similar deviations from the law of mass action.

The cause of these deviations is the complexity of oxidation-reduction reactions which usually involve a number of intermediate stages. The equations for these reactions do not reflect the actual course of the process but serve only as net equations. The rate of such reactions depends on the rates of the individual stages and cannot therefore be predicted on the basis of the overall reaction equation.

That many oxidation-reduction reactions must pass through a series of intermediate steps becomes clear from the following reasoning. Chemical reactions in solutions evidently occur when the reacting ions or molecules collide with one another. The probability of a collision depends, under given conditions, on the number of ions (or molecules) involved in the reaction. In the case of so-called termolecular reactions, which require simultaneous collisions between three particles at a given point in space and in the exact proportions demanded by the reaction equation, e.g.,

$$Sn^{2+} + 2Fe^{3+} \rightarrow Sn^{4+} + 2Fe^{2+}$$

the probability of such collisions must evidently be much less than in the case of bimolecular reactions which involve the collisions of only two particles; for example,

$$Ce^{4+} + Fe^{2+} \rightarrow Ce^{3+} + Fe^{3+}$$

As regards reactions requiring a simultaneous collision of four, five, or more particles, the probability of their occurrence is so low that they may be regarded as practically impossible. But there are very many oxidation-reduction reactions which are represented by equations with large coefficients and might be expected to require collisions between a large number of particles at a given point in space (for example, 5 particles for the reaction between I^- ions and H_2O_2; 14 particles in the reaction between Fe^{2+} and MnO_4^- in acid solution, etc.). It is clear that these final equations are net equations and in fact such reactions proceed through a series of intermediate steps, each of which is a bimolecular or termolecular reaction.

Questions and Problems

1. The standard oxidation potential of the half-cell $Mn^{2+}/Mn\downarrow$ couple is —1.19 volts. What processes will take place in a galvanic cell made up of this half-cell and a standard hydrogen electrode? Write the net equation for this reaction.

2. Do the same for a cell made up of the $Sb^{3+}/Sb\downarrow$ and $2H^+/H_2\uparrow$ couples.

3. What is the equilibrium potential of the $Mn^{2+}/Mn\downarrow$ couple at a concentration of Mn^{2+} equal to: (a) 2 g-ion/litre; (b) 0.005 g-ion/litre?

Answer: (a) $E = -1.181$ volts; (b) $E = -1.256$ volts.

4. A cell is made up of a standard hydrogen electrode and the half-cell $Ni^{2+}/Ni\downarrow$. If the concentration of Ni^{2+} is equal to 0.01 g-ion/litre, the cell emf is 0.288 volt. The nickel electrode is the negative pole. Determine the standard oxidation potential of the $Ni^{2+}/Ni\downarrow$ couple.

Answer: $E^0 = -0.23$ volt.

5. What is the equilibrium potential of the MnO_4^-/Mn^{2+} couple in acid solution if $[MnO_4^-] = [Mn^{2+}]$ and the hydrogen ion concentration is equal to: (a) 1 g-ion/litre; (b) 10^{-5} g-ion/litre?

Answer: (a) 1.51 volts; (b) 1.038 volts.

6. What is the equilibrium potential of the Sn^{4+}/Sn^{2+} couple if the concentration of Sn^{4+} is 0.001 g-ion/litre and that of Sn^{2+} is 0.1 g-ion/litre?

Answer: $E = 0.091$ volt.

7. What is the emf of a cell made up of the Sn^{4+}/Sn^{2+} and Fe^{3+}/Fe^{2+} couples if the concentration of Sn^{4+} and that of Fe^{2+} are equal to 0.01 g-ion/litre, and the concentrations of Sn^{2+} and Fe^{3+} are each equal to 1 g-ion/litre?

Answer: 0.797 volt.

8. Is metallic lead capable of displacing copper from solutions of its salts? Is it possible to displace iron from a solution of $FeSO_4$ by the action of metallic copper?

9. Which of the following metals can displace hydrogen from acids: Sn, Cu, Al, Hg?

10. State whether it is possible to effect the following oxidation reactions in acid solution by the action of $K_2Cr_2O_7$: (a) Fe^{2+} to Fe^{3+}; (b) AsO_2^- to AsO_4^{3-}; (c) Mn^{2+} to MnO_4^-; (d) SO_3^{2-} to SO_4^{2-}; (e) SO_4^{2-} to $S_2O_8^{2-}$.

11. Is it possible to effect the following reduction reactions by the action of Sn^{2+}: (a) Br_2 to Br^-; (b) Fe^{3+} to metallic iron; (c) Fe^{3+} to Fe^{2+}?

12. State whether Fe^{3+} is reduced to Fe^{2+} or to metallic iron if metallic aluminium is placed into an acidified solution of $Fe_2(SO_4)_3$.

13. What will happen if metallic iron is introduced into an acidified solution containing Cu^{2+} and Cd^{2+}?

14. Explain why the action of excess HCl on metallic iron results in the formation of $FeCl_2$, while excess nitric acid gives $Fe(NO_3)_3$.

15. An acid solution containing MnO_4^-, $Cr_2O_7^{2-}$, and Fe^{3+} is acted on by H_2S. What reactions and in what sequence will take place?

16. To a solution of $AgNO_3$, which is one of the electrolytes in a galvanic cell made up of the $Ag^+/Ag\downarrow$ couple and a standard hydrogen electrode, there is added an equivalent amount of HCl. With the value of SP_{AgCl} taken into account, state whether this addition can change the direction of current flow in the cell.

Answer: no, it cannot.

17. To a $0.1M$ solution of $AgNO_3$, which is one of the electrolytes in the same cell as in Problem *16*, there is added as much of solid KCN as required for the concentration of CN^- in solution to become equal to 1 g-ion/litre. Using the value of K_{ins} for the complex $[Ag(CN)_2]^-$ (see Sec. 5.10), figure out whether the direction of current flow is changed in the cell.

Answer: it will be reversed.

18. What reaction will take place in a cell made up of the $Cu^{+}/Cu\downarrow$ and $Cd^{2+}/Cd\downarrow$ half-cells at concentrations of the corresponding ions equal to 1 g-ion/litre? Will the direction of the reaction change if to each of the electrolytes in the cell there is added enough solid KCN to bring the concentration of CN^- to 1 g-ion/litre in solutions containing the $[Cu(CN)_4]^{3-}$ and $[Cd(CN)_4]^{2-}$ ions formed?

Answer: it will be reversed.

19. Derive formulas for calculating the equilibrium constants of the following reactions:

(a) $Sn^{2+} + I_2 \rightleftharpoons Sn^{4+} + 2I^-$

(b) $MnO_4^- + 8H^+ + 5Fe^{2+} \rightleftharpoons Mn^{2+} + 5Fe^{3+} + 4H_2O$ and find their numerical values.

Answer: (a) $\log K = 13.2$; $K = 1.6 \times 10^{13}$; (b) $\log K = 62.7$; $K = 5 \times 10^{62}$.

B. REACTIONS AND ANALYTICAL PROCEDURE FOR GROUP IV AND V CATIONS

6.5. Properties of Group IV Sulphides

As pointed out in Sections 1.5, 5.1, and 5.2, all the cations of Groups IV and V are precipitated by hydrogen sulphide in acid solution (at pH ≈ 0.5). This is the similarity between the cations of these groups.

There is however a distinct difference between the properties of the sulphides of Group IV cations and those of Group V sulphides, and also in the behaviour of these cations towards the action of other reagents. For example, Ag^+, $[Hg_2]^{2+}$, and Pb^{2+} are precipitated by hydrochloric acid as slightly soluble chlorides.* These cations are separated first in the systematic analysis in the form of chlorides and therefore may be included in a special analytical group (Group V). The properties and characteristic reactions of this group and also the procedure used for analysis will be discussed later in the book.

On the other hand, the remaining elements which are precipitated by hydrogen sulphide in acid solution, because of their position in the Periodic System, give sulphides having different properties. They may be subdivided into two subgroups.

(1) The copper subgroup: Cu^{2+}, Cd^{2+}, Hg^{2+}, (Pb^{2+}), Bi^{3+}, Sn^{2+}, and others. The sulphur compounds of the cations of this subgroup do not dissolve in alkalis because of the predominantly basic character of the elements involved.**

---

* Because of the relatively high solubility of $PbCl_2$ in water, lead ions partially remain in solution together with the cations of Group IV.

** As pointed out in Sec. 1.5, mercuric sulphide dissolves partially in sodium sulphide with the formation of a thiosalt. Cupric sulphide dissolves in ammonium polysulphide also with the formation of a thiosalt.

(2) The arsenic subgroup: As^{V}, As^{III}, Sb^{V}, Sb^{III}, Sn^{IV}, and others. The sulphides of the ions of this subgroup are soluble in alkalis and in all the reagents indicated above. An exception is the cation Sn^{2+} whose sulphide, SnS, is insoluble in alkali sulphides, Na_2S and $(NH_4)_2S$, which is accounted for by the fact that Sn^{2+} is more basic than Sn^{IV}. The sulphide SnS can be dissolved only in ammonium polysulphide because this involves the oxidation of Sn^{2+} to Sn^{IV}. Alternatively, the ion Sn^{2+} can be preliminarily oxidized to Sn^{IV} by any suitable reagent.

The elements that form the ions of analytical Groups IV and V, which are considered here, belong to the second half of the fourth, fifth, and sixth long periods of Groups I, II, IV, and V in the Periodic System. The ions of these metals have either the outer electron structure of 18 electrons or an eighteen-plus-two electron configuration of the two outer shells. An exception is the ion Cu^{2+} which has an incomplete outer subshell.

There is a straightforward relation between the outer electron structure of the cations and the properties of the sulphides formed by them.

Indeed, it is known that cations which have an outer multi-electron configuration are strong polarizers and are themselves readily polarized. This is especially pronounced in the case of cations with outer electron structure of 18 and (18 + 2) electrons.

When such cations interact with easily deformable S^{2-} anions, both species of ions undergo strong polarization and their outer electron structures are considerably distorted.

As a result, the bonding between the atoms (ions) approaches the covalent bond. According to this, all cations having 18 and (18 + 2) outer electron structures form, as a rule, sulphides with very small solubility product constants and therefore belong to analytical Groups IV and V.

In contrast, cations with complete 2-electron and 8-electron outer levels are weak polarizers and undergo practically no polarization. When such cations interact with S^{2-} anions, no appreciable distortion of the electron levels takes place. Such cations, as a rule, do not form sulphides slightly soluble in water and are therefore included in analytical Groups I and II. To these groups also belong those cations of Group III, which are precipitated, like Al^{3+}, by ammonium sulphide in the form of hydroxides.

Cations with a transitional electronic structure, i.e., with incomplete outer 18-electron levels, occupy an intermediate position. Being relatively strong polarizers, they themselves undergo noticeable polarization, as a result of which they give slightly soluble sulphides when reacted with S^{-2} anions. The sulphides however differ in solubility, depending on the position of the respective elements in the Periodic System. It is precisely the cations formed by the elements of the second and third triads of the Periodic System (Ru^{3+}, Rh^{3+}, Pd^{2+}, Os^{IV}, Ir^{IV}, Pt^{IV}) that belong to analytical Groups IV and V which also include cations with 18 and (18 + 2) electronic structures. This is accounted for by the fact that the presence of low-lying 18- and 32-electron levels increases their polarizing power and intrinsic polarizability. The increase of polarizing interactions taking place, when these cations combine with S^{2-} anions, leads to a decrease in the solubility of the corresponding sulphides.

Cations formed by the elements of the first triad of the Periodic System (Fe^{2+}, Co^{2+}, Ni^{2+}) and also the Mn^{2+} cation have no low-lying 18- and 32-electron levels; their polarizing power and intrinsic polarizability are relatively low. Accordingly, their sulphides, though slightly soluble in water, have nevertheless higher solubility product constants than the sulphides of Groups IV and V, for which reason they belong to analytical Group III.

All that has been said about the role of outer electronic structures of the cations is illustrated by Table 6.2.

TABLE 6.2. **Electronic Structures of Cations and Analytical Groups**

Outer electronic structure	Cations	Analytical groups
Complete (2 or 8 electrons)	Li^+, K^+, Na^+, Rb^+, Cs^+, Fr^+, (Mg^{2+}), Ca^{2+}, Sr^{2+}, Ba^{2+}, Ra^{2+}	I and II
	Be^{2+}, Mg^{2+}, Al^{3+}, Sc^{3+}, Y^{3+}, La^{3+}, lanthanide ions, Ac^{3+}, actinide ions, Ti^{IV}, Zr^{IV}, Hf^{IV}, Nb^V, Ta^V, (Cr^{3+})	III [subgroup I, precipitated by $(NH_4)_2S$ as hydroxides]
Incomplete ($<$ 18 electrons)	V^{IV}, Cr^{3+}, Mn^{2+}, Fe^{2+}, Fe^{3+}, Co^{2+}, Ni^{2+}, (Zn^{2+}, In^{3+}, Ga^{3+})	III [subgroup II, precipitated by $(NH_4)_2S$ as sulphides]
	Cu^{2+}, Ru^{3+}, Rh^{3+}, Pd^{2+}, Os^{IV}, Ir^{IV}, Pt^{IV}	IV and V
Complete (18 electrons)	Cu^+, Ag^+, Au^+, Zn^{2+}, Cd^{2+}, Hg^{2+}, Ga^{3+}, In^{3+}, Ge^{IV}, Sn^{IV}, Sb^V, As^V	IV and V
(18+2) electrons	Tl^+, Sn^{2+}, Pb^{2+}, As^{III}, Sb^{III}, Bi^{3+}	

Silver always exists in solution in the form of a singly charged cation Ag^+. Likewise, lead, cadmium, and bismuth also form each a single cation of practical importance: Pb^{2+}, Cd^{2+}, Bi^{3+}. Conversely, mercury and copper form two series of ions: $[Hg_2]^{2+}$ and Hg^{2+}; Cu^+ and Cu^{2+}.

The Cu^+ cation, which is very unstable and is of auxiliary importance as compared with the Cu^{2+} ion, is not discussed in this book.

The cation $[Hg_2]^{2+}$, which forms the slightly soluble chloride Hg_2Cl_2, belongs to Group V. The cation Hg^{2+} occupies an intermediate position, between the copper subgroup and the arsenic subgroup in Group IV. It may be included in either subgroup, depending on the method of separation used.

According to the separation method adopted in this book, the cation Hg^{2+} is found to be in the arsenic subgroup in the systematic analysis. Just as in Group III, the hydroxides of the cations in question (with the exception of silver hydroxide) are relatively weak bases, owing to which their salts are more or less strongly hydrolysed and show an acid reaction in solutions.

Most Group IV cations have a distinct tendency to form complexes with ammonia, cyanides, and other substances; only Pb^{2+} and Bi^{3+} form no ammine complexes.

Let us discuss in greater detail the ability of the ions of the arsenic subgroup to form water-soluble thiosalts.

As has already been mentioned, the sulphides of the ions of this subgroup have small solubility product constants: they are insoluble in dilute acids (which have no oxidizing properties) and are precipitated by hydrogen sulphide in acid solution (at pH ≈ 0.5).

In contrast to the sulphides of the copper subgroup, these sulphides are capable of dissolving when acted on by Na_2S, K_2S, or $(NH_4)_2S$ with the formation of so-called thiosalts. For example,

$$SnS_2 \downarrow + Na_2S \rightarrow \underset{\text{sodium thiostannate}}{Na_2SnS_3}$$

$$As_2S_3 \downarrow + 3(NH_4)_2S \rightarrow \underset{\text{ammonium thioarsenite}}{2(NH_4)_3AsS_3}$$

In ionic form this reactions may be represented thus:

$$SnS_2 \downarrow + S^{2-} \rightarrow SnS_3^{2-}$$

$$As_2S_3 \downarrow + 3S^{2-} \rightarrow 2AsS_3^{3-}$$

Thiosalts are the salts of the corresponding thioacids, for example, thiostannic acid, H_2SnS_3, thioarsenous acid, H_3AsS_3. Such acids are similar in composition to oxy-acids of the same elements, with the only difference that the oxygen atoms are replaced in them by the atoms of an analogous element—sulphur. For example,

Orthoarsenous acid . . .	H_3AsO_3	Thioarsenous acid	H_3AsS_3
Orthoarsenic acid . . .	H_3AsO_4	Thioarsenic acid	H_3AsS_4
Orthoantimonous acid . .	H_3SbO_3	Thioantimonous acid . . .	H_3SbS_3
Orthoantimonic acid . .	H_3SbO_4	Thioantimonic acid . . .	H_3SbS_4
Metastannic acid	H_2SnO_3	Thiostannic acid	H_2SnS_3

Thiosalts are unstable and do not exist in acid solutions because they decompose to give hydrogen sulphide and the corresponding sulphide. In contrast to thiosalts proper, their salts are quite stable.

The reactions written above are similar to reactions which result in the formation of salts of oxy-acids through the interaction of acidic oxides with basic ones. For example, the reaction

$$SnS_2 \downarrow + Na_2S \rightarrow Na_2SnS_3$$

is similar to the reaction

$$SnO_2 \downarrow + Na_2O \rightarrow Na_2SnO_3$$

A comparison of the sulphides of the copper subgroup with those of the arsenic subgroup shows that the latter have a sharply pronounced acidic character. They may be referred to as the thioanhydrides of thioacids since thioacids, when liberating hydrogen sulphide, form the corresponding sulphides just as oxy-acids yield anhydrides and water; for example,

$$H_2SnO_3 \rightarrow H_2O + SnO_2 \downarrow \qquad H_2SnS_3 \rightarrow H_2S \uparrow + SnS_2 \downarrow$$

$$2H_3AsO_3 \rightarrow 3H_2O + As_2O_3 \downarrow \qquad 2H_3AsS_3 \rightarrow 3H_2S \uparrow + As_2S_3 \downarrow$$

The acidic character of the sulphides of the arsenic subgroup is attributed to the position of the corresponding elements in the Periodic System. They are in Groups IV (Sn) and V (As and Sb) of the Periodic System, in the second half of the long periods. The elements here show more sharply pronounced metalloid properties which determine the acidic character of their oxides and sulphides. It should be noted that, according to the general rule, the strongest acidic character must be exhibited by the Group V elements, As and Sb, especially in their highest oxidation states, i.e., in the compounds As_2S_5 and Sb_2S_5. The acidic character of the Group IV element Sn is weaker and manifests itself only in a compound of the highest oxidation state, SnS_2. Stannous sulphide, SnS, is a basic sulphide and, accordingly, forms no thiosalts with Na_2S and $(NH_4)_2S$.

Since it is more expedient to detect tin in one rather than in two subgroups, it is preliminarily converted into the highest oxidation state.

The outer electronic structure of the ions of the arsenic subgroup is the same as that of the cations of the copper subgroup. The arsenic subgroup comprises ions having: (a) an outer 18- or (18 + 2)-electron level; (b) an incomplete (transitional) 8- to 18-electron level provided that there is 18- or 32-electron low-lying level, the presence of which enhances both the polarizing power and the intrinsic polarizability of the ions.

It is because of the identical electronic structure that the ions of Groups IV and V show the same behaviour toward the action of hydrogen sulphide in acid solution, i.e., form sulphides with very low solubility product constants. The difference in the chemical nature of the sulphides, on which is based the separation of Group IV, may be ascribed, according to the investigations carried out by N. I. Blok*, to the so-called **ionic potential** of the cations of these groups.

The ionic potential is defined as the ratio of the charge on an ion to its effective radius, *e/r*. The ionic potential may be regarded as a roughly approximate measure of the intensity of the electric field developed by the ion. Sometimes, for example, when comparing the acidic-basic properties of ions with an identical outer electronic structure, the ionic potential can be used as an approximate criterion of these properties. Thus, from Table 6.3 it can be seen that, for the

TABLE 6.3. **Effect of the Ionic Potential on the Chemical Nature of Sulphides Formed by Ions of Groups IV and V**

Ions	Outer electronic structure	Ionic potential, *e/r*	Chemical character of sulphides
Au^+, Ag^+, Cu^+, Hg^{2+}, Cd^{2+}, (Zn^{2+})*	18 electrons	0.7-2.4	Basic
Sn^{IV}, Ge^{IV}, Sb^{V}, As^{V}	18 electrons	5.4-10.6	Acidic
Pb^{2+}, Sn^{2+}, Bi^{3+}	(18+2) electrons	1.5-2.9	Basic
Sb^{III}, As^{III}	(18+2) electrons	3.3-4.4	Acidic

* The cation Zn^{2+}, which is included in analytical Group III from practical considerations, is in fact intermediate between Groups III and IV because of the outer electronic structure, its position in the Periodic System, and also of its tendency to be precipitated by hydrogen sulphide in acid solution (pH 2).

same electronic structure, ions with small values of ionic potential form basic sulphides. Conversely, ions with large values of *e/r* form acid sulphides (thioanhydrides) which dissolve in caustic alkalis and in solutions of the alkali sulphides [Na_2S, $(NH_4)_2S$, etc.] with the formation of thiosalts.

From all that has been said about the properties of the sulphides of the arsenic subgroup, it follows that two operations are to be carried out in order to separate it from other groups, namely:

(a) separation from the cations of Groups III, II, and I by hydrogen sulphide in acid solution (pH $\approx$ 0.5), together with the cations of the copper subgroup (after separation of Group V chlorides);

(b) separation from the copper subgroup by treatment of the sulphides of Group IV (produced after precipitation by H_2S) with any reagent

* Blok, N. I. *Qualitative Chemical Analysis*, Goskhimizdat, Moscow, 1952 (in Russian).

whose action on the sulphides of the arsenic subgroup leads to the formation of thiosalts; the sulphides of the copper subgroup remain in the precipitate and are separated from the arsenic subgroup.

Since the final isolation of the ions of the arsenic subgroup from mixtures with the cations of other groups is achieved only after the second operation, the group reagent of this subgroup is taken to be a reagent that dissolves the sulphides of the arsenic subgroup to give thiosalts.

Various substances are used for this purpose in analytical practice, for example, ammonium polysulphide, $(NH_4)_2S_2$*. An advantage of this reagent is that, being an oxidizing agent, it first oxidizes SnS to SnS_2 which then dissolves to give a thiosalt:

$$SnS\downarrow + (NH_4)_2S_2 \rightarrow SnS_2\downarrow + (NH_4)_2S$$
$$SnS_2\downarrow + (NH_4)_2S_2 \rightarrow (NH_4)_2SnS_3$$

The use of $(NH_4)_2S_2$, however, is fraught with the following drawbacks:

(a) CuS is partially dissolved to give CuS_2^{2-};

(b) the sulphides are heavily contaminated with sulphur formed as a result of the breakdown of excess $(NH_4)_2S_2$ on further acidification of the solution:

$$(NH_4)_2S_2 + 2CH_3COOH \rightarrow 2CH_3COONH_4 + H_2S\uparrow + S\downarrow$$

(c) not infrequently there is formed a mixture which is difficult to separate into its components by centrifugation.

On account of all that has been said, we shall make use of sodium sulphide, Na_2S, (to which NaOH is added to suppress hydrolysis) as the group reagent for the arsenic subgroup.

As compared with ammonium sulphide, which is practically completely hydrolysed in aqueous solutions, sodium sulphide, being the salt formed by the same acid (H_2S) but with a different, stronger base, has a less tendency to undergo hydrolysis (especially in the presence of sodium hydroxide). It is for this reason that the Na_2S solution contains S^{2-} ions in a much higher concentration than solutions of $(NH_4)_2S$ or $(NH_4)_2S_2$.

But in solutions containing S^{2-} ions in a high concentration, HgS also dissolves with the formation of a thiosalt:

$$HgS\downarrow + S^{2-} \rightarrow HgS_2^{2-}$$

Consequently, under the action of Na_2S, the Hg^{2+} ion in the form of HgS_2^{2-} passes completely into solution containing the thiosalts of the elements of the arsenic subgroup. Conversely, when $(NH_4)_2S_2$ is

* The content of sulphur in ammonium polysulphide is variable and therefore the composition of this reagent is more correctly expressed by the formula $(NH_4)_2S_n$.

used as the group reagent, the Hg^{2+} ion remains in the precipitate together with the cations of the copper subgroup.*

It is clear that, just as in the case of oxides, no sharply distinct boundary line can be drawn between the basic and acidic sulphides since some of them behave differently under different conditions.

The elements As, Sb, and Sn exist in compounds in two oxidation states:

As_2O_3 Arsenous oxide	As_2O_5 Arsenic oxide
Sb_2O_3 Antimonous oxide	Sb_2O_5 Antimonic oxide
SnO Stannous oxide	SnO_2 Stannic oxide

In accordance with the position of the elements in the Periodic System, the hydrates of the oxides formed have a close resemblance either to bases (Sn) or to acids (As, Sb), but they always exhibit the properties of both bases and acids, i.e., are more or less amphoteric in character. For example, stannous hydroxide, $Sn(OH)_2$ or H_2SnO_2, dissolves in both acids and bases (in the latter case, stannites are formed):

$$Sn(OH)_2\downarrow + 2H^+ \rightarrow Sn^{2+} + 2H_2O$$
$$H_2SnO_2\downarrow + 2OH^- \rightarrow SnO_2^{2-} + 2H_2O$$

Thus, tin(II) exists predominantly in the form of Sn^{2+} cations in acid solutions and as SnO_2^{2-} anions in alkaline solutions, the latter being due to the dissociation of stannites. Likewise, tin(IV) may exist in solutions either as Sn^{4+} cations, more precisely, in the form of complex ions $[SnCl_6]^{2-}$ (in a solution of $SnCl_4$ acidified with HCl) or as SnO_3^{2-} anions (in a solution of stannates, i.e., the salts of metastannic acid, H_2SnO_3).

The corresponding compounds of arsenic and antimony exhibit almost the same character. Being an element of the odd row of Group V in the Periodic System, arsenic is much closer in properties to metalloids (semimetals) than to metals. To its oxides As_2O_2 and As_2O_5, there correspond arsenous acid, H_2AsO_3, and arsenic acid, H_3AsO_4. These acids and their salts (arsenites and arsenates) are the most common compounds of arsenic, and in the course of analysis we shall have to deal mainly with the anions AsO_3^{3-} and AsO_4^{3-}.

But, depending on the acidity of solutions of As^{III} and As^{V}, ions of different composition may exist in solution, say, AsO^+ and AsO^{3+}. Arsenic(V) is capable of being precipitated by hydrogen sulphide in a

* Use is occasionally made of NaOH or KOH as the group reagent for the arsenic subgroup. These reagents easily dissolve the sulphides of arsenic, antimony, and tin(IV) but do not dissolve the sulphides of the copper subgroup and SnS. The inconvenience caused by their use is that HgS may partially dissolve and partially remain in the precipitate.

strongly acid solution (6-7*N* HCl) in the form of the sulphide As_2S_5. The reaction evidently proceeds through a series of stages involving the formation of various intermediates. The mechanism of this reaction has not yet been established definitely. Therefore, to simplify, the reaction may be represented in the following form:

$$2H_3AsO_4 + 5H_2S \rightarrow As_2S_5\downarrow + 8H_2O \quad (6.8)$$

It should, however, be kept in mind that this is only a simplified schematic representation of a rather complicated multistage process.

For arsenic(III) the reaction proceeds at an even lower acidity of the solution. It may be written thus:

$$2HAsO_2 + 3H_2S \rightarrow As_2S_3\downarrow + 4H_2O \quad (6.9)$$

Considerable amounts of alkali are formed in solutions of the arsenites and arsenates of the alkali metals due to hydrolysis. When hydrogen sulphide is passed into such a solution, Na_2S is formed. The presence in solution of an alkali and an alkali-metal sulphide prevents the formation of the sulphides of arsenic because thiosalts are formed. But if the solution is acidified, conditions are established, under which the occurrence of reactions (6.8) and (6.9) becomes possible.

The behaviour of antimony is analogous to that of arsenic. It is this attitude of the ions of arsenic and antimony towards the action of H_2S that allows us to consider them together with cations rather than with anions.

It thus follows that complete precipitation of the ions of Group IV by hydrogen sulphide occurs only in acid solution. Apart from shifting the equilibrium to the formation of the cations As^{3+}, Sb^{3+}, etc., the addition of acid also favours the coagulation of the Group IV sulphides which have a strong tendency to form colloidal solutions.

In a neutral and, the more so, in alkaline solution, hydrogen sulphide does not precipitate the sulphides of the arsenic subgroup. Nor are they precipitated by an excess of $(NH_4)_2S$ or Na_2S because of the formation of solutions of the corresponding thiosalts.

6.6. Action of the Group Reagent

In acid solutions at pH 0.5, hydrogen sulphide precipitates all the cations of Groups IV and V*; therefore, in spite of the fact that the cations of Group V are separated preliminarily in the form of insoluble chlorides, the properties of the sulphides of both analytical groups (IV and V) may be discussed jointly.

---

* As indicated above, As_2S_5 is precipitated only at a high acidity of the solution (6-7*N* HCl). On prolonged passage of hydrogen sulphide into the heated solution at pH 0.5, arsenic (V) is reduced to arsenic(III) and a precipitate of As_2S_3 is formed.

The method of preparing hydrogen sulphide under semimicroanalysis conditions was described in Sec. 2.2. It will be recalled here that *hydrogen sulphide is very poisonous; all operations with the gas must be conducted in the fume cupboard.* When precipitation is effected by hydrogen sulphide, only that part of it participates in the reaction which has dissolved in the liquid; therefore H_2S should be passed slowly so that the gas bubbles could be easily counted. Otherwise the major portion of hydrogen sulphide will pass uselessly through the solution, thereby contaminating the air.

When the work with the H_2S *generator is completed, do not forget to shut off stopcock!*

With these directions taken into account, study the action of H_2S on the cations of Groups IV and V through experiments.

Take 1-2 drops of a solution of each of the corresponding salts, dilute in a test tube with 5-6 drops of water, and acidify with 1 drop of 2*N* HCl, or CH_3COOH, in the case of the salts of silver, lead, and mercury(I). Pass H_2S into the solutions obtained. After each experiment, it is necessary to wash thoroughly the glass tube through which hydrogen sulphide was passed into the solution.

The action of H_2S on solutions of the salts of silver, lead, copper, bismuth, and mercury(I) and (II) yields black precipitates of Ag_2S, PbS, CuS, Bi_2S_3, and HgS; for example*:

$$2Ag^+ + H_2S \rightarrow Ag_2S\downarrow + 2H^+$$

$$Cu^{2+} + H_2S \rightarrow CuS\downarrow + 2H^+$$

$$2Bi^{3+} + 3H_2S \rightarrow Bi_2S_3\downarrow + 6H^+$$

Solutions of cadmium salts give a characteristic bright-yellow precipitate of CdS, the formation of which is the most important qualitative reaction of Cd^{2+}:

$$Cd^{2+} + H_2S \rightarrow CdS\downarrow + 2H^+$$

Under the action of hydrogen sulphide, the $[Hg_2]^{2+}$ ion gives a precipitate of Hg_2S which immediately breaks down into HgS and Hg:

$$[Hg_2]^{2+} + H_2S \rightarrow HgS\downarrow + Hg\downarrow + 2H^+$$

The resulting mixture of HgS and Hg is coloured black.

In contrast to the sulphides of Group III cations, those of Groups IV and V are insoluble in dilute acids (HCl, H_2SO_4) because their solubility product constants are so small that the S^{2-} ions supplied by them into the solution are not practically bound by hydrogen ions (see Sec. 5.2). They however dissolve readily in dilute HNO_3 which oxidizes S^{2-} to free sulphur; for example,

$$3CuS\downarrow + 2NO_3^- + 8H^+ \rightarrow 3Cu^{2+} + 3S\downarrow + 2NO\uparrow + 4H_2O$$

* In the case of Pb^{2+} and Hg^{2+}, red ($PbCl_2 \cdot PbS$) and white ($HgCl_2 \cdot 2HgS$) precipitates may be formed first. These precipitates are converted by further action of H_2S into black precipitates of PbS and HgS.

The reaction is carried out on heating. To speed up the dissolution, it is useful to add a small amount of KNO_2 or $NaNO_2$. This gives nitrogen oxides which accelerate the reaction catalytically.

The most soluble Group IV sulphide CdS ($SP = 7.9 \times 10^{-27}$) dissolves not only in HNO_3 but also in sufficiently concentrated solutions of HCl. Thus, even a 1*N* solution of HCl dissolves on heating a large amount of CdS, a fact that can be used in confirmatory tests for Cd^{2+} in those cases where the colour of cadmium sulphide is covered up by the black colour of other coprecipitated sulphides, say, CuS and Bi_2S_3.

The dissolution of CdS in HCl is associated not only with the combination of S^{2-} ions with the H^+ ions of the acid but also with the formation of the complex $[CdCl_4]^{2-}$ ion which greatly reduces the concentration of Cd^{2+}. This is supported by the fact that CdS is not precipitated from a moderately acid solution (pH $\approx$ 0.5) if a halide, say, NH_4Cl, is added to the solution until it is saturated. The behaviour of Cd^{2+} towards the action of hydrogen sulphide in the presence of HCl (in a concentration of 1*N* or higher) or other halides can be used for separation of cadmium from copper.

This fact must also be taken into account in the separation of the cations of Group IV from those of Groups III, II, and I. Much care must be taken to see that the pH of the solution does not fall below 0.5 in the precipitation by hydrogen sulphide, otherwise Cd^{2+} may completely remain in solution and be missed.

From the above example it also follows that there may arise complications in analysis due to the complex-forming tendency of Cl^- ions; this is one of the disadvantages of the hydrogen sulphide method.

In distinction to CdS, the least soluble Group IV sulphide—mercuric sulphide, HgS ($SP = 1.6 \times 10^{-52}$)—is insoluble not only in HCl but also in HNO_3; this fact is used for separating mercury from a number of other cations of Group IV. The separation however is incomplete: appreciable amounts of Hg^{2+} may pass into solution.

Complete dissolution of mercuric sulphide is effected by using aqua regia (a mixture of 1 volume of concentrated HNO_3 and 3 volumes of concentrated HCl) which is a more powerful oxidizing agent than nitric acid and, besides, it binds Hg^{2+} ions into the complex $[HgCl_4]^{2-}$:

$$3HgS\downarrow + 2HNO_3 + 12HCl \rightarrow 3H_2[HgCl_4] + 3S\downarrow + 2NO\uparrow + 4H_2O$$

Mercuric sulphide can be dissolved in HCl as well if Hg^{2+} ions are bound into the very stable complex $[HgI_4]^{2-}$ ($K_{ins} = 1.5 \times 10^{-30}$). The reaction proceeds on heating:

$$HgS\downarrow + 2HCl + 4KI \rightarrow K_2[HgI_4] + 2KCl + H_2S\uparrow$$

Let us consider in more detail the properties of the sulphides of the arsenic subgroup.

1. Arsenic. (1) *Arsenic Ion (in the Oxidation State +3)*. To 4-5 drops of a solution of Na_3AsO_3 or K_3AsO_3 (*Caution: very poisonous!*) add HCl until the reaction is acid, and pass H_2S into the solution. A yellow precipitate of As_2S_2 is formed:

$$2AsO_2^- + 8H^+ \rightarrow 2As^{3+} + 4H_2O$$

$$2As^{3+} + 3H_2S \rightarrow As_2S_3\downarrow + 6H^+$$

$$\overline{2AsO_2^- + 3H_2S + 2H^+ \rightarrow As_2S_3\downarrow + 4H_2O}$$

Centrifuge off the As_2S_3 precipitate obtained, and divide it into 4 parts. Keep one for subsequent tests (reaction with HNO_3; see below). Treat the second part with Na_2S, the third with NaOH or NH_4OH, and the fourth one with $(NH_4)_2CO_3$.

In all the three cases the precipitate is dissolved*:

$$As_2S_3\downarrow + 3Na_2S \rightarrow 2Na_3AsS_3$$

$$As_2S_3\downarrow + 6NaOH \rightarrow Na_3AsS_3 + Na_3AsO_3 + 3H_2O$$

$$As_2S_3\downarrow + 3(NH_4)_2CO_3 \rightarrow (NH_4)_3AsS_3 + (NH_4)_3AsO_3 + 3CO_2\uparrow$$

Write the ionic equations for these reactions.

When dissolved in alkalis, along with AsS_3^{3-}, there is formed the anion of orthoarsenous acid, AsO_3^{3-}. A very important property of As_2S_3 and of As_2S_5 is the solubility in $(NH_4)_2CO_3$, which distinguishes them from the other sulphides of the arsenic subgroup. This property can be used for separation of arsenic in the course of analysis. This difference is accounted for by the fact that at a relatively low pH (about 9.19) due to ammonium carbonate (see Table 5.3, Sec. 5.6) thiosalts can give only sulphides with the most sharply pronounced acidic properties. It is the sulphides of arsenic precisely that are the most acidic of all the sulphides considered because arsenic is an element in an odd row of Group V in the Periodic System and, besides, its atomic weight is smaller than that of antimony.

Another important distinctive feature of the sulphides of arsenic(V) and arsenic(III) is their insolubility in concentrated HCl; this property is used for separating arsenic ions from the ions of antimony and tin in the course of analysis. Arsenous sulphide, As_2S_3, can be dissolved in concentrated HNO_3 or in aqua regia. This gives rise to arsenic ion (in the +5 oxidation state). This may be confirmed by an experiment: use the portion of the precipitate of As_2S_3 left and carry out the reaction on heating. The reaction with HNO_3 proceeds according to the equation**

$$As_2S_3\downarrow + 28HNO_3 \rightarrow 2H_3AsO_4 + 3H_2SO_4 + 28NO_2\uparrow + 8H_2O$$

Arsenous sulphide can equally readily be converted into AsO_4^{3-} on heating with a mixture of alkali or ammonium carbonate and hydro-

* Keep the solutions obtained for further tests.

** For setting-up of this equation, see Sec. 5.13.

gen peroxide; for example,

$$As_2S_3\downarrow + 14H_2O_2 + 6(NH_4)_2CO_3 \rightarrow$$
$$\rightarrow 2(NH_4)_3AsO_4 + 3(NH_4)_2SO_4 + 14H_2O + 6CO_2$$

Carbon dioxide reacts with excess $(NH_4)_2CO_3$ to form ammonium bicarbonate:

$$3(NH_4)_2CO_3 + 3CO_2 + 3H_2O \rightarrow 6NH_4HCO_3$$

A very important property of thiosalts is their tendency to decompose under the action of acids with the formation of sulphide precipitates. Carry out this reaction with the solution of Na_3AsS_3 resulting from the action of Na_2S on arsenous sulphide.

The attitude of thiosalts toward acids is accounted for by the instability of free thioacids which decompose instantly, as soon as they are formed from salts, into the corresponding thioanhydride and H_2S. For example,

$$2AsS_3^{3-} + 6H^+ \rightarrow As_2S_3\downarrow + 3H_2S\uparrow$$

When the mixture of arsenite and thioarsenite obtained upon treatment of the As_2S_3 precipitate with alkali or ammonium carbonate is acted on by acids, the following reaction takes place:

$$AsS_3^{3-} + AsO_3^{3-} + 6H^+ \rightarrow As_2S_3\downarrow + 3H_2O$$

(2) *Arsenic Ion* (*in the Oxidation State* $+5$). In moderately acidified solutions of arsenates, say, Na_2HAsO_4 (*Caution*: *a powerful poison*!), hydrogen sulphide does not give a precipitate in the cold for a long time. The liquid gradually becomes turbid and, eventually, yellow As_2S_3 settles down.

This reaction may be explained as follows. The acidic properties of H_3AsO_4 are much more strongly and its basic properties much less strongly pronounced than the corresponding properties of H_3AsO_3. At the same time, according to the position of the $S\downarrow/H_2S$ couple in the table of standard oxidation potentials ($E^0 = +0.14$ volt), hydrogen sulphide is capable of reducing AsO_4^{3-} to AsO_2^- ($E^0 = +0.56$ volt):

$$AsO_4^{3-} + H_2S + 2H^+ \rightarrow AsO_2^- + S\downarrow + 2H_2O$$

We have seen earlier that, in contrast to AsO_4^{3-}, the AsO_2^- ion gives upon acidification a sufficient amount of As^{3+} cations, which means that as the AsO_4^{3-} ions are reduced the following reaction must take place:

$$2AsO_2^- + 2H^+ + 3H_2S \longrightarrow As_2S_3\downarrow + 4H_2O$$

The net equation of the reaction is obtained by multiplying the first equation by 2 and adding up the two equations:

$$2AsO_4^{3-} + 5H_2S + 6H^+ \longrightarrow As_2S_3\downarrow + 2S\downarrow + 8H_2O$$

The reduction of AsO_4^{3-} to AsO_2^- by hydrogen sulphide is a very slow reaction. It can be accelerated by carrying out the precipitation on heating up to about 70°C. Moreover, it is useful to add to the solution 1-2 drops of a catalyst—a solution of NH_4I*. The role of the catalyst is as follows. Under the action of I^- ions, AsO_4^{3-} is rapidly reduced to AsO_2^-:

$$AsO_4^{3-} + 2I^- + 4H^+ \rightarrow AsO_2^- + I_2\downarrow + 2H_2O$$

But since I_2 ($E^0 = +0.54$ volt) is below the $S\downarrow/H_2S$ couple in the table of standard oxidation potentials, it will react with H_2S and will again be reduced to I^-:

$$I_2 + H_2S \rightarrow 2I^- + 2H^+ + S\downarrow$$

Hence, I^- ions are completely recovered and are not used up in the reaction. But in their presence the reaction is considerably hastened because of the stages indicated above, which proceed much faster than the direct interaction between AsO_4^{3-} and H_2S. Arsenic sulphide, As_2O_5, too, can be obtained by the action of H_2S on a salt of H_3AsO_4 if the precipitation is conducted in the cold** using a rapid current of H_2S in the presence of an excess of concentrated HCl. The reaction proceeds relatively rapidly. Arsenic sulphide separates out as a yellow precipitate which reacts with the reagents indicated above in just the same way as As_2S_3. For example,

$$As_2S_5\downarrow + 3Na_2S \rightarrow 2Na_3AsS_4$$

$$4As_2S_5\downarrow + 12(NH_4)_2CO_3 \rightarrow 5(NH_4)_3AsS_4 + 4(NH_4)_3AsO_4 + 12CO_2$$

$$CO_2 + (NH_4)_2CO_3 + H_2O \rightarrow 2NH_4HCO_3$$

Alkalis react with As_2S_5 in an analogous way:

$$4As_2S_5\downarrow + 24NaOH \rightarrow 5Na_3AsS_4 + 3Na_3AsO_4 + 12H_2O$$

2. Antimony. Hydrochloric acid solutions of $SbCl_3$ and $SbCl_5$ contain largely the complex ions $[SbCl_6]^{3-}$ and $[SbCl_6]^-$. In the presence of HCl at pH ≈ 0.5, these compounds react with H_2S:

$$2[SbCl_6]^{3-} + 3H_2S \rightarrow Sb_2S_3\downarrow + 6H^+ + 12Cl^-$$

$$2[SbCl_6]^- - 5H_2S \rightarrow Sb_2S_5\downarrow + 10H^+ + 12Cl^-$$

The orange-red precipitates formed are soluble in Na_2S and caustic alkalis but insoluble in $(NH_4)_2CO_3$ (distinction from As_2S_3 and As_2S_5):

$$Sb_2S_5\downarrow + 3Na_2S \rightarrow 2Na_3SbS_4$$

$$4Sb_2S_5\downarrow + 24NaOH \rightarrow 5Na_3SbS_4 + 3Na_3SbO_4 + 12H_2O$$

The thiosalts of antimony are decomposed by the action of acids:

$$2Na_3SbS_4 + 6CH_3COOH \rightarrow Sb_2S_5\downarrow + 6CH_3COONa + 3H_2S\uparrow$$

---

* If NH_4I is not available, use may be made of an alcoholic solution of iodine. When acted on by hydrogen sulphide, I_2 is instantly reduced to I^-.

** On heating, part of AsO_4^{3-} ions are reduced to AsO_2^- and a mixture of As_2S_5 and As_2S_3 results.

In contrast to the sulphides of arsenic, the sulphides of antimony dissolve when heated with concentrated HCl. In this reaction, Sb^{V} is reduced to Sb^{III}:

$$Sb_2S_5\downarrow + 12HCl \rightarrow 2H_3[SbCl_6] + 3H_2S\uparrow + 2S\downarrow$$

3. Tin. When a solution of $SnCl_2$ is acted on by H_2S in the presence of HCl, a dark-brown precipitate of SnS is formed. In contrast to other sulphides of the arsenic subgroup, SnS is basic in character and is insoluble in alkalis and in ammonium and sodium sulphides.

But SnS dissolves in $(NH_4)_2S_2$ since it is oxidized by the polysulphide to SnS_2 which gives a thiosalt.

The action of H_2S at pH $\approx$ 0.5 on a solution of $SnCl_4$ acidified with HCl and containing complex $[SnCl_6]^{2-}$ ions, yields a yellow precipitate of SnS_2 soluble in Na_2S and alkalis but insoluble in $(NH_4)_2CO_3$ (difference from the sulphides of arsenic):

$$SnS_2\downarrow + Na_2S \rightarrow Na_2SnS_3$$

Thiosalts of tin are decomposed by acids with the formation of SnS_2.

Just as the sulphides of antimony, SnS_2 dissolves on heating with concentrated HCl to form complex stannohydrochloric acid (distinction from the sulphides of arsenic):

$$SnS_2\downarrow + 6HCl \rightarrow H_2[SnCl_6] + 2H_2S\uparrow$$

4. Mercury(II). The ion Hg^{2+} forms with H_2S a black precipitate of mercuric sulphide, HgS. A characteristic feature of this precipitate is its insolubility in HNO_3 (difference from all the other sulphides of Groups IV and V).* Not infrequently this fact is used for separating mercuric sulphide from other sulphides of Group IV.

Mercuric sulphide can be dissolved in aqua regia or in a mixture of concentrated HCl and KI. In the latter case, the reaction proceeds according to the equation

$$HgS\downarrow + 2H^+ + 4I^- \rightarrow [HgI_4]^{2-} + H_2S\uparrow$$

By its properties mercuric sulphide is intermediate between the sulphides of the copper subgroup and those of the arsenic subgroup. Like the sulphides of the copper subgroup**, HgS is insoluble in $(NH_4)_2S$ and $(NH_4)_2S_2$ whose solutions contain relatively little S^{2-} because of the hydrolysis leading to the formation of HS^-. Conversely, sulphide, which is much less hydrolysed, dissolves HgS to form a thiosalt:

$$HgS\downarrow + Na_2S \rightarrow Na_2HgS_2$$

---

* This is because of the very low solubility product constant of HgS ($SP = 1.6 \times 10^{-52}$).

** As has already been indicated, cupric sulphide, CuS, dissolves somewhat in ammonium polysulphide to give CuS_2^{2-}.

An analogous reaction takes place under the action of alkalis since the HgS precipitate always adsorbs hydrogen sulphide. When acted on by alkalis, hydrogen sulphide forms Na_2S which dissolves HgS.

Like the thiosalts of other elements of the arsenic subgroup, the thiosalt of mercury(II) is decomposed by acids, forming a precipitate of HgS.

Since the copper and arsenic subgroups are separated with the aid of Na_2S, the Hg^{2+} ion must be assigned to the arsenic subgroup.

To separate mercury(II) from the other elements of the arsenic subgroup, use may be made of the characteristic tendency of the thiosalt HgS_2^{2-} to be decomposed by ammonium salts (which is not typical of the thiosalts of arsenic, antimony, and tin). This feature of the mercuric thiosalt is attributed, on the one hand, to its very weak acidic properties and, on the other, to the sufficient concentration of hydrogen ions resulting from the hydrolysis of ammonium salts:

$$HgS_2^{2-} + NH_4^+ + 2H_2O \rightarrow HgS\downarrow + HS^- + NH_4OH$$

Thus, HgS is precipitated when solid NH_4Cl is added to solutions of the thiosalts of the arsenic subgroup whereas all the other thiosalts remain in solution.

Conditions for Precipitation of Group IV Cations by Hydrogen Sulphide. (1) Precipitation of Group IV cations must be carried out (see Sec. 5.1) in acid solution, at pH ≈ 0.5, which corresponds to the hydrogen ion concentration 0.3*M*. At a lower acidity (at a higher pH) zinc sulphide may coprecipitate with the sulphides of Group IV cations. Indeed, as pointed out in Sec. 5.21, at pH 2 zinc sulphide is almost completely precipitated by hydrogen sulphide and may thus be missed. On the other hand, at a concentration of 1*M* (i.e., at pH 0) hydrochloric acid dissolves CdS which therefore may also be missed. This may often happen if a large quantity of chlorides is present in the solution. Adjustment of pH in such cases becomes difficult because there is no suitable buffer system available for pH 0.5. Therefore, with the acid present in the solution having been neutralized (since the amount present is unknown), an equal volume of 0.6*N* HCl has to be added. As a result, the concentration of HCl (and hence the hydrogen ion concentration) in the solution becomes equal to about 0.3*M*, which corresponds to pH ≈ 0.5.

In this particular case, the pH must not be determined with the aid of any suitable indicator (for example, methyl violet whose transition range lies between pH 0 and 2). This is accounted for by the fact that strongly acid solutions exert a considerable buffering action, i.e., they change the pH very slowly. For example, in order to lower the pH of a solution from 1 to 0, about 0.9 mole of HCl must be added to one litre of the solution. But the precision of the indicator paper method used for pH measurements lies within one pH unit. It is evident therefore that we cannot achieve the required precision in adjusting the acidity of a solution by using this method. Of course, the accuracy of pH determination can be increased by comparing the colour assumed by methyl violet under the action of the test solution with the colour which it produces under the same conditions with 0.3*N* HCl, but this method is less reliable than the one recommended above and, besides, it requires some skill.

(2) Of the acids used in the laboratory for acidifying solutions before the precipitation by hydrogen sulphide, the most commonly employed one is HCl. Indeed, HNO_3 oxidizes H_2S to sulphur; if H_2SO_4 is used, SO_4^{2-} ions are introduced, which precipitate Ba^{2+}, Sr^{2+}, (Ca^{2+}), and Pb^{2+}, and, as a result, the analysis is complicated; CH_3COOH is a weak acid and does not produce the desired concentration of hydrogen ion in the solution. Acidification with HCl brings about the precipitation of the ions of Group V which is thus separated from the other ions by the action of H_2S and is analysed separately.

(3) To prevent the sulphides from forming colloidal solutions, the precipitation is effected from a hot solution. But, as in the case of any other gas, the solubility of H_2S decreases with rise of temperature, as a result of which the cations of Group IV are separated less completely. Therefore, at the end of precipitation, the solution is diluted with an equal volume of cold distilled water and again saturated with H_2S. This not only cools the solution but also lowers the concentration of hydrogen ions, thus making the precipitation of Cd^{2+} more complete.

(4) As was pointed out in Sec. 5.4, when the precipitated Group IV sulphides are allowed to age in contact with the solution, zinc sulphide is found to be postprecipitated. In such a case, Zn^{2+} may be missed. To avoid this, it is necessary to separate the precipitated sulphides from the solution as soon as complete precipitation is achieved.

6.7. Reactions of the Cupric Ion

Solutions of cupric salts are light blue or green in colour.

1. Alkali Hydroxides, NaOH and KOH. These produce with Cu^{2+} a blue precipitate of $Cu(OH)_2$ which becomes black on heating as a result of its conversion into CuO:

$$Cu^{2+} + 2OH^- \rightarrow Cu(OH)_2 \downarrow$$
$$Cu(OH)_2 \downarrow \rightarrow CuO \downarrow + H_2O$$

Cupric hydroxide is but slightly amphoteric and is therefore partially soluble in an excess of KOH or NaOH (especially on heating) with the formation of cuprites.

In the presence of glycerine (glycerol) and also of tartaric and citric acids and certain other organic compounds which contain hydroxyl groups, the Cu^{2+} ion forms deep-blue complex compounds. These compounds cannot be precipitated by alkalis. Confirm this experimentally, by adding a few drops of a solution of glycerine or tartaric acid to a solution of a cupric salt, followed by alkali.

2. Ammonium Hydroxide, NH_4OH. When added in a small amount, ammonium hydroxide precipitates a basic salt having a greenish colour, say, $(CuOH)_2SO_4$, which dissolves readily in an excess of the reagent. This gives rise (at pH > 9) to a deep-blue copper ammine complex. In ionic form this characteristic reaction of Cu^{2+} may be represented by the following equations:

$$2Cu^{2+} + SO_4^{2-} + 2NH_4OH \rightarrow (CuOH)_2SO_4 \downarrow + 2NH_4^+$$
$$(CuOH)_2SO_4 \downarrow + 8NH_4OH \rightarrow 2[Cu(NH_3)_4]^{2+} + SO_4^{2-} + 2OH^- + 8H_2O$$

The copper ammine complex is destroyed by acidification because of the NH_3 molecule being bound by H^+ to form the more stable complex NH_4^+. This changes the deep-blue colour of the solution to light-blue (the colour of the cupric ion). Such breakdown of the complex $[Cu(NH_3)_4]^{2+}$ is caused by the action of H_2S since the concentration of Cu^{2+} in the solution turns out to be enough for the solubility product of CuS to be exceeded, as a result of which cupric sulphide is precipitated.

3. Potassium Hexacyanoferrate(II), $K_4[Fe(CN)_6]$. At pH $\leqslant 7$ this reagent (also known as potassium ferrocyanide) forms with Cu^{2+} a red-brown precipitate of cupric ferrocyanide:

$$2Cu^{2+} + [Fe(CN)_6]^{4-} \rightarrow Cu_2[Fe(CN)_6] \downarrow$$

The precipitate is insoluble in dilute acids but dissolves in aqueous ammonia (formation of a complex compound). It is also decomposed by alkalis:

$$Cu_2[Fe(CN)_6] \downarrow + 4OH^- \rightarrow 2Cu(OH)_2 \downarrow + [Fe(CN)_6]^{4-}$$

The reaction is more sensitive than the reaction with NH_4OH.

4. Sodium Thiosulphate, $Na_2S_2O_3$. When added to an acidified solution of a cupric salt, sodium thiosulphate decolourizes the solution as the result of the formation of a complex salt.

If the resulting solution is heated, a dark-brown precipitate of a mixture of Cu_2S and sulphur is formed:

$$2Cu^{2+} + 2S_2O_3^{2-} + 2H_2O \rightarrow Cu_2S \downarrow + S \downarrow + 4H^+ + 2SO_4^{2-}$$

Since Cd^{2+} does not form a precipitate of cupric sulphide when reacted with $Na_2S_2O_3$ in acid solution, this reaction may be used for separation of Cu^{2+} from Cd^{2+}.

5. Aluminium, Iron, and Zinc Metals. These metals reduce Cu^{2+} to the metal which precipitates as a red-coloured spongy mass; for example,

$$3Cu^{2+} + 2Al \rightarrow 3Cu \downarrow + 2Al^{3+}$$

This reaction can also be employed for separation of Cu^{2+} from Cd^{2+}

Judging from the position of Cu and Cd in the electromotive series, one should expect that these two metals can be displaced by aluminium, iron, and zinc from solutions of their salts. But in the presence of H_2SO_4 acid-soluble metallic cadmium does not precipitate. Conversely, metallic copper does not dissolve in dilute H_2SO_4, according to its place in the electromotive series (to the right of hydrogen), and is therefore precipitated in the presence of the acid.

Precipitation of copper does not take place in the presence of HNO_3 which dissolves readily the metal:

$$3Cu\downarrow + 2NO_3^- + 8H^+ \rightarrow 3Cu^{2+} + 2NO\uparrow + 4H_2O$$

Thus, before the separation of Cu^{2+} from Cd^{2+}, it is necessary first of all to remove completely HNO_3, which is achieved by evaporating the solution with H_2SO_4.*

6. Potassium Iodide, KI. This reagent is oxidized by cupric salts with the evolution of free iodine. At the same time, a slightly soluble precipitate of cuprous iodide is formed:

$$2Cu^{2+} + 4I^- \rightarrow 2CuI\downarrow + I_2\downarrow$$

7. Potassium Cyanide, KCN. This reagent produces with Cu^{2+} a yellow precipitate of $Cu(CN)_2$ which however is unstable and is instantly reduced to cuprous cyanide, CuCN, white in colour. The CuCN precipitate dissolves in excess of the cyanide, forming the complex $[Cu(CN)_4]^{3-}$.

$$Cu^{2+} + 2CN^- \rightarrow Cu(CN)_2\downarrow$$

$$2Cu(CN)_2\downarrow \rightarrow 2CuCN\downarrow + (CN)_2\uparrow$$

$$CuCN\downarrow + 3CN^- \rightarrow [Cu(CN)_4]^{3-}$$

The complex formed is very stable, this fact being used for separating the ions of copper and cadmium in the course of analysis (see Sec. 6.8, Reaction *3*).

8. Benzidine in the Presence of Potassium Cyanide. Cupric cyanide, $Cu(CN)_2$, is very unstable and is reduced to CuCN as soon as it is formed, the cyanogen $(CN)_2$ being split off (see Reaction 7). Therefore, cupric cyanide is capable of oxidizing benzidine, forming a deep-blue compound.

The drop reaction with benzidine is carried out as follows. Place a drop of the test solution, a drop of benzidine solution, and a drop of KCN solution** on a strip of filter paper (the sequence as written to be followed). In the presence of Cu^{2+}, the spot turns deep blue. The ions Ag^+ and Fe^{2+} must be absent.

---

* This also applies to the separation of Cu^{2+} by $Na_2S_2O_3$ which is oxidized by nitric acid to Na_2SO_4 and sulphur.

** Instead of KCN solution, use may be made of KBr or NH_4SCN solution since $CuBr_2$ and $Cu(SCN)_2$ are unstable, like $Cu(CN)_2$.

8. Zinc Acetate, $Zn(CH_3COO)_2$ and Ammonium Mercuri-Thiocyanate, $(NH_4)_2[Hg(SCN)_4]$. To a drop of the acidified test solution on a spot plate add a drop each of 1-percent solutions of $Zn(CH_3COO)_2$ and $(NH_4)_2[Hg(SCN)_4]$. This gives a precipitate of a complex compound of zinc, mercury, and copper cyanides having a violet colour. For the reaction to be effective, it is necessary to use a very dilute solution of a cupric salt, otherwise a yellow-green precipitate of $Cu[Hg(SCN)_4]$ is formed instead of a violet precipitate. The ions Fe^{3+}, Co^{2+}, and Ni^{2+} must be absent. The dilution limit is 1 in 500,000.

9. Microcrystalloscopic Test. To a drop of the test solution slightly acidified with CH_3COOH add a drop of a solution of $(NH_4)_2[Hg(SCN)_4]$. Characteristic yellow-green crystals of $Cu[Hg(SCN)_4]$ are formed (Fig. 6.2). The dilution (concentration) limit is 1 in 10,000. The Cd^{2+} ion also gives a crystalline precipitate with $(NH_4)_2[Hg(SCN)_4]$.

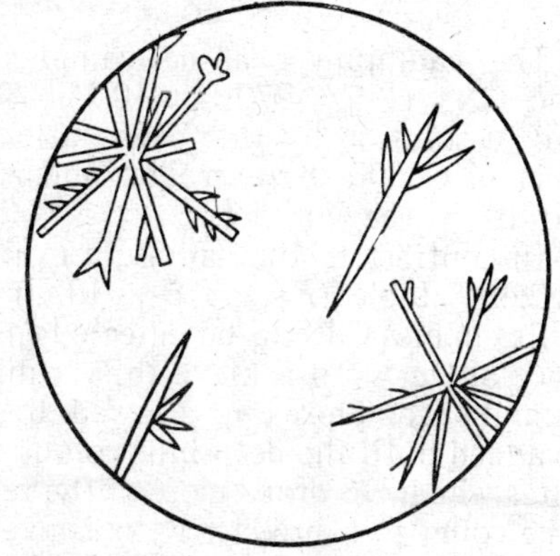

Fig. 6.2. Crystals of $Cu[Hg(SCN)_4]$.

10. Flame Test. Cupric salts impart a deep-blue or green colour to the nonluminous flame of the burner.

6.8. Reactions of the Cadmium Ion

In aqueous solution the Cd^{2+} ion is colourless.

1. Alkali Hydroxides, NaOH and KOH. These alkalis form a white precipitate of $Cd(OH)_2$ which is soluble in acids.

2. Ammonium Hydroxide, NH_4OH. It produces with Cd^{2+} a precipitate of cadmium hydroxide, $Cd(OH)_2$, which dissolves in excess of the precipitant to form an ammine complex:

$$Cd^{2+} + 2OH^{+} \rightarrow Cd(OH)_2\downarrow$$

$$Cd(OH)_2\downarrow + 4NH_4OH \rightarrow [Cd(NH_3)_4]^{2+} + 2OH^- + 4H_2O$$

In distinction to $[Cu(NH_3)_4]^{2+}$, the complex $[Cd(NH_3)_4]^{2+}$ is colourless. The reaction leading to the formation of these complexes is used in the course of analysis to separate Cu^{2+} and Cd^{2+} from Bi^{3+} which forms with NH_4OH a precipitate of a basic salt insoluble in excess ammonia.

3. Potassium Cyanide, KCN. This reagent produces with Cd^{2+} a white amorphous precipitate of $Cd(CN)_2$ which readily dissolves in excess of the reagent to yield the complex ion $[Cd(CN)_4]^{2-}$:

$$Cd^{2+} + 2CN^- \rightarrow Cd(CN)_2 \downarrow$$
$$Cd(CN)_2 \downarrow + 2CN^- \rightarrow [Cd(CN)_4]^{2-}$$

The cadmium cyanide complex dissociates partially into Cd^{2+} and CN^- ($K_{ins} = 7.8 \times 10^{-18}$). The concentration of Cd^{2+} produced in solution by this dissociation is enough for the solubility product of CdS to be exceeded under the action of H_2S and the precipitate to be formed.

In contrast to the cadmium cyanide complex, the copper complex is very stable ($K_{ins} = 5 \times 10^{-31}$) and is not decomposed by H_2S. This enables Cd^{2+} to be detected in the course of analysis in the presence of Cu^{2+}. To achieve this, a mixture of the copper and cadmium ammine complexes is allowed to react with KCN solution which is added until the deep-blue colour of the ion $[Cu(NH_3)_4]^{2+}$ disappears, after which 2-3 drops more of the reagent solution are added in excess. The complex $[Cu(NH_3)_4]^{2+}$ is converted into the more stable complex $[Cu(CN)_4]^{3-}$. If, after KCN has been added, H_2S is passed into the solution, then in the presence of Cd^{2+} a characteristic yellow precipitate of CdS is formed.

In carrying out this reaction it is necessary to remember that KCN is the most powerful poison. *Extreme care should be exercised in working with potassium cyanide!* Acid solutions must not be acted on by KCN because this would result in the formation of volatile and very poisonous hydrocyanic acid, HCN. When the tests with the cyanides are completed, the solutions should be drained into the sink (or into a bottle) in the fume cupboard.

4. Hydrogen Sulphide, H_2S. In not too strongly acid solutions and in neutral and alkaline solutions (at pH $\geqslant$ 0.5), hydrogen sulphide forms a yellow precipitate of CdS which is insoluble in alkali hydroxides and also in sodium sulphide (distinction from SnS_2 and the sulphides of arsenic). Excess of acid and also of halides interferes with this most characteristic reaction of Cd^{2+}. In contrast to CdS, cupric sulphide, which has a considerably lower solubility product, is formed in a strongly acid solution, even in the presence of a large excess of NH_4Cl. This fact can be used for separation of Cu^{2+} ions from Cd^{2+} with which they are found to be present in the course of analysis in the form of ammine complexes. They are separated by various methods.

(1) An ammoniacal solution containing the ions $[Cu(NH_3)_4]^{2+}$ and $[Cd(NH_3)_4]^{2+}$ is made slightly acid with 2*N* HCl, until the dark-blue colour of the copper complex changes to the light-blue colour of the cupric ion, following which an equal volume of 2*N* HCl solution (or a drop of concentrated HCl) is added per every 10-11 drops of the solution. The resulting solution (1*N* with respect to HCl) is heated nearly to boiling and is saturated with hydrogen sulphide before

it cools off. Make certain that the precipitation is complete, separate the CuS precipitate, and discard. Dilute the centrifugate three- or fourfold with water* and pass H_2S into it. A yellow precipitate of CdS indicates the presence of Cd^{2+}.

(2) To the ammoniacal solution add 2*N* HCl until acidic and solid NH_4Cl with shaking until the solution is saturated. Heat the solution and pass H_2S into it. When CuS is completely precipitated, separate the precipitate and discard. Dilute the centrifugate several-fold with water and pass H_2S into it. A yellow precipitate of CdS indicates the presence of Cd^{2+}.

(3) Acidify the ammoniacal solution slightly until the deep-blue colour of the copper ammine complex changes to the light-blue colour of the cupric ion, heat the solution, and pass H_2S into it until the sulphides of copper and cadmium are precipitated completely. Separate the precipitate and heat with a mixture of 5 drops of 2*N* HCl and 5 drops of water. Cadmium sulphide is dissolved, while CuS remains in the precipitate. Centrifuge off the precipitate, dilute the centrifugate threefold with water, and treat with hydrogen sulphide or freshly prepared H_2S water.

5. Separation of Cd^{2+} from Cu^{2+}, Pb^{2+}, and Bi^{3+}. This separation is effected with the aid of glycerol. It is based on the fact that glycerol prevents the precipitation of Cu^{2+}, Pb^{2+}, and Bi^{3+} by alkali hydroxides due to the formation of the corresponding glycerol complexes called glycerates. If, for example, to a solution of a cupric salt there is added glycerol, followed by NaOH, no precipitate of $Cu(OH)_2$ is formed and the solution assumes a dark-blue colour which is proof of the formation of copper glycerate:

$$Cu^{2+} + CH_2OH-CHOH-CH_2OH + 2OH^- \rightarrow CH_2OH-\underset{\lfloor\text{—Cu—}\rfloor}{CHO-CH_2O} + 2H_2O$$

Analogous compounds, though colourless, are also formed by Pb^{2+} and Bi^{3+}; the Cd^{2+} ion does not form complexes with glycerol and is therefore precipitated by alkalis in the presence of glycerol. So, if a solution containing Cu^{2+}, Pb^{2+}, Bi^{3+}, and Cd^{2+} is treated with glycerol and NaOH, the precipitate will be found to consist of $Cd(OH)_2$ and the solution will contain the glycerates of the other cations. Separate and wash the precipitate. Then dissolve in HCl and test for the presence of Cd^{2+}. Look for Pb^{2+} and Cu^{2+} in the centrifugate by fractional reactions after the glycerates have been destroyed by the action of CH_3COOH; Bi^{3+} is to be sought for directly in alkaline solution (through the reaction with $SnCl_2$; see below, Sec. 6.10, Reaction *4*). Upon this principle is based the glycerol-alkali method for analysis of the copper subgroup described in detail in Sec. 6.14.

---

* Without dilution no precipitation of CdS takes place.

6.9. Reactions of the Mercuric Ion

Depending on the course of analysis chosen, Hg^{2+} may be found to be either in the precipitate with the other sulphides of the copper subgroup, if the treatment of the entire precipitate of the Group IV sulphides and the separation of the arsenic subgroup have been carried out with the aid of $(NH_4)_2S_2$, or in the solution of thiosalts together with the cations of the arsenic subgroup, if the precipitated Group IV sulphides have been treated with Na_2S.

Accordingly, the Hg^{2+} ion may be assigned either to the copper or to the arsenic subgroup.

1. Alkali Hydroxides, NaOH and KOH. These form with mercuric salts a yellow precipitate of HgO since mercuric hydroxide is unstable:

$$Hg^{2+} + 2OH^- \rightarrow Hg(OH)_2 \downarrow$$
$$Hg(OH)_2 \downarrow \rightarrow HgO \downarrow + H_2O$$

2. Ammonium Hydroxide, NH_4OH. In solutions of mercuric salts, ammonium hydroxide produces a white precipitate of a complex salt; for example,

$$HgCl_2 + 2NH_4OH \rightarrow [NH_2Hg]Cl \downarrow + NH_4^+ + Cl^- + 2H_2O$$

The precipitate is soluble in acids.

3. Potassium Iodide, KI. Iodide ions produce in solutions of mercuric salts an orange-red precipitate of mercuric iodide, HgI_2:

$$Hg^{2+} + 2I^- \rightarrow HgI_2 \downarrow$$

The precipitate dissolves in an excess of potassium iodide, forming the complex ion $[HgI_4]^{2-}$ which is fairly stable ($K_{ins} = 1.5 \times 10^{-30}$):

$$HgI_2 \downarrow + 2I^- \rightarrow [HgI_4]^{2-}$$

If to the solution obtained there is added a drop of NH_4OH solution or of a solution of an ammonium salt, followed by a few drops of a concentrated solution of alkali, the reaction will produce a characteristic red-brown precipitate of ammonium mercuri-iodide:

$$NH_4^+ + 2[HgI_4]^{2-} + 4OH^- \rightarrow \left[O \begin{smallmatrix} \diagup Hg \diagdown \\ \\ \diagdown Hg \diagup \end{smallmatrix} NH_2 \right] I \downarrow + 7I^- + 3H_2O$$

This reaction known as the Nessler reaction has already been applied for detection of NH_4^+. It can also be used for identification of Hg^{2+}.

4. Potassium Chromate, K_2CrO_4. A yellow precipitate of $HgCrO_4$ is produced, which turns reddish on standing (due probably to the formation of a basic salt).

5. Reduction of Hg^{2+} to $[Hg_2]^{2+}$ and to Metallic Mercury. (1) When solutions of mercuric salts are treated with $SnCl_2$, a white precipitate of Hg_2Cl_2 is first formed, which gradually becomes gray in excess of the reagent, being reduced to metallic mercury.

The course of the reaction may be depicted by the following equations:

$$2HgCl_2 + Sn^{2+} \rightarrow Hg_2Cl_2\downarrow + Sn^{4+} + 2Cl^-$$

$$Hg_2Cl_2\downarrow + Sn^{2+} \rightarrow 2Hg\downarrow + Sn^{4+} + 2Cl^-$$

(2) Metallic copper displaces mercuric ions from solutions of mercuric salts, this resulting in the precipitation of metallic mercury-

$$Hg^{2+} + Cu\downarrow \rightarrow Hg\downarrow + Cu^{2+}$$

The same effect is produced by many other metals which stand to the left of mercury in the electromotive series.

6. Diphenylcarbazide.* In solutions of mercuric salts, diphenylcarbazide gives precipitates coloured violet or deep blue. The composition of the precipitates formed has not yet been established. The sensitivity of the test depends on the pH of the solution, decreasing with increasing acidity.

In neutral and acetic acid solutions, other cations of heavy metals (Cu^{2+}, Fe^{3+}, Co^{2+}, etc.) also produce coloured compounds with diphenylcarbazide but, in solutions containing 0.2 mole of HNO_3 per litre, the reaction is specific for Hg^{2+} (provided chromates and molybdates are absent). The sensitivity of the reaction in nitric acid medium is characterized by the detectable minimum: 1 μg Hg^{2+}.

The presence of large quantities of chlorides makes the identification of mercury difficult because of the formation of the slightly dissociated salt $HgCl_2$.

6.10. Reactions of the Bismuth Ion

In aqueous solutions the Bi^{3+} ion is colourless.

1. Action of Water (Hydrolysis). When a solution of a bismuth salt (say, $BiCl_3$) is poured into a large volume of water, a white precipitate of a basic salt (BiOCl) is produced. The hydrolysis taking place according to the equation

$$BiCl_2 + 2H_2O \rightarrow Bi(OH)_2Cl\downarrow + 2HCl$$

yields first a basic salt containing two hydroxyl groups in the molecule. This salt however is unstable and gives a water molecule as fast as it is formed, producing a new basic salt of the composition BiOCl, in which one oxygen atom has been substituted for two hyd-

* Its formula is

$$O{=}C\begin{matrix}\diagup NH-NH-C_6H_5\\ \diagdown NH-NH-C_6H_5\end{matrix}$$

roxyl groups:

$$Bi(OH)_2Cl\downarrow \rightarrow BiOCl\downarrow + H_2O$$

Hence, as the result of the formation of water, the basic salt may contain no hydroxyl groups in its molecule. The ionic equation of the reaction is

$$Bi^{3+} + Cl^- + H_2O \rightarrow BiOCl\downarrow + 2H^+$$

The compound $Bi(NO_3)_3$, which forms a basic salt of the composition $BiONO_3$, is hydrolysed in just the same way. With bismuth nitrate however the reaction is less effective because $BiONO_3$ is more soluble than BiOCl; that is why it is better to add a chloride, say, NaCl, to the solution of $Bi(NO_3)_3$. The presence of this chloride favours the precipitation of BiOCl.

The radical —Bi=O entering the composition of basic bismuth salts is known as bismuthyl. The compounds BiOCl and $BiONO_3$ are respectively called bismuthyl chloride and bismuthyl nitrate.

The reactions leading to the formation of basic salts are reversible. Therefore, addition of acid to BiOCl and heating of the solution dissolve the precipitate:

$$BiOCl\downarrow + 2H^+ \rightarrow Bi^{3+} + Cl^- + H_2O$$

Addition of water again precipitates the basic salt.

2. Alkali Hydroxides, NaOH and KOH. A white precipitate of bismuth hydroxide, $Bi(OH)_3$, is formed, which is soluble in acids but insoluble in alkalis:

$$Bi^{3+} + 3OH^- \rightarrow Bi(OH)_3\downarrow$$

3. Ammonium Hydroxide, NH_4OH. A white precipitate of a basic salt of variable composition is precipitated. The composition of the salt varies, depending on the concentration of the solution and on temperature.

4. Potassium and Sodium Stannites, K_2SnO_2 and $NaSnO_2$. These salts reduce Bi^{3+} to metallic bismuth (a black-coloured precipitate).

In carrying out this reaction, proceed as follows. To 2 drops of a solution of $SnCl_2$ add (in the cold) 8-10 drops of $2N$ NaOH or KOH solution in order to dissolve the initially precipitated $Sn(OH)_2$ with the formation of a stannite:

$$Sn^{2+} + 2OH^- \rightarrow Sn(OH)_2\downarrow \quad (\text{or } H_2SnO_2)$$

$$H_2SnO_2\downarrow + 2OH^- \rightarrow 2H_2O + SnO_2^{2-}$$

To the alkaline solution obtained add a drop of a solution of a bismuth salt. This yields a black precipitate of metallic bismuth:

$$Bi^{3+} + 3OH^- \rightarrow Bi(OH)_3\downarrow$$

$$2Bi(OH)_3\downarrow + SnO_2^{2-} \rightarrow 2Bi\downarrow + 3SnO_3^{2-} + 3H_2O$$

This reaction is very interesting from the theoretical standpoint. Metallic bismuth finds wide application as a rather powerful reducing agent in acid solution ($E^0_{BiO^+/Bi\downarrow} = +0.32$ volt). As an example, metallic bismuth is capable of reducing iron(III) to iron(II) ($E^0_{Fe^{3+}/Fe^{2+}} = +0.77$ volt).

In the reaction with stannites given above, bismuth(III) is reduced to the metal because in alkaline medium the reducing power of tin(II) is strongly increased ($E^0_{Sn(OH)_6^{2-}/HSnO_2^-,\ OH^-} = -0.93$ volt) and it becomes possible to reduce Bi^{III} to the metal.

Under the stated conditions, the reduction of Pb^{II} to metallic lead is also a slow reaction ($E^0_{Pb^{2+}/Pb\downarrow} = -0.126$ volt). This often introduces complications into the analysis since, with the separation of the groups being not sharp enough, the lead present in the test solution may be mistaken for bismuth.

The compounds Na_2SnO_3 and K_2SnO_3 formed in this reaction are known as stannates. They are the salts of metastannic acid, H_2SnO_3, which corresponds to the oxide SnO_2 (stannic oxide).

This most important reaction of Bi^{3+} may also be carried out by using a solution of a stannite for treating the precipitate of the basic salts of bismuth obtained after the separation of Bi^{3+} from Cu^{2+} and Cd^{2+} through the action of excess NH_4OH.

The reaction is also positive with an alkaline solution of a complex compound of bismuth with glycerol since the dissociation of this compound produces a high enough concentration of Bi^{3+}.

In carrying out this reaction, one must avoid adding concentrated alkali and heating because under these conditions a black precipitate of metallic tin may be formed as a result of the decomposition of the stannite:

$$2SnO_2^{2-} + H_2O \rightarrow SnO_3^{2-} + Sn\downarrow + 2OH^-$$

If the excess of alkali is too small, then, with the solution being allowed to stand, black stannous oxide, SnO, may be precipitated.

5. Potassium Dichromate, $K_2Cr_2O_7$. A yellow precipitate of bismuthyl dichromate, $(BiO)_2Cr_2O_7$, is produced when potassium dichromate is allowed to react with solutions of bismuth salts:

$$2Bi^{3+} + Cr_2O_7^{2-} + 2H_2O \rightarrow (BiO)_2Cr_2O_7\downarrow + 4H^+$$

The precipitate resembles plumbous chromate in external appearance but differs from it by being soluble in acetic acid and insoluble in alkalis.

6. Disodium Hydrogen Phosphate, Na_2HPO_4. This reagent brings down Bi^{3+} in the form of a white finely divided precipitate of bismuth phosphate:

$$Bi^{3+} + 2HPO_4^{2-} \rightarrow BiPO_4\downarrow + H_2PO_4^-$$

The precipitate is practically insoluble in dilute nitric acid. Owing to this fact, the formation of bismuth phosphate can be used for separation of PO_4^{3-} (which interfere with the separation of the cations of Groups III and II by ammonium sulphide) in the systematic cation analysis (see Sec. 6.21).

7. Potassium Iodide, KI. In sufficiently concentrated acidified solutions of bismuth salts, the reagent brings down a black precipitate of bismuth triiodide, BiI_3, which is soluble in excess KI with the formation of orange-coloured complex $[BiI_4]^-$ ions:

$$Bi^{3+} + 3I^- \rightarrow BiI_3\downarrow$$
$$BiI_3\downarrow + I^- \rightarrow [BiI_4]^-$$

When the solution is moderately diluted with water, a black precipitate of BiI_3 appears again:

$$[BiI_4]^- \rightarrow BiI_3\downarrow + I^-$$

On strong dilution, an orange precipitate of a basic salt settles down instead of BiI_3:

$$[BiI_4]^- + H_2O \rightarrow BiOI\downarrow + 3I^- + 2H^+$$

8. Thiourea, $SC(NH_2)_2$. The action of a solution of thiourea on nitric acid solutions of bismuth salts produces a yellow colouration due to the formation of complex $[Bi[SC(NH_2)_2]_3]^{3+}$ ions. The ions Pb^{2+} and Cd^{2+} give slightly soluble compounds with thiourea but do not interfere with the detection of Bi^{3+}.

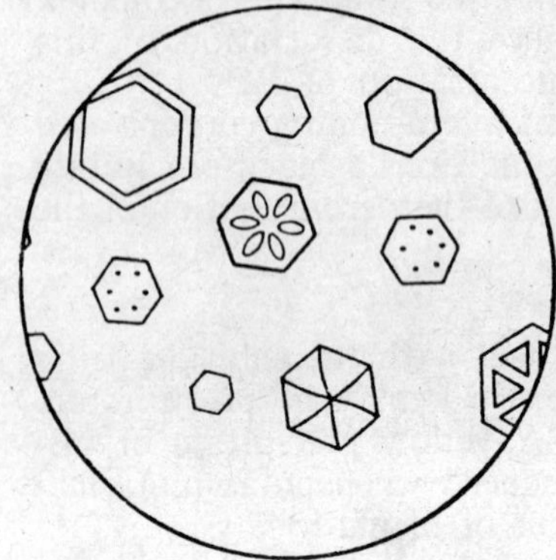

Fig. 6.3. Crystals of $Cs_2[BiI_5]\cdot 2.5H_2O$.

9. 8-Hydroxyquinoline, C_9H_7ON, and Potassium Iodide, KI. In solutions of bismuth salts, these reagents produce an orange-red precipitate (a compound of 8-hydroxyquinoline with the complex acid $HBiI_4$, which has the composition $C_9H_7ON\cdot HBiI_4$). The solution must not contain Ag^+, $[Hg_2]^{2+}$, and Pb^{2+}, which precipitate I^-, and also oxidizing agents, say, Cu^{2+} and Fe^{3+}, which are capable of oxidizing I^- to I_2. The interfering effect of oxidizing agents is eliminated by adding $Na_2S_2O_3$ or $SnCl_2$. This is done as follows. To a few drops of the solution add 2-3 drops of the reagent (a mixture of 8-hydroxyquinoline and KI) and a drop of starch solution. If the solution turns deep-blue in colour (due to the evolution of iodine by the oxidants present in the solution), add, drop by drop, $Na_2S_2O_3$ solution until colourless. In the presence of Bi^{3+}, an orange-red precipitate is formed.

The reaction can also be carried out by the spot-test technique. Moisten a strip of filter paper with a solution of $SnCl_2$ and put on it a drop each of the test solution and the reagent. A test for complete reduction of the oxidants is made by moistening the reaction spot with a drop of starch solution. An orange-red colouration indicates the presence of Bi^{3+}

10. Microcrystalloscopic Test. To a drop of a solution of a bismuth salt add a small drop of 2*N* HCl, a drop of KI solution, and a small crystal of CsCl (or RbCl). This gives orange-red hexagonal crystals (Fig. 6.3) of $Cs_2[BiI_5]\cdot 2.5H_2O$ or $Rb_2[BiI_5]\cdot 2.5H_2O$. The detectable minimum (sensitivity) is 0.13 μg.

Analogous crystals are formed by Sb^{3+} ions.

6.11. Reactions of Arsenous and Arsenic Ions

In solutions of the salts of arsenous acid, arsenic(III) is present mainly in the form of the anion AsO_2^-. But when solutions containing these anions are acidified, As^{3+} cations are formed in concentrations enough for certain reactions to take place.

Arsenic(V) forms almost exclusively AsO_4^{3-} anions (or $HAsO_4^{2-}$). All these anions are colourless in aqueous solutions.

All arsenic compounds are very poisonous!

(a) Reactions of Arsenite Ions (AsO_2^- and AsO_3^{3-})

1. Silver Nitrate, $AgNO_3$. With AsO_2^- silver nitrate gives a yellow precipitate of silver orthoarsenite, Ag_3AsO_3, soluble in nitric acid and ammonium hydroxide. In the latter case, the complex ion $[Ag(NH_3)_2]^+$ is produced:

$$AsO_2^- + H_2O + 3Ag^+ \rightarrow Ag_3AsO_3\downarrow + 2H^+$$

$$Ag_3AsO_3\downarrow + 6NH_3 \rightarrow 3[Ag(NH_3)_2]^+ + AsO_3^{3-}$$

2. Iodine Solution. Iodine oxidizes arsenite to arsenate in a neutral or faintly alkaline solution:

$$AsO_2^- + I_2 + 2H_2O \rightarrow H_2AsO_4^- + 2I^- + 2H^+$$

This is easily recognized by the disappearance of the dark-brown colour characteristic of iodine solutions.

Since the standard potential of the couple $H_3AsO_4/HAsO_2$ ($E^0 = +0.56$ volt) is greater than that of the couple $I_2\downarrow/2I^-$ ($E^0 = +0.54$ volt), the equilibrium of the reaction is shifted from right to left.

To make the reaction proceed in the opposite direction, it is necessary to bind H^+ by addition of $NaHCO_3$; as a result of this, the equilibrium potential of the couple $H_3AsO_4/HAsO_2$ becomes less than that of the couple $I_2\downarrow/2I^-$ (see Sec. 6.3).

To 2-3 drops of an acid solution of $NaAsO_2$ add a little solid $NaHCO_3$* and, after the latter is dissolved, add a drop of iodine solution; the iodine solution becomes colourless. The anion AsO_2^- can reduce other oxidants as well (e.g., chlorine, bromine).

(b) Reactions of the Arsenate Ion (AsO_4^{3-})

1. **Silver Nitrate, $AgNO_3$.** With AsO_4^{3-} silver nitrate forms a characteristic chocolate-brown (distinction from AsO_2^-) precipitate of silver arsenate, Ag_3AsO_4, soluble in HNO_3 and NH_4OH:

$$3Ag^+ + AsO_4^{3-} \rightarrow Ag_3AsO_4 \downarrow$$

2. **Potassium Iodide, KI.** This compound is oxidized by AsO_4^{3-} in acid solution to iodine, as a result of which the solution turns brown:

$$H_2AsO_4^- + 2I^- + 3H^+ \rightarrow HAsO_2 + I_2 \downarrow + 2H_2O$$

The sensitivity of the reaction can be increased by addition of starch (a deep-blue colouration) or by extraction of I_2 with the aid of an organic solvent: benzene, chloroform, etc. (violet colouration of the organic layer).

This reaction is exactly the reverse of the one with arsenate (AsO_2^-) and iodine considered above. In this case a greater increase of the hydrogen ion concentration is required, which, according to the law of mass action, favours the shifting of the equilibrium in the desired direction (i.e., to the right).

According to what has just been said, the reaction in question is to be conducted in the presence of a few drops of concentrated HCl. Add to the solution a little benzene, in which I_2 is more soluble than in water, shake and allow to stand. A greater portion of iodine passes into the benzene layer which assumes a characteristic violet colour.

3. **Magnesium Salts in the Presence of NH_4OH and NH_4Cl.** A white crystalline precipitate of magnesium ammonium arsenate, $MgNH_4AsO_4 \cdot 6H_2O$, is produced, which is soluble in acids but practically insoluble in a 2.5-percent ammonia solution:

$$Mg^{2+} + NH_4OH + HAsO_4^{2-} \rightarrow MgNH_4AsO_4 \downarrow + H_2O$$

The reaction is analogous to the one with Mg^{2+} (see Sec. 3.15), the only difference being that, instead of PO_4^{3-}, the ion AsO_4^{3-}, which is very similar to it in chemical properties, takes part in the reaction.

The reaction is conducted as follows. Mix 2-3 drops of a solution of arsenate with an equal volume of magnesia mixture** and, if required, let it stand.

---

* It must not be replaced with Na_2CO_3 because sodium carbonate decolorizes iodine in the absence of AsO_2^- too

$$Na_2CO_3 + H_2O \rightarrow NaOH + NaHCO_3$$
$$I_2 + 2NaOH \rightarrow NaI + NaIO + H_2O$$

** A solution containing $MgCl_2$, NH_4Cl, and NH_4OH.

4. Ammonium Molybdate Solution. A solution of ammonium molybdate, $(NH_4)_2MoO_4$, in nitric acid, which contains NH_4NO_3, produces with the arsenate ion a yellow crystalline precipitate of ammonium arsenomolybdate of the composition $(NH_4)_3[AsMo_{12}O_{40}] \cdot xH_2O$.

The precipitate is easily soluble in an excess of arsenate with the formation of a complex anion. Consequently, a relatively large excess of the reagent must be added.

To 2-3 drops of the arsenate solution add 10-15 drops of the ammonium molybdate solution. Now add solid NH_4NO_3 (to increase the sensitivity of the reaction through the action of NH_4^+ ions) and heat for a few minutes in the water bath.*

The precipitate is insoluble in nitric acid but dissolves readily in caustic alkalis and ammonium hydroxide. The ion PO_4^{3-} gives a precipitate of $(NH_4)_3[PMo_{12}O_{40}] \cdot xH_2O$ which is analogous to the one mentioned above, except that it settles down in the cold (the precipitation of $(NH_4)_3[AsMo_{12}O_{40}] \cdot xH_2O$ requires that the solution be heated nearly to boiling).

5. Reduction of Arsenous and Arsenic Compounds to AsH_3. To detect small amounts of arsenic in various samples, use is often made of a very sensitive test in which arsenous and arsenic compounds are reduced to gaseous arsenous hydride (arsine, AsH_3).

The formation of AsH_3 can be recognized in various ways, most conveniently by the blackening of the paper moistened with a solution of $AgNO_3$. The blackening occurs as a result of the reduction of Ag^+ to metallic silver by arsine. *Arsine is extremely poisonous. The reactions must be carried out in the fume cupboard.*

(1) *Reduction in Acid Solution* (according to Tananaev). Place a few drops of HCl solution in a micro crucible and put it on a sheet of filter paper. Moisten the paper around the crucible with a solution of $AgNO_3$. Now introduce into the crucible a piece of metallic magnesium (or zinc**) and cover at once the crucible and the moistened spot on the paper with a small funnel whose tube is sealed (or rubber-stoppered). After the lapse of a few minutes examine the paper: if it has not blackened, it means that the reagents are pure and the test for arsenic may be started. Now repeat the test described above, by placing into the crucible, apart from HCl and magnesium, 2-3 drops of the test solution. In the presence of arsenic, the paper is rapidly blackened because of the reduction of the Ag^+ ions by the evolved AsH_3 to the metal. The reactions involved are as follows:

(a) in the crucible

$$AsO_3^{3-} + 9H^+ + 3Mg \downarrow \rightarrow AsH_3 \uparrow + 3Mg^{2+} + 3H_2O$$

---

* In this case, NH_4NO_3 acts also as a coagulant-electrolyte (see Sec. 5.3). Coagulation of the precipitate is also favoured by heating.

** Use concentrated HCl for zinc.

(b) on the paper

$$AsH_3\uparrow + 6Ag^+ + 3H_2O \rightarrow 6Ag\downarrow + H_3AsO_3 + 6H^+$$

Under the given conditions, antimony forms gaseous antimonous hydride, SbH_3, which also blackens the paper moistened with $AgNO_3$. Besides, no sulphides must be present because the hydrogen sulphide evolved when sulphides are acted on by HCl also causes a blackening of the paper due to the formation of Ag_2S.

(2) *Reduction in Alkaline Medium.* Use a concentrated (30-percent) NaOH solution, instead of HCl, and powdered aluminium or zinc dust*, instead of magnesium; other conditions are the same as in the reduction in acid solution. For example,

$$AsO_3^{3-} + 3Zn\downarrow + 3OH^- \rightarrow AsH_3\uparrow + 3ZnO_2^{2-}$$

In alkaline medium, only AsO_3^{3-} ions are reduced to AsH_3 but not AsO_4^{3-} ions. Hence, one must first reduce AsO_4^{3-} to AsO_3^{3-} by the action of KI in the presence of H_2SO_4. The iodine formed is removed by evaporating the solution to dryness (or until H_2SO_4 vapours are evolved), following which the residue left after the evaporation is treated with alkali and aluminium, or zinc dust, as described above. Antimony is not reduced by these metals in alkaline medium and does not interfere with the test.

6. Reduction of As^{III} and As^{V} to Free Arsenic. This reduction can be effected through the action of a saturated solution of $SnCl_2$ in concentrated HCl:

$$2As^{3+} + 3Sn^{2+} \rightarrow 2As\downarrow + 3Sn^{4+}$$

This reaction may be used to detect arsenic in the presence of other cations. Place into a micro crucible a drop or two of the solution obtained after the separation of Group V by HCl, add 2 drops each of solutions of NH_4OH, H_2O_2, and $MgCl_2$. Hydrogen peroxide oxidizes arsenous compounds to AsO_4^{3-} which gives a precipitate of $MgNH_4AsO_4$ with ammonia and a magnesium salt. Evaporate the solution slowly to dryness and ignite strongly to remove $HgCl_2$ which may be present and which can mask the reaction. On ignition, $MgNH_4AsO_4$ is converted into the non-volatile salt $Mg_2As_2O_7$ (magnesium pyroarsenate) which remains in the crucible. After the dry residue is cooled, treat it with a few drops of a solution of $SnCl_2$ in concentrated HCl and heat slightly. In the presence of arsenic, the solution turns brown or a black precipitate of As is formed.

The reagent that must be used is a solution of $SnCl_2$ in concentrated HCl since only the cations As^{V} and As^{III} are reduced and these are formed in a sufficient concentration only in a strongly acid solution.

---

* If zinc is used, the solution in the dish must be heated. Of course, a preliminary test for purity of the reagents is required in this case as well.

7. **Microcrystalloscopic Test.** (1) The above reaction leading to the formation of crystals of $MgNH_4AsO_4 \cdot 6H_2O$ can be carried out as a microcrystalloscopic test. Mix a drop of the test arsenate solution on a glass slide with a drop of magnesia mixture with the aid of a sharpened glass rod and examine the crystals formed under the microscope. They are of the same shape as the crystals of $MgNH_4PO_4 \cdot 6H_2O$

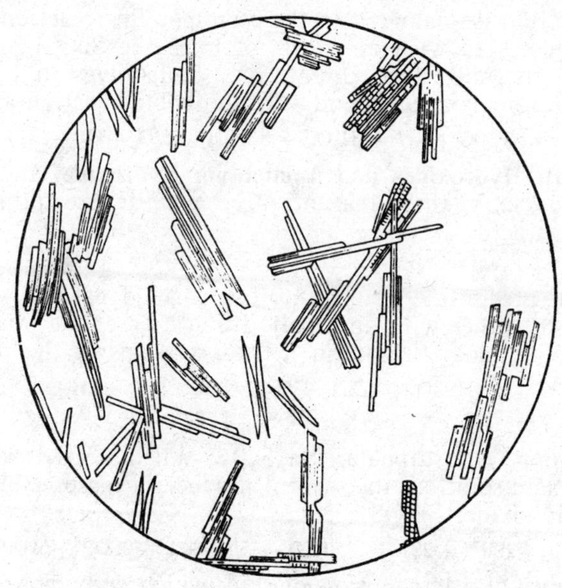

Fig. 6.4. Crystals of $CaNH_4AsO_4 \cdot 6H_2O$.

(see Figs. 3.7 and 3.8). The dilution (concentration) limit is 1 in 20,000. If the concentration of AsO_3^{3-} exceeds 0.5-1.0 per cent, it may give analogous crystals.

(2) Introduce a small crystal of calcium acetate, $Ca(CH_3COO)_2$ into a drop of an ammoniacal solution of arsenate and cover it with a cover glass (to protect against the action of CO_2 present in the air). An amorphous precipitate is formed first, which give rise to crystals of $CaNH_4AsO_4 \cdot 6H_2O$ (Fig. 6.4). The concentration limit is 1 in 30,000.

6.12. Reactions of Antimonous and Antimonic Ions

Antimony, like arsenic, forms two series of compounds in which its oxidation states are +3 and +5. Solutions of antimony salts are colourless.

(a) Reactions of the Antimonous Ion

1. Action of Water (Hydrolysis). When a few drops of a solution of an antimonous salt are diluted with a large amount of water, a white precipitate of the basic salt SbOCl (antimonyl chloride) is formed*:

$$[SbCl_6]^{3-} + H_2O \rightarrow SbOCl\downarrow + 5Cl^- + 2H^+$$

The reaction is analogous to the corresponding reaction of the Bi^{3+} ion. When heated with an excess of HCl, the SbOCl precipitate is dissolved. In contrast to BiOCl, it also dissolves in tartaric acid, forming a complex compound. The probable reaction is

$$SbOCl\downarrow + H_2C_4H_4O_6 \rightarrow Sb(OH)C_4H_4O_6 + H^+ + Cl^-$$

2. Alkali Hydroxides and Ammonium Hydroxide. In solutions of antimonous salts, these reagents give a white precipitate of meta-antimonous acid:

$$[SbCl_6]^{3-} + 3OH^- \rightarrow HSbO_2\downarrow + 6Cl^- + H_2O$$

Divide the precipitate formed into two parts. Treat one with a strong acid and the other with an alkali. In both cases the precipitate dissolves. (What does this mean?) The equations of the reactions are

$$HSbO_2\downarrow + 3H^+ + 6Cl^- \rightarrow [SbCl_6]^{3-} + 2H_2O$$

$$HSbO_2\downarrow + OH^- \rightarrow SbO_2^- + H_2O$$

3. Sodium Thiosulphate, $Na_2S_2O_3$. When heated with Sb^{III} in a faintly acid solution, the reagent produces a red precipitate of antimony sulphoxide, Sb_2OS_2:

$$2Sb^{3+} + 2S_2O_3^{2-} + 3H_2O \rightarrow Sb_2OS_2\downarrow + 2SO_4^{2-} + 6H^+$$

The presence of a large excess of acid, which decomposes the reagent into SO_2 and sulphur, interferes with the reaction.

4. Reduction of Sb^{III} to Metallic Antimony. (1) *Reduction by Tin.* Put a drop of a hydrochloric acid solution of an antimonous salt on a piece of tin foil and let it stand for some time. A black stain (deposit) of metallic antimony is formed:

$$2[SbCl_6]^{3-} + 3Sn\downarrow \rightarrow 2Sb\downarrow + 3Sn^{2+} + 12Cl^-$$

In contrast to arsenic, the antimony stain does not disappear upon treatment with a drop of a freshly prepared solution of NaBrO**. Before this test is started, wash off the test acid solution, remaining on the tin surface, because in the presence of acid the antimony stain may also disappear. The presence of tin ions does not interfere with the detection of antimony by means of this reaction.

* The radical —Sb=O is known as antimonyl.

** The NaBrO solution is prepared by adding, drop by drop, 2*N* NaOH to 2-3 drops of saturated bromine water until colourless (or until the colour becomes pale-yellow):

$$2NaOH + Br_2 \rightarrow NaBrO + NaBr + H_2O$$

(2) *Reduction by Zinc.* Put a drop or two of a solution of an antimonous salt on platinum foil and dip a piece of zinc into it so that the zinc and platinum are brought into contact with each other. This forms a galvanic couple in which Zn is the negative and Pt the positive pole. The platinum gains the electrons lost by zinc and gives them to antimony ions which are reduced to the metal. As a result of this, a black stain of metallic antimony appears on the platinum.* In distinction to tin, antimony is insoluble in HCl but dissolves in HNO_3.

Antimony(III) is reduced in acid solution to metallic antimony by many other metals to the left of Sb in the electromotive series (Mg, Fe, etc.).

5. Methyl Violet. To a drop of the test solution add 3 drops of a concentrated (12*N*) solution of HCl and a drop of 1*N* KNO_2 or $NaNO_2$ solution. This oxidizes Sb^{III} to $[SbCl_6]^-$ in which antimony is pentavalent.** Let it stand for at least one minute (to complete the reaction) and add a drop of a saturated solution of thiourea, $CO(NH_2)_2$, in order to destroy the excess of nitrite. Dilute the test solution to 3 ml with water, add 3-4 drops of a 0.06-percent solution of methyl violet***, and shake.

In the presence of Sb^V, a fine suspension appears, which consists of crystals of a salt formed by the organic cation of the dye and the anion $[SbCl_6]^-$. This salt is violet in transmitted light. If Sb^V is absent, the solution has a yellow-green colour.

The dilution limit is 1 in 10,000,000. Detection of antimony by means of this reaction is possible in the presence of tin, iron, copper, zinc, and many other elements which may be present in amounts more than 1000 times the content of antimony.

The organic dye rhodamine B**** reacts similarly with antimony(V), producing a deep-blue or violet colouration. If antimony(V) is absent,

---

* In a strongly acid solution, Zn reduces partially Sb^{III} to SbH_3 (compare with the reaction of As^{III}).

** It is not always possible to convert antimonic compounds into $[SbCl_6]^-$ by the direct action of HCl; therefore, Sb^V must preliminarily be reduced to Sb^{III} by $SnCl_2$. The reaction is conducted as described, except that a larger amount of the nitrite is taken (to oxidize the excess of $SnCl_2$).

*** See Section 5.21, Reaction *10*.

**** It is also known as tetraethyl-rhodamine:

O

$(C_2H_5)N$ — C — $= \overset{+}{N}(C_2H_5)_2Cl^-$

HOOC —

The composition of the compound formed in the reaction under consideration is unknown.

the colour is pink. The test is performed as described above, except that in this case there is no need to add urea.

6. **Phosphomolybdic Acid*, $H_3[PMo_{12}O_{40}]\cdot xH_2O$.** Molybdenum(VI) entering into the composition of this acid is highly reactive, as is sometimes the case in complex compounds; it can be reduced to a substance known as molybdenum blue or heteropoly blue (a mixture of various molybdenum compounds in the lowest oxidation state) by reducing agents that do not react with molybdic acid and molybdates. This reduction reaction is used for determination of antimony(III).

Place a drop of the test solution on a piece of filter paper which has been impregnated with a 5-percent solution of phosphomolybdic acid, and hold the paper in steam for a few minutes. A more or less deep-blue colouration appears in the presence of Sb^{III}. The concentration limit is 1 in 250,000.

A similar colouration is produced in the presence of Sn^{2+}; therefore Sn^{2+} ions must be absent in the solution to be analysed. Tin(IV) ions do not interfere.

7. **Potassium Mercuri-Iodide, $K_2[HgI_4]$.** In the presence of alkali, the reagent is reduced by antimony(III) to metallic mercury which settles as a black precipitate. Antimony(III) reacts in the same way with solutions of $Hg(CN)_2$. Antimony(V) gives no such reactions. Stannous ions act similarly to antimonous ions and must be absent in the test solution.

8. **8-Hydroxyquinoline, C_9H_7ON.** In the presence of potassium iodide, KI, in a strongly acid solution ($[H^+] = 6\text{-}7N$), the reagent produces a yellow precipitate which is believed to be a compound of 8-hydroxyquinoline with the complex acid $HSbI_4$. The probable composition is $C_9H_7ON\cdot HSbI_4$ (compare the analogous reaction of Bi^{3+}, Sec. 6.10, Reaction *9*). The ions of tin and arsenic do not interfere, but the cations of other groups must be absent. The test may thus be applied only in the systematic analysis, after the arsenic subgroup has been separated from the copper subgroup.

The acid method of separating the ions of the arsenic subgroup (see Sec. 6.14, Procedure *5*) yields a solution which is $6N$ in hydrochloric acid and which may contain $[SnCl_6]^{2-}$ and $[SbCl_6]^{3-}$ ions. To detect antimony in this solution, proceed as follows. Place into a dry test tube 2-3 drops of the test solution and treat with a drop or two of the reagent solution (a mixture of 8-hydroxyquinoline and KI). If the solution contains antimony ions, a yellow precipitate forms instantly at the boundary between the solutions. It dissolves

---

* Phosphomolybdic acid can be prepared by boiling its ammonium salt with aqua regia. When the salt is completely dissolved, the solution is evaporated to dryness; the dry residue is dissolved in water, allowed to recrystallize twice (from water), and stored in a vessel of dark glass.

on heating and reappears on cooling. The precipitate also dissolves upon addition of water. The detectable minimum (sensitivity) is 2 μg Sb.

(b) Reactions of the Antimonic Ion

1. **Action of Water.** When a solution of an antimonic salt is diluted with water, a white precipitate of the basic salt SbO_2Cl is formed:

$$[SbCl_6]^- + 2H_2O \rightarrow SbO_2Cl\downarrow + 5Cl^- + 4H^+$$

The precipitate dissolves when heated with excess HCl.

2. **Alkali Hydroxides and Ammonium Hydroxide.** A white precipitate of metaantimonic acid is produced:

$$[SbCl_6]^- + 5OH^- \rightarrow HSbO_3\downarrow + 6Cl^- + 2H_2O$$

3. **Metallic Sn, Zn, Mg, Al, and Fe.** These metals act on solutions of antimonic salts in just the same way as on solutions of antimonous salts.

6.13. Reactions of Stannous and Stannic Ions

Tin forms two series of compounds: stannous salts in which tin is in the +2 oxidation state, and stannic salts in which its oxidation state is +4. Solutions of tin salts are colourless.

(a) Reactions of the Stannous Ion

1. **Alkali Hydroxides and Ammonium Hydroxide.** These produce with Sn^{2+} a white precipitate of $Sn(OH)_2$:

$$Sn^{2+} + 2OH^- \rightarrow Sn(OH)_2\downarrow \quad (\text{or } H_2SnO_2)$$

The precipitate is amphoteric, i.e., soluble in both acids and alkalis (*Test!*). In the latter case, stannites (SnO_2^{2-}) are formed:

$$H_2SnO_2 + 2OH^- \rightarrow SnO_2^{2-} + 2H_2O$$

The $Sn(OH)_2$ precipitate is insoluble in excess NH_4OH.

2. **Reduction Reactions Effected by Stannous Ion.** The standard oxidation potential of the couple Sn^{4+}/Sn^{2+} is +0.15 volt. Accordingly, the Sn^{2+} ion is a rather strong reducing agent, capable of reducing many oxidants with a high oxidation potential. Let us consider some examples of such reactions.

(1) *Reduction of Mercuric Chloride.* To 2-3 drops of a solution of $SnCl_2$ add a drop of a solution of $HgCl_2$. This yields a white precipitate of Hg_2Cl_2 turning gradually black because of the subsequent reduction to metallic mercury:

$$2HgCl_2 + SnCl_2 \rightarrow Hg_2Cl_2\downarrow + SnCl_4$$
$$Hg_2Cl_2\downarrow + SnCl_2 \rightarrow 2Hg\downarrow + SnCl_4$$

This reaction has already been described as the reaction for Hg^{2+} but it is of no less importance for detection of Sn^{2+}.

(2) *Reduction of* Bi^{3+}. This reaction is also employed for identification of Bi^{3+}:

$$2Bi(OH)_3\downarrow + 3SnO_2^{2-} \rightarrow 2Bi\downarrow + 3SnO_3^{2-} + 3H_2O$$

The procedure for carrying out this reaction was described in Sec. 6.10.

(3) *Reduction of* Fe^{3+} *to* Fe^{2+}. To a drop of a solution of a ferric salt add a drop each of $2N$ solutions of HCl and $K_3[Fe(CN)_6]$. To the resulting solution add a drop of a solution of a stannous salt. A dark-blue precipitate of Turnbull's blue is instantly formed:

$$2Fe^{3+} + Sn^{2+} \rightarrow 2Fe^{2+} + Sn^{4+}$$

$$4Fe^{2+} + 4[Fe(CN)_6]^{3-} \rightarrow Fe_4[Fe(CN)_6]_3 + [Fe(CN)_6]^{4-}$$

(4) *Reduction of* $(NH_4)_3[PMo_{12}O_{40}]\cdot xH_2O$. The antimonous ion is capable of reducing to molybdenum blue only the free phosphomolybdic acid $H_3[PMo_{12}O_{40}]\cdot xH_2O$ or its soluble salts, while the stannous ion, which is a stronger reducing agent, can, in addition, reduce the slightly soluble ammonium salt $(NH_4)_3[PMo_{12}O_{40}]\cdot xH_2O$. Upon this fact is based the test for Sn^{2+}, which allows the ion to be detected in the presence of Sb^{3+}.

Moisten a piece of filter paper with a solution of phosphomolybdic acid and hold it for a short time over the mouth of a flask containing a concentrated ammonia solution. A stain (deposit) of ammonium phosphomolybdate, $(NH_4)_3[PMo_{12}O_{40}]\cdot xH_2O$, is formed on the paper. Dry the paper and place a drop of the test solution on it. If stannous ion is present, a more or less intense colouration appears, the intensity depending on the amount present. The test is very sensitive. The dilution (concentration) limit is 1 in 1,670,000.

The reaction in a test tube is carried out as follows. To a drop of Na_2HPO_4 solution add 2-4 drops of ammonium molybdate solution and heat. After cooling, dilute the contents of the test tube with the precipitated ammonium phosphomolybdate with about 3 ml of water and add the test solution a drop at a time. A blue colouration confirms the presence of Sn^{2+}.

In using this reaction for detection of stannous and stannic tin in the course of analysis, one must remember that it may be present in solution in the form of $[SnCl_6]^{2-}$ anions. These anions must be reduced to Sn^{2+} through the action of metallic magnesium.

(5) *Reduction of Picric Acid,* $C_6H_2(NO_2)_3OH$. The reduction of picric acid to picramic acid, $C_6H_2(NO_2)_2(NH_2)OH$, is very characteristic of the Sn^{2+} ion. It can most conveniently be carried out by the spot-test technique on a strip of newspaper.* Place a drop of an

* Newspaper is used instead of filter paper because it contains lignin whose presence favours the occurrence of the reaction.

aqueous solution of picric acid on the paper, followed by a drop of the test solution. To speed up the impregnation of the paper with the solutions, it is useful to heat the paper. If stannous ion is present, an intense orange colouration appears, which is due to the formation of picramic acid.

The reaction may also be carried out in a test tube in alkaline medium. To 5 drops of the test solution add a drop of the picric acid solution and, drop by drop, NH_4OH solution until a precipitate of $Sn(OH)_2$ separates out, and then heat. The picramic acid formed colours the solution and the precipitate, which become red.

The interferences come only from the ions of silver and mercury which are reduced to the free metals. Their black colour covers up the red colour of picramic acid.

If tin is present in the form of stannic ion, it must first be reduced to stannous ion through the action of metallic aluminium or magnesium. The reaction is conducted in the presence of HCl on heating. The solution is allowed to stand for a few minutes.

3. Metallic Zinc. This metal reduces stannous and stannic ions to the free metal which precipitates as a gray spongy layer. In contrast to antimony, tin is soluble in HCl. Therefore, the solution prepared for the test must not be too acid.

(b) Reactions of the Stannic Ion

1. Alkali Hydroxides and Ammonium Hydroxide. A white gelatinous precipitate of orthostannic acid, $Sn(OH)_4$ or H_4SnO_4, is produced:

$$[SnCl_6]^{2-} + 4OH^- \rightarrow H_4SnO_4\downarrow + 6Cl^-$$

The precipitate dissolves in alkalis to give stannates*:

$$H_4SnO_4\downarrow + 2NaOH \rightarrow Na_2SnO_3 + 3H_2O$$

It also dissolves in concentrated HCl, forming complex stannohydrochloric acid:

$$Sn(OH)_4\downarrow + 6HCl \rightarrow H_2[SnCl_6] + 4H_2O$$

2. Reduction of Stannous to Stannic Tin. In the course of analysis adopted in this book, the dissolution of SnS_2 gives rise to complex $[SnCl_6]^{2-}$ ions containing stannic tin, while tin is commonly detected by making use of the reducing properties of Sn^{2+} ions. Therefore,

* The formula of the stannate is somewhat simplified. I n fact, its structure corresponds to the formula of the complex hexahydroxystannous acid, $H_2[Sn(OH)_6]$. Apart from this acid (also known as α-stannous acid) there is also β-stannous acid which differs from the former by its insolubility in acids. The formula assigned to β-stannous acid is $[H_2SnO_3]_5$. It is formed by the action of concentrated solutions of HNO_3 on metallic tin.

$[SnCl_6]^{2-}$ must be reduced to Sn^{2+}, which is effected through the action of Mg, Al, Fe, and of other metals. For example,

$$Mg\downarrow + [SnCl_6]^{2-} \rightarrow Mg^{2+} + Sn^{2+} + 6Cl^-$$

$$Fe\downarrow + [SnCl_6]^{2-} \rightarrow Fe^{2+} + Sn^{2+} + 6Cl^-$$

Metallic iron reduces stannic tin only to Sn^{2+} but not to the metal. The reduction by magnesium or aluminium with the amount of acid being insufficient may lead to the formation of a gray precipitate of metallic tin. Acidification with a few drops of concentrated HCl dissolves it with the formation of Sn^{2+}:

$$Sn\downarrow + 2H^+ \rightarrow Sn^{2+} + H_2\uparrow$$

3. Reduction of Stannic and Stannous Tin to SnH_4. Mix 2-3 drops of a solution of a stannic or a stannous salt on a watch glass with a few drops of concentrated HCl and introduce a fragment of metallic zinc into the resulting mixture.* Then, grip a micro crucible with the aid of tongs and dip its bottom into the liquid. Hold for some time. Now introduce the crucible into the burner flame. If tin is present, the flame at the bottom of the crucible assumes a characteristic deep-blue colour due to the burning of stannic hydride, SnH_4.

4. Microcrystalloscopic Test. Stannic tin can however be detected without resorting to its reduction to stannous tin if use is made of a microcrystalloscopic test for $[SnCl_6]^{2-}$. Put a small crystal of RbCl or CsCl into a drop of the test hydrochloric acid solution of a stannic salt and let it stand. If stannic tin is present, a white cloudy precipitate of the slightly soluble salt $Rb_2[SnCl_6]$ or $Cs_2[SnCl_6]$ appears around the crystal. When viewed under the microscope, the crystals of these salts have the appearance of small octahedra (Fig. 6.5). Larger crystals are usually formed at the periphery of the precipitate. An increase in the concentration of HCl decreases the rate of crystallization, favouring the growth of crystals.

Very similar crystals are formed by the ammonium ion with $[SnCl_6]^{2-}$ ions. The crystals of $(NH_4)_2[SnCl_6]$ are usually larger and more regular in shape than those of $Rb_2[SnCl_6]$. The test is performed on a slide. To a drop of the hydrochloric acid solution of a stannic salt add a drop of $1N$ ammonia solution. If a turbidity appears due to hydroxides, add a drop of $2N$ HCl, then heat gently the solution until a small rim is formed, and let the liquid cool. After 2 minutes examine the $(NH_4)_2[SnCl_6]$ crystals under the microscope. In most cases, they are relatively large colourless octahedra.

The reaction is not interfered with by all other cations, including the ions of titanium, vanadium, molybdenum, and tungsten. Of

* Zinc must contain no impurity of tin whose absence must be confirmed by an appropriate control test.

course, the cations precipitable by Cl^- ions should be preliminarily removed from the solution. Moreover, before the test is started, stan-

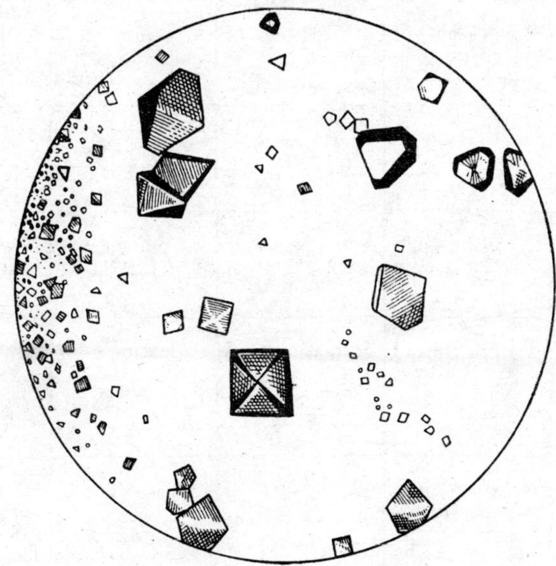

Fig. 6.5. Crystals of $Rb_2[SnCl_6]$.

nous tin must be oxidized to stannic tin by H_2O_2 in the presence of HCl. The test is suitable for fractional detection of tin in a separate portion of the test solution. The dilution limit is 1 in 20,000.

6.14. Analysis of Mixtures of Group IV Ions

The reactions of the most important reagents with the cations of the copper subgroup are listed in Table 6.4.

The effects of the most important reagents on the ions formed by the elements of the arsenic subgroup are compared in Table 6.5.

When the solution containing a mixture of Group IV cations is to be analysed, one has first of all to establish the oxidation state of arsenic. As pointed out in Sec. 6.5, arsenic can exist in solutions (depending on the pH of the medium) either as ions of different composition (AsO_4^{3-}, $HAsO_4^{2-}$, $H_2AsO_4^-$, AsO_3^{3-}, $HAsO_3^{2-}$, $H_2AsO_3^-$) or as undissociated molecules (H_3AsO_4, H_3AsO_3, $HAsO_2$). In order to precipitate arsenic as the sulphide, the solution must be made sufficiently acidic so that the required concentration of the cations As^V and As^{III} is produced. It is especially important in the case of As^V

TABLE 6.4. **Reactions of Cations of Copper Subgroup**

Reagent	Cation			
	Hg^{2+}*	Cu^{2+}	Cd^{2+}	Bi^{3+}
H_2S in 0.3*N* HCl or $(NH_4)_2S$	A black ppt. of HgS insoluble in HNO_3	A black ppt. of CuS	A yellow ppt. of CdS	A black ppt. of Bi_2S_3
		Dissolve in nitric acid		
HCl or chlorides	—	—	—	A white ppt. of BiOCl soluble in excess HCl
KOH or NaOH	A yellow ppt. of HgO	A pale-blue ppt.** of $Cu(OH)_2$	A white ppt.*** of $Cd(OH)_2$	A white ppt.** of $Bi(OH)_3$
NH_4OH (excess)	A white ppt. of NH_2HgCl	Formation of $[Cu(NH_3)_4]^{2+}$ ions (deep-blue)	Formation of colourless $[Cd(NH_3)_4]^{2+}$ ions	A white ppt. of basic salts
$SnCl_2$ + + HCl	A white ppt. gradually turning black: $Hg_2Cl_2 \rightarrow Hg$	—	—	—
Na_2SnO_2 + + NaOH	A yellow ppt. of HgO	A pale-blue ppt. of $Cu(OH)_2$	A white ppt. of $Cd(OH_2)$	A white ppt. of Bi

* For the solubility of HgS in Na_2S, see Sec. 6.6, Reaction 4.
** Does not precipitate in the presence of glycerol.
*** Precipitates in the presence of glycerol.

which is precipitated by hydrogen sulphide only in a strongly acid solution ($\geqslant$6*N* HCl).

1. Detection of As^V. To a few drops of the test solution add an equal volume of concentrated (12*N*) hydrochloric acid. This makes the solution 6*N* in HCl if the original solution was neutral, or more acidic if the pH of the original solution was lower than 7. Pass hydrogen sulphide into the solution for 2-3 min. The formation of a yellow precipitate indicates the presence of As^V in the solution. No other sulphides are formed at such a low value of pH.

If the presence of As^V in the solution has been established, it must be completely removed before the precipitation of all the sulphides of Group IV is started. This is achieved by passing H_2S into the heated solution or by adding a drop of an alcoholic solution of iodine as a catalyst to speed up the reduction of As^{V} to As^{III} (instead of iodine solution, use may be made of a small crystal of NH_4I).

2. Adjustment of the Acidity of Solution and Precipitation of Group IV Cations by Hydrogen Sulphide. To achieve complete precipitation of Group IV cations by hydrogen sulphide, it is necessary that the hydrogen ion concentration be the most favourable (equal to about 0.3*M*, i.e., pH ca. 0.5).

Neglecting the presence of a precipitate (SbOCl, BiOCl), neutralize the acid present in the solution in a very large and not exactly known concentration. For this purpose add, drop by drop, a 25-percent solution of NH_4OH, with thorough stirring, until a further drop makes the solution alkaline (pH 9-10).* Then destroy the excess of ammonia by adding, a drop at a time, 2*N* HCl until pH ca. 7 is achieved. And, finally, add to the solution an equal volume of 0.6*N* hydrochloric acid to achieve the required acidity (0.3*M*).

Place the test tube with the solution into the nearly boiling water bath and pass gase: H_2S into the solution for about 3 minutes (*In the hood!*). Then dilute the contents of the test tube with an approximately equal volume of cold distilled water and again saturate the solution with hydrogen sulphide** in the cold for 2 minutes.

Centrifuge the precipitate and test the centrifugate for complete precipitation with freshly prepared H_2S water.***

If the precipitation is incomplete, dilute the solution with further 20 drops of water and again saturate with H_2S. Having achieved complete precipitation, wash the precipitate with water to which NH_4Cl is added (to prevent peptization of the precipitate).

Of the sulphides only SnS_2 forms a thiosalt. Therefore, Sn^{II} must be oxidized to Sn^{IV} prior to precipitation by hydrogen sulphide. To do this, add to the solution 3-4 drops of a 3-percent solution of hydrogen peroxide and heat for a few minutes to destroy the excess of H_2O_2.

If the subsequent separation of the ions of the copper and arsenic subgroups is accomplished with the aid of $(NH_4)_2S_2$, this operation may be omitted because ammonium polysulphide is itself an oxidizing agent.

3. Separation of the Copper Subgroup from the Arsenic Subgroup. Treat the washed precipitate of the sulphides with 10 drops of an alkaline solution of Na_2S with gentle heating (for 2-3 min) and stir-

* The formation of a precipitate of basic salts and hydroxides in this operation may be neglected since they will be converted into the less soluble sulphides in subsequent tests.

** The importance of the individual conditions in the precipitation by hydrogen sulphide is discussed in detail in Sec. 6.6.

*** Addition of H_2S water decreases the hydrogen ion concentration and increases the concentration of S^{2-} in solution, which ensures a more complete precipitation of the lphides. Use freshly prepared H_2S water because hydrogen sulphide is easily oxidized to sulphur by the oxygen of the air.

TABLE 6.5. **Reactions of Ions**

Reagent	Ion	
	AsO_3^{3-}	AsO_4^{3-}
H_2S in the presence of HCl	A yellow ppt. of As_2S_3	A yellow ppt. of As_2S_3 (or As_2S_5)
Action of Na_2S and $(NH_4)_2S_2$** on sulphides	Sulphides dissolve	
Action of conc. HCl on sulphides	Sulphides do not dissolve	
Action of $(NH_4)_2CO_3$ on sulphides	Sulphides dissolve	
H_2O	—	—
Alkali hydroxides and NH_4OH	—	—
Iron (metallic) in the presence of HCl	—	—
Bismuth(III) salts in alkaline medium	—	—
$HgCl_2$	—	—
$MgCl_2$ in the presence of NH_4OH and NH_4Cl	—	A white crystalline ppt. of $MgNH_4AsO_4$
$(NH_4)_2MoO_4$ in the presence of HNO_3	—	A yellow ppt. of $(NH_4)_3[AsMo_{12}O_{40}] \times xH_2O$
$AgNO_3$	A yellow ppt. of Ag_3AsO_3	A chocolate-brown ppt. of Ag_3AsO_4
Oxidizing agents	AsO_4^{3-}	—

* Though the reactions of Sn^{II} are given here, the properties of its sulphides are

** The reagent Na_2S dissolves the sulphides without changing the oxidation state formation of thiosalts: $(NH_4)_3AsS_4$, $(NH_4)_3SbS_4$, and $(NH_4)_2SnS_3$.

*** The action of $(NH_4)_2S_2$ leads to dissolution since SnS is oxidized by the reagent

**** The reaction proceeds in alkaline medium.

of Arsenic Subgroup*

Ion			
$[SbCl_6]^{3-}$	$[SbCl_6]^{-}$	Sn^{2+}	$[SnCl_6]^{2-}$
An orange-red ppt. of Sb_2S_3	An orange-red ppt. of Sb_2S_5	A brown ppt. of SnS	A yellow ppt. of SnS_2
to form thiosalts		Does not dissolve in Na_2S***	Dissolves to form thiosalts
Sulphides dissolve			
Sulphides do not dissolve			
A white ppt. of SbOCl	A white ppt. of SbO_2Cl	A white ppt. of basic salts	A white ppt. of basic salts
A white ppt. of $HSbO_2$	A white ppt. of $HSbO_3$	A white ppt. of $Sn(OH)_2$	A white ppt. of H_4SnO_4
Dissolve in excess of alkali hydroxides			
A black ppt. of Sb		—	Reduced to Sn^{2+}
—	—	A black ppt. of Bi	—
—	—	A white ppt. turning black: $Hg_2Cl_2 \rightarrow Hg$	—
—	—	—	—
—	—	—	—
—	—	A black ppt. of Ag****	—
Sb^{V}	—	Sn^{IV}	—

closer to those of the sulphides of the copper subgroup.
of the corresponding elem nts. The action of $(NH_4)_2S_2$ leads to their oxidation with the

to SnS_2.

ring. As a result, the sulphides of the arsenic subgroup dissolve to form the corresponding thiosalts while the sulphides of the copper subgroup remain in the precipitate. Dilute the contents of the test tube with 10 drops of water, centrifuge the precipitate and separate it from the supernatant liquid, and then treat with sodium sulphide once again. Combine the centrifugates obtained after the two treatments and examine as directed in Procedure *4*.

Wash the precipitate, which may contain CuS, CdS, and Bi_2S_3, with water containing a little NH_4Cl and examine as directed below (see Procedure *8*).

As has already been noted, the ions of the copper subgroup and those of the arsenic subgroup can be separated with the aid not only of sodium sulphide but also of ammonium polysulphide, $(NH_4)_2S_2$. In this case, the thiosalts and oxysalts of arsenic, antimony, and tin (and also small amounts of CuS_2^{2-}) will pass into solution while mercuric sulphide, HgS, will remain in the precipitate together with the sulphides of the copper subgroup.

For this separation to be achieved, treat the washed sulphide precipitate with 15-20 drops of a solution of $(NH_4)_2S_2$ on gentle heating (40-50°C) and with thorough stirring for 2-3 min. A higher temperature is not tolerated since a considerable amount of copper may dissolve. After centrifuging, separate the centrifugate and repeat the treatment with $(NH_4)_2S_2$ under the same conditions. Combine the centrifugates obtained after the two treatments and examine according to Procedure *4*, taking into account that there is no mercury(II) in the centrifugate.

4. Decomposition of Thiosalts. (1) *Separation of Hg^{II} from the Ions of the Arsenic Subgroup and Detection of Hg^{2+}.* To the combined centrifugates, which may contain AsS_3^{3-} ions (As^V is reduced to As^{III} during the precipitation by hydrogen sulphide), SbS_4^{3-}, SnS_3^{2-}, and HgS_2^{2-}, add solid NH_4Cl together with excess of an alkaline solution of Na_2S. Heat gently. Of all the ions of the arsenic subgroup, the Hg^{2+} ion is the most basic and therefore the pH produced by the hydrolysis of NH_4Cl is found to be enough for the thiosalt of mercury to be decomposed. Mercury is precipitated as the black sulphide HgS. Separate the centrifugate and wash with water containing NH_4Cl.

To detect Hg^{II}, proceed as follows. Treat the precipitate containing HgS and S for 2-3 min, with heating and stirring, with a mixture of 2 drops of concentrated (ca. 12*N*) HCl and 2 drops of KI solution. The mercuric sulphide dissolves to form complex $[HgI_4]^{2-}$ ions. Centrifuge the undissolved precipitate (S) and discard. Remove the H_2S formed from the centrifugate by heating* and add a drop of 2*N* NH_4OH and an excess of a concentrated solution of NaOH. The formation of a red-brown precipitate indicates the presence of Hg^{2+}.

* If H_2S has not been removed, a black precipitate of HgS forms again instead of a brown-red precipitate.

The Hg^{2+} ion may be detected by a different method as well. Heat the precipitate with a few drops of aqua regia. With HgS being dissolved, remove the sulphur by centrifuging and evaporate the centrifugate very cautiously in a crucible to remove the excess of aqua regia (not to dryness, otherwise the $HgCl_2$ formed in the reaction will volatilize). Now dilute the solution with 5-8 drops of water and test a portion of it with $SnCl_2$. A white precipitate, which turns black in excess of the reagent, indicates the presence of Hg.

Other characteristic tests can also be carried out (see Sec. 6.9).

(2) *Decomposition of the Thiosalts of Arsenic, Antimony, and Tin.* Acidify the solution resulting from the separation of HgS (see Procedure *1*), or the original solution of the thiosalts if Hg^{II} is absent, with concentrated CH_3COOH by adding it, drop by drop, with constant stirring until the solution reacts acid (pH $\leqslant$ 5). The thiosalts are decomposed and the corresponding sulphides (As_2S_3, Sb_2S_5, and SnS_2) and the sulphur formed from sodium sulphide* are precipitated.

Heat the test tube with the precipitate for a few minutes in a water bath, centrifuge, and test for complete precipitation by adding a further drop of CH_3COOH. When the precipitation is proved to be complete, separate the centrifugate from the precipitate and discard. Wash the precipitate with water containing NH_4Cl and examine as directed in Procedure *5*.

5. Separation of As from Sb and Sn. Heat the precipitated sulphides for 3-5 minutes with 5-6 drops of concentrated HCl and an equal volume of water. This dissolves the sulphides of tin and antimony to give the ions** $[SbCl_6]^{3-}$ and $[SnCl_6]^{2-}$ while As_2S_3, which is insoluble in HCl, and sulphur remain in the precipitate.*** Wash the precipitate with water containing KNO_3 and examine as directed in Procedure *6*. Test the centrifugate according to Procedure *7*.

6. Dissolution of the As_2S_3 Precipitate and Detection of Arsenic. To the washed precipitate add 5-7 drops of water and an excess of solid**** $(NH_4)_2CO_3$ and heat slightly with stirring for about a minute.

---

* Sulphur results from the interaction of acetic acid with polysulphides (present in the Na_2S solution because of its oxidation by atmospheric oxygen). For example,

$$Na_2S_2 + 2H^+ \rightarrow H_2S\uparrow + S\downarrow + 2Na^+$$

The sulphur precipitate is milky white in colour and the sulphides of the arsenic subgroup are yellow (SnS_2 and As_2S_3) or orange-red (Sb_2S_5). So, if acidification of the CH_3COOH solution gives a white precipitate, it means that the arsenic subgroup is absent.

** It will be recalled that, when Sb_2S_5 is dissolved in HCl, Sb^V is reduced to Sb^{III} to form the complex acid $H_3[SbCl_6]$. Under the same conditions, Sn^{IV} is not reduced but forms the complex acid $H_2[SnCl_6]$.

*** Small amounts (traces) of arsenous sulphide may pass into solution. To precipitate such small amounts, hydrogen sulphide is passed, prior to centrifugation, into the hot mixture (for about one minute).

**** The $(NH_4)_2CO_3$ solution used for precipitation of Group II cations must not be employed in this case since it contains NH_4OH and will therefore

This dissolves As_2S_3 to give $(NH_4)_3AsS_3$ and $(NH_4)_3AsO_3$ while sulphur remains in the precipitate.

Centrifuge off the sulphur and examine the centrifugate as follows.

(1) To the portion of the centrifugate known to contain $(NH_4)_3AsS_3$, $(NH_4)_3AsO_3$, and excess $(NH_4)_2CO_3$ add cautiously* $2N$ HCl until the reaction is acid. A yellow precipitate of As_2S_3 indicates the presence of As.

(2) Heat the remaining portion of the centrifugate for 1-2 min with a drop or two of H_2O_2 solution. This oxidizes AsS_3^{3-} and AsO_3^{3-} to AsO_4^{3-}.

Arsenic is detected as follows. Acidify the solution with HNO_3, add a little solid NH_4NO_3, and then an excess (10-15 drops) of ammonium molybdate solution**. Heat the contents of the test tube for 5-10 min. The formation of a yellow crystalline precipitate of $(NH_4)_3[AsMo_{12}O_{40}] \cdot xH_2O$ proves the presence of As. The ion AsO_4^{3-} can be detected in the alkaline solution after treatment with H_2O_2 by means of the reaction with magnesia mixture or by a microcrystalloscopic test (see Sec. 6.11).

7. Detection of Sn and Sb. Test the centrifugate obtained in Procedure *5*, which is known to contain $[SnCl_6]^{2-}$ and $[SbCl_6]^{3-}$ ions together with excess HCl, for tin and antimony.

(1) *Detection of* Sn^{IV}. To a drop of the solution on a microscope slide add a drop of $2N$ NH_4OH solution. If a precipitate is formed, acidify with a drop of $2N$ HCl and heat gently until a rim appears around the drop. After cooling, examine the crystals under the microscope. The formation of characteristic octahedral crystals of $(NH_4)_2SnCl_6$ (see Fig. 6.5) is proof of the presence of tin. The same test can be applied to a drop of the original solution (made acid with HCl) after the precipitated ions of Group V have been removed and the solution heated with a drop of H_2O_2 to oxidize Sn^{II} to Sn^{IV}.

(2) *Detection of Sb.* Place a drop of the solution on a piece of tin foil and let stand. If a black stain (deposit) is formed, wash thoroughly with water and treat with a drop of a freshly prepared solution of NaBrO. If the stain does not disappear, this is an indication of the presence of antimony.***

---

dissolve the sulphides of tin and antimony if they have not been completely dissolved in HCl. Because of this, As could be detected "in error".

* HCl should be added carefully, otherwise the vigorous evolution of CO_2 may throw the liquid out of the test tube.

** To provide coagulation and to decrease the solubility of $(NH_4)_3[AsMo_{12}O_{40}] \cdot xH_2O$, NH_4NO_3 is added. Ammonium molybdate solution must be taken in a large excess, otherwise a precipitate will not form because of its solubility in excess of arsenate.

*** A black stain is also produced by As if it has dissolved partially by the action of HCl on As_2S_3 (see Procedure *5*). But the stain of metallic arsenic disappears under the action of NaBrO while the stain of Sb remains. In acid solution, the Sb stain disappears too; therefore it is necessary to wash the stain thoroughly with water in order to remove HCl.

Alternatively, the test with 8-hydroxyquinoline and KI, or with methyl violet, or else the drop reaction with phosphomolybdic acid (see Sec. 6.12), may be carried out.

(3) *Reduction of Sn^{IV} to Sn^{II} and Confirmatory Tests for Sn^{II}.* If the reaction of the formation of $(NH_4)_2SnCl_6$ crystals has not given sufficiently distinct results, confirmatory tests must be made. First, Sn^{IV} must be reduced to Sn^{II}. Proceed as follows. To the acid solution obtained in Procedure 7 add a piece of metallic magnesium and let it stand until the metal dissolves completely.* This reduces the $[SnCl_6]^{2-}$ ion to $[SnCl_4]^{2-}$ ion or even to metallic tin (if the amount of acid is insufficient), and the $[SbCl_6]^{3-}$ ion to metallic antimony which settles as a black precipitate. After the Mg metal has completely dissolved, add to the test tube a drop or two of concentrated HCl and warm in order to dissolve metallic tin that might have been formed. Centrifuge off the Sb residue** and look for Sn^{II} in the centrifugate through the action of $HgCl_2$ (a white precipitate of Hg_2Cl_2 turning black in excess of Sn^{II}) by means of the reaction with ammonium phosphomolybdate or with picric acid (see Sec. 6.13).

(4) *Confirmatory Tests for Sn and Sb.* These tests can also be made by using the tendency of the ions toward complex formation.

Dilute a few drops of the solution obtained in Procedure 5 with an equal volume of distilled water and add crystalline potassium oxalate to saturation. The tin is bound into the stable complex $[Sn(C_2O_4)_3]^{2-}$. When hydrogen sulphide is passed into this solution, an orange precipitate of antimonous sulphide is produced.

Dilute another portion of the solution obtained in Procedure 5 with water and add crystalline tartaric acid until the solution is saturated. In this case, a stable antimony complex is obtained.*** When hydrogen sulphide is passed into the solution, a yellow precipitate of stannic sulphide is formed.

8. Dissolution of Precipitated Sulphides of the Cations of the Copper Subgroup. Treat the precipitate, which may contain CuS, Bi_2S_3, and CdS, with 5 drops of water and 10 drops of 6*N* HNO_3. Add a little KNO_2 or $NaNO_2$ to hasten the reaction and heat the con-

* If required, a drop or two of HCl is added.

** It can also be dissolved by heating in HCl and H_2O_2, and then the test for Sb may be carried out.

*** The probable composition of the complex for Sb^{III} is

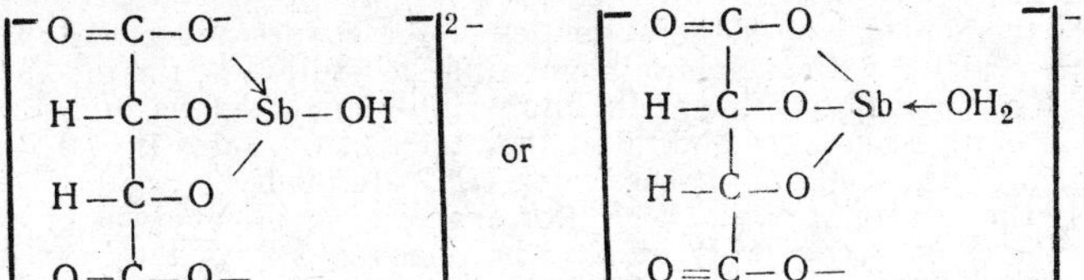

```
┌ O=C—O⁻              ┐2-        ┌ O=C—O               ┐-
│   |      ↘           │          │   |    ＼             │
│ H—C—O—Sb—OH          │    or    │ H—C—O—Sb ← OH₂       │
│   |      /           │          │   |    /              │
│ H—C—O                │          │ H—C—O                 │
│   |                  │          │   |                   │
└ O=C—O—               ┘          └ O=C—O—                ┘
```

The exact composition of the complex for Sb^V is not known.

tents of the test tube in the water bath. Centrifuge off the undissolved sulphur precipitate or collect it with a glass spatula and discard.*

Test part of the solution for Bi^{3+}. To 2-3 drops of the solution in a test tube add 8-10 drops of water. A white turbidity indicates the presence of Bi^{3+} (formation of basic salts).

9. Separation ** **of Bi^{3+} and Detection of Cu^{2+}.** To the centrifugate (from Procedure *8*) add a 25-percent solution of NH_4OH until alkaline. Now add 2-3 drops more of the solution indicated.

The Bi^{3+} ion is precipitated as a basic salt or hydroxide (white-coloured), and the Cu^{2+} and Cd^{2+} ions are bound into the ammine complexes $[Cd(NH_3)_4]^{2+}$ and $[Cu(NH_3)_4]^{2+}$ (deep-blue) which remain in the solution. Heat the precipitate for 2 minutes (to provide better coagulation), centrifuge off, wash, and examine as directed in Procedure *10*. A deep-blue colour of the centrifugate indicates the presence of copper.

If the colouration does not appear or is not sufficiently distinct, acidify a drop of the centrifugate with CH_3COOH and treat with a drop of $K_4[Fe(CN)_6]$ solution. The formation of a red-brown precipitate of $Cu_2[Fe(CN)_6]$ confirms the presence of Cu^{2+}.

Examine the centrifugate as directed in Procedure *11*.

10. Confirmatory Tests for Bi^{3+}. Treat the washed $Bi(OH)_3$ precipitate with a few drops of a solution of sodium stannite, Na_2SnO_2, and mix.***

The blackening of the precipitate due to the formation of metallic bismuth proves the presence of Bi^{3+}.

11. Detection of Cd^{2+}. (1) If Cu^{2+} is absent, acidify slightly (pH $\geqslant$ $\geqslant$ 0.5) the ammoniacal solution with 2*N* HCl and pass hydrogen sulphide into it. The formation of a yellow precipitate of CdS insoluble in 6*N* NaOH solution indicates the presence of Cd^{2+}

Note. If the precipitate dissolves in a concentrated solution of NaOH, it may be regarded as being any of the sulphides of the arsenic subgroup (SnS_2 or As_2S_3) rather than CdS. This happens in the analysis of mixtures of the cations of all the five groups if the separation of the ions of the arsenic subgroup by Na_2S was not sufficiently complete.

(2) If Cu^{2+} is present, it must be separated first. To do this, to the deep-blue ammoniacal solution add, drop by drop, 2*N* HCl until

---

* If the separation of the sulphides of the copper and arsenic subgroups have been effected with the aid of ammonium polysulphide, the precipitate will contain HgS (black) together with sulphur. In this case, the precipitate is to be treated with aqua regia and, after the excess of the acids is boiled off in an evaporating dish, confirmatory tests for Hg^{II} are made.

** If the solution contains Pb^{2+} (for example, in the analysis of a mixture of the cations of Groups V-I), it is to be removed first in the form of $PbSO_4$ by evaporating the solution with 2*N* H_2SO_4 (see Sec. 6.1.9, Procedure *13*).

*** The solution of sodium stannite is prepared by adding an excess (8-10 drops) of 2*N* NaOH to 2 drops of the $SnCl_2$ solution and mixing.

the deep-blue colour changes to light-blue. Now add an equal volume of the same HCl solution (or a drop of 12*N* HCl for every 10-11 drops of the solution). Heat the solution and pass hydrogen sulphide into it. Centrifuge the CuS precipitate, test for complete precipitation, dilute the centrifugate three- or fourfold with water, and treat again with hydrogen sulphide. A yellow precipitate of CdS insoluble in NaOH confirms the presence of Cd^{2+}. The Cu^{2+} ion may also be separated as described in Sec. 6.8 (Reactions *5*).

(a) The Glycerol-Alkali Method for Analysis of the Copper Subgroup

1. Separation of Cd^{2+} from Cu^{2+} and Bi^{3+}. To the nitric acid solution, resulting from the dissolution of the sulphides of the copper subgroup in Procedure *8*, add 5-6 drops of glycerol diluted with water (1 : 1). Now add, a drop at a time, a 30-percent solution of NaOH until alkaline, following which add 8-10 drops more of it. While adding NaOH, mix thoroughly the contents of the test tube.

In this case, Cu^{2+} and Bi^{3+} form* complex compounds with glycerine and remain in solution while Cd^{2+} is precipitated as $Cd(OH)_2$. Heat the solution for a minute, centrifuge the precipitate, and wash twice with a 2*N* solution of alkali and glycerine (2-3 drops of glycerine and 20 drops of alkali) in order to remove small amounts of the hydroxides of other metals (especially, bismuth). And, finally, wash once with water and examine as directed in Procedure *2*.

Test the centrifugate according to Procedure *3*.

2. Detection of Cd^{2+}. Dissolve the washed precipitate in 4-5 drops of 2*N* HCl, dilute the solution with 20 drops of water, and pass H_2S into it. A yellow precipitate of CdS indicates the presence of Cd^{2+}.

Note. If the CdS precipitate is dark-coloured rather than pure-yellow, due to the incomplete removal of the coprecipitated hydroxides of bismuth or copper when washed with NaOH solution, it should be centrifuged off and treated with HCl on heating. The Cd^{2+} ion is detected by the action of H_2S in the resulting solution after strong dilution. The precipitate obtained is lemon-yellow in colour, the colour being characteristic of CdS.

3. Detection of Cu^{2+} and Bi^{3+}. The ions Cu^{2+} and Bi^{3+} are searched for in separate portions of the solution (see Procedure *1*).

(1) *The Cu^{2+} Ion.* The deep-blue colour of the solution indicates Cu^{2+}. If the colour is not sufficiently distinct or is absent, acidify a drop of the centrifugate with CH_3COOH and treat with $K_4[Fe(CN)_6]$ solution. A red-brown precipitate of $Cu_2[Fe(CN)_6]$ indicates the presence of Cu^{2+}.

---

* If Pb^{2+} is present in solution, it also gives a complex compound with glycerine and remains in solution together with Cu^{2+} and Bi^{3+}.

(2) *The Bi^{3+} Ion.* To 2 drops of $SnCl_2$ solution add 8-10 drops of 2N NaOH and mix. To the resulting solution of sodium stannite, Na_2SnO_2, add the alkaline test solution drop by drop.

The formation of a black precipitate of metallic bismuth indicates the presence of Bi^{3+}.

(b) The Cyanide Method of Separation of C[illegible] *Cd^{2+}*

As pointed out in Sec. 6.8 (Reaction *3*), [illegible]anide complexes of copper and cadmium differ markedly in [illegible]bility constants. This fact can be used for detection of cadmium [illegible] the presence of copper after bismuth has been separated as the hydroxide by the action of ammonia.

To the solution of copper and cadmium ammines obtained in the course of analysis (see Procedure *9*) add KCN solution, drop by drop, until the colour disappears completely, and 5-6 drops more (excess) to repress the dissociation of the copper complex. Pass hydrogen sulphide into the solution. The appearance of a yellow precipitate of CdS is evidence of the presence of cadmium. If the precipitate is dark in colour, it means that too small an excess of KCN has been added. In such a case, repeat the precipitation using a new portion of the ammoniacal solution of copper and cadmium and adding an excess of 8-10 drops of KCN solution.

6.15. Group V Cations. Action of the Group Reagent

Group V comprises Ag^+, $[Hg_2]^{2+}$, and Pb^{2+}, and also certain other cations whose reactions are not discussed in the book. The cations of Group V form chlorides slightly soluble in water and dilute acids. Upon this property of the chlorides is based the separation of Group V cations in the systematic analysis from the mixture of all cations.

The behaviour of the chlorides AgCl, Hg_2Cl_2, and $PbCl_2$ toward the action of water, excess HCl, ammonia, and other reagents is essential for the completeness of precipitation and further separation of Group V cations. Therefore we shall consider in more detail the properties of the chlorides of these cations.

Dilute hydrochloric acid (and also solutions of chlorides) bring down Ag^+, $[Hg_2]^{2+}$, and Pb^{2+} in the form of white precipitates of AgCl, Hg_2Cl_2, and $PbCl_2$. For example,

$$Ag^+ + Cl^- \rightarrow AgCl\downarrow$$

$$[Hg_2]^{2+} + 2Cl^- \rightarrow Hg_2Cl_2\downarrow$$

(1) On shaking the contents of the test tube, the AgCl precipitate collects as white curdy flakes. In the light, it becomes violet or even black (decomposition). Silver chloride, being a salt formed by a strong acid, is insoluble in dilute acids (HNO_3, H_2SO_4). It readily dissolves

in NH_4OH because of the Ag^+ ions being bound into complex $[Ag(NH_3)_2]^+$ ions:

$$AgCl\downarrow + 2NH_4OH \rightarrow [Ag(NH_3)_2]^+ + Cl^- + 2H_2O$$

If the resulting solution of the ammine complex is treated with an acid, say, HNO_3, the hydrogen ions introduced with it will combine with NH_3 molecules into the more stable complex NH_4^+ and the equilibrium

$$[Ag(NH_3)_2]^+ \rightleftarrows Ag^+ + 2NH_3$$

will be shifted to the right. This will result in the breakdown of the silver ammine complex and accumulation of Ag^+ ions. But since the solution contains the Cl^- ions formed in the primary dissociation of $[Ag(NH_3)_2]Cl$, the solubility product of AgCl is found to be exceeded and this salt is precipitated. The overall equation of the reaction involved is

$$[Ag(NH_3)_2]^+ + Cl^- + 2H^+ \rightarrow AgCl\downarrow + 2NH_4^+$$

The action of NH_4OH is used in the course of analysis for separation of AgCl from Hg_2Cl_2, and the action of HNO_3 on the resulting solution of $[Ag(NH_3)_2]^+$ is employed for identification of Ag^+. The Ag^+ ion may also be detected through the use of the fact that it forms with I^- a slightly soluble precipitate of AgI whose solubility product constant is extremely small ($SP_{AgI} = 8.3 \times 10^{-17}$). Consequently, for this precipitate to be formed even the low concentration of Ag^+ contained in the solution of the ammine complex $[Ag(NH_3)_2]^+$ ($K_{ins} = 5.8 \times 10^{-8}$) is enough for the precipitation of AgI.

To a few drops of the ammoniacal solution obtained after the dissolution of AgCl in ammonium hydroxide add 2-3 drops of KI solution. A yellow precipitate of AgI is formed:

$$[Ag(NH_3)_2]^+ + I^- + 2H_2O \rightarrow AgI\downarrow + 2NH_4OH$$

The precipitate is soluble to a certain extent in excess KI due to the formation of a complex:

$$AgI\downarrow + I^- \rightarrow [AgI_2]^-$$

A similar complex is formed by Ag^+ also with Cl^- as the ligand (if its concentration is large). Therefore, AgCl dissolves appreciably in concentrated solutions of HCl and chlorides:

$$AgCl\downarrow + Cl^- \rightarrow [AgCl_2]^-$$

From the foregoing it follows that, when the cations of Group V are to be precipitated, the concentration of hydrochloric acid must not be too high, otherwise Ag^+ may precipitate incompletely.

(2) The Hg_2Cl_2 precipitate is white in colour and is soluble to a certain extent in excess HCl with the formation of a complex; in contrast to AgCl, the chloride Hg_2Cl_2 does not dissolve but turns black

by the action of NH_4OH. At first this results in the formation of a complex compound of mercury(I):

$$Hg_2Cl_2 \downarrow + 2NH_4OH \rightarrow [NH_2Hg_2]Cl \downarrow + NH_4^+ + 2H_2O + Cl^-$$

or

$$Hg_2Cl_2 \downarrow + 2NH_3 \rightarrow [NH_2Hg_2]Cl \downarrow + NH_4^+ + Cl^-$$

The compound $[NH_2Hg_2]Cl$ corresponds in composition to NH_4Cl in which two hydrogen atoms are replaced by two Hg^I atoms. It is unstable and rapidly decomposes:

$$[NH_2Hg_2]Cl \downarrow \rightarrow NH_2HgCl \downarrow + Hg \downarrow$$

As seen from the equation, this involves a disproportionation* reaction

$$[Hg_2]^{2+} \rightarrow Hg^{2+} + Hg \downarrow$$

The two amino compounds—$[NH_2Hg_2]Cl$ and NH_2HgCl—are white in colour. The blackening of the precipitate occurs due to the formation of metallic mercury. This is the most important reaction of the $[Hg_2]^{2+}$ ion, which is generally employed for identification of Hg^I.

(3) The $PbCl_2$ precipitate is white, soluble in excess of HCl and alkali chloride. It also dissolves well in hot water. This fact is used for separation of $PbCl_2$ from AgCl and Hg_2Cl_2.

The chloride $PbCl_2$ dissolves rather appreciably in cold water too (about 10 g/litre at 20°C). Owing to this, Pb^{2+} is separated incompletely by hydrochloric acid; part of it remains in solution and is precipitated together with the cations of Group IV in the form of PbS.

6.16. Reactions of the Silver Ion

The Ag^+ ion is colourless in aqueous solutions.

1. Alkali Hydroxides, NaOH and KOH. A brown precipitate of Ag_2O is formed in solutions of silver salts. Silver oxide results from the decomposition of unstable silver hydroxide which forms first:

$$2Ag^+ + 2OH^- \rightarrow 2AgOH \downarrow$$

$$2AgOH \downarrow \rightarrow Ag_2O \downarrow + H_2O$$

The precipitate dissolves in NH_4OH owing to the formation of the complex silver ammine:

$$Ag_2O \downarrow + 4NH_4OH \rightarrow 2[Ag(NH_3)_2]^+ + 2OH^- + 3H_2O$$

The dissolution of Ag_2O in NH_4OH must be conducted in the presence of an ammonium salt, otherwise the silver azide, Ag_3N, may be formed, which is explosive on slight percussion.

* A disproportionation reaction is a self-oxidation reaction in which one ionic species gives rise to two (or more) others by a gain and loss of electrons.—*Tr.*

The action of NH_4OH on solutions of silver salts may also yield a precipitate of Ag_2O which however immediately dissolves in excess of the reagent.

2. **Potassium Chromate, K_2CrO_4.** A brick-red precipitate of silver chromate, Ag_2CrO_4, is produced:

$$2Ag^+ + CrO_4^{2-} \rightarrow Ag_2CrO_4\downarrow$$

The precipitate is soluble in HNO_3 and NH_4OH but almost insoluble in CH_3COOH.

The reaction should be carried out at pH ≈ 7. In alkaline solution, a precipitate of silver oxide is produced. No precipitate forms in ammoniacal solution and in strongly acid solutions.

3. **Disodium Hydrogen Phosphate, Na_2HPO_4.** A yellow precipitate of silver phosphate, Ag_3PO_4, is formed, which is soluble in HNO_3 and NH_4OH:

$$3Ag^+ + 2HPO_4^{2-} \rightarrow Ag_3PO_4\downarrow + H_2PO_4^-$$

4. **Bromide and Iodide Solutions (Br^- and I^- Ions).** These give with Ag^+ a pale-yellow precipitate of silver bromide and yellow silver iodide:

$$Ag^+ + Br^- \rightarrow AgBr\downarrow \quad (SP = 5.3 \times 10^{-13})$$
$$Ag^+ + I^- \rightarrow AgI\downarrow \quad (SP = 8.3 \times 10^{-17})$$

In distinction to silver chloride ($SP = 1.78 \times 10^{-10}$), silver iodide is insoluble in ammonium hydroxide, and silver bromide dissolves but very slightly in it. This is explained by the difference in the solubility product constants of these silver compounds.

If however the precipitates are treated with solutions of KCN or $Na_2S_2O_3$, which form with Ag^+ the complexes $[Ag(CN)_2]^-$ and $[AgS_2O_3]^-$ (less dissociated than $[Ag(NH_3)_2]^+$), the compounds AgBr and AgI will also dissolve. This can be proved by treating the centrifuged precipitate with an excess of a solution of sodium thiosulphate, $Na_2S_2O_3$.*

5. **Reduction of Ag^+ to Metallic Silver.** The Ag^+ ion has a relatively high standard potential ($E^0 = 0.80$ volt) and can therefore be reduced to metallic silver by various reducing agents.

(1) *Reduction by Formaldehyde, HCOH.* Place a few drops of a solution of a silver salt in a test tube that has been purified from grease by washing first with chromic acid mixture and then with water, and dilute with 10 drops of water. Now add 8-10 drops of $2N$ NH_4OH and a few drops of a dilute solution of formaldehyde (formalin). When the test tube is immersed into hot water, a bright mirror of metallic silver forms on the interior of the tube:

$$2Ag^+ + 3NH_4OH + HCOH \rightarrow 2Ag\downarrow + HCOO^- + 3NH_4^+ + 2H_2O$$

* Silver iodide dissolves only in concentrated solutions of $Na_2S_2O_3$ (see Sec. 5.10).

(2) *Reduction by Mn^{2+} Ions.* This reaction has already been used as a spot test for the Mn^{2+} ion (see Sec. 5.20, Reaction 7):

$$2AgCl\downarrow + Mn^{2+} + 4OH^- \rightarrow 2Ag\downarrow + 2Cl^- + MnO(OH)_2\downarrow + H_2O$$

The reaction can also be employed for detection of Ag^+ by the spot-test technique. In this case, a drop of the test solution is used instead of the solution of a silver salt and the Mn^{2+} ion is introduced with a drop of $Mn(NO_3)_2$ solution. The presence of Ag^+ is recognized by the instant blackening of the spot under the action of alkali.* The dilution limit is 1 in 25,000.

6.17. Reactions of the Lead Ion

The Pb^{2+} ion is colourless in aqueous solutions.

1. Alkalis (NaOH, KOH) and Ammonium Hydroxide. A white precipitate of lead hydroxide is formed:

$$Pb^{2+} + 2OH^- \rightarrow Pb(OH)_2\downarrow$$

The precipitate is amphoteric, i.e., dissolves in both acids and bases. In the latter case, the plumbite Na_2PbO_2 or K_2PbO_2 is formed:

$$Pb(OH)_2\downarrow + 2H^+ \rightarrow Pb^{2+} + 2H_2O$$

$$H_2PbO_2\downarrow + 2OH^- \rightarrow PbO_2^{2-} + 2H_2O$$

The $Pb(OH)_2$ precipitate is insoluble in ammonia solution.

2. Sulphuric Acid and Soluble Sulphates (SO_4^{2-}). The sulphate ion from sulphuric acid or soluble sulphates produces in solutions of lead salts a white precipitate of lead sulphate:

$$Pb^{2+} + SO_4^{2-} \rightarrow PbSO_4\downarrow$$

The precipitate dissolves on heating in solutions of alkali hydroxides through the formation of plumbites. For example,

$$PbSO_4\downarrow + 4OH^- \rightarrow PbO_2^{2-} + SO_4^{2-} + 2H_2O$$

Lead sulphate also dissolves on heating with a concentrated (30-per cent) solution of CH_3COONH_4 or $(NH_4)_2C_4H_4O_6$:

$$2PbSO_4\downarrow + 2CH_3COO^- \rightarrow [Pb(CH_3COO)_2\cdot PbSO_4] + SO_4^{2-}$$

$$PbSO_4\downarrow + C_4H_4O_6^{2-} \rightarrow PbC_4H_2O_6^{2-} + SO_4^{2-} + 2H^+$$

The structures of the compounds formed may be depicted as follows:

```
                                               H   H
                                               |   |
CH3COO—Pb—O     O     and     O=C—C—C—C=O
           \   //                          |   |   |   |
            S                              O   O   O   O
           /   \\                          |    \ /    |
CH2COO—Pb—O     O                               Pb
```

---

* The slow darkening is associated with the conversion of Mn^{2+} into $MnO(OH)_2$ or $MnO_2\cdot xH_2O$ by atmospheric oxygen.

The acids HCl and HNO_3 also considerably increase the solubility of $PbSO_4$, which is explained by the fact that the SO_4^{2-} ions supplied by lead sulphate to the solution are bound by hydrogen ions:

$$PbSO_4\downarrow \rightleftharpoons Pb^{2+} + SO_4^{2-}$$
$$+$$
$$HNO_3 \rightarrow NO_3^- + H^+$$
$$\uparrow\downarrow$$
$$HSO_4^-$$

From this it follows that the complete separation of Pb^{2+} in the form of $PbSO_4$ is possible only after the acids indicated have been removed from the solution.

The presence of acids does not cause an appreciable dissolution of $BaSO_4$ because of its much smaller solubility product constant ($SP = 1.1 \times 10^{-10}$) than that of $PbSO_4$ ($SP = 1.6 \times 10^{-8}$).

3. Potassium Chromate, K_2CrO_4, and Potassium Dichromate, $K_2Cr_2O_7$. A slightly soluble yellow precipitate of lead chromate is formed ($SP = 1.8 \times 10^{-14}$):

$$Pb^{2+} + CrO_4^{2-} \rightarrow PbCrO_4\downarrow$$
$$2Pb^{2+} + Cr_2O_7^{2-} + H_2O \rightarrow 2PbCrO_4\downarrow + 2H^+$$

The precipitate is easily soluble in solutions of caustic alkalis:

$$PbCrO_4\downarrow + 4OH^- \rightarrow PbO_2^{2-} + CrO_4^{2-} + 2H_2O$$

The $PbCrO_4$ precipitate is slightly soluble in dilute nitric acid. It is almost insoluble in ammonia, acetic acid, ammonium acetate and tartrate. This is one of the most important reactions of Pb^{2+}.

4. Potassium Iodide, KI. In solutions of lead salts, potassium iodide produces a yellow precipitate of PbI_2:

$$Pb^{2+} + 2I^- \rightarrow PbI_2\downarrow$$

When the precipitate is obtained, a few drops of water and of 2*N* CH_3COOH solution are added to the test tube and the solution is heated. This dissolves the precipitate but, when the tube is dipped into cold water, PbI_2 reprecipitates as bright golden crystals (plates). This test, which is characteristic of Pb^{2+}, is one of the most beautiful analytical reactions.

5. Disodium Hydrogen Phosphate, Na_2HPO_4. A white precipitate of lead phosphate is produced:

$$3Pb^{2+} + 4HPO_4^{2-} \rightarrow Pb_3(PO_4)_2\downarrow + 2H_2PO_4^-$$

Lead phosphate, $Pb_3(PO_4)_2$, is relatively slightly soluble in dilute nitric and acetic acids. This fact is sometimes used for separation of PO_4^{3-} from the cations of Groups III, II, and I. Lead phosphate is soluble in alkalis.

6. Benzidine. Moisten filter paper with an ammoniacal solution of 3-percent H_2O_2 and apply a drop of the solution to be tested for Pb^{2+} onto the moistened area. Hold the paper in the vapour of a boiling water bath for some time. The lead hydroxide formed by the interaction of the lead salt with NH_4OH is oxidized by hydrogen peroxide to lead dioxide (brown in colour):

$$Pb(OH)_2\downarrow + H_2O_2 \rightarrow PbO_2\downarrow + 2H_2O$$

the excess of H_2O_2 being destroyed at the same time.

If the moistened spot on the paper is then treated with a drop of a solution of benzidine acetate, the reagent is oxidized by the PbO_2 formed and the spot turns deep-blue.* The dilution (concentration) limit is 1 in 33,000.

6.18. Reactions of the Mercurous Ion

1. Alkali Hydroxides, NaOH and KOH. On interaction with $[Hg_2]^{2+}$, these alkali hydroxides give a black precipitate of mercurous oxide:

$$[Hg_2]^{2+} + 2OH^- \rightarrow Hg_2O\downarrow + H_2O$$

2. Ammonium Hydroxide, NH_4OH. Ammonium hydroxide reacts with solutions of mercurous salts in various ways, depending on the compound of mercury taken for the reaction.

The reaction with mercurous chloride, Hg_2Cl_2, proceeds as shown above (see Sec. 6.15) with the formation of mercurous aminochloride which decomposes into the corresponding mercuric compound and metallic mercury.

With mercurous nitrate, $Hg_2(NO_3)_2$, the reaction results in the formation of metallic mercury and a basic aminomercuric salt (a salt of Millon's base). The probable structural formula of this compound is

```
 ┌   Hg       H  ┐+
 │  /   \    /   │
 │ O      N      │
 │  \   /    \   │
 └   Hg       H  ┘
```

This reaction serves as a test for NH_4^+ (see Sec. 3.14).

3. Potassium Iodide, KI. With soluble mercurous salts, potassium iodide forms a greenish precipitate of mercurous iodide, Hg_2I_2:

$$[Hg_2]^{2+} + 2I^- \rightarrow Hg_2I_2\downarrow$$

4. Potassium Chromate, K_2CrO_4. With mercury(I), potassium chromate produces a red precipitate of mercurous chromate, Hg_2CrO_4, insoluble in alkalis and in dilute nitric acid:

$$[Hg_2]^{2+} + CrO_4^{2-} \rightarrow Hg_2CrO_4\downarrow$$

5. Reduction of $[Hg_2]^{2+}$ to Metallic Mercury. (1) The action of 2-3 drops of a solution of $SnCl_2$ on a drop of a solution of a mercurous salt precipitates white Hg_2Cl_2 which turns black due to the reduction

* For an analogous test for Mn^{2+}, see Sec. 5.20, Reaction *6*.

of $[Hg_2]^{2+}$ to metallic mercury:

$$[Hg_2]^{2+} + 2Cl^- \rightarrow Hg_2Cl_2\downarrow$$
$$Hg_2Cl_2\downarrow + Sn^{2+} \rightarrow 2Hg\downarrow + Sn^{IV} + 2Cl^-$$

(2) If a drop of a solution of $Hg_2(NO_3)_2$ is put on a bright copper plate (or a coin) and allowed to stand, a gray stain (deposit) of metallic mercury is formed:

$$[Hg_2]^{2+} + Cu\downarrow \rightarrow Cu^{2+} + 2Hg\downarrow$$

If the solution is washed off with water and the stain is rubbed with cotton wool or filter paper, it becomes bright.

The effects of the most important reagents on the cations of Group V are shown in Table 6.6.

TABLE 6.6. **Reactions of Group V Cations**

Reagent	Cation		
	Ag^+	Pb^{2+}	$[Hg_2]^{2+}$
H_2S in 0.3*N* HCl or sulphides	A black ppt. of Ag_2S	A black ppt. of PbS	A black ppt. of HgS+Hg
HCl or chlorides (in a slight excess)	A white ppt. of AgCl	A white ppt. of $PbCl_2$	A white ppt. of Hg_2Cl_2
KOH or NaON (no excess)	A brown ppt. of Ag_2O	A white ppt. of $Pb(OH_2)$	A black ppt. of Hg_2O
NH_4OH (excess)	Formation of $[Ag(NH_3)_2]^+$ ions	A white ppt. of $Pb(OH)_2$	A black ppt. of $[NH_2Hg_2O]NO_3+$ $+Hg$
H_2SO_4 (2*N*)	No ppt. is formed	A white ppt. of $PbSO_4$	A white ppt. of Hg_2SO_4
$SnCl_2$+HCl	A black ppt. of Ag	A white ppt. of $PbCl_2$	A white ppt. gradually turning black: $Hg_2Cl_2 \rightarrow$ $\rightarrow Hg$
Na_2SnO_2+NaOH	A brown ppt. of Ag_2O	A white ppt. gradually turning black: $Pb(OH)_2 \rightarrow Pb$	A black ppt. of Hg_2O

6.19. Analysis of Mixtures of Group V-I Cations

In the general systematic analysis of a mixture of the cations of all the five groups, the test solution often contains the precipitated chlorides and sulphates of a number of cations: AgCl, Hg_2Cl_2, $PbCl_2$,

$PbSO_4$, $BaSO_4$, $SrSO_4$, ($CaSO_4$); the precipitate may also contain the oxychlorides of antimony and bismuth, SbOCl and BiOCl. Let us consider the complete analysis of this precipitate.

First, it is necessary to make sure that the sulphates, if there are any (see Sec. 5.24), have been completely precipitated. To do this, centrifuge a small portion of the solution and divide it into two parts. Acidify one part of the centrifugate with HCl and test for the sulphate ion by adding a few drops of $BaCl_2$ solution. The formation of a precipitate shows that SO_4^{2-} is present in the solution, which is proof of the complete precipitation of Ba^{2+}, Sr^{2+}, and Pb^{2+} (if they are present).

If the test for SO_4^{2-} is negative, to another portion of the centrifugate add a few drops of Na_2SO_4 solution or $2N$ H_2SO_4. The appearance of a precipitate indicates that the solution contains ions capable of forming sulphates insoluble in dilute acids (Ba^{2+}, Sr^{2+}, Pb^{2+}).

If it is found that the initial precipitate contains acid-insoluble sulphates and that they have not been precipitated completely, they should be postprecipitated. These cations must be precipitated from the entire solution in the form of sulphates by adding $2N$ H_2SO_4 until the precipitation is complete (until further addition of sulphuric acid yields no precipitate). Postprecipitation is recommended because the ions Ba^{2+} and Sr^{2+} may be easily "lost" in the course of analysis: they will be distributed between the precipitate and the solution and may come down together with the cations of Group III if ammonium sulphide used for precipitation contains SO_4^{2-} (due to the oxidation of S^{2-} by the oxygen of the air) or if the ammoniacal solution contains CO_3^{2-} (due to the absorption of CO_2 from the air). The cations Ba^{2+}, Sr^{2+}, and Pb^{2+} can easily be detected in the precipitated sulphates.

In a large excess of HCl, the chlorides of Group V dissolve to form complex ions and therefore an excess of HCl must be avoided. But, on the other hand, the concentration of HCl must be sufficient so that the precipitate does not contain impurities of SbOCl and BiOCl. The best conditions for precipitation of chlorides are provided at $pH \approx 2$. At this value of pH, the chlorides of silver, mercury(I), and lead are precipitated practically completely, while the oxychlorides of antimony and bismuth (and also of tin) are found to be contained in small amounts in the precipitate.

On account of all that has been said, the following systematic course is proposed for analysis of a mixture of the cations of Groups V-I, which may be altered if the student fails to find this or that element during his work.

1. Preliminary Observations and Tests. As always, analysis is started with preliminary observations and tests. First, note the colour of the solution, which may sometimes serve as an indication of the presence of Cu^{2+}. To render this colour more intense, it is useful to add a drop or two of a 25-percent ammonia solution to a drop of the test solution in order to bind Cu^{2+} into $[Cu(NH_3)_4]^{2+}$. It must however be kept in mind that a light-blue colouration may develop due to the formation of nickel ammine, $[Ni(NH_3)_6]^{2+}$.

Then use separate portions of the solution to perform tests for Fe^{3+} and Fe^{2+} (since after the passage of H_2S all the iron may be found to be in solution in the form of Fe^{2+}, in which case it will be impossible to determine the oxidation state of iron) and also for NH_4^+ (introduced in adjusting the acidity of the solution before the precipitation of the cations of Group IV by hydrogen sulphide).

Moreover, the pH of the solution must be determined. A low value of pH may be due to the presence of free acids or of especially strongly hydrolysable salts of bismuth and mercury* and also of titanium and iron. If the pH value does not change noticeably after a portion of the solution is *carefully* neutralized until a permanent turbidity just appears, the presence of any of these cations may be assumed. If the solution shows an alkaline reaction, it means that of the cations of Groups IV and V the solution may contain PbO_2^{2-}, $[Ag(NH_3)_2]^+$, $[Cu(NH_3)_4]^{2+}$, $[Cd(NH_3)_4]^{2+}$, and some other complex ions, say, the complex cyanides, tartrates, glycerates. Arsenic, antimony, and tin may in this case be present in solution in the form of the anions of the corresponding oxy-acids.

At the same time the tests for SO_4^{2-} and ions capable of forming insoluble sulphates are carried out in separate portions of the solution. The sulphates can be precipitated from the entire test solution by adding, drop by drop, $2N$ H_2SO_4 until the precipitation is complete.

2. Precipitation of Group V Cations. Place 2 drops of a concentrated ($12N$) solution of HCl into a conical test tube and add, with stirring, 25 drops of the test solution with the suspended precipitate (if there is any).** Centrifuge off the precipitate and wash with cold water containing several drops of $2N$ HCl (to depress the solubility of $PbCl_2$).

The precipitate, which may contain AgCl, Hg_2Cl_2, $PbCl_2$, $PbSO_4$, $BaSO_4$, $SrSO_4$, and (partially) $CaSO_4$ together with the basic salts of bismuth (and also of Sb and Sn), is to be examined according to Procedure *3* and subsequent procedures.

The centrifugate, containing the cations of Groups IV, III, II, and I, is prepared for precipitation of Group IV by hydrogen sulphide in accordance with the course of analysis described in Sec. 6.14 (see Procedure *9*).

---

* According to the data obtained by N. Tananaev, the salts of bismuth and mercury(II) develop a pH' of about 2 in solution (without acidification), except $HgCl_2$ whose solutions have a pH of about 4.

** The reason why the test solution should be added to the concentrated solution of HCl (and not vice versa) is that a lesser number of basic salts are precipitated in this separation of Group V.

3. Detection of Pb^{2+} and Removal of $PbCl_2$. Treat the precipitate with 10 drops of distilled water and heat; part of $PbCl_2$ is dissolved. Separate quickly the hot solution from the precipitate and test for Pb^{2+} by means of the reaction with $K_2Cr_2O_7$ or KI. The formation of a yellow precipitate of $PbCrO_4$ soluble in caustic alkalis or, after the solution is heated and cooled, of a golden precipitate of PbI_2 indicates the presence of Pb^{2+}.

If Pb^{2+} is present, it is to be eliminated from the precipitate. To the test tube containing the precipitate add an excess (2-3 ml) of distilled water and heat nearly to boiling. Centrifuge quickly the undissolved precipitate (AgCl, Hg_2Cl_2, $PbSO_4$, $BaSO_4$, $SrSO_4$, and partially $CaSO_4$) and basic salts (if they have not been washed off completely in Procedure *2*) and examine according to Procedure *4*. The centrifugate containing $PbCl_2$ is discarded.

The precipitate should be washed with hot water several times until the $PbCl_2$ is completely dissolved (the test for the absence of Pb^{2+} ions in the washings).

4. Separation of Ag^+ and Detection of $[Hg_2]^{2+}$. To the precipitate add, depending on its amount, 4-8 drops of a concentrated (25-percent) solution of NH_4OH and mix thoroughly. This dissolves AgCl to give the complex ion $[Ag(NH_3)_2]^+$, while Hg_2Cl_2 is converted into a mixture of $[NH_2Hg]Cl$ and Hg which is black in colour. Hence, the instant blackening of the precipitate under the action of ammonia indicates the presence of $[Hg_2]^{2+}$. This conclusion is valid provided that the ions of Group IV are absent since the blackening may also be due to the reduction of AgCl to metallic silver by a basic stannous salt. So, if the ions of Group IV are likely to be present, the final conclusion as to the presence of $[Hg_2]^{2+}$ can be made only after the preliminary tests for mercury (see Procedure *8.1*). Centrifuge the undissolved residue and examine as directed in Procedure *6*. The centrifugate containing $[Ag(NH_3)_2]^+$ is to be investigated according to Procedure *5*.

5. Detection of Ag^+. To a portion of the centrifugate add a drop of phenolphthalein solution, then, a drop at a time, 5*N* HNO_3 (until the red colour disappears), and a drop or two of excess HNO_3. The formation of a white precipitate (turbidity) of AgCl indicates the presence of Ag^+ in the test solution.

Another portion of the ammoniacal solution may be treated with a drop of a dilute* solution of KI. The formation of a pale-yellow precipitate of AgI proves the presence of Ag^+.

If the precipitate is blackened by NH_4OH (see Procedure *4*) and no silver is found in the ammoniacal solution, the conclusion as to its absence is not reliable since Ag^+ could have been reduced completely to the metal as the result of the interaction with metallic

---

* The KI solution must be sufficiently dilute as an excess of KI dissolves the AgI precipitate due to the formation of the $[AgI_2]^-$ complex.

mercury (formed by the reaction between Hg_2Cl_2 and ammonia) or with a basic stannous salt. In such cases, if no silver has been found in the solution, it must be searched for in the precipitate (see Procedure *8.1*).

If Ag^+ has been detected, it must be removed completely from the precipitate of Group V, for which purpose the precipitate must be treated once again with ammonium hydroxide (according to Procedure *4*) until the test for Ag^+ in the centrifugate is negative.

6. Separation of $PbSO_4$. When AgCl has been removed from the precipitate, wash it once with water and treat on heating with 10-15 drops of a 30-percent solution of CH_3COONH_4. This dissolves $PbSO_4$ and $[NH_2Hg]Cl$*. Centrifuge off the precipitate and look for Pb^{2+} in a portion of the centrifugate with the aid of the reaction with K_2CrO_4 or with H_2SO_4. If Pb^{2+} is present, treat the precipitate once again with CH_3COONH_4 until $PbSO_4$ is completely removed, then wash the undissolved residue once with water, and examine as directed in Procedure *7*.

7. Dissolution of Basic Salts. After the residue has been treated with ammonium acetate according to Procedure *6*, heat it for a few minutes with 1-3 drops of concentrated (12*N*) HCl and a double amount of water (2-6 drops). This will dissolve the basic salts of bismuth (and also of antimony and tin) if they are present in the precipitate. Centrifuge off the undissolved residue, combine the centrifugate with the solution resulting from the separation of the cations of Group V by the action of HCl (according to Procedure *2*), and examine as directed in Procedures *9-15*. Examine the precipitate according to Procedure *8*.

8. Confirmatory Tests for Silver and Mercury. Detection of $BaSO_4$ and $SrSO_4$. (1) If the precipitate turns black by the action of NH_4OH (see Procedure *4*), it may contain, after the treatment with HCl according to Procedure *7*, metallic mercury and silver along with the sulphates of barium and strontium (see Procedure *5*). To separate mercury and silver and to detect them (if silver could not be detected in the solution according to Procedure *5*), heat the precipitate with a mixture of 1 drop of 6*N* HNO_3 and 3 drops of concentrated (12*N*) HCl. The mercury will pass into solution in the form of $HgCl_2$, while the silver will be converted to AgCl. Centrifuge off the precipitate and add a few drops of $SnCl_2$ solution** to a drop of the centrifugate. A white precipitate turning black ($Hg_2Cl_2 \rightarrow Hg$) confirms the presence of mercury.

Wash the residue (AgCl, $BaSO_4$, $SrSO_4$) that has not been dissolved by the action of acids and treat with a few drops of concentrated

* The presence of $[NH_2Hg]Cl$ in the solution obtained can be proved by treating a portion of it with $SnCl_2$ solution. This produces a black (or gray) precipitate containing the basic salts of tin and metallic mercury. If $[Hg_2]^{2+}$ is absent, a white precipitate of basic tin salts is obtained.

** An excess of $SnCl_2$ is required to reduce the HNO_3 present in solution

(25%) NH_4OH solution. Centrifuge off the precipitate and look for Ag^+, as directed in Procedure *5*, in the centrifugate known to contain the complex ion $[Ag(NH_3)_2]^+$.

Investigate the precipitate known to contain $BaSO_4$ and $SrSO_4$ as described earlier (see Sec. 4.13, Procedure *3*).

(2) If the precipitate has not turned black upon addition of NH_4OH, the residue is found, after the treatment according to Procedure *7*, to contain only $BaSO_4$ and $SrSO_4$. It is examined at once as indicated earlier (see Sec. 4.13, Procedure *3*).

9. Adjustment of the Acidity of the Solution and Precipitation of Group IV Cations by Hydrogen Sulphide. Look for As^V in the mixture of centrifugates obtained in Procedures *2* and *7* (see Sec. 6.14, Procedure *1*). Then oxidize Sn^{II} to Sn^{IV}, adjust the solution to the required pH (ca. 0.5), and precipitate Group IV by hydrogen sulphide (see Sec. 6.14, Procedure *2*). After the separation of the sulphides, the centrifugate is treated as directed in Procedure *15*.

10. Separation of the Arsenic Subgroup from the Copper Subgroup. Follow the procedure described in Sec. 6.14, Procedure *3*.

11. Analysis of the Arsenic Subgroup. To be carried out in accordance with Sec. 6.14, Procedures *4-7*.

12. Dissolution of Precipitated Sulphides of the Copper Subgroup. The sulphides CuS, Bi_2S_3, CdS, and PbS are dissolved as directed in Sec. 6.14, Procedure *8*.

13. Precipitation of Pb^{2+} and Removal of HNO_3. Lead is precipitated as $PbCl_2$ and $PbSO_4$ in a strongly acid solution incompletely and it should therefore be separated from the other cations of the copper subgroup. Place the solution into a crucible (or an evaporating dish), add 5-6 drops of $2N$ H_2SO_4, and evaporate until a thick white cloud of sulphuric acid appears (*In the hood!*). This eliminates free HNO_3 which prevents the complete precipitation of Pb^{2+} as $PbSO_4$. Nitric acid interferes also with the separation of Cu^{2+} in the form of metallic copper and with the detection of Cd^{2+} as CdS. Simultaneously, all the nitrates pass into the sulphates due to the low volatility of H_2SO_4 as compared with HNO_3.

The $PbSO_4$ precipitate dissolves in concentrated H_2SO_4 and the solution should therefore be diluted with distilled water for the $PbSO_4$ to be separated completely. After the residue in the crucible is cooled, add 10-12 drops of distilled water cautiously, and transfer the contents of the crucible to a conical test tube. Centrifuge off the precipitated lead sulphate and test it for the content of an impurity of basic bismuth salts. For this purpose, treat the precipitate on heating with a 30-percent solution of CH_3COONH_4. Perform confirmatory tests in the resulting solution for Pb^{2+} by means of the reaction with K_2CrO_4 or H_2SO_4.

If the action of CH_3COONH_4 leaves an undissolved residue, heat it with a few drops of HCl solution, following which dilute the resulting solution strongly (to the top of the test tube) with water. A white turbidity indicates the presence of Bi^{3+}.

14. Detection of Cu^{2+}, Bi^{3+}, and Cd^{2+}. The solution obtained after the separation of $PbSO_4$ according to Procedure *13* is to be examined as directed in Sec. 6.14, Procedures *9-11*.

15. Preparation of the Solution for Detection of Group III-I Cations. Transfer the centrifugate obtained after the separation of the cations of Group IV by H_2S (according to Procedure *9*) to a crucible and evaporate (to remove H_2S) until a volume of about 20 drops is left.* Separate the precipitated sulphur by centrifugation and discard. Examine the centrifugate for the cations of Groups III, II, and I as directed in Sec. 5.24.

The schemes for analysis of a mixture of the cations of Groups V-I are given in Tables 6.7 and 6.8. If the test solution contains PO_4^{3-},

TABLE 6.7. **Scheme of Analysis of Precipitates of Group V Chlorides in the Presence of Sulphates ($PbSO_4$, $BaSO_4$, $SrSO_4$, partly $CaSO_4$) and Oxychlorides (BiOCl and SbOCl)**

1. Test (in solution) for SO_4^{2-} and cations precipitated as sulphates. If such cations are present, they are to be precipitated by 2*N* H_2SO_4 solution.	
2. Separate Group V by the action of HCl (in the cold).	
Precipitate: AgCl, $PbCl_2$ (partially), $PbSO_4$, Hg_2Cl_2, BiOCl, SbOCl, Group II sulphates. 3. Treat the ppt. with hot water.	
Precipitate: AgCl, $PbSO_4$, Hg_2Cl_2, BiOCl, SbOCl, Group II sulphates. 4. Treat the ppt. with NH_4OH solution.	*Solution*: $PbCl_2$. Look for Pb^{2+}.
Precipitate: $[NH_2Hg]Cl + Hg$ (black), $PbSO_4$, BiOCl, SbOCl, Ag, Group II sulphates. 5. Treat the ppt. with CH_3COONH_4 solution.	*Solution*: $[Ag(NH_3)_2]Cl$. Look for Ag^+.
Precipitate: BiOCl, SbOCl, Hg, Ag, Group II sulphates. 6. Dissolve BiOCl and SbOCl in HCl on heating.	*Solution*: Pb^{2+} and Hg^{2+}. Look for Pb^{2+}.
Precipitate: Hg, Ag, Group II sulphates. 7. Test the ppt. for silver and mercury (see Sec. 6.19, Procedure *8.1*) and for Group II sulphates (see Sec. 6.19, Procedure *8.2*).	*Solution*: Bi^{3+} and Sb^{III} (HCl). To be added to the solution obtained by the action of HCl according to Procedure *2*.

* If preliminary tests for the cations of Group III are omitted, there is no need to remove H_2S.

TABLE 6.8. Scheme of Analysis of

1. Preliminary observations and tests: colour solution, solution pH, the presence of the ions NH_4^+, Fe^{2+}, Fe^{3+}, As^{V}

2. Test for SO_4^{2-} and for cations forming acid-insoluble sulphates. If required,

3. Separate Group V chlorides together with sulphates ($PbSO_4$ and Group II impurities of basic salts of bismuth, antimony, and tin) as directed in

4. *Centrifugate*: Cu^{2+}, Cd^{2+}, Pb^{2+} (partially), Bi^{3+}, Hg^{2+}, As^{III}, As^{V}, Sb^{III}, H_2SO.

5. Oxidize Sn^{II} to Sn^{IV} by the action of H_2O_2.

6. Adjust the pH of solution and precipitate Group IV by hydrogen sulphide of I_2 or NH_4I).

7. *Precipitate* contains sulphides: CuS, CdS, PbS, Bi_2S_3, HgS, As_2S_3, Sb_2S_5, $Na_2S+NaOH$.

9. *Precipitate*: PbS, CuS, CdS, Bi_2S_3. Treat the ppt. on heating with dilute HNO_3.

10. *Solution*: Pb^{2+}, Cu^{2+}, Cd^{2+}, Bi^{3+} (HNO_3).

11. Remove HNO_3 and precipitate Pb^{2+} by evaporating the solution with H_2SO_4 until a white cloud of sulphuric acid appears; dilute with water.

<table>
<tr><td rowspan="3">Precipitate: $PbSO_4$. Dissolve in CH_3COONH_4 and look for Pb^{2+}.</td><td colspan="3">Solution: Cu^{2+}, Cd^{2+}, Bi^{3+} (H_2SO_4).
12. Separate Bi^{3+} by NH_4OH. Look for Cu^{2+}</td></tr>
<tr><td>Precipitate: basic bismuth salts.</td><td colspan="2">Solution: $[Cu(NH_3)_4]^{2+}$,$[Cd(NH_3)_4]^{2+}$,(NH_4OH).
14. Separate Cu^{2+} by H_2S in presence of 1N HCl (or saturated solution of NH_4Cl)*.</td></tr>
<tr><td>13. Test for Bi^{3+}.</td><td>Precipitate: CuS.</td><td>Solution: Cd^{2+} (HCl or NH_4Cl).
Dilute and look for Cd^{2+}.</td></tr>
</table>

* The Cd^{2+} ion can also be detected by using the glycerol or cyanide method for separa

Cations of Groups V-I

or absence of ppt., colour and solubility of ppt. in acids and alkalis. Detection

postprecipitate by $2N$ H_2SO_4.

sulphates) by HCl in the cold. Analyse pptd. chlorides and sulphates (containing Table 6.7 and in Sec. 6.19 (Procedures *3-8*).

Sb^V, Sn^{II}, Sn^{IV}, cations of Groups III, II, and I, and also excess of HCl and

(if As^V has been detected, the precipitation is to be effected in the presence

SnS_2 (S). Separate copper subgroup from arsenic subgroup by

15. *Centrifugate*: AsS_3^{3-}, SbS_4^{3-}, SnS_3^{2-}, HgS_2^{2-} (Na_2S, NaOH). Treat with solid NH_4Cl.

16. *Precipitate*: HgS+ +S. Dissolve in solution of KI+HCl and look for Hg^{II}.

17. *Centrifugate*: thiosalts of As^{III}, Sb^V, and Sn^{IV}. Treat with CH_3COOH (decomposition of thiosalts).

18. *Precipitate*: As_2S_3, Sb_2S_5, SnS_2 (S). Treat with conc. HCl.

19. *Precipitate*: As_2S_3+S. Dissolve by $(NH_4)_2CO_3$ or by conc. HNO_3. After As^{III} is oxidized to As^V, look for arsenic in the form of $(NH_4)_3\times[AsMo_{12}O_{40}]\times xH_2O$.

20. *Centrifugate*: $[SbCl_6]^{3-}$, $[SnCl_6]^{2-}$ (HCl).

21. Reduce Sb^{III} to Sb and Sn^{IV} to Sn^{II} by magnesium.

22. *Precipitate*: metallic Sb.

23. *Solution*: Sn^{II} and magnesium. Look for Sn^{II}.

Tests for Sb^{III} and Sn^{IV} may also be carried out in solution according to Procedure *20* by using oxalate and tartrate complexes (Sec. 6.14, Procedure *7.4*.)

8. *Centrifugate*: cations of Groups III, II, and I. Remove H_2S and concentrate the solution by evaporation. Test for cations of Groups III-I according to the scheme in Sec. 5.24 (Tables 5.6 and 5.7).

tion from other cations of the copper subgroup (see Sec. 6.14, Procedure *11.2*).

which is to be established by special tests, it must be eliminated before the precipitation of Group III cations. Some of the methods of detection and separation of PO_4^{3-} and also the method of analysis of a mixture of the cations of Groups III-I without removal of PO_4^{3-} are discussed in Sec. 6.21.

6.20. Systematic Analysis with Thioacetamide Used Instead of Hydrogen Sulphide

Hydrogen sulphide is highly toxic and therefore numerous attempts have been made to replace it in the systematic analysis by other sulphur-containing reagents, for example, by thioacetamide which forms hydrogen sulphide on hydrolysis. Precipitation by thioacetamide is an example of the so-called homogeneous precipitation or, more precisely, the internal-generation method. In this method, the reagent to be used is a product gradually formed as the result of a definite reaction.

Below is given a description of the method of precipitation of the sulphides of Groups III and IV with the aid of thioacetamide.

Thioacetamide, CH_3CSNH_2, is the amide of thioacetic acid. It decomposes in aqueous solutions, forming thioacetic acid, CH_3COSH, and acetamide, CH_3CONH_2, as intermediate products.

With increasing temperature, the rate of decomposition increases and the reaction proceeds eventually according to the equation:

$$CH_3CSNH_2 + 2H_2O \xrightleftharpoons{\text{heat}} CH_3COOH + NH_3 + H_2S$$

The rate of decomposition of thioacetamide increases also in the presence of hydrogen and hydroxyl ions. The overall equations for the reactions may be represented as follows:

in acid solution

$$CH_3CSNH_2 + 2H_2O + H^+ \xrightleftharpoons{\text{heat}} CH_3COOH + NH_4^+ + H_2S \qquad (6.10)$$

and in alkaline solution

$$CH_3CSNH_2 + 3OH^- \xrightleftharpoons{\text{heat}} CH_3COO^- + NH_3\uparrow + H_2O + S^{2-} \qquad (6.11)$$

As can be seen from the above equations, the concentration of S^{2-} in acid solution is considerably lower than in alkaline solution since the dissociation of hydrosulphuric acid is suppressed by the presence of free hydrogen ions in solution. But this concentration of S^{2-} ions resulting from the dissociation of H_2S is enough for the solubility product constants of the sulphides of all the cations of Group IV to be exceeded if the precipitation is carried out on heating the solution up to 80-90°C and at $[H^+] \approx 0.3N$.

Thus, the conditions for precipitation of the sulphides of Group IV by thioacetamide are almost the same as those for precipitation by hydrogen sulphide, i.e., the precipitation is carried out in acid solution at pH $\approx$ 0.5 (0.3*N* in hydrochloric acid) on heating up to 80-90°C. After the sulphides of Group IV are separated, the solution is neutralized by ammonium hydroxide to pH $\approx$ 9, the sulphides and hydroxides of Group III being precipitated (also on heating up to 80-90°C). Treatment of the sulphides of Group IV and III cations is carried out in just the same way as in the hydrogen sulphide method. When the detection of the cations of antimony, cadmium, and zinc requires hydrogen sulphide, use may also be made of thioacetamide.

1. Precipitation of Group IV Sulphides. In the solution obtained after the separation of the precipitated chlorides of Group V and the sulphates of lead and Group II cations, Sn^{II} is preliminarily oxidized to Sn^{IV}. To detect arsenic(V), a preliminary test is made with thioacetamide in a solution which is 6*N* in HCl. In the case of formation of a yellow precipitate of arsenic(V) sulphide, it is useful, in the precipitation of all the sulphides of Group IV cations, to add a drop of iodine water which speeds up the reduction of As^{V} to As^{III}. If the solution is heated for a fairly extended time in the water bath, the precipitation is found to be complete without addition of iodine water.

Place into a beaker 2 ml of the test solution containing all cations of Groups I-IV and neutralize with ammonium hydroxide up to pH $\approx$ 7. Then add an equal volume of 0.6*N* HCl. This provides a pH about 0.5. Neglecting the precipitate formed (the oxychlorides of antimony and bismuth), add thioacetamide to the test solution (*In the hood!*) until the precipitation is complete (about 2 ml of 1*M* solution). Place the solution with the precipitate into two test tubes, heat in the water bath for 3-5 min, and centrifuge. Dilute a portion of the solution twofold and add on heating a few drops of thioacetamide solution. When a yellow precipitate of cadmium sulphide is formed, dilute the entire centrifugate with water and precipitate again; then wash and treat the sulphides in the usual way (see Sec. 6.19). The sulphides of arsenic, antimony, and tin can be dissolved by a mixture of 30-percent sodium hydroxide and thioacetamide. Under these conditions, mercury(II) gives a thiosalt and passes into solution together with the thiosalts of arsenic, antimony, and tin, where it is to be detected in the usual manner (precipitation is effected with the aid of solid ammonium chloride).

2. Precipitation of the Sulphides and Hydroxides of Group III Cations. After the Group IV cations are separated, neutralize the solution known to contain the cations of Groups III-I and excess of thioacetamide with a solution of ammonia up to pH $\approx$ 9 and heat in a water bath up to 80-90°C for 3-5 minutes. Test for completeness of precipitation. If precipitation is incomplete, add a few drops of thioacetamide solution and heat again. Separate the precipitated sulphides and hydroxides of Group III cations and then wash and treat the precipitate as usual (see Sec. 5.24).

Examine the centrifugate for the cations of Groups II and I according to the scheme given in Sec. 4.13.

6.21. Identification and Removal of Phosphate Ions

The presence of PO_4^{3-} in the solution interferes with the separation of the cations of Group III from those of Groups II and I as the phosphates of Group II cations and magnesium phosphate are insoluble in water but soluble in acids and precipitate when the acidic test solution is rendered alkaline or neutral for the precipitation of Group III cations by ammonium sulphide (see Sec. 5.15).

It is clear that, in the presence of PO_4^{3-}, the action of $(NH_4)_2S$ can no longer be used to separate the cations of Group III from those of Group II which may precipitate completely or partially, depending on the amount of PO_4^{3-} present, together with Mg^{2+}. In order to separate them by the method considered in this book, the first thing to do is to remove PO_4^{3-}. There are a number of methods for this separation to be achieved: precipitation of PO_4^{3-} by solutions of stannic chloride or zirconium(IV) nitrate in hydrochloric acid solution; by bismuth(III) ions in nitric acid solution; by ferric ions in the presence of the acetate buffer, etc.

Let us first consider the last-named method. It is based on the fact that in the presence of the acetate buffer ($CH_3COOH + CH_3COONH_4$), which develops pH ≈ 5 in solution, the phosphate ion can be completely precipitated in the form of $FePO_4$. At this pH value, the ions Ba^{2+}, Sr^{2+}, Ca^{2+}, and Mg^{2+} and also Zn^{2+}, Mn^{2+}, Co^{2+}, Ni^{2+}, and Fe^{2+} (in fact, all the doubly charged cations) remain in solution. Conversely, the phosphates of aluminium, chromium, and titanium coprecipitate with $FePO_4$.

Iron(III) is chosen as the precipitant for PO_4^{3-} for a number of reasons: (a) the cations Fe^{2+} and Fe^{3+} can easily be detected in the original solution; (b) of the phosphates of Group III cations, ferric phosphate is the least soluble ($SP_{FePO_4} = 1.3 \times 10^{-22}$); (c) an excess of Fe^{3+} is required for complete precipitation of the phosphate ion (this excess is easily recognized by the red colour of the ferric acetate complex formed, while with the acetate ion aluminium gives a colourless basic acetate and chromium forms a soluble complex compound).

To remove excess of Fe^{3+} and precipitate PO_4^{3-} completely, the solution is boiled. This precipitates Fe^{3+} ions in the form of a basic acetate. The basic acetates of aluminium and chromium coprecipitate with it (in the presence of iron, the soluble chromium acetate complex is decomposed and a basic chromium acetate is precipitated) and so

TABLE 6.9. Scheme of Analysis of Mixtures of Cations of Groups III-I in the Presence of PO_4^{3-} by the Acetate Method

<table>
<tr><td colspan="3">1. Remove H_2S by boiling. Look for PO_4^{3-} and Ti^{IV} (if required, for NH_4^+, Fe^{2+}, and Fe^{3+}) in separate portions of solution. Oxidize Fe^{2+} to Fe^{3+} by conc. HNO_3.</td></tr>
<tr><td colspan="3">2. Separate PO_4^{3-} by the action of $FeCl_3$ in the presence of the acetate buffer (pH ≈ 5). Precipitate excess of Fe^{III}, Al^{III}, and Cr^{III} as oxyacetates and also of titanic acid by boiling.</td></tr>
<tr><td colspan="2">3. *Precipitate*: $FePO_4$, titanic acid, the oxyacetates of Fe^{III}, Cr^{III}, Al^{III}.
Treat the ppt. with 30-percent NaOH and 3-percent H_2O_2.</td><td rowspan="2">*Filtrate*: remaining Group III cations together with cations of Groups II and I.
4. Test the solution for cations of Groups III, II, and I (see Sec. 5.24, Tables 5.6 and 5.7).</td></tr>
<tr><td>*Precipitate*: $FePO_4$, $Fe_2O_3 \times xH_2O$, $TiO_2 \cdot xH_2O$ (not to be examined).</td><td>*Centrifugate*: AlO_2^-, CrO_4^{2-}.
Look for Al^{III} and CrO_4^{2-}.</td></tr>
</table>

does the titanic acid formed by hydrolysis of a titanium salt at the pH indicated above.

The precipitate obtained is to be tested for Al^{3+}, Cr^{3+}, and also Ti^{IV} (if it has not been detected earlier) and the solution is to be analysed for the cations of Groups III, II, and I.

If the test solution is expected to contain the cations of all the five groups and PO_4^{3-}, the separation and analysis of the cations of Groups V and IV is to be conducted as usual (Sec. 6.19). The solution obtained after their separation, which is known to contain the cations of Groups III, II, and I, is to be investigated as directed below.

(a) Separation of Phosphate Ion as Ferric Phosphate

1. Preliminary Tests. (1) If the test solution containing a mixture of the cations of Groups III, II, and I is obtained after the separation of Groups V and IV, there is no need to detect the ions NH_4^+, Fe^{2+}, and Fe^{3+} in it because this has already been done. If the solution under analysis does not contain the cations of Groups V and IV, these tests must be performed.

It is also desirable to separate Ba^{2+} and Sr^{2+} in the form of sulphates before applying the tests.

(2) In all the methods of separation of the phosphate ion described below, titanium is detected in a preliminary test after hydrogen sulphide has been expelled (if the cations of Group IV have been separated as sulphides). Owing to the low solubility of titanium phosphate in

acids, it is found to be practically completely in the precipitate (partially in the form of $TiO_2 \cdot xH_2O$) and can easily be lost.

To detect titanium, the solution resulting from the separation of the ions of Group IV by hydrogen sulphide is evaporated to expel H_2S to a volume of 20-30 drops and, when the precipitated sulphur is centrifuged off, a test is made for Ti^{IV} by using the reaction with H_2O_2 or with chromotropic acid (see Sec. 5.18).

Titanium should be detected before the oxidation of Fe^{II} as the colour of iron(III) may interfere with the identification of titanium.

If Group IV has not been separated by hydrogen sulphide and the Fe^{III} ion has been detected in the solution, it should be reduced by adding a few drops of $SnCl_2$ solution to the hydrochloric acid solution.

(3) If the solution contains Fe^{II}, it should be oxidized to Fe^{III} by heating with a few drops of concentrated HNO_3 solution. The phosphate ion is then detected in a separate portion by the action of ammonium molybdate solution.

This is done as follows. To 2 drops of the test solution add a drop of concentrated HNO_3 solution and evaporate to dryness to drive out HCl which interferes with the test for phosphate ion (*In the hood!*). Treat the dry residue with 2 drops of water and a drop of concentrated HNO_3 solution. If required, heat until the precipitate is completely dissolved, add a small crystal of NH_4NO_3 and 6 drops of ammonium molybdate solution, and heat. The formation of a yellow precipitate of $(NH_4)_3[PMo_{12}O_{40}] \cdot xH_2O$ indicates the presence of phosphate ions in the test solution.

If the phosphate ion (PO_4^{3-}) has been detected, the analysis is carried out as described in Procedure *2*. If it is absent, the solution is to be analysed as directed in Sec. 5.24 (see Tables 5.6 and 5.7).

2. Separation of the Phosphate Ion. To the remainder of the solution add, a drop at a time, NH_4OH solution until a permanent turbidity of phosphates and hydroxides appears or until alkaline if no turbidity is formed.* Now add, drop by drop, 2*N* HCl until the precipitate is dissolved or until the solution is just acid (a large excess of HCl should be avoided). To the resulting faintly acid solution add 15-20 drops of a 30-percent solution of CH_3COONH_4 and check the pH of the solution.** The pH value is brought to about 5 by addition of HCl or CH_3COONH_4.

If Fe^{3+} ions are present in excess as compared with their amount required for precipitation of PO_4^{3-}, the solution is coloured red-brown,

* If no precipitate (turbidity) is formed upon addition of ammonium hydroxide until the solution reacts alkaline, it may be assumed that no cations of Group II and no Mg^{2+} are present; of the cations of Group III, only Zn^{2+}, Co^{2+}, and Ni^{2+} may be present; their phosphates are well soluble in NH_4OH, and may therefore fail to be precipitated by the action of this reagent.

** When excess of CH_3COONH_4 is added to the solution containing free HCl, an equivalent amount of CH_3COOH is formed. Thus, an acetate buffer is produced in the solution, which is required for the separation of PO_4^{3-}.

this colour being characteristic of solutions of the acetate complex of iron(III). If this does not happen or if there is no iron at all in the solution, add a solution of $FeCl_3$, drop by drop, until a red-brown colour develops. Then, pour the solution with the precipitate into a beaker containing 5-10 ml of hot distilled water and boil for 2-3 min. This precipitates the oxyacetates of Al^{III}, Cr^{III}, and Fe^{III} and also titanic acid. Without allowing the liquid to cool off, quickly filter off the precipitate (since it is usually difficult to separate it by centrifugation) through a small filter with the aid of the apparatus for filtration (see Fig. 2.17) and collect the filtrate. Wash the precipitate on the filter with hot water and reject the washings.

The washed precipitate of the phosphates, basic acetates, and titanic acid is to be examined as directed in Procedure *3*, and the filtrate according to Procedure *4*.

3. Detection of Al^{3+} and Cr^{3+}. Transfer the precipitate obtained in Procedure *2* to a test tube and treat with an excess of a 30-percent solution of NaOH and a few drops of 3-percent H_2O_2. Heat the solution together with the precipitate in a boiling water bath for 2-3 min. This oxidizes Cr^{3+} to CrO_4^{2-} which passes into solution together with AlO_2^-. Ferric phosphate, ferric hydroxide, and titanic acid remain in the precipitate. Since the ions Fe^{III} and Ti^{IV} have already been detected, the precipitate is not to be analysed. It is separated and discarded.

Tests for CrO_4^{2-} and Al^{III} are performed in separate portions of the centrifugate as described in the Group III cation analysis (Sec. 5.24).

4. Detection of Cations of Groups III, II and I. The filtrate obtained in Procedure *2* may contain the remaining cations of Group III together with the cations of Groups II and I. Since PO_4^{3-} has been removed, the analysis is carried out by ordinary methods (see Sec. 5.24).

(b) Separation of Phosphate Ion by Coprecipitation with Stannic Acid

The method is based on the formation of an adsorption compound of phosphate with stannic acid and also on the low solubility of the precipitate in dilute (about 0.5*N*) hydrochloric acid. Stannic acid is produced by the hydrolysis of Sn^{IV} at the stated acidity of the solution and on heating.

The reagent is prepared by shaking at room temperature 1 g of $SnCl_4 \cdot 3H_2O$ with 1 ml of water. Before the reagent is added, it is recommended to add 2-3 drops of 0.5*M* $(NH_4)_2HPO_4$ solution to the test solution since in the case of a low concentration of phosphate ions stannic acid is badly separated upon centrifugation and the supernatant liquid becomes turbid.

The phosphate ion is removed by this method as follows. First of all, all the necessary preliminary tests are carried out in the test solution as described above (see Sec. 6.21, Procedure *1*). After this, place the test solution in a test tube and add 3 drops of $(NH_4)_2HPO_4$ solution. Then add, drop by drop, a 10-percent ammonia solution until the solution shows a faintly acid reaction toward litmus paper (the precipitate formed upon addition of ammonia solution must gradually dissolve on mixing). Now add 4 drops of 6*N* hydrochloric acid and water to a volume of 2 ml, heat to boiling, and precipitate phosphate ions by gradually adding 3-4 drops of the reagent solution. Mix the solution with the precipitate thoroughly and centrifuge. Using two drops of the clear centrifugate obtained, test for completeness of precipitation of phosphate ions, making use of the reaction with the ammonium molybdate reagent for the purpose.

If the phosphate ions have been precipitated incompletely, the precipitation must be repeated, by adding a drop more of the reagent and heating the solution to boiling. If the test for PO_4^{3-} with ammonium molybdate solution gives a negative result, separate the precipitate of stannic acid with the adsorbed PO_4^{3-} and discard.

Adjust the pH of the medium to the required value (about 0.5) and precipitate the Sn^{IV} ions left in solution in the form of SnS_2 by hydrogen sulphide. Centrifuge off the precipitate, boil off the hydrogen sulphide from the solution, evaporate to a volume of 1-2 ml, and test for the cations of Groups III, II, and I in the usual way (see Sec. 5.24, Tables 5.6 and 5.7).

(c) Separation of Phosphate Ion as Bismuth Phosphate

This method is based on the fact that bismuth phosphate is practically insoluble in dilute (0.5*N*) HNO_3. In hydrochloric acid, however, it dissolves much better and Cl^- ions must therefore be absent in solution. These ions can be removed by twofold evaporation of the solution with HNO_3 almost to dryness.

Phosphate ions are separated by this method as follows. Place the test solution into a micro crucible (or a dish) and evaporate to a volume of a few drops. Add 5-6 drops of HNO_3 solution (6*N*) and evaporate the solution almost to dryness (do not ignite). Add a few drops more of HNO_3 and evaporate again. Then add a few drops of water, transfer the contents of the crucible to a conical test tube having a capacity of not less than 4 ml, and dilute to a volume of about 2 ml. Neutralize the solution with ammonia until it is neutral to litmus, and add HNO_3, drop by drop with stirring, until a further drop renders the solution acid. This being done, add such a number of drops of the acid, by calculation, that its concentration in the solution is about 0.5*N*.*

---

* The calculation is based on the normality of HNO_3 and it is assumed that the total volume of the solution during precipitation is equal to 3 ml.

Dilute the solution to a volume of 3 ml, heat, and precipitate PO_4^{3-}, adding 0.1*M* $Bi(NO_3)_3$ solution by 5-drop portions until the precipitation is complete. Add a small excess of bismuth nitrate, heat the contents of the test tube for several minutes (for better coagulation of the precipitate), centrifuge, and again test for complete precipitation. If precipitation is complete, transfer the centrifugate to another test tube and remove the excess of Bi^{3+} ions by precipitating them with hydrogen sulphide. With the precipitation being complete, separate the precipitated Bi_2S_3 by centrifugation and discard. Then remove H_2S from the centrifugate, concentrate the solution by evaporation to a volume of 1 ml, and examine for the cations of Groups III, II, and I as indicated earlier (see Sec. 5.24, Tables 5.6 and 5.7).

(d) *Separation of Phosphate Ion as Zirconium Phosphate*

This method is based on the fact that zirconium phosphate, $Zr(HPO_4)_2$, is almost insoluble in 0.3*N* HCl and the excess of Zr^{IV} introduced can be eliminated by the action of ammonia.

The PO_4^{3-} ion is precipitated as follows. First, boil off hydrogen sulphide from the solution remaining after the separation of Group IV and then add a little solid NH_4Cl. Then add 10-15 drops of a solution* of $Zr(NO_3)_4$ and an ammonia solution, a drop at a time with thorough stirring using a glass rod, until the solution reacts basic to litmus, heat for about 5 min in the water bath, and acidify with 2*N* HCl. The acid should be added dropwise until the solution reacts acid to litmus, and then add an excess of it (a drop of HCl for every 10 drops of the solution).

Heat the test tube again in the water bath to dissolve the hydroxides of the cations of Group III. Zirconium phosphate and metazirconic acid, H_2ZrO_3, remain in the precipitate. The acid is produced when zirconium hydroxide (precipitated under the action of ammonia solution) is heated. It is almost insoluble in dilute HCl.

Centrifuge off the precipitate formed and discard it. Evaporate a few drops of the centrifugate to dryness with HNO_3, dissolve the dry residue in 2-3 drops of HNO_3, and test for the presence of a phosphate ion by using ammonium molybdate solution (10-15 drops) on heating. If the precipitation of the phosphate ion is incomplete, add more of the $Zr(NO_3)_4$ solution, and repeat the precipitation.

When complete precipitation has been achieved and the precipitate has been separated from the solution, it is to be analysed for the cations of Groups III, II, and I.

The precipitation of Zr^{IV} as H_2ZrO_3 is incomplete and a small amount of it passes into solution. On further separation of the cations of Group III by the action of $(NH_4)_2S$ in an ammoniacal solution, zirconium is completely precipitated. When Group III is separated

* One millilitre of this solution must contain 50 mg of zirconium nitrate.

into subgroups, zirconium is found to be present in the precipitate together with the hydroxides of iron(III), aluminium, and chromium(III). On further treatment of the precipitate with excess NaOH and H_2O_2, a very small amount of zirconium may pass into solution together with CrO_4^{2-} and AlO_2^-. The presence of zirconium has to be considered only in the detection of Al^{3+} since after the solution is acidified and acted on by ammonia solution, a precipitate of zirconium hydroxide, which is similar to aluminium hydroxide, coprecipitates with $Al(OH)_3$.

To avoid errors, before the acidified solution is allowed to react with ammonium hydroxide, the acidity of the solution must be adjusted to the same value as in the precipitation of PO_4^{3-} (i.e., an excess of 2*N* HCl must be equal to 1 drop for every 10 drops of the solution). Then add a drop or two of Na_2HPO_4 solution and heat for 3-5 min. If a precipitate of $Zr(HPO_4)_2$ is formed, separate it and discard. Precipitate Al^{3+} in the centrifugate by addition of ammonia solution until alkaline.*

Apart from the methods described above, use is also made of a method for separating PO_4^{3-} in the form of lead or titanium phosphates.

(e) Separation of Phosphate Ion by Chromatography

The chromatographic method** of separation of phosphate ions and other interfering anions is based on the absorption of all the cations present in the solution by passing it through a column of any H^+-cation exchanger. All the anions (including PO_4^{3-}) remain in solution in the form of the corresponding free acids. When the column is washed with water, the absorbed cations are eluted by HCl solution and, after the column is washed again, the resulting solution of chlorides together with the washings is tested for cations. A somewhat modified version of this method is described below.

1. Preparation of Cation-Exchanger. The test is carried out in a glass column with a stopcock (a burette may also be used for this purpose), with a capacity of 25 ml (Fig. 6.6). A pad of glass wool is placed in the constricted end of the column and 10 g of a cation-exchanger (e.g. Dowex 50), which has been allowed to swell in water for at least 24 hours, is put over it. The upper surface of the cation-exchanger is covered with a pad of glass wool. Before the test is started, it is necessary to prepare the cation-exchanger. This is done by passing 15-20 ml of 6*N* HCl through the cation-exchanger to displace all the cations that it may contain by hydrogen ions. The acid solution must flow out of the column at a rate of 1-2 drops per second. When

* Aluminium is precipitated as $Al_2O_3 \cdot xH_2O$ since the solubility product of $Al(OH)_3$ is 1×10^{-32} and that of $AlPO_4$ is 5.75×10^{-19}.

** The chromatographic method has been described in Sec. 2.6.

not more than 1 ml of the solution is left over the cation-exchanger bed in the column, the cation-exchanger is washed with water until a drop of the wash liquid ceases to impart a pink colour to the paper impregnated with methyl orange (i.e., until the pH of the wash liquid is equal to or more than 4).

2. Absorption of Cations and Their Separation from Phosphate Ion and Other Anions. The pH of the test solution is to be determined: it must not be lower than 1.5-2. More strongly acid solutions (1-2 ml) are neutralized with ammonia until a turbidity appears, which is dissolved in a drop or two of 6*N* HCl, and the solution is diluted with water to a volume of 10 ml. This solution is passed at the same rate (1-2 drops per second) through a bed of the cation-exchanger, whereby the cations present in the solution are absorbed, displacing an equivalent amount of hydrogen ions from the cation-exchanger. To remove the solution, which contains H_3PO_4 and other acids, from the column, the latter is again washed thoroughly (until the pH of the washings becomes approximately equal to 4).*

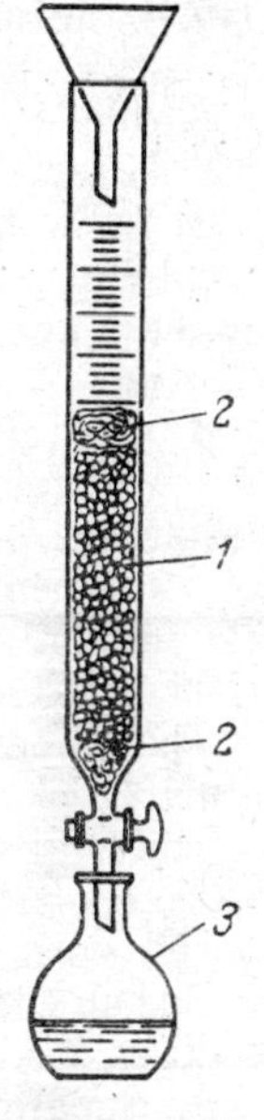

Fig. 6.6. Apparatus for separation of the phosphate ion by chromatography:
1 – cation-exchanger; 2 – glass wool wads; 3 – flask for collecting eluate.

3. Isolation of Absorbed Cations and Recovery of Cation-Exchanger. After the cation-exchanger is washed with water, a flask or beaker is placed under the column and 15-20 ml of 6*N* HCl is again passed at the same rate. Then the column is washed with water to remove the displaced cations, the washings being collected in the same vessel. When the total volume of the liquid is 40-50 ml, it is poured into a porcelain dish, evaporated in the water bath to the initial volume (1-2 ml), and examined for cations as usual. The cation-exchanger may be used for new determinations.

The separation of cations by this method is somewhat complicated by the necessity to wash the column three times and to evaporate a large volume of the solution.

(f) Method of Analysis of a Mixture of Cations of Groups III-I without Separation of Phosphate Ion

To analyse a mixture of the cations of Groups III-I, use may be made of a procedure based on the different solubility of the phosphates of the cations of these groups in acids and alkalis. If the solution

* It is necessary to check by the reaction with the molybdate solution whether the phosphate ion has been completely washed off.

does not contain phosphate ions, it is introduced deliberately. Thus, the phosphate ion is a precipitant which can be used for separation of the cations of the groups indicated by changing the acidity of the solution.

All the preliminary tests described above (see Sec. 6.21 (a), Procedure *1*) are made in the solution remaining after the separation of Group IV by hydrogen sulphide. The Fe^{2+} ions are oxidized to Fe^{3+} by the action of concentrated nitric acid.

If no iron ions are present, they should be introduced for further complete precipitation of chromium phosphate which is difficult to separate out in the absence of ferric ions. Further separation is effected by using the following scheme.

1. Precipitation of Phosphates Insoluble in Ammonium Hydroxide. A 1-ml volume of the mixture of cations being analysed is added (after the removal of H_2S) gradually, with constant stirring, to 2-3 ml of a 25-percent ammonia solution. Then, the phosphates insoluble in ammonium hydroxide are precipitated by adding a solution of $(NH_4)_2HPO_4$.

The precipitate contains $FePO_4$, $AlPO_4$, $CrPO_4$, $Ti(HPO_4)_2$, $MnNH_4PO_4$, $MgNH_4PO_4$, $Ba_3(PO_4)_2$, $Sr_3(PO_4)_2$, and $Ca_3(PO_4)_2$.

The solution contains the ions $[Zn(NH_3)_6]^{2+}$, $[Co(NH_3)_6]^{2+}$, $[Ni(NH_3)_6]^{2+}$, K^+, Na^+, and NH_4^+; the ammonium ion has been introduced with the precipitant and ammonia solution.

The precipitate is centrifuged off and examined according to Procedures *2-4*. The centrifugate is analysed as directed in Procedure *5*.

2. Separation of Phosphates Soluble in Acetic Acid from Those Insoluble in It. Wash thoroughly the precipitate obtained in Procedure *1* with a 10-percent solution of NH_4OH and treat on heating with a 30-percent solution of CH_3COOH. This dissolves the phosphates of magnesium, barium, strontium, calcium, and manganese. The phosphates $FePO_4$, $AlPO_4$, $CrPO_4$, and $Ti(HPO_4)_2$ are retained in the precipitate. Centrifuge off the precipitate, wash with distilled water, and analyse as directed in Procedure *3*. Examine the centrifugate according to Procedure *4*.

3. Examination of Precipitated $FePO_4$, $AlPO_4$, $CrPO_4$, and $Ti(HPO_4)_2$. Treat the washed precipitate with a 30-percent solution of NaOH and a 3-percent solution of H_2O_2. This dissolved $AlPO_4$, and the aluminium passes into solution in the form of AlO_2^-. Chromium(III) is oxidized to CrO_4^{2-} and also passes into solution. Both these ions can easily be detected by ordinary methods (see Sec. 5.24). The phosphates $FePO_4$ and $Ti(HPO_4)_2$ do not dissolve under these conditions. Since the ions of iron and titanium were detected earlier, there is naturally no need to examine the precipitate.

4. Separation of Mn^{2+} from Ba^{2+}, Sr^{2+}, Ca^{2+}, and Mg^{2+}. To the centrifugate obtained in Procedure *2* add 5 drops of concentrated HNO_3 and several minute crystals of $KBrO_3$. Heat the mixture. The

Mn^{II} ions are oxidized to the tetravalent manganese which separates out as a brown precipitate of $MnO(OH)_2$. Separate the precipitate, wash, and test for manganese (see Sec. 5.20).

Neutralize the solution known to contain Ba^{2+}, Sr^{2+}, Ca^{2+}, and Mg^{2+} with a 10-percent alkali solution and precipitate the cations of Group II and Mg^{2+} by using a saturated solution of an alkali carbonate.* Wash the precipitate well and dissolve on heating in a dilute solution of acetic acid. Then separate the cations of Group II from Mg^{2+} and look for them in the usual way (see Sec. 4.13). The Mg^{2+} ion is detected in the filtrate.

After the carbonates of Group II cations are separated from the centrifugate, look for magnesium with the aid of characteristic reactions (see Sec. 3.15).

5. Separation of the Sulphides of Cobalt, Nickel, and Zinc from Group I Cations. To the solution obtained in Procedure *1* add $(NH_4)_2S$ until the sulphides CoS, NiS, and ZnS are completely precipitated. Heat the mixture in the water bath up to 60-70°C and allow to stand for 20-30 min. Centrifuge off the precipitated sulphides, wash, and analyse in the usual way (see Sec. 5.24).

After the removal of S^{2-} in the centrifugate (by acidifying with HCl and boiling to expel H_2S and NH_4^+ completely), look for K^+, and Na^+ in it, for which purpose a suitable medium is produced (see Sec. 3.16).

* If Ba^{2+} and Sr^{2+} were precipitated as the sulphates at the beginning of the analysis of the mixture of Group V-I cations, only $CaCO_3$ is precipitated here.

Chapter 7

Vanadium, Molybdenum, and Tungsten

7.1. General Characteristics

The elements vanadium, molybdenum, and tungsten were formerly considered to belong to the so-called "rarer" elements. The term "rarer" elements as applied to the elements described in this chapter and to some other elements must now be regarded as a misnomer since they are fairly widespread in nature. These elements find wide application in the production of special grades of steel and ferroalloys.

In the Periodic System, the elements in question are in the first half of large periods (IV, V, and VI) and differ markedly in their chemical properties.

The chemistry of vanadium, molybdenum, and tungsten is extremely complicated. These elements have variable oxidation states and form a large number of compounds in which they are contained as a cation or anion.

Especially complicated is the question of the state of these elements in aqueous solutions. Depending on the conditions (primarily, the acidity of the medium), they are capable of forming a series of complex ions: aqua-complexes, oxo- and hydroxy-complexes, polymeric ions, etc.

These questions cannot be discussed in any appreciable length in this book. Detailed information can be found in special textbooks.* Here we shall consider the properties of only those compounds of vanadium, molybdenum, and tungsten, which are produced in the dissolution of the alloys of these elements in acids.

Since the dissolution of these alloys requires the use of concentrated nitric acid in admixture with 2*N* sulphuric acid or with concentrated hydrochloric acid (aqua regia), the elements in question pass into their highest oxidation states: vanadium(V), molybdenum(VI), and tungsten(VI). Therefore, we shall deal with the properties of these elements in their highest oxidation states.

Vanadium in the +5 oxidation state can give rise in solution to ions of different composition: VO_4^{3-}, $[VO_3(OH)]^{2-}$, $[V_2O_6(OH)]^{3-}$, $[VO_2(OH)_2]^-$, $[V_3O_9]^{3-}$, etc. To simplify the reaction equations, we shall write VO_3^- since solid salts are known, which contain this anion,

---

* Cotton, F. A. and G. Wilkinson, *Advanced Inorganic Chemistry: A Comprehensive Text*, 3rd ed., John Wiley and Sons, New York, 1972.

for example, NH_4VO_3, though in aqueous solution this salt, being hydrated, forms the complex ions indicated, depending on the conditions.

For molybdenum(VI) and tungsten(VI) we know natural compounds and solid salts containing the anions MoO_4^{2-} and WO_4^{2-}. For instance, $PbMoO_4$, $MgMoO_4$, $(NH_4)_2MoO_4$, $FeWO_4$, $MnWO_4$, $CaWO_4$, $PbWO_4$. Therefore, although, as in the case of vanadium(V), various complex ions can exist in solution, we shall use in the reaction equations the formulas of the anions MoO_4^{2-} and WO_4^{2-}.

As regards the hydroxides of Mo^{VI} and W^{VI}, they correspond to the formula $MeO_3 \cdot H_2O$ or $MeO_3 \cdot 2H_2O$, where $Me = Mo^{VI}$ or W^{VI}.

The ions of the elements considered, corresponding to the highest oxidation state, namely, V^V (the ion VO_3^-), Mo^{VI} (MoO_4^{2-}), and W^{VI} (WO_4^{2-}), are not included in a separate analytical group.

The action of $(NH_4)_2S$ on the ions formed by vanadium, molybdenum, and tungsten is the most important factor in deciding to which analytical groups these ions should be assigned.

The ions VO_3^-, MoO_4^{2-}, and WO_4^{2-} are not precipitated by $(NH_4)_2S$ but they form thiosalts. For example, when $(NH_4)_2S$ is added to a solution of ammonium molybdate, $(NH_4)_2MoO_4$, the solution turns brown due to the formation of the thiomolybdate:

$$MoO_4^{2-} + 8NH_4^+ + 4S^{2-} + 4H_2O \rightarrow MoS_4^{2-} + 8NH_3OH$$

Tungstates and vanadates react with $(NH_4)_2S$ in an analogous way:

$$WO_4^{2-} + 8NH_4^+ + 4S^{2-} + 4H_2O \rightarrow WS_4^{2-} + 8NH_4OH$$

$$VO_3^- + 6NH_4^+ + 3S^{2-} + 3H_2O \rightarrow VS_3^- + 6NH_4OH$$

The thiosalts resulting from these reactions* are similar in their properties to the thiosalts of arsenic, antimony, and tin. They are all well soluble in water and are decomposed by acids. For instance:

$$MoS_4^{2-} + 2H^+ \rightarrow MoS_3 \downarrow + H_2S \uparrow$$

Molybdenum(VI) sulphide, MoS_3, separates out as a dark-brown precipitate. If this precipitate is treated with an excess of $(NH_4)_2S$ or Na_2S, it redissolves with the formation of a thiosalt:

$$MoS_3 \downarrow + S^{2-} \rightarrow MoS_4^{2-}$$

The same behaviour is shown by the ions WO_4^{2-} and VO_3^-. Thus, the thiosalts produced by the action of ammonium sulphide (or sodium sulphide) on Na_2WO_4 and $NaVO_3$ are soluble in water and decompose on boiling with acids, forming tungsten(VI) sulphide (light-brown) and vanadium(V) sulphide (brown).

* This may also result in the formation of hydroxy thiosalts (i.e., the products of partial replacement of oxygen in the molecules of salts of oxy-acids by sulphur). It is because of their formation that the colour of the corresponding solution is gradually changed.

Like MoS_3, these sulphides redissolve when acted on by $(NH_4)_2S$ or Na_2S, giving rise to thiosalts. In other words, just like the sulphides of arsenic, antimony, and tin, the sulphides of vanadium, tungsten, and molybdenum are acidic in character, this being explained by the position of the corresponding elements in the Periodic System.

Based on these properties, we should include the ions VO_3^-, MoO_4^{2-}, and WO_4^{2-} in the arsenic subgroup of Group IV cations. They, however, differ from the ions of the arsenic subgroup by their behaviour towards the action of hydrogen sulphide. Whereas the ions formed by arsenic, antimony, and tin are precipitated by hydrogen sulphide in acid solution, of the ions considered in this chapter only the ion MoO_4^{2-} can be precipitated by this reagent:

$$MoO_4^{2-} + 2H^+ + 3H_2S \rightarrow MoS_3\downarrow + 4H_2O$$

Moreover, owing to the formation of a colloidal solution of MoS_3, the precipitation is complete only under pressure. The ions VO_3^- and WO_4^{2-} are not precipitated at all by hydrogen sulphide.

For this reason, only molybdenum is found in the systematic cation analysis to be in the group to which it belongs. The ion WO_4^{2-} is precipitated by HCl as slightly soluble tungstic acid, $WO_3 \cdot H_2O$, together with the silver group, and is thus separated from most of the cations. In contrast to WO_4^{2-}, the ion VO_3^- is precipitated neither by HCl nor by hydrogen sulphide which reduces it to the vanadyl:

$$2VO_3^- + H_2S + 6H^+ \rightarrow 2VO^{2+} + S\downarrow + 4H_2O$$

The action of $(NH_4)_2S$ on the resulting solution (in an ammoniacal medium) precipitates the vanadyl ion in the form of the sulphide VOS, together with the cations of Group III.

As has already been mentioned, all these elements can exist in several states of oxidation. Consequently, oxidation-reduction reactions should be of importance among their analytical reactions, especially as such reactions are usually accompanied by a colour change.

It must be borne in mind that oxidation-reduction reactions proceed very slowly. Therefore, if they do not occur at once, the reaction mixture must be allowed to stand for an extended period of time.

The ions VO_3^-, MoO_4^{2-}, and WO_4^{2-} are colourless.

7.2. Reactions of the Vanadate Ion

1. Acids. When solutions of vanadates are acted on by acids, they acquire an orange-yellow colour due to the formation of polyvanadates of different composition (see Sec. 7.1).

2. Ammonium Sulphide, $(NH_4)_2S$. When being added cautiously to a metavanadate solution, the reagent produces a brown precipitate which slowly dissolves in excess of the reagent to form a thiosalt (the ion VS_3^-).

3. Hydrogen Sulphide, H_2S. In acid solution, hydrogen sulphide does not precipitate VO_3^- but merely reduces it to the vanadyl which is blue in colour:

$$2VO_3^- + H_2S + 6H^+ \rightarrow 2VO^{2+} + S\downarrow + 4H_2O$$

The reaction proceeds slowly in the cold, but more rapidly on heating. Some other reducing agents (e.g., SO_2 or Fe^{2+}) show a similar effect.

4. Reduction of VO_3^- to V^{III}. This reaction takes place when an acidified vanadate solution is acted on by KI:

$$VO_3^- + 2I^- + 6H^+ + 6H_2O \rightarrow [V(H_2O)_6]^{3+} + I_2 + 3H_2O$$

The hydrated vanadium(III) ions are green-coloured. The green colour becomes noticeable however only after the iodine formed in the reaction is removed by boiling, extraction with an organic solvent, or by reduction with the thiosulphate $Na_2S_2O_3$.

5. Reduction of VO_3^- to V^{II}. This reaction is accomplished by the action of metallic zinc on an acidified vanadate solution; a violet colour appears, which is characteristic of $[V(H_2O)_6]^{2+}$:

$$2VO_3^- + 3Zn + 12H^+ + 6H_2O \rightarrow 2[V(H_2O)_6]^{2+} + 3Zn^{2+}$$

6. Hydrogen Peroxide, H_2O_2. To an acidified solution of a vanadate there is added a drop of H_2O_2 solution. A cherry-red or pinkish-brown colour develops due to the formation of peroxy compounds of vanadium(V). Probably, first the VO_3^- ion is hydrated, forming VO_4^{3-}:

$$VO_3^- + H_2O \rightarrow VO_4^{3-} + 2H^+$$

Then two or four oxygen atoms in the VO_4^{3-} ion are displaced by the peroxide groups O_2^{2-}:

$$VO_4^{3-} + 2H_2O_2 \rightarrow [VO_2(O_2)_2]^{3-} + 2H_2O$$

or

$$VO_4^{3-} + 4H_2O_2 \rightarrow [V(O_2)_4]^{3-} + 4H_2O$$

Excess of hydrogen peroxide weakens the colour and reduces the sensitivity of the reaction.

7. Aniline, $C_6H_5NH_2$. Compounds of vanadium(V) are reduced in a nitric acid solution by aniline hydrochloride, $C_6H_5NH_2 \cdot HCl$, to compounds of vanadium(IV), the aniline being simultaneously oxidized to the so-called aniline black. The reaction proposed by Tananaev is carried out as follows.

Place a drop of a solution of aniline in concentrated HCl on a strip of filter paper, followed by a drop of the test solution which has been preliminarily acidified with concentrated HNO_3. In the presence of VO_3^-, depending on its concentration, a black or blue-green spot or ring appears. The detectable minimum (sensitivity) is 3 μg of vanadium.

8. Lignin. When a drop of a metavanadate solution and a drop of an orthophosphoric acid solution are applied to a strip of newspaper (which contains lignin), a black-green colouration instantly appears due to the reduction of vanadium(V) to its lower oxides. Instead of orthophosphoric acid, use may be made of nitric acid, in which case the reaction is less effective. If filter paper, which does not contain lignin, is used instead of newspaper, no black-green colouration is obtained.

7.3. Reactions of the Molybdate Ion

1. Acids. When molybdate solutions are acted on by acids, a white precipitate of molybdic acid is formed. The precipitate is soluble in excess of acid (distinction from tungsten). When solutions of ammonium molybdate, $(NH_4)_2MoO_4$, are evaporated with a few drops of concentrated H_2SO_4, a dark-blue precipitate is produced, whose composition may be expressed by the formula $Mo_3O_8 \cdot xH_2O$.

2. Hydrogen Sulphide, H_2S. When hydrogen sulphide is allowed to react with an acidified molybdate solution, a blue colouration is first produced due to the formation of molybdenum blue, a mixture of molybdenum dioxide and trioxide.

Then a brown precipitate of MoS_3 separates out:

$$MoO_4^{2-} + 3H_2S + 2H^+ \rightarrow MoS_3\downarrow + 4H_2O$$

Complete precipitation is achieved only when H_2S acts under pressure. The MoS_3 precipitate dissolves in $(NH_4)_2S$ to form the thiosalt $(NH_4)_2MoS_4$. This thiosalt can also be obtained by the action of $(NH_4)_2S$ on $(NH_4)_2MoO_4$. When a solution of the thiosalt is acidified, it decomposes into MoS_3 and H_2S (the reaction equations are given in Sec. 7.1).

3. Ammonium Thiocyanate, NH_4SCN, in the Presence of Stannous Chloride, $SnCl_2$. When an acidified solution of stannous chloride is allowed to react with a molybdate solution, molybdenum(VI) is reduced to the quinquevalent state. On further addition of ammonium thiocyanate, molybdenum(V) forms the complex $[MoO(SCN)_5]^{2-}$ which has a characteristic deep-red colour:

$$2MoO_4^{2-} + Sn^{2+} + 12H^+ + 10SCN^- \rightarrow 2[MoO(SCN)_5]^{2-} + Sn^{4+} + 6H_2O$$

As known, the Fe^{3+} ion gives a similar colour when acted on by thiocyanates. Therefore, Fe^{3+} must be bound into the complex $[Fe(PO_4)_2]^{3-}$ by orthophosphoric acid. The colour of the molybdenum complex is not destroyed by H_3PO_4.

The reaction under consideration can be carried out in a test tube or on paper. In the latter case, proceed as follows. Place a drop of an acid solution of $SnCl_2$ on a strip of filter paper and treat the moistened spot formed with a drop of the test solution, such as $(NH_4)_2MoO_4$, followed by a drop of NH_4SCN solution. A red spot appears on the paper due to the formation of $(NH_4)_2[MoO(SCN)_5]$. The colour is destroyed by the action of concentrated HCl. The presence of Fe^{3+} ions does not interfere with the reaction* since they are reduced by $SnCl_2$ to Fe^{2+} ions which do not react with NH_4SCN. The dilution limit is 1 in 500,000.

4. Disodium Hydrogen Phosphate, Na_2HPO_4. Acidify 5-6 drops of a solution of $(NH_4)_2MoO_4$ with nitric acid, heat the solution to about 50°C, and add a drop or two of Na_2HPO_4 solution. A yellow crystalline precipitate of ammonium phosphomolybdate is obtained:

$$12MoO_4^{2-} + 3NH_4^+ + PO_4^{3-} + 24H^+ \rightarrow (NH_4)_3[PMo_{12}O_{40}]\downarrow + 12H_2O$$

It will be recalled that the ion AsO_4^{3-} forms an analogous precipitate of $(NH_4)_3[AsMo_{12}O_{40}]$; the reaction is used in the systematic analysis for detection of AsO_4^{3-}.

5. Potassium Xanthate.** This compound is the most sensitive and specific reagent for molybdenum(VI). It gives with molybdates an intense-red colouration. With large quantities of molybdenum(VI), almost black oily drops separate out.

To a drop of the neutral or faintly acid test solution on a spot plate add a small grain of potassium xanthate and 2 drops of 2*N* HCl solution. The dilution limit is 1 in 250,000.

In contrast to MoO_4^{2-}, the xanthate gives a yellow precipitate with Cu^{2+}.

7.4. Reactions of the Tungstate Ion

1. Acids. When solutions of tungstates are allowed to react with concentrated mineral acids, say, HCl, a white precipitate of tungstic acid is formed:

$$WO_4^{2-} + 2H^+ + H_2O \rightarrow WO_3 \cdot 2H_2O$$

When the solution is boiled, $WO_3 \cdot 2H_2O$ is converted into $WO_3 \cdot H_2O$ which is yellow in colour. Tungstic acid is insoluble in concentrated acids (distinction from $MoO_3 \cdot H_2O$). The reaction of its formation is used for separation of WO_4^{2-} from other ions.

2. Hydrogen Sulphide, H_2S. In acid solution no precipitate of WO_4^{2-} is formed.

3. Ammonium Sulphide, $(NH_4)_2S$. With tungstates this reagent gives water-soluble thiosalts. For example,

$$WO_4^{2-} + 8NH_4^+ + 4S^{2-} + 4H_2O \rightarrow WS_4^{2-} + 8NH_4OH$$

---

* But if the concentration of Fe^{3+} in the solution is high, say, in the analysis of steels, the amount of stannous chloride may turn out to be insufficient for Fe^{3+} to be reduced completely. Therefore, the red spot formed should be moistened once more with $SnCl_2$ solution. A conclusion as to the presence of molybdenum can be made provided that the colour does not disappear.

** Its formula is $KS-(S)C-OC_2H_5$. The compound formed in the reaction with molybdenum(VI) has the formula $MoO_3[SC(SK)-(OC_2H_5)]_2$.

Upon acidification the thiosalt formed decomposes with the formation of a light-brown precipitate of WS_3.

4. **Reduction of WO_4^{2-}.** A tungstate solution acidified with HCl or H_2SO_4 is allowed to react with a piece of metallic zinc in an evaporating dish or in a crucible and is left to stand for some time. A precipitate of tungstic acid, which forms first, turns dark-blue due to the formation of products of variable composition, containing compounds of tungsten(VI) and (V).

The same compound is produced when a solution of $SnCl_2$ is used instead of metallic zinc.

The reduction can also be effected by using the precipitate of tungstic acid obtained as described above (see Reaction *1*), instead of a solution of Na_2WO_4.

7.5. Analysis of Mixtures of VO_3^-, MoO_4^{2-} and WO_4^{2-} Ions

We shall not consider the general scheme of analysis of cation mixtures in the presence of the ions of vanadium, molybdenum, and tungsten. Instead, we shall discuss only the case of greatest practical interest where these ions are present together with the cations of Group III (with the exception of Zn^{2+}) and Cu^{2+}, which happens in the analysis of special grades of steel and ferroalloys.*

The analytical procedures described below are based on the following properties of the elements under study: (a) the insolubility of $WO_3 \cdot H_2O$ in water and in excess of HCl; (b) the tendency of MoO_4^{2-} and Cu^{2+} to be precipitated by hydrogen sulphide in acid solution; and (c) the tendency of most of the cations of Group III to be precipitated by excess of caustic alkali on heating, while VO_3^- remains in solution together with AlO_2^-.**

The scheme of analysis is given in Table 7.1

1. **Separation of WO_4^{2-}.** Place into a test tube 20-25 drops of the test solution with the precipitate*** (if there is any), add 4-5 drops of concentrated HCl,**** and heat for 2-3 min.

This brings down WO_4^{2-} in the form of $WO_3 \cdot 2H_2O$ (white) which is gradually converted on heating into $WO_3 \cdot H_2O$ (yellow). The acid-soluble constituents of the precipitate are dissolved.

---

* Since the elements in question are determined in alloys, the oxidation state of iron need not be established.

** In the cold, Cr^{III} would also pass into solution in the form of CrO_2^-. On heating, however, the chromates are hydrolysed with the formation of a precipitate of $Cr_2O_3 \cdot xH_2O$.

*** The precipitate may contain tungstic acid and also various compounds soluble in HCl, such as the hydroxides, carbonates, or sulphides of the cations of Group III, etc.

**** Addition of HCl is required also for the subsequent precipitation of molybdenum and copper by hydrogen sulphide.

TABLE 7.1. **Scheme of Analysis of Mixtures of VO_3^-, MoO_4^{2-}, WO_4^{2-}, Cu^{2+} with Group III Cations (in the Absence of Zn^{2+})**

<table>
<tr><td colspan="4">1. Separate WO_4^{2-} and dissolve some of the constituents of the ppt. by conc. HCl.</td></tr>
<tr><td>Precipitate: $WO_3 \cdot 2H_2O$ (white) or $WO_3 \cdot H_2O$ (yellow).</td><td colspan="3">Centrifugate: V^V, MoO_4^{2-}, Cu^{2+}, Fe^{3+}, Al^{3+}, Cr^{3+}, Ti^{IV}, Co^{2+}, Ni^{2+}, Mo^{2+}.</td></tr>
<tr><td rowspan="6">2. Detect WO_4^{2-}.</td><td colspan="3">3. Precipitate Cu^{2+} and Mo^{VI} by the action of H_2S in the presence of $(NH_4)_2S_2O_8$.</td></tr>
<tr><td>Precipitate: CuS, MoS_3 (S).</td><td colspan="2">Centrifugate: Ti^{IV}, V^{IV}, (MoO_4^{2-}), Fe^{2+}, Mn^{2+}, Cr^{3+}, Al^{3+}, CO^{2+}, Ni^{2+}.</td></tr>
<tr><td>4. Dissolve the ppt. in HNO_3. Remove excess HNO_3 by evaporation with H_2SO_4.</td><td colspan="2">6. Oxidize V^{IV} and Fe^{2+} by HNO_3. Remove excess acids (HCl and HNO_3) by evaporation with H_2SO_4.</td></tr>
<tr><td rowspan="3">Solution: Cu^{2+}, MoO_4^{2-}.
5. Look for Cu^{2+} and Mo^{VI} in separate portions of the solution.</td><td colspan="2">Solution: Ti^{IV}, V^V, (MoO_4^{2-}), Fe^{3+}, Mn^{2+}, Cr^{3+}, Al^{3+}, Ni^{2+}, Co^{2+}.</td></tr>
<tr><td colspan="2">7. Separate V^V and Al^{3+} by a 30-per-cent solution of NaOH.</td></tr>
<tr><td>Precipitate: $TiO_2 \cdot xH_2O$, $Fe_2O_3 \cdot xH_2O$, $Cr_2O_3 \times \times xH_2O$, $Mn(OH)_2$, $Co(OH)_2$, $Ni(OH)_2$, partially, $MnO_2 \cdot xH_2O$.
8. Dissolve the ppt. in H_2SO_4. Look for Ti^{IV}, Mn^{2+}, Cr^{3+}, Ni^{2+}, Fe^{3+}, Co^{2+} in separate portions of the solution.</td><td>Centrifugate: AlO_2^-, VO_3^-, (MoO_4^{2-}).
9. Look for V^V and Al^{3+} ions in separate portions of the solution.</td></tr>
</table>

Separate the precipitate by centrifugation, wash with hot water containing a few drops of 2*N* HCl, and examine as directed in Procedure *2*.

Examine the centrifugate in accordance with Procedure *3*.

2. Detection of WO_4^{2-}. To the precipitate add a drop or two of concentrated HCl, followed by a piece of metallic zinc or 2-3 drops of $SnCl_2$ solution. Let it stand for some time. The formation of molybdenum blue indicates the presence of WO_4^{2-}.

3. Precipitation of Cu^{2+} and MoO_4^{2-}. To the solution remaining after the separation of WO_4^{2-} add a drop or two of $(NH_4)_2S_2O_8$ solution*, heat the mixture, and pass H_2S into it for 3 min. Then dilute the solution with an equal volume of cold water and again saturate with hydrogen sulphide for 2-3 min. Test for completeness of precipitation, centrifuge the precipitate, and evaporate the centrifugate in an evaporating dish or crucible to about half the initial volume. If more of the precipitate (MoS_3) settles down, separate it by centrifugation and combine it with the precipitate obtained earlier.

Wash the combined precipitates and examine as directed in Procedure *4*.

The centrifugate is to be tested as described in Procedure *6*.

4. Dissolution of Precipitate and Removal of HNO_3. Dissolve the precipitate, which may contain CuS, MoS_3, and S, on heating in 5-6 drops of 6*N* HNO_3. With the sulphur residue having been removed, transfer the solution to a crucible.

Then add 4-5 drops of 2*N* H_2SO_4 and evaporate the solution to remove HNO_3 until white fumes of sulphuric acid appear.** Cool the solution, dilute it with 5 drops of water, stir well, transfer to a test tube, and examine as directed in Procedure *5*.

5. Detection of Cu^{2+} and MoO_4^{2-}. (1) *Detection of Cu^{2+}*. To 2-3 drops of the solution add an excess of a 25-percent solution of NH_4OH. The blue colour developed by the ion $[Cu(NH_3)_4]^{2+}$ indicates the presence*** of Cu^{2+}.

(2) *Detection of MoO_4^{2-}*. Put a drop each of an acid solution of $SnCl_2$, the test solution, and NH_4SCN solution on a strip of filter paper. The appearance of a red colour due to the formation of $[MoO(SCN)_5]^{2-}$ indicates MoO_4^{2-}. The test for MoO_4^{2-} can also be made with potassium xanthate using another drop of the solution (see Sec. 7.3, Reaction *5*).

6. Oxidation of VO^{2+} and Fe^{2+} and Removal of Excess Acids. The solution obtained in Procedure *3* may contain Ti^{IV}, Mn^{2+}, Ni^{2+}, Al^{3+}, Cr^{3+}, and Co^{2+} along with excess HCl, and also VO^{2+} and Fe^{2+} which have been formed as the result of the reduction of vanadium(V) and ferric ion by hydrogen sulphide.

To this solution add 2 drops of 6*N* HNO_3 and 4 drops of 2*N* H_2SO_4 and evaporate it in a porcelain dish or in a crucible until white fumes

* The role of $(NH_4)_2S_2O_8$ consists in oxidizing H_2S to free sulphur which serves as a collector (carrier), i.e., while precipitating, it entrains MoS_3 from the colloidal solution. Molybdenum is not precipitated completely, but this does not interfere with the analysis.

** The presence of HNO_3 as a strong oxidizing agent would interfere in subsequent analysis with the detection of MoO_4^{2-} by the action of the reducing agent $SnCl_2$ or Zn.

*** The reaction with $K_4[Fe(CN)_6]$ cannot be applied here since the reagent gives with molybdenum(VI) a precipitate similar to $Cu_2[Fe(CN)_6]$.

of sulphuric acid appear (not to dryness*). This oxidizes the ions VO^{2+} and Fe^{2+} and eliminates excess HCl and HNO_3.

Allow the solution to cool off, dilute it with 5 drops of water, transfer to a test tube, and examine according to Procedure *7*.

7. Separation of VO_3^- and Al^{3+}. To the solution add a 30-percent solution of NaOH until alkaline, after which add 4-5 drops more of the reagent solution, and heat for a minute or two.** Most of the cations of Group III are precipitated, but the ions VO_3^- and AlO_2^- remain in solution. Separate the precipitate by centrifuging, wash twice with water, and examine as directed in Procedure *8*.

Examine the centrifugate as directed in Procedure *9*.

8. Detection of Group III Cations. Dissolve the washed precipitate on heating in 6-8 drops of $2N$ H_2SO_4. If brown flakes of $MnO_2 \cdot xH_2O$ remain undissolved, dissolve them on heating by adding a drop or two of KNO_2 or $NaNO_2$ solution.***

Look for the cations of Group III in separate portions of the hydrosulphuric acid solution obtained.

(1) *The Ti^{IV} Ion.* To 2 drops of the resultant solution add 2 drops of H_3PO_4 solution to bind Fe^{3+} into the colourless complex**** $[Fe(PO_4)_2]^{3-}$ and add cautiously, down the side of the test tube, a few drops of H_2O_2 so that hydrogen peroxide does not mix with the test solution.

The appearance of a yellow or orange ring of a peroxy-compound of titanium at the interface between the solutions indicates the presence of Ti^{IV} (see Sec. 5.18, Reaction *5*).

The Ti^{IV} ion can also be detected by a spot test with chromotropic acid (see Sec. 5.18, Reaction *6*).

(2) *The Mn^{2+} Ion.* The ion is detected by the reaction of oxidation to MnO_4^- (see Sec. 5.20, Reaction *5*).

(3) *The Cr^{3+} Ion.* This ion is detected by the reaction leading to the formation of $Cr_2O_7^{2-}$ through the action of $KMnO_4$ in the presence of HNO_3 with subsequent conversion into perchromic acid, and by other reactions with preliminary oxidation of Cr^{III} to Cr^{VI} (see Sec. 5.17, Reaction *5*).

(4) *The Ni^{2+} Ion.* Use is made of a spot test with dimethyl-glyoxime on a substrate of Na_2HPO_4 (see Sec. 5.23, Reaction *5*).

(5) *The Fe^{3+} and Co^{2+} Ions.* To 2 drops of the solution add 8 drops of a saturated solution of NH_4SCN (or the solid salt). The red colour

---

* It must not be evaporated to dryness since oxides and basic sulphates may be formed, which often become insoluble in acids after ignition.

** The solution must not be heated for a longer time, otherwise, because of the dehydration, $TiO_2 \cdot xH_2O$ will pass into metatitanic acid slightly soluble in mineral acids.

*** Under the action of the nitrite, Mn^{IV} is reduced to Mn^{II}.

**** The Fe^{3+} ion, which has a yellow colour, would interfere with the subsequent detection of Ti^{IV} in the form of a peroxy-compound of titanium.

of $[Fe(SCN)_5]^{2-}$ (or complexes of a different composition) indicates Fe^{3+}. The colour is eliminated by adding solid NH_4F or NaF, followed by a few drops of amyl alcohol (or a mixture of amyl alcohol and ether), and shaking (extraction).

The deep-blue colour of the alcohol layer indicates Co^{2+}.

9. Detection of VO_3^- and Al^{3+}. To the solution obtained as indicated in Procedure 7 add HCl until faintly acid (pH $\approx$ 3-5)* and look for VO_3^- and Al^{3+} in separate portions of the solution.

(1) *The VO_3^- Ion.* To 6-8 drops of the faintly acid solution add cautiously a few drops of H_2O_2 solution down the side of the test tube taking care that the two solutions do not mix. The formation of a reddish-brown ring of peroxy-compounds of vanadium(V) indicates the presence of VO_3^- (see Sec. 7.2, Reaction *6*).

Alternatively, mix a drop of the solution with a drop of 6*N* HNO_3 and look for VO_3^- in the resultant mixture using a spot test with aniline (see Sec. 7.2, Reaction *7*) or the reaction with lignin in the presence of H_3PO_4 (see Sec. 7.2, Reaction *8*).

(2) *The Al^{3+} Ion.* To the remaining (acid) solution add a 30-percent solution of NH_4OH until alkaline (pH $\approx$ 9) and heat. A white flocculent precipitate of $Al_2O_3 \cdot xH_2O$ indicates the presence of Al^{3+}. To check, separate the precipitate by centrifuging, wash with water, dissolve on heating in 2*N* HCl, add solid CH_3COONa, and perform confirmatory tests for Al^{3+} with alizarin, aluminon, or morin (see Sec. 5.16, Reactions *6*, *7*, *8*).

---

* An acid solution is required for detection of VO_3^- and Al^{3+}, but the test solution shows a strongly alkaline reaction. The success of further tests depends on how correctly the acidification is carried out at this point.

Chapter 8

Anions

8.1. Classification of Anions

The classification of anions is mostly based on the different solubilities of the barium and silver salts of the corresponding acids. This classification, however, is one of the various classifications proposed by different authors and based on other properties.

Tananaev, for example, proposed the following classification: *Group I*—oxidizing anions (MnO_4^-, CrO_4^{2-}, NO_3^-, NO_2^-, AsO_4^{3-}, $[Fe(CN)_6]^{3-}$, etc.); *Group II*—reducing anions (Cl^-, Br^-, I^-, S^{2-}, SO_3^{2-}, $S_2O_3^{2-}$, SCN^-, AsO_3^{3-}, $[Fe(CN)_6]^{4-}$, $C_2O_4^{2-}$, etc.); *Group III*—indifferent anions (SO_4^{2-}, PO_4^{3-}, BO_2^-, SiO_3^{2-}, CO_3^{2-}, F^-, CH_3COO^-, etc.).

We shall classify the anions studied into three groups as shown in Table 8.1.

There are not many analytical interferences between the anions. Therefore, elimination tests are resorted to in relatively rare cases. More often the anions are detected by fractional precipitation, i.e., in separate portions of the test solution. Accordingly, in the anion analysis, as a rule, the group reagent is used to show the presence or absence of certain ions, and not for their separation. If it has been found that a particular group is absent, it is pointless to carry out tests for the individual anions of this group. Thus, the use of group reactions considerably facilitates the analysis.

We have already encountered most of the reactions of the anions in our study of the cations. For example, the ions Ba^{2+} and Pb^{2+} are detected with the aid of the anions SO_4^{2-} and CrO_4^{2-}, which can be identified by using barium and lead salts. The Cl^- ion is the reagent for Ag^+ and, vice versa, Cl^- can be detected by means of the Ag^+ ion. Many other examples could be cited.

A. GROUP I ANIONS

The anions of Group I are listed in Table 8.1. With Ba^{2+} these anions form salts slightly soluble in water, which (except $BaSO_4$) are soluble in dilute mineral acids. Therefore, the anions of Group I are not precipitated (except SO_4^{2-}) from acid solution in the form of barium salts.

TABLE 8.1. **Classification of Anions**

Group No.	Group characteristics	Anions	Group reagent
I	Barium salts slightly soluble in water	SO_4^{2-} (sulphate ion) SO_3^{2-} (sulphite ion) $S_2O_3^{2-}$ (thiosulphate ion) CO_3^{2-} (carbonate ion) PO_4^{3-} (phosphate ion) AsO_4^{3-} (arsenate ion) AsO_3^{3-} (arsenite ion) BO_2^- (borate ion) or $B_4O_7^{2-}$ (tetraborate ion) CrO_4^{2-} (chromate ion) $Cr_2O_7^{2-}$ (dichromate ion) F^- (fluoride ion) SiO_3^{2-} (silicate ion) $C_2O_4^{2-}$ (oxalate ion), etc.	$BaCl_2$ in a neutral or faintly alkaline solution
II	Silver salts slightly soluble in water and in dilute HNO_3*	Cl^- (chloride ion) Br^- (bromide ion) I^- (iodide ion) S^{2-} (sulphide ion) SCN^- (thiocyanate ion) $[Fe(CN)_6]^{4-}$ (hexacyanoferrate (II) or ferrocyanide ion) $[Fe(CN)_6]^{3-}$ (hexacyanoferrate (III) or ferricyanide ion) CN^- (cyanide ion) BrO_3^- (bromate ion) IO_3^- (iodate ion) ClO^- (hydrochlorite ion), etc.	$AgNO_3$ in the presence of 2*N* HNO_3
III	Barium salts and silver salts soluble in water	NO_3^- (nitrate ion) NO_2^- (nitrite ion) CH_3COO^- (acetate ion) ClO_3^- (chlorate ion) MnO_4^- (permanganate ion), etc.	No group reagent

* Silver sulphide dissolves in HNO_3 on heating.

In keeping with what has been said, the group reagent for Group I anions is $BaCl_2$ in neutral or weakly alkaline (pH 7-9) solutions.

In contrast to the salts formed by the anions of Group II with Ag^+, those produced by the anions of Group I with Ag^+ are soluble in dilute acids and even in water (Ag_2SO_4 and AgF).

The reactions of the anions AsO_4^{3-} and AsO_3^{3-} (or AsO_2^-) have already been discussed in studying the properties of As^V and As^{III} (see Sec. 6.11). The reactions of the ions CrO_4^{2-} and $Cr_2O_7^{2-}$ have been described in Sec. 5.17 where the properties of Cr^{III} were considered.

8.2. Reactions of the Sulphate Ion

The sulphate ion, SO_4^{2-}, is the anion of sulphuric acid, H_2SO_4 Sulphuric acid is a strong acid, being inferior in strength only to the halide acids HCl, HBr, HI and to nitric acid.

Of the salts of sulphuric acid, only the salts of barium, strontium, lead, and calcium are slightly soluble in water. Other sulphates are well soluble. Solutions of Na_2SO_4 and K_2SO_4 are neutral. Conversely, $0.1N$ solutions of $NaHSO_4$ and $KHSO_4$ give an acid reaction (pH $\approx$ $\approx$ 2). The SO_4^{2-} ion is colourless.

1. Barium Chloride, $BaCl_2$. On interaction with solutions containing SO_4^{2-}, the Ba^{2+} ion forms a white precipitate of $BaSO_4$. Being a slightly soluble salt of a strong acid, barium sulphate is insoluble in acids. This distinguishes it from the barium salts of all other anions, which is used in the detection of SO_4^{2-}. A certain complication arises because of the presence in the test solution of $S_2O_3^{2-}$ or of a mixture of SO_3^{2-} and S^{2-} since in both cases upon acidification a white precipitate of sulphur is formed, which, like $BaSO_4$, is insoluble in acids:

$$S_2O_3^{2-} + 2H^+ \rightarrow S\downarrow + SO_2\uparrow + H_2O$$

$$2S^{2-} + SO_3^{2-} + 6H^+ \rightarrow 3S\downarrow + 3H_2O$$

The compound $BaSO_4$ can be distinguished from sulphur by making use of the tendency of barium sulphate to form mixed crystals with $KMnO_4$, which have a pink colour (see Sec. 5.4).

To 2-3 drops of the test solution acidified with HCl there are added 5 drops of $0.01N$ $KMnO_4$ and a solution of $BaCl_2$. Then, the excess of $KMnO_4$ is reduced by H_2O_2, as a result of which the solution is decolourized and the precipitate assumes a pink colour. This proves the presence of $BaSO_4$. The coprecipitated $KMnO_4$ does not react with hydrogen peroxide because it is contained in the interior of the crystals of $BaSO_4$. This reaction however cannot be used in the presence of SO_3^{2-} and $S_2O_3^{2-}$ since they are oxidized by the permanganate to SO_4^{2-} which may thus be detected in error.

2. Strontium Salts (Sr^{2+}). These salts give a white precipitate of $SrSO_4$ which is almost insoluble in acids

3. Lead Salts (Pb^{2+}). A white precipitate of $PbSO_4$ is produced, which is soluble in caustic alkalis and in ammonium acetate solution.

4. Silver Nitrate, $AgNO_3$. With dilute solutions of sulphates, silver nitrate gives no precipitate since Ag_2SO_4 dissolves fairly well in water. But in concentrated solutions of sulphates a precipitate of Ag_2SO_4 (silver sulphate) may be formed (white in colour).

5. Microcrystalloscopic Test. To a drop of the solution to be tested for SO_4^{2-} add a small drop of a solution of $Ca(CH_3COO)_2$ or $Ca(NO_3)_2$ and heat until a rim appears. Crystals of gypsum, $CaSO_4 \cdot 2H_2O$, are formed (see Figs. 4.5 and 4.6). The dilution limit is 1 in 86,000.

8.3. Reactions of the Sulphite Ion

The sulphite ion SO_3^{2-} is the anion of sulphurous acid, H_2SO_3. Sulphurous acid does not exist in the free state. When displaced from its salts, it decomposes into its anhydride and water. A solution of SO_2 in water shows an acid reaction and contains H_2SO_3 in small amounts. Sulphurous acid belongs to the category of moderately strong acids.

Of the salts of sulphurous acid, only alkali sulphites are soluble in water. The remaining sulphites are more or less insoluble in water but dissolve in acids, say, in HCl. Aqueous solutions of Na_2SO_3 and K_2SO_3 give an alkaline reaction ($pH \approx 9$) due to hydrolysis. The acid salts of the same metals, however, have a weakly acid reaction (pH 4-5) in solutions.

In aqueous solution the sulphites are slowly oxidized to sulphates. The anion SO_3^{2-} is colourless.

1. Barium Chloride, $BaCl_2$. A white precipitate of $BaSO_3$ is produced with SO_3^{2-}:

$$Ba^{2+} + SO_3^{2-} \rightarrow BaSO_3 \downarrow$$

The precipitate is soluble in acids.

2. Silver Nitrate, $AgNO_3$. In solutions of sulphites, a white precipitate of Ag_2SO_3 is formed, which is soluble in HNO_3 and in excess of SO_3^{2-} ions.

3. Strontium Salts (Sr^{2+}). These salts produce in solutions of sulphites a white precipitate of $SrSO_3$ (distinction from $S_2O_3^{2-}$) which is also soluble in acids.

4. Mineral Acids. These acids decompose all water-soluble and insoluble sulphites with the evolution of sulphur dioxide, SO_2, which is easily recognized by the odour of burning sulphur:

$$BaSO_3 \downarrow + 2H^+ \rightarrow Ba^{2+} + SO_2 \uparrow + H_2O$$
$$SO_3^{2-} + 2H^+ \rightarrow H_2O + SO_2 \uparrow$$

Sulphur dioxide, SO_2, can be detected by its ability to decolourize iodine or $KMnO_4$ solutions; for example,

$$SO_2 \uparrow + I_2 + 2H_2O \rightarrow 4H^+ + SO_4^{2-} + 2I^-$$

Potassium permanganate, $KMnO_4$, oxidizes SO_2 partly to sulphuric acid and partly to dithionic acid, $H_2S_2O_6$:

$$5SO_2\uparrow + 2MnO_4^- + 2H_2O \rightarrow 2SO_4^{2-} + 2Mn^{2+} + 4H^+$$

$$6SO_2\uparrow + 2MnO_4^- + 2H_2O \rightarrow H_2S_2O_6 + 4SO_4^{2-} + 2Mn^{2+} + 2H^+$$

The reaction is carried out in one of the devices used for detection of gases (see, for example, Fig. 2.22).

A few drops of a very dilute iodine solution coloured dark-blue by starch are placed in one leg of the apparatus and a few drops of the test solution of a sulphite are introduced into the other leg. To the test sulphite solution is added a solution of HCl or H_2SO_4. Then the apparatus is tightly stoppered and the decolourization of the iodine solution caused by the action of the evolving SO_2 is observed. To hasten the reaction, the leg with the sulphite is heated. Other devices designed for detection of gases may also be employed (see Figs. 8.1, 8.2, and 8.3).

In the experiment just described, a strongly dilute solution of $KMnO_4$ may be used instead of iodine solution.

Thiosulphates also evolve SO_2 upon acidification. Sulphides give off H_2S when acted on by acids, the evolved hydrogen sulphide exerting a decolourizing effect on iodine and $KMnO_4$ solutions. Thus, in the presence of $S_2O_3^{2-}$ and S^{2-}, the test in question is inapplicable.

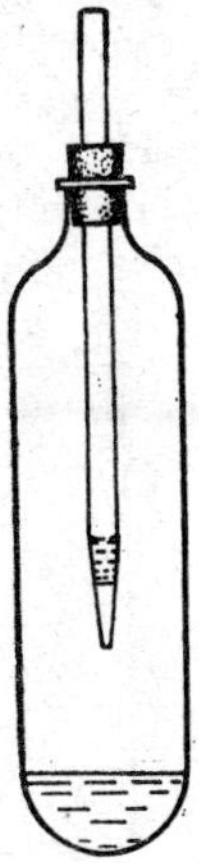

Fig. 8.1. Apparatus for detection of CO_2 and SO_2.

5. Reduction Reactions Effected by the Sulphite Ion. Iodine is reduced not only by gaseous SO_2 but also by solutions of sulphites:

$$SO_3^{2-} + I_2 + H_2O \rightarrow SO_4^{2-} + 2I^- + 2H^+$$

As a result, the solution shows an acid reaction, which is an additional evidence of the presence of SO_3^{2-}. The ions $S_2O_3^{2-}$ and S^{2-} form no acid upon reduction of iodine (see Sec. 8.4, Reaction 4).

Potassium permanganate is also reduced by sulphites; the reaction can be carried out in both acid and alkaline solutions. For example, if to a solution of Na_2SO_3 acidified with sulphuric acid there is added, drop by drop, a solution of $KMnO_4$, the latter is decolourized (reduction to Mn^{2+}):

$$5SO_3^{2-} + 2MnO_4^- + 6H^+ \rightarrow 5SO_4^{2-} + 2Mn^{2+} + 3H_2O$$

The reaction proceeds differently in alkaline solution. First, the crimson colour of MnO_4^- changes to green due to the formation of the anion of manganic acid, MnO_4^{2-}, which corresponds to manganese in the +6 oxidation state. If the resulting green solution is heated, it becomes brown and gives a brown precipitate of manganous acid, $MnO(OH)_2$:

$$3SO_3^{2-} + 2MnO_4^- + 3H_2O \rightarrow 2MnO(OH)_2\downarrow + 3SO_4^{2-} + 2OH^-$$

Many other reducing anions also reduce permanganate to $MnO(OH)_2$ in alkaline solution. Examples are S^{2-}, $S_2O_3^{2-}$, AsO_3^{3-}, SCN^-, $[Fe(CN)_6]^{4-}$, and I^-.

6. Oxidation Reactions Effected by the Sulphite Ion. The sulphur in SO_3^{2-} has an intermediate oxidation state (+4) which therefore may not only increase but also decrease. In other words, the SO_3^{2-} ion can be reduced as well as oxidized. The reduction products of SO_3^{2-} may be either free sulphur or H_2S.

(1) Mix 2 drops each of solutions of Na_2SO_3 and Na_2S and acidify the mixture with hydrochloric or sulphuric acid. This precipitates sulphur as a result of the oxidation of hydrogen sulphide by sulphurous acid which is reduced to sulphur:

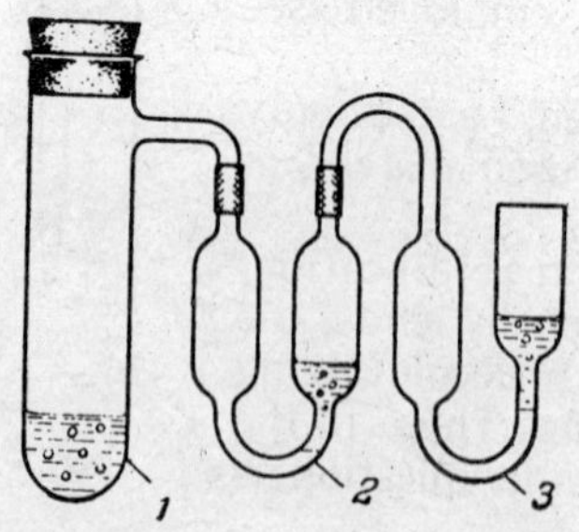

Fig. 8.2. Apparatus for detection of CO_2 and SO_2:
1 — test tube; 2 — absorption tube with lime water; 3 — tubing with alkali.

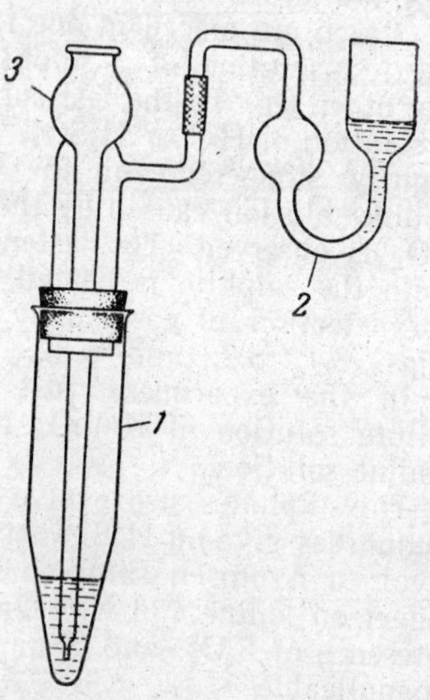

Fig. 8.3. Apparatus for detection of CO_2 and SO_2:
1 — test tube; 2 — absorption tube containing lime water; 3 — funnel.

$$2S^{2-} + SO_3^{2-} + 6H^+ \rightarrow 3S\downarrow + 3H_2O$$

(2) To a sulphite solution add an excess of 2*N* HCl and introduce a piece of metallic zinc. Heat. This will evolve H_2S which can be identified by the odour or by the blackening of the paper moistened with a solution of $Pb(CH_3COO)_2$ or Na_2PbO_2.

The reaction in the test tube proceeds according to the equation

$$3Zn + 8H^+ + SO_3^{2-} \rightarrow 3Zn^{2+} + H_2S\uparrow + 3H_2O$$

The reaction on the paper takes place as follows:

$$H_2S\uparrow + Pb^{2+} + 2CH_3COO^- \rightarrow PbS\downarrow + 2CH_3COOH$$

7. Decolourization of Fuchsin or Malachite Green. Sulphites decolourize solutions of certain organic dyes, say, fuchsin or malachite green. This fact can be used for detection of SO_3^{2-}.

To a drop of a solution of fuchsin or malachite green (or a mixture of both) placed into the cavity of a spot plate, there is added a drop of the neutral test solution. In the presence of SO_3^{2-}, the solution is decolourized.

Acid solutions are preliminarily neutralized with $NaHCO_3$, and alkaline solutions are saturated with CO_2 until the colour fades away upon addition of a drop of phenolphthalein solution. The ion S^{2-} interferes with the reaction since it decolourizes these dyes too.

8. Sodium Nitroprusside, $Na_2[Fe(CN)_5NO]$. The reagent colours sulphite solutions pink-red. When a solution of $ZnSO_4$ is added, the colour becomes more intense. If a drop of a solution of $K_4[Fe(CN)_6]$ is added, a red precipitate is formed. Acids destroy the colour and therefore acid solutions must first be neutralized. The ion $S_2O_3^{2-}$ gives no such reaction. The ion S^{2-} gives with sodium nitroprusside a violet colour and therefore interferes with the detection of SO_3^{2-}.

8.4. Reactions of the Thiosulphate Ion

The thiosulphate ion ($S_2O_3^{2-}$) is the anion of thiosulphuric (or hydrosulphurous) acid, $H_2S_2O_3$. This acid is even more unstable than H_2SO_3 and, when evolved in the free state, it decomposes into SO_2 and S. The thiosulphates of the alkali metals and also of strontium, zinc, cadmium, nickel, cobalt, manganese, and iron(II) are well soluble in water. The remaining thiosulphates are slightly soluble. The $S_2O_3^{2-}$ ion is colourless. The reaction of aqueous solutions of the alkali thiosulphates is close to neutral since $H_2S_2O_3$ is a rather strong but unstable acid.

1. Barium Chloride, $BaCl_2$. When added in excess, barium chloride forms with the anion $S_2O_3^{2-}$ a white precipitate of BaS_2O_3. Barium thiosulphate is capable of forming supersaturated solutions; therefore, in carrying out the reactions, the sides of the test tubes should be scratched with a glass rod.

2. Silver Nitrate, $AgNO_3$. When added in excess, silver nitrate produces a white precipitate of $Ag_2S_2O_3$ which quickly turns first yellow, then brown, and, finally, black due to the conversion into silver sulphide*:

$$Ag_2S_2O_3 + H_2O \rightarrow Ag_2S\downarrow + 2H^+ + SO_4^{2-}$$

* This reaction proceeds especially rapidly on heating.

This is one of the most characteristic reactions of $S_2O_3^{2-}$. In carrying out this reaction, one must bear in mind that the $Ag_2S_2O_3$ precipitate dissolves in an excess of thiosulphate with the formation of complex ions $[AgS_2O_3]^-$ and $[Ag(S_2O_3)_3]^{5-}$. Therefore, the precipitate separates out only in an excess of Ag^+.

3. **Acids.** Free $H_2S_2O_3$ is evolved from thiosulphate solutions upon addition of acids. The thiosulphuric acid decomposes into H_2O, SO_2, and S:

$$S_2O_3^{2-} + 2H^+ \rightarrow H_2S_2O_3$$

$$H_2S_2O_3 \rightarrow H_2O + SO_2\uparrow + S\downarrow$$

The sulpnur formed in this reaction makes the solution turbid; the higher the concentration of the thiosulphate ion and the temperature of the solution, the sooner the solution becomes turbid. Since under analogous conditions sulphites liberate SO_2 without the formation of a sulphur turbidity, the reaction enables the detection of $S_2O_3^{2-}$ in the presence of SO_3^{2-} and is one of the most important reactions of the thiosulphate ion.

4. **Iodine Solution.** Iodine solutions are decolourized by solutions of thiosulphates with the formation of tetrathionates (i.e., the salts of tetrathionic acid, $H_2S_4O_6$); for example,

$$2S_2O_3^{2-} + I_2 \rightarrow S_4O_6^{2-} + 2I^-$$

The reaction between iodine and a thiosulphate is widely employed in quantitative analysis.

5. **Potassium Permanganate, $KMnO_4$.** The permanganate is reduced by thiosulphates to Mn^{2+} in acid solution and to $MnO(OH)_2$ in alkaline solution, the latter coming down as a brown precipitate. This involves the formation of sulphates and tetrathionates.

6. **Ferric Chloride.** In solutions of thiosulphates, a dark-violet colour is produced due to the formation of the complex compound $Na[Fe(S_2O_3)_2]$; this colour rapidly disappears. This happens as the result of the reduction of iron(III) to iron(II):

$$2Fe^{3+} + 2S_2O_3^{2-} \rightarrow 2Fe^{2+} + S_4O_6^{2-}$$

8.5. Reactions of the Carbonate Ion

The carbonate ion, CO_3^{2-}, is the anion of carbonic acid, H_2CO_3, which does not exist in the free state. When formed from its salts, it decomposes almost completely into the anhydride CO_2 and water. Since H_2CO_3 is a very week acid, alkali carbonates show an alkaline reaction. The only normal carbonates soluble in water are those of potassium, sodium, and ammonium. The cations of the other groups are precipitated by the ion CO_3^{2-} in the form of normal and basic carbonates, and some others (e.g., Al^{3+}, Cr^{3+}, Ti^{IV}) as the hydroxides.

As a result of hydrolysis, 0.1N solutions of Na_2CO_3 and K_2CO_3 have a strongly alkaline reaction (pH $\approx$ 11.6). The CO_3^{2-} ion is colourless.

1. Barium Chloride, $BaCl_2$. The carbonate ion separates out as a white precipitate of $BaCO_3$ soluble in dilute mineral acids and even in CH_3COOH with the evolution of CO_3. The action of sulphuric acid results simultaneously in the dissolution of $BaCO_3$ and the formation of slightly soluble barium sulphate.

2. Silver Nitrate, $AgNO_3$. A white precipitate of Ag_2CO_3 is formed, which is soluble in acids. The action of HCl gives AgCl. On boiling Ag_2CO_3 decomposes into Ag_2O and CO_2.

3. Acids. The action of acids decomposes carbonates with the liberation of CO_2:

$$CO_3^{2-} + 2H^+ \rightarrow H_2CO_3$$

$$H_2CO_3 \rightarrow H_2O + CO_2\uparrow$$

The evolution of CO_2 can be recognized by the fact that lime or baryta water, i.e., saturated solutions of $Ca(OH)_2$ or $Ba(OH)_2$, becomes turbid. This turbidity caused by the action of CO_2 is due to the formation of precipitates of $CaCO_3$ or $BaCO_3$; for example,

$$Ca(OH)_2 + CO_2 \rightarrow CaCO_3\downarrow + H_2O$$

The reaction can be carried out in the gas-testing devices shown in Figs. 2.22 and 8.1. If the apparatus shown in Fig. 8.1 is used, a small amount of a solid substance or 5-8 drops of the test solution are placed into the vessel and an approximately equal volume of 2N HCl is added. A pipette with a drop or two of lime water* is inserted at once into the vessel (not tightly). The rubber stopper with the pipette is then forced farther into the mouth of the vessel until the drop of lime water rises a little in the pipette (indicating that the vessel is tightly sealed).

The vessel is held against the light and the bottom tapped slightly with the finger. The lime water becomes turbid. This turbidity quickly disappears because of the formation of a water-soluble acid salt:

$$CaCO_3 + CO_2\uparrow + H_2O \rightarrow Ca(HCO_3)_2$$

The lime water can however be made turbid again by addition of $Ca(OH)_2$ solution to the pipette.

Instead of the simplest apparatus shown in Fig. 8.1, use may be made of more complicated devices which provide more reliable results. An example is a device shown in Fig. 8.2. A few drops of the test

---

* Lime water must be freshly prepared and absolutely clear. The suitability of lime water for the test must be checked as follows. Air is blown from the mouth with the aid of a glass tube through a few drops of lime water. Use may also be made of baryta water. But it must be carefully protected from contact with air, otherwise it may become strongly turbid, even in the absence of CO_3^{2-} in the test solution.

solution and $2N$ HCl are put into a test tube *1*. The tube is at once tightly stoppered and heated in the water bath. The CO_2 formed in the reaction is passed through a few drops of lime water placed in an absorption tube *2*. The tubing *3* connected with it contains a caustic alkali which protects the solution of $Ca(OH)_2$ against the action of the CO_2 of the air.

The apparatus proposed by V. Alexeyev is shown in Fig. 8.3. The formation of CO_2 in this device takes place in test tube *1* on heating for 1-2 min in the water bath. If the lime water in tubing *2* does not become turbid, water is poured into tube *1* through funnel *3* up to the top of the funnel. In this way, it is very easy to detect the evolution of CO_2 even if its volume is very small. The carbon dioxide must be displaced without removing the apparatus from the water bath; otherwise, on cooling, the surrounding air containing CO_2 will get through the $Ca(OH)_2$ solution. To avoid this, an additional tubing with alkali is connected to the apparatus, just as in the device in Fig. 8.2. The water used for displacement of CO_2 (to reduce its solubility) should be hot.

Before using either of the devices shown in Figs. 8.2. and 8.3, it is necessary to wash thoroughly their individual parts. The absorption tube *2* is washed first with dilute HCl and then with distilled water.

4. Detection of CO_3^{2-} in the Presence of SO_3^{2-} and $S_2O_3^{2-}$. The presence of SO_3^{2-} or $S_2O_3^{2-}$ interferes with the detection of CO_3^{2-} because the sulphur dioxide gas, evolved by the sulphites and thiosulphates acted on by acids, forms a white precipitate of $CaSO_3$ when absorbed by the $Ca(OH)_2$ solution. If the interfering anions are detected, they must be oxidized. Add an excess (4-6 drops) of dilute H_2O_2 to the solution before the addition of H_2SO_4. Then proceed as described above (Procedure *3*). Instead of hydrogen peroxide, use may be made of other oxidizing agents (e.g., $KMnO_4$ or $K_2Cr_2O_7$).

8.6. Reactions of the Phosphate Ion

The phosphate ion, PO_4^{3-}, is the anion of orthophosphoric acid, H_3PO_4 (commonly called phosphoric acid), which is a moderately strong acid. Being a tribasic acid, it forms three series of salts:

Phosphates: Na_3PO_4, $Ca_3(PO_4)_2$, $AlPO_4$, etc.
Hydrogen phosphates: Na_2HPO_4, $CaHPO_4$, etc.
Dihydrogen phosphates: NaH_2PO_4, $Ca(H_2PO_4)_2$, etc.

The only salts of H_3PO_4 that are soluble in water are those of the alkali metals and ammonium, and also the dihydrogen phosphates of the alkaline earths, such as $Ca(H_2PO_4)_2$. All the other metal phosphates are insoluble in water but are dissolved by mineral acids. Many of them are also dissolved by acetic acid. The phosphates insoluble in acetic acid are $FePO_4$ and $AlPO_4$. The bismuth salt $BiPO_4$ is slightly soluble in dilute (0.5*N*) HNO_3. The phosphates of the alkali metals give a strongly alkaline reaction. Thus, the pH value calculated for a 0.1*M* solution of Na_3PO_4 is about 12. At the same concentration, a solution of Na_2HPO_4 has a pH of about 9.8, and that of NaH_2PO_4 has a pH ≈ 4.7 (see Table 5.3). The PO_4^{3-} ion is colourless.

1. Barium Chloride, $BaCl_2$. In a solution of Na_2HPO_4 barium chloride forms a white precipitate of $BaHPO_4$ soluble in acids:

$$Ba^{2+} + HPO_4^{2-} \rightarrow BaHPO_4\downarrow$$

If the reaction is carried out in the presence of caustic alkalis or NH_4OH, the ion HPO_4^{2-} is converted to PO_4^{3-} and a precipitate of $Ba_3(PO_4)_2$

$$3Ba^{2+} + 2HPO_4^{2-} + 2OH^- \rightarrow Ba_3(PO_4)_2\downarrow + 2H_2O$$

The action of acids on the precipitate is the same as on $BaHPO_4$.

2. Silver Nitrate, $AgNO_3$. A yellow precipitate of Ag_3PO_4 is produced, which is soluble in HNO_3 and NH_4OH. Let us recall that the AsO_4^{3-} ion, which is similar to PO_4^{3-}, forms with Ag^+ a chocolate-brown precipitate of Ag_3AsO_4.

3. Magnesia Mixture. This is a mixture of $MgCl_2$, NH_4OH, and NH_4Cl. It gives a white crystalline precipitate of $MgNH_4PO_4$:

$$HPO_4^{2-} + NH_4OH + Mg^{2+} \rightarrow MgNH_4PO_4\downarrow + H_2O$$

This reaction was considered in studying Mg^{2+} (see Sec. 3.15). Under the same conditions, a similar precipitate is formed by the ion AsO_4^{3-} (see Sec. 6.11). The ion AsO_3^{3-} gives no precipitate with magnesia mixture if its concentration is not too high.

4. Ammonium Molybdate Solution. This is a solution of ammonium molybdate, $(NH_4)_2MoO_4$, in HNO_3. It forms with phosphates a characteristic yellow crystalline precipitate of ammonium phosphomolybdate or, according to modern nomenclature, ammonium 12-molybdophosphate:

$$PO_4^{3-} + 3NH_4^+ + 12MoO_4^{2-} + 24H^+ \rightarrow (NH_4)_3[PMo_{12}O_{40}]\downarrow + 12H_2O$$

The precipitate formed is an ammonium salt of 12-molybdophosphoric heteropoly acid, $H_3[PMo_{12}O_{40}]$.

The precipitated ammonium phosphomolybdate dissolves in excess of a phosphate. Therefore, the reagent must be added in a large excess. It will be recalled that arsenomolybdates also dissolve in excess of AsO_4^{3-}.

To a drop or two of a phosphate solution there are added 8-10 drops of ammonium molybdate solution and the mixture is slightly heated (up to 40-50°C). If required, the solution is allowed to stand for some time.

The sensitivity of the reaction is increased by addition of solid NH_4NO_3 (the effect of the common NH_4^+ ions which depress the solubility of the precipitate).

The presence in solution of reducing agents, say, SO_3^{2-}, $S_2O_3^{2-}$, S^{2-}, etc., interferes with the test because they reduce Mo^{VI} to molybdenum blue (a mixture of molybdenum dioxide and trioxide). These reducing agents are also capable of reducing the heteropoly anions $[PMo_{12}O_{40}]^{3-}$ to heteropoly acids in which molybdenum is in the lowest oxidation state. Also, the solution must not contain large amounts of chlorides. For the reducing agents to be oxidized and part of Cl^- ions to be removed, a drop or two of the solution should be boiled with 2-3 drops of 6*N* HNO_3 and the above test performed on the solution obtained.

5. Benzidine, $C_{12}H_8(NH_2)_2$. As was pointed out in discussing complex compounds, the ion of molybdenum(VI) bound into a complex with AsO_4^{3-} or PO_4^{3-} is highly reactive and can oxidize substances that cannot be oxidized by molybdates. One such substance is benzidine which has been repeatedly used in the cation analysis and which forms an intense-blue oxidation product; the resulting heteropoly acid has also a deep-blue colour. This and also the fact that one molecule of a phosphate binds and hence activates 12 molecules of MoO_3 renders the reaction with benzidine highly sensitive.

The heteropoly anion of molybdenum(VI) exhibits an increased oxidizing power as compared with other complex ions of molybdenum(VI).

The reaction is most conveniently carried out by the spot-test technique. A drop of the acid solution to be tested for PO_4^{3-} is placed on a piece of ashless filter, followed by a drop each (one after the other) of a nitric acid solution of ammonium molybdate and of benzidine. To decrease the acidity of the solution, the moist spot is held over ammonia vapour or is treated with a drop of a solution of CH_3COONa. In the presence of PO_4^{3-}, a blue colour appears, the intensity of the colouration depending on the concentration of PO_4^{3-}.

Under such conditions, the AsO_4^{3-} ion does not react since the compound $(NH_4)_3[AsMo_{12}O_{40}]$ is formed in the cold very slowly. An analogous (though water-soluble) salt is also formed by SiO_3^{2-} which may always be present in small amounts as a result of the extraction by solutions of traces of silicates from the glass of the vessels used. Therefore, it is better to perform the test in the presence of tartaric acid, $H_2C_4H_4O_6$, which prevents the formation of silicomolybdic and arsenomolybdic acids. Molybdophosphoric acid is also formed in the presence of tartaric acid.

The test for PO_4^{3-} by this method is made as described above, the only difference being that the reagent used is ammonium molybdate solution which contains tartaric acid. Before the addition of CH_3COONa (or treatment with NH_3), it is useful to hold the moist spot over a heated wire gauze in order to speed up the reaction.

The reaction enables detection of 1.5 μg of PO_4^{3-} in the presence of a 500-fold amount of silicic acid. The dilution limit is 1 in 33,000.

Procedures for Mixtures of PO_4^{3-}, AsO_4^{3-}, and AsO_3^{3-}

If the anions PO_4^{3-}, AsO_4^{3-}, and AsO_3^{3-} are present together, the tests for their detection can be carried out as described below.

1. Separation of AsO_4^{3-} and PO_4^{3-} from AsO_3^{3-}. To 5-8 drops of the test solution add 3 drops of 2*N* NH_4Cl and NH_4OH until alkaline (pH $\approx$ 9), followed by 4-5 drops of a solution of a magnesium salt. If no precipitate separates at once, scratch the sides of the test tube with a glass rod. After the lapse of 15 min, centrifuge off the precipitate consisting of $MgNH_4AsO_4$, $MgNH_4PO_4$, and other magnesium salts*, wash with water containing a few drops of NH_4OH (to depress the solubility of the precipitate), and examine as directed in Procedure *2*.

Examine the centrifugate according to Procedure *3*.

2. Detection of AsO_4^{3-} and PO_4^{3-}. Dissolve the precipitate in a few drops of 2*N* CH_3COOH and look for the ions AsO_4^{3-} and PO_4^{3-} in separate portions of the solution.

(1) *Test for AsO_4^{3-}.* To one portion of the solution made acid with an equal volume of concentrated (12*N*) HCl add a drop of KI solution and a few drops of a solution of benzene, C_6H_6. The appearance of a violet colour in the benzene layer (evolution of iodine) indicates AsO_4^{3-}**.

(2) *Test for PO_4^{3-}.* Use a drop of the solution for carrying out a spot test for PO_4^{3-} with $(NH_4)_2MoO_4$ in the presence of tartaric acid (see Sec. 8.6, Reaction *5*).

In the absence of AsO_4^{3-}, the PO_4^{3-} ion may also be tested for by the action of ammonium molybdate solution in the presence of HNO_3 (see Sec. 8.6, Reaction *4*).

3. Detection of AsO_3^{3-}. Acidify the solution, remaining after the separation of AsO_4^{3-} and PO_4^{3-} in Procedure *1*, with a few drops of 2*N* HCl, heat, and pass H_2S into it. The prompt formation*** of a yellow precipitate of As_2S_3, which dissolves in NH_4OH and reprecipitates upon acidification of the solution, indicates the presence of AsO_3^{3-}.

8.7. Reactions of Borate Ions (Tetraborate and Metaborate)

Free boric (orthoboric) acid, H_3BO_3, is a white crystalline solid which is moderately soluble in water and is an extremely weak acid.

* Formed as the result of the presence of other anions in the solution.

** In the presence of oxidizing agents (NO_2^-, CrO_4^{2-}, MnO_4^-), which may coprecipitate with $MgNH_4AsO_4$, this reaction is not reliable. In such cases, a portion of the solution must be acidified with an equal volume of concentrated HCl and hydrogen sulphide passed into it. A yellow precipitate of As_2S_5 indicates AsO_4^{3-}.

*** Very slow precipitation of As_2S_3 may be due to the incomplete precipitation of AsO_4^{3-} in the form of $MgNH_4AsO_4$ (slow reduction of AsO_4^{3-} to AsO_3^{3-} and precipitation of the latter by hydrogen sulphide).

The salts of boric acid (borates) correspond in composition not to orthoboric but to metaboric acid, HBO_2, or to tetraboric acid, $H_2B_4O_7$.

Only the alkali borates are soluble in water. The most important of these is sodium tetraborate, $Na_2B_4O_7 \cdot 10H_2O$, which is commonly known as borax. Being the salts of a very weak acid, all the soluble metaborates are strongly hydrolysed in aqueous solution and give an alkaline reaction (pH $\approx$ 11). The water-insoluble borates are dissolved by acids. Solutions of borates are colourless.

1. Barium Chloride, $BaCl_2$. In a moderately dilute solution of $Na_2B_4O_7$, barium chloride forms a white precipitate of barium metaborate, $Ba(BO_2)_2$, which is soluble in acids.

The formation of barium metaborate may be represented by the following equations:

$$B_4O_7^{2-} + 3H_2O \rightarrow 2BO_2^- + 2H_3BO_3$$

$$Ba^{2+} + 2BO_2^- \rightarrow Ba(BO_2)_2 \downarrow$$

$$B_4O_7^{2-} + Ba^{2+} + 3H_2O \rightarrow Ba(BO_2)_2 \downarrow + 2H_3BO_3$$

As can be seen from the reaction equation, only half of the metaborate ion present in solution is precipitated. Complete precipitation can be achieved if the reaction is conducted in alkaline solution:

$$2H_3BO_3 + 2OH^- + Ba^{2+} \rightarrow Ba(BO_2)_2 \downarrow + 4H_2O$$

2. Silver Nitrate, $AgNO_3$. A white precipitate of $AgBO_2$ is formed, which is soluble in HNO_3 and NH_4OH:

$$B_4O_7^{2-} + 2Ag^+ + 3H_2O \rightarrow 2AgBO_2 \downarrow + 2H_3BO_3$$

3. Turmeric-Paper Test. Moisten a strip of turmeric paper* with a drop of an acidified solution of a borate and dry it. A brown-red spot appears. This colour is not changed by acids. Alkalis change the colour to greenish-black or (in very dilute solutions) to violet.

4. Flame Test. Volatile compounds of boron impart a green colour to the non-luminous burner flame. Since borates are not volatile, the test is carried out by either of the following two methods.

(1) Evaporate** 5-6 drops of a solution of $Na_2B_4O_7$ to dryness in a crucible. Let it cool and scrape off the dry residue using a glass rod with a rubber policeman, mix thoroughly with powdered CaF_2, and moisten the mixture with 2-3 drops of concentrated H_2SO_4. The boron compound formed (BF_3) is so volatile that if a small amount of the resulting mixture is brought near the flame (not touching), the edge of the flame is coloured green after a while.

---

* Also called curcuma paper.—*Tr.*

** Free H_3BO_3 volatilizes together with water vapours and therefore acid solutions of borates must not be evaporated; they must first be made basic with NaOH.

(2) Evaporate 4-5 drops of $Na_2B_4O_7$ solution in a crucible. Let the dry residue cool and treat with 3-4 drops of concentrated H_2SO_4 and 5-6 drops of ethyl (or methyl) alcohol. Mix well and set fire to it. The alcohol burns with a green flame. The reaction proceeds according to the equation

$$B_4O_7^{2-} + 2H^+ + 5H_2O \rightarrow 4H_3BO_3 \quad [\text{or } B(OH)_3]$$

$$B(OH)_3 + 3C_2H_5OH \rightarrow B(OC_2H_5)_3\uparrow + 3H_2O$$

Ethyl borate, $B(OC_2H_5)_3$, which belongs to a class of compounds known as esters, is highly volatile and imparts a characteristic green colour to the flame.

8.8. Reactions of the Fluoride Ion

The fluoride ion (F^-) is the anion of hydrofluoric acid*, HF, which is a volatile liquid boiling at 19.4°C. Its vapour has a sharp, penetrating odour and is *highly poisonous.* In contrast to other halogen acids, HF is a moderately strong acid. Its distinctive feature is that the F^- ion is incapable of being oxidized by the action of chemical reagents to free F_2 since fluorine is an exceedingly powerful oxidizing agent. Another feature distinguishing it from the other halogen acids is that silver fluoride, AgF, is well soluble in water whereas such compounds as AgCl, AgBr, and AgI are almost insoluble in water.

A characteristic feature of HF is its tendency to react with SiO_2 and silicates with the evolution of gaseous silicon fluoride, SiF_4 For example,

$$SiO_2 + 4HF \rightarrow 2H_2O + SiF_4\uparrow$$

$$CaSiO_3\downarrow + 6HF \rightarrow CaF_2\downarrow + 3H_2O + SiF_4\uparrow$$

The fluorides of the alkali metals, ammonium, and also of silver, aluminium, tin, and mercury are soluble in water. The fluorides of the alkaline-earth metals, lead, copper, and zinc are practically insoluble. The reaction of fluoride solutions is close to neutral. The fluoride ion is colourless.

1. Barium Chloride, $BaCl_2$. With the fluoride ion barium chloride gives a voluminous white precipitate of BaF_2 soluble in excess of mineral acids and in solutions of ammonium salts.**

2. Calcium Chloride, $CaCl_2$. A white precipitate of CaF_2 is produced on interaction with the fluoride ion. The precipitate is slightly soluble in mineral acids and insoluble in acetic acid.

3. Formation of SiF_4. Place a small amount (the size of a rice grain) of a dry fluoride (e.g., CaF_2) in the dry vessel shown in Fig. 2.21 and treat it with a few drops of concentrated H_2SO_4. Then introduce

* This is a solution of hydrogen fluoride in water.

** The dissolution takes place under the action of the acid formed as the result of the hydrolysis of the ammonium salt.

a drop of water in the loop of a platinum wire into the apparatus and let it stand. The water becomes turbid due to the formation of a precipitate of silicic acid.

The fluoride reacts first with H_2SO_4:

$$CaF_2\downarrow + H_2SO_4 \rightarrow CaSO_4\downarrow + 2HF\uparrow$$

The hydrogen fluoride formed attacks the silicates of the glass, forming the salts of silicofluoric acid, H_2SiF_6, and gaseous silicon fluoride, SiF_4. The composition of ordinary glass corresponds approximately to the formula $Na_2SiO_3 \cdot CaSiO_3 \cdot 4SiO_2$. So, the reaction involved may be formulated as follows:

$$SiO_2\downarrow + 4HF \rightarrow SiF_4\uparrow + 2H_2O$$

The more extended form of the equation is:

$$Na_2SiO_3 \cdot CaSiO_3 \cdot 4SiO_2 + 28HF \rightarrow 14H_2O + Na_2SiF_6 + CaSiF_6\downarrow + 4SiF_4\uparrow$$

The SiF_4 formed reacts with the water in the platinum wire loop and is thus hydrolysed, forming a gel of orthosilicic acid:

$$SiF_4 + 4H_2O \rightarrow H_4SiO_4\downarrow + 4HF\uparrow$$

The formation of orthosilicic acid is the reason why the water drop becomes turbid. This is the most characteristic reaction of the fluoride ion.

In this test for the fluoride ion, the solution should first be evaporated to dryness. Alternatively, the fluoride ion may be precipitated first by the action of $CaCl_2$ in the presence of CH_3COOH. In this case, separate the CaF_2 precipitate by centrifugation, dry thoroughly, and test for the fluoride ion as suggested above.

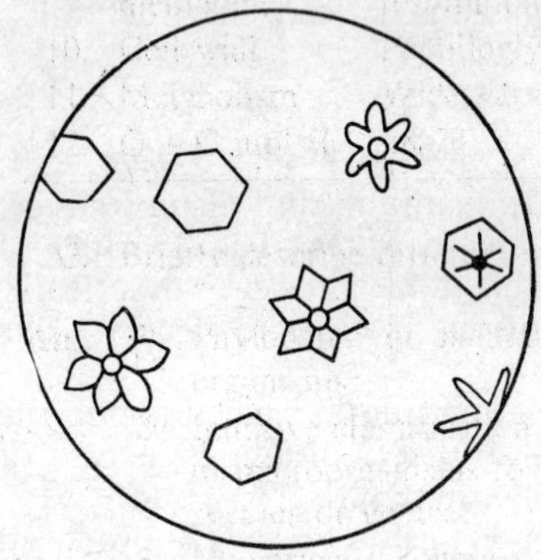

Fig. 8.4. Crystals of $Na_2[SiF_6]$.

4. Alizarin and Zirconium (IV) Nitrate. If a drop of alizarin solution is mixed with a drop of a solution of zirconium(IV) nitrate, a violet-coloured zirconium-alizarin lake is precipitated.

If to the resulting mixture, there is then added a little of a fluoride, the zirconium(IV) ions will be bound into a more stable complex $[ZrF_6]^{2-}$. As a result, the violet colour of the zirconium-alizarin lake disappears and, instead, a greenish-yellow colour is formed due to alizarin. The reaction may be depicted schematically as follows:

$$\underset{\text{(a violet lake)}}{\text{Zr-alizarin}} + 6F^- \rightarrow [ZrF_6]^{2-} + \underset{\text{(greenisn-yellow)}}{\text{alizarin}}$$

Phosphates and oxalates interfere with the test.

5. Microcrystalloscopic Test. To a drop of the test solution acidified with hydrochloric acid add a little solid silicic acid, followed by a small grain of NaCl. The crystals of $Na_2[SiF_6]$ are formed (Fig. 8.4), which have a pale-pink colour. The dilution limit is 1 in 2500.

8.9. Reactions of the Metasilicate Ion

A large variety of silicon compounds known as silicates occur in nature and can be produced artificially. They are the salts of various silicic acids whose composition may be expressed by the general formula $m\text{SiO}_2 \cdot n\text{H}_2\text{O}$.

In most cases, silicic acids cannot be isolated in the free state and their existence can be deduced only from the presence of the corresponding salts. Of these, soluble in water are only the alkali silicates of metasilicic acid, Na_2SiO_3 and K_2SiO_3, which are called "soluble glasses". Aqueous $0.1N$ solutions of these compounds give a very strongly alkaline reaction ($pH > 12$) due to hydrolysis. Some of the water-insoluble silicates are decomposed by mineral acids to give free silicic acids; others do not decompose in this way. Silicates that are insoluble in water and non-decomposable by acids can be dissolved by fusion with a mixture of Na_2CO_3 and K_2CO_3, the soluble silicates of the alkali metals being formed.

Solutions of silicates are colourless.

1. Barium Chloride, $BaCl_2$. With SiO_3^{2-} barium chloride gives a white precipitate of $BaSiO_3$. The compound formed is decomposed by acids with the formation of a gelatinous precipitate which is a mixture of various silicic acids.

2. Silver Nitrate, $AgNO_3$. In solutions of silicates, silver nitrate produces a yellow precipitate of Ag_2SiO_3 soluble in HNO_3:

$$SiO_3^{2-} + 2Ag^+ \rightarrow Ag_2SiO_3 \downarrow$$

3. Dilute Acids. When being gradually added to moderately dilute solutions of silicates, these acids give a white gelatinous precipitate (gel) of silicic acids. Occasionally, the entire solution forms a semisolid gel. On rapid addition of an excess of acids, especially to dilute solutions of silicates, a precipitate either fails to separate at once or is formed after a very long time. The silicic acids remain in solution in the form of a colloid (sol). A colloidal solution also results partially in the formation of a gel. The precipitation of silicic acids by the action of acids cannot therefore be complete.

For complete precipitation of silicic acids to be achieved, the solution is evaporated repeatedly to dryness with concentrated hydrochloric acid. In this case, the silicic acids are dehydrated and become practically insoluble. They can be separated by centrifugation after the precipitate is treated with acidulated water.

4. Ammonium Salts. These salts precipitate silicic acids from solutions of silicates more completely than acids. The factors responsible for the occurrence of this reactions are the disturbance of the hydrolysis equilibrium of the silicate and the coagulation of the silicic acid by the NH_4^+ ions of an ammonium salt:

$$SiO_3^{2-} + 2H_2O \rightarrow H_2SiO_3\downarrow + 2OH^-$$
$$2OH^- + 2NH_4^+ \rightarrow 2NH_4OH$$

$$SiO_3^{2-} + 2H_2O + 2NH_4^+ \rightarrow H_2SiO_3\downarrow + 2NH_4OH$$

If no precipitate forms in the cold, the solution is heated for several minutes in the water bath.

It should be kept in mind that aluminates, when heated with NH_4Cl, give a precipitate of $Al(OH)_3$ which is very similar in appearance to the precipitate of silicic acids. The corresponding compounds of tin and antimony, when allowed to react with NH_4Cl, give analogous precipitates of hydroxides. For this reason, to avoid errors, it is recommended that a confirmatory test be made for SiO_3^{2-}. Evaporate a portion of the test solution to dryness, transfer the dry residue to a filter, and treat it with a 0.5-percent acetic acid solution of methylene blue. Then wash with cold water. This leaves a gelatinous precipitate of silicic acid coloured deep-blue by methylene blue.

5. Microcrystalloscopic Test. Dissolve a small grain of the substance under test in hydrofluoric acid (or treat with a solution of NH_4F in HCl) on a microscope slide covered with paraffin wax (or on photographic film, in which case the light-sensitive emulsion should be removed). Introduce a small grain of NaCl into the resulting solution. If silicon is present, crystals of $Na_2[SiF_6]$ are formed (see Fig. 8.4). The dilution limit is 1 in 20,000.

8.10. Reactions of the Oxalate Ion

The oxalate ion, $C_2O_4^{2-}$ is the anion of oxalic acid, $H_2C_2O_4$.

Free oxalic acid is a white substance which crystallizes with two molecules of water. It is well soluble in water and is a moderately strong dibasic acid. When warmed, oxalic acid readily sublimes. It is decomposed by strong ignition into H_2O, CO_2, and CO (carbon monoxide). Of the salts of $H_2C_2O_4$ (oxalates), only the oxalates of the alkali metals and magnesium are soluble. The other oxalates are slightly soluble in water but are dissolved by mineral acids.

Oxalic acid has a tendency to form complex anions with Cr^{3+}, Fe^{3+}, Ni^{2+}, and some others (e.g., $[Cr(C_2O_4)_3]^{3-}$, $[Fe(C_2O_4)_3]^{3-}$, $[Ni(C_2O_4)_3]^{4-}$).

The anion $C_2O_4^{2-}$ is colourless.

1. Barium Chloride, $BaCl_2$. With $C_2O_4^{2-}$ barium chloride gives a white precipitate of BaC_2O_4 soluble in mineral acids and (on boiling) in acetic acid.

2. Silver Nitrate, $AgNO_3$. A white curdy precipitate of $Ag_2C_2O_4$ is formed. It is soluble in HNO_3 and NH_4OH.

3. Calcium Salts (Ca^{2+}). These salts bring down $C_2O_4^{2-}$ in the form of a white precipitate of CaC_2O_4 soluble in mineral acids but insoluble in acetic acid. This behaviour toward acetic acid distinguishes CaC_2O_4 from the calcium salts of all the other anions discussed in this book, except CaF_2 (and also $CaSO_4$ which may form if the concentration of SO_4^{2-} in the solution is high).

4. Potassium Permanganate, $KMnO_4$. In acid solution, the permanganate oxidizes $C_2O_4^{2-}$ to CO_2:

$$2MnO_4^- + 5C_2O_4^{2-} + 16H^+ \rightarrow 2Mn^{2+} + 8H_2O + 10CO_2\uparrow$$

The reaction is carried out on heating the solution up to 70-80°C. The occurrence of this reaction is easily recognized by the decolourization of permanganate which is added, a drop at a time, to the heated acid test solution. Permanganate becomes colourless not at once. Addition of a drop or two of a solution of $MnSO_4$ accelerates the reaction catalytically. Since other reducing agents (e.g., sulphides, sulphites, thiosulphates, nitrites) are also capable of decolourizing (reducing) $KMnO_4$, before the test for $C_2O_4^{2-}$ is started, the permanganate is separated from the reducing agents present by precipitation with $CaCl_2$ in the presence of CH_3COOH.

This is done as follows. Place a few drops of the test solution in a test tube and make it slightly alkaline with NaOH in order to neutralize strong acids which may be present in solution; otherwise, they will inhibit the precipitation of CaC_2O_4. This done, acidify the solution with acetic acid until a pH of 4-5 is reached. To the acetic acid solution add an excess of $CaCl_2$ solution and heat for several minutes. Centrifuge off the precipitate of calcium salts, wash thoroughly with hot water (to remove the remainder of the solution which may contain the reducing agents to be separated), and treat on heating with 2*N* H_2SO_4. Having separated the precipitate, to the solution obtained add a drop of $MnSO_4$ solution, heat in the water bath, and add dropwise 0.01*N* permanganate solution. Its decolourization indicates the presence of $C_2O_4^{2-}$. The oxidation of $C_2O_4^{2-}$ by permanganate in alkaline solution is very slow, so that a negative result in the test does not yet mean that $C_2O_4^{2-}$ is absent.

5. Concentrated Sulphuric Acid. The acid abstracts water from oxalic acid on heating, equal volumes of CO_2 and CO being evolved. Carbon monoxide burns with a blue flame:

$$H_2C_2O_4 \rightarrow H_2O + CO_2\uparrow + CO\uparrow$$

6. Resorcinol*, $C_6H_4(OH)_2$. When dissolved in a hundredfold amount (mass) of concentrated H_2SO_4 and heated with dry oxalates until fumes of SO_3 appear, resorcinol gives a characteristic blue colouration.

7. Ignition. Strong heating (ignition) decomposes oxalates with slight charring. The oxalates of the alkali metals and barium decompose on ignition with formation of the corresponding carbonates and evolution of carbon monoxide. For example,

$$Na_2C_2O_4 \rightarrow Na_2CO_3 + CO\uparrow$$

The oxalates of iron, nickel, cobalt, copper, and noble metals produce on ignition free metals, and the other oxalates form metal oxides.

Oxalates are capable of forming complexes with a number of cations which may thus not be found during the analysis. Therefore, oxalates must be removed from the substance to be analysed for cations. This removal can be effected either by heating the corresponding solid substance on exposure to the air or by repeated treatment with a mixture of concentrated sulphuric and nitric acids. This treatment is done on heating until fumes of SO_3 appear and is repeated several times until the test for $C_2O_4^{2-}$ is negative. The ions Ba^{2+}, Sr^{2+}, Pb^{2+}, and partially Ca^{2+} are, of course, precipitated in this operation in the form of slightly soluble sulphates; they are dissolved and the solution is examined.

B. GROUP II ANIONS

Group II includes the following anions: Cl^- (chloride ion), Br^- (bromide ion), I^- (iodide ion), S^{2-} (sulphide ion), SCN^- (thiocyanate ion), $[Fe(CN)_6]^{4-}$ (ferrocyanide ion), $[Fe(CN)_6]^{3-}$ (ferricyanide ion), CN^- (cyanide ion), BrO_3^- (bromate ion), IO_3^- (iodate ion), ClO^- (hypochlorite ion), etc.

These anions are precipitated by Ag^+ in a solution made faintly acid with nitric acid since the silver salts** formed by them are insoluble in dilute (2*N*) nitric acid.

In contrast to the anions of Group II, the silver salts formed by the anions of Group III and also by SO_4^{2-} and F^- are soluble in water, and those formed by the other anions of Group I are soluble in dilute HNO_3.

Hence, the group reagent for the anions of Group II is $AgNO_3$ in the presence of HNO_3.

---

* The structural formula of resorcinol is

OH (at position 1), OH (at position 3) on a benzene ring

** Silver sulphide, Ag_2S, dissolves in HNO_3 on heating.

Sulphide ions are oxidized by nitric acid to sulphur (or even to SO_4^{2-}); therefore, when the test for the presence of Group II anions is to be carried out by the group reagent, not the test solution but the silver nitrate solution must be acidified.

The barium salts of Group II anions are soluble in water. It is for this reason that, in contrast to the anions of Group I, those of Group II are not precipitated by $BaCl_2$. In aqueous solutions, the anions of Group II are colourless, except $[Fe(CN)_6]^{4-}$ and $[Fe(CN)_6]^{3-}$

8.11. Reactions of the Chloride Ion

Th chloride ion, Cl^-, is the anion of hydrochloric acid, HCl. It is a solution of gaseous hydrogen chloride in water and is one of the most powerful mineral acids. Solutions of the chlorides resulting from the interaction of HCl with strong bases, say, NaCl, $CaCl_2$, give a neutral reaction, whereas solutions of the salts formed with weak bases show an acid reaction. This applies equally well to bromides and iodides since HBr and HI are as strong as HCl.

The chlorides slightly soluble in water are AgCl, Hg_2Cl_2, $PbCl_2$, and also the basic salts of bismuth, antimony, and tin. The other chlorides are well soluble.

1. Silver Nitrate, $AgNO_3$. With Cl^-, silver nitrate gives a white curdy precipitate of AgCl. The precipitate turns gray-violet by the action of light and then becomes black (decomposition).

Being a slightly soluble salt of a strong acid, AgCl is insoluble in acids, say, in HNO_3; but it readily dissolves under the action of substances capable of binding Ag^+ ions into a complex (e.g., NH_4OH, $Na_2S_2O_3$, KCN). In the case of ammonium hydroxide, the reaction proceeds according to the equation

$$AgCl + 2NH_4OH \rightarrow [Ag(NH_3)_2]^+ + Cl^- + 2H_2O$$

When the solution is acidified, AgCl reprecipitates (see Sec. 5.10):

$$[Ag(NH_3)_2]^+ + Cl^- + 2H^+ \rightarrow AgCl\downarrow + 2NH_4^+$$

In distinction to AgCl ($SP = 1.78 \times 10^{-10}$), AgI has a much smaller solubility product constant (8.3×10^{-17}) and is practically insoluble in ammonium hydroxide. Silver bromide, AgBr ($SP = 5.3 \times 10^{-13}$), and silver thiocyanate, AgSCN ($SP = 1.1 \times 10^{-12}$), dissolve but slightly in ammonium hydroxide and even this slight solubility of silver bromide and thiocyanate can be further reduced to a very negligible value by using a solution of ammonium carbonate* (12-percent) as the reagent. This solution actually becomes an ammonium buffer with pH $\approx$ 9.2 due to hydrolysis.

---

* No solution of $(NH_4)_2CO_3$ containing ammonia must be used for this reaction.

When this solution is shaken thoroughly with the precipitated silver salts, silver chloride dissolves, whereas silver bromide and thiocyanate remain practically unaffected. If the undissolved precipitate is centrifuged off and a few drops of KBr solution are added to the clear centrifugate*, then, in the presence of Cl^- in the original solution, the concentration of Ag^+ in the ammoniacal solution will be sufficient for the solubility product of AgBr to be considerably exceeded. Therefore, an abundant yellowish-white turbidity of AgBr appears in the solution; its formation is proof of the presence of Cl^-. If Cl^- is absent, the solution remains almost clear (more precisely, a very slight turbidity appears) because of the low solubility of silver bromide and thiocyanate in ammonium carbonate solution.

To make sure that the results obtained are reliable, the test should be carried out with: (a) a mixture of AgCl, AgBr, and AgSCN; and (b) a mixture of AgBr and AgSCN (without AgCl). Then the turbidities obtained should be compared.

Silver chloride can be dissolved by the action of a special mixture containing 0.25*M* NH_4OH, 0.25*M* KNO_3, and 0.01*M* $AgNO_3$ in a litre. The ions $[Ag(NH_3)_2]^+$, which are contained in this mixture along with NH_4OH, are formed when silver halides are dissolved in ammonium hydroxide:

$$AgBr + 2NH_4OH \rightarrow [Ag(NH_3)_2]^+ + Br^- + 2H_2O$$

These ions therefore reduce the solubility of silver halides, i.e., displace the equilibrium to the left. This mixture dissolves AgCl fairly well, but it does not practically dissolve AgBr. Hence, if a solution of AgCl, AgBr, and AgI is treated with the mixture indicated, only AgCl passes into solution and can be detected there (after the undissolved residue is separated) by acidifying the centrifugate with nitric acid. If the precipitate contains AgSCN, it is better to use $(NH_4)_2CO_3$ as a solvent in order to detect Cl^-.

2. **Concentrated H_2SO_4.** When this acid is allowed to react with dry chlorides, say, with NaCl or KCl, gaseous HCl is liberated (recognized by the odour or by the reddening of the moist blue litmus paper held over the mouth of the tube):

$$NaCl + H_2SO_4 \rightarrow NaHSO_4 + HCl\uparrow$$

3. **Action of Oxidizing Agents.** Free Cl_2 is a very powerful oxidizing agent. The value of the standard potential of the couple $Cl_2/2Cl^-$ is +1.36 volts. Chloride ions are a weak reducing agent. But when acted on by oxidizing agents with higher standard potentials, the Cl^- ions are oxidized to Cl_2.

Such oxidizing agents include, for example, $KMnO_4$, PbO_2, and $KClO_3$ in acid solution. It is sometimes possible to oxidize Cl^- by using oxidizing agents whose standard potentials are lower than the

* If the centrifugate is turbid, it must be made clear by filtration.

potential for chlorine, provided that suitable concentrations of the reacting substances are chosen. Thus, if to a solution of a chloride there are added a few drops of concentrated H_2SO_4 and a little solid MnO_2 ($E^0_{MnO_2/Mn^{2+}} = 1.23$ volts) and the solution is heated, free chlorine will be evolved (detected by the odour or by the formation of a blue colour on the iodide-starch paper* held over the mouth of the test tube). The equation for the reaction is

$$MnO_2\downarrow + 2Cl^- + 4H^+ \rightarrow Cl_2\uparrow + Mn^{2+} + 2H_2O$$

That the reaction proceeds as written is explained as being due to a large concentration of hydrogen ions in solution, which considerably increases the equilibrium potential of the MnO_2/Mn^{2+} system.

Since the standard potentials of the couples $Br_2/2Br^-$ (+1.09 volts) and $I_2/2I^-$ (+0.54 volt) are lower than the potential of the $Cl_2/2Cl^-$ system, it follows that all oxidizing agents that oxidize Cl^- ions to free chlorine must oxidize I^- and Br^- even more vigorously to free I_2 and Br_2. It should be noted that the Cl^- anion is a very weak reducing agent and is incapable of reducing $KMnO_4$ to $MnO(OH)_2$ in alkaline solution.

4. Potassium Dichromate, $K_2Cr_2O_7$, and Sulphuric Acid. Place a small amount (equal to half the size of a rice grain) of solid sodium chloride, NaCl, mixed thoroughly with an approximately triple amount of finely powdered $K_2Cr_2O_7$ in a dry apparatus (see Fig. 2.21). Moisten the mixture with 4-5 drops of concentrated H_2SO_4; put a drop of 2*N* NaOH in the loop of a platinum wire and cautiously close the apparatus with a rubber stopper; avoid touching the sides of the vessel with the wire. Then heat the contents of the vessel by placing it into a beaker or a dish with hot water. The reactions that take place are as follows.

(1) The action of $K_2Cr_2O_7$ on the chloride, in the presence of H_2SO_4, results in the formation of a volatile chromium compound known as chromyl chloride, CrO_2Cl_2:

$$4NaCl + K_2Cr_2O_7 + 6H_2SO_4 \rightarrow 2CrO_2Cl_2\uparrow + 2KHSO_4 + 4NaHSO_4 + 3H_2O$$

(2) The evolved CrO_2Cl_2 reacts with the alkaline solution to form a chromate:

$$CrO_2Cl_2\uparrow + 4OH^- \rightarrow CrO_4^{2-} + 2Cl^- + 2H_2O$$

After several minutes take the wire cautiously out of the vessel and wash the drop off the loop with 2-3 drops of 2*N* H_2SO_4 into a test tube. Add a few drops of amyl alcohol and a drop of H_2O_2 and shake at once. The formation of a blue ring of H_2CrO_6 indicates the presence of CrO_4^{2-}.

Alternatively, the drop of the solution in the loop may be transferred to a test tube with 2-3 drops of 2*N* CH_3COOH and treated with a

* This is filter paper moistened with solutions of starch and KI. The interaction of KI with Cl_2 results in the formation of I_2 which gives a blue colour with starch.

drop of $BaCl_2$ solution. A yellow precipitate (turbidity) of $BaCrO_4$ indicates CrO_4^{2-}.

If fluorides are absent, chromium may volatilize only in the form of CrO_2Cl_2; therefore, the presence of CrO_4^{2-} ions in the drop is evidence that the substance being analysed contains chlorides (Cl^- ions).

Under the conditions of the experiment, fluorides form liquid (but volatile) chromyl fluoride, CrO_2F_2, and therefore interfere with the test. The ions Br^- and I^- do not interfere.

The test described requires strict observation of the procedure since, if $K_2Cr_2O_7$ gets into the drop from the reaction mixture or from the sides of the vessel, the Cl^- ion will be detected in error. The reaction under consideration, along with the one described in Procedure *1*, is used for detection of Cl^- in the presence of Br^- and I^-.

8.12. Reactions of the Bromide Ion

The ion Br^- is the anion of hydrobromic acid, HBr, which is a solution of hydrogen bromide in water. Hydrobromic acid is just as strong as HCl. The salts of hydrobromic acid (bromides) insoluble in water are AgBr, Hg_2Br_2, and $PbBr_2$, i.e., the salts formed by the same cations which give insoluble chlorides. Bromides are in general very similar in properties to chlorides, the main difference being that they are somewhat less stable to the action of oxidizing agents.

1. Silver Nitrate, $AgNO_3$. In solutions of bromides, silver nitrate gives a pale-yellow precipitate of AgBr insoluble in HNO_3. The precipitate is poorly soluble in NH_4OH and practically insoluble in $(NH_4)_2CO_3$ and an ammoniacal solution of $AgNO_3$.

Silver bromide is easily decomposed by zinc dust in the presence of water or 2*N* H_2SO_4:

$$2AgBr\downarrow + Zn\downarrow \rightarrow Zn^{2+} + 2Br^- + 2Ag\downarrow$$

The silver formed separates out as a black precipitate and the Br^- ions pass into solution where they can be detected.

2. Concentrated H_2SO_4. When concentrated sulphuric acid is allowed to react with solid bromides, say, with NaBr, hydrogen bromide gas is evolved

$$NaBr + H_2SO_4 \rightarrow NaHSO_4 + HBr\uparrow$$

In contrast to hydrogen chloride, hydrogen bromide is partly oxidized to free bromine (Br_2) by sulphuric acid, which can be recognized by the brownish colour of the escaping gases:

$$2HBr + H_2SO_4 \rightarrow Br_2\uparrow + 2H_2O + SO_2\uparrow$$

3. Action of Oxidizing Agents. It has already been pointed out that, since the standard oxidation potential of the couple $Br_2/2Br^-$ ($E^0 =$ $= +1.09$ volts) is lower than that of $Cl_2/2Cl^-$ ($E^0 = +1.36$ volts), the bromide ion is a stronger reducing agent, i.e., is oxidized more readily, than Cl^- ions. Therefore, all oxidizing agents, like $KMnO_4$, $KClO_3$, PbO_2, MnO_2, which oxidize (in acid solution) Cl^- to free chlorine (Cl_2), must oxidize Br^- to Br_2. Oxidizing agents are known to exist, which are incapable of oxidizing Cl^- but can oxidize Br^-. An example of such an oxidizing agent is concentrated H_2SO_4, as can be seen from the above reaction.

Sodium hypochlorite, NaClO, and chlorine water (an aqueous solution of Cl_2) exert the same effect on solutions of bromides. To carry out this reaction, which is characteristic of Br^-, proceed as follows. Place a drop or two of NaBr solution in a test tube, acidify with a few drops of $2N$ H_2SO_4, and add a drop or two of chlorine water or of NaClO solution. The solution turns brown due to the evolution of free bromine. If a few drops of chloroform, $CHCl_3$, or benzene*, C_6H_6, in which Br_2 is more soluble than in water, are added to the liquid, and the solution is shaken and allowed to stand, the solvent layer is coloured orange, the colour being characteristic of bromine. With an excess of chlorine water, the colour changes to yellowish because of the formation of BrCl. This is one of the most important reactions of Br^-

Since, like the Cl^- ion, the anion Br^- is a very weak reducing agent, it is incapable of reducing $KMnO_4$ to $MnO(OH)_2$ in alkaline solution.

4. Fuchsin Test. The organic dye fuchsin forms a colourless addition product with $NaHSO_3$; if this product is allowed to react with free bromine, a new bromo-substituted dye with a bright red-violet colour is produced. Since elemental chlorine or iodine gives no colour, the reaction makes it possible to detect even very small amounts of Br^- in the presence of Cl^- and I^-.

Place a few drops of 0.1-percent aqueous fuchsin solution in a test tube and add solid $NaHSO_3$ and a drop or two of concentrated ($12N$) HCl to decolourize the fuchsin solution. Moisten a piece of filter paper with the solution and attach it to the inner surface of the upper watch glass of the gas chamber (see Fig. 2.6). Now put 2-3 drops of the test solution and 4-5 drops of a 25-percent solution of perchromic acid on the lower glass of the chamber, and then heat the chamber for 10 min over the boiling water bath.

The free bromine formed as the result of the oxidation of Br^- ions by perchromic acid volatilizes, reacting with the reagent applied to the paper which acquires more or less rapidlv a red-violet colour. The test is very sensitive.**

---

* Benzene must contain no impurities of unsa urated compounds which decolourize solutions of Br_2. Instead of chloroform and benzene, use may be made of carbon tetrachloride, CCl_4, or carbon sulphide, CS_2.

** Since this reaction is highly sensitive, the substances used for preparing solutions to be tested must be of high purity. For example, potassium iodide usually contains a small amount (traces) of bromides which cannot be detected by other reactions but give a noticeable colour in the fuchsin test.

8.13. Reactions of the Iodide Ion

The iodide ion, I^-, is the anion of hydriodic acid, HI, which is a solution of hydrogen iodide in water. This acid is as strong as HCl and HBr. The only common iodides that are insoluble in water are those of silver, mercury, lead, and copper(I). Hydriodic acid and its salts are oxidized to I_2 more readily than HBr or bromides since the standard potential of the couple $I_2/2I^-$ (+0.54 volt) is even lower than that of the $Br_2/2Br^-$ couple (+1.09 volts).

1. Silver Nitrate, $AgNO_3$. In iodide solutions, silver nitrate gives a yellow precipitate of AgI insoluble in nitric acid and also in ammonium hydroxide (distinction from AgCl). Silver iodide dissolves in $Na_2S_2O_3$ and KCN solutions which form with Ag^+ the complexes $[AgS_2O_3]^-$, $[Ag(S_2O_3)_3]^{5-}$, and $[Ag(CN)_2]^-$ which are less dissociated than $[Ag(NH_3)_2]^+$. The reaction with $Na_2S_2O_3$ requires a large excess of this reagent:

$$AgI\downarrow + 3S_2O_3^{2-} \rightarrow [Ag(S_2O_3)_3]^{5-} + I^-$$

Like AgBr, silver iodide is decomposed by the action of zinc dust in the presence of water or 2*N* H_2SO :

$$2AgI\downarrow + Zn\downarrow \rightarrow Zn^{2+} + 2I^- + 2Ag\downarrow$$

2. Lead Salts (Pb^{2+}). In iodide solutions, lead salts give a yellow or golden precipitate of PbI_2 (see Sec. 6.17, Reaction *4*).

3. Concentrated H_2SO_4. The acid attacks iodides in just the same way as in the case of bromides, the only difference being that the oxidation of the HI formed to I_2 proceeds even faster than the oxidation of HBr. Sulphuric acid is reduced here to SO_2 or to H_2S. For example,

$$8I^- + SO_4^{2-} + 10H^+ \rightarrow 4I_2\downarrow + 4H_2O + H_2S\uparrow$$

The iodine formed separates out in the cold in the form of a dark-gray precipitate or colours the solution brown. On heating, violet-coloured vapours of iodine are liberated.

4. Action of Chlorine and NaClO. Many other oxidizing agents (e.g., H_2O_2, $KMnO_4$, and $K_2Cr_2O_7$ in acid solution, Cl_2 and Br_2, NaClO and NaBrO, ferric salts, etc.) have the same effect on iodides as concentrated H_2SO_4. In contrast to Cl^- and Br^- ions, the iodide ions are oxidized by $KMnO_4$ not only in acid but also in alkaline solution, the permanganate being reduced to $MnO(OH)_2$. The iodide ion is especially often detected by using chlorine water or a solution of sodium hypochlorite since this reaction enables identification of both I^- and Br^- ions. To carry out this reaction, proceed as follows. Mix solutions of KI and KBr (a drop of KI solution with a drop of KBr solution, diluted two- or three-fold with water) and add 2-3 drops of 2*N* H_2SO_4 and a few drops of benzene, C_6H_6, or chloroform, $CHCl_3$. Then add, drop by drop, chlorine water or a solution of sodium hypochlorite, NaClO, and shake thoroughly after each drop. Iodide ions, which are a stronger reducing agent than Br^-, are oxidized first:

$$2I^- + Cl_2 \rightarrow I_2 + 2Cl^-$$

In this reaction, the benzene or chloroform layer is coloured violet, this colour being characteristic of iodine. Upon further addition of chlorine water, or NaClO solution, the colour disappears because of the oxidation of I_2 to iodic acid, HIO_3:

$$I_2 + 5Cl_2 + 6H_2O \rightarrow 12H^+ + 2IO_3^- + 10Cl^-$$

In the presence of Br^-, as soon as the violet colour of I_2 disappears, it is replaced by the orange colour of Br_2, which in turn changes to the yellow colour of BrCl. In carrying out this reaction, a very small amount of the solution to be tested for I^- and Br^- should be taken, and chlorine water or NaClO solution added a drop at a time. Otherwise, a large amount of chlorine water or NaClO solution will be required for oxidizing all the iodine and there is a risk of missing Br^- if the amount of chlorine (or NaClO) added is insufficient. And if we use too much chlorine water (or NaClO) all at once, the colour due to iodine may disappear instantly and I^- will be missed.

If the solution contains S^{2-}, $S_2O_3^{2-}$, or SO_3^{2-} which are stronger reducing agents than I^- and Br^-, their oxidation will begin only after the ions S^{2-}, $S_2O_3^{2-}$, and SO_3^{2-} have been completely oxidized. These anions can be oxidized, for example, by the action of $KMnO_4$ in acid solution. The iodide and bromide ions are more conveniently identified by using the precipitate remaining after the precipitation of Group II anions by $AgNO_3$ in the presence of HNO_3.

5. Nitrite Ion, NO_2^-. In the presence of sulphuric, hydrochloric, and even acetic acids, the nitrite ion oxidizes I^- to elemental iodine:

$$2I^- + 2NO_2^- + 4H^+ \rightarrow I_2\downarrow + 2H_2O + 2NO\uparrow$$

The oxidation of iodide ion in the presence of acetic acid is a characteristic feature of NO_2^- and is used for its detection. The iodine formed in this reaction separates out as a dark-gray precipitate or colours the solution brown. It can be identified with the aid of benzene or starch emulsion. In the latter case, the resulting brown-coloured solution is added dropwise to 2-3 drops of starch emulsion diluted with about 1 ml of water. The solution assumes an intense-blue colour due to the formation of a compound of iodine and starch. The colour disappears on heating and, when the solution is allowed to cool off, it appears again.

In contrast to I^-, bromide ions are not oxidized by the nitrite ion. This is accounted for by the fact that the standard potential of the couple NO_2^-/NO (+0.99 volt) is higher than that of $I_2/2I^-$ (+0.54 volt) but lower than that of the system $Br_2/2Br^-$ (+1.09 volts).

The same effect on I^- and Br^- is produced by Fe^{3+} ($E^0 = +0.77$ volt).

As free iodine is volatile, I^- ions can be removed from solution as follows. Acidify the solution with $2N$ H_2SO_4, add a few drops of a nitrite solution, and boil until the brown colour fades away. As the liquid is evaporated, add water dropwise in order to prevent the volatilization of HBr.

6. Cupric Salts (Cu^{2+}). Cupric sulphate, for example, reacts with I^- to form a precipitate of cuprous iodide, CuI, which has an ivory colour:

$$2Cu^{2+} + 4I^- \rightarrow 2CuI\downarrow + I_2$$

It may seem at first sight that the direction of this reaction does not correspond to the standard potentials of the systems $I_2/2I^-$ ($E^0 = +0.54$ volt) and Cu^{2+}/Cu^+ ($E^0 = +0.15$ volt). This is accounted for by the low solubility of CuI ($SP \approx 10^{-12}$), as a result of which the concentration of Cu^+ in the solution is strongly reduced and the equilibrium potential of the couple Cu^{2+}/Cu^+ is increased. It becomes larger than that of the couple $I_2/2I^-$, which is what forces the reaction to the oxidation of I^- by Cu^{2+} ions ($E^0_{Cu^{2+}/CuI\downarrow} = +0.86$ volt).

8.14. Reactions of the Sulphide Ion

The sulphide ion, S^{2-}, is the anion of hydrosulphuric acid which is a solution of hydrogen sulphide, H_2S, in water. Hydrosulphuric acid is one of the weakest acids. Accordingly, the sulphide formed by strong bases show a strongly alkaline reaction in solutions. Thus, calculation shows that $0.1M$ solutions of Na_2S and K_2S have a pH of about 13. A solution of $(NH_4)_2S$ of the same concentration has a pH of 9.25. Some of the salts formed by very weak bases (Al_2S_3, Cr_2S_3, TiS_2) cannot exist in the presence of water because they undergo hydrolysis, giving rise to the corresponding hydroxides.

The behaviour of sulphides toward the action of water, acids, alkalis, and alkali sulphides, which is used for separation of cations into subgroups, has been considered in the previous chapters.

1. Silver Nitrate, $AgNO_3$. In sulphide solutions, silver nitrate produces a black precipitate of silver sulphide, Ag_2S. The precipitate is insoluble in NH_4OH but is dissolved by heating in dilute HNO_3.

2. Acids. Dilute solutions of H_2SO_4 and HCl, for example, decompose many sulphides with the formation of hydrogen sulphide gas:

$$Na_2S + H_2SO_4 \rightarrow Na_2SO_4 + H_2S\uparrow$$

$$FeS + 2HCl \rightarrow FeCl_2 + H_2S\uparrow$$

The evolution of hydrogen sulphide can be identified by its nauseating odour of rotten eggs and also by its ability to blacken a piece of paper moistened with a solution of lead acetate, $Pb(CH_3COO)_2$, or sodium plumbite*, Na_2PbO_2:

$$H_2S + Pb(CH_3COO)_2 \rightarrow PbS\downarrow + 2CH_3COOH$$

$$H_2S + Na_2PbO_2 \rightarrow PbS\downarrow + 2NaOH$$

---

* It will be recalled that sodium plumbite, Na_2PbO_2, is produced by the action of an excess of NaOH on a lead salt solution.

The reaction is most conveniently carried out in a gas chamber (see Fig. 2.6). Place the sulphide to be tested and a few drops of 2*N* H_2SO_4 or HCl on the lower glass and attach the paper moistened with a solution of the lead salt to the upper glass.

Sulphides that are non-decomposable by acids can be decomposed by mixing them with zinc dust* and adding HCl. This liberates hydrogen sulphide along with hydrogen; the former can be detected as described earlier.

3. Sodium Nitroprusside, $Na_2[Fe(CN)_5NO]$. In sulphide solutions (with S^{2-} and not with HS^-), the reagent produces a characteristic red-violet colour due to the formation of the complex compound $Na_4[Fe(CN)_5NOS]$. Hydrosulphuric acid, which forms almost exclusively HS^- ions, gives no colouration. Conversely, upon addition of alkali, the colour develops because a sufficient amount of S^{2-} ions s formed:

$$H_2S + 2OH^- \rightarrow S^{2-} + 2H_2O$$

4. Cadmium Salts (Cd^{2+}). In sulphide solutions (witl. S^{2-}), cadmium salts give a characteristic bright-yellow precipitate of CdS. If the precipitate (after it is separated from the solution) is treated with a drop or two of a solution of $CuSO_4$, it turns black due to the formation of CuS:

$$CdS\downarrow + Cu^{2+} \rightarrow Cd^{2+} + CuS\downarrow$$

The formation of CdS is used for separation of S^{2-} from other sulphur containing anions. But since, in the case of soluble cadmium salts, $CdSO_3$ would also precipitate together with CdS, solid cadmium carbonate, $CdCO_3$, should be used as the reagent. Cadmium carbonate is slightly soluble in water and therefore creates in solution a very small concentration of Cd^{2+} ions. This concentration however turns out to be quite enough for the solubility product of the less soluble cadmium sulphide to be exceeded, and CdS is therefore precipitated completely:

$$CdCO_3\downarrow + S^{2-} \rightarrow CdS\downarrow + CO_3^{2-}$$

Conversely, the solubility product of $CdSO_3$ is not reached and SO_3^{2-} remains in solution.

5. Reduction Reactions Effected by Sulphide Ions. Since the standard potential of the couple S/S^{2-} is low ($E^0 = -0.48$ volt), the S^{2-} ion is easily oxidized to free sulphur and is therefore an active reducing agent. Thus, like SO_3^{2-} and $S_2O_3^{2-}$ ions, it decolourizes iodine solution, reducing I_2 to I^-:

$$S^{2-} + I_2 \rightarrow S\downarrow + 2I^-$$

In just the same way, it reduces $KMnO_4$ in acid and alkaline solutions, a brown precipitate of $MnO(OH)_2$ being formed in the latter

* First it is necessary to make sure that the zinc used does not contain an impurity of a sulphide, i.e., does not liberate H_2S when brought into contact with HCl.

case, as usual. Other reactions of reduction of oxidizing agents by S^{2-} have already been considered. For example, S^{2-} ions reduce Fe^{3+} to Fe^{2+}, and AsO_4^{3-} to AsO_3^{3-}. Apart from free sulphur*, the oxidation of S^{2-} may give SO_2 or even H_2SO_4.

Procedures for Mixtures of S^{2-}, $S_2O_3^{2-}$, SO_3^{2-}, and SO_4^{2-} (in the Absence of Other Anions)

If the anions S^{2-}, $S_2O_3^{2-}$, and SO_3^{2-} are present together, they must be detected in the systematic analysis because of their mutual interferences. Indeed, if the test solution contains S^{2-} and SO_3^{2-}, hydrogen sulphide and sulphur dioxide are evolved upon acidification and react with each other at once:

$$SO_2 + 2H_2S \rightarrow 3S\downarrow + 2H_2O$$

As a result of this reaction, only the gas present in excess will be detected. If there is an excess of SO_2, the acidification of the solution will probably be accompanied by the same phenomena (formation of SO_2 and S) as would be observed if the solution contained only $S_2O_3^{2-}$. This anion would thus be detected in error. It is clear that a mixture of SO_3^{2-} and $S_2O_3^{2-}$ forms upon acidification the same products as does $S_2O_3^{2-}$. And, finally, if S^{2-} is present, the tests for SO_3^{2-} using fuchsin and also sodium nitroprusside with addition of $ZnSO_4$ and $K_4[Fe(CN)_6]$, etc., cannot be used.

Similarly, the detection of SO_3^{2-} is complicated by the presence of $S_2O_3^{2-}$ or of a mixture of S^{2-} and SO_3^{2-}. In both cases, when the solution is made acid in the reaction with $BaCl_2$, a white acid-insoluble precipitate of sulphur is formed, which may be mistaken for $BaSO_4$.

The systematic analysis of a mixture of S^{2-}, $S_2O_3^{2-}$, SO_3^{2-}, and SO_4^{2-} (described in detail below) is based on: (a) the precipitation of S^{2-} by the action of $CdCO_3$, the ions SO_3^{2-}, SO_4^{2-}, and $S_2O_3^{2-}$ remaining in solution; (b) the precipitation of SO_3^{2-} and SO_4^{2-} by a strontium salt which does not precipitate $S_2O_3^{2-}$; (c) the fact that $SrSO_4$ is practically insoluble in dilute acids, whereas $SrSO_3$ dissolves in them.

1. **Detection and Separation of S^{2-}.** To a drop of the alkaline test solution add a drop of sodium nitroprusside solution. A red-violet colour indicates S^{2-}. If S^{2-} has been detected, to 5 drops of the solution add a little solid $CdCO_3$** and shake. Centrifuge off the precipitate and test for completeness of precipitation by adding sodium nitroprusside to a drop of the solution. If required, add more $CdCO_3$ and again shake the contents of the test tube. When you are sure that the precipitation is complete, centrifuge off the precipitate containing CdS and other cadmium salts along with excess $CdCO_3$, and wash. A yellow colour of the

* It is very characteristic that the iodate ion, IO_3^-, oxidizes S^{2-} to S even in neutral and weakly alkaline solutions.

** Cadmium carbonate, $CdCO_3$, must contain no impurity of sulphates.

precipitate confirms the presence of S^{2-}*. If the colour is not distinct, add a drop of $CuSO_4$ solution and 2-3 drops of 2*N* HCl to the test tube. Cadmium carbonate will be dissolved and the yellow CdS precipitate will be replaced by the less soluble black precipiate of CuS.

The centrifugate containing $S_2O_3^{2-}$, SO_3^{2-} and SO_4^{2-} (and other anions) is to be examined according to Procedures *2* and *3*.

2. Detection of $S_2O_3^{2-}$. To 2 drops of the centrifugate (or of the original solution if S^{2-} is absent) add 3-4 drops of 2*N* HCl and heat. The formation of a white or yellow turbidity (sulphur) indicates $S_2O_3^{2-}$.

3. Detection of SO_3^{2-} and SO_4^{2-}. To the remainder of the centrifugate add a solution of $SrCl_2$ or $Sr(NO_3)_2$ until the precipitation is complete. Wash thoroughly the resultant precipitate (consisting of $SrSO_3$, $SrSO_4$, and other strontium salts) to remove $S_2O_3^{2-}$, add 3-4 drops of water, shake, and divide the resulting turbid liquid into two parts.

(1) *Detection of* SO_3^{2-}. To one part add 2-3 drops of 2*N* HCl and a solution of I_2, a drop at a time. The decolourization of the solution indicates SO_3^{2-}.

(2) *Detection of* SO_4^{2-}. Treat the other portion with 2 drops of a solution of $BaCl_2$ (to convert $SrSO_4$ into a less soluble salt—$BaSO_4$) and with an excess of 2*N* HCl. If the precipitate does not dissolve in it, this means that SO_4^{2-} is present. It can also be detected in a separate portion of the original solution by the action of $BaCl_2$ in the presence of HCl and $KMnO_4$ (see Sec. 8.2, Reaction *1*).

8.15. Reactions of the Thiocyanate Ion

The thiocyanate ion, SCN^-, is the anion of thiocyanic acid, HSCN. This acid is a colourless liquid with a pungent, penetrating odour. In aqueous solutions, it behaves as a strong monobasic acid. Most of the salts of thiocyanic acid are well soluble in water. Lead thiocyanate is slightly soluble and the thiocyanates of silver, copper(I), and mercury(II) are insoluble in water. The SCN^- ion is colourless.

1. Silver Nitrate, $AgNO_3$. In solutions of thiocyanates, silver nitrate gives a white precipitate of silver thiocyanate:

$$Ag^+ + SCN^- \rightarrow AgSCN\downarrow$$

The precipitate is insoluble in dilute nitric acid and, according to the value of the solubility product (1.1×10^{-12}) which is close to that of AgBr (5.3×10^{-13}), is only slightly soluble in NH_4OH. Like AgBr, it is practically insoluble in a solution of $(NH_4)_2CO_3$.

2. Mercuric Nitrate, $Hg(NO_3)_2$. This reagent brings down a white precipitate of mercuric thiocyanate which dissolves in excess of thio-

* The ion S^{2-} can also be detected in a gas chamber by the evolution of H_2S when a solution of HCl diluted with water (1 : 1) is allowed to react with the precipitate.

cyanate to form a complex compound:

$$Hg^{2+} + 2SCN^- \rightarrow Hg(SCN)_2\downarrow$$
$$Hg(SCN)_2\downarrow + 2SCN^- \rightarrow [Hg(SCN)_4]^{2-}$$

3. Ferric Salts (Fe^{3+}). In thiocyanate solutions, a red or pink (at small concentrations of SCN^-) colouration is produced because of the formation of ferric thiocyanate*:

$$Fe^{3+} + 3SCN^- \rightarrow Fe(SCN)_3$$

While carrying out this reaction, the solution should be acidified in order to suppress the hydrolysis of $Fe(SCN)_3$; otherwise, the red colour of the solution will change to red-brown or even to yellow (in very dilute solutions). At low concentrations of SCN^-, the faint colour can be intensified by extraction of ferric thiocyanate with organic solvents, say, amyl alcohol or diethyl ether, in which it dissolves more readily than in water. This is the most important test for detection of SCN^-. The test, however, is interfered with by many substances and ions which bind Fe^{3+} into fairly stable complexes (e.g., oxalic, tartaric, citric, and acetic acids, the ions PO_4^{3-}, F^-, and $[Fe(CN)_6]^{4-}$). Iodide ions must also be absent because they are oxidized to free iodine by Fe^{3+}, as a result of which a brownish-red colour appears, which is similar to the colour of $Fe(SCN)_3$.

To detect SCN^- in the presence of all the compounds indicated, the complex $[Fe(CN)_6]^{4-}$ ion must be removed from the solution by the action of $ZnSO_4$ or $Zn(CH_3COO)_2$ and the resultant centrifugate treated with $AgNO_3$ solution in the presence of HNO_3. This precipitates SCN^- ions which will thereby be separated from most of the interfering substances remaining in solution. To bring the SCN^- ions back into solution and to separate them from I^- ions, the precipitate, which may contain AgSCN, AgCl, AgBr, AgI, and Ag_2S, is shaken with KBr solution. Most of the AgSCN is converted into the less soluble silver bromide and SCN^- ions appear in solution:

$$AgSCN\downarrow + Br^- \rightarrow AgBr\downarrow + SCN^-$$

In contrast to silver thiocyanate, silver iodide is not converted into AgBr to any appreciable extent by the action of KBr because the solubility product of AgI is lower than that of AgBr, and, consequently, I^- ions do not pass into solution. Thus, SCN^- is found to have been separated from I^- and can be identified through the addition of $FeCl_3$ to the solution.

Alternatively, a separate portion of the solution placed in a depression of a spot plate may be treated with a drop of an acidified solution of $FeCl_3$. In the presence of SCN^-, the precipitate becomes noticeably pink.

---

* This reaction in fact proceeds in a more complicated way: a range of complexes, from $Fe[SCN]^{2+}$ to $[Fe(SCN)_6]^{3-}$, are formed.

4. Action of H_2SO_4. A dilute (1:1) solution of H_2SO_4 decomposes dry thiocyanates with the evolution of carbon sulphoxide, COS (*poisonous!*), which burns with a blue flame:

$$SCN^- + 2H^+ + H_2O \rightarrow NH_4^+ + COS\uparrow$$

Concentrated sulphuric acid decomposes dry thiocyanates vigorously with the evolution of vapours and gases with a pungent odour (COS, HCOOH*, CS_2, CO_2, SO_2) and formation of sulphur.

5. Oxidizing Agents. The sulphur in SCN^- is oxidized according to the scheme

$$S^{-II} - 8e \rightarrow S^{+VI}$$

Nitric acid, even if dilute, decomposes thiocyanates:

$$3SCN^- + 10H^+ + 13NO_3^- \rightarrow 16NO\uparrow + 3CO_2\uparrow + 3SO_4^{2-} + 5H_2O$$

Solutions of thiocyanates are also oxidized by permanganate in acid and in alkaline (or neutral) solution.

In an acid solution, the reaction proceeds according to the equation

$$6MnO_4^- + 5SCN^- + 13H^+ \rightarrow 6Mn^{2+} + 5HCN\uparrow + 5SO_4^{2-} + 4H_2O$$

A very small amount (1-2 drops) of thiocyanate solution must be used and the test carried out in a fume cupboard because *the HCN vapour is extremely poisonous!* The occurrence of the reaction can be recognized by the decolourization of a strongly dilute permanganate solution when it is added dropwise to the acidified thiocyanate solution.

6. Cupric Salts (Cu^{2+}). When added drop by drop, a cupric salt first produces an emerald-green colour. If the reagent is present in excess, a black precipitate of $Cu(SCN)_2$ is formed. In the presence of sulphurous acid, a white precipitate of the corresponding cuprous salt separates out:

$$2SCN^- + 2Cu^{2+} + SO_2 + 2H_2O \rightarrow 2CuSCN\downarrow + SO_4^{2-} + 4H^+$$

7. Cobaltous Salts (Co^{2+}). Complex compounds are formed with thiocyanates. The resulting compounds dissolve in amyl alcohol to give a blue ring (see Sec. 5.22, Reaction *5*). A dilute solution of $Co(NO_3)_2$ must be used and added drop by drop because the complex $[Co(SCN)_4]^{2-}$ is formed only in the presence of a large excess of SCN^- ions.

8. Sodium Nitrite, $NaNO_2$. In a neutral medium, sodium nitrite does not react with thiocyanates. But when the solution is acidified with a mineral acid, a dark colouration is produced, which disappears after a while. The colour is due to the oxidation of SCN^-.

9. Metallic Zinc. In an acid solution, zinc metal decomposes thiocyanates (including AgSCN) to form hydrogen sulphide and HCN:

$$Zn\downarrow + SCN^- + 3H^+ \rightarrow Zn^{2+} + H_2S\uparrow + HCN\uparrow$$

The reaction must be carried out in the fume cupboard!

* Formic acid.

8.16. Reactions of the Cyanide Ion

The cyanide ion, CN^-, is the anion of a very weak acid—hydrocyanic acid ($K = 6.2 \times 10^{-10}$). Hydrocyanic acid (prussic acid) is extremely poisonous. The salts formed by this acid with the alkali or alkaline-earth metals are well soluble in water. Solutions of these salts give an alkaline reaction due to hydrolysis (the pH of 0.1*M* KCN solution is 11.10).

Most of the cyanides of other metals are insoluble in water but dissolve in excess of potassium cyanide or sodium cyanide with the formation of complex compounds.

Aqueous solutions of cyanides are unstable and decompose on prolonged standing:

$$KCN + 2H_2O \rightarrow HCOOK + NH_3\uparrow$$

On exposure to the air (the action of CO_2 and moisture), solid soluble cyanides are converted into carbonates.

1. Dilute Mineral Acids. These acids decompose in the cold all water-soluble cyanides, except $Hg(CN)_2$, with the evolution of hydrogen cyanide, HCN (*hood!*).

2. Concentrated Sulphuric Acid. Insoluble cyanides and cyanide complexes are decomposed by concentrated sulphuric acid. The reaction proceeds through a number of intermediate stages. The overall equation of the reaction is:

$$Zn(CN)_2\downarrow + 2H_2SO_4 \rightarrow ZnSO_4 + (NH_4)_2SO_4 + 2CO_2\uparrow$$

3. Silver Nitrate, $AgNO_3$. When an excess of silver nitrate is added to a solution of potassium cyanide, a white curdy precipitate of $Ag[Ag(CN)_2]$ is produced.

The reaction proceeds as follows. At first, when the solution contains an excess of CN^- ions, a stable complex is formed, which is soluble in water:

$$2CN^- + Ag^+ \rightarrow [Ag^+(CN)_2]^-$$

The addition of more silver ions brings down a precipitate:

$$[Ag(CN)_2]^- + Ag^+ \rightarrow Ag[Ag(CN)_2]\downarrow$$

If, conversely, potassium cyanide is added to a solution of $AgNO_3$, the cyanide is precipitated as silver cyanide:

$$Ag^+ + CN^- \rightarrow AgCN\downarrow$$

The addition of more cyanide dissolves the precipitate:

$$AgCN\downarrow + CN^- \rightarrow [Ag(CN)_2]^-$$

4. Salts of Zinc (Zn^{2+}) and Lead (Pb^{2+}). These salts give white precipitates of the corresponding cyanides. Zinc cyanide dissolves in excess of KCN to form a complex compound:

$$Zn^{2+} + 2CN^- \rightarrow Zn(CN)_2 \downarrow$$

$$Zn(CN)_2 \downarrow + 2KCN \rightarrow K_2[Zn(CN)_4]$$

5. Formation of Ferric Thiocyanate. To a concentrated solution of a cyanide there are added a few drops of ammonium polysulphide and the solution is evaporated cautiously in the water bath (*hood!*). To the dry residue there are added a few drops of dilute HCl and a drop or two of a solution of ferric chloride. A blood-red colour appears in the solution because silver thiocyanate is formed:

$$CN^- + S_2^{2-} \rightarrow SCN^- + S^{2-}$$

$$S^{2-} + 2H^+ \rightarrow H_2S \uparrow$$

$$3SCN^- + Fe^{3+} \rightarrow Fe(SCN)_3$$

Otherwise thiocyanate complexes of different composition may be produced.

8.17. Reactions of the Ferrocyanide Ion

The ferrocyanide ion, or hexacyanoferrate(II) ion, is the anion of ferrocyanic acid, $H_4[Fe(CN)_6]$. This strong acid is a white crystalline solid which is readily soluble in water and alcohol. In the solid state, it is stable; in the presence of moisture, it gradually turns deep-blue because of the oxidation to Prussian blue. The salts of ferrocyanic acid—ferrocyanides—are more stable. Of these, only the ferrocyanides of the alkali and alkaline-earth metals are soluble in water.* The reaction of their solutions is close to neutral. Most of the other ferrocyanides are slightly soluble or practically insoluble in water and in cold dilute acids. The ferrocyanide ion has a greenish-yellow colour.

1. Silver Nitrate, $AgNO_3$. In solutions of ferrocyanides, silver nitrate forms a white precipitate of silver ferrocyanide:

$$4Ag^+ + [Fe(CN)_6]^{4-} \rightarrow Ag_4[Fe(CN)_6] \downarrow$$

The precipitate is almost insoluble in dilute nitric acid and in ammonium hydroxide.

2. Ferric Salts (Fe^{3+}). In a neutral or faintly acid solution, ferric salts give a precipitate of Prussian blue:

$$4Fe^{3+} + 3[Fe(CN)_6]^{4-} \rightarrow Fe_4[Fe(CN)_6]_3 \downarrow$$

The precipitate is insoluble in dilute and soluble in concentrated HCl. It dissolves also in oxalic acid to yield the complex compound $H_3[Fe(C_2O_4)_3]$. This most important reaction of the ion $[Fe(CN)_6]^{4-}$ is interfered with by SCN^- which forms with Fe^{3+} ferric thiocyanate,

---

* The double salt of calcium and ammonium (calcium ammonium ferrocyanide) is relatively slightly soluble; its formation is used as a qualitative test for Ca^{2+}.

$Fe(SCN)_3$, having a blood-red colour. The interfering effect of this compound can easily be overcome by using the spot-test technique for carrying out the reaction.*

Place on a strip of filter paper, successively, a drop of a saturated solution of lead nitrate, a drop of the test solution, again a drop of $Pb(NO_3)_2$ solution, and a drop of water. A precipitate of lead ferrocyanide is formed in the centre of the spot, while the SCN^- ions, which are not precipitable by Pb^{2+}, are displaced to the periphery of the spot and are thus separated from the $[Fe(CN)_6]^{4-}$ ions. Then moisten the central part of the spot with a solution of $FeCl_3$ and dilute HCl. Lead ferrocyanide is converted into Prussian blue and the spot centre turns deep-blue.

3. **Cupric Salts (Cu^{2+}).** In a neutral or acetic acid solution, cupric salts produce a red-brown precipitate of cupric ferrocyanide:

$$2Cu^{2+} + [Fe(CN)_6]^{4-} \rightarrow Cu_2[Fe(CN)_6] \downarrow$$

The precipitate is soluble in mineral acids and in ammonium hydroxide.

4. **Zinc Salts (Zn^{2+}).** A white precipitate is formed, which is soluble in alkalis:

$$3Zn^{2+} + 2K^+ + 2[Fe(CN)_6]^{4-} \rightarrow K_2Zn_3[Fe(CN)_6]_2 \downarrow$$

$$K_2Zn_3[Fe(CN_6)]_2 \downarrow + 12OH^- \rightarrow 2K^+ + 2[Fe(CN)_6]^{4-} + 3ZnO_2^{2-} + 6H_2O$$

5. **Ferrous Sulphate, $FeSO_4$.** In solutions of ferrocyanides, the reagent produces a white precipitate of $Fe_2[Fe(CN)_6]$ which gradually turns dark-blue as the result of the oxidation to Prussian blue.

6. **Dilute Sulphuric Acid.** In the cold, ferrocyanides are not decomposed by dilute H_2SO_4. Prolonged boiling causes their decomposition with the formation of hydrocyanic acid:

$$[Fe(CN_6)]^{4-} + 6H^+ \rightarrow Fe^{2+} + 6HCN \uparrow$$

The work should be done in the hood!

Concentrated H_2SO_4 decomposes all ferrocyanides on heating:

$$[Fe(CN)_6]^{4-} + 12H^+ + 6H_2O \rightarrow Fe^{2+} + 6NH_4^+ + 6CO \uparrow$$

In this reaction, all the Fe^{2+} ions are oxidized by sulphuric acid which is reduced to SO_2:

$$2Fe^{2+} + SO_4^{2-} + 4H^+ \rightarrow 2Fe^{3+} + 2H_2O + SO_2 \uparrow$$

7. **Alkali Hydroxides (Caustic Alkalis).** These decompose all insoluble ferrocyanides with the formation of an alkali ferrocyanide and the corresponding hydroxide. For example,

$$Fe_4[Fe(CN)_6]_3 \downarrow + 12OH^- \rightarrow 4Fe(OH)_3 \downarrow + 3[Fe(CN)_6]^{4-}$$

$$Cu_2[Fe(CN)_6] \downarrow + 4OH^- \rightarrow 2Cu(OH)_2 \downarrow + [Fe(CN)_6]^{4-}$$

---

* Tananaev, N. A., *The Drop-Reaction Technique*, Goskhimisdat, Moscow, 1954 (in Russian).

8. Oxidizing Agents. Potassium permanganate, potassium dichromate, chlorine, bromine, and other oxidizing agents oxidize $[Fe(CN)_6]^{4-}$ to $[Fe(CN)_6]^{3-}$ in acid solution. For example,

$$MnO_4^- + 5[Fe(CN]_6]^{4-} + 8H^+ \rightarrow Mn^{2+} + 5[Fe(CN)_6]^{3-} + 4H_2O$$

The $KMnO_4$ solution is decolourized in the reaction. In alkaline solution, permanganate is reduced by $[Fe(CN)_6]^{4-}$ to $MnO(OH)_2$.

8.18. Reactions of the Ferricyanide Ion

The ferricyanide ion, or the hexacyanoferrate(III) ion, is the anion of ferricyanic acid, $H_3[Fe(CN)_6]$. In the free state, this is a brown crystalline solid which is well soluble in water. On storage in the dark, the acid is stable but on exposure to light it decomposes to give hydrocyanic acid, HCN.

Ferricyanic acid is a strong tribasic acid and an oxidizing agent. The salts of this acid (ferricyanides) formed by the alkali and alkaline-earth metals and also by iron(III), chromium(III), and aluminium are soluble in water. Other ferricyanides are insoluble even in dilute acids. Solutions of ferricyanides are yellowish-brown in colour. Their reaction is close to neutral.

1. Silver Nitrate, $AgNO_3$. In ferricyanide solutions, silver nitrate forms an orange precipitate of silver ferricyanide:

$$3Ag^+ + [Fe(CN)_6]^{3-} \rightarrow Ag_3[Fe(CN)_6] \downarrow$$

The precipitate is insoluble in nitric acid but dissolves in ammonium hydroxide. The solubility in ammonium hydroxide distinguishes the $Ag_3[Fe(CN)_6]$ precipitate from $Ag_4[Fe(CN)_6]$.

2. Ferrous Salts (Fe^{2+}). In neutral and acid solutions, ferrous salts produce a deep-blue precipitate of Turnbull's blue. (For the composition of this compound, see Sec. 5.19.)

The use of this most important reaction of $[Fe(CN)_6]^{3-}$ is complicated by the fact that ferrous salts are easily oxidized on exposure to air and therefore contain Fe^{3+} along with Fe^{2+}. In the presence of $[Fe(CN)_6]^{4-}$, the Fe^{3+} ions give a precipitate of Prussian blue indistinguishable in appearance from the precipitate formed by Fe^{2+}. Thus, to avoid detecting $[Fe(CN)_6]^{3-}$ in error in those cases when the ion $[Fe(CN)_6]^{4-}$ has been detected in the solution, it is necessary to test the reagent ($FeSO_4$) for purity. It must give no colour with NH_4SCN solution. If the reagent is found to be contaminated,* the $[Fe(CN)_6]^{3-}$ ion should be detected

* The reagent $FeSO_4$ can be purified from an impurity of Fe^{3+} by boiling with metallic iron and also with reducing agents such as hydrazine sulphate or hydroxylamine sulphate. Boiling must be continued until the test for Fe^{3+} with NH_4SCN is negative. The purified solution of $FeSO_4$ containing no Fe^{3+} must be used immediately for detection of $[Fe(CN)_6]^{3-}$.

by carrying out the reaction with $MnSO_4$ in the presence of alkali (see Reaction *6 (1)* below).

3. Ferric Salts (Fe^{3+}). These salts give no precipitate but develop a brown colour (distinction from $[Fe(CN)_6]^{4-}$).

4. Zinc Salts (Zn^{2+}). In ferricyanide solutions, zinc salts produce a yellow precipitate of $Zn_3[Fe(CN)_6]_2$. The reaction can be used for separation of $[Fe(CN)_6]^{3-}$ and $[Fe(CN)_6]^{4-}$ from SCN^-, Cl^-, Br^-, and I^-. This can also be achieved through the use of cadmium salts which do not precipitate SCN^-, Cl^-, Br^-, and I^-.

5. Cupric Sulphate, $CuSO_4$. In ferricyanide solutions, cupric sulphate produces a green precipitate of $Cu_3[Fe(CN)_6]_2$.

6. Oxidation Reactions. The anion $[Fe(CN)_6]^{3-}$, in which iron is in the oxidation state +3, is easily reduced to $[Fe(CN)_6]^{4-}$ and is therefore an oxidizing agent. It can oxidize not only such readily oxidizable inorganic substances as sulphides (S^{2-}), sulphites (SO_3^{2-}), etc., but also organic compounds (oxalic acid, cellulose, sugar) and ammonia. The oxidation by ferricyanides usually proceeds most readily in alkaline medium.

(1) *Oxidation of Mn^{2+} to $MnO(OH)_2$.* Place a drop of $MnSO_4$ solution and a drop of 2*N* NaOH into a depression of a spot plate. Treat immediately the manganous hydroxide formed with a drop of the ferricyanide test solution. The precipitate assumes a dark-brown colour due to the formation of $MnO(OH)_2$:

$$Mn(OH)_2\downarrow + 2[Fe(CN)_6]^{3-} + 2OH^- \rightarrow MnO(OH)_2\downarrow + 2[Fe(CN)_6]^{4-} + H_2O$$

Since $Mn(OH)_2$ is oxidized by the oxygen of the air, the reaction mixture gradually turns brown even in the absence of $[Fe(CN)_6]^{3-}$. To avoid errors, it is necessary to place a drop each of $MnSO_4$ solution and 2*N* NaOH in a depression next to the one in which the reaction is taking place, and then to compare the colour developed in the reaction with that in the control test.

The reaction under consideration is very important since it enables one to distinguish the ion $[Fe(CN)_6]^{3-}$ from the $[Fe(CN)_6]^{4-}$ ion (just as the reaction with Fe^{2+}). Apart from $[Fe(CN)_6]^{3-}$, only MnO_4^- and CrO_4^{2-} among the anions considered in this book give this reaction.

(2) *Oxidation of I^- to I_2.* Acidify a solution of KI with sulphuric acid and add a few drops of benzene or chloroform. Then treat the solution with a drop or two of the ferricyanide test solution, shake, and allow to stand. The organic layer is coloured violet by the iodine formed:

$$2[Fe(CN)_6]^{3-} + 2I^- \rightarrow 2[Fe(CN)_6]^{4-} + I_2$$

Most other oxidizing anions, such as MnO_4^-, CrO_4^{2-}, NO_2^-, AsO_4^{3-}, etc., give this reaction.

(3) *Benzidine and Diphenylamine.* These are oxidized by $[Fe(CN)_6]^{3-}$ ions just as by many other oxidizing agents, as a result of which a blue precipitate or colour is formed.

7. Alkali Hydroxides and Carbonates. All insoluble ferricyanides are decomposed by alkali hydroxides and carbonates (e.g., Na_2CO_3) on boiling, and the $[Fe(CN)_6]^{3-}$ ion passes into solution where it can be detected:

$$Fe_3[Fe(CN)_6]_2\downarrow + 6OH^- \rightarrow 3Fe(OH)_2\downarrow + 2[Fe(CN)_6]^{3-}$$

8.19. Reactions of the Bromate Ion

The bromate ion, BrO_3^-, is the anion of bromic acid, $HBrO_3$, which is a rather strong acid. The only bromates well soluble in water are those of the alkali metals. Bromic acid is a strong oxidizing agent and this is used for its detection.

1. Silver Nitrate, $AgNO_3$. In concentrated solutions of bromates, silver nitrate gives a pale-yellow precipitate of $AgBrO_3$ which readily dissolves in dilute nitric and sulphuric acids. Silver bromate dissolves also in ammonium hydroxide and potassium cyanide solution, forming complex compounds:

$$AgBrO_3\downarrow + 2NH_4OH \rightarrow [Ag(NH_3)_2]^+ + 2H_2O + BrO_3^-$$
$$AgBrO_3\downarrow + 2CN^- \rightarrow [Ag(CN)_2]^- + BrO_3^-$$

2. Barium Chloride, $BaCl_2$. In concentrated solutions of bromates, barium chloride gives a white crystalline precipitate of barium bromate, which is readily soluble in dilute mineral acids.

3. Bromides (Br^-) and Iodides (I^-). In acid solution, bromides and iodides are oxidized by bromates to free bromine and free iodine:

$$BrO_3^- + 5Br^- + 6H^+ \rightarrow 3Br_2 + 3H_2O$$
$$BrO_3^- + 6I^- + 6H^+ \rightarrow 3I_2 + Br^- + 3H_2O$$

4. Sulphide (S^{2-}), Sulphite (SO_3^{2-}), and Thiosulphite ($S_2O_3^{2-}$). These are oxidized by bromates in acid solution to sulphate ions. The reaction proceeds in stages. Free bromine is evolved first and the solution turns yellow. The yellow colour disappears in excess of the reducing agent because bromine is reduced to bromide ion:

$$5S^{2-} + 8BrO_3^- + 8H^+ \rightarrow 5SO_4^{2-} + 4Br_2 + 4H_2O$$
$$S^{2-} + 4Br_2 + 4H_2O \rightarrow SO_4^{2-} + 8Br^- + 8H^+$$
$$\overline{6S^{2-} + 8BrO_3^- \rightarrow 6SO_4^{2-} + 8Br^-}$$

$$5SO_3^{2-} + 2BrO_3^- + 2H^+ \rightarrow 5SO_4^{2-} + Br_2 + H_2O$$
$$SO_3^{2-} + Br_2 + H_2O \rightarrow SO_4^{2-} + 2Br^- + 2H^+$$
$$\overline{6SO_3^{2-} + 2BrO_3^- \rightarrow 6SO_4^{2-} + 2Br^-}$$

$$5S_2O_3^{2-} + 8BrO_3^- + H_2O \rightarrow 10SO_4^{2-} + 4Br_2 + 2H^+$$
$$S_2O_3^{2-} + 4Br_2 + 5H_2O \rightarrow 2SO_4^{2-} + 8Br^- + 10H^+$$
$$\overline{6S_2O_3^{2-} + 8BrO_3^- + 6H_2O \rightarrow 12SO_4^{2-} + 8Br^- + 12H^+}$$

8.20. Reactions of the Iodate Ion

The iodate ion, IO_3^-, is the anion of iodic acid, HIO_3, which is a fairly strong acid ($K = 1.6 \times 10^{-1}$). In aqueous solutions, iodic acid is strongly associated: it exists in the form of the associations $(HIO_3)_x$, where x is usually equal to 2 or 3. Monobasic iodic acid forms two series of salts—normal and acid ($KIO_3 \cdot HIO_3$). The iodates of the alkali metals are well soluble in water.

1. **Silver Nitrate, $AgNO_3$.** Silver nitrate produces in iodate solutions a white curdy precipitate of $AgIO_3$ which is soluble in all reagents that dissolve silver chloride to yield complex compounds, for example, in ammoniacal and cyanide solutions:

$$AgIO_3 \downarrow + 2NH_4OH \rightarrow [Ag(NH_3)_2]^+ + 2H_2O + IO_3^-$$
$$AgIO_3 \downarrow + 2CN^- \rightarrow [Ag(N)_2]^+ + IO_3^-$$

When sulphite ions are allowed to react in acid medium with a solution containing IO_3^- ions and a silver ammine complex, a precipitate of silver iodide is formed:

$$[Ag(NH_3)_2]^+ + IO_3^- + 3H_2SO_3 \rightarrow AgI \downarrow + 3SO_4^{2-} + 4H^+ + 2NH_4^+$$

Silver iodate also dissolves in a mixture of sulphuric and nitric acids (1 : 4), forming iodic acid which is strongly polymerized in solution.

2. **Barium Chloride, $BaCl_2$.** In iodate solutions, barium chloride gives a white precipitate of barium iodate which is soluble (on heating) in dilute hydrochloric and nitric acids. Barium iodate easily forms supersaturated solutions from which a precipitate separates out when the sides of the test tube are scratched with a glass rod:

$$2IO_3^- + Ba^{2+} \rightarrow Ba(IO_3)_2 \downarrow$$

Therefore, in the course of the analysis, barium iodate will not be precipitated when present alone in the solution. When present together with other insoluble barium compounds, the iodate separates out relatively easily.

3. **Lead Salts (Pb^{2+}).** In iodate solutions, lead salts produce a white precipitate of lead iodate which is less soluble in mineral acids than barium iodate:

$$Pb^{2+} + 2IO_3^- \rightarrow Pb(IO_3)_2 \downarrow$$

4. **Iodides (I^-).** In acid solution (even in acetic acid solution), iodides are oxidized by iodates to free iodine and water:

$$IO_3^- + 5I^- + 6H^+ \rightarrow 3I_2 + 3H_2O$$

Therefore, in the course of the analysis, in the precipitation of the anions of Group II by silver nitrate, the solution should be acidified after the addition of the reagent.

5. Sodium Thiosulphate, $Na_2S_2O_3$. Iodic acid is reduced by sodium thiosulphate to free iodine. In excess of the reagent, the reaction pro ceeds as far as the reduction to iodide ion:

$$10S_2O_3^{2-} + 2IO_3^- + 12H^+ \rightarrow I_2 + 5S_4O_6^{2-} + 6H_2O$$
$$I_2 + 2S_2O_3^{2-} \rightarrow 2I^- + S_4O_6^{2-}$$
$$\overline{12S_2O_3^{2-} + 2IO_3^- + 12H^+ \rightarrow 2I^- + 6S_4O_6^{2-} + 6H_2O}$$

The same effect is produced by sulphur dioxide:

$$3SO_2 + IO_3^- + 3H_2O \rightarrow I^- + 3H_2SO_4$$

6. Hydrogen Sulphide, H_2S. Hydrogen sulphide reduces iodic acid to free iodine and then to iodide ion:

$$5H_2S + 2IO_3^- + 2H^+ \rightarrow 5S\downarrow + I_2 + 6H_2O$$
$$I_2 + H_2S \rightarrow 2H^+ + S\downarrow + 2I^-$$
$$\overline{2IO_3^- + 6H_2S \rightarrow 2I^- + 6S\downarrow + 6H_2O}$$

7. Ferrous Salts (Fe^{2+}). In acid solution, ferrous salts reduce iodic acid to iodine:

$$10Fe^{2+} + 2IO_3^- + 12H^+ \rightarrow 10Fe^{3+} + I_2 + 6H_2O$$

8.21. Reactions of the Hypochlorite Ion

The hypochlorite, ClO^-, is the anion of hypochlorous acid, HClO, which does not exist in the free state because it decomposes rapidly,

$$2HClO \rightarrow 2HCl + O_2\uparrow$$

Hypochlorous acid is a very weak acid ($K = 5.0 \times 10^{-8}$). The hypochlorites of the alkali metals are well soluble in water and are strong oxidizing agents both in acid and alkaline medium. In solutions, hypochlorites decompose (slowly in the cold and more rapidly on heating) to give chlorates and chlorides:

$$3ClO^- \rightarrow ClO_3^- + 2Cl^-$$

Aqueous solutions of hypochlorites show an alkaline reaction due to hydrolysis.

1. Concentrated Acids. Concentrated acids react violently with hypochlorites with the liberation of oxygen. If chlorides are present, chlorine is evolved:

$$2ClO^- + 2H^+ \rightarrow 2HCl + O_2\uparrow$$
$$ClO^- + Cl^- + 2H^+ \rightarrow Cl_2\uparrow + H_2O$$

In aqueous solution, dilute sulphuric acid displaces hypochlorous acid from hypochlorites. The same effect is produced by other acids stronger than hypochlorous acid.

2. Iodides (I^-). In a feebly alkaline solution, iodides are oxidized by a hypochlorite to iodine:

$$ClO^- + 2I^- + H_2O \rightarrow I_2 + Cl^- + 2OH^-$$

As hydroxyl ions accumulate in the solution, the following reaction begins to proceed:

$$3I_2+6OH^- \rightarrow IO_3^-+5I^-+3H_2O$$

3. **Organic Dyes.** Indigo, indicator solutions, and other organic dyes are decolourized by hypochlorites.

4. **Silver Nitrate, $AgNO_3$.** When allowed to react with ClO^-, silver nitrate partially precipitates chlorine in the form of silver chloride, one third of the chlorine being converted to chlorate ion:

$$3ClO^-+2Ag^+ \rightarrow 2AgCl\downarrow+ClO_3^-$$

5. **Manganous Sulphate, $MnSO_4$.** In alkaline solution (in the cold), manganous sulphate reduces hypochlorites to chloride ions with the precipitation of metamanganous acid:

$$Mn^{2+}+ClO^-+2OH^- \rightarrow MnO(OH)_2\downarrow+Cl^-$$

Other oxidizing agents (NO_2^-, NO_3^-, ClO_3^-) do not give this reaction in alkaline solution.

C. GROUP III ANIONS

Group III includes the following anions: NO_3^- (nitrate ion), NO_2^- (nitrite ion), CH_3COO^- (acetate ion), MnO_4^- (permanganate ion), ClO_3^- (chlorate ion), etc. A characteristic feature of this group is the solubility of the corresponding salts of barium and silver, owing to which neither $AgNO_3$ nor $BaCl_2$ precipitates the anions of Group III. The least soluble silver salts are $AgNO_2$ ($SP = 1.6 \times 10^{-4}$), CH_3COOAg ($SP = 4 \times 10^{-3}$), and $AgClO_3$ ($SP = 5 \times 10^{-2}$), which may therefore precipitate from rather concentrated solutions. But on dilution with water and on heating they readily dissolve. There is no group reagent for these anions.

The anions ClO_3^-, NO_2^-, NO_3^-, and CH_3COO^- are colourless and the MnO_4^- ion is crimson-violet.

The principal distinguishing feature of the permanganate ion is that it is one of the most powerful oxidizing agents available. The oxidation reactions characteristic of the permanganate ion were considered in Sections 5.13 and 5.20.

In the systematic analysis, the permanganate ion is reduced by any suitable reagent to Mn^{2+} and detected in analytical Group III of cations. That is why the reactions of the permanganate ion are not discussed here.

8.22. Reactions of the Nitrate Ion

The nitrate ion, NO_3^-, is the anion of nitric acid, HNO_3, which is one of the most powerful mineral acids. At the same time, nitric acid is a rather strong oxidizing agent, being capable of oxidizing a large number of various reducing agents. This is why nitric acid is employed in analysis for dissolving many metals and alloys which cannot be dissolved by HCl and H_2SO_4, for dissolving certain sulphides, etc. The reduction of HNO_3 most often yields nitrogen dioxide, NO_2, or nitrogen oxide, NO. The first results from the oxidation of various compounds by concentrated HNO_3. The second is formed when dilute nitric acid is used. In some cases, the reduction of nitric acid may go even further with the formation of either molecular nitrogen, N_2, or even ammonia, NH_3.

All the nitrates, with the exception of the basic salts of bismuth and mercury and also the salts of certain organic bases, are soluble in water. Solutions of the nitrates formed by strong bases show a neutral reaction; solutions of the nitrates formed by weak bases give an acid reaction. Since most nitrates are soluble, the NO_3^- ion is detected by using almost exclusively the oxidation reactions induced by this ion.

1. Copper and Sulphuric Acid. To 2-3 drops of a nitrate solution add a few drops of concentrated H_2SO_4 and a fragment of metallic copper. Heat the contents of the test tube. A brown-coloured gas is evolved (NO_2), which can easily be recognized if the test tube is placed against a sheet of white paper.

Before this is done, it is necessary to make certain that the reaction is not too violent. The equations for the reactions involved are:

$$2NO_3^- + 8H^+ + 3Cu\downarrow \rightarrow 3Cu^{2+} + 4H_2O + 2NO\uparrow$$
$$2NO\uparrow + O_2 \rightarrow 2NO_2\uparrow$$

2. Ferrous Sulphate, $FeSO_4$. Place a drop of the test solution on a spot plate or on a watch glass, introduce a small crystal of $FeSO_4$ (the size of a pinhead) into it, and add a drop of concentrated H_2SO_4. If NO_3^- is present, a brown ring appears around the crystal due to the formation of a complex compound, $[Fe(NO)SO_4]$:

$$6Fe^{2+} + 2NO_3^- + 8H^+ \rightarrow 6Fe^{3+} + 2NO + 4H_2O$$
$$NO + Fe^{2+} + SO_4^{2-} \rightarrow [Fe(NO)SO_4]$$

Iodides and bromides interfere with this reaction because in their presence I_2 and Br_2 are liberated, which form similar rings. The ions CrO_4^{2-}, MnO_4^-, $[Fe(CN)_6]^{3-}$, $[Fe(CN)_6]^{4-}$, SCN^-, $S_2O_3^{2-}$, and SO_3^{2-} must be absent.

The NO_2^- ion gives the same reaction, but this reaction takes place in dilute H_2SO_4 or even in CH_3COOH.

3. Reduction to Ammonia. To a test tube containing 3-4 drops of the nitrate test solution add a few drops of 2*N* NaOH and introduce one or two pieces of metallic aluminium.* Close the tube (not too tightly) with a cotton plug about 1 cm thick to retain the spray of alkali entrained by the gases evolved. Put a piece of moist litmus paper or phenolphthalein paper over the cotton plug and leave it there for

* An alternative metal is zinc but in this case the reaction is slower.

several minutes. To accelerate the reaction, heat gently the contents of the tube but, as soon as hydrogen begins to evolve violently, heating should be stopped immediately.

Under these conditions, the nitrate ion is reduced to NH_3 which turns the litmus paper blue or imparts a red colour to the phenolphthalein paper:

$$3NO_3^- + 8Al\downarrow + 5OH^- + 2H_2O \rightarrow 3NH_3\uparrow + 8AlO_2^-$$

The detection of NO_3^- by this reaction is interfered with by NH_4^+ which must therefore be removed from the solution by boiling with alkali before the test is started. The reaction is also interfered with by all other nitrogen-containing anions, i.e., NO_2^-, SCN^-, $[Fe(CN)_6]^{4-}$, and $[Fe(CN)_6]^{3-}$, which are also reduced by aluminium to NH_3.

4. Reduction to NO_2^-. The action of zinc reduces nitrates to nitrites in the presence of CH_3COOH:

$$NO_3^- + Zn\downarrow + 2CH_3COOH \rightarrow NO_2^- + Zn^{2+} + 2CH_3COO^- + H_2O$$

The formation of NO_2^- can be recognized by its characteristic reactions.

5. Diphenylamine, $(C_6H_5)_2NH$. Place 4-5 drops of a solution of diphenylamine in concentrated H_2SO_4 on a watch glass (washed thoroughly and wiped dry).* Then introduce a very small amount of the nitrate test solution placed on the end of a glass rod, and mix. In the presence of NO_3^-, there appears an intense-blue colour due to the oxidation of diphenylamine by the nitric acid formed.

The same reaction is also shown by NO_2^-, CrO_4^{2-}, MnO_4^-, Fe^{3+}, $[Fe(CN)_6]^{3-}$, and other oxidizing agents having rather high standard potentials.

If iodide ions are present, sulphuric acid oxidizes them to free iodine and this may occasionally mask the colour developed by NO_3^-.

6. Manganous Chloride, $MnCl_2$. When the nitrate test solution is heated with a double volume of a saturated solution of $MnCl_2$, in concentrated HCl, the solution takes on a dark-brown colour as a result of the formation of complex $[MnCl_6]^{2-}$ ions:

$$3MnCl_2 + 12HCl + 2HNO_3 \rightarrow 3H_2[MnCl_6] + 4H_2O + 2NO\uparrow$$

A similar reaction is given by certain other oxidizing agents, for example, MnO_4^- and CrO_4^{2-}; the NO_2^- ion produces a yellow colour when gaseous NO_2 is dissolved in a liquid.

7. Microcrystalloscopic Test. Treat a drop of the nitrate test solution with a drop of a 10-percent solution of the organic base nitron ($C_{20}H_{16}N_4$) in 5-percent acetic acid solution. Characteristic needle bundles of nitron nitrate ($C_{20}H_{16}N_4 \cdot HNO_3$) are formed (Fig. 8.5). The dilution limit is 1 in 120,000. The following anions (of those

* If the solution turns blue, it means that the glass used was not clean enough.

considered in this book) also form crystalline precipitates with nitron: NO_2^- (see Fig. 8.6), SCN^-, and I^-, but the shape of the crystals formed is different from that of the crystals produced by NO_3^-.

8. Brucine Solution. A solution of brucine in concentrated sulphuric acid (0.2 g of brucine in 100 ml of concentrated H_2SO_4) gives with nitrates a bright-pink colour which changes to golden after some time.

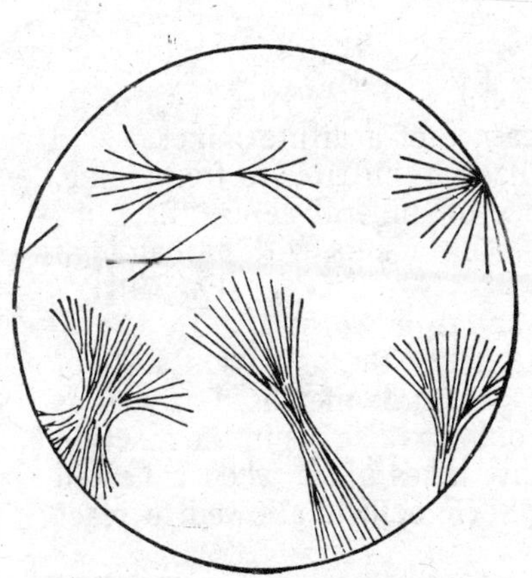

Fig. 8.5 Crystals of nitron nitrate, $C_{20}H_{16}N_4 \cdot HNO_3$.

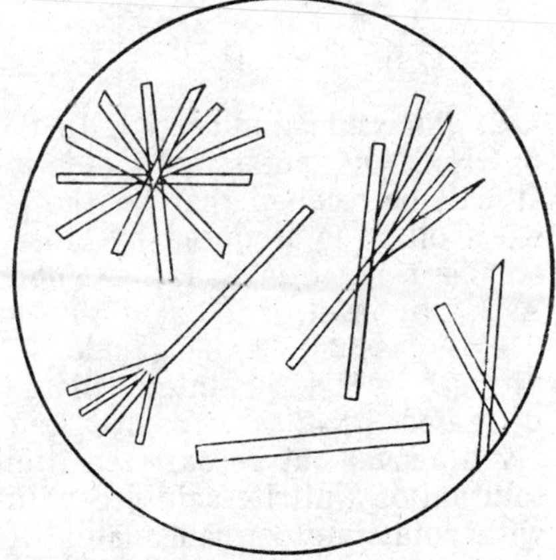

Fig. 8.6. Crystals of nitron nitrite, $C_{20}H_{16}N_4 \cdot HNO_2$.

To the test solution add a tripled volume of concentrated sulphuric acid and brucine solution, and mix. The appearance of a bright-pink colour indicates NO_3^-. The NO_2^- ion does not give this reaction.

8.23. Reactions of the Nitrite Ion

The nitrite ion, NO_2^-, is the anion of nitrous acid, HNO_2, which exists only in cold dilute aqueous solutions: it decomposes very readily into nitrous anhydride (nitrogen trioxide) and water:

$$2HNO_2 \rightarrow N_2O_3 + H_2O$$

Nitrogen trioxide is stable only at low temperatures. At room temperature it decomposes instantly:

$$N_2O_3 \rightarrow NO\uparrow + NO_2\uparrow$$

Nitrites are much more stable than HNO_2. They all are well soluble in water. The least soluble is silver nitrite, $AgNO_2$, but it dissolves readily on heating. Certain complex salts containing NO_2^- are slightly soluble. Examples are $K_3[Co(NO_2)_6]$ and $(NH_4)_3[Co(NO_2)_6]$.

In contrast to nitric acid, nitrous acid is a relatively weak acid ($K = 5.1 \times 10^{-4}$).

All the reactions described for NO_3^- (except Reactions *4* and *8*) are also given by NO_2^-. The NO_2^- ion can be distinguished from NO_3^- by means of the following reactions.

1. **Acids.** All the nitrites are decomposed by acids with the formation of brown nitrogen dioxide gas:

$$2NO_2^- + 2H^+ \rightarrow 2HNO_2$$
$$2HNO_2 \rightarrow NO_2\uparrow + NO\uparrow + H_2O$$

2. **Potassium Iodide, KI.** In the presence of a dilute mineral acid or CH_3COOH, potassium iodide is oxidized by nitrites to free iodine. It will be recalled that, in the presence of mineral acids, the same effect on I^- is produced by other oxidizing agents with rather high standard potentials, for example, by MnO_4^-, CrO_4^{2-}, $[Fe(CN)_6]^{3-}$, AsO_4^{3-} (in a strongly acid solution), and some others.

3. **Potassium Permanganate, $KMnO_4$.** Since the oxidation state of nitrogen in HNO_2 is intermediate (+3), it can be increased as well as decreased. In other words, NO_2^- can not only oxidize (as in the reaction given above) but be oxidized itself. This takes place when a heated solution of a nitrite acidified with sulphuric acid is allowed to react with potassium permanganate:

$$5NO_2^- + 2MnO_4^- + 6H^+ \rightarrow 5NO_3^- + 2Mn^{2+} + 3H_2O$$

The reaction is accompanied by the decolourization of the permanganate solution. We know that $KMnO_4$ is decolourized also by many other reducing agents which have rather low standard oxidation potentials (e.g., S^{2-}, SO_3^{2-}, $S_2O_3^{2-}$, $C_2O_4^{2-}$, SCN^-, AsO_3^{3-}, $[Fe(CN)_6]^{4-}$, I^-, Br^-). In alkaline solution, permanganate is not reduced by NO_2^- ions.

4. **Sulphanilic Acid and α-Naphthylamine.** This is a very sensitive test for NO_2^-. It is based on the formation of a brightly coloured azo dye. The test is carried out as follows. To a drop of a neutral or acetic acid test solution on a spot plate (or on a watch glass) add a drop each of solutions of sulphanilic acid and α-naphthylamine. If NO_2^- is present, a characteristic red colouration appears at once or after some time, depending on the amount of NO_2^- present.*

If no NO_2^- ion is present, this reaction can be used successfully for detection of NO_3^- after it is reduced by zinc dust in the presence of CH_3COOH.

5. **Removal of NO_2^-.** This is achieved by heating cautiously the solution with solid NH_4Cl or $(NH_4)_2SO_4$:

$$NH_4^+ + NO_2^- \rightarrow N_2\uparrow + 2H_2O$$

---

* The formula of sulphanilic acid is $H_2N-C_6H_4-SO_3H$ and that of α-naphthylamine is $C_{10}H_7-NH_2$; the formula of the reaction product is $H_2N-C_{10}H_6-N{=}N-C_6H_4-SO_3H$.

Urea, $CO(NH_2)_2$, reacts similarly in an acid solution:

$$2NO_2^- + 2H^+ + CO(NH_2)_2 \rightarrow 3H_2O + CO_2\uparrow + 2N_2\uparrow$$

Dissolve about 0.1 g of urea in 4 drops of a solution of a nitrite. Add, drop by drop, the resulting solution to 2-4 drops of 2*N* H_2SO_4. Every further drop is added only when the violent reaction with the previous drop is complete. With the entire solution being added, allow the liquid to stand for a while, and test a separate portion of it for NO_2^- by means of the reaction with KI in the presence of starch. It should be noted that when NO_2^- is separated, especially by the action of ammonium salts, a small amount of NO_2^- ions is usually found to have been oxidized to NO_3^-.

If a nitrite ion is present, the nitrate ion can be detected only after the removal of NO_2^- from solution by the methods described above.

It should be kept in mind that nitrites are readily oxidized and, if the solution contains considerable amounts of nitrites, it will almost invariably contain NO_3^- ions as well. Therefore, there is hardly a specific test for NO_3^-.

Only large quantities of NO_3^- can be detected with a sufficient degree of reliability in the presence of small amounts of NO_2^-. This is done with the aid of a relatively insensitive test for NO_3^- (for example, by the reaction with Cu and H_2SO_4) after the removal of NO_2^- by the action of $CO(NH_2)_2$ or of an ammonium salt. The presence of NO_3^- in the test solution may be inferred with certainty only if the test is very distinct.

8.24. Reactions of the Acetate Ion

The acetate ion, CH_3COO^-, is the anion of acetic acid, CH_3COOH, which is a relatively weak monobasic acid ($K = 1.74 \times 10^{-5}$). Nearly all the salts of acetic acid (acetates) dissolve well in water. The least soluble is silver acetate. Certain basic salts, e.g., $[Fe_3(CH_3COO)_6O]OH$, are slightly soluble.

1. Sulphuric Acid, H_2SO_4. When sulphuric acid is allowed to react with acetates, free CH_3COOH is expelled from them. While volatilizing on heating, acetic acid can be identified by its characteristic odour. The same reaction takes place when a solid acetate is ground in a small mortar together with solid $NaHSO_4$ or $KHSO_4$.

2. Formation of Ethyl and Amyl Acetates. To a few drops of a solution of an acetate, or preferably to several small crystals of the solid salt, add 3-4 drops each of concentrated H_2SO_4 and ethyl alcohol, C_2H_5OH. Heat the mixture for a minute or two in the water bath and then pour the contents of the test tube into a beaker with cold water. This results in the formation of ethyl acetate, $CH_3COOC_2H_5$, which has a characteristic pleasant odour (vinegar-like odour):

$$2CH_3COONa + H_2SO_4 \rightarrow Na_2SO_4 + 2CH_3COOH$$
$$CH_3COOH + C_2H_5OH \rightarrow CH_3COOC_2H_5 + H_2O$$

If amyl alcohol, $C_5H_{11}OH$, is used in place of ethyl alcohol, amyl acetate, $CH_3COOC_5H_{11}$, is formed, which is called also banana oil or pear oil; it also has a pleasant odour.

3. Ferric Chloride, $FeCl_3$. To a solution of CH_3COONa add a little $FeCl_3$. A red-brown colour appears due to the formation of a complex. When the solution is diluted with water and heated, a precipitate of the basic salt $[Fe_3(CH_3COO)_6O]OH$ separates out.

The solution must be neutral and must contain no anions capable of precipitating iron(III): CO_3^{2-}, SO_3^{2-}, PO_4^{3-}, $[Fe(CN)_6]^{4-}$, etc. No SCN^- and I^- ions must be present either; iodide is oxidized to free iodine by Fe^{3+}, as a result of which the solution turns red-brown. All the ions indicated can be removed by adding $AgNO_3$ or Ag_2SO_4 to the neutral solution. But this involves the use of considerable amounts of silver salts, and therefore the test should not be applied unless absolutely necessary.

4. Lanthanum Nitrate, $La(NO_3)_3$. Place on a spot plate a drop each of the test solution, a 5-percent solution of $La(NO_3)_3$, and 0.01*N* iodine solution. After a minute add a drop of 1*N* NH_4OH and let stand for several minutes.* If CH_3COO^- is present, a blue or bluish-brown ring forms around the ammonia spot. The test is interfered with by SO_4^{2-} and PO_4^{3-} which give slightly soluble salts with lanthanum. These ions must therefore be precipitated first by the action of $Ba(NO_3)_2$ or $BaCl_2$.

8.25. Reactions of the Chlorate Ion

The chlorate ion, ClO_3^-, is the anion of an extremely weak unstable acid—chloric acid, $HClO_3$, which decomposes rapidly into perchloric acid, chlorine, water, and oxygen:

$$3HClO_3 \rightarrow H_2O + 2ClO_2\uparrow + HClO_4$$

$$2ClO_2\uparrow \rightarrow Cl_2\uparrow + 2O_2\uparrow$$

Therefore, chlorates are extremely strong oxidants in acid solutions.

The salts of chloric acid are well soluble in water and very stable in the solid state and in aqueous solutions.

The principal tests for the chlorate ion are based on its oxidizing properties. A chloride ion is formed in solution as a result of an oxidation-reduction reaction and is detected by means of its characteristic reactions.

1. Dilute Sulphuric Acid. Chlorates are decomposed by dilute sulphuric acid with the formation of chloric acid which decomposes to give chlorine and oxygen. If a few drops of KI solution are added to an acidified solution of a chlorate, iodine is formed:

$$ClO_3^- + 6I^- + 6H^+ \rightarrow 3I_2 + Cl^- + 3H_2O$$

This reaction does not take place in an acetic acid solution.

---

* After NH_4OH has been added, the solution should not be stirred.

2. Concentrated Sulphuric Acid. Concentrated sulphuric acid decomposes chlorates with the liberation of yellowish-green chlorine dioxide gas. If the reaction is carried out at a high temperature, the chlorine dioxide *explodes violently* and is decomposed into its constituent elements:

$$3ClO_3^- + 3H_2SO_4 \rightarrow 3HSO_4^- + HClO_4 + 2ClO_2\uparrow + H_2O$$

$$2ClO_2\uparrow \rightarrow Cl_2\uparrow + 2O_2\uparrow$$

3. Reducing Agents. Sulphurous acid, ferrous salts, and aluminium and zinc metals (in alkaline solution) reduce chlorates to chlorides:

$$ClO_3^- + 3H_2SO_3 \rightarrow Cl^- + 3SO_4^{2-} + 6H^+$$

$$ClO_3^- + 6Fe^{2+} + 6H^+ \rightarrow Cl^- + 6Fe^{3+} + 3H_2O$$

$$ClO_3^- + 2Al\downarrow + 2OH^- \rightarrow Cl^- + 2AlO_2^- + H_2O$$

The chloride ion formed in the reaction is detected in the usual manner (for example, by the formation of a precipitate of AgCl soluble in NH_4OH).

The salts of sulphurous acids should be used in the solid state. A blank experiment must be carried out. Place a solid salt of sulphurous acid (e.g., Na_2SO_3) in a test tube, and add a few drops of distilled water, 2-3 drops of 2*N* H_2SO_4, and 2-3 drops of $AgNO_3$ solution. Then carry out the reaction with the test solution in another tube. If a much larger amount of AgCl is precipitated in the test solution than in the blank experiment (the turbidity is appreciably stronger), the presence of chlorate ions may be deduced.

4. Manganous Sulphate, $MnSO_4$. In the presence of orthophosphoric acid, manganous sulphate gives with chlorates, on heating, a complex compound of Mn^{III} which has a red-violet colour:

$$ClO_3^- + 6Mn^{2+} + 12PO_4^{3-} + 6H^+ \rightarrow Cl^- + 6[Mn(PO_4)_2]^{3-} + 3H_2O$$

To a small amount of the test solution add a saturated solution of manganous sulphate and a few drops of phosphoric acid, and heat. If chlorates are present, a violet colouration appears.

5. Diphenylamine, $(C_6H_5)_2NH$. The reagent is oxidized by a chlorate ion in acid solution (just as by all other strong oxidizing agents) with the formation of a blue oxidation product.

8.26. Analysis of Mixtures of Group I, II, and III Anions

In analytical practice, the systematic analysis of anions has to be resorted to in relatively rare cases. This is accounted for by the fact that natural and commercial materials contain a limited number of anions.

Anions are most often identified by the fractional technique, i.e., in separate portions of the test solution in an arbitrary sequence. Only in comparatively seldom cases, has one to resort to separation tests.

Though no strict sequence of reactions must necessarily be followed in the anion analysis, it nevertheless is recommended that a definite scheme of work be adhered to. It is most expedient to start the analysis with preliminary tests in order to establish the absence of certain anions and thus to simplify the subsequent analysis.

It must be borne in mind that the only cations that may be present in the solution to be analysed for anions are K^+, Na^+, or NH_4^+. All the other cations, which form slightly soluble salts with many anions, interfere with the detection of anions and must therefore be preliminarily removed from the test solution by boiling with sodium carbonate.

The procedures used for removal of interfering cations are discussed in more detail later in the book, where the analysis of a solid substance is considered. They consist briefly in the following. When a solid is boiled with a saturated solution of sodium carbonate, all the cations of heavy metals are converted into carbonates or hydrous oxides and become insoluble in water. Only the water-soluble carbonates of K^+, Na^+, and NH_4^+ remain in the solution together with some anions (AsO_3^{3-}, AsO_4^{3-}, CrO_4^{2-}, $Cr_2O_7^{2-}$, MnO_4^-, etc.) which contain elements capable of forming anions.

To simplify the course of analysis, we shall not consider the relatively rare and practically unimportant cases of the presence in the solution of incompatible oxidizing and reducing agents. Thus, the hypochlorite ion, ClO_4^-, is such a strong oxidizing agent that it can oxidize all reducing anions even in alkaline solution.

The iodate ion, IO^-, oxidizes S^{2-} to S both in acid and alkaline solutions.

The simplest cases are discussed below. In more involved cases, the student should consult special books.

Here we suggest procedures for anions, which require the following operations:

(a) preliminary tests;

(b) the detection of those ions which can be identified by the fractional technique in separate portions of the test solution;

(c) the systematic separation of anions through the use of strontium and barium salts (in a neutral or a feebly-alkaline solution) and silver salts (in a nitric acid solution).

In systematic separations, no identification tests are made for anions that can easily be detected by the fractional method; such anions introduce complications into the course of the analysis and must therefore be removed from the solution prior to group precipitation and detection of other anions. Such anions include: Group I—SiO_3^{2-}, F^-, and $C_2O_4^{2-}$; Group II—S^{2-}, CN^-, $[Fe(CN)_6]^{3-}$, $[Fe(CN)_6]^{4-}$, BrO_3^-, and ClO^-. Also, oxidizing anions that are readily reduced and converted into the corresponding cations (e.g., CrO_4^{2-}, $Cr_2O_7^{2-}$, and MnO_4^-) are not included in the systematic course of analysis.

(1) FRACTIONAL DETECTION OF ANIONS NOT DETERMINED IN THE SYSTEMATIC ANALYSIS

(a) Group I Anions

1. **Detection of SiO_3^{2-}.** To 4-5 drops of the test solution add 2 drops of 2*N* ammonia solution, 3 drops of a saturated solution of NH_4Cl, and head in a water bath for several minutes. The appearance of a white gelatinous precipitate of silicic acid indicates the presence of SiO_3^{2-}. The reaction is unambiguous provided that the solution contains no cations of the elements having amphoteric properties (Al, Sn, Sb). In the presence of cations of these elements, it is necessary to perform confirmatory tests with methylene blue or a microcrystalloscopic reaction of the formation of $Na_2[SiF_6]$ crystals (see Sec. 8.9).

2. **Detection of CrO_4^{2-} (or $Cr_2O_7^{2-}$).** This test is made only if the test solution is coloured yellow (or orange).

To 2-3 drops of the test solution add a drop of phenolphthalein solution. If a red colour appears, add 2*N* H_2SO_4, drop by drop, until the colour is removed. Then add a further drop of the acid to the same test tube, followed by a few drops of amyl alcohol, a drop of hydrogen peroxide, and mix immediately. The formation of a deep-blue ring due to perchromic acid (H_2CrO_6) indicates the presence of CrO_4^{2-} (or $Cr_2O_7^{2-}$).

3. **Detection of $C_2O_4^{2-}$ and F^-.** (1) Acidify a separate portion (3-4 drops) of the test solution with CH_3COOH (to pH $\approx$ 4-5) and treat with an excess of $CaCl_2$ solution on heating. The precipitation of CaC_2O_4 and CaF_2 points to the possible presence of $C_2O_4^{2-}$ and F^- ions. Centrifuge off the precipitate, wash thoroughly (to remove the adsorbed reducing agents), and treat on heating with 2*N* H_2SO_4. If no complete dissolution takes place, separate the undissolved precipitate and add a dilute (0.01*N*) solution of $KMnO_4$, a drop at a time, to the centrifugate heated to 60-70°C*. If the permanganate is decolourized, this confirms the presence of $C_2O_4^{2-}$.

(2) If the previous test indicates the possibility of the anion F^- being present, it can be detected as follows. Place a small amount of the test solid substance** in a dry vessel (see Fig. 2.21) and treat it with a few drops of concentrated H_2SO_4. Then introduce into it a drop of water in the loop of a platinum wire and let stand. If a fluoride

---

* In order to speed up the reaction, it is useful to add a drop of a solution of $MnSO_4$.

** If the solution is to be analysed, it is necessary to evaporate a portion of it to dryness in a crucible or in an evaporating dish.

is present, gaseous SiF_4 is formed, which when absorbed by the water makes it turbid due to the formation of a precipitate of silicic acid.

If the test solution contains no phosphates and oxalates, the F^- ion can be detected by the spot-test technique using alizarin and zirconyl nitrate (see Sec. 8.8).

(b) Group II Anions

1. **Detection of S^{2-}.** Place 5-6 drops of the test solution in a test tube and add HCl solution (1:1) until decidedly acid (pH $\leqslant$ 2). Bring up a strip of filter paper moistened with a solution of lead nitrate or acetate or of sodium plumbite to the mouth of the test tube. A blackening of the paper indicates a sulphide ion.

The reaction may also be carried out with sodium nitroprusside (see Sec. 8.14, Reaction *3*).

2. **Detection of $[Fe(CN)_6]^{4-}$.** To 2-3 drops of the test solution add HCl solution until just acid, followed by a drop or two of $FeCl_3$ solution. The formation of a deep-blue precipitate of Prussian blue indicates the presence of $[Fe(CN)_6]^{4-}$.

3. **Detection of $[Fe(CN)_6]^{3-}$.** To 2-3 drops of the acidified test solution add a drop or two of $FeSO_4$ solution. A dark-blue precipitate of Turnbull's blue indicates the presence of $[Fe(CN)_6]^{3-}$.

4. **Detection of CN^-.** To detect CN^- (cyanide ion), use is made of the fact that hydrocyanic acid is extremely weak and is expelled from its salts even by carbonic acid. To 3-5 drops of the test solution placed in a test tube add a little dry sodium dicarbonate, $NaHCO_3$, and 30-percent CH_3COOH until just acid.

Close the test tube quickly with a stopper into which a bent delivery tubing is inserted. Dip the other end of the tubing into another test tube with a solution of $AgNO_3$ acidified with nitric acid. A white precipitate of AgCN indicates the presence of CN^-.

5. **Detection of BrO_3^-.** To 3-4 drops of the test solution add a solution of silver nitrate acidified with nitric acid until precipitation is complete. Centrifuge off the precipitate, wash two or three times with distilled water, and treat with a few drops of concentrated HCl. Add 2-3 drops of chloroform or carbon tetrachloride to the test tube with the precipitate, shake, and allow to stand. If bromate ions are present, the organic layer assumes a yellow colour.

6. **Detection of ClO^-.** As has already been pointed out, ClO^- liberates free iodine from solutions of KI even in alkaline solution (all other oxidizing agents produce the same effect only in acid solution). To 2-3 drops of the alkaline test solution add an excess of KI solution and a few drops of starch suspension. If the solution becomes deep-blue in colour, this is an indication of the presence of ClO^-. The presence of hypochlorite ions means that no reducing agents are contained in the solution.

(c) *Group III Anions*

1. Detection of MnO_4^-. An indication of the presence of MnO_4^- is the crimson-violet, or pink (at low concentrations of MnO_4^-), colour of the solution. To check, add the test solution, drop by drop, to a solution of $(NH_4)_2C_2O_4$ acidified with H_2SO_4 and heated up to 70°C. The decolourization of the solution proves the presence of MnO_4^-.

The detection of MnO_4^- is interfered with by the presence of CrO_4^{2-}; it must therefore be precipitated first as $BaCrO_4$ by the action of $Ba(OH)_2$.

2. Removal of SiO_3^{2-}, S^{2-}, $[Fe(CN)_6]^{4-}$, $[Fe(CN)_6]^{3-}$, and CN^- by Precipitation as Zinc Salts. First, a solution of hexaamminezinc hydrate must be prepared. To do this, to a solution of a zinc salt cautiously add, drop by drop, a 10-percent solution of NH_4OH until a precipitate of $Zn(OH)_2$ is formed; avoid using an excess of NH_4OH. Wash thoroughly the precipitate of zinc hydroxide with water and dissolve in a 10-percent solution of NH_4OH, avoiding an excess of ammonium hydroxide. It is this solution containing $[Zn(NH_3)_6]^{2+}$ and an equivalent amount of OH^- ions that serves as a precipitant.

To 1-2 ml of the test solution in a neutral or faintly alkaline medium add, drop by drop, a freshly prepared solution of hexaamminezinc hydrate until precipitation is complete. Apart from the anions indicated, the precipitate may also contain PO_4^{3-}, CO_3^{2-}, and some others. But these anions are preliminarily detected in the general solution by the fractional method.

The precipitate of zinc salts obtained can be tested for the presence of CN^- and also of other anions by means of their characteristic reactions.

After the precipitated zinc salts are separated from the centrifugate, the latter is subjected to systematic examination for anions according to the procedures described below.

(2) SYSTEMATIC ANALYSIS OF MIXTURES OF GROUP I-III ANIONS

The solution may contain only the salts of the alkali metals and those anions which are included in the systematic analysis:

Group I: SO_4^{2-}, SO_3^{2-}, $S_2O_3^{2-}$, CO_3^{2-}, BO_2^-, PO_4^{3-}, AsO_3^{3-}, AsO_4^{3-}
Group II: Cl^-, Br^-, I^-, IO_3^-, SCN^-
Group III: NO_3^-, NO_2^-, ClO_3^-, CH_3COO^-

(a) *Preliminary Tests*

1. Determination of Solution pH. (1) If the reaction of the test solution is acid ($pH \leqslant 2$), it means that the following anions of volatile and unstable acids are absent: CO_3^{2-}, SO_3^{2-}, $S_2O_3^{2-}$, and NO_2^-.

Moreover, the following pairs of anions cannot coexist in acid solution: I^- and NO_2^-; I^- and IO_3^-; I^- and ClO_3^-; I^- and AsO_4^{3-}; NO_2^- and ClO_3^-; NO_2^- and AsO_3^{3-}; SO_3^{2-} and AsO_4^{3-}; $S_2O_3^{2-}$ and AsO_4^{3-}.

(2) If the reaction of the test solution is alkaline, all the anions being considered may be present in it.

2. Test for Simultaneous Presence of Certain Oxidizing and Reducing Agents. (1) When the test solution is acidified with 2*N* CH_3COOH in the presence of NO_2^- and I^-, or IO_3^- and I^-, free iodine is liberated.

(2) If the previous test gives a negative result, the solution is made acid with 2*N* HCl. The formation of free iodine points to the possible presence of the following pairs: I^- and ClO_3^-; I^- and AsO_4^{3-}.

If the solution becomes turbid, this is an indication of the presence of $S_2O_3^{2-}$.

3. Test for Evolution of Gases. Treat a few drops of the test solution (preferably, a small grain of a solid substance) with 2*N* H_2SO_4 and shake gently the contents of the tube by tapping cautiously the bottom part with your finger. If no gas bubbles are seen, heat the test tube. The formation of gases (CO_2, SO_2, NO_2) is evidence for the possible presence of CO_3^{2-}, SO_3^{2-}, $S_2O_3^{2-}$, and NO_2^-.

In this test, the composition of the gases evolved should be established as exactly as possible. Let us recall some of their properties:

(a) CO_2 is a colourless gas without odour, making lime water turbid (lime water can also be made turbid by SO_2);

(b) SO_2 is a gas with the odour of burning sulphur; it renders dilute solutions of $KMnO_4$ and I_2 colourless;

(c) NO_2 is a red-brown gas with a characteristic pungent odour; when it is passed into a solution of KI, free iodine is liberated (starch test).

4. Test for Group I Anions. To 2 drops of a neutral or faintly alkaline solution (pH ≈ 7-9)* add 2-3 drops of $BaCl_2$ solution. The formation of a precipitate indicates the presence of Group I**.

The precipitate may contain $BaSO_4$, $BaSO_3$, $Ba_3(AsO_3)_2$, $Ba_3(AsO_4)_2$, $Ba_3(PO_4)_2$, BaS_2O_3, $Ba(IO_3)_2$, $Ba(BO_2)_2$, $BaCO_3$.

5. Test for Group II Anions. To 2-3 drops of the test solution add 2-3 drops of $AgNO_3$ solution. If a precipitate is formed, add a few drops of HNO_3. The insolubility of the precipitate is evidence that the following salts may be present: AgCl, AgBr, AgI, $AgIO_3$, AgSCN, Ag_2S (the last-named one being formed by the reaction between $S_2O_3^{2-}$ and Ag^+).

---

* Acid solutions must be preliminarily neutralized with a solution of $Ba(OH)_2$.

** If no precipitate is formed, the anions $S_2O_3^{2-}$ and IO_3^- may nevertheless be present as they can be precipitated by Ba^{2+} ions only in rather concentrated solutions.

6. Tests for Presence of Reducing Anions. (1) To 3 drops of the test solution add a drop of 2*N* alkali solution and a drop or two of a dilute solution of $KMnO_4$. If no precipitate is formed, heat the solution in a water bath for a minute. The formation of a brown precipitate of $MnO_2 \cdot xH_2O$ indicates the presence of reducing anions (SO_3^{2-}, $S_2O_3^{2-}$, I^-, AsO_3^{3-}, SCN^-, and others).

(2) If the previous test has shown the presence of reducing agents, acidify the other portion of the test solution with H_2SO_4 and add, drop by drop, a dilute iodine solution coloured dark-blue by starch. The decolourization of the solution indicates the presence of $S_2O_3^{2-}$ or SO_3^{2-}.

7. Test for Presence of Oxidizing Anions. To 2-3 drops of the test solution acidified with 2*N* H_2SO_4 add a few drops of benzene or chloroform and a drop or two of KI solution, shake and allow to stand. The appearance of a violet colour in the organic layer indicates the presence of oxidizing anions (NO_2^-, IO_3^-, AsO_4^{3-}, ClO_3^-, etc.).

Note. 1. If the solution contains both oxidizing and reducing agents, no final conclusion as to the absence of the anions listed in Procedures 6 and 7 can be made from the results of the preliminary tests.

The total effect will depend on which anions are present in a larger amount—oxidizing or reducing anions.

2. The AsO_4^{3-} ion liberates iodine from potassium iodide only at a sufficiently high concentration of the hydrogen ion. If the test according to Procedure 7 is negative, additional tests for AsO_4^{3-} should be made in the course of analysis.

(b) Detection of Certain Anions by the Fractional Method in Separate Portions of the Solution

1. Detection of $S_2O_3^{2-}$, SO_3^{2-}, and SO_4^{2-}. If the test for reducing agents by the use of iodine solution, according to Procedure *6(2)*, gives a positive result, the anions $S_2O_3^{2-}$, SO_3^{2-}, and SO_4^{2-} are to be detected in a separate portion of the solution as described in Sec. 8.14. But if the result of the test with I_2 solution is negative, the ion SO_4^{2-} is to be detected by the action of Ba^{2+} on a separate portion of the solution. A white precipitate of $BaSO_4$ insoluble in dilute* HCl indicates the presence of SO_4^{2-}.

2. Detection of CO_3^{2-}. Place a small amount of the test solid substance (or 5-8 drops of the test solution) in a vessel (see Fig. 8.1).

If the previous tests have revealed the presence of SO_3^{2-} or $S_2O_3^{2-}$ in the substance, they are oxidized by adding 5-6 drops of 3-percent H_2O_2. Then treat with 6-8 drops of 2*N* HCl proceed as directed in Sec. 8.5**.

---

* Concentrated HCl must not be used as it can induce the precipitation of $BaCl_2$ (the effect of common Cl^- ions—see Sec. 4.2).

** It is better to detect CO_3^{2-} in the washed precipitate of strontium salts obtained in the systematic analysis.

3. Detection of PO_4^{3-}, AsO_4^{3-} and AsO_3^{3-}. (1) If no AsO_4^{3-} and AsO_3^{3-} may be present in the test solution, PO_4^{3-} is to be detected by the reaction with ammonium molybdate solution. Heat 2 drops of the test solution for several minutes with 4 drops of 6*N* HNO_3 to oxidize reducing agents, then add excess molybdate solution, and heat up to 40-50°C. If no precipitate is formed, add several small crystals of NH_4NO_3 and let stand for a while. The formation of a yellow precipitate of $(NH_4)_3[PMo_{12}O_{40}]$ indicates the presence of PO_4^{3-}.

(2) If the presence of AsO_4^{3-} and AsO_3^{3-} is possible, examine as described above (see Sec. 8.6).

4. Detection of BO_2^-. Evaporate 5-8 drops of the test solution to dryness in a crucible or in an evaporating dish and treat the residue with a few drops of concentrated H_2SO_4. Add a little ethyl (or methyl) alcohol and set fire to the solution. If the flame of the burning alcohol is coloured green (along the edges), this is evidence for the presence of BO_2^-. The BO_2^- ion can also be identified with the aid of the reactions with CaF_2 and sulphuric acid or turmeric paper test (see Sec. 8.7).

5. Detection of NO_2^-. The presence of NO_2^- is recognized by the evolution of brown NO_2 gas when the solution is acidified with H_2SO_4. A confirmatory test is made as follows. To a drop of the neutral or acetic acid (pH 4-7) test solution on a spot plate add a drop each of solutions of sulphanilic acid and α-naphthylamine (see Sec. 8.23, Reaction *4*). If a red colour appears at once or after some time, this is proof of the presence of NO_2^-.

Alternatively, use may be made of the ability of NO_2^- to displace iodine from KI solution in an acetic acid medium (see under Preliminary Tests).

6. Detection of NO_3^-. (1) *No NO_2^- Ion Is Present.* In this case, NO_3^- can be identified, after its conversion into NO_2^- (by heating with metallic zinc in acetic acid—see Sec. 8.22, Reaction *4*), by means of the colour reaction with sulphanilic acid and α-naphthylamine described above. If NO_2^- has been detected by this test, it will be evidence for the presence of NO_3^- in this particular case.

If no SCN^-, $[Fe(CN)_6]^{4-}$, and $[Fe(CN)_6]^{3-}$ are present, the nitrate ion can be detected by its reduction to ammonia (see Sec. 8.22, Reaction *3*). First, the NH_4^+ ion is tested for and, if the result is positive, the ion is removed by evaporating a few drops of the test solution to incipient dryness in a crucible with excess NaOH (*Caution: avoid spattering*). After the contents of the crucible is cooled, dilute with 5-6 drops of water, transfer to a test tube, add 8-10 drops of 2*N* NaOH, pour a little powdered aluminium, and heat slightly. Close the mouth of the tube (to retain the spray) with a cotton plug, put a piece of moist phenolphthalein paper on it, and let stand for a while. If the reaction proceeds too violently, cool the contents of the tube by dipping it into cold water. The reddening of the paper (formation of NH_3) indicates the presence of NO_3^-.

If no other oxidizing anions (MnO_4^-, CrO_4^{2-}, $[Fe(CN)_6]^{3-}$, and NO_2^-) are present, the ion NO_3^- can more conveniently be identified by the diphenylamine test. To carry out this test, place 4-5 drops of the reagent (a solution of diphenylamine in concentrated H_2SO_4) on a thoroughly washed and dry-wiped watch glass, introduce a very small amount of the test solution on the end of a clean glass rod, and mix. The appearance of an intense-blue colour indicates NO_3^-. A large amount of I^- may interfere with the test.

And, finally, the nitrate ion can also be identified by a microcrystalloscopic test with nitron (see Sec. 8.22, Reaction *7*).

(2) *NO_2^- Is Present.* In this case, the presence of considerable quantities of NO_3^- can be detected after the removal of NO_2^- by using the methods described in Sec. 8.23 (Reaction *5*).

7. Detection of CH_3COO^-. (1) To 3-5 drops of the test solution add 3-5 drops of ethyl alcohol, C_2H_5OH, and 3-5 drops of concentrated H_2SO_4. The appearance of a characteristic odour due to ethyl acetate, $CH_3COOC_2H_5$, indicates the presence of CH_3COO^-.

(2) Grind thoroughly a small amount of the test solid substance in a small mortar with solid $NaHSO_4$ or $KHSO_4$. The appearance of an odour due to acetic acid indicates the CH_3COO^- ion*.

(3) If lanthanum nitrate is available in the laboratory, use it for the test in the presence of I_2 and NH_4OH. The ions SO_4^{2-} and PO_4^{3-} must be precipitated first by Ba^{2+} ions (see Sec. 8.24, Reaction *4*).

(c) Systematic Separation of Anions into Groups and Analysis of Precipitates Obtained

1. Action of Strontium Nitrate and Analysis of the Precipitate Obtained. Place a portion of the test solution (15-20 drops) having a neutral or faintly alkaline reaction in a centrifuge tube and heat in a water bath up to 60-70°C. To the hot solution add a saturated solution of $Sr(NO_3)_2$ until complete precipitation is achieved and let it stand for 25-30 min. The precipitate may be found to contain the following salts: $SrSO_4$, $SrSO_3$, $SrCO_3$, $Sr_3(PO_4)_2$, $Sr_3(AsO_3)_2$.

If PO_4^{3-} and AsO_3^{3-} have been detected in the preliminary tests, analyse the precipitated strontium salts for the presence of SO_4^{2-}, SO_3^{2-}, and CO_3^{2-}. Centrifuge off the precipitate and wash thoroughly with 2*N* $Sr(NO_3)_2$. Keep the centrifugate for further analysis.

Wash the precipitate until a negative test for reducing agents is obtained (otherwise they may interfere with the detection of SO_3^{2-}). Discard the washings and proceed to the following tests on the precipitate.

---

* If the action of concentrated H_2SO_4 or grinding with solid hydrosulphate liberates gases or vapours with a pungent odour (Cl_2, I_2, Br_2, HCl, SO_2, etc.), the reaction may be carried out only after the precipitation of the anions of Groups I and II.

1. If the precipitate does not dissolve completely in 2*N* HCl, this is evidence for the presence of SO_4^{2-}.

2. Use the following procedures for detection of SO_3^{2-} ions.

(a) To a part of the precipitate in the test tube add 2*N* HCl, quickly close it with a stopper fitted with a delivery tubing, set the test tube in the water bath, and pass the evolving gas into a weak solution of permanganate (2-3 ml). Then place an equal volume of a permanganate solution as a control in another test tube. After the lapse of 5-8 min, compare the colours of the permanganate solutions in the two test tubes. A complete or partial decolourization of the permanganate indicates the presence of the sulphite ion.

(b) To the other part of the precipitate in the test tube add 2*N* HCl and hold starch-iodide paper over the mouth of the tube. Heat the solution in the test tube cautiously. If SO_2 is liberated, the starch-iodide paper turns blue.

(3) To detect CO_3^{2-}, place a few drops of an oxidizing agent (H_2O_2, $KMnO_4$, or $K_2Cr_2O_7$) in the test tube with a portion of the precipitate, followed by a few drops of 2*N* HCl. Quickly close the tube with a stopper with a bent delivery tubing. Insert the other end of the gas delivery tubing into another test tube containing baryta or lime water. The test tube with the precipitate may be heated gently in a water bath. If the baryta or lime water becomes turbid, this is evidence of the presence of carbonate ion in the precipitate.

2. Action of Barium Nitrate and Analysis of the Precipitate Obtained. If the arsenate ion has been detected in the preliminary tests, the precipitation should be carried out by barium salts. If the solution does not contain AsO_4^{3-}, the anions of Group II may be precipitated by silver salts in the centrifugate (obtained in Procedure *1*), after the separation of the precipitated strontium salts.

Barium salts are to be precipitated as follows. Heat the centrifugate remaining after the separation of strontium salts up to 60-70°C and add a saturated solution of $Ba(NO_3)_2$ until precipitation is complete. Allow the solution with the precipitate to stand for 50-60 min.

The precipitate may contain the following salts: BaS_2O_3, $Ba(IO_3)_2$, $Ba(AsO_4)_2$, $Ba(BO_2)_2$. It should be kept in mind that barium nitrate is capable of precipitating completely the anions specified only from relatively concentrated solutions. Thiosulphate and iodate ions are precipitated, as a rule, only if they are present in the same solution. Therefore, the negative results of the tests for these ions do not necessarily serve as proof of their absence. They may be detected later in the precipitate formed by silver nitrate.

Centrifuge the precipitated barium salts and reserve the centrifugate for subsequent analysis. Wash the precipitate with cold distilled water and examine. Since AsO_4^{3-} and BO_2^- could have been detected in the preliminary tests, look for $S_2O_3^{2-}$ and IO_3^- in the precipitated barium salts.

To the precipitate in the centrifuge tube add 5-6 drops of 2*N* HCl and 2-3 drops of a clear starch suspension. The following effects may be observed.

(a) If the ions $S_2O_3^{2-}$ and IO_3^- are present together, free iodine is liberated and the solution is turned deep-blue.

(b) If only $S_2O_3^{2-}$ is present, the solution remains turbid because of the separation of free sulphur.

(c) If the acidified solution remains clear and no iodine is evolved, add the solid salt Na_2SO_3 to this solution; if iodine is liberated and the solution turns intense-blue, this is evidence for the presence of iodate ion.

3. Action of Silver Nitrate and Analysis of the Precipitate.* To the centrifugate obtained in Procedure *2* or *1* (depending on whether barium salts were used for precipitation) add, drop by drop in the cold, a solution of $AgNO_3$ acidified with 2*N* HNO_3 until precipitation is complete.

The following salts may be found to be present in the precipitate: AgCl, AgBr, AgI, AgSCN, $AgIO_3$, Ag_2S.**

Centrifuge off the precipitate and wash twice with water. Discard the centrifugate. Analyse the precipitate as follows.

(1) If after being separated and washed, the precipitate darkens (Ag_2S), this is evidence for the presence of $S_2O_3^{2-}$.

(2) To detect SCN^-, place a small amount of the precipitate on a porcelain plate (or a cover of a porcelain crucible), add 2-3 drops of H_2SO_4 solution (1:4), and (cautiously, without disturbing the precipitate) a drop or two of $FeCl_3$ solution. A pink colour indicates SCN^-.

(3) To the bulk of the precipitate in the centrifuge tube add 5-6 drops of H_2SO_4 solution (1:4), stir thoroughly the solution with the precipitate using a glass rod, warm in a water bath for 5-10 min, and centrifuge. Decant the supernatant into another test tube, dilute it with an equal volume of water, and add several pieces of granular zinc or a little zinc dust. After a lapse of 5-10 min, decant the solution into a clean test tube and look for the I^- ion, which has resulted from the reduction of IO_3^-, by the action of chlorine water in the presence of an organic solvent.

If the presence of IO_3^- in the barium salt precipitate has been definitely established, it need not be tested for at this point. The only thing to be done is to wash the precipitated silver salts with H_2SO_4 solution (1:4) until a negative test for IO_3^- is obtained.

(4) After the IO_3^- ions have been washed out of the precipitate, the latter may contain AgCl, AgBr, AgI, and AgSCN.

---

* All solutions and precipitates containing silver should be collected in special containers for subsequent recovery of metallic silver.

** The compounds $AgIO_3$ and Ag_2S (resulting from the decomposition of $Ag_2S_2O_3$) will be in the precipitate only if the precipitation by barium nitrate was incomplete or if the precipitation of barium salts was omitted.

4. Analysis of the Precipitate. The following scheme may be used for the purpose. (1) To the precipitate in the centrifuge tube add a 12-percent solution of $(NH_4)_2CO_3$ (prepared without addition of NH_4OH) and shake vigorously. Centrifuge off the precipitate (AgBr, AgI, and AgSCN) and keep for further tests. To the centrifugate add 3-4 drops of KBr solution and mix. The appearance of an abundant turbidity of AgBr indicates the presence of Cl^- in the test solution. If no Cl^- is present, only a slight opalescence may be formed.

It is more reliable to carry out the reaction of formation of chromyl chloride (the solution must be first evaporated to dryness).

(2) Test the precipitate remaining in Procedure *4(1)* for I^- and Br^-. Before starting the identification test, it is necessary to put them into solution by the action of zinc and sulphuric acid.

If the precipitate contains AgSCN, it forms H_2S and volatile hydrocyanic acid, HCN (*Extremely poisonous!*). On account of this, *the test described must be carried out only in a faultlessly operating fume cupboard!* Treat the precipitate with 5-6 drops of H_2SO_4, add a little zinc dust, and heat in a water bath until no gases (H_2, H_2S, HCN) are evolved. Then heat the test tube again for several minutes. Make certain with the aid of a piece of paper moistened with a solution of $Pb(CH_3COO)_2$ or Na_2PbO_2 that no hydrogen sulphide gas is evolved. Now centrifuge off the precipitate (consisting of silver and excess zinc) and discard. Then test the centrifugate (2-3 drops) for I^- and Br^- by means of the reaction with chlorine water or a solution of NaClO in the presence of benzene or chloroform. The I^- and Br^- ions may also be identified by means of the following tests (simultaneously with the above reaction):

(a) the test for iodide ion using KNO_2 in the presence of starch (see Sec. 8.13, Reaction *5*);

(b) the fuchsin test for bromide ion after Br^- has been oxidized to Br_2 (see Sec. 8.12, Reaction *4*).

Note. The tests for iodide and bromide ions should be performed by using very small amounts of the test solution, otherwise too much chlorine water or NaClO solution will be required to displace all the iodine, and the Br^- ion may easily be lost.

The conversion of SCN^- into HCN by the action of zinc and H_2SO_4 may occasionally prove to be incomplete. The presence of these ions in the solution, however, will interfere with the detection of iodide and bromide ions since the chlorine water added will be consumed for oxidation of SCN^-. Therefore, if the chlorine-water tests for I^- and Br^- are found to be negative, a drop of KI or KBr solution should be added to the unknown solution. If the organic layer becomes coloured after shaking, this may be regarded as conclusive evidence for the absence of interfering effects and also of iodide and bromide ions.

Conversely, if after iodide and bromide ions have been introduced into the solution they cannot be detected, it means that the solution contains a reducing agent (SCN^- or H_2S). In this case, it is necessary to take another portion of the centrifugate remaining after the separation of the metallic silver precipitate and to oxidize the reducing agents by the action of $KMnO_4$ in the presence of H_2SO_4. Permanganate should be added in the hood until the solution is turned crimson or yellowish-brown by a further drop of $KMnO_4$ solution. The colour is removed by adding a drop or two of a dilute solution of $Na_2S_2O_3$ (avoid adding an excess of it) and then look for I^- and Br^- in the resulting solution in the usual way.

5. Analysis of Group III Anions. To a portion of the original test solution showing a neutral or faintly alkaline reaction add, with stirring, a saturated solution of Ag_2SO_4 until precipitation is complete. All the anions of Group I and II are precipitated. Centrifuge off the precipitate and wash two times with water. Combine the washings with the centrifugate and examine for Group III anions after evaporation in a porcelain dish.

(1) *Test for* NO_3^-, NO_2^-, ClO_3^-. Place a drop of the test solution on a porcelain plate or a cover of a porcelain crucible and add a drop of diphenylamine solution. The appearance of a deep-blue colour indicates the presence of oxidizing agents—NO_3^-, NO_2^-, ClO_3^-.

(2) *Test for* NO_2^-. To 2-3 drops of the test solution add (until just acid) $2N$ H_2SO_4 and 1-2 drops of a dilute solution of $KMnO_4$ ($0.05N$). If NO_2^- is present, the solution remains colourless or takes on a pale-pink colour. For comparison, place an equal volume of distilled water and a drop or two of $KMnO_4$ solution ($0.05N$) in another test tube.

(3) *Test for* ClO_3^-. To 2-3 drops of the test solution add several small crystals of Na_2SO_3 and $2N$ HNO_3 until just acid. Heat the solution. The ions ClO_3^- (if they are present) are reduced to Cl^-; chloride ions give a precipitate with the Ag^+ ions added previously in excess. The formation of a noticeable precipitate of AgCl indicates the presence of ClO_3^-. For comparison, a blank experiment should be carried out with all the components specified, except the test solution, and the amounts of the precipitates formed are compared (the reagents used almost invariably contain chlorides as an impurity).

(4) *Test for* NO_3^-. If NO_2^- was detected in Procedure *2*, it should be destroyed by one of the methods indicated in Sec. 8.23, Reaction *5*. A certain amount of NO_3^- is always formed in the solution and therefore the highly sensitive tests described in (b), Procedure *6* should not be used for detection of NO_3^- after the breakdown of NO_2^-.

Alternatively, the brucine test may be carried out for detection of NO_3^- in the presence of NO_2^- (see Sec. 8.22, Reaction *8*).

(5) *Test for* CH_3COO^-. Examine the centrifugate remaining after the separation of the anions of Groups I and II in the form of silver precipitates for CH_3COO^- as directed in (b), Procedure *7*.

Chapter 9

General Course of Analysis (Detection of Cations and Anions)

9.1. Preliminary Observations and Preparation of a Substance for Analysis

Before the complete analysis of any sample is started, a physical examination is to be carried out. Since the procedures for examination of solutions and also of solutions with a precipitate have already been discussed, here we shall deal largely with the analysis of solid substances.

Solid non-metallic substances should be carefully examined with a magnifying glass. Not infrequently it is possible in this way not only to establish whether the sample is homogeneous or heterogeneous but also to obtain some indication of the composition. In this examination, attention should be directed to the colour, shape, and size of the various particles in the mixture. An expecially important characteristic, which may yield information on the presence of certain elements, is the colour of the particles. For example, blue crystals in the sample suggest the presence of copper salts; pink crystals may indicate cobalt or manganese salts, etc.

Of course, no final conclusions can be made on the basis of such observations. But when used in combination with the analytical results obtained, these observations increase the confidence in the analysis.

After the visual observation, the sample must be prepared for analysis. If the test substance is a solid, it must be ground to the finest possible powder because small particles are more rapidly dissolved by the solvents. This operation is required for another reason too: only when the heterogeneous sample to be analysed is thoroughly ground, is it possible to attain uniform mixing of its constituents. The importance of this operation in semimicroanalysis, in which use is made of a very small amount of substance, is obvious.

Easily breakable solids are ground in a porcelain or an agate mortar until a fine powder is obtained. Harder solids (e.g., various minerals) are first broken into small pieces in a steel mortar and then ground in an agate mortar. The required fineness is often not achieved at once. Therefore, the particles have to be passed from time to time through a fine screen and the residue left on the screen ground again in the mortar. This operation is continued until all the solid to be analysed passes through the screen. This is important because the harder particles may differ in chemical composition from the readily breakable ones. So, if the grinding is not taken to the end, there is a risk of obtaining erroneous analytical results.

Metals or alloys must be converted into filings or shavings with the aid of suitable tools.

The solid substance prepared for analysis (unless it is a metal or an alloy) is divided into three parts: one is used for detection of cations, the second for detection of anions, and the third for the preliminary tests, a portion of it being kept as a reserve for a check analysis.

The amount of substance to be used for analysis is very important in the semimicromethod. On no account should too large an amount be used since the apparatus and the semimicro technique are intended for work with small quantities and are inapplicable if large quantities are used. Too small an amount should not be taken either since some of the ions to be identified may escape detection. Since semimicroanalysis, as a rule, makes use of highly sensitive reactions which enable detection of thousandths of a milligram (and, sometimes, even smaller quantities), it may seem at first glance that very small amounts of substance may also be suitable for carrying out the analysis as a whole. In fact, the amount taken has to be a little larger than required if the sensitivity of the reactions is considered. This is explained as being due to the insufficient specificity of the tests and to their sensitivity being decreased in the presence of foreign ions. This complicates the detection of individual ions by the fractional method. Separation reactions used in the systematic analysis are invariably associated with considerable losses of the substance.

Accordingly, the amount used for the systematic analysis of cations is about 25 mg (0.02-0.03 g) of a substance and, when it is dissolved, the volume of the solution is brought to about 1 ml. In those cases when traces of impurities are to be detected in a substance, such an amount may turn out to be insufficient, and a larger quantity must be used. The solution to be used for detection of anions should be prepared with a certain reserve (2-3 ml).

In order to avoid crude errors in the weighing of a substance to be subjected to analysis, it must be weighed on a sufficiently sensitive balance.

9.2. Preliminary Tests

Before the systematic analysis of the sample is started, it is expedient to perform a number of preliminary tests on its separate portions. These tests provide some indications which may be helpful in choosing the most efficient method of putting the sample into solution and may facilitate the analysis. If the sample to be analysed is a solution, a portion of it must be evaporated to dryness and separate parts the residue obtained must be used for preliminary tests. The preliminary tests used for detection of anions (see Sec. 8.26) have already been studied. Therefore, here we shall consider the tests employed for detection of cations.

1. Fusion with Sodium Carbonate on Charcoal. Make a small cavity in a charcoal block with the aid of a knife, put into it a small amount of the test solid substance mixed with a double quantity of anhydrous sodium carbonate, Na_2CO_3, and heat it in the reducing flame of a blowpipe (Fig. 9.1). A reducing flame is produced if the tip of the

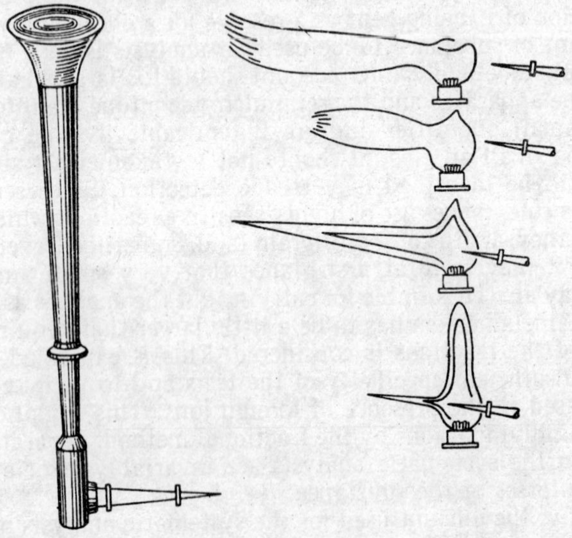

Fig. 9.1. Blowpipe.

blowpipe is placed at the outer edge of a luminous burner flame, and the flame is blown gently in a sloping downward direction towards the cavity in the charcoal. The flame is merely deflected, remaining luminous. This flame reduces substances, which come in contact with it, owing to the presence of incandescent carbon particles. On the other hand, if we put the tip of the blowpipe inside the burner flame and blow more strongly, a nonluminous oxidizing flame is obtained.*

When the mixture of the test substance with sodium carbonate is heated in the reducing flame on charcoal, the following takes place. First, the salt taken is converted into the corresponding carbonate as the result of a double replacement reaction; for example,

$$PbCl_2 + Na_2CO_3 \rightarrow PbCO_3 + 2NaCl$$

* In carrying out this test, it must be kept in mind that good results are obtained only when blowing is uniform. Therefore, it is very important to inhale through the nose while blowing.

Most carbonates, however, decompose upon ignition into a metal oxide and carbon dioxide. In this case,

$$PbCO_3 \rightarrow PbO + CO_2\uparrow$$

The oxide formed is almost invariably reduced immediately by the incandescent carbon particles to the free metal; for example,

$$PbO + C \rightarrow Pb + CO\uparrow$$

As a result, a bright bead (fused metal) appears in the cavity, or an infusible spongy mass if the metal is refractory. If the metal formed is volatile, part of it is converted into vapour which is oxidized by the oxygen of the air on leaving the flame. Therefore, various coloured oxide deposits may appear on the cold regions of the charcoal near the cavity.

For example, a yellow deposit of PbO is formed in the presence of lead; zinc gives a white (yellow when hot) deposit of ZnO. The most volatile metals (zinc, cadmium, and arsenic) do not form beads at all, because they volatilize completely and form deposits of the corresponding oxides.

Sometimes the metal particles formed during the test are so small that they are difficult to see. In such cases, the contents of the cavity should be cut out with a knife and ground cautiously in a mortar. Then water is poured over the powdered substance; the heavier metal grains remain on the bottom of the mortar while the light carbon particles are carried away by the flowing water.

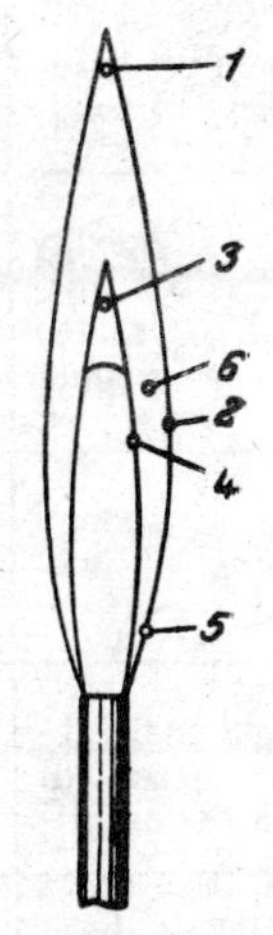

Fig. 9.2. The structure of the flame of a gas burner:
1, *2* — oxidizing flame; *3*, *4* — reducing flame; *5* — lower-temperature region; *6* — hottest point.

Metals are classified as malleable and brittle. A bead of a malleable metal is flattened out with a pestle or hammer, whereas a bead of a brittle metal is ground to a powder (in an agate mortar).

The results usually obtained in such tests are given in Table 9.1.

2. Borax Bead Test. The technique of this test and the chemistry of the process were described in Sec. 5.17. Let us note here that not infrequently the beads formed in the oxidizing and reducing flames differ in colour. Therefore, heat the bead first in the oxidizing flame (flame zones *1* and *2* in Fig. 9.2), and then, its colour noted, transfer to the reducing flame (*3* and *4* in Fig. 9.2) and reheat. In both cases, note the colour after the bead cools off.

The characteristics of the beads formed are given in Table 9.2.

3. Flame Test. Clean the platinum or nichrome wire to be used in the test by dipping it into HCl and igniting strongly, pick up a small amount of the unknown solid substance, introduce it into the burner

TABLE 9.1. Fusion with Sodium Carbonate on Charcoal

Appearance of mass formed in charcoal cavity	Presence of deposit on cold charcoal around cavity	Metal probably present	Appearance of mass formed in charcoal cavity	Presence of deposit on cold charcoal around cavity	Metal probably present
White infusible mass glowing strongly on ignition	No deposit	Mg, Sr, Ca, Al*	Bright malleable bead	Yellow	Pb
Green infusible mass of oxide	No deposit	Cr	Bright brittle bead	Brown-yellow	Bi
Gray spongy mass of metal, attracted by a magnet	No deposit	Fe, Co, Ni	Bright brittle bead	Blue-edged white oxide deposit	Sb
Red spongy mass of metal	No deposit	Cu	No bead	Oxide deposit yellow when hot and white when cold	Zn
Bright malleable bead of white metal	No deposit	Ag	No bead	Brown-red	Cd
Malleable bead coated with white oxide when removed from flame	No deposit	Sn	No bead	White. Garlic-like odour appears (*caution!*)	As

* Aluminium can be detected as follows. Moisten the mass formed in the charcoal cavity with a few drops of dilute $Co(NO_3)_2$ and reheat. If aluminium is present, blue cobalt aluminate is formed (Thenard's blue, see Sec. 5.16, Reaction *10*). Avoid using a large amount of cobalt since this will result in the formation of black cobalt oxide which may cover up the colour formed.

flame, and observe the colour. If the sample is a mixture, it is necessary to ensure that its constituents are volatilized separately according to their different volatilities and interfere as little as possible with one another. Proceed as follows. First, place the wire in the relatively cool region near the base of the flame (zone *5* in Fig. 9.2). After some time, transfer the wire to the hottest point in the flame (just above the tip of the inner blue cone, zone *6* in Fig. 9.2). Finally, remove the wire from the flame, dip it into HCl to convert non-volatile salts into vola-

TABLE 9.2. **Borax Bead Colours**

Colour of bead (cold) formed in oxidizing flame	Metal probably present	Colour of bead (cold) formed in reducing flame
Blue	Co	Not characteristic (indefinite)
Blue*	Cu	Red-brown
Green	Cr	Green
Brown-yellow or greenish-yellow	Fe	Dingy-green
Red-brown	Ni	Gray with red or violet tinge
Colourless	Sn**	Carmine-red (in presence of traces of Cu)
Colourless or indefinite (not characteristic)	Zn, Ag, Pb, Bi, Cd, Sb	Gray

* In the presence of a large amount of copper, the colour of the bead in the oxidizing flame is green.

** To test for tin, mix a very small amount of a copper salt to borax (1 : 1200). With large amounts, a turbid brown-red bead is obtained.

tile chlorides, and repeat the above procedure. Different cations impart different colours to the flame (Table 9.3).

4. Evolution of Gases and Sublimation on Ignition. To carry out this test, place a small amount of a solid on the bottom of a test tube

TABLE 9.3. **Flame Colouration**

Flame colouration	Metal probably present	Flame colouration	Metal probably present
Yellow (invisible through indigo solution or blue cobalt glass)	Na	Brick-red	Ca
		Carmine-red	Sr
		Yellow-green	Ba
Violet (appears purple through indigo solution or blue cobalt glass)	K	Green	Cu, Bi
		Blue	Cu, Pb, As, Sb

made of refractory glass. Use may be made of special tubes with a bulb at the end or in the middle. Such a tube enables one to identify the volatile decomposition products with great ease.

The following effects may be observed when the substance is heated in a bulb-tube (or in a test tube).

(1) The appearance of water droplets (condensation) on the inner sides of the tube indicates that the test substance either contains water of crystallization or liberates it upon thermal decomposition.

(2) The formation of a white deposit is evidence for the possible presence of ammonium salts, mercury salts, and arsenic or antimony oxides; a yellow deposit may be caused by the presence of arsenic sulphide, compounds of mercury and sulphur (for example, thiosulphate). If a gray or dark deposit with a metallic tint is formed, this means that the substance contains mercury or arsenic compounds in the presence of reducing agents, or iodides in the presence of oxidizing agents.

(3) Occasionally, heating results in a colour change; this is accounted for by the fact that the ability of a substance to absorb light changes with increasing temperature.

Thus, when heated, yellow chromates become red and white zinc oxide turns yellow.

(4) Certain substances when heated liberate volatile products that can be identified by the colour, odour, or by their characteristic reactions. For example, the following gases may evolve: CO_2 (from carbonates); SO_2 (from sulphites or thiosulphates of heavy metals); SO_3 (from corresponding sulphates); NO_2 (from nitrates and nitrites). The following fumes may also evolve: brown fumes of Br_2 (from bromides in the presence of oxidizing agents); violet fumes of I_2 (from iodides in the presence of oxidizing agents); ammonium salts decompose rapidly with the liberation of NH_3.

5. **Action of Acids on Dry Solid.** (1) The action of 2*N* hydrochloric or sulphuric acid may liberate the following gases: HCN, H_2S, NO_2 SO_2, CO_2.

(2) When a dry solid substance is acted on by concentrated sulphuric acid, colourless gases (HCl, SO_2, CO_2, NO, CO) and coloured gases or fumes (NO_2, Cl_2, Br_2, I_2) may evolve.

On the basis of this test, conclusions can be made concerning the presence of certain components in the test substance.

It should be kept in mind that all the preliminary tests described above give correct results if pure substances are being studied. In the analysis of mixtures, some metals may mask the reactions of other metals. Therefore, one has to be careful in interpreting the results obtained; each conclusion must be confirmed by other reactions of the ions involved.

After the preliminary tests have been completed, proceed to the systematic analysis of the test sample, using one of its portions for detection of cations and another for identification of anions.

9.3. Detection of Cations

As will be explained later, it is preferable to start the analysis with detection of cations since the presence of certain cations is evidence for the absence of a number of anions; this facilitates the work. Since the analysis is largely performed by the wet method, the first step in the analysis is to put the solid into solution. Many variants are possible. We shall consider cases where the sample is not similar in appearance to a metal or an alloy (the analysis of metals and their alloys are discussed in Sec. 9.5).

(a) Sample Is a Solid

In this case, a suitable solvent is chosen by carrying out special tests with small portions of the substance. After the substance has been dissolved, the resulting solution is to be analysed for cations. The solid is treated first with water (see below, Procedure *1*). If it does not dissolve in water (or dissolves partially), its behaviour with acids (HCl, HNO_3, or a mixture of these—aqua regia) is tested (see Procedure *2*). If the substance does not dissolve in acids either, dissolve and analyse as suggested in Procedure *3*.

1. Dissolving and Analysis of Substances Soluble in Water. Treat a very small amount of the test substance (about the size of a match head) with 15-20 drops of distilled water in a conical test tube with constant stirring. If required, heat the liquid in a water bath for 2-3 min.

If the substance does not dissolve completely, it is necessary to find out whether it dissolves partially. Centrifuge off the undissolved residue and evaporate 2-3 drops of the centrifugate to dryness on a watch glass. The appearance of a deposit on the glass is an indication of the partial solubility of the substance in water.

If the substance dissolves completely in water, prepare 25-30 drops of a solution by using the portion (0.02-0.03 g) to be tested for cations, and analyse the resulting solution as indicated in Table 6.8.

If the substance is partially soluble in water, it is expedient, in many cases, to dissolve, using the above method, all the water-soluble components and to analyse the solution separately from the precipitate. Dissolve the precipitate and examine as described below (see Procedure *2*). Of course, the total time required to analyse the various solutions separately is a little longer than is required to make one analysis of the sample dissolved in acid. But the analysis will be much simpler to carry out and the results will be more reliable than if all the solutions were mixed and just a single analysis done on the whole sample. The more ions there are in the test solution the greater chance there is for interference and, consequently, the greater possibility of making an error.

Extraction of the water-soluble components from the mixture and separate analysis of the solution and of the insoluble residue are advantageous also because complications, that may arise due to the simultaneous presence of oxidizing and reducing agents, are sometimes avoided; this procedure enables one to obtain a more correct idea of the actual composition of the mixture being analysed. For example, if the mixture contains $BaCrO_4$ and KI, these two substances can easily be separated because the first is insoluble and the second readily soluble in water. Separate analysis of the solution and of the undissolved residue gives a quite correct indication of the composition of the mixture. The dissolution of the mixture in acid involves the following reaction:

$$2BaCrO_4\downarrow + KI + 16H^+ \rightarrow 2Ba^{2+} + 2Cr^{3+} + 3I_2 + 6K^+ + 8H_2O$$

Therefore, if a sufficiently large amount of KI is present, Cr^{3+} ions will be detected instead of CrO_4^{2-} ions, and free I_2 along with I^-, i.e., quite wrong conclusions will be arrived at about the composition of the mixture.

Before starting the systematic analysis of the solution obtained, its colour should be noted and the pH determined.

The colour may yield information on the presence or absence of coloured ions (Cu^{2+}, Co^{2+}, Ni^{2+}, Fe^{3+}, Cr^{3+}, CrO_4^{2-}, $Cr_2O_7^{2-}$, MnO_7^-) and sometimes a direct indication of the presence of some of them.

Different coloured ions when present in the same solution mutually interfere and therefore it is very convenient to observe the colours after a chromatographic separation. Introduce 5 drops of the test solution into a column with alumina (see Sec. 2.6) and elute with 4-5 drops of water.

The following coloured bands or zones are formed in the column:

Ion present	*Zone colour*
Fe^{3+}	Dark-brown
Cu^{2+}	Pale-blue
Co^{2+}	Pink

If no Fe^{3+} ions are present, Cr^{3+} can be identified by the formation of a gray-blue zone above the blue zone of Cu^{2+}. In just the same way, in the absence of Co^{2+}, the ion Ni^{2+} can be detected by the formation of a pale-green zone below the zone of Cu^{2+}.*

Determination of the pH of the solution also gives preliminary indications of the presence of certain ions (see Sections 6.19 and 8.26). This determination is made either directly in separate portions of the test solution (if the substance is completely soluble in water) or after the undissolved residue has been separated by centrifugation (if the substance is partially soluble in water). If the substance has to be dissolved in acids (see below, Procedure *2*), then before the pH is determined it is necessary to neutralize cautiously a separate portion of the solution with alkali until a slight but persistent turbidity is formed, which is removed by centrifuging.

---

* The sorbability of Co^{2+} and Ni^{2+} ions by alumina is the same. Therefore, if they are present in the same solution, they form a single zone; the brighter colour of Co^{2+} covers up the pale-green colour of Ni^{2+}. The same applies to the ions Fe^{3+} and Cr^{3+} which are also equally sorbable.

The results of pH determinations must not be regarded as final. They must be checked during the analysis by the tests for individual ions. In such cases, they are very useful since they increase the confidence in the results or else indicate that an error has been made.

2. Dissolving and Analysis of Substances Insoluble in Water but Soluble in Acids. If the sample is found to be insoluble in water, it is tested for solubility in acids.

1. Treat a small portion of the substance in a test tube with 2*N* hydrochloric acid with stirring. Observe to determine whether any gases (CO_2, SO_2, H_2S, NO_2) are evolved, and take this into account in the subsequent analysis for anions. If the substance does not dissolve in the cold, heat the liquid. If it does not dissolve completely on heating either, treat another portion of the substance with concentrated acid instead of the dilute one.

2. Regardless of the results obtained in the preceding test, heat another portion of the substance with 6*N* HNO_3.

3. Finally, if the substance dissolves neither in HCl nor in HNO_3, try to dissolve it in a mixture of the two (i.e., aqua regia) on heating.

Most water-insoluble substances are dissolved by the action of even dilute hydrochloric acid. Acting as a strong oxidant, nitric acid dissolves a number of substances insoluble in hydrochloric acid. Such are, for example, the sulphides of the cations of Groups IV and V: CuS, PbS, Bi_2S_3, etc. Certain of the compounds of bismuth, antimony, and tin insoluble in nitric acid are however dissolved in hydrochloric acid. Aqua regia, which is a more powerful oxidizing agent than nitric acid, dissolves some compounds that do not dissolve in hydrochloric and in nitric acid, taken separately (e.g., HgS). There are substances, however, which dissolve neither in these acids nor in aqua regia. They are dissolved and examined as described in Procedure *3*.

If the substance is soluble in hydrochloric and in nitric acid, preference should be given to nitric acid because nitrates are less volatile than chlorides and when the excess acid is removed by evaporation there is less risk of losing any of the constituents of the sample*.

Evaporation is required because the precipitation of certain of the sulphides of Group IV (especially, CdS) is incomplete at high hydrogen ion concentrations. Excess of nitric acid or aqua regia is intolerable also because, in the precipitation of Group IV by hydrogen sulphide, they would oxidize H_2S to sulphur, thereby preventing the precipitation of this group. At the same time, most of the anions, which might interfere with detection of cations, are removed from solution during evaporation with acids (as the result of volatilization as the corresponding free acids, or of oxidation).

---

* For example, hydrochloric acid solutions containing arsenic must not be evaporated since $AsCl_3$ is volatile. Because of the volatility of $HgCl_2$, mercury may also be lost by incautious evaporation.

This procedure does not remove PO_4^{3-} ions. In their presence, the cations of Group II and magnesium are precipitated together with the cations of Group III by the action of $(NH_4)_2S$, which complicates the analysis.

The systematic procedures for analysis of cations in the presence of PO_4^{3-} were discussed in Sec. 6.21*

Having found the most suitable solvent-acid (e.g., 6*N* HNO_3) as the result of the preliminary tests, treat 0.02-0.03 g of the test substance (or the residue left after all the water-soluble components have been extracted from it) with 25-30 drops of the acid in a porcelain dish or in a crucible. Boil the liquid (a drop or two of concentrated HCl being added if necessary) and evaporate cautiously to incipient dryness, avoiding strong heating** (*Hood*!). Then add to the crucible 25-30 drops of water, mix to speed up the dissolution, and transfer the contents of the crucible to a conical test tube. If the substance does not dissolve completely in acids, separate the undissolved residue by centrifuging and examine as directed in Procedure *3* (see below).

Analyse the solution for cations as usual (see Table 6.8).***

3. Dissolving and Analysis of Substances Insoluble in Acids. Substances insoluble in acids include: silver halides**** (AgCl, AgBr, AgI); slightly soluble sulphates***** ($BaSO_4$, $SrSO_4$, ($CaSO_4$), $PbSO_4$); some natural oxides or oxides formed on ignition of precipitates (Al_2O_3, Cr_2O_3, Fe_2O_3, and also SnO_2 and SiO_2). Many silicates, chromite $Fe(CrO_2)_2$, fluorspar CaF_2, sulphur, carbon, and certain other substances also belong to this class.

Methods of dissolving the halides and sulphates indicated have already been considered.

For example, AgCl dissolves in NH_4OH with the formation of the complex salt $[Ag(NH_3)_2]Cl$ which decomposes when the solution is acidified with HNO_3; AgI and AgBr can be decomposed by adding a

* The detection of cations may also be interfered with by the anions F^-, SiO_3^{2-}, $C_2O_4^{2-}$. The directions concerning the removal of F^- and SiO_3^{2-} are given in Procedure *3* of this section, and those concerning the removal of $C_2O_4^{2-}$ are given in Sec. 8.10, Reaction 7.

** The residue must not be ignited after evaporation because the salts contained in the residue decompose on ignition to form the corresponding anhydrous oxides which sometimes become insoluble in acids after the ignition. To avoid this, the evaporation is best carried out (or at least completed) in a water or air bath.

*** It is obvious that the possibility of the ions Sn^{2+}, Fe^{2+}, and Hg^+ being present in solution after the action of HNO_3 or aqua regia must be excluded since they would have been completely oxidized to Sn^{IV}, Fe^{3+}, and Hg^{2+}. This should be taken into account in the analysis.

**** In concentrated HCl, silver chloride dissolves with the formation of a complex compound $H[AgCl_2]$ but it precipitates again when the solution is diluted with water.

***** Lead sulphate is noticeably soluble in concentrated acids with the formation of the acid salt $Pb(HSO_4)_2$.

fragment of metallic zinc and H_2SO_4 on heating, e.g.,

$$2AgI\downarrow + Zn\downarrow \rightarrow 2Ag\downarrow + Zn^{2+} + 2I^-$$

The black precipitate of metallic silver is dissolved by heating in 6*N* HNO_3 and then Ag^+ is detected by its characteristic reactions.

The anions I^- and Br^- are detected in the resulting solution with the aid of chlorine water.

Of the slightly soluble sulphates, lead sulphate, $PbSO_4$, dissolves in a 30-percent solution of ammonium acetate and also in excess of alkali hydroxides. The ion Pb^{2+} is detected in the resulting solution by the action of K_2CrO_4 (after the solution is acidified with CH_3COOH if the solvent used is an alkali). The sulphates of barium, strontium, and calcium* are dissolved by reboiling with a concentrated solution of Na_2CO_3 and are examined as indicated in Sec. 4.6. Alternatively, they may be converted into the corresponding carbonates by fusing with a mixture of sodium carbonate, Na_2CO_3, and potassium carbonate, K_2CO_3 (see Sec. 4.13, Procedure *3*).

Silicon dioxide and silicates non-decomposable by acids are dissolved also by fusing finely divided solids with an approximately sixfold amount of a mixture of Na_2CO_3 and K_2CO_3. This results in the formation of soluble sodium and potassium silicates:

$$SiO_2 + Na_2CO_3 \rightarrow Na_2SiO_3 + CO_2\uparrow$$

Treat cautiously the cooled molten flux with dilute hydrochloric acid and evaporate the solution (to convert the silicic acid to an insoluble form) 2-3 times to dryness with a concentrated solution of HCl. Heat the residue left after the last evaporation with a few drops of a concentrated solution of HCl, add 20-30 drops of water, separate the undissolved silicic acid by centrifugation, and examine the solution for cations in the usual way.

It is clear that K^+ and Na^+ cannot be detected if silicates are decomposed by this methods because these cations are introduced in the fusion. If K^+ and Na^+ are to be identified, silicates must be decomposed by the action of a mixture of hydrofluoric acid (HF) and sulphuric acid. The reaction is carried out in a platinum crucible. Place a small amount of the finely divided test substance in the crucible and treat it with 15-20 drops of pure hydrofluoric acid and a drop or two of concentrated sulphuric acid. Then cautiously evaporate the liquid to dryness, avoiding strong heating, over a low flame (*In the hood!*). Moisten the dry residue again with a few drops of HF and again evaporate. As a result,

---

* As indicated earlier, calcium sulphate dissolves noticeably in mineral acids, especially in concentrated acids. Therefore, calcium often cannot be detected in the sulphate precipitate after the precipitate is treated with a saturated solution of sodium carbonate and the sulphates are converted into the carbonates.

the SiO_2 is converted into gaseous silicon fluoride, SiF_4, following which the silicic acid is found to have been completely removed from the substance after two or three evaporations with HF. After the last evaporation, dissolve the residue in HCl and examine it for cations.

The insoluble oxides (Al_2O_3 and Fe_2O_3) can be dissolved by fusion with $KHSO_4$ or with $K_2S_2O_7$, the corresponding sulphates being formed; for example,

$$Al_2O_3 + 6KHSO_4 \rightarrow Al_2(SO_4)_3 + 3K_2SO_4 + 3H_2O$$

The fusion is carried out as follows. Mix thoroughly a small amount of the finely divided test substance with an approximately 12-fold amount of $KHSO_4$, place the mixture in a crucible, and heat, increasing the temperature gradually, until the mixture ceases to swell and is converted to a homogeneous liquid. Then allow the melt to cool off, treat it on heating with a small amount of water, and examine the resulting solution for cations.*

Chromium oxide, Cr_2O_3, and chromite, $Fe(CrO_2)_2$, can also be put into solution by fusion with an approximately sixfold quantity of a mixture of Na_2CO_3 (2 parts) and $NaNO_3$ or KNO_3 (1 part). Trivalent chromium is oxidized to the sexivalent state:

$$Cr_2O_3 + 2Na_2CO_3 + 3KNO_3 \rightarrow 2Na_2CrO_4 + 3KNO_2 + 2CO_2 \uparrow$$

The chromate formed can be extracted from the melt with water and detected (see Sec. 5.17). After this treatment of chromite, the iron remains in the residue in the form of an acid-soluble modification—Fe_2O_3.

Stannic oxide, SnO_2, and also metaantimonic acid, $HSbO_3$, can be brought into solution by fusion with an approximately sixfold amount of a mixture of sodium carbonate and sulphur (1:1). This yields water-soluble thiosalts; for example,

$$2SnO_2 + 2Na_2CO_3 + 9S \rightarrow 2Na_2SnS_3 + 2CO_2 \uparrow + 3SO_2 \uparrow$$

The fusion should be carried out over a low flame in a covered crucible, the heating being continued until all the excess sulphur has been driven off and burnt. Allow the melt to cool off, treat it with warm water, centrifuge off, discard the precipitate, and analyse the solution for antimony and tin as suggested in Table 6.8 (starting with the decomposition of thiosalts).

It should also be noted that freshly precipitated metastannic (H_2SnO_3) and metaantimonic ($HSbO_3$) acids, which are obtained, for example, by dissolving the melts in HNO_3, are noticeably soluble in hot concentrated HCl; this fact can conveniently be used in the detection of tin and antimony in alloys.

* It should be kept in mind that the amphoteric oxides Al_2O_3 and Cr_2O_3 can also be dissolved by fusion with Na_2CO_3, in which case they are converted into the aluminate and chromite.

Insoluble fluorides (e.g., CaF_2) are identified by the reaction leading to the formation of SiF_4 (see Sec. 8.8, Reaction *3*). They are decomposed by heating the substance with concentrated H_2SO_4 in a platinum crucible. The heating is continued until H_2SO_4 has been evaporated almost completely (*In the hood!*). Free HF, which volatilizes on heating, is expelled from the fluoride, and the residue is found to contain the corresponding sulphate (e.g., $CaSO_4$) which is dissolved and examined by the methods that have already been discussed.

Carbon is identified by the black colour of the precipitate formed and by combustion when fused with a mixture of Na_2CO_3 and $NaNO_3$ with the formation of CO_2 which makes a drop of lime water turbid.

Sulphur is identified by the blue colour of the flame when it burns with formation of SO_2 (recognized by its acrid odour), and also by its ability to sublime when the sample is heated in a narrow tubing sealed at one end.*

(b) Sample Is a Solution (or a Solution with a Precipitate)

In those cases when the test substance is a clear solution, it is to be analysed as indicated in Table 6.8.

If the test sample is a solution with a precipitate, the latter is to be separated and tested for solubility in acids. All the constituents of the sample that are dissolved by acids are either added (after the removal of excess acid by evaporation and the addition of water) to the rest of the solution and analysed for cations in the usual way (see Table 6.8) or, according to what has been said above, both solutions are analysed separately. The undissolved residue is to be decomposed and examined as described above (see Procedure *3*).

9.4. Detection of Anions

As has already been stated, the tests for anions should be carried out only after the sample has been analysed for cations. This sequence is preferred for the following reasons.

1. The presence or absence of anions is established concurrently with the detection of cations. For example, PO_4^{-3} must be identified before the precipitation of the cations of Group III by ammonium sulphide. If no As^{III} and As^{V} have been detected during the analysis of Group IV, it is pointless to look for the ions AsO_3^{3-} and AsO_4^{3-}. If no Cr^{III} has been identified in Group III, the absence of CrO_4^{2-} and $Cr_2O_7^{2-}$ may be reported since these anions should have been reduced

* It should be remembered that sulphur may form upon treatment of the substance with acids, which results in decomposition of thiosulphates or oxidation of sulphides.

to Cr^{III} in the precipitation of Group IV by hydrogen sulphide. If no Mn^{2+} has been detected, the absence of MnO_4^- may be deduced.*

The presence of certain anions (S^{2-}, SO_3^{2-}, $S_2O_3^{2-}$, CO_3^{2-}, and NO_2^-) is usually established concurrently with the detection of cations since, when the solution is acidified, gases may evolve (H_2S, SO_2, CO_2, NO_2) and, in the case of $S_2O_3^{2-}$, the separation of sulphur may take place (if the solution contains no oxidizing agents).

2. Some valuable conclusions may also be drawn concerning the absence of certain anions on the basis of the solubility of the salt of the cations detected in the test solution. Suppose, we have detected cations in the test substance, which form slightly soluble salts with the anions S^{2-}, CO_3^{2-}, PO_4^{3-}, etc. If this substance is soluble in water, these anions must be absent. If Ba^{2+}, Sr^{2+}, or Pb^{2+} are present and the substance is soluble in dilute HCl or HNO_3, no SO_4^{2-} must be present. Likewise, if Ag^+ has been shown to be present in the test solution (provided that there is no precipitate), the absence of Group II anions may be reported. If the solution shows no acid reaction at that, no anions of Group I may be present either, except SO_4^{2-} and F^-, because the silver salts formed by these anions are insoluble in water (with the exception of Ag_2SO_4 and AgF).

After conclusive evidence for the absence of certain anions has been obtained, proceed to the detection of anions that may be present.

First, it is necessary to prepare the solution properly. This preparation consists in removing so-called heavy-metal cations which include in this particular case the cations of Group II-V together with Mg^{2+}. Many of these cations introduce interfering effects because they are coloured, exhibit oxidizing and reducing properties, and are capable of forming precipitates with a number of anions. The removal of heavy-metal cations is achieved by boiling the test solution with Na_2CO_3 solution. This puts the anions of the substance into solution in the form of corresponding sodium salts, whereas the cations remain in the precipitate as normal or basic carbonates or (some of them) as hydroxides; for example,

$$BaSO_3\downarrow + CO_3^{2-} \rightarrow BaCO_3\downarrow + SO_3^{2-}$$

$$2CuSO_4 + 2CO_3^{2-} + H_2O \rightarrow (CuOH)_2CO_3\downarrow + 2SO_4^{2-} + CO_2\uparrow$$

$$2AlCl_3 + 3CO_3^{2-} + 3H_2O \rightarrow 2Al(OH)_3\downarrow + 6Cl^- + 3CO_2\uparrow$$

The resulting solution, called the sodium carbonate extract, is used for detection of anions. It should however be kept in mind that certain substances decompose with difficulty when boiled with a solution of sodium carbonate. Such are, for example, certain phosphates, fluorides,

* No reverse conclusions, however, can be made in such cases: if Cr^{III} has been detected, it does not yet mean that the substance under study contains CrO_4^{2-} or $Cr_2O_7^{2-}$; likewise, the detection of Mn^{II} cannot be taken as proof of the presence of MnO_4^-.

and sulphides, and also silver halides. Therefore, if no corresponding anions have been detected in the prepared solution, they are looked for in the precipitate obtained by the action of Na_2CO_3 or occasionally in separate portions of the test solid substance.

Of course, if the test substance contains only K^+, Na^+, or NH_4^+, the preparation of the sodium carbonate extract may be dispensed with. In such a case, about 0.1 g of the substance is simply dissolved in 2-3 ml of distilled water and the anions are tested for in separate portions of the resulting solution as described in Sec. 8.26.

If heavy-metal cations are present, CO_3^{2-} cannot evidently be detected in the sodium carbonate extract since it is deliberately introduced into the solution. Therefore, CO_3^{2-} is to be detected in a separate portion of the test substance (in a reserve portion).

A sodium carbonate extract is produced as follows.

1. If the test substance is a solid, mix about 0.1 g of its fine powder in a small beaker (or in a crucible) with 0.4 g of pure anhydrous sodium carbonate (Na_2CO_3), add about 2.5 ml (50-60 drops) of distilled water, and heat with stirring using a glass rod. Bring the liquid to boiling and maintain at the boiling point for about 5 min, adding water drop by drop as the liquid is evaporated. Now transfer the contents of the beaker (or crucible) to a conical test tube and centrifuge. Reserve the precipitate for further tests for anions that have not been brought into solution (see below).

2. If the substance is a solution (or a solution with a precipitate), make 25-30 drops of its alkaline with freshly prepared $3N$ Na_2CO_3 solution and then add 25-30 drops of excess Na_2CO_3 and proceed as directed above (see Procedure *1*).

3. In those cases when the substance has to be fused with a mixture of Na_2CO_3 and K_2CO_3 (see Sec. 9.3, Procedure *3*), in order to put it into solution for detection of anions, the anions are identified in the solution resulting from the treatment of the molten flux with water.

Before any attempt is made to detect anions, a portion of the sodium carbonate extract obtained should be neutralized to pH ≈ 7 with acetic acid to remove the excess Na_2CO_3 added. For the acetate ion to be detected too, another portion of the centrifugate is neutralized with HNO_3 or H_2SO_4.

The neutralization should be carried out very cautiously, stirring thoroughly the liquid after each drop of acid and avoiding excess of acid; otherwise, some of the anions (e.g., S^{2-}, NO_2^-) may be lost. Furthermore, one has to remember that some cations are precipitated incompletely when boiled with Na_2CO_3. For example, cations that form amphoteric hydroxides may be partially carried over into the sodium carbonate extract in the form of corresponding anions (e.g., AlO_2^-).

Copper, cobalt, and nickel form complex compounds and also remain partly in solution, thereby imparting to it a corresponding characteristic colour. If the subsequent neutralization is carried out accurately, all these compounds are decomposed and the cations are precipitated. For instance,

$$AlO_2^- + CH_3COOH + H_2O \rightarrow Al(OH)_3\downarrow + CH_3COO^-$$

$$2[Cu(CO_3)_2]^{2-} + 4H^+ \rightarrow (CuOH)_2CO_3\downarrow + 3CO_2\uparrow + H_2O$$

Conversely, if the neutralization has not been done accurately, the precipitates formed will dissolve partially or completely in acid, and the aim of the entire procedure will not be achieved.* If the neutralization of the prepared solution leads to the formation of a precipitate, the latter is centrifuged off and discarded, and the centrifugate is examined for anions in the usual manner.

As already pointed out, certain anions, namely, PO_4^{3-}, F^-, S^{2-}, Br^-, and I^-, may not be carried into solution by Na_2CO_3 and may remain in the precipitate. If these anions have not been detected in the prepared solution, they must be looked for in the precipitate.

To do this, proceed as follows.

1. Treat a portion of the precipitate on heating with a few drops of $6N$ HNO_3; if the undissolved residue is left, separate it by centrifugation, wash, and examine for the presence of I^- and Br^- Test the centrifugate for PO_4^{3-} if it has not been detected earlier.

(2) Test another portion of the precipitate for S^{2-} by using dilute H_2SO_4 and metallic zinc**. Conduct the reaction in a gas chamber (see Fig. 2.6); use, as an indicator, a paper moistened with a solution of $Pb(CH_3COO)_2$ or Na_2PbO_2.

(3) Dry a portion of the precipitate and test for F^- by means of the reaction of formation of SiF_4. If borates are absent, this reaction may be carried out with a separate portion of the test (solid) substance.

Finally, if no PO_4^{3-} is present, the F^- ion can be detected by using the zirconium-alizarin reaction, for which purpose a part of the precipitate should be dissolved with dilute HCl solution.

9.5. Analysis of Metals and Their Alloys

In conclusion, let us consider the case where the sample to be analysed is a technically pure metal or an alloy.

The purity of a technically pure metal (the content of the base component) is usually 95-97 per cent; it may sometimes reach 99.99 per cent. It is evident that the content of the various impurities (which may be more than 10 in number) in such metals must be low, in some cases only thousandths of a per cent. The detection of such vanishingly

---

* Copper can be precipitated by boiling the solution with a few drops of the alkaline solution.

** Zinc must be preliminarily checked for the absence of an impurity of a sulphide.

small amounts (traces) of impurities is sometimes a rather formidable task and requires the use of highly sensitive methods of investigation, such as spectrum or luminescence analysis.

Alloys are systems consisting of one or several phases, each of which may in turn be composed of an individual element, a chemical compound, or a solid solution of variable composition. The properties of alloys depend on their composition and on their structure.

We give here only schemes of analysis for the most commonly used types of alloy. Four principal types may be distinguished (Table 9.4).

Especially important is the behaviour of alloys with different solvents, in most cases, it depends on the properties of the base metal component.

1. Iron Alloys. These alloys, just like iron itself, which is to the left of hydrogen in the electromotive series, usually dissolve in dilute and concentrated HCl, and also in dilute H_2SO_4 and HNO_3. Concentrated nitric acid not infrequently passivates iron and its alloys. The metal becomes coated with a thin oxide film which protects it from contact with the acid and greatly retards its dissolution.

Certain other oxidizing agents act similarly on metals. It must however be kept in mind that certain iron alloys can be dissolved only by boiling with aqua regia or with other mixtures of acids, and in some cases only by fusion with any suitable fluxing agents, such as Na_2O_2 or NaOH.*

Oxidizing agents, such as HNO_3 and $(NH_4)_2S_2O_8$, have to be resorted to also in those cases when an alloy dissolves in dilute HCl or H_2SO_4 but leaves a residue of the carbides (i.e., compounds with carbon) of certain metals (Cr, W, V, Mo, etc.); they must be oxidized in order to put them into solution. An oxidizing agent should be added only after the process of dissolution is complete (recognized by the cessation of hydrogen evolution); otherwise iron will be passivated by the oxidizing agent and will dissolve slowly or not at all.

2. Copper Alloys. Like copper itself, which is to the right of hydrogen in the electromotive series, copper alloys are insoluble in dilute HCl and H_2SO_4. On the other hand, they are readily dissolved by concentrated and dilute HNO_3 (on heating).

The reaction of copper with dilute HNO_3 may be represented by the equation

$$3Cu\downarrow + 2NO_3^- + 8H^+ \rightarrow 3Cu^{2+} + 2NO\uparrow + 4H_2O$$

When copper is brought into contact with concentrated HNO_3, NO_2 is liberated instead of NO:

$$Cu\downarrow + 2NO_3^- + 4H^+ \rightarrow Cu^{2+} + 2NO_2\uparrow + 2H_2O$$

* The fusion is carried out in steel or nickel crucibles; the crucible material passes partially into the alloy. The resulting alloy is dissolved in sulphuric acid and the solution is examined in the usual manner (see below).

TABLE 9.4. **Composition and Some Characteristics of Most Important Alloys**

Type of alloy	Name	Base metal	Composition	Colour	Density g/cm³	Solvent
Iron alloys (ferrous alloys)	Cast iron, steel, ferroalloys, certain superhard alloys, etc.	Iron	Fe, Mn, Ni, Cr, V, Co, Mo, Cu, W, Al, Ti (C, P, Si, S)	Dark-gray (steel), sometimes nearly black	ca. 8	H_2SO_4, HCl, HCl + HNO_3, etc.
Copper alloys (non-ferrous alloys)	Bronzes, brasses, German silver, argentan, etc.	Copper	Cu, Sn, Zn, Pb, Cd, Sb, Al, Mn, Fe, Ni (P)	Golden, yellow, silvery-white	7-9	HNO_3
Aluminium alloys and magnesium alloys (light alloys)	Duralumin, magnalium, etc.	Aluminium, magnesium	Al, Mg, Ca, Cu, Mn, Ni, Fe, Zn, Sn (Cd, Si)	Silvery-white	1.8-2.7	NaOH, HCl, H_2SO_4
Lead alloys (heavy alloys)	Babbitts, solders, printing metals, etc.	Lead, tin	Pb, Sn, Cu, Fe, Cd, Zn, Sb (Ca, Na)	Gray, bluish-gray, silvery-white	8.5-11.5	HNO_3

When the dissolution takes place in nitric acid (if the concentration is high enough), the tin and antimony of an alloy form β-stannic, H_2SnO_3, and metaantimonic acid, $HSbO_3$, which separate out as white precipitates*:

$$Sn\downarrow + 4HNO_3 \rightarrow H_2SnO_3\downarrow + H_2O + 4NO_2\uparrow$$

$$2Sb\downarrow + 10HNO_3 \rightarrow 2HSbO_3\downarrow + 4H_2O + 10NO_2\uparrow$$

The precipitates are appreciably soluble in hot concentrated hydrochloric acid, a fact that can be used for detection of tin and antimony in it.

TABLE 9.5. **Preliminary Tests for Determination of Alloy Type**

1. Treat a grain of alloy in the cold with 2-3 drops of 30-percent NaOH solution. The abundant evolution of hydrogen (effervescence) taking place after 1-5 min indicates an aluminium alloy.*

2. Heat another grain of alloy with 2-3 drops of concentrated HCl for a minute. While heating, add (to oxidize Fe^{2+} to Fe^{3+}) a drop of $6N$ HNO_3. After a minute, treat a drop of the solution (if required, after the ppt. has been centrifuged off) with a drop of NH_4SCN solution. An *intense* blood-red colour indicates an iron alloy.

3. Heat a grain or two of alloy with 2 drops of $6N$ HNO_3 and 2 drops of water. If a ppt. (H_2SnO_3 and $HSbO_3$) is formed, centrifuge it off. Carry out the tests indicated below on separate portions of the centrifugate.

4. The blue colour of the solution indicates a copper alloy. Check as follows. Treat a drop of the solution with excess (2-3 drops) of conc. NH_4OH. An *intense* blue colour due to the formation of $[Cu(NH_3)_4]^{2+}$ is formed.	5. Dilute another portion of the solution two- or threefold with water and add a drop or two of $2N$ H_2SO_4. An *abundant* white ppt. of $PbSO_4$ indicates a lead alloy.**

* If on the basis of physical characteristics (colour, density) it may be assumed that the alloy is a light one (aluminium-magnesium) and does not react with NaOH, a grain of the alloy should be treated (in the cold) with $2N$ CH_3COOH. The effervescence of hydrogen allows one to assume that the base metal is magnesium. In such a case, dissolve 0.02-0.03 g of the alloy in dilute HCl and analyse, as usual, the solution for the cations of metals which may be present in light alloys (see Table 9.4).

** The presence of large amounts of tin in the alloy is indicated by the formation of an abundant white precipitate of H_2SnO_3 upon treatment of the sample with HNO_3 (see Procedure 3). Such alloys are best dissolved not in nitric but in dilute (1 : 1) hydrochloric acid.

---

* It has been found that the compounds formed have no definite composition, they are the hydrous oxides $SnO_2 \cdot xH_2O$ and $Sb_2O_5 \cdot xH_2O$.

Such alloys are best dissolved by aqua regia. In this case, soluble chloride complexes of tin and antimony are formed at once:

$$3Sn\downarrow + 4HNO_3 + 18HCl \rightarrow 3[SnCl_6]^{2-} + 6H^+ + 4NO\uparrow + 8H_2O$$
$$3Sb\downarrow + 5HNO_3 + 18HCl \rightarrow 3[SbCl_6]^- + 3H^+ + 5NO\uparrow + 10H_2O$$

3. Aluminium and Magnesium Alloys. These alloys are well soluble in acids. Besides, aluminium alloys and aluminium itself dissolve readily in alkali hydroxides which are the most efficient solvents for them:

$$2Al\downarrow + 2OH^- + 2H_2O \rightarrow 2AlO_2^- + 3H_2\uparrow$$

Zinc and tin (and also silicon) also dissolve in alkalis.

The other components of an alloy are unaffected by the action of alkali and remain in the precipitate in the form of free metals (or partly in the form of hydroxides). This precipitate should be dissolved in HNO_3 because of the possible presence of copper.

4. Lead and Tin Alloys. Lead stands to the left of hydrogen in the electromotive series, but it dissolves with difficulty in dilute HCl and H_2SO_4, because the slightly soluble salts formed, $PbCl_2$ and $PbSO_4$,

TABLE 9.6. **Analysis of Iron and Its Alloys**

1. Heat about 0.02-0.03 g of alloy in an evaporating dish or a crucible with 12-15 drops of 2*N* H_2SO_4. After the evolution of hydrogen stops completely, add, a drop at a time, conc. HNO_3 (to destroy the carbides) until frothing ceases, and heat until black specks disappear completely.* Transfer the contents of the dish to a conical test tube and centrifuge. Wash the precipitate with water containing HCl.	
Precipitate: $H_2WO_4(H_2SiO_3)$.	Centrifugate: Fe^{3+}, Mn^{2+}, Cr^{3+}, Ni^{2+}, Co^{2+}, Al^{3+}, Ti^{IV}, Cu^{2+}, MoO_4^{2-}, VO_3^-.
2. Detection of W (see Sec. 7.5, Procedure *2*).	3. Detection of Group III cations and the ions** Cu^{2+}, MoO_4^{2-}, and VO_3^- as suggested in Table 7.1 (Procedures *3-9*) after the removal of excess HNO_3 by evaporating with H_2SO_4 until white fumes of H_2SO_4 appear.***

* In certain cases, iron alloys can be dissolved by boiling with aqua regia or even by fusion. In the case of boiling with aqua regia, further analysis is carried out in the same way. The fusion is not discussed in this book.

** If V, Mo, and W are known to be absent, the analysis is to be carried out as indicated in Sec. 7.5, with omission of the procedures for detection of WO_4^{2-}, MoO_4^{2-}, and VO_3^-. In this case, there is need to add $(NH_4)_2S_2O_8$ before the precipitation by hydrogen sulphide (Procedure *3*) and to evaporate with H_2SO_4 after the dissolution of the CuS ppt. in HNO_3 (Procedure *4*).

*** The removal of excess HNO_3 (or aqua regia) is required for subsequent precipitation by hydrogen sulphide. The repeated evaporation with H_2SO_4 indicated in Table 7.1 (Procedure *6*) may be omitted.

TABLE 9.7. Analysis of Copper and Its Alloys

<table>
<tr><td colspan="4">1. Treat about 0.02—0.03 g of alloy in a conical test tube with 12-15 drops of $6N\ HNO_3$, by adding it in portions. After the violent reaction stops, add* 5-6 drops of NH_4NO_3 solution and heat in a water bath until the alloy is completely decomposed. If a ppt. is formed, separate it by centrifuging and wash it with hot water containing NH_4Cl.</td></tr>
<tr><td rowspan="5">Precipitate: H_2SnO_3, $HSbO_3$.
2. Look for Sb and Sn. Heat the precipitate for several minutes with conc. HCl. Look for the following elements in the resulting solution (if required, after the undissolved residue has been centrifuged off):
(a) Sb—by placing a drop of the solution on a tin foil (see Sec. 6.12, Reaction 4.1);
(b) Sn—by reactions with NH_4OH and HCl (see Sec. 6.13) or with $HgCl_2$. In the latter case, it is necessary first to reduce tin (IV) to tin (II) by the action of metallic iron or magnesium.</td><td colspan="3">Centrifugate: Cu^{2+}, Pb^{2+}, Cd^{2+}, Al^{3+}, Zn^{2+}, Mn^{2+}, Fe^{3+}, Ni^{2+} (HNO_3).
3. Removal of HNO_3 and separation of Pb^{2+}. Evaporate the solution in an evaporating dish or a crucible with 5-6 drops of $2N\ H_2SO_4$ until white fumes of H_2SO_4 appear. After cooling, add 20 drops of water, transfer the contents of the dish to a test tube, and centrifuge off the ppt.</td></tr>
<tr><td>Precipitate: $PbSO_4$</td><td colspan="2">Centrifugate: Cu^{2+}, Cd^{2+}, Al^{3+}, Zn^{2+}, Mn^{2+}, Fe^{3+}, Ni^{2+} (H_2SO_4)</td></tr>
<tr><td rowspan="2">May be dissolved by heating with 30-percent CH_3COONH_4 and the solution tested for Pb^{2+} (a confirmatory test), by the action of K_2CrO_4 or H_2SO_4.</td><td colspan="2">4. Separation of Cu^{2+} and Cd^{2+}. Precipitate Cu^{2+} and Cd^{2+} in the usual way by H_2S. When the precipitation is complete, centrifuge off the ppt. and wash.</td></tr>
<tr><td>Precipitate: CuS, CdS.</td><td>Centrifugate: Al^{3+}, Zn^{2+}, Mn^{2+}, Fe^{3+}, Ni^{2+} (H_2S).</td></tr>
<tr><td colspan="2">5. Heat with $1N$ HCl, separate the CuS residue, dilute the centrifugate threefold with water, and look for Cd^{2+} by the action of H_2S. Dissolve CuS in HNO_3 and look for Cu^{2+}.</td><td>6. Look for the cations of Group III, as indicated in Table 5.6, after the removal of H_2S by evaporation and the oxidation of Fe^{2+} by heating with 2-3 drops of $6N\ HNO_3$.</td></tr>
</table>

* For better coagulation of H_2SnO_3 and $HSbO_3$. If the precipitate indicated is not formed, NH_4NO_3 need not be added.

are deposited on the metal surface, thereby protecting it against the action of the acid. Therefore, HNO_3 should be used (with heating) to dissolve lead and tin alloys. The Sb and Sn contained in the alloy will then be precipitated in the form of H_2SnO_3 and $HSbO_3$.

TABLE 9.8. **Analysis of Aluminium and Its Alloys**

<table>
<tr><td colspan="3">1. Treat about 0.02-0.03 g of alloy in a conical test tube with 8-10 drops of 30-percent alkali solution and with 2-3 drops of Na_2CO_3 solution (if Ca is present), first in the cold and then on heating in a water bath. After the gas bubbles cease to evolve, separate the ppt. by centrifuging and wash once with water.</td></tr>
<tr><td rowspan="2">Precipitate: hydroxides, basic carbonates or the carbonates of copper, iron, manganese, nickel, magnesium, and calcium.
2. Look for Cu^{2+}, Fe^{3+}, Mn^{2+}, Ni^{2+} and separate them from Mg^{2+} and Ca^{2+}. Dissolve the ppt. by heating with a few drops of 6N HNO_3.* Look for Cu^{2+}, Fe^{3+}, Mn^{2+}, and Ni^{2+} by appropriate reactions in separate portions of the solution.** Then precipitate them from the remainder of the solution by $(NH_4)_2S$ in the presence of NH_4OH and NH_4Cl. Centrifuge off the ppt. of sulphides.</td><td colspan="2">Centrifugate: AlO_2^-, ZnO_2^{2-}, SnO_2^{2-} (NaOH).
4. Look for Sn^{2+} by reaction with $HgCl_2$ or $Bi(OH)_3$.
5. Separate Sn^{2+}. Acidify the alkaline centrifugate with HCl and precipitate by hydrogen sulphide. Centrifuge off the ppt.</td></tr>
<tr><td>Precipitate: SnS.
Not to be examined.</td><td>Centrifugate: Al^{3+}, Zn^{2+} (HCl, H_2S).
6. Look for Al^{3+} and Zn^{2+}.</td></tr>
<tr><td>Precipitate: CuS, Fe_2S_3, MnS, NiS.
Not to be examined.</td><td colspan="2">Centrifugate: Ca^{2+}, Mg^{2+} and excess $(NH_4)_2S$.
3. Look for Ca^{2+} and Mg^{2+}.</td></tr>
</table>

* Instead of fractional analysis of the solution, use may be made of the systematic analysis according to the usual scheme for analysis of a mixture of the cations of Groups V-I; the HNO_3 must be preliminarily removed by evaporation.

** The ion Cu^{2+} can be detected through the action of excess of NH_4OH (concentrated); if required, first centrifuge off the ppt. It is convenient to use the following spot-test technique based on the combined action of NH_4OH and $K_4[Fe(CN)_6]$. Place a drop of conc. NH_4OH in the periphery of a strip of filter paper and then a drop of the test solution in the middle of the paper, followed by a drop of conc. NH_4OH. Dry the moist spot to remove ammonium hydroxide, and draw a capillary containing $K_4[Fe(CN)_6]$ solution along the periphery of the spot. A reddish-brown ring due to $Cu_2[Fe(CN)_6]$ indicates the presence of copper. Other spot-test technique may also be used (see Sec. 6.7).

Since alloys of each given type may contain only metals characteristic of that type, the analysis is simplified if we first determine the type to which the alloy belongs and then conduct the analysis accordingly.

TABLE 9.9. **Analysis of Lead and Tin and Their Alloys**

<table>
<tr><td colspan="3">1. Treat about 0.02-0.03 g of alloy in a conical test tube with 12-15 drops of 6N HNO_3. Heat the contents of the tube for 2-3 min in a water bath, then dilute with 10 drops of water, add 5-6 drops of NH_4NO_3 solution and continue heating until the alloy is completely decomposed. Centrifuge off the ppt. and wash once with hot water containing NH_4Cl.</td></tr>
<tr><td rowspan="2">Precipitate: H_2SnO_3 and $HSbO_3$.
2. Test for Sb and Sn as indicated in Table 9.7 (Procedure 2).</td><td colspan="2">Centrifugate: Pb^{2+}, Cu^{2+}, Fe^{3+}, Cd^{2+}, Zn^{2+}(HNO_3).
3. Add 10-12 drops of 2N H_2SO_4. Evaporate the solution until white fumes of H_2SO_4 appear; then dilute with water (twofold) and centrifuge off the $PbSO_4$ ppt.</td></tr>
<tr><td>Precipitate: $PbSO_4$.
4. Carry out confirmatory tests for Pb^{2+}.</td><td>Centrifugate: Cu^{2+}, Fe^{3+}, Cd^{2+}, Zn^{2+}(H_2SO_4).
5. Analyse the solution as directed in the analysis of copper alloys (see Table 9.7, Procedures 4-6).</td></tr>
</table>

To do so, we have not only to determine the colour and specific gravity* (density) of alloys (see Table 9.4) but also to perform the preliminary tests described in Table 9.5.

In carrying out the tests given in Table 9.5, one should observe closely the intensity of the colour or the amount of the precipitate formed in the reactions.

For example, iron is present not only in ferrous metals but in many other alloys as well. But only in the analysis of iron alloys can the extremely intense colour (blood-red) be produced through the action of NH_4SCN; in other alloys, it will be incomparably paler (pale-red).

As a second example, a precipitate of $PbSO_4$ may be formed in the analysis of copper alloys, but it will be much more abundant when lead alloys are subjected to analysis.

If no definite conclusion can be arrived at from all the characteristics listed in Tables 9.4 and 9.5 concerning the type of an alloy (which may be the case with metals or alloys that do not fit into this general classification), a suitable solvent should be chosen in accordance with the general rules given in Sec. 9.3; this being done, dissolve about 0.02-0.03 g of the alloy in the solvent and analyse the resulting solution for

* What is meant here is merely a qualitative assessment (and not a quantitative determination) of the density, which enables one to distinguish, for example, a light alloy from a heavy one. It may however be used if the amount of the alloy is not too small.

cations in the usual way. The detection of non-metals that may be present in alloys is not discussed in this book.

If the type of an alloy has been ascertained, the alloy is dissolved and analysed according to one of the schemes presented in Tables 9.6-9.9.

9.6. Qualitative Analysis Without Hydrogen Sulphide

The classical method of qualitative cation analysis considered in the book has a drawback: the separation of groups is associated with the use of hydrogen sulphide. The work with this substance is very unpleasant, harmful to the health, and must therefore be carried out in a properly operating fume cupboard; in macroanalysis, a special room is required. The method has certain other disadvantages.

1. In the first place, the separation of analytical groups by the hydrogen-sulphide method is not sufficiently distinct because of the coprecipitation and postprecipitation. For example, as has already been pointed out, when the precipitated sulphides of Group IV are allowed to stand in contact with the solution, zinc sulphide postprecipitates and therefore Zn^{2+} may sometimes be lost. Likewise, certain cations of Group III (e.g., Ni^{2+} and Co^{2+}) may coprecipitate with the sulphides of the cations of Group IV.

2. As stated in Sec. 6.5, the separation of Group IV cations into two subgroups (the copper and arsenic subgroups) is rather arbitrary. Depending on the reagent employed—NaOH, Na_2S, $(NH_4)_2S$, or $(NH_4)_2S_2$, the cations of the copper subgroup may form thiosalts (CuS_2^{2-} and HgS_2^{2-}). When use is made of NaOH, Na_2S, or $(NH_4)_2S$, it is necessary first to oxidize Sn^{II} to Sn^{IV}; otherwise tin will remain in the precipitate of the sulphides of the copper subgroup in the form of SnS.

3. The use of solutions of H_2S and $(NH_4)_2S$ which often contain SO_4^{2-} (due to the oxidation of S^{2-} by atmospheric oxygen) leads not infrequently to the loss of Group II cations which coprecipitate as slightly soluble sulphates with the sulphides of Groups III and IV. The coprecipitation of the cations of Group II with those of Group III may also take place because ammonium sulphide and ammonium hydroxide (used for separation of Group III) not infrequently contain an impurity of ammonium carbonate (absorption of CO_2 from the air).

4. Detection of Group I cations in the solution remaining after the precipitation of the cations of all the other groups often yields wrong results due to the considerable decrease of their concentration in the solution caused by adsorption and other kinds of coprecipitation, and the reactions of some of them are not sensitive enough. As has already been indicated, the cations K^+ and Na^+ are therefore more conveniently detected by the fractional method.

5. Adjustment of the pH of the solution is very important in the precipitation of sulphides. But this is rather difficult to accomplish because of the absence of suitable buffer systems for producing a pH of about 0.5. The acids present in the solution have to be preliminarily neutralized by adding NH_4OH until the solution reacts basic, after which its excess is neutralized by hydrochloric acid. As a result, a large amount of ammonium salts accumulate in the solution, which enhances the salting effect. Furthermore, the complex-forming tendency of Cl^- has also to be considered since it may result in an incomplete precipitation of Ag^+, Cd^{2+}, and other ions.

Some other disadvantages could also be cited. Numerous attempts have been and are being made by many authors to find a substitute for hydrogen sulphide and to develop a fairly satisfactory method of systematic cation analysis without the use of hydrogen sulphide.

Some authors suggest separations of groups based on the properties of sulphides, just as in the classical method, but with the use of sulphur-containing reagents other than hydrogen sulphide for their preparation: sodium thiosulphate, $Na_2S_2O_3$, which precipitates many cations in the form of sulphides in acid solution; ammonium polysulphide, $(NH_4)_2S_2$; sodium sulphide, Na_2S; thioacetamide, CH_3CSNH_2; ammonium dithiocarbonate, $(NH_4)_2COS_2$; and some other compounds. Since the sulphides precipitated by the action of these reagents have to be dissolved by acids in subsequent procedures (usually with the evolution of H_2S), all these methods enable avoiding only some of the disadvantages of the hydrogen-sulphide method.

According to other authors, it is better to separate cations into groups on the basis of the properties of any other compounds of metals, for example, hydroxides, phosphates, carbonates, etc., or even of free metals.

For instance, according to the method suggested by I. Sakhiev*, the separation of groups is accomplished as follows:

Group I: the ions of antimony and tin—precipitated as $HSbO_3$ and H_2SnO_3 by evaporation of the solution with HNO_3;

Group II: the cations Ag^+, Hg_2^{2+}, Pb^{2+}—precipitated by hydrochloric acid**;

Group III: the cations Fe^{3+}, Bi^{3+}, Cr^{3+}, Al^{3+}, Pb^{2+}, and Mn^{2+}—precipitated as hydroxides or basic salts by ammonium hydroxide in the presence of an ammonium salt;

* Sakhiev, I. S., "Qualitative Analysis without the Use of Hydrogen Sulphide and Ammonium Sulphide", in *Qualitative Analysis without Hydrogen Sulphide*, edited by N. E. Orlov, Goskhimizdat, Moscow, 1934 (in Russian).

** The ion Hg^{2+} is reduced to $[Hg_2]^{2+}$ by the action of orthophosphorous acid, H_3PO_3; arsenic is driven off in the form of the volatile chloride $AsCl_3$. Therefore, Hg^{2+} and also As^{III} and As^{V} are not included in this classification.

Group IV: the cations Ba^{2+}, Sr^{2+}, Ca^{2+}—precipitated as carbonates by ammonium carbonate;

Group V: the cations Cu^{2+}, Cd^{2+}, Zn^{2+}, Ni^{2+}, Co^{2+}, Mg^{2+}—precipitated as hydroxides by barium hydrous oxide;

Group VI: the cations Na^+, K^+, NH_4^+—having no group reagent.

As another example, let us consider the method of analysis without the use of hydrogen sulphide, which is based on the differences in the properties of metal hydroxides, and also on the different solubility of the chlorides and sulphates formed by them.* The separation of cations into groups is achieved in this case as follows.

Group I: Ag^+, Hg_2^{2+}, Pb^{2+}, which form slightly soluble chlorides;

Group II: Ba^{2+}, Sr^{2+}, Pb^{2+}, and Ca^{2+}, which form slightly soluble sulphates;

Group III: the cations Al^{3+}, Cr^{3+}, Zn^{2+}, and the ions of arsenic and tin, which form amphoteric hydroxides and therefore remain in solution upon addition of excess NaOH and H_2O_2 (the cations of Groups IV and V are precipitated under these conditions);

Group IV: the cations Fe^{3+}, Mn^{2+}, Mg^{2+}, Bi^{3+}, Sb^{3+}, which form hydroxides having no amphoteric properties and insoluble in excess of ammonium hydroxide;

Group V: the cations Cu^{2+}, Co^{2+}, Ni^{2+}, Cd^{2+}, Hg^{2+}, which give hydroxides soluble in excess of ammonium hydroxide with the formation of ammines;

Group VI: the cations K^+, Na^+, NH_4^+.

In this method of analysis, first HCl and then H_2SO_4 are used to precipitate the cations of Groups I and II from the solution (according to the above classification). Then the ions NH_4^+, K^+, and Na^+ are detected by the fractional technique in separate portions of the solution obtained after the removal of the cations of the groups indicated. The remainder of the solution is treated with excess NaOH and H_2O_2. The cations of Group III remain in solution, whereas those of Groups IV and V are precipitated. The precipitate is treated with HNO_3, all the hydroxides being dissolved, with the exception of $HSbO_3$ which remains unaffected. After the $HSbO_3$ is separated, an excess of ammonium hydroxide is added to the solution (after HNO_3 is neutralized with Na_2CO_3). In this procedure, the cations of Group V remain in solution in the form of complex ammines, whereas the hydroxides of Group IV cations are precipitated.

The method of separation into groups, proposed by V. I. Petrashen, is based on the different solubility of phosphates in ammonium hydroxide, acetic acid, and mineral acids. N. A. Tananaev suggested a method according to which all the cations are divided into two large

---

* Beskov, S. D. and O. A. Slizkovskaya, Analytical Chemistry, Uchpedgiz, Moscow, 1956 (in Russian).

groups by treatment of the test solution (chlorides or nitrates) with powdered zinc metal. Some of the cations are reduced to the corresponding metals, while others remain in solution.

Numerous methods involving no hydrogen sulphide are discussed in detail in special literature. None of them have however won wide recognition up to now. This is evidently explained by the fact that they all have no less, and sometimes even more, shortcomings than the hydrogen-sulphide method which may be noted for the undoubted advantage of having all its details thoroughly studied for a period of more than 120 years of its existence. Besides, the classical method of qualitative analysis is of great pedagogical importance since it allows the study of theory to be connected in the best way with analytical practice. The analytical classification of cations adopted in this method is closely associated with the position of the elements in the Periodic System and with the outer electronic structure of cations.

Name Index

Page numbers in italic type refer to bibliography

N

O

P

R

S

T

V

W

Y

Subject Index

B

C

T

U

V

W

Z

TO THE READER

Mir Publishers would be grateful for your comments on the content, translation, and design of this book. We would also be pleased to receive any other suggestions you may wish to make.

Our address is: USSR, 129820, Moscow I-110,
GSP Pervy Rizhsky Pereulok 2
Mir Publishers